SON

ARTHUR C. GUYTON, M.D.

Professor and Chairman of the
Department of Physiology and
Biophysics, University of Mississippi
School of Medicine

BASIC HUMAN PHYSIOLOGY: NORMAL FUNCTION AND MECHANISMS OF DISEASE

W. B. SAUNDERS COMPANY
Philadelphia · London · Toronto

W. B. Saunders Company: West Washington Square
Philadelphia, Pa. 19105

12 Dyott Street
London, WC1A 1DB

1835 Yonge Street
Toronto 7, Ontario

BASIC HUMAN PHYSIOLOGY: Normal Function and Mechanism of Disease

SBN 0-7216-4382-5

Print No.:   9   8   7   6   5   4   3

Dedicated to
the modern student,
often a quixotic being,
but far brighter than his forebears
despite what his forebears often think

PREFACE

I have written this book for those students who wish more than an elementary introduction to the physiology of the human body and yet who cannot afford the time or perhaps might not have the background to study one of the more formidable textbooks. A special attempt has been made to give wide coverage to all aspects of human physiology but still to present the material at a level that is acceptable to the previous training of essentially all college and early professional students. And throughout the text I have attempted to present bodily mechanisms in the light of well known physical and chemical laws, not merely to describe physiological function as if it were unrelated to other scientific disciplines.

I have also given a strong "human" flavor to the book. Many textbooks of human physiology are in reality animal physiology placed in a human setting. Obviously, a major share of the information used to develop the present book also came from basic experiments in animals. Yet, another large body of knowledge has come from human experimentation, both planned experiments in normal human beings and unplanned experiments caused by disease. Indeed, one of the reasons for discussing the physiology of some diseases in this text has been to give it its human flavor. For instance, when one discusses the regulation of blood glucose or of carbohydrate metabolism or of fat metabolism, it is almost ridiculous not to discuss simultaneously the basic physiology of diabetes mellitus, a disease that affects all these physiological functions and that is widespread among the human population, one that affords a constant example of a ubiquitous and important physiological experiment. By the same token, literally thousands of human "experiments" proceed each day in the fields of high blood pressure, congestive heart failure, gastrointestinal disturbances, respiratory diseases, and so forth. The physiology of these abnormalities should be the property of every student of human physiology rather than being reserved in a special niche for only the medical student or physician.

I hope that I can leave the student with the knowledge that the human body is one of the most complex and yet most beautiful of all functional mechanisms. Think for a moment that the human brain carries within it a computer with capabilities and functions that all the electronic computers of the world put together cannot at present achieve. Think of the individual living cell, the basic structural component of the body, that has within it all the genetic components of the entire human being, in reality a myriad of control systems for literally thousands of chemical reactions

v

within each cell. I could go on detailing the miracles of the human body. That, indeed, is the purpose of this entire text. The success that I have in exciting the student to further study in the field of physiology or to a lifetime of physiological thinking will determine whether or not I have been successful in my own goal in writing this text.

A vast amount of actual labor always goes into the development and publication of a text. For the majority of the figures in the text I am especially indebted to Mrs. Carolyn Hull, and for the great quantity and quality of secretarial services I owe my gratitude to Mrs. Billie Howard and Mrs. Judy Bass. Finally, I extend my appreciation to the staff of the W. B. Saunders Company for its continued excellence in all publication matters, particularly for the novel and, to me at least, enticing format that has been achieved.

ARTHUR C. GUYTON
Jackson, Mississippi

CONTENTS

PART VII   ENDOCRINOLOGY AND REPRODUCTION

THE CELL
AND
GENERAL
PHYSIOLOGY

PART I

FUNCTIONAL ORGANIZATION OF THE HUMAN BODY AND CONTROL OF THE "INTERNAL ENVIRONMENT"

In human physiology we attempt to explain such effects as the chemical reactions in the cells, the transmission of nerve impulses from one part of the body to another, contraction of the muscles, reproduction, and even the minute details of transformation of light energy into chemical energy to excite the eyes, thus allowing us to see the world. The very fact that we are alive is almost beyond our own control, for hunger makes us seek food and fear makes us seek refuge. Sensations of cold make us provide warmth, and other forces cause us to seek fellowship and reproduce. Thus, the human being is actually an automaton, and the fact that we are sensing, feeling, and knowledgeable beings is part of this automatic sequence of life; these special attributes allow us to exist under widely varying conditions, which otherwise would make life impossible.

Cells as the Basic Living Units of the Body. The basic living unit of the body is the cell, and each organ is actually an aggregate of many different cells held together by intercellular supporting structures. Each type of cell is specially adapted to perform one particular function. For instance, the red blood cells, twenty-five trillion in all, transport oxygen from the lungs to the tissues. Though this type of cell is perhaps the most abundant of any in the whole body, there are approximately another 75 trillion cells. The entire body, then, contains about 100 trillion cells.

Cells are automatons that are capable of living, growing, and providing their special functions so long as the proper concentrations of oxygen, glucose, the different electrolytes, amino acids, and fatty substances are available in the tissue fluids of the body.

3

Almost all cells also have the ability to reproduce, and whenever cells of a particular type are destroyed for one cause or another, the remaining cells of this type usually divide again and again until the appropriate number is replenished.

THE EXTRACELLULAR FLUID — THE INTERNAL ENVIRONMENT

About 56 per cent of the adult human body is fluid. Some of this fluid is inside the cells and is called, collectively, the *intracellular fluid*. The fluid in the spaces outside the cells is called the *extracellular fluid*. Among the dissolved constituents of the extracellular fluids are the ions and the nutrients needed by the cells for maintenance of life. The extracellular fluid is in constant motion throughout the body and is constantly mixed by the blood circulation and by diffusion between the blood and the tissue spaces. Therefore, all cells live in essentially the same environment, for which reason the extracellular fluid is often called the *internal environment* of the body.

Differences Between Extracellular and Intracellular Fluids. The extracellular fluid contains large amounts of sodium, chloride, and bicarbonate ions, plus nutrients for the cells, such as oxygen, glucose, fatty acids, and amino acids. It also contains carbon dioxide, which is being transported from the cells to the lungs, and other cellular excretory products, which are being transported to the kidneys.

The intracellular fluid is much the same from one cell to another, but it differs significantly from the extracellular fluid; particularly, the intracellular fluid contains large amounts of potassium, magnesium, and phosphate ions instead of the sodium and chloride ions found in the extracellular fluid. Special mechanisms for transporting ions through the cell membranes maintain these differences. These mechanisms will be discussed in detail in Chapter 4.

HOMEOSTASIS

The term *homeostasis* is used by physiologists to mean *maintenance of static*, or *constant, conditions in the internal environment*. Essentially all the organs and tissues of the body perform functions that help to maintain these constant conditions. For instance, the lungs provide new oxygen as it is required by the cells, the kidneys maintain constant electrolyte concentrations, and the gut provides nutrients. A large segment of this text is concerned with the manner in which each organ or tissue contributes to homeostasis. To begin this discussion, the different functional systems of the body and their homeostatic mechanisms will be outlined briefly; then the basic theory of the control systems that cause the functional systems to operate in harmony with each other will be discussed.

THE FLUID TRANSPORT SYSTEM

Extracellular fluid is transported to parts of the body in two different stages. The first stage entails movement of blood around and around the circulatory system, and the second, movement of fluid between the blood capillaries and the cells. Figure 1-1 illustrates the overall circulation of blood, showing the heart is actually two separate pumps, one of which propels blood through the lungs and the other through the systemic circulation. All the blood in the circulation traverses the entire circuit of the circulation an average of once each minute when a person is at rest and as many as five times each minute when he becomes extremely active.

As blood passes through the capillaries, continual exchange occurs between the plasma portion of the blood and the interstitial fluid in the spaces surrounding the capillaries. This is illustrated in Figure 1-2. Note that the capillaries are porous so that large amounts of fluid can *diffuse* back and forth between the blood and the

FUNCTIONAL ORGANIZATION OF THE HUMAN BODY AND CONTROL OF THE "INTERNAL ENVIRONMENT"

In human physiology we attempt to explain such effects as the chemical reactions in the cells, the transmission of nerve impulses from one part of the body to another, contraction of the muscles, reproduction, and even the minute details of transformation of light energy into chemical energy to excite the eyes, thus allowing us to see the world. The very fact that we are alive is almost beyond our own control, for hunger makes us seek food and fear makes us seek refuge. Sensations of cold make us provide warmth, and other forces cause us to seek fellowship and reproduce. Thus, the human being is actually an automaton, and the fact that we are sensing, feeling, and knowledgeable beings is part of this automatic sequence of life; these special attributes allow us to exist under widely varying conditions, which otherwise would make life impossible.

Cells as the Basic Living Units of the Body. The basic living unit of the body is the cell, and each organ is actually an aggregate of many different cells held together by intercellular supporting structures. Each type of cell is specially adapted to perform one particular function. For instance, the red blood cells, twenty-five trillion in all, transport oxygen from the lungs to the tissues. Though this type of cell is perhaps the most abundant of any in the whole body, there are approximately another 75 trillion cells. The entire body, then, contains about 100 trillion cells.

Cells are automatons that are capable of living, growing, and providing their special functions so long as the proper concentrations of oxygen, glucose, the different electrolytes, amino acids, and fatty substances are available in the tissue fluids of the body.

Almost all cells also have the ability to reproduce, and whenever cells of a particular type are destroyed for one cause or another, the remaining cells of this type usually divide again and again until the appropriate number is replenished.

THE EXTRACELLULAR FLUID— THE INTERNAL ENVIRONMENT

About 56 per cent of the adult human body is fluid. Some of this fluid is inside the cells and is called, collectively, the *intracellular fluid*. The fluid in the spaces outside the cells is called the *extracellular fluid*. Among the dissolved constituents of the extracellular fluids are the ions and the nutrients needed by the cells for maintenance of life. The extracellular fluid is in constant motion throughout the body and is constantly mixed by the blood circulation and by diffusion between the blood and the tissue spaces. Therefore, all cells live in essentially the same environment, for which reason the extracellular fluid is often called the *internal environment* of the body.

Differences Between Extracellular and Intracellular Fluids. The extracellular fluid contains large amounts of sodium, chloride, and bicarbonate ions, plus nutrients for the cells, such as oxygen, glucose, fatty acids, and amino acids. It also contains carbon dioxide, which is being transported from the cells to the lungs, and other cellular excretory products, which are being transported to the kidneys.

The intracellular fluid is much the same from one cell to another, but it differs significantly from the extracellular fluid; particularly, the intracellular fluid contains large amounts of potassium, magnesium, and phosphate ions instead of the sodium and chloride ions found in the extracellular fluid. Special mechanisms for transporting ions through the cell membranes maintain these differences. These mechanisms will be discussed in detail in Chapter 4.

HOMEOSTASIS

The term *homeostasis* is used by physiologists to mean *maintenance of static, or constant, conditions in the internal environment*. Essentially all the organs and tissues of the body perform functions that help to maintain these constant conditions. For instance, the lungs provide new oxygen as it is required by the cells, the kidneys maintain constant electrolyte concentrations, and the gut provides nutrients. A large segment of this text is concerned with the manner in which each organ or tissue contributes to homeostasis. To begin this discussion, the different functional systems of the body and their homeostatic mechanisms will be outlined briefly; then the basic theory of the control systems that cause the functional systems to operate in harmony with each other will be discussed.

THE FLUID TRANSPORT SYSTEM

Extracellular fluid is transported to all parts of the body in two different stages. The first stage entails movement of blood around and around the circulatory system, and the second, movement of fluid between the blood capillaries and the cells. Figure 1–1 illustrates the overall circulation of blood, showing that the heart is actually two separate pumps, one of which propels blood through the lungs and the other through the systemic circulation. All the blood in the circulation traverses the entire circuit of the circulation an average of once each minute when a person is at rest and as many as five times each minute when he becomes extremely active.

As blood passes through the capillaries, continual exchange occurs between the plasma portion of the blood and the interstitial fluid in the spaces surrounding the capillaries. This process is illustrated in Figure 1–2. Note that the capillaries are porous so that large amounts of fluid can *diffuse* back and forth between the blood and the tissue spaces, as illustrated by the arrows. This process of diffusion is caused by kinetic motion of the molecules in both the plasma and the extracellular fluid. That is, all fluid and dissolved molecules are continually moving and bouncing in all directions, through the pores, through the tissue spaces, and so forth. Almost no cell is located more than 25 to 50 microns from a capillary, which allows movement of any substance from the capillary to the cell within a few seconds. Thus, the extracellular fluid throughout the body is continually mixed and thereby maintains almost complete homogeneity.

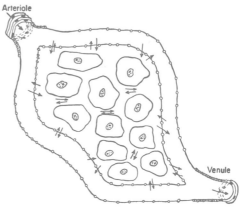

FIGURE 1–2 Diffusion of fluids through the capillary walls and through the interstitial spaces.

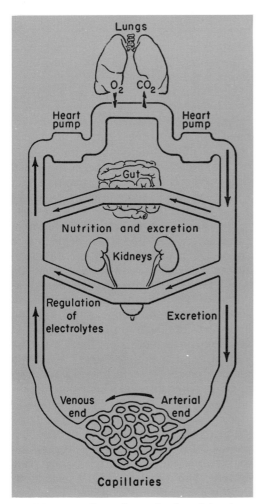

FIGURE 1–1 General organization of the circulatory system.

ORIGIN OF NUTRIENTS IN THE EXTRACELLULAR FLUID

The Respiratory System. Figure 1–1 shows that each time the blood passes through the body it also flows through the lungs. The blood picks up oxygen in the alveoli, thus acquiring the oxygen needed by the cells. The membrane between the alveoli and the lumen of the pulmonary capillaries is only 1 to 4 microns in thickness, and oxygen diffuses through this membrane into the blood in exactly the same manner that water, nutrients, and excreta diffuse through the tissue capillaries.

The Gastrointestinal Tract. Figure 1–1 also shows that a large portion of the blood pumped by the heart passes through the walls of the gastrointestinal organs. Here, different dissolved nutrients, including carbohydrates, fatty acids, amino acids, and others, are absorbed into the extracellular fluid.

The Liver and Other Organs that Perform Primarily Metabolic Functions. Not all substances absorbed from the gastrointestinal tract can be used in their absorbed form by the cells. The liver alters the chemical compositions of many of these to more usable forms, and other tissues of the body — such as the fat cells, the gastrointestinal mucosa, the kidneys, and the endocrine glands — help to modify the absorbed substances or store them until they are needed at a later time.

Musculoskeletal System. Sometimes the question is asked: How does the musculoskeletal system fit into the homeostatic functions of the body? Were it not for this system, the body could not move to the appropriate place at the appropriate time to obtain the foods required for nutrition. The musculoskeletal system also provides motility for protection against adverse surroundings, without which the entire body, and along with it all the homeostatic mechanisms, could be destroyed instantaneously.

REMOVAL OF METABOLIC END-PRODUCTS

Removal of Carbon Dioxide by the Lungs. At the same time that blood picks up oxygen in the lungs, carbon dioxide is released from the blood into the alveoli, and the respiratory movement of air into and out of the alveoli carries the carbon dioxide to the atmosphere. Carbon dioxide is the most abundant of all the end-products of metabolism.

The Kidneys. Passage of the blood through the kidneys removes most substances from the plasma that are not needed by the cells. These substances include especially different end-products of metabolism and excesses of electrolytes or water that might have accumulated in the extracellular fluids. The kidneys perform their function by, first, filtering large quantities of plasma through the glomeruli into the tubules and then reabsorbing into the blood those substances needed by the body, such as glucose, amino acids, large amounts of water, and many of the electrolytes. However, substances not needed by the body generally are not reabsorbed but, instead, pass on through the renal tubules into the urine.

REGULATION OF BODY FUNCTIONS

The Nervous System. The nervous system is composed of three major parts: the *sensory portion*, the *central nervous system*, or *integrative portion*, and the *motor portion*. Sensory nerves detect the state of the body or the state of the surroundings. For instance, nerves present everywhere in the skin apprise one every time an object touches him at any point. The eyes are sensory organs that give one a visual image of the surrounding area. The ears also are sensory organs. The central nervous system is comprised of the brain and spinal cord. The brain can store information, generate thoughts, create ambition, and determine reactions that the body should perform in response to the sensations. Appropriate signals are then transmitted through the motor portion of the nervous system to carry out the person's desires.

A large segment of the nervous system is called the *autonomic system*. It operates at a subconscious level and controls many functions of the internal organs, including the action of the heart, the movements of the gastrointestinal tract, and the secretion by different glands.

The Hormonal System of Regulation. Located in the body are eight major endocrine glands that secrete chemical substances called *hormones*. Hormones are transported in the extracellular fluids to all parts of the body to help regulate function. For instance, thyroid hormone increases the rates of almost all chemical reactions in all cells. In this way thyroid hormone helps to set the tempo of bodily activity. Likewise, insulin controls glucose metabolism; adrenocortical hormones, electrolyte and protein metabolism; and parathormone, bone metabolism. Thus, the hormones represent a system of regulation that complements that of the nervous system. The nervous system, in general, regulates rapid muscular and secretory activities of the body, whereas the hormonal system regulates mainly the slowly reacting metabolic functions.

REPRODUCTION

Reproduction is another function of the body that sometimes is not considered to be a homeostatic function. But it does help to maintain a static situation by generating new beings to take the place of ones that are dying. This perhaps sounds like a farfetched usage of the term homeostasis, but it does illustrate that, in the final analysis, essentially all structures of the body are so organized that they help to maintain continuity of life.

THE CONTROL SYSTEMS OF THE BODY

The human body has literally thousands of control systems in it. Some of these operate within the cell to control intracellular function, a subject that will be discussed in detail in Chapter 3. Others operate within the organs to control functions of the individual parts of the organs, while others operate throughout the entire body to control the interrelationships between the different organs. For instance, the respiratory system regulates the concentration of carbon dioxide in the extracellular fluids. The liver and the pancreas regulate the concentration of glucose in the extracellular fluids. And the kidneys regulate the concentrations of hydrogen, sodium, potassium, phosphate, and other ions in the extracellular fluids.

An Example of a Control Mechanism: Regulation of Arterial Pressure. Several different systems contribute to the regulation of arterial pressure. One of these, the *baroreceptor system*, is very simple and is an excellent example of a control mechanism. In the walls of most of the great arteries of the upper body, especially the bifurcation region of the carotids and the arch of the aorta, are many nerve stretch receptors, called *baroreceptors*, which are stimulated by stretch of the arterial wall. When the arterial pressure becomes great, these baroreceptors are stimulated excessively, and impulses are transmitted to the medulla of the brain. Here the impulses inhibit the *vasomotor center*, which in turn decreases the number of impulses transmitted through the sympathetic nervous system to the heart and blood vessels. Lack of these impulses causes diminished pumping activity by the heart and increased ease of blood flow through the peripheral vessels, both of which lower the arterial pressure back toward normal. Conversely, a fall in arterial pressure relaxes the stretch receptors, allowing the vasomotor center to become more active than usual and thereby causing the arterial pressure to rise back toward normal.

Negative Feedback Nature of Control Systems. The control systems of the body act by a process of *negative feedback*, which can be explained best by analyzing the baroreceptor pressure regulating mechanism that was just explained. In this mechanism, a high pressure causes a series of reactions that promote a lowered pressure, or a low pressure causes a series of reactions that promote an elevated pressure. In both instances these effects are negative with respect to the initiating stimulus.

Essentially all other control mechanisms of the body also operate by the process of negative feedback. For instance, if the oxygen concentration in the body fluids falls too low, the mechanism for controlling oxygen automatically returns the oxygen back to a higher level. Thus, the effect is *negative* to the initiating stimulus. Likewise, elevated carbon dioxide concentration in the body fluids causes increased respiration, which then removes the excess carbon dioxide. Again, the response is negative to the stimulus. Also, essentially all of the endocrine control systems operate in this manner. For instance, when the concentration of sodium rises too high in the body fluids, the adrenal glands decrease their secretion of the hormone aldosterone, and lack of this hormone allows the kidneys to excrete excessive amounts of sodium into the urine. This is still another example of negative feedback.

Therefore, in general, if some factor becomes excessive or too little, a control system initiates *negative feedback*, which consists of a series of changes that returns the factor toward a certain mean value, thus maintaining homeostasis.

Amplification, or gain, of a control system. The degree of effectiveness with which a control system maintains constant conditions is called the *amplification*, or *gain*, of the system.

For instance, let us assume that a large volume of blood is suddenly transfused into a person and that this immediately raises the arterial pressure from a normal value of 100 mm. Hg up to 160 mm. Hg. However, within 15 to 30 seconds the baroreceptor control mechanism becomes fully operative, and the arterial pressure is reduced back to 120 mm. Hg. Thus, the pressure is corrected 40 mm. Hg, while the final abnormality is only 20 mm. Hg instead of the 60 mm. Hg that would have occurred without the control system. The gain of the mechanism is calculated by the following equation:

$$\text{Gain} = \frac{\text{Amount of correction}}{\text{Amount of abnormality still remaining}}$$

In the above example the correction is 40 mm. Hg, and the amount of abnormality still remaining is 20 mm. Hg; therefore, the gain of the baroreceptor system for control of arterial pressure is approximately 2.

The gains of different control systems of the body vary markedly, with gains as low as 1 to 2 for control of arterial pressure by a hormone called renin that is released from the kidney, and as high as 50 for control of body temperature in the face of changing atmospheric temperature. In other words, the pressure controlling ability of the renin mechan-ism is weak, while the temperature controlling ability of the temperature feedback system is very great.

AUTOMATICITY OF THE BODY

The purpose of this chapter has been to point out, first, the overall organization of the body and, second, the means by which the different parts of the body operate in harmony. To summarize, the body is actually an *aggregate of about 100 trillion cells* organized into different functional structures, some of which are called *organs*. Each functional structure provides its share in the maintenance of homeostatic conditions in the extracellular fluid, which is often called the *internal environment*. As long as normal conditions are maintained in the internal environment, the cells of the body will continue to live and function properly. Thus, each cell benefits from homeostasis, and in turn each cell contributes its share toward the maintenance of this state. This reciprocal interplay provides continuous automaticity of the body until one or more functional systems lose their ability to contribute their share of function. When this happens, all the cells of the body suffer. Extreme dysfunction of any integrated system leads to death, while moderate dysfunction leads to sickness.

REFERENCES

Adolph, E. F.: Origins of Physiological Regulations. New York, Academic Press, Inc., 1968.

Cannon, W. B.: The Wisdom of the Body, New York, W. W. Norton & Company, Inc., 1932.

Milhorn, H. T.: The Application of Control Theory to Physiological Systems. Philadelphia, W. B. Saunders Company, 1966.

Reeve, E. B., and Guyton, A. C.: Physical Bases of Circulatory Transport: Regulation and Exchange. Philadelphia, W. B. Saunders Company, 1967.

Sherrington, C. S.: Man on His Nature. New York, The Macmillan Company, 1941.

Wolf, W.: Rhythmic functions in the living system. *Ann. N.Y. Acad. Sci.,* 98:753, 1962.

CHAPTER 2

THE CELL AND ITS FUNCTION

Each of the 100 trillion cells in the human being is a living structure that can survive indefinitely and in most instances can even reproduce itself provided its surrounding fluids simply remain constant. To understand the function of organs and other structures of the body, it is essential that we first understand the basic organization of the cell and the functions of its component parts.

PHYSICAL STRUCTURE OF THE CELL

The cell is not merely a bag of fluid, enzymes, and chemicals; it also contains highly organized physical structures called *organelles*, which are as important to the function of the cell as are the cell's chemical constituents. For instance, without one of the organelles, the *mitochondria*, the energy supply of the cell would cease almost entirely. Some principal organelles of the cell are the *cell membrane, nuclear membrane, endoplasmic reticulum, mitochondria*, and *lysosomes*, as illustrated in Figure 2–1. Others not shown in the figure are the *Golgi complex, centrioles, cilia*, and *microtubules*.

MEMBRANES OF THE CELL

Essentially all physical structures of the cell are lined by a membrane composed primarily of lipids and proteins. All of the different membranes are similar in structure, and this common type of structure is called the "unit membrane," which will be described below in relation to the cell membrane. The different membranes include the *cell*

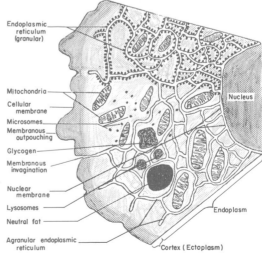

FIGURE 2–1 Organization of the cytoplasmic compartment of the cell.

9

membrane, the *nuclear membrane*, the *membrane of the endoplasmic reticulum*, and the *membranes of the mitochondria, lysosomes, Golgi complexes*, and so forth.

The Cell Membrane. The cell membrane is thin (approximately 75 to 100 Ångstroms) and elastic. It is composed almost entirely of proteins and lipids (fats), with a percentage composition approximately as follows: proteins, 55 per cent; lipids, 40 per cent; polysaccharides, perhaps 5 per cent. The proteins in the cell are mainly a type of protein called *stromatin*, an insoluble structural protein having elastic properties. The lipids are approximately 65 per cent phospholipids, 25 per cent cholesterol, and 10 per cent other lipids.

The precise molecular organization of the cell membrane is unknown, but many experiments point to the structure illustrated in Figure 2–2A. This shows a central layer of lipids covered by protein layers and a thin mucopolysaccharide layer on the outside surface. The presence of protein and mucopolysaccharide on the surfaces supposedly makes the membrane *hydrophilic*, which means that water adheres easily to the membrane. The lipid center of the membrane makes the membrane mainly impervious to lipid-insoluble substances. The small knobbed structures lying at the bases of the protein molecules in Figure 2–2A are phospholipid molecules; the fat portion of the phospholipid molecule is attracted to the central lipid phase of the cell membrane, and the polar (ionized) portion of the molecule protrudes toward either surface where it is bound electrochemically with the inner and outer layers of proteins. The thin layer of mucopolysaccharide on the outside of the cell membrane makes the outside different from the inside, thus polarizing the membrane so that the chemical reactivities of the cell's inner surface are different from those of the outer surface.

Pores in the membrane. The membrane is believed to have many minute pores that pass from one side to the other as shown in Figure 2–2B. These pores have never been demonstrated, even with an electron microscope; but functional experiments to study the movement of molecules of different sizes between the extra- and intracellular fluids have demonstrated free diffusion of molecules up to a size of approximately 8 Ångstroms. It is through these pores that lipid-insoluble substances of very small sizes, such as water and urea molecules, are believed to pass with relative ease between the interior and exterior of the cell.

The Nuclear Membrane. The nuclear membrane is actually a double membrane having a wide space between two

CELL MEMBRANE

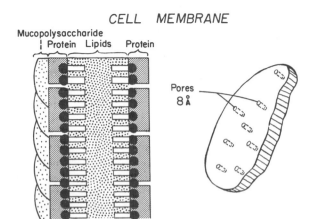

FIGURE 2–2 (*A*) Postulated molecular organization of the cell membrane. (*B*) Pores in the cell membrane.

"unit" membranes. Each unit membrane is almost identical to the cell membrane, having lipids in its center and protein on its two surfaces, but having no mucopolysaccharide layer. The nuclear membrane probably has characteristics similar to those of the cell membrane, except that very large holes, or "pores," of several hundred Ångstroms diameter are present in the membrane, so that almost all dissolved substances can move with ease between the fluids of the nucleus and those of the cytoplasm.

The Endoplasmic Reticulum. Figure 2-1 illustrates in the cytoplasm a network of tubular and vesicular structures, constructed of a system of unit membranes, called the *endoplasmic reticulum*. The detailed structure of this organelle is illustrated in Figure 2-3. The space inside the tubules and vesicles is filled with *endoplasmic matrix*, a fluid medium that is different from the fluid outside the endoplasmic reticulum.

Electron micrographs show that the space inside the endoplasmic reticulum is connected with the space between the two membranes of the double nuclear membrane. Also, in some instances the endoplasmic reticulum connects directly through small openings with the exterior of the cell. Substances formed in different parts of the cell are believed to enter the space between the two nuclear membranes and then to be conducted to other parts of the cell through the endoplasmic reticular tubules.

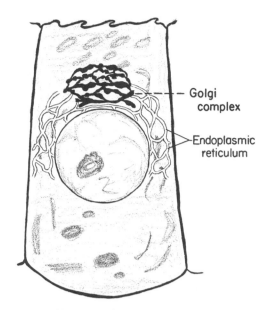

FIGURE 2-4 A typical Golgi complex.

Ribosomes. Attached to the outer surfaces of many parts of the endoplasmic reticulum are large numbers of small granular particles called ribosomes. Where these are present, the reticulum is frequently called the *granular endoplasmic reticulum*, or the *ergastoplasm*. The ribosomes are composed mainly of ribonucleic acid, which functions in the synthesis of protein in the cells, as is discussed in the following chapter.

Part of the endoplasmic reticulum has no attached ribosomes. This part is called the *agranular*, or *smooth*, *endoplasmic reticulum*. It is believed that the agranular reticulum helps in the synthesis of lipid substances and probably also acts as a medium for transporting secretory substances to the exterior of the cell, as is discussed later in the chapter.

Golgi Complex. The Golgi complex, illustrated in Figure 2-4, is probably a specialized portion of the endoplasmic reticulum. It has membranes similar to those of the agranular endoplasmic reticulum and is composed of four or more layers of thin vesicles. Electron micrographs show direct connections

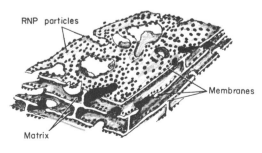

FIGURE 2-3 Structure of the membranes of the endoplasmic reticulum. (Redrawn from De Robertis, Nowinski, and Saez: Cell Biology, 4th Ed.)

between the endoplasmic reticulum and parts of the Golgi complex.

The Golgi complex is very prominent in secretory cells; in these, it is located on the side of the cell from which substances will be secreted. Its function is believed to be temporary storage of secretory substances and preparation of these substances for final secretion.

PHYSICAL NATURE OF THE CYTOPLASM

The cytoplasm is filled with both minute and large dispersed particles ranging in size from a few Ångstroms to 1 micron in diameter. The clear fluid portion of the cytoplasm in which the particles are dispersed is called *hyaloplasm*; this contains mainly dissolved proteins, electrolytes (mainly potassium ions, magnesium ions, phosphate ions, and organic ions), glucose, and small quantities of phospholipids, cholesterol, and esterified fatty acids.

The portion of the cytoplasm immediately beneath the cell membrane is frequently gelled into a semi-solid called the *cortex*, or *ectoplasm*. The cytoplasm between the cortex and the nuclear membrane is liquefied and is called the *endoplasm*.

Among the large dispersed particles in the cytoplasm are neutral fat globules, glycogen granules, ribosomes, secretory granules, and two important organelles — the *mitochondria* and *lysosomes* — which are discussed below.

Mitochondria. Mitochondria are present in the cytoplasm of all cells, as illustrated in Figure 2–1, but the number per cell varies from a few hundred to many thousand, depending on the amount of energy required by each cell to perform its functions. Mitochondria are also very variable in size and shape; some are only a few hundred millimicrons in diameter and globular in shape while others are as large as 1 micron in diameter, as long as 7 microns, and filamentous in shape.

The basic structure of a mitochondrion is illustrated in Figure 2–5, which shows it to be surrounded by a double-unit membrane that is similar in structure to the nuclear membrane. Many infoldings of the inner unit membrane form shelves on which almost all the oxidative enzymes of the cell are believed to be adsorbed. When nutrients and oxygen come in contact with these enzymes in the mitochondrion, they combine to form carbon dioxide and water, and the liberated energy is used to synthesize a substance called *adenosine triphosphate (ATP)*. The ATP then diffuses throughout the cell and releases its stored energy wherever it is needed for performing cellular functions. The function of ATP is so important to the cell that it is discussed in detail later in the chapter.

Lysosomes. Another structure recently discovered in cells is the lysosome. The lysosome is 250 to 750 millimicrons in diameter, and is surrounded by a lipoprotein unit membrane. It is filled with large numbers of small granules 55 to 80 Ångstroms in diameter which are protein aggregates of hydrolytic (digestive) enzymes. A hydrolytic enzyme is capable of splitting an organic compound into two or more parts by combining hydrogen from a water molecule with part of the compound and by combining the hydroxyl portion of the water molecule with the other part of the

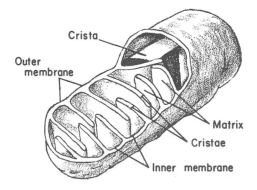

FIGURE 2–5 Structure of a mitochondrion. (Redrawn from De Robertis, Nowinski, and Saez: Cell Biology, 4th Ed.)

compound. For instance, protein is hydrolyzed to form amino acids, and glycogen is hydrolyzed to form glucose. More than a dozen different *acid hydrolases* have been found in lysosomes, and the principal substances that they digest are proteins, nucleic acids, mucopolysaccharides, and glycogen.

Ordinarily, the membrane surrounding the lysosome prevents the enclosed hydrolytic enzymes from coming in contact with other substances in the cell. However, many different conditions of the cell will break the membranes of some of the lysosomes, allowing release of the enzymes. These enzymes then split the organic substances with which they come in contact into small, highly diffusible substances, such as amino acids and glucose. Some of the more specific functions of lysosomes are discussed later in the chapter.

Microtubules. Located in many cells are fine tubular structures that are approximately 250 Å in diameter and that have lengths of 1 to many microns. These structures, called *microtubules*, are very thin in relation to their length, but they are usually arranged in bundles, which gives them, en masse, considerable structural strength. Furthermore, microtubules are usually stiff structures that break if bent too severely. Figure 2–6 illustrates some typical microtubules that have been teased from the flagellum of a sperm. Another example of microtubules is the tubular filaments that give cilia their structural strength, radiating upward from within the cell cytoplasm to the tip of the cilium.

The primary function of microtubules appears to be to act as a *cytoskeleton*, providing rigid physical structures for certain parts of cells such as the cilia just mentioned. However, the tubular nature of their structure also suggests that substances might be transported through the tubules. Indeed, cytoplasmic streaming has been observed in the vicinity of microtubules, indicating that these tubular structures might play a role in causing movement of cytoplasm.

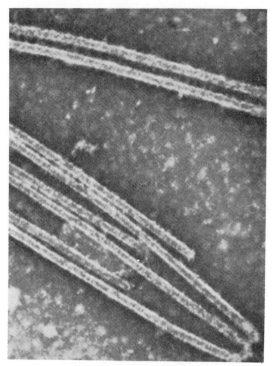

FIGURE 2–6 Microtubules teased from the flagellum of a sperm. (From Porter, K. R.: Ciba Foundation Symposium: Principles of Biomolecular Organization. Little, Brown and Co., 1966.)

Other Organelles of the Cytoplasm. The cytoplasm of each cell contains two *centrioles*, which are small cylindrical structures that play a major role in cell division, as will be discussed in Chapter 3. Also, most cells contain small *lipid droplets* and *glycogen granules* that play important roles in energy metabolism of the cell. And certain cells contain highly specialized structures such as the *cilia* of ciliated cells which are actually outgrowths from the cytoplasm, and the *myofibrils* of muscle cells. All of these are discussed in detail at different points in this text.

THE NUCLEUS

The nucleus is the control center of the cell. It controls both the chemical reactions that occur in the cell and the reproduction of the cell. Briefly, the nucleus contains large quantities of *deoxyribonucleic acid*, which we have called *genes* for many years. The genes control the characteristics of the protein enzymes of the cytoplasm, and in this way control cytoplasmic activities. To control reproduction, the genes reproduce themselves, and after this is accomplished the cell splits by a special process called *mitosis* to form two daughter cells, each of which receives one of the two sets of genes. These activities of the nucleus are considered in detail in the following chapter.

The appearance of the nucleus under the microscope does not give much of a clue to the mechanisms by which it performs its control activities. Figure 2–7 illustrates the interphase nucleus (period between mitoses), showing darkly staining, granular *chromatin material* throughout the *nuclear sap*. During mitosis, the chromatin material becomes readily identifiable as part of the highly structured *chromosomes*, which can be seen easily with a light microscope. Even during the interphase of cellular activity the granular chromatin material is still organized into chromosomal structures, but this is impossible to see except in a few types of cells.

Nucleoli. The nuclei of many cells contain one or more lightly staining structures called nucleoli. The nucleolus, unlike most of the organelles that

we have discussed, does not have a limiting membrane. Instead, it is simply an aggregate of loosely bound granules composed mainly of *ribonucleic acid*. It usually becomes considerably enlarged when a cell is actively synthesizing proteins. The genes of the chromosomes are believed to synthesize the ribonucleic acid and then to store it in the nucleolus; this ribonucleic acid later disperses from the nucleolus into the cytoplasm where it controls cytoplasmic function. The details of these mechanisms are discussed in the following chapter.

FUNCTIONAL SYSTEMS OF THE CELL

In the remainder of this chapter some of the more important functional systems of the cell are discussed. However, one of the most important functions of all, control of protein synthesis and of other cellular functions by the genes of the nucleus, is so important that the following entire chapter is devoted to it.

FUNCTIONS OF THE ENDOPLASMIC RETICULUM

The endoplasmic reticulum exists in many different forms in different cells, sometimes highly granular with a large number of ribosomes on its surface, sometimes agranular, sometimes tubular, sometimes vesicular with large shelf-like surfaces, and so forth. Therefore, simply from anatomical considerations alone, it is certain that the endoplasmic reticulum performs a very large share of the cell's functions. Yet, understanding of its functions has been extremely slow to develop because of the difficulty of studying function with the electron microscope.

On the vast surfaces of the endoplasmic reticulum are adsorbed many of the protein enzymes of the cell, and per-

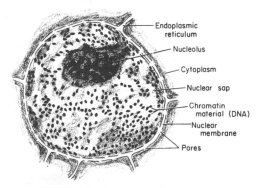

FIGURE 2–7 Structure of the nucleus.

haps still other enzymes are actually integral parts of the reticulum itself. Many of these enzymes synthesize substances in the cells, and others undoubtedly act to transport substances through the membrane of the endoplasmic reticulum from the hyaloplasm of the cytoplasm into the matrix of the reticulum or in the opposite direction. Some of the proved functions of the endoplasmic reticulum are the following:

Secretion of Proteins by Secretory Cells. Many cells, particularly cells in the various glands of the body, form special proteins that are secreted to the outside of the cells. The mechanism for this involves the endoplasmic reticulum and Golgi complex as illustrated in Figure 2–8. The ribosomes on the surface of the endoplasmic reticulum synthesize the protein that is to be secreted. This protein either is discharged directly into the tubules of the endoplasmic reticulum by the ribosomes or is immediately transported into the tubules to form small protein granules. These granules then move slowly through the tubules toward the Golgi complex, arriving there a few minutes to an hour or more later. In the Golgi complex the granules are condensed into coalesced granules that then evaginate outward through

the membrane of the Golgi complex into the cytoplasm of the cell to form *secretory granules.* Each of these granules carries with it part of the membrane of the Golgi apparatus which provides a membrane around the secretory granule and prevents it from dispersing in the cytoplasm. Gradually, the secretory granules move toward the surface of the cell where their membranes become miscible with the membrane of the cell itself, and in some way not completely understood they expel their substances to the exterior. It is in this way that protein enzymes, for instance, are secreted by the exocrine glands of the gastrointestinal tract.

Lipid Secretion. Almost exactly the same mechanism applies to lipid secretion. The one major difference is that lipids are synthesized by the agranular portion of the endoplasmic reticulum. The Golgi complex also provides much the same function in lipid secretion as in protein secretion, for here lipids are stored for long periods of time before finally being extruded into the cytoplasm as lipid droplets and thence to the exterior of the cells.

Release of Glucose from Glycogen Stores of Cells. Electron micrographs show that glycogen is stored in cells as minute granules lying in close apposition to agranular tubules of the endoplasmic reticulum. The actual chemical reactions that cause polymerization of glucose to form glycogen granules probably occur in the hyaloplasm and not in the wall of the endoplasmic reticulum, but the reticulum seems to play a transport role in bringing glucose to the site of glycogen formation and in carrying glucose away from the site when the glycogen is later broken down.

Other Possible Functions of the Endoplasmic Reticulum. The functions presented thus far for the endoplasmic reticulum have involved, first, synthesis of substances and, second, transport of these substances to the exterior of the cell. It is probable that many other sub-

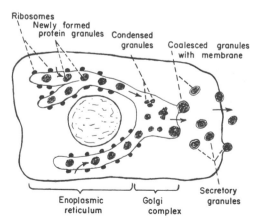

FIGURE 2–8 Function of the endoplasmic reticulum and Golgi complex in secreting proteins.

stances are synthesized by the endoplasmic reticulum and are then secreted to the interior of the cell rather than to the exterior. It is likewise probable that the Golgi complex plays a role in these internal secretory processes, for large Golgi complexes are found in some cells, such as nerve cells, that do not perform any external secretory function.

Because of the multitude of different anatomical forms of the endoplasmic reticulum, it is certain that we will discover many more functional roles that it plays besides those few that have been discussed here.

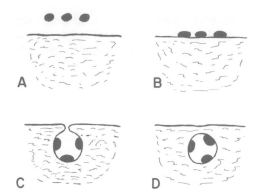

FIGURE 2–9 Mechanism of pinocytosis.

INGESTION BY THE CELL— PINOCYTOSIS

If a cell is to live and grow, it must obtain nutrients and other substances from the surrounding fluids. Substances can pass through a cell membrane in three separate ways: (1) by *diffusion* through the pores in the membrane or through the membrane matrix itself; (2) by *active transport* through the membrane, a mechanism in which enzyme systems and special carrier substances "carry" the substances through the membrane; and (3) by *pinocytosis*, a mechanism by which the membrane actually engulfs some of the extracellular fluid and its contents. Transport of substances through the membrane is such an important subject that it will be considered in detail in Chapter 4, but one of these mechanisms of transport, pinocytosis, is such a specialized cellular function that it deserves mention here as well.

Figure 2–9 illustrates the mechanism of pinocytosis. Figure 2–9*A* shows three molecules of protein in the extracellular fluid approaching the surface of the cell. In Figure 2–9*B* these molecules have become attached to the surface, presumably by the simple process of adsorption. The presence of these proteins causes the surface tension proper-

ties of the cell surface to change in such a way that the membrane invaginates as shown in Figure 2–9*C*. Immediately thereafter, the invaginated portion of the membrane breaks away from the surface of the cell, forming a *pinocytic vesicle* which then penetrates deep into the cytoplasm away from the cell membrane.

Pinocytosis occurs only in response to certain types of substances that contact the cell membrane, the two most important of which are *proteins* and *strong solutions of electrolytes*. It is especially significant that proteins cause pinocytosis, because pinocytosis is the only means by which proteins can pass through the cell membrane.

Phagocytosis. Phagocytosis means the ingestion of large particulate matter by a cell, such as the ingestion of (a) a bacterium, (b) some other cell, or (c) particles of degenerating tissue. Pinocytosis was not discovered until the advent of the electron microscope because pinocytic vesicles are smaller than the resolution of the light microscope. However, phagocytosis has been known to occur from the earliest studies using the light microscope.

The mechanism of phagocytosis is almost identical with that of pinocytosis. The particle to be phagocytized must have a surface that can become adsorbed to the cell membrane. Many bacteria have membranes that, on contact with the cell membrane, actually become miscible with the cell membrane so that the cell membrane simply spreads around the bacterium and invaginates to form a *phagocytic vesicle* containing the bacterium, a vesicle that is essentially the same as the pinocytic vesicle shown in Figure 2–9 but much larger. Phagocytosis will be discussed at further length in Chapter 8 in relation to function of the white blood cells.

THE DIGESTIVE ORGAN OF THE CELL — THE LYSOSOMES

Almost immediately after a pinocytic or phagocytic vesicle appears inside a cell, one or more lysosomes become attached to the vesicle and empty their hydrolases into the vesicle, as illustrated in Figure 2–10. Thus, a *digestive vesicle* is formed in which the hydrolases begin hydrolyzing the proteins, glycogen, nucleic acids, mucopolysaccharides, and other substances in the vesicle. The products of digestion are small molecules of amino acids, glucose, phosphates, and so forth that can then diffuse

through the membrane of the vesicle into the cytoplasm. What is left of the digestive vesicle, called the *residual body*, is then either excreted or undergoes dissolution inside the cytoplasm. Thus, the lysosomes may be called the *digestive organ* of the cells.

EXTRACTION OF ENERGY FROM NUTRIENTS

The principal nutrients from which cells extract energy are oxygen and one or more of the foodstuffs — carbohydrates, fats, and proteins. In the human body essentially all carbohydrates are converted into glucose before they reach the cell, the proteins are converted into amino acids, and the fats are converted into fatty acids. Figure 2–11 shows oxygen and the foodstuffs — glucose, fatty acids, and amino acids — all entering the cell. Inside the cell, the foodstuffs react chemically with the oxygen under the influence of various enzymes that control their rates of reactions and channel the energy that is released in the proper direction.

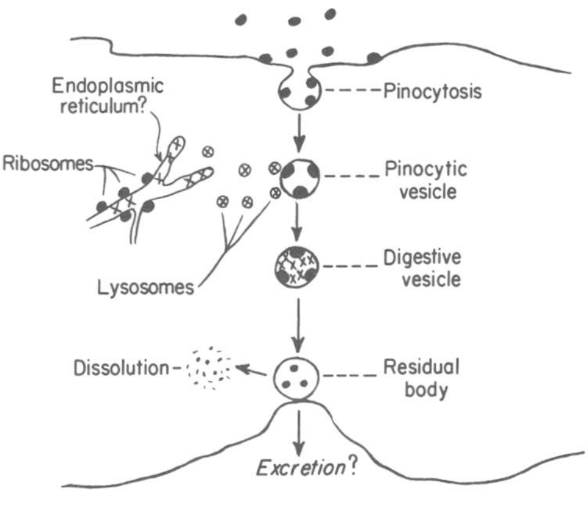

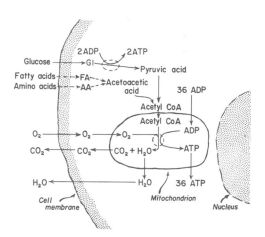

FIGURE 2–10 Digestion of substances in pinocytic vesicles by enzymes derived from lysosomes. (Modified from C. De Duve, in Lysosomes, Ed. by Reuck and Cameron, Little, Brown and Co., 1963.)

FIGURE 2–11 Formation of adenosine triphosphate in the cell, showing that most of the ATP is formed in the mitochondria.

Formation of Adenosine Triphosphate (ATP). The energy released from the nutrients is used to form adenosine triphosphate, generally called ATP, the formula for which is shown at the bottom of this page.

Note that ATP is a nucleotide composed of the nitrogenous base *adenine,* the pentose sugar *ribose,* and three *phosphate radicals.* The last two phosphate radicals are connected with the remainder of the molecule by so-called *high energy phosphate bonds,* which are represented by the symbol ~. Each of these bonds contains about 8000 calories of energy per mole of ATP under the physical conditions of the body (7000 calories under standard conditions), which is many times the energy stored in the average chemical bond, thus giving rise to the term "high energy" bond. Furthermore, the high energy phosphate bond is very labile so that it can be split instantly on demand whenever energy is required to promote other cellular reactions.

When ATP releases its energy, a phosphoric acid radical is split away, and *adenosine diphosphate (ADP)* is formed. Then, once again, energy derived from the cellular nutrients causes the ADP and phosphoric acid to recombine to form new ATP, the entire process continuing over and over again. For these reasons, ATP has been called the *energy currency* of the cell, for it can be spent and remade again and again.

By far the major portion of the ATP formed in the cell is formed in the mitochondria. The pyruvic and acetoacetic acids are both converted into the compound *acetyl co-A* in the cytoplasm, and this is transported along with oxygen through the mitochondrial membrane into the matrix of the mitochondria. Here this substance is acted upon by a series of enzymes and undergoes dissolution in a sequence of chemical reactions called the *tricarboxylic acid cycle,* or *Krebs cycle.* These chemical reactions will be explained in detail in Chapter 45.

In the tricarboxylic acid cycle, acetyl co-A is split into its component parts, hydrogen atoms and carbon dioxide. The carbon dioxide in turn diffuses out of the mitochondria and eventually out of the cell. The hydrogen atoms combine with carrier substances and are carried to the surfaces of the shelves that protrude into the mitochondria, as shown in Figure 2–5. Attached to these shelves are so-called *oxidative enzymes* which, by a series of additional reactions, cause the hydrogen atoms to combine with oxygen. The enzymes are arranged on the surfaces of the shelves in such a way that the products of one chemical reaction are immediately relayed to the next enzyme, then to the next, and so on until the complete sequence of reactions has taken place. During the course of these reactions, the energy released from the combination of hydrogen with oxygen is used to manufacture tremendous quantities of ATP from ADP. The ATP then diffuses out of the mitochondria into all parts of the cytoplasm and nucleoplasm where its energy is used to energize the functions of the cell.

Uses of ATP for Cellular Function. ATP is used to promote three major categories of cellular functions: (1) *membrane transport*, (2) *synthesis of chemical compounds* throughout the cell, and (3) *mechanical work*. These three different uses of ATP are illustrated in Figure 2–12: (a) to supply energy for the transport of glucose through the membrane, (b) to promote protein synthesis by the ribosomes, and (c) to supply the energy needed during muscle contraction.

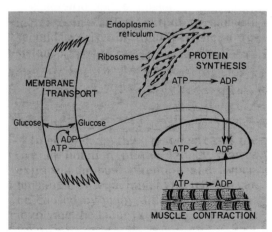

FIGURE 2–12 Use of adenosine triphosphate to provide energy for three major cellular functions: membrane transport, protein synthesis, and muscle contraction.

In addition to membrane transport of glucose, energy from ATP is required for transport of sodium ions, potassium ions, and, in certain cells, calcium ions, phosphate ions, chloride ions, uric acid, hydrogen ions, and still many other special substances. Membrane transport is so important to cellular function that some cells utilize as much as 30 per cent of the ATP formed in the cells for this purpose alone.

In addition to synthesizing proteins, cells also synthesize phospholipids, cholesterol, purines, pyrimidines, and a great host of other substances. Synthesis of almost any chemical compound requires energy. For instance, a single protein molecule might be composed of as many as several thousand amino acids attached to each other by peptide linkages, and the formation of each of these linkages requires the breakdown of an ATP molecule to ADP; thus several thousand ATP molecules must release their energy as each protein molecule is formed. Indeed, cells often utilize as much as 75 per cent of all the ATP formed in the cell simply to synthesize new chemical compounds; this is particularly true during the growth phase of cells.

The final major use of ATP is to supply energy for special cells to perform mechanical work. We shall see in Chapter 7 that each contraction of a muscle fibril requires expenditure of tremendous quantities of ATP. This is true whether the fibril is in skeletal muscle, smooth muscle, or cardiac muscle. Other cells perform mechanical work in two additional ways, by *ciliary* or *ameboid motion*, both of which will be described in the following section. The source of energy for all these types of mechanical work is ATP.

In summary, therefore, ATP is always available to release its energy rapidly and almost explosively wherever in the cell it is needed. To replace the ATP used by the cell, other but much slower chemical reactions break down carbohydrates, fats, and proteins and use the energy derived from these to form new ATP.

CELL MOVEMENT

By far the most important type of cell movement that occurs in the body is that of the specialized muscle cells in skeletal, cardiac, and smooth muscle, which comprise almost 50 per cent of the entire body mass. The specialized functions of these cells will be discussed in Chapter 7. However, two other types of movement occur in other cells, *ameboid movement and ciliary movement.*

Ameboid Motion. Ameboid motion means movement of an entire cell in relation to its surroundings, such as the movement of white blood cells through tissues. Typically ameboid motion begins with protrusion of a pseudopodium from one end of the cell. The pseudopodium projects far out, away from the cell body; and then the remainder of the cell moves toward the pseudopodium. Formerly it was believed that the protruding pseudopodium attached itself far away from the cell and then pulled the remainder of the cell toward it. However in recent years, new studies have changed this idea to a "streaming" concept, as follows: Figure 2–13 illustrates diagrammatically an elongated cell, the right-hand end of which is a protruding pseudopodium. The membrane of this end of the cell is continually expanding forward while the membrane at the left-hand end of the cell is continually con-tracting and following along as the cell moves. It has been postulated that this movement is caused by "streaming" of the endoplasmic portion of the cytoplasm through the center of the cell from left to right. On reaching the right-hand end, the endoplasm turns toward the outside wall of the cell, becomes gelled, and therefore becomes ectoplasm that is fixed in position. Simultaneously, at the back end of the cell, the gelated ectoplasm becomes dissoluted and streams forward as endoplasm to make the pseudopodium move forward still farther. The continuous repetition of this process presumably makes the cell move in the direction in which the pseudopodium projects. One can readily see that this streaming movement inside the cell is analogous to the revolving movement of the track of a Caterpillar tractor.

Control of ameboid motion— "chemotaxis." At the onset of ameboid motion, an electrochemical change occurs at the cell membrane and causes an *action potential* similar to that which occurs in nerves for the transmission of nerve impulses, as explained in Chapter 5. It is presumably this action potential that initiates the forward projection of the pseudopodium.

The most important factor that usually initiates ameboid motion, presumably by causing action potentials in the cell membrane, is the appearance of certain chemical substances in the tissues. This phenomenon is called *chemotaxis*, and the chemical substance causing it to occur is called a *chemotaxic substance.* Most ameboid cells move toward the source of the chemotaxic substance—that is, from an area of lower concentration toward an area of higher concentration—which is called *positive chemotaxis.* However, some cells move away from the source, which is called *negative chemotaxis.*

Movement of Cilia. A second type of cellular motion, *ciliary movement*, occurs along the surface of cells in the respiratory tract and in some portions of the reproductive tract. As illustrated

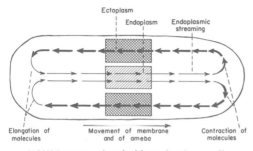

FIGURE 2–13 Ameboid motion by a cell.

in Figure 2–14, a cilium looks like a minute, sharp-pointed hair that projects 3 to 4 microns from the surface of the cell. Many cilia can project from a single cell.

The cilium is covered by an outcropping of the cell membrane, and it is supported by 11 microtubular filaments, 9 double tubular filaments located around the periphery of the cilium and 2 single tubular filaments down the center, as shown in the cross-section illustrated in Figure 2–14. Each cilium is an outgrowth of a structure that lies immediately beneath the cell membrane called the *basal body* of the cilium.

In the inset of Figure 2–14 movement of the cilium is illustrated. The cilium

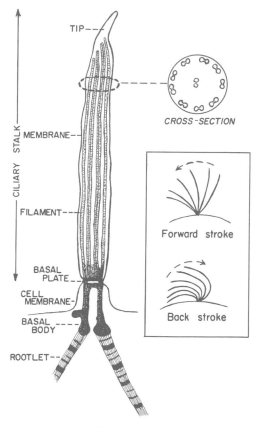

FIGURE 2–14 Structure and function of the cilium. (Modified from Satir, *Sci. Amer., 204*[2]: 108, 1961.)

moves forward with a sudden rapid stroke, bending sharply where it projects from the surface of the cell. Then it moves backward very slowly in a whiplike manner. The rapid forward movement pushes the fluid lying adjacent to the cell in the direction that the cilium moves, then the slow whiplike movement in the other direction has almost no effect on the fluid. As a result, fluid is continually propelled in the direction of the forward stroke. Since most ciliated cells have large numbers of cilia on their surfaces, and since ciliated cells on a surface are oriented in the same direction, this is a very satisfactory means for moving fluids from one part of the surface to another; for instance, for moving mucus out of the lungs or for moving the ovum along the fallopian tube.

Mechanism of ciliary movement. The mechanism of ciliary movement is unknown, though we do know some facts about it. First, the movement requires energy supplied by ATP. Second, an electrochemical impulse similar to that transmitted over nerves is transmitted over the membrane of the cilium to produce the movement. Third, the two filaments in the center of the cilium are necessary for movement to occur.

The filaments of the cilium are believed to be nothing more than elastic cytoskeletal microtubules that give the cilium its physical rigidity. These filaments are not known to have the capability of contracting, even though various theories have postulated that contraction occurs. It is more likely that some cytoplasmic element surrounding the filaments causes the forward movement of the cilium, while the filaments, by virtue of their elastic rigidity, cause the slower whiplike return movement of the cilium. It is possible that the tubular structure of the filaments plays a role in transporting necessary substances into the cilium for the contractile process to take place.

REFERENCES

Bourne, G. H.: Cytology and Cell Physiology. 3rd ed., New York, Academic Press, Inc., 1964.

De Robertis, E. D. P., Nowinski, W. W., and Saez, F. A.: Cell Biology. 4th ed., Philadelphia, W. B. Saunders Company, 1965.

Fawcett, D. W.: The Cell. Philadelphia, W. B. Saunders Company, 1966.

Giese, A. C.: Cell Physiology. 3rd ed., Philadelphia, W. B. Saunders Company, 1968.

Hayashi, T., and Szent-Gyorgy: Molecular Architecture in Cell Physiology. Englewood Cliffs, New Jersey, Prentice-Hall, Inc., 1966.

Porter, K. R., and Franzini-Armstrong, C.: The sarcoplasmic reticulum. *Sci. Amer.*, *212*:72(3), 1965.

Threadgold, L. T.: The Ultrastructure of the Animal Cell. Oxford, England, Pergamon Press, Inc., 1967.

CHAPTER 3

GENETIC CONTROL OF PROTEIN SYNTHESIS, CELL FUNCTION, AND CELL REPRODUCTION

Almost everyone knows that the genes control heredity from parents to children, but most persons do not realize that the same genes control the day-by-day function of all cells. The genes control function of the cell by determining what substances will be synthesized within the cell—what structures, what enzymes, what chemicals. Figure 3–1 illustrates the general schema by which the genes perform this function. The gene, which is a nucleic acid called *deoxyribonucleic acid* (*DNA*), automatically controls the formation of another nucleic acid, *ribonucleic acid* (*RNA*), which spreads throughout the cell and controls the formation of the different proteins. Some of these proteins are *structural proteins*, which, in association with various lipids, form the structures of the various organelles that were discussed in the preceding chapter. But by far the majority of the proteins are *enzymes* that promote the different chemical reactions in the cells. For instance, enzymes promote all the oxidative reactions that supply energy to the cell, and they promote the synthesis of various chemicals such as lipids, glycogen, adenosine triphosphate, and so forth.

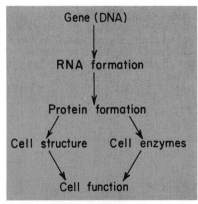

FIGURE 3–1 General schema by which the genes control cell function.

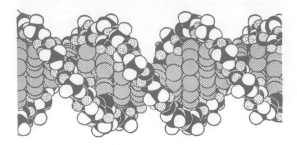

FIGURE 3–2 The helical, double-stranded structure of the gene. The outside strands are composed of phosphoric acid and the sugar deoxyribose. The internal molecules connecting the two strands of the helix are purine and pyrimidine bases; these determine the "code" of the gene.

THE GENE

The gene is a long, double-stranded, helical molecule of *deoxyribonucleic acid (DNA)* having a molecular weight usually measured in the millions. A very short segment of such a molecule is illustrated in Figure 3–2. This molecule is composed of several simple chemical compounds arranged in a regular pattern explained in the following few paragraphs.

The Basic Building Blocks of DNA. Figure 3–3 illustrates the basic chemical compounds involved in the formation of DNA. These include *phosphoric acid*, a sugar called *deoxyribose*, and four nitrogenous bases (two purines, *adenine* and *guanine*, and two pyrimidines, *thymine* and *cytosine*). The phosphoric acid and deoxyribose form the two helical strands of DNA, and the bases lie between the strands and connect them.

The Nucleotides. The first stage in the formation of DNA is the combination of one molecule of phosphoric acid, one molecule of deoxyribose, and one of the four bases to form a nucleotide. Four separate nucleotides are thus formed: *adenylic, thymidylic, guanylic,* and *cytidylic acids.* Figure 3–4*A* illustrates the chemical structure of adenylic acid, and Figure 3–4*B* illustrates simple symbols for all the four basic nucleotides that form DNA.

Note also in Figure 3–4*B* that the nucleotides are separated into two pairs. Adenylic acid and thymidylic acid form one pair. Guanylic acid and cytidylic

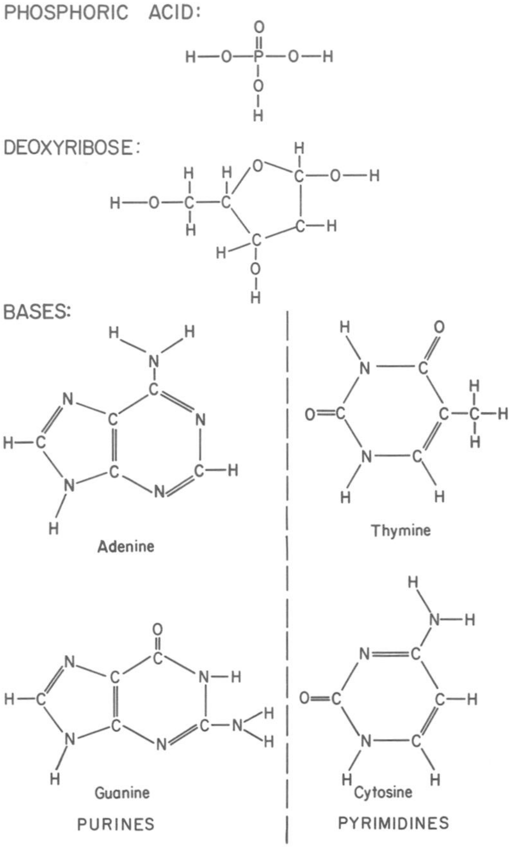

PHOSPHORIC ACID:

DEOXYRIBOSE:

BASES:

Adenine

Thymine

Guanine

Cytosine

PURINES

PYRIMIDINES

FIGURE 3–3 The basic building blocks of DNA.

acid form the other pair. The pairs of nucleotides provide the means by which the two strands of the DNA helix are bound together: one nucleotide of a pair is on one strand of DNA, the other nucleotide of that pair is in a corresponding position on the other strand, and these are bound together by loose and reversible bonds.

A. NUCLEOTIDE

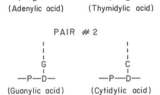

Adenylic acid

B. THE FOUR BASIC NUCLEOTIDES OF DNA

PAIR #1

A   T
—P—D—   —P—D—
(Adenylic acid)   (Thymidylic acid)

PAIR #2

G   C
—P—D—   —P—D—
(Guanylic acid)   (Cytidylic acid)

FIGURE 3–4 Combinations of the basic building blocks of DNA to form nucleotides. (*A*, adenine; *C*, cytosine; *D*, deoxyribose; *G*, guanine; *P*, phosphoric acid; *T*, thymine.)

Organization of the Nucleotides to Form DNA. Figure 3–5 illustrates the manner in which multiple numbers of nucleotides are bound together to form DNA. Note that these are combined in such a way that phosphoric acid and deoxyribose alternate with each other in the two separate strands, and these strands are held together by the respective pairs of bases. Thus, in Figure 3–5

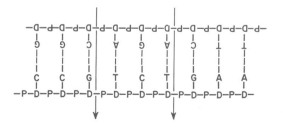

FIGURE 3–5 Combination of deoxyribose nucleotides to form DNA.

the respective pairs of bases are CG, CG, GC, TA, CG, TA, GC, AT, and AT. However, the base pairs are bound together by very loose chemical bonds, represented in the figure by dashed lines. Because of the looseness of these bonds, the two strands can pull apart with ease, and they do so many times during the course of their function in the cell.

Now, to put the DNA of Figure 3–5 into its proper physical perspective, one needs merely to pick up the two ends and twist them into a helix. Ten pairs of nucleotides are present in each full turn of the helix in the DNA molecule as illustrated in Figure 3–2.

THE GENETIC CODE

The importance of the gene lies in its ability to control the formation of other substances in the cell. It does this by means of the so-called genetic code. When the two strands of a DNA molecule are split apart, this exposes a succession of purine and pyrimidine bases projecting to the side of each strand. It is these projecting bases that form the code.

Research studies in the past few years have demonstrated that the so-called *code words* consist of "triplets" of bases —that is, each three successive bases are a code word. And the successive code words control the sequence of amino acids in a protein molecule to be synthesized in the cell. Note in Figure 3–5 that each of the two strands of the DNA molecule carries its own genetic code. For instance, the top strand, reading from left to right, has the genetic code GGC, AGA, CTT, the code words being separated from each other by the arrows. As we follow this genetic code through Figures 3–6 and 3–7, we shall see that the code words are responsible for placement of the three amino acids, *proline*, *serine*, and *glutamic acid*, in a molecule of protein. Furthermore, these three amino acids will be lined up in the protein molecule in exactly the same way that the genetic code is lined up in this strand of DNA.

It is also important that some code words do not cause amino acids to incorporate into proteins but, instead, perform other functions in the synthesis of protein molecules, such as initiating and stopping the formation of a protein molecule. For instance, some DNA molecules of viruses have molecular weights of 40,000,000 or more and cause the formation of many more than a single protein molecule — usually 8, 10, or 20 molecules. It is the code words that do not cause incorporation of amino acids into a protein molecule that signal when to stop the formation of one protein and to start the formation of another.

RIBONUCLEIC ACID (RNA)

Since almost all DNA is located in the nucleus of the cell and yet most of the functions of the cell are carried out in the cytoplasm, some means must be available for the genes of the nucleus to control the chemical reactions of the cytoplasm. This is done through the intermediary of another type of nucleic acid, ribonucleic acid (RNA), the formation of which is controlled by the DNA of the nucleus. The RNA is then transported into the cytoplasmic cavity where it controls protein synthesis.

At least three separate types of RNA are important to protein synthesis: *messenger RNA*, *transfer RNA*, and *ribosomal RNA*. Before we describe the functions of messenger and transfer RNA's in the synthesis of proteins, let us see how DNA controls the formation of RNA.

Synthesis of RNA. One strand of the DNA molecule, which constitutes the gene, acts as a template for synthesis of each type of RNA molecule. (The other strand of the DNA has no genetic function but does function for replication of the gene itself, which will be discussed later in the chapter.) The code words in DNA cause the formation of *comple-*

mentary code words (or *codons*) in RNA. The stages of RNA synthesis are as follows:

The basic building blocks of RNA. The basic building blocks of RNA are almost the same as those of DNA except for two differences. First, the sugar deoxyribose is not used in the formation of RNA. In its place is another sugar of very slightly different composition, *ribose*. The second difference in the basic building blocks is that thymine is replaced by another pyrimidine, *uracil*.

Formation of RNA Nucleotides. The basic building blocks of RNA first form nucleotides exactly as described above for the synthesis of DNA. Here again, four separate nucleotides are used in the formation of RNA. These nucleotides contain the bases *adenine*, *guanine*, *cytosine*, and *uracil*, respectively, the uracil replacing the thymine found in the four nucleotides that make up DNA. Also, uracil takes the place of thymine in pairing with the purine adenine, as we shall see in the following paragraphs.

Activation of the Nucleotides. The next step in the synthesis of RNA is activation of the nucleotides. This occurs by addition to each nucleotide of two phosphate radicals to form triphosphates of the nucleotides. These last two phosphates are combined with the nucleotide by *high energy phosphate bonds* derived from adenosine triphosphate in the cell.

The result of this activation process is that large quantities of energy are made available to each of the nucleotides, and it is this energy that is used in promoting the chemical reactions that eventuate in the formation of the RNA chain.

Combination of the Activated Nucleotides with the DNA Strand. The next stage in the formation of RNA is the splitting apart of the two strands of the DNA molecule. Then, activated nucleotides become attached to the bases on the DNA strand that is the gene, as illustrated by the top panel in Figure 3–6. Note that a ribose nucleotide base

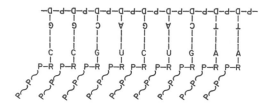

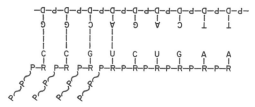

FIGURE 3–6 Combination of ribose nucleotides with a strand of DNA to form a molecule of ribonucleic acid (RNA) that carries the DNA code from the gene to the cytoplasm.

always combines with the deoxyribose bases in the following combinations:

DNA base	RNA base
guanine	cytosine
cytosine	guanine
adenine	uracil
thymine	adenine

Polymerization of the RNA Chain. Once the ribose nucleotides have combined with the DNA strand as shown in the top panel of Figure 3–6, they are then lined up in proper sequence to form code words (codons) that are complementary to those in the DNA molecule. At this time an enzyme called *RNA polymerase* causes the two extra phosphates on each nucleotide to split away and, at the same time, to liberate enough energy to cause bonds to form between the successive ribose and phosphoric acid radicals. As this happens, the RNA strand automatically separates from the DNA strand and becomes a free molecule of RNA. This bonding of the ribose and phosphoric acid radicals and the simultaneous splitting of the RNA from the DNA is illustrated in the lower part of Figure 3–6.

Once the RNA molecules are formed, they are transported, by means yet unknown, into all parts of the cytoplasm where they perform further functions.

MESSENGER RNA—THE PROCESS OF TRANSCRIPTION

The type of RNA that carries the genetic code to the cytoplasm for formation of proteins is called messenger RNA. Molecules of messenger RNA are usually composed of several hundred to several thousand nucleotides in a single strand containing codons that are exactly complementary to the code words of the genes. Figure 3–7 illustrates a small segment of a molecule of messenger RNA. Its codons are CCG, UCU, and GAA. These are the codons for proline, serine, and glutamic acid. If we now refer back to Figure 3–6, we see that the events recorded in this figure represent transfer of this particular genetic code from the DNA strand to the RNA strand. This process of transferring the genetic code from DNA to messenger RNA is called *transcription.*

Messenger RNA molecules are long straight strands that are suspended in the cytoplasm. These molecules migrate to the ribosomes where protein molecules are manufactured, which is explained below.

RNA Codons. Table 3–1 gives the RNA codons for the 20 common amino acids found in protein molecules. Note that several of the amino acids are represented by more than one codon; and, as was pointed out before, some codons represent such signals as "start manufacturing a protein molecule" or "stop

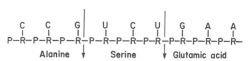

FIGURE 3–7 Portion of a ribonucleic acid molecule, showing three "code" words, CCG, UCU, and GAA, which represent the three amino acids *proline, serine, and glutamic acid.*

TABLE 3–1 RNA CODONS FOR THE DIFFERENT AMINO ACIDS AND FOR START AND STOP

Amino Acid	RNA Codons				
Alanine	GCU	GCC	GCA	GCG	
Arginine	CGU	CGC	CGA	CGG	AGA AGG
Asparagine	AAU	AAC			
Aspartic acid	GAU	GAC			
Cysteine	UGU	UGC			
Glutamic acid	GAA	GAG			
Glutamine	CAA	CAG			
Glycine	GGU	GGC	GGA	GGG	
Histidine	CAU	CAC			
Isoleucine	AUU	AUC	AUA		
Leucine	CUU	CUC	CUA	CUG	
Lysine	AAA	AAG			
Methionine	AUG				
Phenylalanine	UUU	UUC			
Proline	CCU	CCC	CCA	CCG	
Serine	UCU	UCC	UCA	UCG	
Threonine	ACU	ACC	ACA	ACG	
Tryptophan	UGG				
Tyrosine			UAU	UAC	
Valine	GUU	GUC	GUA	GUG	
Start (CI)	AUG	GUG			
Stop (CT)	UAA	UAG	UGA		

manufacturing a protein molecule." In Table 3–1, these two codons are designated CI for "chain-initiating" and CT for "chain-terminating."

TRANSFER RNA

Another type of RNA that plays a prominent role in protein synthesis is called transfer RNA (also called soluble RNA) because it transfers amino acid molecules to protein molecules as the protein is synthesized. There is a sepa-

rate type of transfer RNA for each of the 20 amino acids that are incorporated into proteins. Furthermore, each type of transfer RNA combines with only one type of amino acid—one and no more. Transfer RNA then acts as a *carrier* to transport its specific type of amino acid to the ribosomes where protein molecules are formed. In the ribosomes, each specific type of transfer RNA recognizes a particular code word on the messenger RNA, as is described below, and thereby delivers the appropriate amino acid to the appropriate place in the chain of the newly forming protein molecule.

Transfer RNA, containing only about 75 nucleotides, is a relatively small molecule in comparison with messenger RNA. It is a folded chain of nucleotides believed to have a cloverleaf appearance similar to that illustrated in Figure 3–8. At one end of the molecule is always an adenylic acid; it is to this that the transported amino acid attaches to a hydroxyl group of the ribose in the adenylic acid. A specific enzyme causes this attachment for each specific type of transfer RNA; this enzyme also determines the type of amino acid that will attach to the respective type of transfer RNA.

Since the function of transfer RNA is to cause attachment of a specific amino acid to a forming protein chain, it is essential that each type of transfer RNA also has specificity for a particular

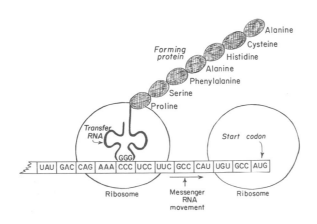

FIGURE 3–8 Postulated mechanism by which a protein molecule is formed in ribosomes in association with messenger RNA and soluble RNA.

codon in messenger RNA. The specific prosthetic group in the transfer RNA that allows it to recognize a specific codon is called an *anticodon*, and this is believed to be located approximately in the middle of the transfer RNA molecule (at the bottom of the cloverleaf configuration illustrated in Fig. 3–8). During formation of a protein molecule, the anticodon bases combine loosely by hydrogen bonding with the codon bases of the messenger RNA. In this way the respective amino acids are lined up one after another along the messenger RNA chain, thus establishing the appropriate sequence of amino acids in the protein molecule.

RIBOSOMAL RNA

The third type of RNA in the cell is that found in the ribosomes; it constitutes between 40 and 50 per cent of the ribosome. The remainder of the ribosome is protein.

The function of ribosomal RNA is mainly unknown, but it is known that it exists in particles of several different sizes. The transfer RNA, along with its attached amino acid, first binds with one of the ribosomal particles. Then as the messenger RNA passes across or through the ribosome, the amino acid is released to the forming protein while the transfer RNA is released back into the cytoplasm to combine again with another molecule of amino acid of the same type. Thus the ribosome acts as a physical structure upon which the protein is formed, and its RNA is in some way necessary for this process to take place.

FORMATION OF PROTEINS ON THE RIBOSOMES — THE PROCESS OF TRANSLATION

When a molecule of messenger RNA comes in contact with a ribosome, it travels through or across the surface of the ribosome beginning at a predetermined end specified by one of the "start" (or "chain-initiating") codons. Then, as illustrated in Figure 3–8, while the messenger RNA travels past the ribosome, a protein molecule is formed. The exact events in the formation of this protein molecule are still a mystery, though it is presumed that the ribosome reads the code of the messenger RNA in much the same way that a tape is "read" as it passes through the playback head of a tape recorder, a process called *translation*. As each codon slips past the ribosome, the specific transfer RNA corresponding to each specific codon supposedly attaches to the codon of the messenger RNA, and the ribosome in some unexplained way causes the successive amino acids to combine with each other by the process of peptide linkage. At the same time, the transfer RNA is freed from the amino acid, and returns to the cytoplasm to pick up another amino acid of its specific type.

Thus, as the messenger RNA passes through the ribosome, a protein molecule is formed as shown in Figure 3–8. Then, when a "stop" (or "chain-terminating") codon slips past the ribosome, the end of a protein molecule is signaled, and the entire molecule is freed into the cytoplasm.

It is especially important that a messenger RNA can cause the formation of a protein molecule in any ribosome, and that there is no specificity of ribosomes for given types of protein. The ribosome seems to be simply the physical structure in which or on which chemical reactions take place.

SYNTHESIS OF OTHER SUBSTANCES IN THE CELL

Many hundred or perhaps a thousand or more protein enzymes formed in the manner just described control essentially all the other chemical reactions that take place in cells. These enzymes promote synthesis of lipids, glycogen, purines, pyrimidines, and hundreds of other substances. We will discuss many of these synthetic processes in relation to carbohydrate, lipid, and protein metabolism in Chapters 45 and 46. It is by means of all these different substances that the many functions of the cells are performed.

CONTROL OF GENETIC FUNCTION AND BIOCHEMICAL ACTIVITY IN CELLS

There are basically two different methods by which the biochemical activities in the cell are controlled. One of these can be called *genetic regulation*, in which the activities of the genes themselves are controlled, and the other can be called *enzyme regulation*, in which the rates of activities of the enzymes within the cell are controlled.

GENETIC REGULATION

Figure 3–9 illustrates the general plan by which function of the genes is regulated. At the top of the figure is a *regulatory gene* that has the ability to regulate the degree of activity of other genes. It does this by controlling the formation of

a *repressor substance*, a compound of small molecular weight, which in turn represses the activities of other genes. The repressor substance does not act directly on these other genes but instead on a small part of the DNA helix called an *operator*, which lies adjacent to the genes that are to be controlled. The genetic operator excites the genes, but when the operator is repressed by the repressor substance, the genes become inactive.

The genes controlled by the operator are called *structural genes*; it is they that cause the eventual formation of enzymes in the cell. The group of structural genes under the influence of each genetic operator is known as an *operon*. In Figure 3–9, the structural genes initiate the formation of enzymes A, B, and C, which in turn control biochemical reactions in the cell leading to the formation of some specific chemical product.

As noted in Figure 3–9, the repressor substance from the regulatory gene can be either *inhibited* or *activated*. If inhibited, the genetic operator is no longer repressed and therefore becomes very active. As a result the biochemical synthesis proceeds unabated. On the other hand, if the repressor substance is activated, it immediately represses the genetic operator, and the events leading to the synthesized product cease within minutes or hours.

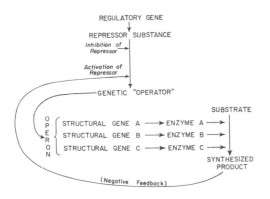

FIGURE 3–9 Control system that regulates function of genes. See text for detailed explanation.

Control of Cellular Constituents by Feedback Repression of the Genes. The importance of the genetic regulatory system of Figure 3–9 is that it allows the concentrations of different cellular substances to be controlled. The mechanism of this is the following: The synthesized product often can activate the repressor substance, and the repressor substance then inactivates the genetic operator, leading to decrease or cessation of production of the enzymes required for forming the synthesized product. Thus, negative feedback occurs in which the concentration of the synthesized product controls the product's rate of synthesis. When its concentration becomes too great, the rate of synthesis falls; when its concentration falls too low, its rate of synthesis increases.

Genetic control systems of this type are especially important in controlling the intracellular concentrations of amino acids, amino acid derivatives, and some of the intermediary substances of carbohydrate, lipid, and protein metabolism.

Induction of Enzyme Formation. Note in Figure 3–9 that the repressor substance can be inhibited as well as activated. When it is inhibited, the repressor no longer holds the operon in abeyance, and the structural genes operate to their full capacity in forming enzymes. Therefore, a simple way to "induce" the formation of a particular group of enzymes is to inhibit the repressor. This is probably one of the means by which hormones act to increase or decrease specific cellular activities. For instance, the steroid hormone, aldosterone, increases the quantities of enzymes required for transport of sodium through membranes, and the steroid sex hormones induce the formation of enzymes required for protein synthesis in the sex organs. These effects will be discussed later at greater length in relation to the functions of these hormones.

Induction also occurs in response to substances inside the cell itself. For instance, cells grown in tissue culture in the absence of lactose do not have the ability to utilize lactose when first exposed to it. However, within a few hours after exposure, because of inhibition of the appropriate repressor substance, the enzyme required for splitting lactose into its monosaccharides appears in the cell. Thus, lactose induces a specific enzyme synthesis and, in so doing, makes it possible for the cell to utilize the lactose.

CONTROL OF ENZYME ACTIVITY

In the same way that inhibitors and activators can affect the genetic regulatory system, so also can the enzymes themselves be directly controlled by other inhibitors or activators. This, then, represents a second mechanism by which cellular biochemical functions can be controlled.

Enzyme Inhibition. A great many of the chemical substances formed in the cell have a direct feedback effect on the respective enzyme systems that synthesize them. Thus, in Figure 3–9, the synthesized product is shown to act directly back on enzyme A to inactivate it. If enzyme A is inactivated, then none of the substrate will begin to be converted into the synthesized product. Almost always in enzyme inhibition, the synthesized product acts on the first enzyme in a sequence, rather than on the subsequent enzymes. One can readily recognize the importance of inactivating this first enzyme; it prevents buildup of intermediary products that will not be utilized.

This process of enzyme inhibition is another example of negative feedback control; it is responsible for controlling the intracellular concentrations of some of the amino acids not controlled by the genetic mechanism, of the purines, the pyrimidines, vitamins, and other substances.

Enzyme Activation. Enzymes that are normally inactive or those that have been inactivated by some inhibitor substance can often be activated. An example of this is the action of cyclic adenosine monophosphate (AMP) in causing glycogen to split so that the released glucose molecules can be used to form high energy ATP, as discussed in the previous chapter. When most of the adenosine triphosphate has been depleted in the cell, a considerable amount of cyclic AMP begins to be formed as a breakdown product of the ATP; the presence of cyclic AMP in the cell indicates that the reserves of ATP have approached a low ebb. The cyclic AMP immediately activates the glycogen-splitting enzyme phosphorylase, liberating glucose molecules that are rapidly used for replenishment of the ATP stores. Thus, in this case the cyclic AMP acts as an enzyme activator and thereby helps to control intracellular ATP concentration.

Another interesting instance of both enzyme inhibition and enzyme activation occurs in the formation of the purines and the pyrimidines. These substances are needed by the cell in approximately equal quantities for formation of DNA and RNA. When purines are formed, they inhibit the enzymes that are required for formation of additional purines, but they activate the enzymes for formation of the pyrimidines. Conversely, the pyrimidines inhibit their own enzymes but activate the purine enzymes. In this way there is continual cross-feed between the synthesizing systems for these two substances, resulting in almost exactly equal amounts cf the two substances in the cells at all times.

In summary, there are two different methods by which the cells control proper proportions and proper quantities of different cellular constituents: the mechanism of genetic regulation and the mechanism of enzyme regulation. The genes can be either activated or inhibited, and, likewise, the enzymes can be either activated or inhibited. Furthermore, it is principally the substances formed in the cells that cause activation or inhibition; but on occasion, substances from without the cell (especially some of the hormones which will be discussed later in this text) also control the intracellular biochemical reactions.

CELL REPRODUCTION

Most cells are continually growing and reproducing. The new cells take the place of the old ones that die, thus maintaining a complete complement of cells in the body.

Also, one can remove most types of cells from the human body and grow them in tissue culture where they will continue to grow and reproduce so long as appropriate nutrients are supplied and so long as the end-products of the cells' metabolism are not allowed to accumulate in the nutrient medium. Thus, the life lineage of cells is indefinite.

As is true of almost all other events in the cell, reproduction also begins in the nucleus itself. The first step is *replication (duplication) of all genes and of all chromosomes*. The next step is division of the two sets of genes between two separate nuclei. And the final step is splitting of the cell itself to form two new daughter cells, a process called *mitosis*.

The complete life cycle of a cell that is not inhibited in some way is about 10 to 30 hours from reproduction to reproduction, and the period of mitosis lasts for approximately one-half hour. The period between mitoses is called *interphase*. However, in the body there are almost always inhibitory controls that slow or stop the uninhibited life cycle of the cell and give cells life cycle periods that vary from as little as 10 hours for stimulated bone marrow cells to an entire lifetime of the human body for nerve cells.

REPLICATION OF THE GENES

The genes are reproduced several hours before mitosis takes place, and the duration of gene replication is only about one hour. When replication begins, it occurs for all genes at the same time and not for only part of them. Furthermore, the genes are duplicated only once. The net result is two exact duplicates of all genes, which respectively become the genes in the two new daughter cells that will be formed at mitosis. Following replication of the genes, the nucleus continues to function normally for several hours before mitosis begins abruptly.

Chemical and Physical Events. The genes are duplicated in almost exactly the same way that RNA is formed from DNA. First, the two strands of the DNA helix of the gene pull apart. Second, each of these strands combines with deoxyribose nucleotides of the four types described early in the chapter as the basic building blocks of DNA. Each of the bases on each strand of DNA in the chain attracts a nucleotide containing the appropriate *complementary* base. In this way, the appropriate nucleotides are lined up side by side. Third, appropriate enzyme mechanisms then provide energy and cause polymerization of the nucleotides to form a new DNA strand. The only difference between this formation of the new strand of DNA and the formation of an RNA strand is that the new strand of DNA remains attached to the old strand that has formed it, thus forming a new double-stranded DNA helix. At the same time the other strand of the original helix forms its complementary DNA strand and thereby also forms a new double-stranded helix.

THE CHROMOSOMES AND THEIR REPLICATION

Unfortunately, we know much less about replication of the chromosomes than about replication of the genes themselves. The primary reason for this is that the structure of the chromosomes itself is almost unknown.

The chromosomes consist of two major parts: the genes, consisting of DNA, and protein. The DNA is bound loosely with the protein and sometimes during the life cycle of the cell seems to become separated from the protein. The combination of the two is known as a *nucleoprotein.*

Recent experiments indicate that all the DNA of a particular chromosome is arranged in one long double helix and that the genes are attached end-on-end with each other. Such a molecule, if spread out linearly, would be approximately 2 cm. long, or several hundred times as long as the diameter of the nucleus itself; but the experiments also indicate that this long double helix is folded like a spring and is held in this folded position by its linkages to protein molecules. The protein has nothing to do with the genetic potency of the chromosome, and the protein molecules can be replaced by new protein molecules without any alteration in the functions of the genes.

Replication of the chromosomes follows as a natural result of replication of the DNA strand. When the new double helix separates from the original double helix, it presumably carries some of the old protein with it or combines with new protein, the DNA acting as the backbone of the newly replicated chromosome and the protein acting only as an accessory to the chromosomal structure.

Number of Chromosomes in the Human Cell. Each human cell contains 46 chromosomes arranged in 23 pairs. In general, the genes in the two chromosomes of each pair are almost identical with each other, so that it is usually stated that the different genes exist in pairs, though occasionally this is not the case. Also, most genes have several duplicates even on the same chromosome, so that loss of any one usually will not completely remove a particular characteristic from a cell.

MITOSIS

The actual process by which the cell splits into two new cells is called mitosis. Once the genes have been duplicated and each chromosome has split to form two new chromosomes, each of which is now called a *chromatid*, mitosis follows automatically, almost without fail, within a few hours.

Replication of the Centrioles. The first event of mitosis takes place in the cytoplasm, occurring during the latter part of interphase in the small structures called *centrioles*. As illustrated in Figure 3–10, two pairs of centrioles lie close to each other near one pole of the nucleus. Each centriole is a small cylindrical body about 0.4 micron long and about 0.15 micron in diameter, consisting mainly of nine parallel, tubular-like structures arranged around the inner wall of the

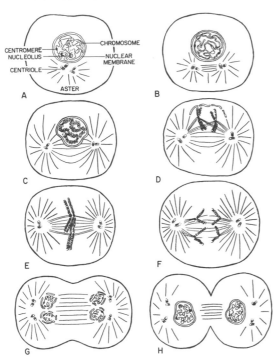

A

B

C

D

E

F

G

H

FIGURE 3–10 Stages in the reproduction of the cell. *A* and *B*, late interphase; *C* and *D*, prophase; *E*, metaphase; *F*, anaphase; *G* and *H*, telophase. (Redrawn from Mazia: *Sci. Amer., 205*[3]:102, 1961.)

cylinder. The two centrioles of each pair lie at right angles to each other.

In so far as is known, the two pairs of centrioles remain dormant during interphase until shortly before mitosis is to take place. At that time, the two pairs begin to move apart from each other. This is caused by protein microtubules growing between the respective pairs and actually pushing them apart. At the same time, microtubules grow radially away from each of the pairs. Some of these penetrate the nucleus. The set of microtubules connecting the two centriole pairs is called the *spindle*, and the entire set of microtubules plus the two centrioles is called the *mitotic apparatus*.

While the spindle is forming, the *chromatin material* of the nucleus (the DNA) becomes condensed into well-defined but disoriented chromosomes, as shown in Figures 3–10A and B.

Prophase. The first stage of mitosis, called prophase, is shown in Figures 3–10C and D. The nuclear envelope dissolutes, and some of the microtubules from the forming mitotic apparatus become attached to the chromosomes. This attachment always occurs at the same point on each chromosome, at a small condensed portion called the *centromere*.

Metaphase. During metaphase (Fig. 3–10E) the centriole pairs are pushed far apart by the growing spindle, and the chromosomes are thereby pulled tightly by the attached microtubules to the very center of the cell, lining up in the equatorial plane of the mitotic spindle.

Anaphase. With still further growth of the spindle, each pair of chromosomes is now broken apart, a stage of mitosis called anaphase (Fig. 3–10F). A microtubule connecting with one pair of centrioles pulls one chromatid and a microtubule connecting with the other centriole pair pulls the opposite chromatid. Thus, all 46 pairs of chromatids are separated, forming 46 daughter chromosomes that are pulled toward one end of the mitotic spindle and another 46 *complementary* chromosomes that are pulled toward the other end of the mitotic spindle.

Telophase. In telophase (Figs. 3–10*G* and *H*) the mitotic spindle grows still longer, pulling the two sets of daughter chromosomes completely apart. Then a new nuclear membrane develops around each set of chromosomes, this membrane perhaps being formed from portions of the endoplasmic reticulum that are already present in the cytoplasm. Simultaneously, the mitotic apparatus undergoes dissolution, and the cell pinches in two midway between the two nuclei, for reasons totally unexplained at present.

Note, also, that each of the two pairs of centrioles is replicated during telophase, the mechanism of which is not understood. These new pairs of centrioles remain dormant through the next interphase until a mitotic apparatus is required for the next cell division.

CONTROL OF CELL GROWTH AND REPRODUCTION

Cell growth and reproduction usually go together, growth occurring simultaneously with replication of the genes of the nucleus, followed a few hours later by mitosis.

In the normal human body, regulation of cell growth and reproduction is mainly a mystery. We know that certain cells grow and reproduce all the time, such as the blood-forming cells of the bone marrow, the germinal layers of the skin, and the epithelium of the gut. However, many other cells, such as some muscle cells, do not reproduce for many years. And a few cells, such as the neurons, do not reproduce during the entire life of the person.

Thus, most cells of the human body do have the ability to reproduce continually, though the rate of reproduction usually remains greatly suppressed. Yet if there is an insufficiency of a given type of cell in the body, this type of cell will grow and reproduce very rapidly until appropriate numbers of them are again available. For instance, seven-eighths of the liver can be removed surgically, and the cells of the remaining one-eighth will grow and divide until the liver mass returns almost to normal. The same effect occurs for almost all glandular cells, for cells of the bone marrow, the subcutaneous tissue, the intestinal epithelium, and almost any other tissue except highly differentiated cells, such as nerve cells.

We know very little about the mechanisms that maintain proper numbers of the different types of cells in the body. It is assumed that control substances are secreted by the different cells and cause feedback effects to stop or slow their growth when too many of them have been formed, though only a few such substances have been found. We know that cells of any type removed from the body and grown in tissue culture can grow and reproduce rapidly and indefinitely if the medium in which they grow is continually replenished. Yet they will stop growing when even small amounts of their own secretions are allowed to collect in the medium, which supports the idea that control substances limit cellular growth.

CANCER

Cancer can occur in any tissue of the body. It results from a change in certain cells that allows them to disrespect normal growth limits, no longer obeying the feedback controls that normally stop cellular growth and reproduction after a given number of such cells have developed. As pointed out above, even normal cells when removed from the body and grown in tissue culture can grow and proliferate indefinitely if the growth medium is continually changed. Therefore, in tissue culture, normal tissue cells behave exactly as cancer cells, but in the body, normal tissue cells behave differently, for they are subject to limits, whereas cancer cells are not.

What is the difference between the cancer cell and the normal tissue cell that allows the cancer cell to grow and reproduce unabated? The answer to this question is not known, but researchers have found the genetic make-up of some cancer cells to be different from those of normal cells. This has led to the idea that cancer usually results from mutation of part of the genetic system in the nucleus (or from a change of this genetic system induced by the presence of a virus in the cell). The mutated or otherwise changed "genome" eliminates the feedback mechanisms that normally limit growth and reproduction of the cell. Once even a single such cell is formed, it obviously will grow and proliferate indefinitely, its number increasing exponentially.

Cancerous tissue competes with normal tissues for nutrients, and because cancer cells continue to proliferate in-definitely, their number multiplying day-by-day, one can readily understand that the cancer cells will soon demand essentially all the nutrition available to the body. As a result, the normal tissues gradually suffer nutritive death.

REFERENCES

Brenner, S.: Theories of gene regulation. *Brit. Med. Bull., 21*:244, 1965.

Crick, F. H. C.: The genetic code. *Sci. Amer., 215*:55(4), 1966.

Gaze, R. M.: Growth and differentiation. *Ann. Rev. Physiol., 29*:59, 1967.

Gurdon, J. B.: Transplanted nuclei and cell differentiation. *Sci. Amer., 219*:24(6), 1968.

Marifield, R. B.: The automatic synthesis of proteins. *Sci. Amer., 218*:56(3), 1968.

Siddiqi, O.: The mechanism of gene action and regulation. *In* Bittar, E. Edward, and Bittar, Neville (eds.): The Biological Basis of Medicine. New York, Academic Press, Inc., 1969, Vol. 4, p. 43.

Watson, J. D.: Molecular Biology of the Gene. New York, Benjamin Company, Inc., 1965.

TRANSPORT THROUGH THE CELL MEMBRANE

The fluid inside the cells of the body, called *intracellular fluid*, is very different from that outside the cells, called *extracellular fluid*. The extracellular fluid circulates in the spaces between the cells and also mixes freely with the fluid of the blood through the capillary walls. Thus, it is the extracellular fluid that supplies the cells with nutrients and other substances needed for cellular function. But before the cell can utilize these substances, they must be transported through the cell membrane.

Note in Figure 4–1 that the extracellular fluid contains large quantities of *sodium* but only small quantities of *potassium*. Exactly the opposite is true of the intracellular fluid. Also, the extracellular fluid contains large quantities of chloride, while the intracellular fluid contains very little. On the other hand, the concentration of phosphates in the intracellular fluid is considerably greater than that in the extracellular fluid. All these differences between the components of the intracellular and extracellular fluids are extremely important to the life of the cell. It is the purpose of this chapter to explain how these dif-

ferences are brought about by the transport mechanisms in the cell membrane.

Substances are transported through the cell membrane by two major processes, *diffusion* and *active transport*. Though there are many different variations of these two basic mechanisms,

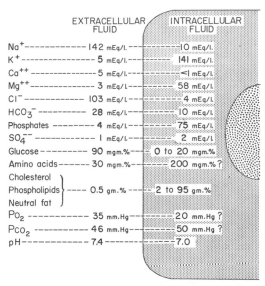

	EXTRACELLULAR FLUID	INTRACELLULAR FLUID
Na^+	142 mEq/l.	10 mEq/l.
K^+	5 mEq/l.	141 mEq/l.
Ca^{++}	5 mEq/l.	<1 mEq/l.
Mg^{++}	3 mEq/l.	58 mEq/l.
Cl^-	103 mEq/l.	4 mEq/l.
HCO_3^-	28 mEq/l.	10 mEq/l.
Phosphates	4 mEq/l.	75 mEq/l.
SO_4^{--}	1 mEq/l.	2 mEq/l.
Glucose	90 mgm.%	0 to 20 mgm.%
Amino acids	30 mgm.%	200 mgm.%?
Cholesterol Phospholipids Neutral fat	0.5 gm.%	2 to 95 gm.%
Po_2	35 mm.Hg	20 mm.Hg?
Pco_2	46 mm.Hg	50 mm.Hg?
pH	7.4	7.0

FIGURE 4–1 Chemical compositions of extracellular and intracellular fluids.

37

as we shall see later in this chapter, basically, diffusion means movement of substances in a random fashion caused by the normal kinetic motion of matter, whereas active transport means movement of substances as a result of chemical processes that impart energy to cause the movement.

DIFFUSION

All molecules and ions in the body fluids, including both water molecules and dissolved substances, are in constant motion, each particle moving its own separate way. Motion of these particles is what physicists call heat — the greater the motion, the higher is the temperature — and motion never ceases under any conditions except absolute zero temperature. When a moving particle, A, bounces against a stationary particle, B, its electrostatic forces repel particle B, momentarily adding some of its energy of motion to particle B. Consequently, particle B gains kinetic energy of motion while particle A slows down, losing some of its kinetic energy. Thus, as shown in Figure 4–2, a single molecule in solution bounces among the other molecules first in one direction, then another, then another, and so forth, bouncing hundreds to millions of times each second. At times it travels a far distance before striking the next molecule, but at other times only a short distance.

This continual movement of molecules among each other in liquids, or in gases, is called *diffusion*. Ions diffuse in exactly the same manner as whole molecules, and even suspended colloid particles diffuse in a similar manner, except that because of their very large sizes they diffuse far less rapidly than molecular substances.

KINETICS OF DIFFUSION — THE CONCENTRATION DIFFERENCE

When a large amount of dissolved substance is placed in a solvent at one end of a chamber, it immediately begins to diffuse toward the opposite end of the chamber. If the same amount of substance is placed in the opposite end of the chamber, it begins to diffuse toward the first end, the same amount diffusing in each direction. As a result, the *net rate of diffusion* from one end to the other is zero. If, however, the concentration of the substances is greater at one end of the chamber than at the other end, the net rate of diffusion from the area of high concentration to low concentration is directly proportional to the larger concentration minus the lower concentration. This concentration change along the axis of the chamber is called a *concentration difference* or a *diffusion gradient*.

If we consider all the different factors that affect the rate of diffusion of a substance from one area to another, they are the following: (1) The greater the concentration difference, the greater the rate of diffusion. (2) The less the molecular weight, the greater the rate of diffusion. (3) The shorter the distance, the greater the rate. (4) The greater the cross-section of the chamber in which diffusion is taking place, the greater is the rate of diffusion. (5) The greater the temperature, the greater is the molecular motion and also the greater is the rate of diffusion. All these can be placed

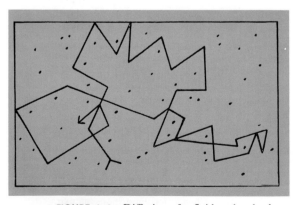

FIGURE 4–2 Diffusion of a fluid molecule during a fraction of a second.

in an approximate formula, found at the bottom of this page, for diffusion in solutions.

DIFFUSION THROUGH THE CELL MEMBRANE

The cell membrane is essentially a sheet of lipid material, called the *lipid matrix*, covered on each surface by a layer of protein, the detailed structure of which was discussed in Chapter 2 and shown in the diagram of Figure 2–2. The fluids on each side of the membrane are believed to penetrate the protein portion of the membrane with ease so that any dissolved substances can diffuse into this portion of the cell membrane without any impediment. However, the lipid portion of the membrane is an entirely different type of fluid medium, acting as a limiting membrane between the extracellular and intracellular fluids.

Two different methods by which the substances can diffuse through the membrane are: (a) by becoming dissolved in the lipid and diffusing through it in the same way that diffusion occurs in water, and (b) by diffusing through minute pores that pass directly through the membrane at wide intervals over its surface.

Effect of Lipid Solubility on Diffusion Through the Lipid Matrix. The primary factor that determines how rapidly a substance can diffuse through the lipid matrix of a cell membrane is its solubility in lipids. If it is very soluble, it becomes dissolved in the membrane very easily and therefore passes on through. On the other hand, a substance that dissolves very poorly in lipids will be greatly retarded. For instance, oxygen, carbon dioxide, alcohol, and fatty acids, which are very soluble, pass through the lipid matrix with ease, as shown in Figure 4–3; whereas water, which is almost completely insoluble in lipids, passes through the lipid matrix almost not at all.

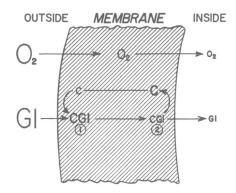

FIGURE 4–3 Diffusion of substances through the lipid matrix of the membrane. The upper part of the figure shows *free diffusion* of oxygen through the membrane and the lower part shows *facilitated diffusion* of glucose.

Facilitated diffusion. Some substances are very insoluble in lipids and yet can still pass through the lipid matrix by a process called facilitated diffusion. This is the means by which different sugars in particular cross the membrane. The most important of these sugars is glucose, the membrane transport of which is illustrated in the lower part of Figure 4–3. This shows that glucose G1 combines with a *carrier* substance C at point 1 to form the compound CG1. This combination is soluble in the lipid so that it can diffuse (or simply move by rotation of the large carrier molecule) to the other side of the membrane, where the glucose breaks away from the carrier (point 2) and passes to the inside of the cell, while the carrier moves back to the outside surface of the membrane to pick up still more glucose and transport it also to the inside. Thus, the effect of the carrier is to make the glucose soluble in the membrane; without it, glucose cannot pass through the membrane. The carrier itself is probably a small protein or lipoprotein.

Diffusion Through the Membrane Pores. Some substances, such as water and

$$\text{Diffusion rate} \propto \frac{\text{Concentration difference} \times \text{Cross-sectional area} \times \text{Temperature}}{\text{Molecular weight} \times \text{Distance}}$$

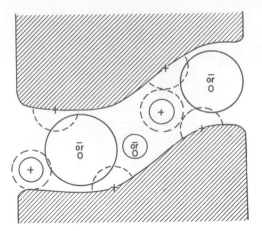

FIGURE 4-4 Postulated structure of the pore in the mammalian red cell membrane, showing the sphere of influence exerted by charges along the surface of the pore. (Modified from Solomon: *Sci. Amer., 203*[6]:146, 1960.)

many of the dissolved ions, seem to go through holes in the cell membrane called *membrane pores*. The nature of these pores is unknown. They might be round holes, small slits in the cell membrane, or possibly some large hydrophilic molecule penetrating all the way through the cell membrane and providing a pathway for movement of water-soluble substances along the axis of the molecule. At any rate, these so-called "pores" in the cell membrane behave as if they were minute round holes approximately 8 A in diameter and as if the total area of the pores should equal approximately 1/1600 of the total surface area of the cell. Despite this very minute total area of the pores, molecules and ions diffuse so rapidly that the entire volume of fluid in a cell can easily pass through the pores within a few hundredths of a second.

Figure 4-4 illustrates a postulated structure of a pore, indicating that its surface is probably lined with positively charged prosthetic groups. This figure shows several small particles passing through the pore and also shows that the maximum size of the particle that can pass through is approximately equal to the size of the pore itself.

Effect of pore size on diffusion through the pore. Table 4-1 gives the effective diameters of various substances in comparison with the diameter of the pore. Note that some substances, such as the water molecule, urea molecule, and chloride ion, are considerably smaller in size than the pore. All these pass through the pore with great ease. For instance, the rate per second of diffusion of water in each direction through the pores of a cell is about one hundred times as great as the volume of the cell itself. It is fortunate that an identical amount of water normally diffuses in each direction, which keeps the cell from either swelling or shrinking despite the rapid rate of diffusion. The rates of diffusion of urea and chloride ions through the membrane are somewhat less than that of water, which is in keeping with the fact that their effective diameters are slightly greater than that of water.

Table 4-1 also shows that most of the sugars, including glucose, have effective diameters that are slightly greater than that of the pores. Obviously, not even a single molecule as large as these could go through a pore that is smaller than its size. For this reason essentially none of

TABLE 4-1 RELATIONSHIP OF EFFECTIVE DIAMETERS OF DIFFERENT SUBSTANCES TO PORE DIAMETER*

Substance	Diameter Å	Ratio to Pore Diameter	Approximate Relative Diffusion Rate
Water molecule	3	0.38	50,000,000
Urea molecule	3.6	0.45	40,000,000
Hydrated chloride ion			
(red cell)	3.86	0.48	36,000,000
(nerve membrane)	–	–	200
Hydrated potassium ion	3.96	0.49	100
Hydrated sodium ion	5.12	0.64	1
Lactate ion	5.2	0.65	?
Glycerol molecule	6.2	0.77	?
Ribose molecule	7.4	0.93	?
Pore size	8 (Ave.)	1.00	–
Galactose	8.4	1.03	?
Glucose	8.6	1.04	0.4
Mannitol	8.6	1.04	?
Sucrose	10.4	1.30	?
Lactose	10.8	1.35	?

*These data have been gathered from different sources but relate primarily to the red cell membrane. Other cell membranes have different characteristics.

the sugars can pass through the pores; instead most of these pass through the lipid matrix by the process of facilitated diffusion.

Effects of electrical charge on transport of ions through the membrane. Positively charged ions, such as sodium and potassium ions, pass through the cell membrane with extreme difficulty, as shown in Table 4–1. The reason for this is believed to be the presence of positive charges of proteins or adsorbed positive ions, such as calcium ions, lining the pores. Figure 4–4 illustrates that each positive charge causes a sphere of electrostatic space charge to protrude into the lumen of the pore. A positive ion attempting to pass through a pore also exerts a sphere of positive electrostatic charge so that the two positive charges repel each other. This repulsion, therefore, blocks or greatly impedes movement of the positive ion through the pore.

In contrast to the positive ions, negatively charged ions pass through mammalian cell membrane pores much more easily. Thus, for the nerve membrane, chloride ions permeate the membrane about two times as easily as potassium ions and 100 to 200 times as easily as sodium ions. Moreover, for the red cell membrane, chloride ions permeate the membrane still another half million times as easily. It is believed that these differences are caused by lack of, or few numbers of, negative charges lining the pores in contrast to the large numbers of positive charges.

Effect of different factors on pore permeability. Pore permeability does not always remain exactly the same under different conditions. For instance, excess *calcium* in the extracellular fluid causes the permeability to decrease, and diminished calcium causes considerably increased permeability. This is extremely important in the function of nerves, for the enlarged pores that occur in extracellular fluid calcium deficiency cause excessive diffusion of ions, which results in spurious nerve discharge throughout the body.

Another factor that has an important effect on pore permeability of many cells is *antidiuretic hormone*, which is secreted in the hypothalamus. This hormone has an especially important effect on the membranes of the cells lining the kidney tubules. Increased quantities of the hormone increase the pore diameter, which allows water and other substances to diffuse out of the tubules and back into the blood with ease.

EFFECT OF CONCENTRATION DIFFERENCE ON NET DIFFUSION RATE

Figure 4–5 illustrates a membrane with a substance in high concentration on the outside and low concentration on the inside. The rate at which the substance diffuses *inward* is proportional to the concentration of molecules on the outside, for this concentration determines how many of the molecules strike the pore each second. On the other hand, the rate at which the molecules diffuse *outward* is proportional to their concentration inside the membrane. Obviously, therefore, the rate of net diffusion is proportional to the concentration on the outside *minus* the concentration on the inside, or

$$\text{Net diffusion} \propto P(C_1 - C_2)$$

in which C_1 is the concentration on the outside, C_2 is the concentration on the inside, and P is the permeability of the membrane for the substance.

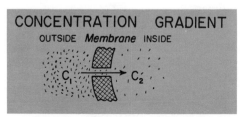

CONCENTRATION GRADIENT

OUTSIDE *Membrane* INSIDE

C_1 → C_2

FIGURE 4–5 Effect of concentration difference on diffusion of molecules and ions through a cell membrane.

NET DIFFUSION OF WATER ACROSS CELL MEMBRANES— OSMOSIS

By far the most abundant substance to diffuse through the cell membrane is water, but it should be recalled again that enough water ordinarily diffuses in each direction through the red cell membrane per second to equal about *100 times the volume of the cell itself.* Yet, *normally*, the amount that diffuses in the two directions is so precisely balanced that not even the slightest *net* diffusion of water occurs. Therefore, the volume of the cell remains constant. However, under certain conditions, a *concentration difference for water* can develop across a membrane, just as concentration differences for other substances can also occur. When this happens, net diffusion of water does occur across the cell membrane, causing the cell either to swell or to shrink, depending on the direction of the net diffusion. This process of net diffusion of water caused by a concentration difference is called *osmosis*.

To give an example of osmosis, let us assume that we have the conditions shown in Figure 4–6, with pure water on one side of the cell membrane and a solution of sodium chloride on the other side. Water molecules pass through the cell membrane with extreme ease while sodium ions pass through only with extreme difficulty. And chloride ions cannot pass through the membrane because the positive charge of the sodium ions holds the negatively charged chloride ions back. Therefore, sodium chloride solution is actually a mixture of diffusible water molecules and nondiffusible sodium ions and chloride ions. Yet, the presence of the sodium and chloride has reduced the concentration of water molecules below that of pure water. As a result, in the example of Figure 4–6, more water molecules strike the pores on the left side where there is pure water than on the right side where the water concentration has been reduced. Thus, osmosis occurs from left to right.

Osmotic Pressure. If in Figure 4–6 pressure were applied to the sodium chloride solution, osmosis of water into this solution could be slowed or even stopped. The amount of pressure required to stop osmosis completely is called the osmotic pressure of the sodium chloride solution.

The principle of a pressure difference opposing osmosis is illustrated in Figure 4–7, which shows a semipermeable membrane separating two separate columns of fluid, one containing water and the other containing a solution of water and some solute that will not penetrate the membrane. Osmosis of water from chamber B into chamber A causes the levels of the fluid columns to become farther and farther apart, until eventually a pressure difference is developed that is great enough to oppose the osmotic effect. The pressure difference across the membrane at this time is the osmotic pressure of the solution containing the nondiffusible solute.

Osmolality. The amount of osmotic pressure caused by a solute is proportional to the concentration of the solute in numbers of molecules or ions, because each of these particles decreases the molecular concentration of water by

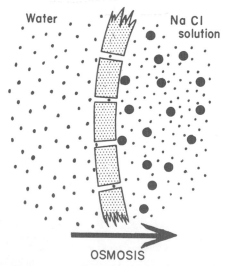

FIGURE 4–6 Osmosis at a cell membrane when a sodium chloride solution is placed on one side of the membrane and water on the other side.

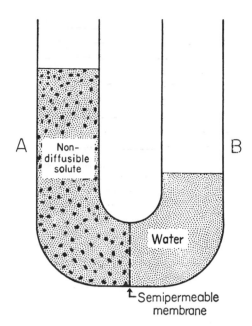

FIGURE 4-7 Demonstration of osmotic pressure on the two sides of a semipermeable membrane.

exactly the same amount regardless of size of the particles. Because of this effect, expressing the solute concentration in terms of mass is of no value in determining osmotic pressure. To express the concentration in terms of numbers of particles, the unit called the *osmol* is used in place of grams.

One osmol is the number of particles in 1 gram molecular weight of undissociated solute. Thus, 180 grams of glucose is equal to 1 osmol of glucose, because glucose does not dissociate. On the other hand, if the solute dissociates into two ions, 1 gram molecular weight of the solute equals 2 osmols, because the number of osmotically active particles is now twice as great as is the case in the undissociated solute. Therefore, 1 gram molecular weight of sodium chloride, 58.5 gm., is equal to 2 osmols.

A solution that has 1 osmol of solute dissolved in each liter of water is said to have a concentration of 1 osmol per liter, and a solution that has $1/1000$ osmol dissolved per liter has a concentration of 1 milliosmol per liter. The normal concentration of the extracellular and intracellular fluids is about 300 milliosmols per liter.

Relationship of osmolality to osmotic pressure. At normal body temperature, a concentration of 1 osmol per liter will cause 19,300 mm. Hg osmotic pressure in the solution. Likewise, 1 milliosmol per liter concentration is equivalent to 19.3 mm. Hg osmotic pressure.

ACTIVE TRANSPORT

Often only a minute concentration of a substance is present in the extracellular fluid, and yet a large concentration of the substance is required in the intracellular fluid. For instance, this is true of potassium ions. Conversely, other substances frequently enter cells and must be removed even though the concentration inside is far less than that outside. This is true of sodium ions.

From the discussion thus far it is evident that *no substances can diffuse against a concentration difference*, or, as is often said, "uphill." To cause movement of substances uphill, energy must be imparted to the substance. This is analogous to the compression of air by a pump. After compression, the concentration of the air molecules is far greater than before compression, but to create this greater concentration, energy must be expended by the piston of the pump as it compresses the air molecules. Likewise, as molecules are transported through a cell membrane from a dilute solution to a concentrated solution, energy must be expended. The process of moving molecules uphill against a concentration difference (or uphill against an electrical or pressure difference) is called *active transport*. Among the different substances that are actively transported through the cell membrane in at least some parts of the body are sodium ions, potassium ions, calcium ions, iron ions, hydrogen ions, chloride ions, iodide ions, ureate ions, several different sugars, and the amino acids.

BASIC MECHANISM OF ACTIVE TRANSPORT

The mechanism of active transport is believed to be similar for all substances and to be dependent on transport by carriers. Figure 4–8 illustrates the basic mechanism, showing a substance S entering the outside surface of the membrane where it combines with carrier C. At the inside surface of the membrane, S separates from the carrier and is released to the inside of the cell. C then moves back to the outside to pick up more S.

One will immediately recognize the similarity between this mechanism of active transport and that of facilitated diffusion discussed earlier in the chapter. The difference, however, is that energy is imparted to the system in the course of transport, so that transport can occur against a concentration difference (or against an electrical or pressure difference).

Unfortunately, little is known about the mechanisms by which energy is utilized in the transport mechanism. The energy could cause combination of the carrier with the substance, or it could cause splitting of the substance away from the carrier. Regardless of which of

these is true, the energy is probably supplied to the transport system at the inside surface of the cell membrane, because it is inside the cell that large quantities of the energy-giving substance ATP are available for promoting chemical reactions.

It is also evident from Figure 4–8 that the chemical reactions are believed to be promoted by specific enzymes which catalyze the chemical reactions. Indeed, the carrier might itself be an enzyme as well as a carrier.

Chemical Nature of the Carrier. Carrier substances are believed to be either proteins or lipoproteins, the protein moiety providing a specific site for attachment of the substance to be transported and the lipid moiety providing solubility in the lipid phase of the cell membrane.

It has been suggested that the carrier might transport substances by a simple process of thermal motion, such as by rotating to expose its carrier site first to the outside surface of the cell membrane and then to the inside, or by sliding a substance from one reactive site of the carrier to another until the substance enters the cell. Whether or not a "shuttling" type of carrier exists, nevertheless, essentially the same principles of transport through the membrane would still apply.

Specificity of Carrier Systems. Several different carrier systems exist in cell membranes, each of which transports only certain specific substances. One carrier system, for instance, transports sodium to the outside of the membrane and probably transports potassium to the inside at the same time. Another system actively transports sugars through the membranes of certain cells, whereas still other specific carrier systems transport different ones of the amino acids.

The specificity of active transport systems for substances is determined either by the chemical nature of the carrier, which allows it to combine only with certain substances, or by the nature of the enzymes that catalyze the specific

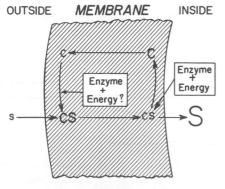

FIGURE 4–8 Basic mechanism of active transport.

chemical reactions. Unfortunately, very little is known about the precise function of these two components of the carrier systems.

Energetics of Active Transport. In terms of calories, the amount of energy required to concentrate 1 osmol of substance 10-fold is about 1400 calories. One can see that the energy expenditure for concentrating substances in cells or for removing substances from cells against a concentration gradient can be tremendous. Some cells, such as those lining the renal tubules, expend as much as 50 per cent of their energy for this purpose alone.

ACTIVE TRANSPORT OF SODIUM, POTASSIUM, AND OTHER ELECTROLYTES

Because minute quantities of sodium and potassium can diffuse through the pores of the cell, the concentrations of the two ions would eventually become equal inside and outside the cell unless there were some means to remove the sodium from the inside and to transport potassium back in.

Fortunately, a system for active transport of sodium and potassium is present probably in all cells of the body. The mechanism has been postulated to be that illustrated in Figure 4–9, which shows sodium (Na) inside the cell combining with carrier Y at the membrane surface to form large quantities of the combination NaY. This then moves to the outer surface where sodium is released and the carrier Y changes its chemical composition slightly to become carrier X. This carrier then combines with potassium K to form KX, which moves to the inner surface of the membrane where energy is provided to split K from X under the influence of the enzyme ATPase, the energy being derived from MgATP.

The transport mechanism in Figure 4–9 is believed to be more effective in transporting sodium than in transporting

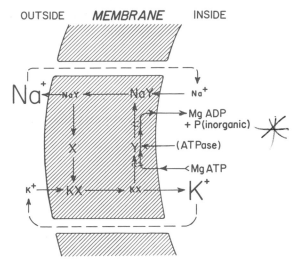

FIGURE 4–9 Postulated mechanism for active transport of sodium and potassium through the cell membrane, showing coupling of the two transport mechanisms and delivery of energy to the system at the inner surface of the membrane.

potassium, usually transporting about three sodium ions for every two potassium ions. The carrier is probably a lipoprotein, and it is likely that this same lipoprotein molecule acts as the enzyme ATPase to release the energy required for transport.

The sodium transport mechanism is so important to many different functioning systems of the body — such as to nerve and muscle fibers for transmission of impulses, various glands for the secretion of different substances, and all cells of the body to prevent cellular swelling — that it is frequently called the *sodium pump*. We will discuss the sodium pump at many places in this text.

Other Electrolytes. Calcium and magnesium are probably transported by all cell membranes in much the same manner that sodium and potassium are transported, and certain cells of the body have the ability to transport still other ions. For instance, the glandular cell membranes of the thyroid gland can transport large quantities of iodide ion; the epithelial cells of the intestine can transport sodium, chloride, calcium, iron, bicarbonate, and probably many other ions; and the epithelial cells of the renal tubules can transport hydrogen, calcium, sodium, potassium, and a number of other ions.

ACTIVE TRANSPORT OF SUGARS

Facilitated diffusion of glucose and certain other sugars occurs in essentially all cells of the body, but active transport of sugars occurs in only a few places in the body. For instance, in the intestine and renal tubules, glucose and several other monosaccharides are continually transported into the blood even though their concentrations be minute in the lumens. Therefore, essentially none of these sugars is lost either in the intestinal excreta or the urine.

Though not all sugars are actively transported, almost all monosaccharides that are important to the body *are* actively transported, including *glucose, galactose, fructose, mannose, xylose, arabinose*, and *sorbose*. On the other hand, the disaccharides such as sucrose, lactose, and maltose are not actively transported at all.

As is true of essentially all other active transport mechanisms, the precise carrier system and chemical reactions responsible for transport of the monosaccharides are yet unknown. The one common denominator in transport of sugars seems to be the necessity for an intact —OH group on the C_2 carbon of the monosaccharide molecule. It is presumed that the monosaccharide attaches to the carrier at this point.

Transport of the monosaccharides will not occur when active transport of sodium is blocked. Because of this, it is believed that active transport of sodium ions, by means of coupled reactions with the glucose transport mechanism, provides the energy required to move the monosaccharides through the membrane. Therefore, it is stated that monosaccharide transport is *secondary active transport*, in contrast to *primary active*

transport represented by the transport of sodium through the membrane.

ACTIVE TRANSPORT OF AMINO ACIDS

Amino acids and the closely related substances, the amines, are actively transported through the membranes of all cells that have been studied. Active transport through the epithelium of the intestine, the renal tubules, and many glands is especially important. Probably at least four different carrier systems exist for transporting different groups of amino acids. However, the carrier systems for transport of amino acids are just as poorly understood as the systems for transporting other substances. One of the few known features of the amino acid carrier system is that transport of at least some amino acids depends on pyridoxine (vitamin B_6). Therefore, deficiency of this vitamin causes protein deficiency.

ACTIVE TRANSPORT THROUGH CELLULAR SHEETS

In many places in the body substances must be transported through an entire *cellular layer* instead of simply through the cell membrane itself. Transport of this type occurs through the intestinal epithelium, the epithelium of the renal tubules, the epithelium of all exocrine glands, the membrane of the choroid plexus of the brain, and many other membranes.

The general mechanism of active transport through a sheet of cells is illustrated in Figure 4–10. This figure shows two adjacent cells in a typical epithelial membrane as found in the intestine, in the gallbladder, and in certain other areas of the body. On the luminal surface of the cells is a brush border that is highly permeable to both water and solutes, allowing both of these to diffuse readily from the lumen of the

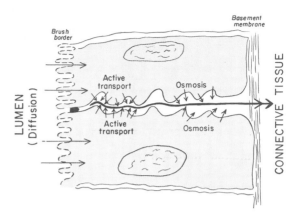

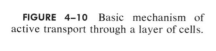

FIGURE 4–10 Basic mechanism of active transport through a layer of cells.

intestine to the interior of the cell. Once inside the cell, some of the solutes are actively transported into the space between the cells. This space is closed at the brush border of the epithelium but is wide open at the base of the cells where they rest on the basement membrane. Furthermore, the basement membrane is extremely permeable. Therefore, substances transported into this channel between the cells flow toward the connective tissue.

One of the most important substances actively transported in this manner is sodium; and when sodium ions are transported into the space between the cells, their positive electrostatic charges pull negatively charged chloride ions through the membrane as well. Then, when the concentration of sodium and chloride ions increases, this in turn causes osmosis of water out of the cell and into the intercellular space. In consequence, the sodium and chloride flow along with the water into the connective tissue behind the basement membrane

where the water and ions diffuse into the blood capillaries. These same principles of transport apply generally wherever such transport occurs through cellular sheets, whether in the intestine, gallbladder, kidneys, or elsewhere.

REFERENCES

Biological Membranes (Symposium). *Ann. N.Y. Acad. Sci., 137*:403, 1966.

Diamond, J. M., and Tormey, J. M.: Studies on the structural basis of water transport across epithelial membranes. *Fed. Proc., 25*:1458, 1968.

Gamble, J. L.: Chemical Anatomy, Physiology and Pathology of Extracellular Fluid: A Lecture Syllabus. 6th ed., Cambridge, Massachusetts, Harvard University Press, 1954.

Rothstein, A.: Membrane phenomena. *Ann. Rev. Physiol., 30*:15, 1968.

Skou, J. C.: Enzymatic basis for active transport of Na^+ and K^+ across cell membrane. *Physiol. Rev., 45*:596, 1965.

Structure and Function of Membranes (Symposium). *Brit. Med. Bull., 24*:99, 1968.

Whittam, R., and Wheeler, K. P.: Transport across cell membranes. *Ann. Rev. Physiol., 32*:21, 1970.

CHAPTER 5

MEMBRANE POTENTIALS, ACTION POTENTIALS, EXCITATION, AND RHYTHMICITY

Electrical potentials exist across the membranes of essentially all cells of the body, and some cells, such as nerve and muscle cells, are "excitable" — that is, capable of transmitting electrochemical impulses along their membranes. And in still other types of cells, such as glandular cells, macrophages, and ciliated cells, changes in membrane potentials probably play significant roles in controlling many of the cell's functions. However, the present discussion is concerned with membrane potentials generated both at rest and during action by nerve and muscle cells.

BASIC PHYSICS OF MEMBRANE POTENTIALS

Before beginning this discussion, let us first recall that the fluids both inside and outside the cells are electrolytic solutions containing approximately 155 mEq./liter of anions and the same concentration of cations. Generally, an excess number of negative ions (anions) accumulates immediately inside the cell membrane along its inner surface, and an equal number of positive ions (cations) accumulates immediately outside the membrane. The result of this is the development of a *membrane potential*.

The two basic means by which membrane potentials can develop are: (1) active transport of ions through the membrane, thus creating an imbalance of negative and positive charges on the two sides of the membrane, and (2) diffusion of ions through the membrane as a result of a concentration difference between the two sides of the membrane, also creating an imbalance of charges.

Membrane Potentials Caused by Active Transport. Figure 5–1*A* illustrates how active transport can create a membrane potential. This shows the sodium

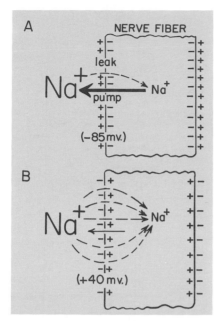

FIGURE 5-1 (*A*) Establishment of a membrane potential as a result of active transport of sodium ions. (*B*) Establishment of a membrane potential of opposite polarity caused by rapid diffusion of sodium ions from an area of high concentration to an area of low concentration.

"pump," which was explained in Chapter 4, transporting *positively charged* sodium ions to the exterior.

Membrane Potentials Caused by Diffusion. Figure 5-1*B* illustrates the nerve fiber under another condition in which the permeability of the membrane to sodium has increased so much that diffusion of sodium through the membrane is now great in comparison with the transport of sodium by active transport. Since the sodium concentration is great outside the cells and slight inside, the positively charged sodium ions now move rapidly to the inside of the membrane, obviously causing the inside to become positively charged in relation to the outside. The magnitude of the potential is determined by the difference in tendency of the ions to diffuse in one direction versus the other direction, in accord with the following formula (at body temperature, 38°C.):

$$\text{EMF (in millivolts)} = 61 \log \frac{\text{Conc. inside}}{\text{Conc. outside}} \quad (1)$$

Thus, when the concentration of ions on one side of a membrane is 10 times that on the other, the log of 10 is 1, and the potential difference calculates to be 61 millivolts.

However, two conditions are necessary for a membrane potential to develop as a result of diffusion: (1) The membrane must be semipermeable, allowing ions of one charge to diffuse through the pores more readily than ions of the opposite charge. (2) The concentration of the diffusible ions must be greater on one side of the membrane than on the other side.

Calculation of the membrane potential when the membrane is permeable to several different ions. When a membrane is permeable to several different ions, the diffusion potential that will develop depends on three factors: (1) the polarity of the electrical charge of each ion, (2) the permeability of the membrane (*P*) to each ion, and (3) the concentrations of the respective ions on the two sides of the membrane. Thus, the following formula, called the *Nernst equation*, gives the calculated membrane potential when two cations (*C*) and two anions (*A*) are involved.

$$\text{EMF (in millivolts)} = \\ -61 \log \frac{C_{1_i}P_1 + C_{2_i}P_2 + A_{3_o}P_3 + A_{4_o}P_4}{C_{1_o}P_1 + C_{2_o}P_2 + A_{3_i}P_3 + A_{4_i}P_4} \quad (2)$$

Note that a cation gradient from the inside (*i*) to the outside (*o*) of a membrane causes electronegativity inside the membrane, while an anion gradient in *exactly the opposite direction* also causes electronegativity on the inside.

ORIGIN OF THE CELL MEMBRANE POTENTIAL

Role of Sodium and Potassium Pumps and of Nondiffusible Anions in the Origin of the Cell Membrane Potential. Before attempting to explain the origin of the cell membrane potential, several basic facts need to be understood as follows:

1. The nerve membrane is endowed with a sodium and a potassium pump, sodium being pumped to the exterior and potassium to the interior. These pumps were discussed in the previous chapter.

2. The resting nerve membrane is normally 50 to 100 times as permeable to potassium as to sodium. Therefore, potassium diffuses with relative ease through the resting membrane, whereas sodium diffuses only with difficulty.

3. Inside the nerve fiber are large numbers of anions that cannot diffuse through the nerve membrane at all or that diffuse very poorly. These anions include especially organic phosphate ions, organic sulfate ions, and protein ions.

Now let us put the above facts together to see how the resting nerve membrane potential comes about. First, the sodium pump decreases the concentration of sodium ions inside the nerve fiber to a very low level, only 10 mEq./liter in contrast to the very high level in the extracellular fluids of 142 mEq./liter. Second, the potassium pump causes a high concentration of potassium ions inside the nerve fiber, 140 mEq./liter in contrast to only 5 mEq./liter in the extracellular fluids. These concentration differences are illustrated in Figure 5–2. Third, also illustrated in Figure 5–2 is a very high concentration of nondiffusible anions, 150 mEq./liter, inside the nerve fiber in contrast to a very low concentration of only 5 mEq./liter outside the fiber.

With the above distributions of sodium, potassium, and nondiffusible anions on the two sides of the membrane, a membrane potential develops in accordance with the Nernst equation discussed previously. Potassium ions diffuse through the membrane relatively easily while the sodium ions and nondiffusible anions diffuse very poorly. Therefore, the main tendency for diffusion of electrical charges will be

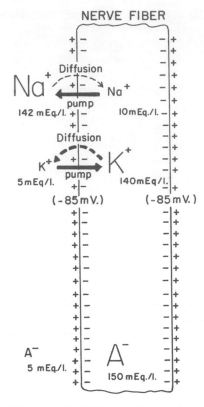

NERVE FIBER

FIGURE 5–2 Establishment of a membrane potential of −85 millivolts in the normal resting nerve fiber, and development of concentration differences of sodium, potassium, and chloride ions between the two sides of the membrane. The dashed arrows represent diffusion and the solid arrows represent active transport (pumps).

caused by diffusion of the positive potassium ions from inside the nerve membrane to the outside. This effect obviously removes positive charges from inside the membrane while building up positive charges outside. As a result, charges line up on the two sides of the membrane as shown in Figure 5–2, giving negativity on the inside and positivity on the outside.

Importance of potassium in determining the magnitude of the membrane potential. Since potassium can diffuse through the membrane very easily in comparison with either sodium or the nondiffusible anions, the Nernst equation predicts that it is almost entirely the concentration difference of potassium across the membrane that determines the magnitude of the membrane potential. Using Equation 1 discussed earlier in the chapter, one can calculate that the diffusion potential created by the concentration difference of potassium shown in Figure 5-2 will be approximately 90 millivolts. This is very near to the usually measured potential across the resting nerve membrane, which averages about 85 millivolts.

Membrane Potentials Measured in Nerve and Muscle Fibers. Figure 5-3 illustrates a method that has been used for measuring the resting membrane potential. A micropipet is made from a minute capillary glass tube so that the tip of the pipet has a diameter of only 0.5 to 2 microns. Inside the pipet is a strong solution of potassium chloride, which acts as an electrical conductor. The fiber whose membrane is to be measured is pierced with the pipet and electrical connections are made from the pipet to an appropriate meter, as illustrated in the figure. The resting membrane potential measured in many different nerve and muscle fibers of mammals has usually ranged between −75 and −95 millivolts, with −85 millivolts as a reasonable average of the different measurements.

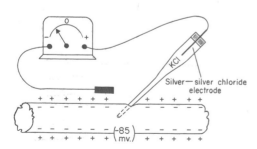

FIGURE 5-3 Measurement of the membrane potential of the nerve fiber using a microelectrode.

THE ACTION POTENTIAL

Any factor that suddenly increases the permeability of the membrane to sodium is likely to elicit a *sequence of rapid changes* in membrane potential lasting a minute fraction of a second, followed immediately by return of the membrane potential to its resting value. This sequence of potential changes is called the *action potential.*

Some of the factors that can elicit an action potential are *electrical stimulation* of the membrane, *application of chemicals* to the membrane to cause increased permeability to sodium, *mechanical damage* to the membrane, *heat, cold,* or almost any other factor that momentarily disturbs the normal resting state of the membrane.

Depolarization and Repolarization. The action potential occurs in two separate stages called depolarization and repolarization, which may be explained by referring to Figure 5-4. Figure 5-4A illustrates the resting state of the membrane, with negativity inside and positivity outside. When the permeability of the membrane to sodium ions suddenly increases, many of the sodium ions rush to the inside of the fiber, carrying enough positive charges to the inside to cause complete disappearance of the normal resting potential and usually enough charges actually to develop a positive state inside the fiber instead of the normal negative state, as illustrated in Figure 5-4B. This positive state is called a *reversal potential.*

Almost immediately after depolarization takes place, the pores of the membrane again become almost totally impermeable to sodium ions. Because of this, the reversal potential inside the fiber disappears, and the normal resting membrane potential returns. This is called *repolarization* (Fig. 5-4C).

Now, let us explain in more detail *how* this sequence of events occurs.

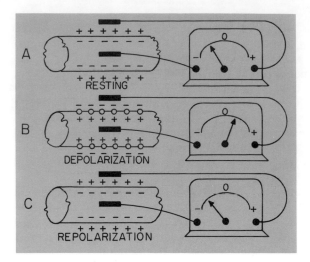

FIGURE 5–4 Sequential events during the action potential, showing: (*A*) the normal resting potential, (*B*) development of a reversal potential during depolarization, and (*C*) re-establishment of the normal resting potential during repolarization.

CONDUCTANCE CHANGES IN THE MEMBRANE DURING THE ACTION POTENTIAL

Sudden Increase in Sodium Conductance at the Onset of the Action Potential —"Activation" of the Membrane. The first event in the action potential is a sudden increase in membrane permeability to sodium. It is postulated that under normal circumstances the membrane pores are lined with calcium ions bound to the walls of the pores. Because of their positive charges, the calcium ions repel sodium and other positive ions and resist their passage through the pores. Yet if some calcium ions can be dislodged from their binding sites, a few sodium ions can begin to move inward. The inward movement of the inrushing sodium ions then theoretically can displace more and more calcium ions from their binding sites; as a result, still more sodium ions rush inward; and the process accelerates until essentially no calcium ions remain to block the movement of sodium through the pores.

The change in sodium conductance is illustrated by the lowermost curve of Figure 5–5, which shows that at the onset of the action potential the conductance of the membrane for sodium ions increases several hundred- to several thousand-fold in only a small fraction of a millisecond. This tremendous increase in sodium conductance is called *activation of the membrane.*

Subsequent Decrease in Sodium Conductance. As sodium ions diffuse to the inside of the membrane, they carry positive charges also to the inside, causing a *reversal potential* with the development of positivity inside the membrane and negativity outside. The positivity inside the membrane now repulses further inflow of the positively charged sodium ions. This obviously stops the movement of the sodium ions in the pores and theoretically allows calcium ions to begin rebinding with the binding sites along the walls. Now, another vicious cycle develops, operating in the opposite direction: as the calcium ions bind, sodium conductance decreases, which allows still more calcium ions to bind, which causes sodium conductance to decrease still more, and so forth. This cycle continues until the membrane becomes almost totally impermeable to sodium once again. (It must be remembered that this is only a theory at present, not fact.)

Thus, the lowermost curve in Figure 5–5 shows that immediately after the sodium conductance increases several hundred- or thousand-fold at the onset of the action potential, it decreases back

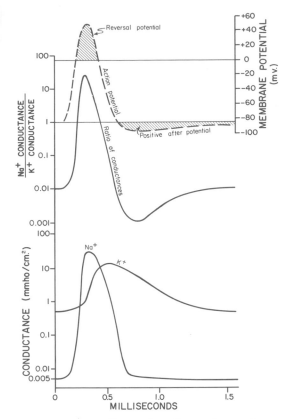

FIGURE 5-5 Changes in sodium and potassium conductances during the course of the action potential. Note that sodium conductance increases several thousand-fold during the early stages of the action potential, while potassium conductance increases only about 30-fold during the latter stages of the action potential and for a short period thereafter. (Curves constructed from data in Hodgkin and Huxley papers but transposed from squid axon to large mammalian nerve fibers.)

with the tremendous increase in sodium conductance, but it nevertheless plays a significant role in the membrane potential changes that occur during the course of the action potential.

The mechanism of the increased potassium conductance during the stages of the action potential, like that for changes in sodium conductance, is also almost completely hypothetical. However, the loss of negativity inside the membrane allows the highly concentrated potassium ions to begin moving *outward* through so-called *potassium "channels."* These channels are believed to be pores separate from the sodium pores, but this might not be true. It is assumed that this movement of potassium through the channels opens them to potassium in the same manner that sodium movement opens the sodium pores. Regardless of the mechanism that causes the increased potassium conductance, the potassium curve in Figure 5–5 shows that potassium conductance remains far above normal for almost 1 millisecond in large nerve fibers.

Return of the Membrane Potential to the Resting Level — Repolarization. The reversal potential lasts for only a fraction of a millisecond in large nerve fibers, only for that short period of time during which sodium conductance is greater than potassium conductance; this can be seen by comparing the top two curves of Figure 5–5. As the membrane becomes "inactivated" — that is, as sodium conductance returns toward its normal state — the membrane potential also returns toward its resting level.

Potassium diffusion as the cause of return of the resting potential. On first thought, one might suspect that return of the resting membrane potential to its original resting level would be caused by movement of sodium ions out of the fiber by the sodium pump. However, the sodium pump plays almost no role in this process because it operates much too slowly to recharge the membrane in a fraction of a millisecond. Instead, the return of the resting membrane

to its original resting level in another small fraction of a millisecond.

Changes in Potassium Conductance During the Action Potential. Referring once again to Figure 5–5, we see that the normal resting membrane conductance for potassium is about 100 times that for sodium. Also, this figure shows that *potassium conductance does not change significantly during the first half of the action potential.* Yet toward the end of the action potential the potassium conductance increases about 30- to 40-fold. This increase is minute compared potential is caused almost entirely by diffusion of potassium ions outward through the membrane. The potassium conductance curve of Figure 5–5 shows that at the same time that the sodium conductance returns toward its resting level, the potassium conductance increases to approximately 30 times its normal level. Thus, potassium ions now diffuse outward through the membrane extremely rapidly, while essentially no sodium ions diffuse inward through the membrane. Thus, we now have a situation in which a diffusion potential develops across the membrane caused almost entirely by potassium ions and almost none at all by the other ions. With an intracellular potassium concentration of 140 mEq./liter and an extracellular concentration of 5 mEq./liter, this is a 28-fold concentration ratio. Referring again to Equation 1, we see that a 28-fold ratio in potassium concentration calculates to give a membrane potential of −88 millivolts. Thus, one can readily understand that the rapid diffusion of potassium to the outside is more than capable of returning the membrane potential back to its resting level of −85 mv.

PROPAGATION OF THE ACTION POTENTIAL

In the preceding paragraphs we have discussed the action potential as if it occurs all at one spot on the membrane. However, an action potential elicited at any one point on an excitable membrane usually excites adjacent portions of the membrane, resulting in propagation of the action potential. The mechanism of this is illustrated in Figure 5–6. Figure 5–6A shows a normal resting nerve fiber, and Figure 5–6B shows a nerve fiber that has been excited in its midportion— that is, the midportion has suddenly developed increased permeability to sodium. The arrows illustrate a *local circuit* of current flow between the depolarized and the resting membrane

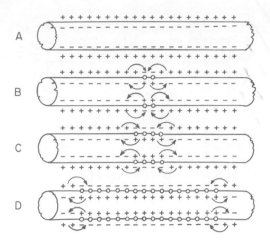

FIGURE 5–6 Propagation of action potentials in both directions along a conductive fiber.

areas; current flows inward through the depolarized membrane and outward through the resting membrane, thus completing a circuit. In some way not understood, *the current flow through the resting membrane now increases the membrane's permeability to sodium*, which immediately allows sodium ions to diffuse inward through the membrane, thus setting up the vicious cycle of membrane activation discussed earlier in the chapter. As a result, depolarization occurs at this area of the membrane as well. Therefore, as illustrated in Figure 5–6C and *D*, successive portions of the membrane become depolarized. And these newly depolarized areas cause local circuits of current flow still farther along the membrane, causing progressively more depolarization. Thus, the depolarization process travels in both directions along the entire extent of the fiber. The transmission of the depolarization process along a nerve or muscle fiber is called a *nerve* or *muscle impulse*.

The All-or-Nothing Law. It is obvious that, once an action potential has been elicited at any point on the membrane of a normal fiber, the depolarization process will travel over the entire membrane, even backwards and outwards along all branches of the nerve fiber.

This is called the all-or-nothing law, and it applies to all normal excitable tissues.

Propagation of Repolarization. The action potential normally lasts almost exactly the same length of time at each point along a fiber. Therefore, repolarization occurs at the point of stimulus first and then spreads progressively along the membrane, moving in the same direction that depolarization had previously spread, but a few ten-thousandths of a second later.

Importance of Energy Metabolism for "Recharging" the Fiber Membrane. Since transmission of large numbers of impulses along the nerve fiber reduces the ionic concentration differences across the membrane, these differences must be re-established. This is achieved by the action of the sodium and potassium pumps in exactly the same way as that described in the first part of the chapter for establishment of the original resting potential. That is, the sodium ions that have diffused during the action potentials to the interior of the cell and the potassium ions that have diffused to the exterior are returned to their original state by the sodium and potassium pumps. Since both the sodium and potassium pumps require energy for operation, this process of "recharging" the nerve fiber is an active metabolic one, utilizing energy derived from the adenosine triphosphate energy "currency" system of the cell.

THE SPIKE POTENTIAL AND THE AFTER-POTENTIAL

Figure 5–7 illustrates an action potential recorded with a much slower time scale than that illustrated in Figure 5–5; many milliseconds of recording are shown in comparison with only the first 1.5 milliseconds of the action potential in Figure 5–5.

The Spike Potential. The initial very large change in membrane potential shown in Figure 5–7 is called the spike potential. In large type A nerve fibers it

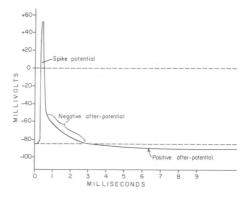

FIGURE 5–7 An idealized action potential, showing: the initial spike followed by a negative after-potential and a positive after-potential.

lasts for about 0.4 millisecond. The spike potential is analogous to the action potential that has been discussed in the preceding paragraphs, and it is the spike potential that is called the *nerve impulse.*

The Positive After-Potential. Once the membrane potential has returned to its resting value, it becomes a little more negative than its normal resting value; this excess negativity is called the *positive after-potential.* It is a fraction of a millivolt to, at most, a few millivolts more negative than the normal resting membrane potential, but it lasts from 50 milliseconds to as long as many seconds.

This positive after-potential is caused principally by the pumping of sodium outward through the nerve fiber membrane, which is the recharging process that was discussed previously. If the active transport processes are poisoned, the positive after-potential is lost, though the action potential continues to occur.

(The student might wonder why excess negativity of the resting membrane potential is called positive rather than negative. The reason for this is that these potentials were first measured *outside* the nerve fibers rather than inside, and all potential changes on the outside are of exactly opposite polarity, whereas modern terminology expresses membrane potentials in terms of the inside

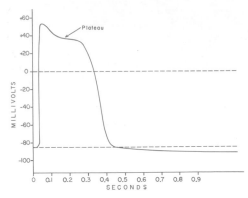

FIGURE 5–8 An action potential from cardiac muscle, showing a "plateau."

potential rather than the outside potential.)

PLATEAU IN THE ACTION POTENTIAL

In some instances the excitable membrane does not repolarize immediately after depolarization, but, instead, the potential remains on a plateau near the peak of the spike sometimes for many milliseconds before repolarization begins. Such a plateau is illustrated in Figure 5–8, from which one can readily see that the plateau greatly prolongs the period of depolarization. It is this type of action potential that occurs in the heart, where the plateau lasts for as long as two- to three-tenths second and causes contraction of the heart muscle during this entire period of time.

The cause of the action potential plateau is not known, but it has been postulated that the efflux of potassium at the end of the spike potential is too low in these membranes to initiate early membrane inactivation. Yet when the process does begin, the inactivation process proceeds unabated, and the membrane becomes repolarized rapidly, as illustrated in Figure 5–8 by the rapid decline of the potential at the end of the plateau.

SPECIAL ASPECTS OF IMPULSE CONDUCTION IN NERVE FIBERS

"Myelinated" and Unmyelinated Nerve Fibers. Figure 5–9 illustrates a cross-section of a typical small nerve trunk, showing a few very large nerve fibers that comprise most of the cross-sectional area and many more small fibers lying between the large ones. The large fibers are *myelinated* and the small ones arc *unmyelinated.* The average nerve trunk contains about twice as many unmyelinated fibers as myelinated fibers.

Figure 5–10 illustrates a typical myelinated fiber. The central core of the fiber is the *axon*, and the membrane of the axon is the actual *conductive membrane.* The axon is filled in its center with *axoplasm*, which is a viscid intracellular fluid. Surrounding the axon is a *myelin sheath* that is approximately as thick as the axon itself, and about once every millimeter along the extent of the axon the myelin sheath is interrupted by a *node of Ranvier.*

The myelin sheath is deposited around the axon by Schwann cells in the following manner: The membrane of a Schwann cell first envelopes the axon. Then it rotates around the axon several times, laying down multiple layers of cellular membrane containing the lipid substance

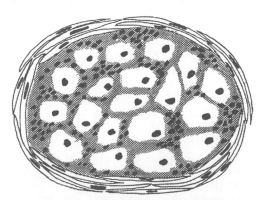

FIGURE 5–9 Cross-section of a small nerve trunk containing myelinated and unmyelinated fibers.

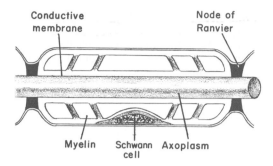

FIGURE 5–10 A myelinated nerve fiber. (Redrawn from Ramon y Cajal: Histology, William Wood and Co.)

sphingomyelin. This substance is an excellent insulator that prevents almost all flow of ions. However, at the juncture between each two successive Schwann cells along the axon, a small uninsulated area remains where ions can flow with ease between the extracellular fluid and the axon. This is the node of Ranvier.

Saltatory Conduction in Myelinated Fibers. Even though ions cannot flow to a significant extent through the thick myelin sheaths of myelinated nerves, they can flow with considerable ease through the nodes of Ranvier. Indeed, the membrane at this point is 500 times as permeable as the membranes of some unmyelinated fibers. Impulses are conducted from node to node by the myelinated nerve rather than, as occurs in unmyelinated fibers, continuously along the entire fiber. This process, illustrated in Figure 5–11, is called *saltatory con-*

duction. That is, electrical current flows through the surrounding extracellular fluids and through the axoplasm from node to node, exciting successive nodes one after another. Thus, the impulse jumps down the fiber, which is the origin of the term "saltatory."

Saltatory conduction is believed to be of value for two reasons: First, by causing the depolarization process to jump long intervals along the axis of the nerve fiber, it is reasonable to believe that this mechanism helps to explain the high velocity of nerve transmission in myelinated fibers. Second, saltatory conduction conserves energy for the axon, for only the nodes depolarize, allowing very little loss of ions and requiring little extra metabolism for retransporting the ions across the membrane.

VELOCITY OF CONDUCTION IN NERVE FIBERS

The velocity of conduction in nerve fibers varies from as little as 0.5 meter per second in very small fibers (0.3 μ diameter) up to as high as 130 meters per second (the length of a football field) in very large fibers (17 μ diameter). Thus the velocity increases with the fiber diameter.

ELECTRICAL STIMULATION OF NERVE FIBERS

Electrical stimulation of a nerve will initiate an action potential. An electrical charge artificially induced across the membrane can cause excess flow of ions through the membrane; this in turn initiates an action potential. For reasons not yet understood, when two electrodes, one positive and one negative, are placed on a nerve, the negative electrode stimulates the nerve while the positive electrode actually inhibits it.

Excitability Curve of Nerve Fibers. A so-called "excitability curve" of a nerve fiber is shown in Figure 5–12. To obtain

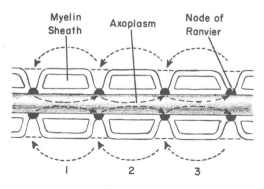

FIGURE 5–11 Saltatory conduction along a myelinated axon.

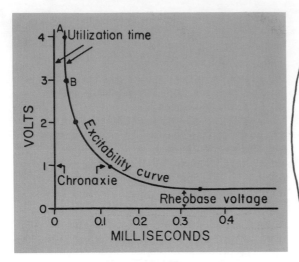

FIGURE 5–12 Excitability of a large myelinated nerve fiber.

this curve, a high voltage stimulus (4 volts in this instance) is first applied to the fiber, and the minimum duration of stimulus required to excite the fiber is found. The voltage and minimal time are plotted as point A. Then a stimulus voltage of 3 volts is applied, and the minimal time required is again determined; the results are plotted as point B. The same is repeated at 2 volts, 1 volt, 0.5 volt, and so forth, until the least voltage possible at which the membrane is stimulated has been reached. On connecting these points, the so-called *excitability curve* is determined.

The excitability curve of Figure 5–12 is that of a large myelinated nerve fiber. For other types of nerve, and also for muscle, more voltage and longer times of stimulation are usually required for excitation, thus shifting the curve upward and to the right.

THE REFRACTORY PERIOD

A second action potential cannot occur in an excitable fiber as long as the membrane is still depolarized from the preceding action potential. Therefore, even an electrical stimulus of maximum strength applied before the first spike potential is almost over will not elicit a second one. This interval of inexcitability is called the *absolute refractory period*. The absolute refractory period of large myelinated nerve fibers is about $1/2500$ second. Therefore, one can readily calculate that such a fiber can carry a maximum of about 2500 impulses per second. Following the absolute refractory period is a *relative refractory period* lasting about one quarter as long. During this period, stronger than normal stimuli are required to excite the fiber. In some types of fibers, a short period of supernormal excitability follows the relative refractory period.

RHYTHMICITY OF CERTAIN EXCITABLE TISSUES — REPETITIVE DISCHARGE

All excitable tissues can discharge repetitively if the threshold for stimulation is reduced low enough. For instance, even nerve fibers and skeletal muscle fibers, which normally are highly stable, discharge repetitively when the calcium ion concentration falls below a critical value and increases the membrane permeability. Repetitive discharges, or rhythmicity, occur normally in the heart, in most smooth muscle, and probably also in some of the neurons of the central nervous sytem.

The Re-excitation Process Necessary for Rhythmicity. For rhythmicity to occur, the membrane, even in its natural state, is already permeable enough to allow automatic depolarization. That is, (a) sodium ions flow inward, (b) this further increases the membrane permeability, (c) still more sodium ions flow inward, (d) the permeability increases more, and so forth, until the spike potential is generated. Then the spike declines until the resting membrane potential is re-established. Shortly thereafter a new action potential occurs spontaneously.

Yet why does the membrane not remain depolarized all the time instead of re-establishing a membrane potential, only to become depolarized again shortly thereafter? The answer to this can be found by referring back to Figure 5–5, which shows that toward the end of the action potential, and continuing for a short period thereafter, the membrane becomes excessively permeable to potassium. The excessive outflow of potassium carries tremendous numbers of positive charges to the outside of the membrane, creating inside the fiber considerably more negativity than would otherwise occur. This is a state called *hyperpolarization*. As long as this state exists, re-excitation will not occur; but gradually the excess potassium conductance (and the state of hyperpolarization) disappears, thereby allowing the onset of a new action potential.

Figure 5–13 illustrates this relationship between repetitive action potentials and potassium conductance. The state of hyperpolarization is established immediately after each preceding action potential; but it gradually recedes, and the membrane potential correspondingly increases until it reaches the threshold for excitation; then suddenly a new action potential results, the process occurring again and again.

RECORDING MEMBRANE POTENTIALS AND ACTION POTENTIALS

The Cathode Ray Oscilloscope. In this chapter we have noted that the membrane potential changes very rapidly

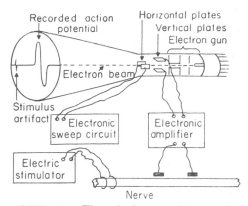

FIGURE 5–14 The cathode ray oscilloscope for recording transient action potentials.

throughout the course of an action potential. Indeed, most of the action potential complex of large nerve fibers takes place in less than $1/1000$ second. In some figures of this chapter a meter has been shown recording these potential changes. However, it must be understood that any meter capable of recording them must be capable of responding extremely rapidly. For practical purposes the only type of meter that is capable of responding accurately to the very rapid membrane potential changes of most excitable fibers is the cathode ray oscilloscope.

Figure 5–14 illustrates the basic components of a cathode ray oscilloscope. The cathode ray tube itself is composed basically of an *electron gun* and a *fluorescent surface* against which electrons are fired. Where the electrons hit the surface, the fluorescent material glows. If the electron beam is moved across the surface, the spot of glowing light also moves.

In addition to the electron gun and fluorescent surface, the cathode ray tube is provided with two sets of plates: one set, called the horizontal deflection plates, positioned on either side of the electron beam; and the other set, called the vertical deflection plates, positioned above and below the beam. If a negative charge is applied to the left-hand plate

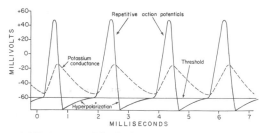

FIGURE 5–13 Rhythmic action potentials, and their relationship to potassium conductance and to the state of hyperpolarization.

and a positive charge to the right-hand plate, the electron beam will be repelled away from the left plate and attracted toward the right plate, thus bending the beam toward the right, and this will cause the spot of light on the fluorescent surface of the cathode ray screen to move to the right. Likewise, positive and negative charges can be applied to the vertical deflection plates to move the beam up or down.

Since electrons travel at extremely rapid velocity and since the plates of the cathode ray tube can be alternately charged positively or negatively within less than a millionth of a second, it is obvious that the spot of light on the face of the tube can also be moved to almost any position in less than a millionth of a second. For this reason, the cathode ray tube oscilloscope can be considered to be an inertialess meter capable of recording with extreme fidelity almost any change in membrane potential.

To use the cathode ray tube for recording action potentials, two electrical circuits must be employed. These are (1) an *electronic sweep circuit* that controls the voltages on the horizontal deflection plates and (2) an *electronic amplifier* that controls the voltages on the vertical deflection plates. The sweep circuit automatically causes the spot of light to begin at the left-hand side and move slowly toward the right. When the spot reaches the right side it jumps back immediately to the left-hand side and starts a new trace.

The electronic amplifier amplifies signals that come from the nerve. If a change in membrane potential occurs while the spot of light is moving across the screen, this change in potential will be amplified and will cause the spot to rise above or fall below the mean level of the trace, as illustrated in the figure. In other words, the sweep circuit provides the lateral movement of the electron beam while the amplifier provides the vertical movement in direct proportion to the changes in membrane potentials picked up by appropriate electrodes.

REFERENCES

Caldwell, P. C.: Factors governing movement and distribution of inorganic ions in nerve and muscle. *Physiol. Rev., 48*:1, 1968.

Cole, K. S.: Electrodiffusion models for the membrane of squid giant axon. *Physiol. Rev., 45*: 340, 1965.

Hodgkin, A. L.: The Conduction of the Nervous Impulse. Springfield, Ill., Charles C Thomas, 1963.

Huxley, A. F.: Ion movements during nerve activity. *Ann. N.Y. Acad. Sci., 81*:221, 1959.

Katz, B.: Nerve, Muscle, and Synapse. New York, McGraw-Hill Book Company, 1968.

Martin, A. R., and Veale, J. L.: The nervous system at the cellular level. *Ann. Rev. Physiol., 29*:401, 1967.

Noble, D.: Applications of Hodgkin-Huxley equations to excitable tissues. *Physiol. Rev., 46*:1, 1966.

Shanes, A. M.: Electrochemical aspects of physiological and pharmacological action in excitable cells. *Pharmacol. Rev., 10*:59, 1958; *10*:165, 1958.

FUNCTION OF NEURONAL SYNAPSES AND NEUROMUSCULAR JUNCTIONS

Every medical student is aware that information is transmitted in the central nervous system through a succession of neurons, one after another. However, it is not immediately apparent that the impulse may be (a) blocked in its transmission from one neuron to the next, (b) changed from a single impulse into repetitive impulses, or (c) integrated with impulses from other neurons to cause highly intricate patterns of impulses in successive neurons. All these functions can be classified as *synaptic functions of neurons*.

PHYSIOLOGIC ANATOMY OF THE SYNAPSE

The juncture between one neuron and the next is called a *synapse*. Figure 6–1 illustrates a typical *motoneuron* in the anterior horn of the spinal cord. It is comprised of three major parts, the *soma*, which is the main body of the neuron; the *axon*, which extends from the soma into the peripheral nerve; and the *dendrites*, which are relatively short projections of the soma into the surrounding areas of the cord.

It should be noted, also, that literally hundreds to thousands of small *presynaptic terminals* lie on the surfaces of the dendrites and soma. These terminals are the ends of nerve fibrils that originate

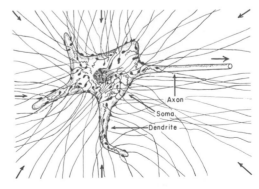

FIGURE 6–1 A typical motoneuron, showing synapses with presynaptic terminals that originate from other neurons. Note the knobbed type of terminals which are believed to be excitatory and the wavy type which are believed to be inhibitory.

61

in many other neurons, and, usually, not more than a few of the terminals are derived from any single previous neuron. Many of these presynaptic terminals are *excitatory* and still others are *inhibitory*.

EXCITATORY FUNCTIONS OF THE SYNAPSES

MECHANISM OF EXCITATION AND THE EXCITATORY TRANSMITTER SUBSTANCE

The mechanism by which the presynaptic terminals excite the postsynaptic neuron is by secretion of an excitatory substance, called the *excitatory transmitter*. Figure 6–2 illustrates the basic structure of the presynaptic terminal. It is separated from the neuronal soma by a *synaptic cleft* having an average width of approximately 200 Ångstroms. In the terminal are many synaptic vesicles containing the *excitatory transmitter*, which, when released into the synaptic cleft, excites the membrane of the neuronal soma.

When an action potential spreads over a presynaptic terminal, the membrane

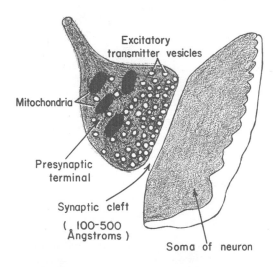

FIGURE 6–2 Physiologic anatomy of the synapse.

depolarization causes automatic emptying of a small number of these vesicles into the cleft; and the released excitatory transmitter in turn causes an immediate increase in permeability of the subsynaptic somal membrane, which allows sodium ions to flow freely from the extracellular fluids to the inside of the neuron, causing decreased electronegativity on the inside. If sufficient numbers of sodium ions diffuse inward, the neuron will discharge. This is discussed in detail in the following paragraphs.

Chemical Nature of the Excitatory Transmitter. It is almost certain that the excitatory transmitter at the synapses in the autonomic ganglia is *acetylcholine*, because measurable quantities of acetylcholine can be recovered from fluids perfusing the autonomic ganglia after strong stimulation of the preganglionic neurons, and injection of acetylcholine will stimulate the postganglionic neurons.

In the central nervous system, injected acetylcholine will stimulate some of the neurons. Therefore, it is believed that at least some of the central nervous system excitatory presynaptic terminals secrete acetylcholine, and perhaps this is true of most of them.

However, it is probable that more than one type of excitatory transmitter substance is secreted in the central nervous system, because some excitatory nerve endings do not seem to have the necessary enzymes to synthesize acetylcholine. On the other hand, some do have the requisite enzymes to synthesize *norepinephrine*, which is the transmitter substance of the peripheral sympathetic nerve endings, as will be discussed later in the chapter. Also, this transmitter can excite at least some of the neurons in the central nervous system. Therefore it, too, is probably an excitatory transmitter at some synapses, particularly those on the intermediolateral horn cells of the spinal cord and perhaps some in the hypothalamus and other basal regions of the brain. Other substances that have been suggested as possible excitatory transmitters include serotonin and L-glutamic acid.

The Excitatory Postsynaptic Potential and its Relationship to the Threshold for Excitation of the Neuron. When an excitatory presynaptic terminal releases its transmitter, the neuronal membrane becomes highly permeable to sodium ions for about 1 millisecond. During this time sufficient sodium diffuses to the interior of the cell to neutralize part of the normal resting potential. That is, the potential inside the cell becomes less negative. This change in resting membrane potential is called the *excitatory postsynaptic potential* (EPSP), and it persists for as long as 15 milliseconds. If the postsynaptic potential is very slight, the neuron will not discharge at all, but, if the potential rises above a certain level, called the *threshold for excitation of the neuron*, action potentials begin to appear in the neuron. The average threshold for excitation of the anterior motoneuron is 11 mv. above the resting membrane potential—that is, when the potential rises from its normal value of −70 to −59 mv., an action potential occurs.

Summation of Excitatory Postsynaptic Potentials. *Spatial summation and the low threshold for excitation of the initial segment of the axon.* Simultaneous excitation of successively greater numbers of excitatory presynaptic terminals causes progressive increase in the postsynaptic potential. This is called *spatial summation*. It should be noted particularly that spatial summation does not occur only in the immediate vicinity of the excited terminals but, instead, *occurs over the entire membrane of the soma and initial segment of the axon, all at the same time.* Also, postsynaptic potentials generated in the dendrites will summate partially with those of the soma despite the wide spatial separation. The reason for all this summation is that the intracellular electrical resistance is very low, which allows a change in potential anywhere in the cell to spread to all areas of the inner surface of the cell membrane almost instantaneously.

Electrical recordings made directly from large neurons show that the first part of the neuron to initiate an action potential, regardless of the positions of the excited presynaptic terminals on the neuron, is the *initial segment of the axon*, that is, the axon hillock and first 50 to 100 microns of the axon, which is non-myelinated. This part of the neuron discharges first because its threshold for excitation (in the anterior motoneuron) is only 11 mv. in contrast to 25 mv. for the soma. After the initial segment discharges, the action potential then travels forward in the axon and backwards over the soma.

This ability of the initial segment of the axon to respond to discharges from widely separated presynaptic terminals means that all the presynaptic discharges can summate together and not simply the discharges of presynaptic terminals that lie near each other on the surface of the cell.

Temporal summation. Not only can discharges from separate presynaptic terminals summate with each other, but rapidly successive discharges *from the same presynaptic terminal* can also summate. Each excitatory postsynaptic potential lasts as long as 15 milliseconds, and, if the terminal discharges a second time before the 15 milliseconds is over, the new postsynaptic potential simply adds to what is left of the old. This effect is called *temporal summation.*

Facilitation of the Neuron; Excitation versus Facilitation. The postsynaptic potential caused by discharge of a single presynaptic terminal is almost always very slight—almost never sufficient to cause *excitation* of the neuron. Instead, either spatial summation or temporal summation, or both, must occur for the potential to become great enough to excite the neuron. However, even subthreshold levels of postsynaptic potential cause a neuron to become *facilitated*. This means that the neuron can then be excited much more easily by additional presynaptic discharges than would normally be true.

In summary, if the summated postsynaptic potential is greater than the threshold for excitation of the neuron, an action potential occurs, but if it is less than the threshold, then the neuron becomes facilitated but not excited.

Relationship of Summated Postsynaptic Potential to Rate of Discharge of the Neuron. A barely threshold excitatory potential causes the neuron to fire approximately 5 to 20 times per second; but as the postsynaptic potential rises progressively above threshold, the repetitive discharges appear closer and closer together, rising to as many as 1000 or more impulses per second in some neurons.

Figure 6-3 demonstrates these relationships between level of summated postsynaptic potential and rate of firing of the neuron. Note that no action potential occurs until the postsynaptic potential rises to threshold level. As it rises somewhat above threshold, discharges then occur every 25 milliseconds, or at a firing rate of 40 per second. Shortly thereafter the postsynaptic potential falls back exactly to threshold level, and 90 milliseconds then lapse between successive action potentials. Finally, toward the end of the

record the postsynaptic potential rises to a very high level, and action potentials then occur about 8 milliseconds apart, giving a repetitive firing rate of 125 discharges per second.

INHIBITORY FUNCTION OF SYNAPSES

The Inhibitory Postsynaptic Potential and the Mechanism of "Postsynaptic" Inhibition. Two different types of inhibition occur at the synapse—*postsynaptic inhibition* and *presynaptic inhibition*. Postsynaptic inhibition results from secretion by certain presynaptic terminals of an *inhibitory transmitter* instead of an *excitatory* transmitter. These terminals derive from special types of neurons that secrete only inhibitory transmitter at all of their endings, as will be discussed later.

The mechanism of postsynaptic inhibition is not certain, but it has been postulated that the inhibitory transmitter increases the permeability of the neuronal membrane to potassium and chloride ions but not to sodium ions, possibly by opening a second set of pores that are too small or otherwise indisposed to the passage of sodium ions. At any rate, potassium ions move outward through the membrane rapidly because there is an excess of potassium inside the cell in relation to the resting membrane potential that is attempting to hold them inside. Movement of potassium ions out of the neuron causes positive charges to move from inside the membrane to the outside, leaving a greater degree of negativity in the neuron. In the spinal motoneuron, this negativity can increase from the normal value of approximately −70 mv. to a value as great as −80 mv., which is 10 mv. more negative than the resting somal membrane potential. This state, called *hyperpolarization*, makes the neuron far less excitable than usual.

The inhibitory transmitter. The precise chemical nature of the inhibitory transmitter, like that of the excitatory

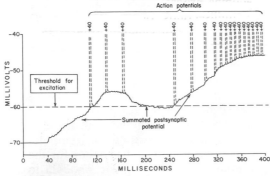

FIGURE 6–3 Relationship between the level of summated excitatory postsynaptic potential and the rate of discharge of the neuron.

transmitter, is unknown. A substance called *gamma aminobutyric acid* (GABA), which is present in the cerebral cortex, is known to inhibit the excitatory action of excitatory presynaptic terminals in some lower animals. Therefore, this substance has been postulated to be an inhibitory transmitter. Another substance that will inhibit neurons and that might possibly act as an inhibitory transmitter in some areas of the nervous system is the amino acid glycine.

Inhibitory Neurons. The inhibitory presynaptic terminals all derive from special types of neurons called inhibitory neurons, and all terminal fibers branching from these neurons secrete only inhibitory transmitter and never excitatory transmitter. (Likewise, excitatory neurons secrete only excitatory transmitter.) When we study the organization of the nervous system in Chapters 31 through 41, it will become apparent that the mechanism by which most inhibition occurs in the nervous system is the passage of a signal first through one of these inhibitory neurons, which in turn secretes the inhibitory transmitter at the next synapse, resulting in inhibition.

"Presynaptic" Inhibition. The second type of inhibition, called presynaptic inhibition, is believed to be caused by excitatory terminals lying on the surfaces of excitatory presynaptic nerve fibrils. When these terminals release the excitatory transmitter, the presynaptic fibrils and their terminals are believed to become partially depolarized, and this prevents their releasing normal quantities of excitatory transmitter. Though the exact mechanism of presynaptic inhibition is still not clear, it is known that the amount of excitatory transmitter released is decreased.

Function of Inhibition in the Central Nervous System. Many nervous pathways are routed through inhibitory synapses, thereby causing inhibition rather than excitation of successive neurons. This inhibition prevents nervous activity that otherwise would occur and thereby helps to determine the overall function of the nervous system. For instance, all parts of the central nervous system are continually barraged by large numbers of incoming nerve impulses from peripheral nerves. If it were not for adequate inhibition of most of these input signals, the brain would remain in a state of such continual excitation that it would be almost useless. Thus, inhibition plays a role in *selecting* those signals that are important while blocking those that are unimportant. Such effects will be discussed much more fully in Chapters 31 through 41 in relation to central nervous system function.

SOME SPECIAL CHARACTERISTICS OF SYNAPTIC TRANSMISSION

Synaptic Delay. The minimal period of time required for transmission of an impulse from presynaptic terminals to a postsynaptic neuron is approximately 0.5 millisecond. This is called the *synaptic delay*. It is important for the following reason: Neurophysiologists can measure the minimal delay time between an input volley of impulses and an output volley and from this can estimate the number of neurons in the circuit.

Fatigue of Synaptic Transmission. When the presynaptic terminals are repetitively stimulated at a rapid rate, the number of discharges by the postsynaptic neuron is at first very great but becomes progressively less in succeeding milliseconds or seconds. This is called fatigue of synaptic transmission. Fatigue is an exceedingly important characteristic of synaptic function, for when areas of the nervous system become overexcited, fatigue causes them to lose this excess excitability after a while. For example, fatigue is probably the means by which the excess excitability of the brain during an epileptic fit is finally subdued so that the fit ceases. Thus, the development of fatigue is a protective mechanism against excess neuronal activity. This will be discussed further in the description of reverberating neuronal circuits in Chapter 31.

The mechanism of fatigue has been presumed to be simply exhaustion of the stores of transmitter substance in the presynaptic terminals, particularly since it has been calculated that the excitatory presynaptic terminals can store enough excitatory transmitter for only 10,000 normal synaptic transmissions, an amount that can be exhausted in only a few seconds.

Effect of Acidosis and Alkalosis on Synaptic Transmission. The neurons are highly responsive to changes in pH of the surrounding interstitial fluids. *Alkalosis greatly increases neuronal excitability.* For instance, a rise in arterial pH from the normal of 7.4 to about 7.8 often causes cerebral convulsions because of increased excitability of the neurons.

On the other hand, *acidosis greatly depresses neuronal activity*; a fall in pH from 7.4 to about 7.0 usually causes a comatose state. For instance, in very severe diabetic or uremic acidosis, coma always develops.

Effect of Hypoxia on Synaptic Transmission. Neuronal excitability is also highly dependent on an adequate supply of oxygen. Cessation of oxygen supply for only a few seconds can cause complete inexcitability of the neurons. This is often seen when the cerebral circulation is temporarily interrupted, for within less than 3 to 5 seconds the person becomes unconscious.

TRANSMISSION OF IMPULSES FROM NERVES TO SKELETAL MUSCLE FIBERS: THE NEUROMUSCULAR JUNCTION

Physiologic Anatomy of the Neuromuscular Junction. Figure 6–4A illustrates the neuromuscular junction between a large myelinated nerve fiber and a skeletal muscle fiber. The nerve fiber branches at its end to form the compli-

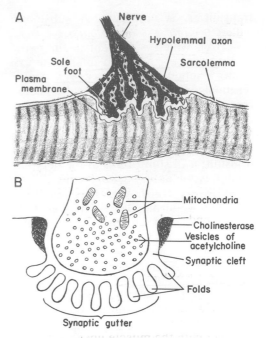

FIGURE 6–4 (*A*) The neuromuscular junction. (*B*) Invagination of a sole foot into the membrane of the muscle fiber.

cated structure called the *end-plate*, which invaginates into the muscle fiber but lies entirely outside the muscle fiber membrane. At the tips of the many nerve branches in the end-plate are *sole feet*.

Figure 6–4B shows an electron micrographic sketch of the juncture between a sole foot and the muscle fiber membrane. The invagination of the membrane is called the *synaptic gutter*, and the space between the sole foot and the fiber membrane is called the *synaptic cleft*. The synaptic cleft is filled with a gelatinous "ground" substance through which diffuses extracellular fluid. At the bottom of the gutter are numerous *folds* of the muscle membrane which greatly increase the surface area at which the synaptic transmitter can act. In the sole foot are many mitochondria that supposedly synthesize the excitatory transmitter *acetylcholine* or that at least supply the energy for this synthesis. The acetylcholine is stored in the sole foot in large numbers of small *vesicles*. Around the rim of the synaptic gutter are large aggregates of the enzyme *cholinesterase*, which is capable of destroying acetylcholine, as is explained in further detail below.

Secretion of Acetylcholine by the Sole Feet, and its Destruction by Cholinesterase. When a nerve impulse reaches the neuromuscular junction, some 50 to 100 vesicles of acetylcholine are released by the sole feet into the synaptic clefts between the sole feet and the muscle fiber membrane. Within approximately 2 milliseconds after acetylcholine is released, it diffuses out of the synaptic gutter and no longer acts on the muscle fiber membrane. The cholinesterase accumulated around the borders of the gutters begins immediately to destroy the acetylcholine, and this destruction is probably complete within another few milliseconds. The very short period of time that the acetylcholine remains in contact with the muscle fiber membrane, about 2 milliseconds, is almost always sufficient to excite the muscle fiber, and yet the rapid removal of the acetylcholine prevents re-excitation after the muscle fiber has recovered from the first action potential.

The "End-Plate Potential" and Excitation of the Skeletal Muscle Fiber. Although the acetylcholine released into the space between the end-plate and the muscle membrane lasts for only a minute fraction of a second, even during this period of time it can nevertheless affect the muscle membrane sufficiently to make it very permeable to sodium ions, allowing rapid influx of sodium into the muscle fiber. As a result, the membrane potential rises a few millivolts, creating a potential called the end-plate potential, which is entirely analogous to the excitatory postsynaptic potential that excites a neuron, as explained earlier in the chapter. Ordinarily, each impulse that arrives at the neuromuscular junction creates an end-plate potential about three to four times that required to stimulate the muscle fiber. Therefore, the normal neuromuscular junction is said to have a very high *safety factor*.

Fatigue of the Neuromuscular Junction. Stimulation of the nerve fiber at rates greater than 150 times per second for many minutes often diminishes the quantity of acetylcholine released with each impulse, so that impulses then often fail to pass into the muscle fiber. This is called *fatigue* of the neuromuscular junction, and it is analogous to fatigue of the synapse. However, under normal functioning conditions, fatigue of the neuromuscular junction almost never occurs, because it is rare that more than 150 impulses per second reach even the most active neuromuscular junctions.

MYASTHENIA GRAVIS

A type of paralysis known as myasthenia gravis occurs in human beings possibly or probably as a result of the inability of the end-plate to secrete adequate acetylcholine. If the disease is intense enough, the patient dies of paralysis of the respiratory muscles. However, the disease can usually be ameliorated by treatment with the drug neostigmine, which is capable of inactivating or destroying cholinesterase. If a sequence of nerve impulses arrives at the end-plate, the quantity of acetylcholine present at the membrane then increases progressively until finally the end-plate potential caused by the acetylcholine rises above threshold value for stimulating the muscle fiber. Thus, it is possible by diminishing the quantity of cholinesterase in the muscles of a patient with myasthenia gravis to allow even the inadequate quantities of acetylcholine secreted at the end-plates to effect almost normal muscular activity.

EXCITATION OF SMOOTH MUSCLE

Physiologic Anatomy of Smooth Muscle Neuromuscular Junctions. Figure 6–5 illustrates a nerve fiber arborizing into a fine reticulum of terminal fibrils and spreading among smooth muscle fibers.

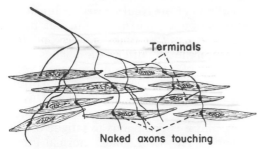

Terminals

Naked axons touching

FIGURE 6-5 Junction of terminal nerve fibrils with smooth muscle fibers, showing a rare discrete terminal but many points where naked axons touch smooth muscle fibers, which points are also believed to act as transmission junctions.

It is believed that most transmission occurs at the points of contact between the naked fibrils and the smooth muscle fibers. Also, some terminal nerve fibers probably release transmitter substance diffusely among the smooth muscle fibers, and the transmitter substance then diffuses in all directions to excite the smooth muscle.

Excitatory and Inhibitory Transmitter Substances at the Smooth Muscle Neuromuscular Junction. Two different transmitter substances known to be secreted by the autonomic nerves innervating smooth muscle are *acetylcholine* and *norepinephrine*. Acetylcholine is an excitatory substance for smooth muscle fibers in some organs but an inhibitory substance for smooth muscle in other organs. And when acetylcholine excites a muscle fiber, norepinephrine ordinarily inhibits it, or when acetylcholine inhibits a fiber norepinephrine excites it.

It is believed that *receptor substances* in the membranes of the different smooth muscle fibers determine which will excite them, acetylcholine or norepinephrine. These receptor substances will be discussed in more detail in Chapter 40 in relation to function of the autonomic nervous system.

Transmission of Action Potentials to Smooth Muscle Fibers—The Junctional Potential. Transmission of impulses from terminal nerve fibers to smooth muscle fibers occurs in very much the same manner as transmission at the neuromuscular junction of skeletal muscle fibers except for temporal differences. That is, depolarization of the smooth muscle fiber is slow to develop, and once it does develop it lasts for a very long period of time. To be more exact, when an action potential reaches the terminal of an excitatory nerve fibril, there is a latent period of about 50 milliseconds before any change in the membrane potential of the smooth muscle fiber can be detected. Then the potential rises to a maximal level in approximately 100 milliseconds. If an action potential does not occur, this potential gradually disappears at a rate of approximately one-half every 200 to 500 milliseconds. This complete sequence of potential changes is called the *junctional potential*; it is analogous to both the excitatory postsynaptic potential of a neuron and the end-plate potential of the skeletal muscle fibers except that its duration is 50 to 200 times as long.

If the junctional potential rises to the threshold level for discharge of the smooth muscle membrane, an action potential will occur in the smooth muscle fiber in exactly the same way that an action potential occurs in a skeletal muscle fiber. A typical smooth muscle fiber has a normal resting membrane potential of −50 to −60 mv., and the threshold potential at which the action potential occurs is about −30 to −40 mv.

Inhibition at the smooth muscle neuromuscular junction. When an inhibitory transmitter substance is released instead of an excitatory transmitter, the membrane potential of the muscle fiber becomes more negative than ever; that is, it becomes *hyperpolarized* and therefore becomes much more difficult to excite than is usually the case. Thus, inhibition occurs at the smooth muscle fiber in identically the same way that inhibition occurs at the neuronal synapse.

TRANSMISSION TO HEART MUSCLE

Autonomic nerve fibers that enter the heart arborize and synapse with cardiac muscle fibers in much the same manner as with smooth muscle fibers, and transmission to the cardiac fibers probably also occurs in much the same manner. Norepinephrine acts as an excitatory transmitter, and acetylcholine acts as an inhibitory transmitter.

REFERENCES

Bradley, P. B.: Pharmacology of the central nervous system. *Brit. Med. Bull., 21*:1, 1965.

Eccles, J.: The Physiology of Synapses. New York, Academic Press, Inc., 1964.

Hebb, C.: CNS at the cellular level: identity of transmitter agents. *Ann. Rev. Physiol., 32*:165, 1970.

Kandel, E. R., and Kupfermann, I.: The functional organization of invertebrate ganglia. *Ann. Rev. Physiol., 32*:193, 1970.

Martin, A. R.: Quantal nature of synaptic transmission. *Physiol. Rev., 46(1)*:51, 1966.

McLennan, H. D.: Synaptic Transmission. Philadelphia, W. B. Saunders Company, 1963.

Tauc, L.: Transmission in invertebrate and vertebrate ganglia. *Physiol. Rev., 47(3)*:521, 1967.

CHAPTER 7

CONTRACTION OF MUSCLE

Approximately one half of the body is skeletal, smooth, or cardiac muscle. Many of the same principles of contraction apply to all these different types of muscle, but in the present chapter the function of skeletal and smooth muscle is considered mainly, whereas the specialized functions of cardiac muscle will be discussed in Chapter 11.

PHYSIOLOGIC ANATOMY OF SKELETAL MUSCLE

The Skeletal Muscle Fiber. All skeletal muscles of the body are made up of numerous muscle fibers ranging between 10 and 100 microns in diameter. Each muscle fiber contains several hundred to several thousand myofibrils, which are illustrated by the small dots in the cross-sectional view of Figure 7-1. Each myofibril in turn has, lying side-by-side, about 1500 myosin filaments and two times this many actin filaments. These filaments are large polymerized protein molecules that are responsible for muscle contraction. They can be seen in the electron micrograph of Figure 7-2, and are represented diagrammatically in Figure 7-3. The thick

70

filaments are *myosin* and the thin filaments are *actin*. Note that the myosin and actin filaments interdigitate and thus the myofibrils have alternate light and dark bands. The light bands, which contain only actin filaments, are called *I bands* because they are *isotropic* to polarized light. The dark bands, which contain the myosin filaments as well as the ends of the action filaments where they overlap the myosin, are called *A bands* because they are *anisotropic* to polarized light. The combination of an A and an I band is called a sarcomere,

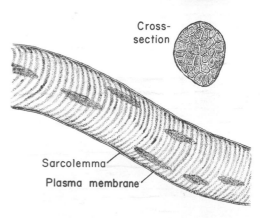

Cross-section

Sarcolemma

Plasma membrane

FIGURE 7-1 Side and cross-sectional views of a skeletal muscle fiber.

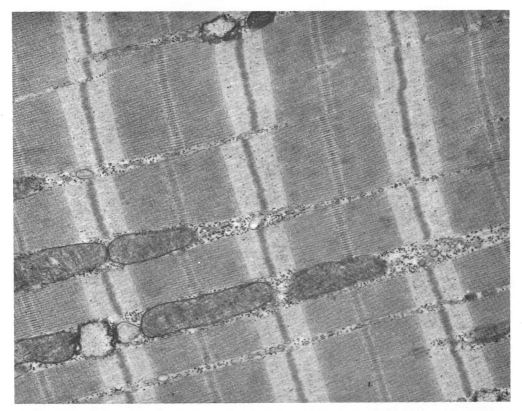

FIGURE 7–2 Electron micrograph of muscle myofibrils, showing the detailed organization of actin and myosin filaments in the fibril. Note the mitochondria lying between the myofibrils. (From Fawcett: The Cell.)

the total length of which in the resting state is about 2 microns.

Note in Figure 7–3 that the actin filaments are attached to each other at the so-called *Z line* or *Z membrane*. The Z membrane passes from one myofibril

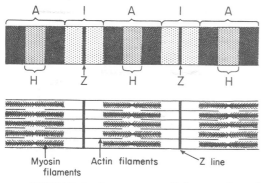

FIGURE 7–3 Arrangement of the myosin and actin filaments in the sarcomeres.

to the next, attaching them to each other all the way across the muscle fiber and causing the respective sarcomeres of adjacent myofibrils to lie side by side.

The actin filaments are longer than the myosin filaments, 2.05 microns in comparison with 1.60 microns. Therefore, as the actin filaments are pulled toward the center of the myosin filaments, their ends overlap. Indeed, the normal muscle functions with the ends of the actin filaments usually slightly overlapped, as will be discussed later.

The sarcoplasm. The myofibrils are suspended in a matrix called *sarcoplasm*, which is composed of usual intracellular constituents. The fluid of the sarcoplasm contains large quantities of potassium, magnesium, phosphate, and protein enzymes. Also present are large numbers of mitochondria that lie mainly in close apposition to the actin filaments of the I bands, suggesting that the actin filaments play a major role in utilizing the ATP formed by the mitochondria.

The sarcoplasmic reticulum. Also in the sarcoplasm is an extensive endoplasmic reticulum, which in the muscle fiber is called the *sarcoplasmic reticulum.* This reticulum has a special organization that is extremely important in the control of muscle contraction, which will be discussed later in the chapter. Figure 7–4 illustrates the arrangement of this sarcoplasmic reticulum and shows how extensive it can be.

The sarcoplasmic reticulum is composed of two distinct and separate types of tubules called the *transverse*, or *T, tubules* and the *longitudinal tubules.* Figure 7–4 illustrates primarily the longitudinal tubules, which lie parallel to the myofibrils. However, it will be noted that each end of each longitudinal tubule terminates in a bulbous cistern abutting against a very small T tubule cut in cross-section. This area of contact between the longitudinal system and transverse system is called the *triad* because it is composed of a central small tubule and, on its sides, two bulbous cisterns of the longitudinal tubules. One triad occurs at each point where the actin and myosin filaments overlap, thus making two triads to each sarcomere.

The T tubules collectively make up the *T system*, which provides a means of communication from the outside of the muscle fiber to its innermost portions. When an action potential spreads over

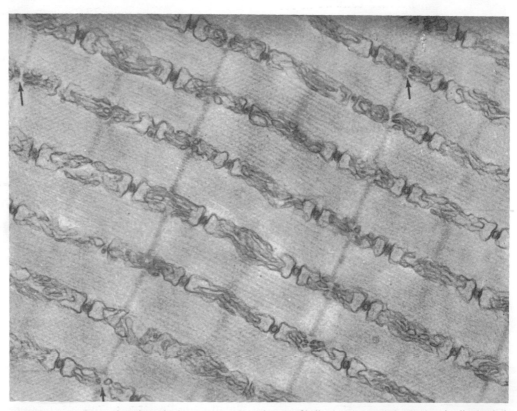

FIGURE 7–4 Sarcoplasmic reticulum surrounding the myofibril, showing mainly the longitudinal tubule system paralleling the myofibrils but also showing in cross-section the small T tubules that lead to the exterior of the fiber membrane and that contain extracellular fluid (arrows). (From Fawcett: The Cell.)

the muscle fiber membrane, electrical currents are transmitted to the interior of the muscle fiber by way of the T tubule system.

ACTION POTENTIALS IN SKELETAL MUSCLE

The normal resting membrane potential of skeletal muscle fibers is essentially the same as that of nerve fibers, about 80 to 90 millivolts. Likewise, the spike potential generated by excitation of a skeletal muscle fiber has essentially the same voltage as that in large nerve fibers, but it lasts 5 to 10 milliseconds instead of 0.5 milliseconds—that is, about 15 times as long.

Velocity of Conduction in Skeletal Muscle Fibers. The average velocity of conduction in skeletal muscle fibers is approximately 5 meters per second, about the same as that in very small myelinated nerve fibers.

SKELETAL MUSCLE CONTRACTION

INITIATION OF CONTRACTION BY THE ACTION POTENTIAL

When an action potential spreads along the muscle fiber, the fiber begins to contract after an initial latent period of about 3 milliseconds. To cause contraction of the myofibrils deep in the muscle fiber, the action potential causes electrical current to flow deep into the interior of the muscle fiber by way of the T tubules. Thus, a local circuit of ionic current flow occurs throughout the entire muscle fiber with each action potential.

Calcium Ions in the Initiation of Contraction. The presence of very low concentrations of calcium ions in the myofibrils, only 2×10^{-4} molar, will cause maximum muscle contraction, but

the normal concentration of calcium ions is only 10^{-7} molar, too little to elicit contraction. Yet, the flow of electrical current through the walls of the longitudinal tubules causes them to release calcium ions into the surrounding sarcoplasm. Within a few ten-thousandths of a second, these ions diffuse to the interior of the myofibrils and there initiate the contractile process.

However, the calcium ions do not remain in the myofibrils for more than a few milliseconds, because once the electrical current caused by the action potential is over the longitudinal tubules almost immediately reabsorb the calcium ions out of the sarcoplasm. Therefore, in effect, the action potential causes a short pulse of calcium ions in the myofibril, and it is during this time that the contractile process is activated. At the termination of this pulse of calcium ions, the muscle immediately relaxes.

THE CONTRACTILE PROCESS

Figure 7–5 illustrates the most likely method of skeletal muscle contraction, showing in the upper drawing a relaxed state of the fiber and in the lower drawing a contracted state. It will be noted from this figure that neither the myosin nor the actin filaments decrease in length during contraction; instead the actin filaments simply slide like pistons inward among the myosin filaments. In the resting, elongated state of the muscle, the ends of the actin filaments are usually barely overlapping. Then during contraction, these ends overlap considerably while the two Z membranes approach the ends of the myosin filaments.

Basic Structures of Myosin and Actin Filaments. Electron micrographs show cross-bridges projecting outward from the myosin filaments toward the actin filaments every 435 Å along the axis of the myosin. On the other hand, the actin filament does not have actual projections in the form of cross-bridges, but it does

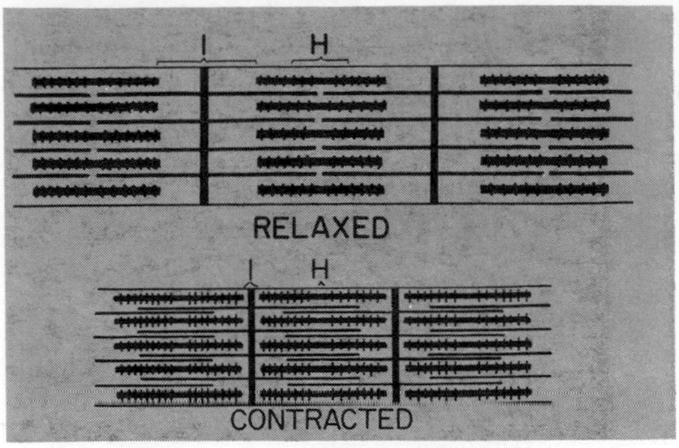

FIGURE 7–5 The sliding model for contraction of skeletal and cardiac muscle.

have a reactive site along its axis occurring once every 405 Å. It is believed that these reactive sites in some way interact with the ends of the cross-bridges to provide the force required for pulling the actin filaments inward among the myosin filaments.

The Ratchet Theory for Muscle Contraction. It was only 15 years ago that the sliding mechanism of the filaments was discovered, and it was suggested then by Huxley, one of the discoverers of this mechanism, that the cross-bridges provide a possible "ratchet" mechanism for causing forceful movement of the actin filaments along the myosin filaments. The theory has been expanded with time and is now essentially the following: First, it is believed that, in the resting state, ATP binds with the cross-bridges of the myosin filaments. The negative charges of the ATP, acting in opposition to the negative charges of the actin filaments, supposedly cause the actin and myosin filaments to remain separated from each other, with no attractive forces between them. Second, on the appearance of calcium ions the following events occur: (a) The calcium ions bind with the negative reactive sites of the ATP on the myosin cross-bridges and, at the same time, with the negative reactive sites on the actin filaments, pulling these filaments together. (b) It is assumed that the cross-bridges, having

strong electronegativity in the resting state because of the presence of ATP, normally project straight outward from the myosin filaments because the shank of the filament also is negatively charged. That is, repulsion of the two charges makes the cross-bridges project outward. But when calcium ions bind with the ATP on the cross-bridges, the negativity of the cross-bridges becomes neutralized. Therefore, the bridges now bend inward toward the axis of the myosin filament. This also pulls the actin filament, thus shortening the muscle. (c) When the cross-bridges fold in against the shank of the myosin filament, the ATPase activity of the myosin filament causes the ATP to split immediately to ADP. This breaks the calcium-linked connections between the myosin cross-bridges and the actin, but in the meantime the actin filament has been pulled along the axis of the myosin filament. (d) Subsequent similar reactions occur at other cross-bridges, and the actin filament is pulled another step. (e) Energy from other sources, such as from high energy creatine phosphate, causes almost immediate reconstitution of the ADP to ATP. Therefore, the cross-bridges that had folded inward now bend outward again and bind with other calcium ions to pull the actin filament another step.

Thus, by a series of "ratchets" — that

is, bonding and pulling, bonding and pulling—the actin filaments theoretically are pulled inward among the myosin filaments. However, one can readily see that this is strictly a theory—indeed, one with so many possible flaws that it will be very surprising if many of its details turn out in the end to be correct. It merely shows how movement of the actin filaments along the myosin filaments could possibly occur.

Function of ATP in contraction of the myofibril. One can see from the theory of myofibrillar contraction just explained that ATP is postulated to perform two principal functions: first, to keep the myosin and actin filaments from bonding during the resting state; and second, to provide the energy for successive folding and unfolding of the cross-bridges as they pull the actin filaments forward.

Relationship Between Actin and Myosin Filament Overlap and Tension Developed by the Contracting Muscle. Figure 7–6 illustrates the relationship between the length of sarcomere and the tension developed by the contracting muscle fiber. To the right are illustrated the degrees of overlap of the myosin and actin filaments at different stages of contraction. At point D on the diagram, the actin filament has pulled all the way out

to the end of the myosin filament with no overlap at all. At this point, the tension developed by the activated muscle is zero. Then as the sarcomere shortens and the actin filament overlaps the myosin filament more and more, the tension increases progressively until the sarcomere length decreases to about 2.2 microns. At this point the actin filament has already overlapped all the cross-bridges of the myosin filament but has not yet reached the center of the myosin filament. Upon further shortening, the sarcomere maintains full tension until point B at a sarcomere length of approximately 2.0 microns. It is at this point that the ends of the two actin filaments begin to overlap. As the sarcomere length falls from 2 microns down to about 1.65 microns at point A, the strength of contraction decreases.

This diagram illustrates that maximum contraction occurs when there is complete overlap between the actin filaments and the cross-bridges of the myosin filaments, but not overlap of the two ends of the actin filaments. The decrease in contractile strength when the actin filaments overlap each other has been postulated to be caused by backward pull of the myosin cross-bridges on the overlapped portions of the actin filaments.

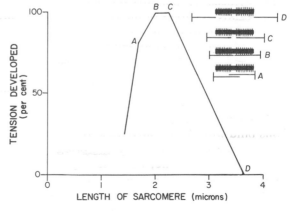

FIGURE 7–6 Length-tension diagram for a single sarcomere, illustrating maximum strength of contraction when the sarcomere is 2.0 to 2.2 microns in length. At the upper right are shown the relative positions of the actin and myosin filaments at different sarcomere lengths from point A to point D. (Modified from Gordon, Huxley, and Julian: *J. Physiol., 171*:28P, 1964.)

Relation of Force of Contraction of the Intact Muscle to Muscle Length. Figure 7–7 illustrates a diagram similar to that in Figure 7–6, but this time for the intact whole muscle rather than for the isolated muscle fiber. The whole muscle has a large amount of connective tissue in it, and the sarcomeres in different parts of the muscle do not necessarily contract exactly in unison. Therefore, the curve has somewhat different dimensions from those illustrated for the individual muscle fiber, but it nevertheless exhibits the same form.

Note in Figure 7–7 that when the muscle is at its normal resting *stretched* length and is then activated, it contracts with maximum force of contraction. If the muscle is stretched to greater than normal length prior to contraction, a large amount of *resting tension* develops in the muscle even before contraction takes place. That is, the two ends of the muscle are pulled toward each other by the elastic forces of the connective tissue, of the sarcolemma, the blood vessels, the nerves, and so forth. However, the *increase* in tension during contraction, called *contractile tension*, decreases as the muscle is stretched beyond its normal length.

Note also in Figure 7–7 that when the resting muscle is shortened to less than its normal fully stretched length, the maximum tension of contraction decreases progressively and reaches zero when the muscle has shortened to approximately 60 per cent of its maximum resting length.

Relation of Muscle Energy Expenditure to Work Performed—the "Fenn" Effect. The shortening process of muscles can lift objects or move objects against force and thereby perform *work*. The amounts of oxygen and other nutrients consumed by the muscle increase greatly when the muscle performs work rather than simply contracting without causing work. This is called the "Fenn" effect. Though this seems to be an obvious effect that one would expect, nevertheless, the chemical basis for it has not been discovered. In some way the contraction of a muscle against a load causes the rate of breakdown of ATP to ADP to increase. This results possibly from the fact that increased numbers of reactive sites must be activated to overcome the load.

EFFICIENCY OF MUSCLE CONTRACTION

The "efficiency" of an engine or a motor is calculated as the percentage of energy input that is converted into work instead of heat. The percentage of the input energy to a muscle (the chemical

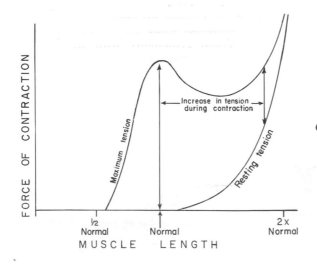

FIGURE 7–7 Relation of muscle length to force of contraction.

energy in the nutrients) that can be converted into work is less than 20 to 25 per cent, the remainder becoming heat. Maximum efficiency can be realized only when the muscle contracts at a moderate velocity.

CHARACTERISTICS OF A SINGLE MUSCLE TWITCH

Many features of muscle contraction can be especially well demonstrated by eliciting single *muscle twitches*. This can be accomplished by instantaneously exciting the nerve to a muscle or by passing a short electrical stimulus through the muscle itself, giving rise to a single, sudden contraction lasting for a fraction of a second.

Isometric versus Isotonic Contraction. Muscle contraction is said to be *isometric* when the muscle does not shorten during contraction, and *isotonic* when it shortens but the tension on the muscle remains constant.

In comparing the rapidity of contraction of different types of muscles, isometric recordings such as those illustrated in Figure 7–8 are usually used instead of isotonic recordings, because the duration of an isotonic recording is almost as dependent on the inertia of the recording system as upon the contraction itself, and this makes it difficult to compare time relationships of contractions from one muscle to another.

Muscles can contract both isometrically and isotonically in the body, but most contractions are actually a mixture of the two. When a person stands, he tenses his quadriceps muscles to tighten the knee joints and to keep the legs stiff. This is isometric contraction. On the other hand, when a person lifts a weight using his biceps, this is mainly an isotonic contraction. Finally, contractions of leg muscles during running are a mixture of isometric and isotonic contractions — isometric mainly to keep the limbs stiff when the legs hit the ground and isotonic mainly to move the limbs.

Characteristics of Isometric Twitches Recorded from Different Muscles. Figure 7–8 illustrates isometric contractions of three different types of skeletal muscles: an ocular muscle, which has a duration of contraction of less than $1/100$ second; the gastrocnemius muscle, which has a duration of contraction of about $1/30$ second; and the soleus muscle, which has a duration of contraction of about $1/10$ second. It is interesting that these durations of contraction are adapted to the function of each of the respective muscles, for ocular movements must be extremely rapid to maintain fixation of the eyes upon specific objects, the gastrocnemius muscle must contract moderately rapidly to provide sufficient velocity of limb movement for running and jumping, while the soleus muscle is concerned principally with slow reactions for continual support of the body against gravity.

CONTRACTION OF SKELETAL MUSCLE IN THE BODY

THE MOTOR UNIT

Each motor neuron that leaves the spinal cord usually innervates many different muscle fibers, the number de-

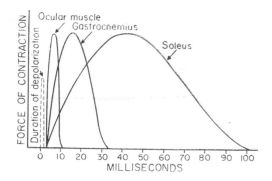

FIGURE 7–8 Duration of isometric contractions of different types of mammalian muscles, showing also a latent period between the action potential and muscle contraction.

pending on the type of muscle. All the muscle fibers innervated by a single motor nerve fiber are called a *motor unit*. In general, muscles that react rapidly and whose control is exact have few muscle fibers (as few as 10 to 15 in ocular muscles) in each motor unit and have a large number of nerve fibers going to each muscle. On the other hand, the slowly acting postural muscles, which do not require a very fine degree of control, may have as many as 300 to 800 muscle fibers in each motor unit. An average figure for all the muscles of the body can be considered to be about 180 muscle fibers to the motor unit.

SUMMATION OF MUSCLE CONTRACTION

Summation means the adding together of individual muscle twitches to make strong and concerted muscle movements. In general, summation occurs in two different ways: (1) by increasing the number of motor units contracting simultaneously and (2) by

increasing the rapidity of contraction of individual motor units. These are called, respectively, *multiple motor unit summation* and *wave summation*, or spatial summation and temporal summation.

Multiple Motor Unit Summation. As more and more motor units are stimulated simultaneously, the greater becomes the strength of contraction. Even within a single muscle, the numbers of muscle fibers and their sizes vary tremendously, so that one motor unit may be as much as 50 times as strong as another. The smaller motor units are far more easily excited than are the larger ones because they are innervated by smaller nerve fibers whose cell bodies in the spinal cord have a naturally high level of excitability. This effect causes the gradations of muscle strength during weak muscle contraction to occur in very small steps, while the steps become progressively greater as the intensity of contraction increases.

Wave Summation. Figure 7–9 illustrates the principles of wave summation, showing in the lower left-hand corner a single muscle twitch followed by suc-

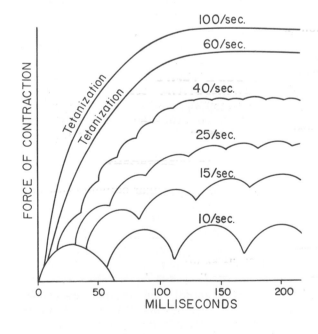

FIGURE 7–9 Wave summation and tetanization.

cessive muscle twitches at various frequencies. When the frequency of twitches is 10 per second, the first muscle twitch is not completely over by the time the second one begins. Therefore, since the muscle is already in a partially contracted state when the second twitch begins, the degree of muscle shortening this time is slightly greater than that which occurs with the single muscle twitch. The third, fourth, and additional twitches add still more shortening.

At more rapid rates of contraction, the degree of summation of successive contractions becomes greater and greater, because the successive contractions appear at earlier times following the preceding contraction.

Tetanization. When a muscle is stimulated at progressively greater rates, a frequency of stimulation is finally reached at which the successive contractions fuse together and cannot be distinguished one from the other. This state is called *tetanization*, and the lowest frequency at which it occurs is called the *critical frequency*.

Tetanization results partly from the viscous properties of the muscle and partly from overlap of successive periods of muscle activation. Once the critical frequency for tetanization is reached, further increase in rate of stimulation increases the force of contraction only a few more per cent, as shown in Figure 7–9.

Asynchronous Summation of Motor Units. Actually it is rare for either multiple motor unit summation or wave summation to occur separately from each other in normal muscle function. Instead, special neurogenic mechanisms in the spinal cord normally increase both the impulse rate and the number of motor units firing at the same time. And even when tetanization of individual motor units of a muscle is not occurring, the tension exerted by the whole muscle is still continuous and nonjerky because *the different motor units fire asynchronously.* That is, while one is contracting another is relaxing; then another fires, followed by still another, and so forth.

Maximum Strength of Contraction. The maximum strength of tetanic contraction of a muscle operating at a normal muscle length is about 3 kilograms per square centimeter of muscle, or 42 pounds per square inch. Since a quadriceps muscle can at times have as much as 16 square inches of muscle belly, as much as 600 to 700 pounds of tension may at times be applied to the patellar tendon. One can readily understand, therefore, how it is possible for muscles sometimes to pull their tendons out of the insertions in bones.

Changes in Muscle Strength at the Onset of Contraction—The Staircase Effect (Treppe). When a muscle begins to contract after a long period of rest, its initial strength of contraction may be as little as one-half its strength 30 to 50 muscle twitches later. That is, the strength of contraction increases to a plateau, a phenomenon called the *staircase effect*, or *treppe*. This phenomenon is believed to be caused primarily by net increase in calcium ions inside the muscle fiber because of movement of calcium ions inward through the membrane with each action potential.

SPECIAL FEATURES AND ABNORMALITIES OF SKELETAL MUSCLE FUNCTION

MUSCULAR HYPERTROPHY

Forceful muscular activity causes the muscle size to increase, a phenomenon called hypertrophy. The diameters of the individual muscle fibers increase, and the fibers gain in total numbers of myofibrils as well as in various nutrient and intermediary metabolic substances, such as adenosine triphosphate, phosphocreatine, glycogen, and so forth. Briefly, muscular hypertrophy increases both the motive power of the muscle and the nutrient mechanisms for maintaining increased motive power.

Weak muscular activity, even when sustained over long periods of time, does not result in significant hypertrophy. Instead, hypertrophy results mainly from *very* forceful muscle activity, though it might occur for only a few minutes each day. For this reason, strength can be developed in muscles much more rapidly when "resistive" or "isometric" exercise is used rather than simply prolonged mild exercise. Indeed, essentially no new myofibrils develop unless the muscle contracts to at least 75 per cent of its maximum tension.

MUSCULAR ATROPHY

Muscular atrophy is the reverse of muscular hypertrophy; it results any time a muscle is not used or even when a muscle is used only for very weak contractions. Atrophy is particularly likely to occur when limbs are placed in casts, thereby preventing muscular contraction. As little as one to two months of disuse can sometimes decrease the muscular size to one-half normal.

Reaction of Muscle to Denervation. When a muscle is denervated it immediately begins to atrophy, and the muscle continues to decrease in size for several years. If the muscle becomes re-innervated during the first three to four months, full function of the muscle usually returns, but after four months of denervation some of the muscle fibers usually will have degenerated. Re-innervation after two years rarely results in return of any function at all. Pathological studies show that the muscle fibers have by that time been replaced by fat and fibrous tissue.

FAMILIAL PERIODIC PARALYSIS

Occasionally, a hereditary disease called familial periodic paralysis occurs. In persons so afflicted, the extracellular fluid potassium concentration periodically falls to very low levels, causing various degrees of paralysis. It is believed that the paralysis is caused in the following manner: A great decrease in extracellular fluid potassium increases the muscle fiber membrane potential to a very high value. This results in strong hyperpolarization of the membrane, making the fiber almost totally inexcitable; that is, the membrane potential is so high that the normal end-plate potential at the neuromuscular junction is incapable of reducing it to the threshold level for excitation.

CONTRACTION OF CARDIAC MUSCLE

The basic mechanism of cardiac muscle contraction is very nearly the same as that in skeletal muscle except that (1) the heart is excited by a rhythmical self-excitation process, and (2) the duration of each single contraction is about 10 times as long as that in skeletal muscle. However, the heart is such an important organ, and the details of its conductive and contractile processes are so important, that these will be presented fully in Chapters 11 and 12.

CONTRACTION OF SMOOTH MUSCLE

Though the smooth muscle in each organ is usually different from that in other organs, smooth muscle can be divided into two major types as illustrated in Figure 7–10: *multiunit smooth muscle* and *visceral smooth muscle*.

Multiunit Smooth Muscle. This type of smooth muscle is composed of discrete smooth muscle fibers, usually having lengths of 100 to 200 microns and diameters averaging about 10 microns. This is the type of smooth muscle found in the blood vessels, in the iris of the eye, and in the nictitating membrane that covers the eyes of some lower ani-

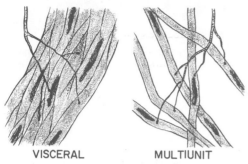

VISCERAL   MULTIUNIT

FIGURE 7–10 Visceral and multiunit smooth muscle fibers.

mals. In general, the multiunit type of smooth muscle contracts only when directly stimulated by a nerve.

Visceral Smooth Muscle. Visceral smooth muscle fibers are very similar to multiunit fibers except that they are crowded together and lie in such close contact with each other that it is sometimes difficult to see the limits between the adjacent cells. For this reason, this type of smooth muscle is also known as *unitary smooth muscle.* This type of muscle is found in most of the organs of the body, especially in the walls of the gut, the bile ducts, the ureters, the uterus, and so forth.

When even one fiber in a visceral muscle tissue is stimulated, the impulse ordinarily is conducted to the surrounding fibers by *ephaptic conduction.* This means that the action potential generated in one of the fibers is strong enough electrically to excite the adjacent fibers without secretion of any excitatory substance. Visceral smooth muscle fibers can transmit impulses one to another in this manner because the fiber membranes where the fibers contact each other develop enhanced permeability, thus allowing easy flow of current from the interior of one cell to the next and, therefore, ease of transmission of impulses from one fiber to the next.

Membrane Potentials in Smooth Muscle. In the normal resting condition, the membrane potential of smooth muscle is usually about −50 millivolts. This is

greatly increased by inhibitory transmitters, which cause *hyperpolarization*, resulting in membrane potentials of −70 or more millivolts and making the fibers very inexcitable. The potential can be decreased to lower values by excitatory transmitters, by stretching the smooth muscle fiber, by hypoxia, by various types of hormones and many other factors. When decreased to about −40 millivolts, action potentials usually occur.

Characteristics of Action Potentials in Smooth Muscle. Because of the extreme variability of smooth muscle in different parts of the body, one would expect many variations in the characteristics of the action potentials. These appear in three different forms as follows:

Spike potentials. Typical spike action potentials, such as those seen in skeletal muscle, occasionally occur in smooth muscle fibers, such as in uterine muscle fibers that have been exposed to large quantities of estrogen. However, this is not a very common form of smooth muscle action potential.

Action potentials with plateaus. One of the most common types of action potentials is that illustrated in Figure 7–11, which shows a typical plateau type of action potential lasting in this case for 0.3 seconds. Such action potentials may last from 0.05 second to as long as 0.5 second, depending on the type of smooth muscle. The muscle remains contracted as long as this action potential persists.

Small spikes and ripples. The resting membrane potential also frequently shows small spikes of only a few millivolts instead of the 70 or more millivolt spikes one sees in the usual action potentials. And many small ripples of 1 to 3 millivolts frequently appear in the resting membrane potential. The small spikes and ripples probably result from local discharges along the surface of the fiber, discharges that do not develop into a full action potential.

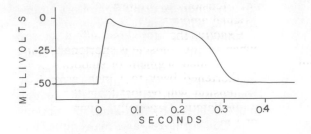

FIGURE 7–11 Monophasic action potential from a smooth muscle fiber of the rat uterus.

Rhythmicity of Visceral Muscle. Most types of visceral muscle exhibit rhythmicity. This results from the presence of at least some smooth muscle fibers within the muscle mass that are self-excitable — that is, their membranes are excessively permeable to sodium, which allows spontaneous generation of action potentials as was explained in Chapter 5. After recovery from each action potential, self-excitation occurs again a few seconds later. As an example, the muscle in the pelvis of the kidney is self-excitable and transmits impulses along the ureter, causing a peristaltic wave to travel along this tubular structure, thus forcing urine into the bladder. Similar effects occur in the bile ducts, in other gland ducts, in the vas deferens, in the gut, and in many other tubular structures of the body.

THE CONTRACTILE PROCESS IN SMOOTH MUSCLE

The Chemical Basis for Contraction. Smooth muscle contains exactly the same chemical substances as skeletal muscle, including both actin and myosin. Typical actin filaments can be seen in electron micrographs; the tips of some of these are apparently attached to the cell membrane, thereby providing a method for mechanically coupling the contractile process inside the fiber with the tissues outside the fiber. On the other hand, typical myosin filaments as seen in striated muscle are rarely seen in smooth muscle, so that there is actually some doubt that myosin fila-

ments of the type known to occur in skeletal muscle are necessary for smooth muscle contraction. Yet, chemically there is approximately the same ratio of myosin molecules to actin molecules in smooth muscle as in skeletal muscle. It is possible that some of the smaller filaments that have tentatively been identified as actin filaments are instead myosin filaments, or it could be that unpolymerized myosin molecules too small to be seen with the electron microscope could provide binding forces between the actin filaments during contraction.

Nevertheless, the contractile process appears in many other ways to be the same as in skeletal muscle: that is, it is activated by calcium ions; ATP is converted to ADP, just as occurs in skeletal muscle; and contraction occurs during depolarization of the membrane and stops following repolarization.

One significant difference between smooth muscle contraction and skeletal muscle contraction is the timing of the process; contraction develops only one-fourth to one-twentieth as rapidly in smooth muscle as in skeletal muscle. Also, relaxation at the end of the action potential is equally slow. This difference probably results from lack of a T tubule system in smooth muscle, a system that greatly accelerates contractile events in skeletal muscle.

Tonus of Smooth Muscle. Smooth muscle can also maintain tonic contraction over long periods of time, and the degree of tonic contraction may be entirely distinct from superimposed rhythmic contractions. Tonic contraction probably results in most instances from summation of repetitive action potentials in the same way that summation occurs in skeletal muscle. If the repetitive action potentials are infrequent, a rhythmic contraction results, but if the repetitive action potentials occur in rapid succession, one can obtain typical tetanization.

Stress-Relaxation of Smooth Muscle. A very important characteristic of smooth muscle is its ability to change length greatly without marked changes in tension. This results from a phenomenon called stress-relaxation, which may be explained as follows:

If a segment of smooth muscle 1 inch long is suddenly stretched to 2 inches, the tension between the two ends increases tremendously at first, but the extra tension begins to disappear immediately, and within a few minutes the tension has returned almost to its level prior to the stretch, even though the muscle is now twice as long. This probably results from the fact that the actin and myosin filaments are arranged randomly in smooth muscle. Over a period of time, the filaments of the stretched muscle presumably rearrange their bonds and gradually allow the sliding process to take place, thus allowing the tension to return almost to its original amount.

Exactly the converse effect occurs when smooth muscle is shortened. Thus, if the 2 inch segment of smooth muscle is shortened back to 1 inch, essentially all tension will be lost from the muscle immediately. Gradually, over a period of 1 minute or more, the tension returns, this again presumably resulting from slow sliding of the filaments. This is called *reverse stress-relaxation*.

REFERENCES

Astrand, P., and Rodahl, K.: Textbook of Work Physiology. New York, McGraw-Hill Book Company, 1968.

Ekelund, L.: Exercise, including weightlessness. *Ann. Rev. Physiol., 31*:85, 1969.

Falls, Harold B.: Exercise Physiology. New York, Academic Press, Inc., 1968.

Hill, A. V., et al.: Physiology of voluntary muscle. *Brit. Med. Bull., 12*(Sept.): 1956.

Huxley, A. F.: Muscle. *Ann. Rev. Physiol., 26*:131, 1964.

Mommaerts, W. F. H. M.: Energetics of muscular contraction. *Physiol. Rev., 49*:427, 1969.

New York Heart Association: The Contractile Process. Boston, Little, Brown and Company, 1967.

Sandow, A.: Skeletal muscle. *Ann. Rev. Physiol., 32*:87, 1970.

CHAPTER 8

RED BLOOD CELLS, WHITE BLOOD CELLS, AND RESISTANCE OF THE BODY TO INFECTION

With this chapter we begin a discussion of blood, which is the vehicle for transport of nutrients, cellular excreta, hormones, electrolytes, and other substances from one part of the body to another. First, we will describe the cells of the blood, and in later chapters we will discuss blood plasma and the relationships of blood to the extravascular fluids.

THE RED BLOOD CELLS

The major function of red blood cells is to transport hemoglobin, which in turn carries oxygen from the lungs to the tissues.

Normal red blood cells are biconcave disks having a mean diameter of approximately 8 microns and a thickness at the thickest point of 2 microns and in the center of 1 micron or less. The shapes of red blood cells can change remarkably as the cells pass through capillaries.

Actually, the red blood cell is a "bag" that can be deformed into almost any shape. Furthermore, because the normal cell has a great excess of cell membrane for the quantity of material inside, deformation does not stretch the membrane, and consequently does not rupture the cell as would be the case with many other cells.

The average number of red blood cells per cubic millimeter of blood is about 5,000,000; women are a few per cent below this value, and men are a few per cent above it. The altitude at which the person lives and his degree of exercise affect the number of red blood cells; these are discussed later.

Quantity of Hemoglobin in the Cells and Transport of Oxygen. Normal red blood cells contain approximately 34 grams of hemoglobin per 100 ml. of cells. However, when hemoglobin formation is deficient in the bone marrow, the percentage of hemoglobin in the cells may fall to as low as 15 grams per cent or less.

When the hematocrit (the percentage of the blood that is cells—normally about 40 per cent) and the quantity of hemoglobin in each respective cell are normal, the whole blood contains an average of 15 grams of hemoglobin per 100 ml. As will be discussed in connection with the transport of oxygen in Chapter 28, each gram of hemoglobin is capable of combining with approximately 1.33 ml. of oxygen. Therefore, 20 ml. of oxygen can normally be carried in combination with hemoglobin in each 100 ml. of blood.

GENESIS OF THE RED BLOOD CELL

The blood cells are produced in the bone marrow and are derived from a cell known as the *hemocytoblast*, illustrated in Figure 8–1. New hemocytoblasts are continually being formed from primordial *stem* cells located throughout the bone marrow.

As illustrated in Figure 8–1, the hemocytoblast first forms the *basophil erythroblast* which begins the synthesis of hemoglobin. The erythroblast then becomes a *polychromatophil erythroblast*, so called because of a mixture of basophilic material and the red hemoglobin. Following this, the nucleus of the cell shrinks while still greater quantities of hemoglobin are formed, and the cell becomes a *normoblast*. During all these stages the different cells continue to divide so that greater and greater numbers of cells are formed. Finally, after the cytoplasm of the normoblast has become filled with hemoglobin to a concentration of approximately 34 per cent, the nucleus is autolyzed and absorbed. The cell that is finally formed, the *erythrocyte,* usually contains no nuclear material when it passes by the process of

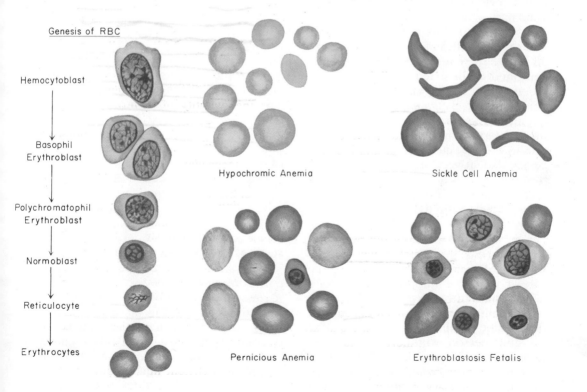

Genesis of RBC

Hemocytoblast

Basophil Erythroblast

Polychromatophil Erythroblast

Normoblast

Reticulocyte

Erythrocytes

Hypochromic Anemia

Sickle Cell Anemia

Pernicious Anemia

Erythroblastosis Fetalis

FIGURE 8–1   Genesis of red blood cells, and red blood cells in different types of anemias.

diapedesis (squeezing through the pores of the membrane) into the blood capillaries.

Some, if not most, of the erythrocytes entering the blood contain small amounts of basophilic reticulum interspersed among the hemoglobin in the cytoplasm. This reticulum is chiefly the remains of the endoplasmic reticulum that produces hemoglobin in the early cell, and hemoglobin continues to be produced as long as the reticulum persists, a period of up to two to three days. In this stage the cells are known as *reticulocytes*. Ordinarily, the total proportion of reticulocytes in the blood is less than 0.5 per cent.

REGULATION OF RED BLOOD CELL PRODUCTION

The total mass of red blood cells in the circulatory system is regulated within very narrow limits, so that an adequate number of red cells is always available to provide sufficient tissue oxygenation and, yet, so that the cells are not concentrated to the extent that they impede blood flow. Thus, when a person becomes extremely *anemic* as a result of hemorrhage or any other condition, the bone marrow immediately begins to produce large quantities of red blood cells. Also, at very *high altitudes*, where the quantity of oxygen in the air is greatly decreased, insufficient oxygen is transported to the tissues, and red cells are produced so rapidly that their number in the blood increases considerably. Finally, the degree of physical activity of a person determines to a great extent the rate at which he will produce red blood cells. Thus, the athlete will have a red blood cell count as high as 6.5 million per cubic millimeter, whereas the asthenic person will have a count as low as 3 million per cubic millimeter.

Erythropoietin, its Response to Hypoxia, and its Function in Regulating Red Blood Cell Production. Despite the very marked effect of hypoxia on red blood cell production, hypoxia does not have a direct effect on the bone marrow. Instead, the hypoxia first causes formation of a hormone called *erythropoietin*. This is a glycoprotein having a molecular weight of about 35,000. Especially large quantities of erythropoietin are formed in the kidneys during hypoxia, but lesser amounts seem to be formed also in the liver and possibly even in other tissues.

Though erythropoietin begins to be formed by the kidneys almost immediately upon placing an animal or person in an atmosphere of low oxygen, very few new red blood cells appear in the circulating blood within the first two days; and it is only after three to four days that the maximum rate of new red cell production is reached. Thereafter, cells continue to be produced as long as the person remains in the low oxygen state or until he has produced enough red blood cells to carry adequate amounts of oxygen to his tissues despite the low oxygen. Erythropoietin seems to increase both maturation rate and numbers of mitotic cells, this effect occurring at all stages in the genesis of the red cell but mainly *at the stem cell* itself to increase the rate of conversion of stem cells into hemocytoblasts.

In the complete absence of erythropoietin, very few red blood cells are formed by the bone marrow. At the other extreme, when extreme quantities of erythropoietin are formed, the rate of red blood cell production can rise to as high as 25 to 50 times normal.

VITAMINS NEEDED FOR FORMATION OF RED BLOOD CELLS

The Maturation Factor—Vitamin B_{12} (Cyanocobalamin). Vitamin B_{12} is an essential nutrient for all cells of the body, and growth of tissues in general is greatly depressed when this vitamin is lacking. This results from the fact that vitamin B_{12} is required for conversion of ribose nucleotides into deoxyribose nucleotides, one of the essential steps in DNA formation. Therefore, lack of this vitamin causes failure of nuclear maturation and division, and greatly inhibits the rate of red blood cell production.

Maturation failure caused by poor absorption of vitamin B_{12} — pernicious anemia. The usual cause of maturation failure is not a lack of vitamin B_{12} in the diet but instead failure to absorb vitamin B_{12} from the gastrointestinal tract. This most commonly occurs in the disease called *pernicious anemia*, in which the basic abnormality is an *atrophic gastric mucosa* that fails to secrete normal gastric secretions. In the mucus secreted by the fundus and the body of the stomach is a mucopolysaccharide or mucopolypeptide (molecular weight about 50,000) called *intrinsic factor*, which combines with vitamin B_{12} of the food and makes the B_{12} available for absorption by the gut. It does this by protecting the B_{12} from digestion by the gastrointestinal enzymes.

Once vitamin B_{12} has been absorbed from the gastrointestinal tract, it is stored in large quantities in the liver and then released slowly as needed to the bone marrow and other tissues of the body. The total amount of vitamin B_{12} required each day to maintain normal red cell maturation is less than 1 microgram, and the normal store in the liver is about 1000 times this amount.

Relationship of folic acid (pteroylglutamic acid) to vitamin B_{12}. Occasionally a patient with maturation failure anemia responds as well to folic acid as to vitamin B_{12}, so that it has become apparent that this vitamin is also concerned with the maturation of red blood cells. Folic acid, like B_{12}, is required for formation of DNA but in a different way. It promotes the methylation of deoxyuridylate to form deoxythymidylate, one of the nucleotides required for DNA synthesis.

FORMATION OF HEMOGLOBIN

Synthesis of hemoglobin begins in the erythroblasts and continues through the

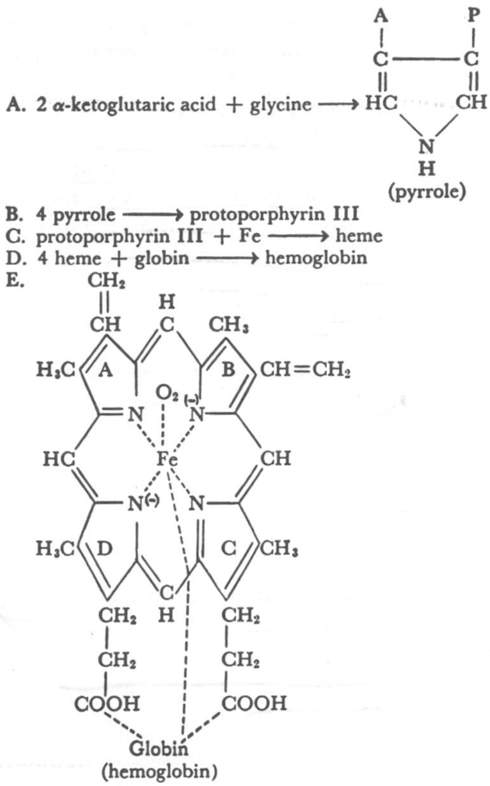

A. 2 α-ketoglutaric acid + glycine ⟶

(pyrrole)

B. 4 pyrrole ⟶ protoporphyrin III
C. protoporphyrin III + Fe ⟶ heme
D. 4 heme + globin ⟶ hemoglobin
E.

Globin
(hemoglobin)

FIGURE 8–2 Formation of hemoglobin.

normoblastic stage. Even when young red blood cells leave the bone marrow and pass into the blood stream, they continue to form hemoglobin for a few days.

Figure 8–2 gives the basic chemical steps in the formation of hemoglobin. From tracer studies with isotopes it is known that hemoglobin is synthesized mainly from acetic acid and glycine, which together form a pyrrole compound. In turn, four pyrrole compounds combine to form a protoporphyrin compound. One of the protoporphyrin compounds, known as protoporphyrin III, then combines with iron to form the heme molecule. Finally, four heme molecules combine with one molecule of globin, a globulin, to form hemoglobin, the formula for which is shown in Figure 8–2E. Hemoglobin has a molecular weight of 68,000.

Combination of Hemoglobin with Oxygen. The most important feature of the hemoglobin molecule is its ability to combine loosely and reversibly with oxygen. This ability will be discussed in detail in Chapter 28 in relation to respiration, for the primary function of hemoglobin in the body depends upon its ability to combine with oxygen in the lungs and then to release this oxygen readily in the tissue capillaries where the gaseous tension of oxygen is much lower than in the lungs.

Oxygen *does not* combine with the two positive valences of the ferrous iron in the hemoglobin molecule. Instead, it binds loosely with two of the six "coordination" valences of the iron atom. This is an extremely loose bond, so that the combination is easily reversible.

IRON METABOLISM

Because iron is important for formation of hemoglobin, myoglobin, and other substances such as the cytochromes, cytochrome oxidase, peroxidase, and catalase, it is essential to understand the means by which iron is utilized in the body.

The total quantity of iron in the body averages about 4 grams, approximately 65 per cent of which is present in the form of hemoglobin. About 4 per cent is present in the form of myoglobin, 1 per cent in the form of the various heme compounds that control intracellular oxidation, 0.1 per cent in the form of transferrin in the blood plasma, and 15 to 30 per cent stored mainly in the form of ferritin.

Transport and Storage of Iron. Transport, storage, and metabolism of iron in the body is illustrated in Figure 8-3, which may be explained as follows: When iron is absorbed from the small intestine, it immediately combines with a beta globulin to form the compound *transferrin*, in which form it is transported in the blood plasma. The iron in this compound is very loosely combined with the globulin molecule and, consequently, can be released to any of the tissue cells at any point in the body. Excess iron in the blood is deposited in all cells of the body *but especially in the liver cells*, where more than 60 per cent of the excess is stored. There it combines with the protein *apoferritin* to form *ferritin*. Apoferritin has a molecular weight of approximately 460,000, and varying quantities of iron can combine in clusters of iron radicals with this large molecule; therefore, ferritin may contain only a small amount of iron or a relatively large amount. This iron stored in ferritin is called *storage iron*.

When the quantity of iron in the plasma falls very low, iron is absorbed

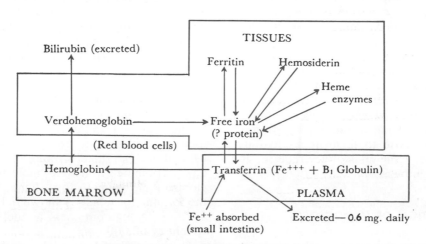

FIGURE 8-3 Iron transport and metabolism.

from ferritin quite easily. The absorbed iron is then transported to the portions of the body where it is needed.

When red blood cells have lived their life span and are destroyed, the hemoglobin released from the cells is ingested by the reticuloendothelial cells. There free iron is liberated, and it can then either be stored in the ferritin pool or be reused for formation of hemoglobin.

Absorption of Iron from the Gastrointestinal Tract. Iron is absorbed almost entirely in the upper part of the small intestine, primarily in the duodenum. It is absorbed mainly in the ferrous form by an active absorptive process, though the precise mechanism of this active absorption is unknown.

Regulation of Total Body Iron by Alteration of Rate of Absorption. When essentially all the apoferritin in the body has become saturated with iron, it becomes difficult for transferrin to release iron to the tissues. As a consequence, the transferrin, which is normally only one-third saturated with iron, now becomes almost fully bound with iron so that the transferrin accepts almost no new iron from the mucosal cells. Then, as a final stage of this process, the buildup of excess iron in the mucosal cells themselves depresses active absorption of iron from the intestinal lumen and at the same time enhances the rate of excretion of iron from the mucosa.

Though the details of the method by which excess iron depresses active absorption of iron by the mucosa are yet mostly unknown, some are beginning to emerge. Most important, if there is excess iron already in the blood and tissues, small vesicles filled with ferritin, called *ferritin bodies*, appear in the newly formed gastrointestinal epithelial cells. For reasons that are not understood, the presence of such bodies in the cells prevents or markedly reduces the active absorption of iron by the cells. On the other hand, if the blood is deficient in iron, ferritin bodies do not appear in the epithelial cells, so that the rate of iron absorption through the intestinal mucosa becomes maximal.

DESTRUCTION OF RED BLOOD CELLS

When red blood cells are delivered from the bone marrow into the circulatory system they normally circulate 100 to 120 days before being destroyed. Even though red cells do not have a nucleus, they do still have cytoplasmic enzymes for metabolizing glucose and other substances and for the utilization of oxygen, but these metabolic systems become progressively less active with time. As the cells become older they become progressively more fragile, presumably because their life processes simply wear out.

Once the red cell membrane becomes very fragile, it ruptures during passage through some tight spot of the circulation. Many of the red cells fragment in the spleen, perhaps where the cells squeeze through the red pulp of the spleen. When the spleen is removed, the number of abnormal cells and old cells circulating in the blood increases considerably.

Destruction of Hemoglobin. The hemoglobin released from the cells when they burst is phagocytized and digested almost immediately by reticuloepithelial cells, releasing iron back into the blood to be carried by transferrin either to the bone marrow for production of new red blood cells or to the liver and other tissues for storage in the form of ferritin. The heme portion of the hemoglobin molecule is converted through a series of stages into the pigment called *bilirubin,* which is excreted by the liver into the bile.

Excretion of Bilirubin in the Bile, and Jaundice. When bilirubin is released into the blood from the reticuloendothelial cells, it combines immediately with some of the plasma proteins, mainly albumin, and is transported in this combination throughout the blood and interstitial fluids. This protein-bound bilirubin eventually is absorbed into the liver cells, and the bilirubin is removed from the protein and conjugated with other substances, mainly glucuronic acid, that make it become highly soluble. It is in this soluble form that bilirubin is excreted into the bile.

Formation and fate of urobilinogen. Once in the intestine, bilirubin is converted by bacterial action mainly into the substance *urobilinogen*, which is highly soluble. Some of the urobilinogen is reabsorbed through the intestinal mucosa into the blood and is re-excreted by the liver back into the bile, or the urobilinogen may be excreted by the kidneys into the urine. After exposure to air in the urine, the urobilinogen becomes oxidized to *urobilin*, or it becomes altered and oxidized in the feces to form *stercobilin*. These interrelationships of bilirubin and the other bile pigments are illustrated in Figure 8–4.

Jaundice. The word "jaundice" means a yellowish tint to the body tissues, including yellowness of the skin and also of the deep tissues. The usual cause of jaundice is large quantities of bilirubin in the extracellular fluids, either protein-bound bilirubin or soluble bilirubin.

The common causes of jaundice are:

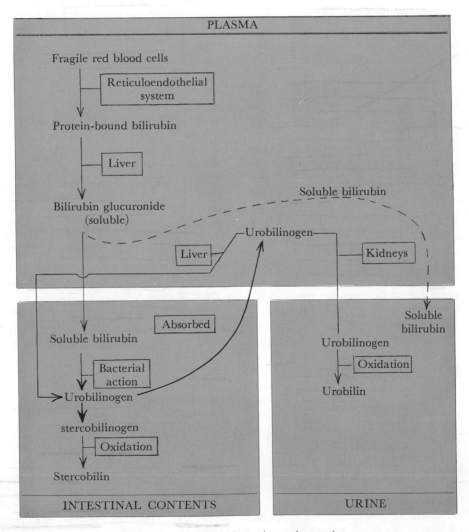

FIGURE 8–4 Bilirubin formation and excretion.

(1) increased destruction of red blood cells with rapid release of bilirubin into the blood or (2) obstruction of the bile ducts or damage to the liver cells so that even the usual amounts of bilirubin cannot be excreted into the gastrointestinal tract. These two types of jaundice are called, respectively, *hemolytic jaundice* and *obstructive jaundice*. They differ from each other in the following ways:

Hemolytic Jaundice. In hemolytic jaundice, red blood cells are hemolyzed rapidly and the hepatic cells simply cannot excrete the bilirubin as rapidly as it is formed. Therefore, the plasma concentration of *protein-bound bilirubin* rises especially high.

Obstructive Jaundice. In obstructive jaundice, caused either by obstruction of the bile ducts or by damage to the liver cells, the rate of bilirubin formation is normal, but the excretion of bilirubin is blocked at the level of the bile ducts. Instead, the bilirubin is returned to the blood mainly by rupture of congested bile canaliculi and direct emptying of bile into the lymph leaving the liver. However, since this bilirubin has already been through the hepatic cells, *most of the bilirubin in the plasma becomes the soluble type* rather than the protein-bound type.

Diagnostic Differences Between Hemolytic and Obstructive Jaundice. A simple test called the *van den Bergh test* can be used to differentiate between protein-bound and soluble bilirubin in the plasma. If the bilirubin is of the soluble type, then the diagnosis is obstructive jaundice; if of the protein-bound type, the diagnosis is hemolytic jaundice.

ANEMIA

Anemia means a deficiency of red blood cells, which can be caused either by too rapid loss or by too slow production of red blood cells. Some of the causes of anemia are:

1. *Blood loss.*

2. *Bone marrow aplasia,* in which the bone marrow is destroyed. Common causes of this are drug poisoning or gamma ray irradiation.

3. *Maturation failure* because of lack of vitamin B_{12} or folic acid, as was previously explained.

4. *Hemolysis of red cells* resulting from many possible causes, such as (a) drug poisoning, (b) hereditary diseases that make the red cell membranes friable, and (c) erythroblastosis fetalis, a disease of the newborn in which antibodies from the mother destroy red cells in the baby (This will be discussed in Chap. 9).

Effects of Anemia on the Circulatory System. The viscosity of the blood, which will be discussed in detail in Chapter 14, is dependent almost entirely on the concentration of red blood cells. In severe anemia the blood viscosity may fall to as low as one and one-half times that of water rather than the normal value of approximately three times the viscosity of water. The greatly decreased viscosity decreases the resistance to blood flow in the peripheral vessels so that far greater than normal quantities of blood return to the heart. As a consequence, the cardiac output increases to as much as two or more times normal.

Moreover, hypoxia due to diminished transport of oxygen by the blood causes the tissue vessels to dilate, allowing further increased return of blood to the heart, increasing the cardiac output to a higher level than ever. Thus, one of the major effects of anemia is greatly *increased work load on the heart*.

The increased cardiac output in anemia offsets many of the symptoms of anemia, for, even though each unit quantity of blood carries only small quantities of oxygen, the rate of blood flow may be increased to such an extent that almost normal quantities of oxygen are delivered to the tissues.

As long as an anemic person's rate of activity is low, he can live without fatal hypoxia of the tissues, even though his concentration of red blood cells may be reduced to one-fourth normal. However, when he begins to exercise, his heart is not capable of pumping much greater quantities of blood than it is already pumping. Consequently, during exercise, which greatly increases the demands for oxygen, extreme tissue hypoxia results, and acute, cardiac failure often ensues.

RESISTANCE OF THE BODY TO INFECTION

The body is constantly exposed to bacteria, these occurring especially in the mouth, the respiratory passageways, the colon, the mucous membranes of the eyes, and even the urinary tract. Many of these bacteria are capable of causing disease if they invade the deeper tissues. In addition, a person is intermittently exposed to highly virulent bacteria and viruses from outside the body which can cause specific diseases such as pneumonia, streptococcal infections, and typhoid fever.

On the other hand, a group of tissues including the *reticuloendothelial system* and the leukocytes constantly combat any infectious agent that tries to invade the body. These tissues function in two different ways to prevent disease: (1) by actually destroying invading agents by the process of phagocytosis and (2) by forming antibodies against the invading agent, the antibodies in turn destroying the invader. The present chapter is concerned with the first of these methods, while the following chapter is concerned with the second.

THE RETICULOENDOTHELIAL SYSTEM

The reticuloendothelial system is comprised of those cells that line many of the vascular and lymph channels and are capable of phagocytizing bacteria, viruses, or other foreign agents or are

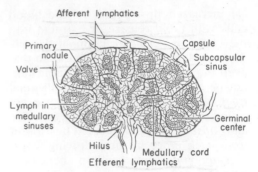

FIGURE 8–5 Functional diagram of a lymph node. (Redrawn from Ham: Histology. J. B. Lippincott Co.)

capable of forming immune bodies against these. They include the phagocytic cells of the bone marrow, spleen, liver, and lymph nodes. All these cells are closely related to each other, and are called *reticulum cells.*

The Lymph Nodes. The lymphatic system is discussed in Chapter 16, where it is pointed out that essentially no particulate matter or even large molecular compounds can be absorbed directly through the capillary membranes into the blood stream. Instead, any invading organisms or large molecules that enter the tissue spaces must pass by way of the lymphatic system into the blood. Lymph nodes are located intermittently along the course of the lymphatics. Figure 8–5 illustrates the general organization of a lymph node, showing lymph entering by way of the *afferent lymphatics,* flowing through the outer capsule of the lymph node into *subcapsular sinuses,* then gradually through the *medullary sinuses,* and finally out the *hilus* into the *efferent lymphatics.* Large numbers of *reticulum cells* line the sinuses, and if any particles or proteins foreign to the person's own body are in the lymph, these cells phagocytize them and prevent general dissemination throughout the body.

The Reticular Structures of the Circulatory System. If invading agents succeed in getting past the lymph nodes or if they should by chance enter the circulation directly without going through the lymph nodes, there are still other lines of defense for their removal even after they enter the circulation. In addition to the circulating granulocytes, which can phagocytize the invaders, three special structures of the circulation contain large numbers of reticulum cells—the spleen, the liver, and the bone marrow.

Figure 8-6 illustrates the general structure of the *spleen*, with an artery entering the splenic capsule, then passing into the *splenic pulp*, and probably connecting through small capillaries with the large *venous sinuses*. The capillaries seem to be highly porous, allowing large quantities of whole blood cells to pass out of the capillaries into the *red pulp* and gradually to *squeeze* through the pulp and then back through the sinus walls into the venous system. The red pulp is loaded with reticulum cells, and, in addition, the venous sinuses are also lined with reticulum cells. This peculiar passage of cells through the red pulp provides an exceptionally effective means for phagocytosis of unwanted debris in the blood, especially old and abnormal red blood cells.

The *sinuses of the liver* are also lined with reticulum cells, called *Kupffer cells*, which are illustrated in Figure 8-7. The reticulum system of the liver is especially valuable because large numbers of

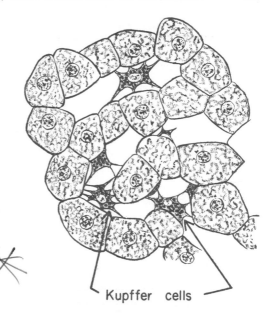

Kupffer cells

FIGURE 8-7 Kupffer cells lining the liver sinusoids, showing phagocytosis of India ink particles. (Redrawn from Copenhaver and Johnson: Bailey's Textbook of Histology. The Williams and Wilkins Co.)

bacteria constantly pass by diapedesis through the gastrointestinal mucosa into the portal blood. The liver filtration system is so effective that essentially none of these bacteria succeeds in passing from the portal blood into the general systemic circulation.

The reticulum cells of the *bone marrow* are particularly capable of removing very fine particles such as protein toxins.

The Tissue Histiocyte and Defense of the Tissues Against Infection. In virtually all tissues of the body are many cells called *histiocytes* or *clasmatocytes* that have properties similar to those of reticulum cells. They can also become mobile and move to points of infection in the tissues, in which case they are called *macrophages*. These are always available to remove foreign substances, and they can even phagocytize necrotic tissue cells and any other debris that might enter the tissues.

Two of the white blood cells, the monocytes and lymphocytes, upon migrating into the tissues, can also become

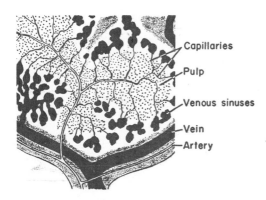

Capillaries

Pulp

Venous sinuses

Vein

Artery

FIGURE 8-6 Functional structures of the spleen. (Modified from Maximow and Bloom: Textbook of Histology.)

histiocytes or macrophages. Histiocyte function is discussed further in connection with the functions of the white blood cell.

WHITE BLOOD CELLS (LEUKOCYTES)

The white blood cells are the *mobile units* of the body's protective system. They are formed partially in the bone marrow (the *granulocytes*) and partially in the lymph nodes (*lymphocytes* and *monocytes*), but after formation they are transported in the blood to the different parts of the body where they are to be used.

The Types of White Blood Cells. Five different types of white blood cells are normally found in the blood. These are *polymorphonuclear neutrophils, polymorphonuclear eosinophils, polymorphonuclear basophils, monocytes,* and *lymphocytes.* In addition, there are large numbers of *platelets,* which are fragments of a sixth type of white blood cell, the *megakaryocyte.* The three types of polymorphonuclear cells have a granular appearance, as illustrated in Figure 8–8, for which reason they are called *granulocytes,* or in clinical terminology they are often called simply "polys."

The granulocytes and the monocytes protect the body against invading organisms by ingesting them—that is, by the process of *phagocytosis.* One of the functions of the lymph glands and lymphocytes is formation of monocytes which in turn destroy invading organisms. Finally, the function of platelets is to activate the blood clotting mechanism. All these functions are protective mechanisms of one type or another.

Concentrations of the Different White Blood Cells in the Blood. The adult human being has approximately 7000 white blood cells per cubic millimeter of blood. The normal percentages of the different types of white blood cells are approximately the following:

Polymorphonuclear neutrophils	62.0%
Polymorphonuclear eosinophils	2.3%
Polymorphonuclear basophils	0.4%
Monocytes	5.3%
Lymphocytes	30.0%

The number of platelets in each cubic millimeter of blood is normally about 300,000.

GENESIS OF THE LEUKOCYTES

Figure 8–8 illustrates the stages in the development of white blood cells. The polymorphonuclear cells are normally formed only in the bone marrow. On the other hand, lymphocytes and monocytes are produced in the various lymphogenous organs, including the lymph glands, the spleen, the thymus, the tonsils, and various lymphoid rests in the gut and elsewhere.

Some of the white blood cells formed in the bone marrow are stored within the marrow until they are needed in the circulatory system. Then when the need arises, various factors that are discussed later cause them to be released.

As illustrated in Figure 8–8, megakaryocytes are also formed in the bone marrow and are part of the myelogenous group of white blood cells. It is generally taught that these megakaryocytes fragment in the bone marrow, the small fragments known as *platelets* passing then into the blood, but it is possible that the megakaryocytes, which are very large cells, are released directly into the blood and fragment later as they attempt to traverse the capillaries.

PROPERTIES OF WHITE BLOOD CELLS

Diapedesis. White blood cells can squeeze through the pores of the blood vessels by the process of diapedesis. That is, even though a pore is much smaller than the size of the cell, a small portion of the cell slides through the pore at a time, the portion sliding

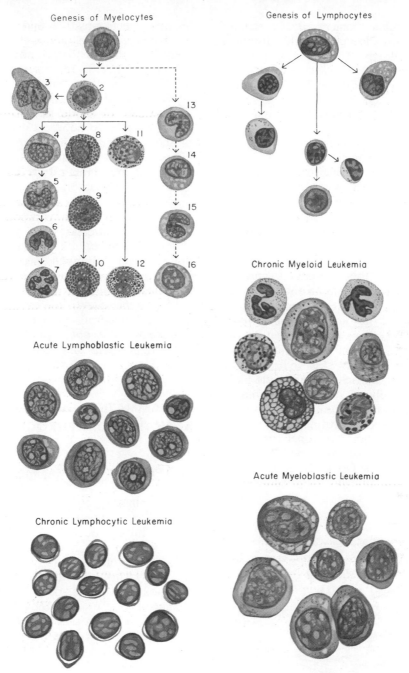

FIGURE 8–8 Genesis of the white blood cells, and the blood picture in different types of leukemia. The different cells of the myelogenous series are: *1*, myeloblast; *2*, promyelocyte; *3*, megakaryocyte; *4*, neutrophil myelocyte; *5*, young neutrophil metamyelocyte; *6*, "band" neutrophil metamyelocyte; *7*, polymorphonuclear neutrophil; *8*, eosinophil myelocyte; *9*, eosinophil metamyelocyte; *10*, polymorphonuclear eosinophil; *11*, basophil myelocyte; *12*, polymorphonuclear basophil; *13–16*, stages of monocyte formation. (Redrawn in part from Piney: A Clinical Atlas of Blood Diseases. The Blakiston Co.)

through being momentarily constricted to the size of the pore, as illustrated in Figure 8–9.

Ameboid Motion. Once the cells have entered the tissue spaces, the polymorphonuclear leukocytes especially, and the large lymphocytes and monocytes to a lesser degree, move through the tissues by ameboid motion, which was described in Chapter 2. Some of the cells can move through the tissues at a rate as great as 40 microns per minute — that is, they can move at least three times their own length each minute.

Phagocytosis. The most important function of the neutrophils and monocytes is phagocytosis.

Obviously, the phagocytes must be selective in the material that is phagocytized, or otherwise some of the structures of the body itself would be ingested. Whether or not phagocytosis will occur depends especially upon three selective procedures. First, if the surface of a particle is rough, the likelihood of phagocytosis is increased. Second, most natural substances of the body have electronegative surface charges and therefore repel the phagocytes, which also carry electronegative surface charges. On the other hand, dead tissues and foreign particles are fre-quently electropositive and are therefore subject to phagocytosis. Third, the body has a means for promoting phagocytosis of specific foreign materials by first combining them with globulin molecules called *opsonins*. After the opsonin has combined with the particle, it allows the phagocyte to adhere to the surface of the particle, which promotes phagocytosis. The special relationship of opsonization to immunity is discussed in the following chapter.

The *macrophages*, which are derived mainly from the monocytes of the blood, are much more powerful phagocytes than the neutrophils. They have the ability to engulf much larger particles and often five or more times as many particles as the neutrophils. And they can even phagocytize whole red blood cells or malarial parasites, whereas neutrophils are not capable of phagocytizing particles much larger than bacteria.

INFLAMMATION AND THE FUNCTION OF LEUKOCYTES

THE PROCESS OF INFLAMMATION

Inflammation comprises the changes in the tissues in response to injury. When tissue injury occurs, whether it be caused by bacteria, trauma, chemicals, heat, or any other phenomenon, the substance *histamine*, along with other humoral substances, is liberated by the damaged tissue into the surrounding fluids. This increases the local blood flow and also increases the permeability of the capillaries, allowing large quantities of fluid and protein, including fibrinogen, to leak into the tissues. Local extracellular edema results, and the extracellular fluid and lymphatic fluid both clot because of the coagulating effect of tissue exudates on the leaking fibrinogen. Thus, *brawny edema* develops in the spaces surrounding the injured cells.

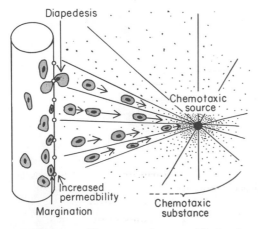

FIGURE 8–9 Movement of neutrophils by the process of *chemotaxis* toward an area of tissue damage.

The "Walling Off" Effect of Inflammation.

The net result of inflammation is to "wall off" the area of injury from the remaining portions of the body. The tissue spaces and the lymphatics are blocked by fibrinogen clots so that fluid flows through the spaces very slowly. Therefore, walling off the area of injury delays the spread of bacteria or toxic products.

Attraction of Neutrophils to the Area of Inflammation.

When tissues are damaged, several effects occur to cause movement of neutrophils into the damaged area. First, the neutrophils stick to the walls of the damaged capillary, resulting in a process known as "margination," which is shown in Figure 8–9. Gradually the cells pass by diapedesis into the tissue spaces.

The second effect is chemotaxis of the neutrophils toward the damaged area; this is caused by bacterial or cellular products that attract the neutrophils. Thus, within a few hours after tissue damage begins, the area becomes well supplied with neutrophils.

Neutrophilia During Inflammation— Leukocytosis-Promoting Factor.

The term *neutrophilia* means an increase above normal in the number of neutrophils in the circulatory system, and the term *leukocytosis* means an excess total number of white blood cells.

A globulin substance known as *leukocytosis-promoting factor* is liberated by most inflamed tissues. This factor diffuses into the blood and finally to the bone marrow where it has two actions: First, it causes large numbers of granulocytes, especially neutrophils, to be released within a few minutes to a few hours into the blood from the storage areas of the bone marrow, thus increasing the total number of neutrophils per cubic millimeter of blood sometimes to as high as 20,000 to 30,000. Second, it increases the rate of granulocyte production by the bone marrow. Within a day or two after onset of the inflammation, the bone marrow becomes hyperplastic and then continues to produce large numbers of granulocytes as long as leukocyte-promoting factor is formed in the inflamed tissues.

THE MACROPHAGE RESPONSE IN CHRONIC INFLAMMATION

The monocytic cells—including the tissue histiocytes, the blood monocytes, and the lymphocytes—also play a major role in protecting the body against infection. First, the tissue histiocytes change into macrophages within minutes, develop ameboid motion, and migrate chemotactically toward the area of inflammation. These cells provide the first line of defense against the infection within the first hour or so, but their numbers are not very great. Within the next few hours, the neutrophils become the primary defense, reaching their maximum effectiveness in about 6 to 12 hours. By that time large numbers of monocytes have begun to enter the tissues from the blood. They change their characteristics drastically during the first few hours; they start to swell, to exhibit increased ameboid motion, and to move chemotactically toward the damaged tissues.

Finally, the largest source of all for developing macrophages is the lymphocytes of the blood, for tremendous numbers of lymphocytes invade the damaged tissue by about the 10th to 12th hour. Then within the next hour or so, these undergo progressive changes first into monocytes and then into macrophages.

Therefore, in summary, very large numbers of macrophages are derived from three different sources: the tissue histiocytes, the monocytes of the blood, and the lymphocytes of the blood.

Formation of Pus.

When the neutrophils and macrophages engulf large amounts of bacteria and necrotic tissue they themselves eventually die. After several days, a cavity is often excavated in the inflamed tissues containing varying portions of necrotic tissue, dead neutrophils, and dead macrophages.

Such a mixture is commonly known as *pus*.

THE EOSINOPHILS

The function of the eosinophils is almost totally unknown even though they normally comprise 1 to 3 per cent of all the leukocytes. Eosinophils are weak phagocytes, and they exhibit chemotaxis, but in comparison with the neutrophils, it is doubtful that the eosinophils are of significant importance in protection against usual types of infection.

Eosinophils enter the blood in large numbers following foreign protein injection, during allergic reactions, and during infection with parasites. The function of eosinophils in these conditions is unknown, though it is supposed that they might remove toxic substances from the tissues.

THE BASOPHILS

The basophils in the circulating blood are very similar to the large *mast* cells located immediately outside many of the capillaries in the body. The mast cells liberate *heparin* into the blood, a substance that can prevent blood coagulation and that can also speed the removal of fat particles from the blood after a fatty meal. Therefore, it is probable that the basophils in the circulating blood perform similar functions within the blood stream, or it is even possible that the blood simply transports basophils to tissues where they then become mast cells and perform the function of heparin liberation.

THE LYMPHOCYTES

Until recently it was believed that the lymphocytes represent a homogeneous group of cells, with a possible distinction between the so-called small lymphocytes and large lymphocytes. However, it is now known that the cell known as the lymphocyte comprises a number of different types of cells, all of which have essentially the same staining characteristics. A few of these cells, probably produced primarily in the bone marrow, seem to be multipotential cells that are similar, if not identical, to the stem cell from which almost any other type of cell can be formed. These multipotential cells can be changed into erythroblasts, myeloblasts, monocytes, plasma cells, other types of small lymphocytes, fibroblasts, and so forth.

Many of the lymphocytes play special roles in the process of immunity. These will be discussed in detail in Chapter 9.

AGRANULOCYTOSIS

A clinical condition known as "agranulocytosis" occasionally occurs, in which the bone marrow stops producing white blood cells, leaving the body unprotected against bacteria and other agents that might invade the tissues. The cause is usually drug poisoning or irradiation following nuclear bomb blast. Within two days after the bone marrow stops producing white blood cells, ulcers appear in the mouth and colon, or the person develops some form of severe respiratory infection. Bacteria from the ulcers then rapidly invade the surrounding tissues and the blood. Without treatment, death usually ensues three to six days after agranulocytosis begins.

THE LEUKEMIAS

Uncontrolled production of white blood cells can be caused by cancerous mutation of a myelogenous or a lymphogenous cell, causing leukemia, which means greatly increased numbers of white blood cells in the circulating blood. Ordinarily, leukemias are divided into two general types: the *lymphogenous leukemias* and the *myelogenous leukemias.*

The first effect of leukemia is metastatic growth of leukemic cells in abnormal areas of the body. The leukemic cells of the bone marrow may reproduce so greatly that they invade the surrounding bone, causing a tendency to easy fracture. Almost all leukemias spread to the spleen, the lymph nodes, the liver, and other especially vascular regions, regardless of whether the origin of the leukemia is in the bone marrow or in the lymph nodes. In each of these areas the rapidly growing cells invade the surrounding tissues, utilizing the metabolic elements of these tissues and consequently causing tissue destruction.

Perhaps the most important effect of leukemia on the body is the excessive use of metabolic substrates by the growing cancerous cells. The leukemic tissues reproduce new cells so rapidly that tremendous demands are made on the body fluids for foodstuffs, especially the amino acids and vitamins. Consequently, the energy of the patient is greatly depleted, and the excessive utilization of amino acids causes rapid deterioration of the normal protein tissues of the body. Thus, while the leukemic tissues grow, the other tissues are debilitated. Obviously, after metabolic starvation has continued long enough, this alone is sufficient to cause death.

REFERENCES

Biology and function of the lymphocyte (Symposium). *Fed. Proc.,* p. 1711, 1966.

Boggs, D. R.: Homeostatic regulatory mechanisms of hematopoiesis. *Ann. Rev. Physiol.,* 28:39, 1966.

Crosby, W. H.: Iron absorption. *In* Handbook of Physiology. Baltimore, The Williams & Wilkins Co., 1968, Vol. III, Sec. 6, p. 1553.

Fisher, J. W.: The structure and physiology of erythropoietin. *In* Bittar, E. Edward, and Bittar, Neville (eds.): The Biological Basis of Medicine. New York, Academic Press, Inc., 1969, Vol. 3, p. 41.

Heller, J. H.: Host defence and the reticulo-endothelial system. *In* Bittar, E. Edward, and Bittar, Neville (eds.): The Biological Basis of Medicine. New York, Academic Press, Inc., 1969, Vol. 4, p. 181.

Weissbach, H., and Dickerman, H.: Biochemical role of vitamin B_{12}. *Physiol. Rev., 45*:80, 1965.

Zweifach, B.: The Inflammatory Process. New York, Academic Press, Inc., 1965.

CHAPTER 9

IMMUNITY; BLOOD GROUPS; TRANSFUSION; AND TRANSPLANTATION

The body has a large amount of *natural immunity*, which makes it resistant to many bacteria, viruses, and toxins even without any specific immune process. For instance, the human body is entirely resistant to distemper virus that results in death of 25 to 30 per cent of all dogs that are afflicted with it. Likewise, the body has native resistance against herpes virus, which is almost always lethal in the rabbit but which rarely causes more than fever blisters or chicken pox in the human being.

However, in addition to its natural immunity, the human body also has the ability to develop immunity against specific invading agents such as lethal bacteria, viruses, toxins, and even foreign tissues from other animals.

Unfortunately, the body can become immune also to cells and tissues from other human beings, which can cause serious difficulties in transfusion of blood or transplantation of organs. This problem will be discussed later in the chapter.

THE IMMUNE PROCESS

The immune process is initiated by exposure to an invading agent. Once exposed, the body begins developing in its lymphoid tissues (1) specific globulin molecules called *antibodies* or (2) *sensitized lymphocytes* that are capable of reacting with and destroying the invading agent. However, it takes a week to several months to develop immunity against the invading agent. Therefore, the person is likely to receive considerable harm from the first exposure before immunity has developed, but upon subsequent exposure the invading agent is often totally blocked before any damage can be done. Exposure to some types of invading agents, such as tetanus toxin, in very minute doses (not enough to cause death) can initiate such intense immunity against the toxin that the person can subsequently withstand 100,000 times the normally lethal dose, thus illustrating the extreme capability of the immune process to protect the body against foreign agents.

Antigens. Any invading agent that can elicit an immune response is called an *antigen.* The usual antigen to which the human body is exposed is a large protein, a large polysaccharide, or a combination of these. For instance, essentially all toxins elaborated by bacteria are proteins, large polysaccharides, or mucopolysaccharides. Also, bacteria and viruses themselves contain mainly these same substances.

Haptens. Though substances with molecular weights less than 10,000 can almost never act as antigens, immunity can nevertheless be developed against substances of low molecular weight in a very special way, as follows: If the low molecular weight compound, which is called a *hapten*, first combines with a substance that *is* antigenic, such as a protein, then the combination will elicit an immune response. The antibodies that develop against the combination can then react either against the protein or against the hapten.

The haptens that elicit immune responses of this type are usually drugs, chemical constituents in dust, breakdown products of dandruff from animals, degenerative products of scaling skin, various industrial chemicals, and so forth.

HUMORAL IMMUNITY — ANTIBODIES

Humoral immunity depends upon the development of *antibodies* in the circulating body fluid. These antibodies are formed primarily by plasma cells in the lymph nodes; this mechanism will be discussed later. The vast majority of the antibodies are gamma globulin molecules, having a mean molecular weight of approximately 160,000.

Specific Chemical Reactivity of Antibodies with Specific Antigens. Gamma globulins formed in response to a particular antigen can react specifically only with that one type of antigen and rarely with any other type.

The exact reason for the specificity of antibodies for specific antigens is not known, but a logical theory to explain this is based on spatial orientation of electrostatic (or hydrogen bond) forces at the surface of the molecule, as illustrated in Figure 9–1. Figure 9–1*A* illustrates the spatial orientation of "polar forces" on the surface of an antigen molecule. By some means yet obscure, the antibodies supposedly develop polar forces of opposite electrostatic charge and arranged as a mirror image to the forces on the antigen molecule, as illustrated in Figure 9–1*B*.

Each antibody has either one or two reactive sites—that is, areas on its molecular surface that have the appropriate image polar forces to react with an antigen—but never more than two reactive sites appear on the same antibody. Therefore, antibodies are said to be either *univalent or bivalent.* Antigens, in contrast to antibodies, can have large numbers of reactive sites: three, five, ten, or more.

Mechanism of Antibody Formation—Function of Plasma Cells

Almost all antibodies are formed in *plasma cells*, which are found in the

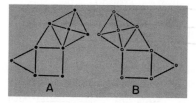

FIGURE 9–1   Polar forces: (*A*) on an antigen and (*B*) on an antibody.

lymphoid tissue of lymph nodes, the spleen, the gastrointestinal tract, and, to a slight extent, other tissues.

The plasma cells, before exposure to an antigen, remain essentially in a dormant state and are called *plasmablasts*. However, upon entry of a foreign antigen into the lymphoid tissue, some of the plasmablasts begin to divide at a rate of approximately once every 10 hours for about nine divisions, giving in four days a total population of about 500 cells for each originally stimulated plasmablast. During this division process, the maturing plasma cells develop an extensive granular endoplasmic reticulum that produces gamma globulin antibodies at an extremely rapid rate, a rate of at least 100 molecules per second. Indeed, radioactive amino acids injected intravenously can become embodied in an active antibody within 15 to 20 minutes, illustrating the rapidity with which antibodies are formed by plasma cells.

Mechanisms for Producing Specific Antibodies. Even though an antigen causes the plasma cells to divide and to form antibodies, the mature plasma cells do not contain any of the antigen. Therefore, the antigen must cause the plasma cells to produce their specific antibodies in some indirect way. Though we do not know the whole mechanism of this, it is known that the antigen is first entrapped by the *fixed reticulum cells* (*macrophages*) of the lymphoid tissue and probably remains there until it is destroyed. The macrophages have long tentacles that extend outward into the surrounding lymphoid tissue, forming a *reticulum* throughout the tissue's meshwork. This reticulum surrounds the plasmablasts and supposedly determines the type of antibody that will be formed.

Mechanism by which Antibodies Inactivate Antigens—The Antigen-Antibody Reaction

Reaction of antibodies with antigens can either destroy or inactivate the antigens. If the antigen is part of a bacterial body, the antibodies can either inactivate the bacterium or destroy it. There are several different mechanisms by which antibodies perform these functions:

1. They can combine with the antigen and thereby *neutralize* its toxic qualities.

2. They can combine with the antigen and cause it to *precipitate*.

3. They can combine with two separate bacteria simultaneously, causing them to *agglutinate*.

4. They can combine with bacteria and cause *rupture* of their cell membranes.

5. They can combine with bacteria or viruses and thereby make them susceptible to phagocytosis, a phenomenon that is called *opsonization*.

FUNCTION OF LYMPHOCYTES IN IMMUNITY (CELLULAR OR LYMPHOCYTE IMMUNITY)

Often immunity can be developed against a foreign agent even without formation of humoral antibodies. In these instances, large numbers of small lymphocytes take on the characteristics of whole cell "antibodies"; that is, specific lymphocytes become sensitized against specific antigens in the same way that specific antibodies are formed against specific antigens in humoral immunity.

Characteristics of the Lymphocyte Type of Immunity. The lymphocyte type of immunity is different from the humoral type in several important ways. First, once lymphocytes have been sensitized to a particular antigen, they continue to circulate in the blood and body tissues sometimes for years. Moreover, these lymphocytes can even divide and form new lymphocytes having the same specific capability. Therefore, the lymphocyte type of immunity usually lasts far longer than does the humoral type of immunity.

Second, the lymphocyte type of immunity will frequently develop upon exposure to extremely minute amounts of antigens, even those amounts too minute to activate the plasmablasts. Therefore, immunity can be developed against some antigens by this mechanism when it cannot be developed by the humoral mechanism.

Formation of Sensitized Lymphocytes. Much less is known about the manner in which specifically sensitized lymphocytes are created than about the mechanism for formation of humoral antibodies. However, most of the normal resting lymphocytes of the lymph nodes are actually dormant *lymphoblasts*. Upon exposure to a foreign antigen, some of these lymphoblasts become sensitized to the antigens and proliferate in the same manner as plasmablasts. Once the lymphocytes are sensitized in this manner they are called *committed lymphocytes*, because thereafter they cannot be changed into lymphocytes sensitized against other types of foreign antigens.

The mechanism by which lymphocytes become sensitized to foreign antigens is probably very similar to the mechanism for formation of antibodies by the plasma cells. Presumably, the small lymphocytes develop antibodies within their cytoplasm or on their cell surfaces, and it is these antibodies that allow the lymphocyte thereafter to attack foreign antigens.

Mechanism by which Sensitized Lymphocytes Attack Antigens. When an organism to which sensitized lymphocytes have been formed invades a tissue, the sensitized lymphocytes (1) swell in size, (2) take on characteristics similar to those of monocytes, and (3) soon thereafter attach side by side to the invading organism and destroy it.

Certain types of antigens seem especially to promote this type of immunity rather than the humoral type. For instance, some bacteria, such as the tubercle bacillus, cause little of the humoral type of immunity but elicit important immune responses of the lymphocyte type. Thus, invading tubercle bacilli become surrounded by swollen lymphocytes, giving rise to a walling off process that prevents further invasion by the tubercle bacillus. Fibrous tissue grows into this walled off area, thereby developing the so-called *tubercle.*

Another type of antigen that is specifically attacked by the lymphocytic type of immunity is foreign animal cells such as those in a grafted heart or other tissue. Beginning approximately a week after the graft, sensitized lymphocytes against the graft begin to appear in the blood, and these rapidly invade the graft itself. Within another four days to a few weeks the lymphocytes gradually cause lysis of the cells in the graft and, subsequently, complete destruction.

FUNCTION OF THE THYMUS GLAND IN IMMUNITY

In the fetus and newborn animal, the thymus gland is a large organ lying in the upper mediastinum beneath the sternum. In the very young fetus it is composed primarily of epithelioid cells, but it has a reticular structure that entraps stem cells circulating in the blood. For reasons not understood, these trapped stem cells are then stimulated to proliferate extremely rapidly, and the cells that are formed are the first lymphocytes to appear in the fetus. These lymphocytes migrate into the blood and thence throughout the body, seeding the lymph nodes and other lymphoid tissue. The seeded cells then become either plasmablasts, the progenitors of the plasma cells, or lymphoblasts, the progenitors of the small lymphocytes formed in the lymphoid tissue. Thus, the thymus plays an important role in the original genesis of the entire lymphoid system of the body.

Failure of the Immune System to Develop in the Absence of the Thymus Gland. If the thymus gland is removed early in fetal life, the lymphoid tissue of the entire body fails to become seeded with lymphocytes, and consequently, essentially no immune system develops. Even if the thymus is removed shortly before birth of the baby or within the first few days after birth, the immune system is still greatly impaired. Furthermore, the lymphocytic type of immunity seems to be much more impaired than the humoral type, indicating that plasmablast seeding of the lymphoid tissue occurs earlier than lymphoblast seeding.

IMMUNOLOGIC TOLERANCE OF ONE'S OWN PROTEINS

Obviously, if one should develop immunity against his own body proteins, he would soon wreak severe damage within his body or actually cause his own death. Fortunately, during development of the immune system, the system learns to recognize the body's own proteins and not to form antibodies against them, while nevertheless retaining the ability to form antibodies against any foreign invader. This specific ability of the immune system to recognize the body's own proteins is called *immunologic tolerance*. Its mechanism has always been extremely perplexing, and even the theories that attempt to explain it are still vague and questionable. Yet, one of these theories deserves comment, as follows:

The Clonal Theory for Immunologic Tolerance. When the thymus forms the first early lymphocytes, it does so at an extremely rapid rate, and it has been postulated that this rapid rate of formation causes tremendous numbers of genetic mutations so that literally thousands of different types of lymphoblasts and plasmablasts seed the lymphoid tissue, each type of lymphoblast and plasmablast having the capability of forming immunity against one particular type of antigen—one and no more. Presumably, some of these early lymphoblasts and plasmablasts can form immunity against the body's own proteins. However, it is postulated that those that are capable of acting against the body's own proteins are quickly bound with the body's proteins and are either inactivated or destroyed before they can seed the lymphoid tissue. This theoretically explains why the immune system thereafter will not react with the body's own tissues but nevertheless will react with foreign antigens that enter the body.

AUTOIMMUNITY

Though in the normal person the immune system fails to develop immunity against the body's own tissue proteins, on occasion such immunity does occur, and the result is called *autoimmunity*. Autoimmunity is believed to be caused most frequently by release of abnormal protein products into the body during infectious diseases, the protein products having antigenicities similar to those of the body's own tissues. The immune system then develops antibodies or sensitized lymphocytes against the body's tissues. At any rate, a large number of the nonspecific diseases that occur in middle and late life are now believed to be autoimmune diseases. These include rheumatoid arthritis, rheumatic fever, excessive fibrosis of tissues, and many others.

TRANSFUSION REACTIONS AND THE O–A–B BLOOD GROUPS

When blood transfusions from one person to another were first attempted, the transfusions were successful in some instances; but, in many more instances, immediate or delayed agglutination and hemolysis of the red blood cells occurred. Soon it was discovered that the bloods of different persons usually have different antigenic and immune properties so that antibodies in the plasma of one blood react with antigens in the cells of another. Furthermore, the antigens and the antibodies are almost never precisely the same in one person as in another. For this reason, it is easy for blood from a donor to be mismatched with that of a recipient, resulting in a typical transfusion reaction that can lead to death.

Two particular groups of antigens are more likely than the others to cause blood transfusion reactions. These are the so-called *O-A-B* system of antigens and the *Rh* system. Bloods are divided into different *groups* and *types* in accordance with the types of antigens present in the cells.

THE A AND B ANTIGENS — THE AGGLUTINOGENS

Two different but related antigens — type A and type B — occur in the cells of different persons. Because of the way these antigens are inherited, a person may have neither of them in his cells, or he may have one, or he may have both simultaneously. The type A and type B antigens in the cells make the cells susceptible to agglutination, for which reason these antigens are called *agglutinogens*; it is on the basis of the presence or absence of agglutinogens in the blood that different types of blood for transfusion are grouped.

The Four Major O-A-B Blood Groups. In transfusing blood from one person to another, the bloods of donors and recipients are normally classified into four major O-A-B groups, as illustrated in Table 9–1, depending on the presence or absence of the two agglutinogens. When neither A nor B agglutinogen is present, the blood group is *group O*. When only type A agglutinogen is present, the blood is *group A*. When only type B agglutinogen is present, the blood is *group B*. And when both A and B agglutinogens are present, the blood is *group AB*.

TABLE 9–1   THE BLOOD GROUPS, WITH THEIR GENOTYPES AND THEIR CONSTITUENT AGGLUTINOGENS AND AGGLUTININS

Genotypes	Blood Groups	Agglutinogens	Agglutinins
OO	O	—	Anti-A and Anti-B
OA or AA	A	A	Anti-B
OB or BB	B	B	Anti-A
AB	AB	A and B	—

Relative frequency of the different blood types. The prevalence of the different blood types among white persons is approximately as follows:

Type	Per cent
O	47
A	41
B	9
AB	3

It is obvious from these percentages that the O and A genes occur frequently but the B gene is infrequent.

THE AGGLUTININS

When type A agglutinogen *is not present* in a person's red blood cells, antibodies known as "anti-A" agglutinins develop in his plasma. Also, when type B agglutinogen *is not present* in the red blood cells, antibodies known as "anti-B" agglutinins develop in the plasma.

Thus referring once again to Table 9–1, it will be observed that group O blood, though containing no agglutinogens, does contain both *anti-A* and *anti-B agglutinins*, whereas group A blood contains type A agglutinogens and *anti-B agglutinins*, and group B blood contains type B agglutinogens and *anti-A agglutinins*. Finally, group AB blood contains both A and B agglutinogens but no agglutinins at all.

Origin of the Agglutinins in the Plasma. The agglutinins are gamma globulins, as are other antibodies, and are produced by the same cells that produce antibodies to infectious diseases.

It is difficult to understand how agglutinins are produced in individuals who do not have the respective antigenic substances in their red blood cells. However, small amounts of group A and B antigens enter the body in the food, in bacteria, or by other means, and these substances, it is believed, initiate the development of anti-A or anti-B agglutinins. One of the reasons for believing this is that injection of group A or group B antigen into a recipient of another blood type causes a typical immune response with formation of greater quantities of agglutinins than ever. Also, the newborn baby has few if any agglutinins, showing that agglutinin formation occurs after birth.

THE AGGLUTINATION PROCESS IN TRANSFUSION REACTIONS

When bloods are mismatched so that anti-A or anti-B agglutinins are mixed with red blood cells containing A or B agglutinogens respectively, the agglutinins attach themselves to the red blood cells. Because the agglutinins are bivalent, a single agglutinin can attach to two different red blood cells at the same time, thereby causing the cells to adhere to each other. This causes the cells to clump and then plug up small blood vessels throughout the circulatory system. During the ensuing few hours to few days, the reticuloendothelial system destroys the cells, releasing hemoglobin into the blood.

Hemolysis in Transfusion Reactions. Sometimes *anti-A* or *anti-B hemolysins*, as well as agglutinins, develop in the plasma. Usually hemolysins are present only when the titer of agglutinins is very high, but the hemolysins are actually different antibodies from the agglutinins. The hemolysins, instead of causing the blood cells to clump, cause lysis of the red blood cells, as described earlier in the chapter.

BLOOD TYPING

Prior to giving a transfusion, it is necessary to determine the blood group of the recipient and the group of the donor blood so that the bloods will be appropriately matched. The usual method of blood typing is the slide technique. In using this technique a drop or more of blood is removed from the person to be typed. This is then diluted approximately 10 times with saline so that clotting will not occur. This leaves essentially a suspension of red blood cells in saline. Two separate drops of this suspension are placed on a microscope slide, and a drop of anti-A agglutinin serum is mixed with one of the drops of cell suspension while a drop of anti-B agglutinin serum is mixed with the second drop of cell suspension. After allowing several minutes for the agglutination process to take place, the slide is observed under a microscope to determine whether or not the cells have clumped. If they have clumped, one knows that an immune reaction has resulted between the serum and the cells.

Table 9–2 illustrates the reactions that occur with each of the four different types of blood. Group O red blood cells have no agglutinogens and therefore do not react with either the anti-A or the anti-B serum. Group A blood has A

TABLE 9–2 BLOOD TYPING — SHOWING AGGLUTINATION OF CELLS OF THE DIFFERENT BLOOD GROUPS WITH ANTI-A AND ANTI-B AGGLUTININS

Red Blood Cells	Anti-A	Anti-B
O	−	−
A	+	−
B	−	+
AB	+	+

agglutinogens and therefore agglutinates with anti-A agglutinins. Group B blood has B agglutinogens and agglutinates with the anti-B serum. Group AB blood has both A and B agglutinogens and agglutinates with both types of serum.

THE Rh BLOOD TYPES

In addition to the O-A-B blood group system, several other systems are sometimes important in the transfusion of blood, the most important of which is the Rh system. The one major difference between the O-A-B system and the Rh system is as follows: In the O-A-B system, the agglutinins responsible for causing transfusion reactions develop spontaneously, whereas in the Rh system spontaneous agglutinins almost never occur. Instead, the person must first be massively exposed to some antigen of the system, usually by transfusion of blood into him, before he will develop enough agglutinins to cause significant transfusion reaction.

The Rh Agglutinogens — Rh Positive and Rh Negative Persons. There are at least eight different types of Rh agglutinogens, but only four of these are especially likely to cause transfusion reactions. Any time a person's blood contains one of these Rh factors that can cause a serious transfusion reaction, he is said to be *Rh positive*. When his blood does not contain one of these factors, he is said to be *Rh negative*; however, his blood might still contain some less antigenic Rh factor that on occasion can cause a mild reaction.

Approximately 85 per cent of all Caucasoids are Rh positive and approximately 15 per cent are Rh negative.

THE Rh IMMUNE RESPONSE

Formation of Anti-Rh Agglutinins. When red blood cells containing Rh factor, or even protein breakdown products of such cells, are injected into an Rh negative person, anti-Rh agglutinins develop very slowly, the maximum concentration of agglutinins occurring approximately two to four months later. This immune response occurs to a much greater extent in some people than in others. On multiple exposure to the Rh factor, the Rh negative person eventually becomes strongly "sensitized" to the Rh factor—that is, he develops a progressively higher titer of anti-Rh agglutinins.

Characteristics of Rh Agglutination. The anti-Rh agglutinins are similar to the anti-A and anti-B agglutinins discussed previously; they attach to the Rh positive red blood cells and cause them to agglutinate. However, the linkages of the agglutination are frequently much weaker than those for the A and B agglutination.

Anti-Rh antibodies do not cause hemolysis directly, but whenever agglutination of the cells occurs, the agglutinated cells plug the capillaries in the peripheral circulatory system, and the cells are gradually destroyed by phagocytes during the next few hours to few days, so that the final effect of the agglutination reaction is hemolysis.

Erythroblastosis Fetalis. Erythroblastosis fetalis is a disease of the newborn infant characterized by progressive agglutination and subsequent phagocytosis of the red blood cells. In most instances of erythroblastosis fetalis the mother is Rh negative, and the father is Rh positive; the baby has inherited the Rh positive characteristic from the father, and the mother has developed anti-Rh agglutinins that have diffused into the fetus to cause red blood cell agglutination.

Effect of the mother's antibodies on the fetus. After anti-Rh antibodies have formed in the mother, they diffuse very slowly through the placental membrane into the fetus' blood. There they cause slow agglutination of the fetus' blood, and clumps of blood cells occlude small blood vessels. The red blood cells of these clumps gradually hemolyze, releasing hemoglobin into the blood. The reticuloendothelial system then converts the hemoglobin into bilirubin, which causes yellowness (jaundice) of the skin. The antibodies probably also attack and damage many of the other cells of the body.

Severe erythroblastosis fetalis is almost always fatal. However, patients with mild cases can be saved by transfusing Rh negative blood into the body while removing its Rh positive blood, a process called *exchange transfusion.*

OTHER BLOOD FACTORS

Many antigenic proteins besides the O, A, B, and Rh factors are present in red blood cells of different persons, but these other factors very rarely cause transfusion reactions and, therefore, are mainly of academic and legal importance. Some of these different blood factors are the M, N, S, s, P, Kell, Lewis, Duffy, Kidd, Diego, and Lutheran factors.

TRANSFUSION REACTIONS RESULTING FROM MISMATCHED BLOOD GROUPS

All transfusion reactions resulting from mismatched blood groups eventually cause hemolysis of the red blood cells. Occasionally, specific *hemolysins* are present in the plasma of the recipient along with the agglutinins and therefore cause immediate hemolysis, but more frequently the cells agglutinate first and then are entrapped in the peripheral vessels. Over a period of hours to days the entrapped cells degenerate and hemolyze, liberating hemoglobin into the circulatory system.

When the rate of hemolysis is rapid, the concentration of hemoglobin in the blood can rise to extremely high values.

A small quantity of hemoglobin, up to 100 mg./100 ml. of plasma, becomes attached to one of the plasma proteins, *haptoglobin,* and continues to circulate in the blood without causing any harm. However, above the threshold value of 100 mg./100 ml. of plasma, the excess hemoglobin remains in the free form and diffuses out of the circulation into the tissue spaces or through the renal glomeruli into the kidney tubules, as is discussed below. The hemoglobin remaining in the circulation or passing into the tissue spaces is gradually ingested by reticuloendothelial cells and converted into bilirubin, which was discussed in Chapter 8. The concentration of bilirubin in the body fluids sometimes rises high enough to cause *jaundice* — that is, the person's tissues become tinged with yellow pigment. But jaundice usually does not appear unless 300 to 500 ml. of blood is hemolyzed in less than a day.

Acute Kidney Shutdown Following Transfusion Reactions. One of the most lethal effects of transfusion reactions is acute kidney shutdown, which can begin within a few minutes to a few hours and continue until the person dies of renal failure.

The kidney shutdown seems to result from two different causes: First, the antigen-antibody reaction of the transfusion reaction releases toxic substances from the hemolyzing blood that cause powerful renal vasoconstriction. Second, if the total amount of free hemoglobin in the circulating blood is greater than that quantity which can bind with haptoglobin, much of the excess leaks through the glomerular membranes into the kidney tubules. If this amount is still slight, it can be reabsorbed through the tubular epithelium into the blood and will cause no harm, but, if it is great, then only a small percentage is reabsorbed. Yet water continues to be reabsorbed, causing the tubular hemoglobin concentration to rise so high that it precipitates and blocks many of the tubules. Thus, renal vasoconstriction and tubular blockage add together to cause acute renal shutdown. If the shutdown is complete, the patient dies within a week to 12 days, as explained in Chapter 26.

TRANSFUSION REACTIONS OCCURRING WHEN Rh FACTORS ARE MISMATCHED

If the recipient has previously received a blood transfusion containing a strongly antigenic Rh factor not present in his own blood, he often will have built up appropriate anti-Rh antibodies that can cause a transfusion reaction on subsequent exposure to the same type of Rh blood. However, since anti-Rh agglutination is usually weak, the intensity of the reaction is usually less severe than when there is an anti-A or anti-B reaction.

OTHER TYPES OF TRANSFUSION REACTIONS

Pyrogenic Reactions. Pyrogenic reactions occur more frequently than either agglutination or hemolysis. These reactions cause fever in the recipient but do not destroy red blood cells. Most pyrogenic reactions probably result from the presence in the donor plasma of proteins to which the recipient is allergic, or they result from the presence of breakdown products in old, deteriorating donor blood.

Transfusion Reactions Resulting from Anticoagulants. The usual anticoagulant used for transfusion is a citrate salt. As discussed in the following chapter, citrate operates as an anticoagulant by combining with the calcium ions of the plasma so that these become nonionizable. Without the presence of ionizable calcium the coagulation process cannot take place. Normal nerve, muscle, and heart function also cannot occur in the absence of calcium ions. Therefore, if blood containing large quantities of citrate anticoagulant is administered rapidly, the recipient is likely to experience typical tetany due to low calcium, which is discussed in detail in Chapter 53. Such tetany can kill the patient within a few minutes because of respiratory muscle spasm.

TRANSPLANTATION OF TISSUES AND ORGANS

Relation of Genotypes to Transplantation. In this modern age of surgery, many attempts are being made to transplant tissues and organs from one person to another, or, occasionally, from lower animals to the human being. All the different antigenic proteins of red blood cells that cause transfusion reactions, and still many more, are present in the other cells of the body. Consequently, any foreign cells transplanted into a recipient can cause immune responses and immune reactions. In other words, most recipients are just as able to resist invasion by foreign cells as to resist invasion by foreign bacteria. The greater the difference in antigenic structure, the more rapid and the more severe are the immune reactions to the graft. However, with proper "tissue typing"—that is, matching donor tissues with recipient tissues by preliminary immunologic tests —it is occasionally possible to transplant tissues and organs from one person to another.

Transplants from one identical twin to another are an example in which homologous transplants of cellular tissues are almost always successful. The reason for this is that the antigenic proteins of both twins are determined by identical genes derived originally from the single fertilized ovum.

Some of the different cellular tissues and organs that have been transplanted either experimentally or for temporary benefit from one person to another are skin, kidney, heart, liver, glandular tissue, bone marrow, and lung. A few kidney homologous transplants have been successful for as long as five years, a rare liver and heart transplant for one to two years, and lung transplants for one month.

Transplantation of Noncellular Tissues. Certain tissues that have no cells or in which the cells are unimportant to the purpose of the graft, such as the cornea, tendon, fascia, and bone, can usually be grafted from one person to another with considerable success. In these instances the grafts act merely as a supporting latticework into which or around which the surrounding living tissues of the recipient grow.

ATTEMPTS TO OVERCOME THE ANTIGEN-ANTIBODY REACTIONS IN TRANSPLANTED TISSUE

Because of the extreme potential importance of transplanting certain tissues and organs, such as skin, kidneys, and lungs, serious attempts have been made to prevent the antigen-antibody reactions associated with transplants. The following specific procedures have met with certain degrees of success.

Use of Anti-Lymphocyte Serum. It has already been pointed out that grafted tissues are usually destroyed by lymphocytes that have become sensitized against the graft. These cells invade the graft and presumably secrete substances that gradually destroy the grafted cells. Upon contact with the lymphocytes, the cells of the graft begin to swell, their membranes become very permeable, and finally they rupture. Within a few days to a few weeks after this process begins, the tissue is completely destroyed, even though the graft had been completely viable and functioning normally for the few days to few weeks after its original transplantation.

Therefore, one of the most effective procedures in preventing rejection of grafted tissues has been to inoculate the recipient with *anti-lymphocyte serum.*

This serum is made in horses by injecting human lymphocytes into them; the antibodies that develop in these animals will then attack and destroy the human lymphocytes.

Glucocorticoid Therapy (Cortisone, Hydrocortisone, and ACTH). The glucocorticoid hormones from the adrenal gland greatly suppress the formation of antibodies and immunologically competent lymphocytes. Therefore, administration of large quantities of these, or of ACTH, which causes the adrenal gland to produce glucocorticoids, often allows transplants to persist much longer than usual in the recipient. However, this is only a delaying process rather than a preventive one.

Suppression of Antibody Formation. Irradiative destruction of most of the lymphoid tissue by either x-rays or gamma rays renders a person much more susceptible than usual to a homologous transplant. Also, treatment with certain drugs, such as azathioprine (Imuran), which suppresses antibody formation, increases the likelihood of success with transplants. Unfortunately, these procedures also leave the person unprotected from disease.

Removal of the thymus gland to prevent antibody formation. It was pointed out earlier in the chapter that the thymus gland during fetal and early postnatal life forms thymic cells that migrate throughout the body and become the precursors of the immunologic lymphocytes and the plasma cells. Therefore, if the thymus gland is removed during fetal life or immediately after birth, the immune system is greatly impaired throughout the remainder of the animal's life. Transplants are frequently successful in such animals, but the loss of immunity to disease makes it difficult for the animal to live.

To summarize, transplantation of living tissues in human beings up to the present has been hardly more than an experiment. But when someone succeeds in blocking the immune response of the recipient to a donor organ without at the same time destroying the recipient's specific immunity for disease, this story will undoubtedly change overnight.

REFERENCES

Gatti, R. A., Stutman, O., and Good, R. A.: The lymphoid system. *Ann. Rev. Physiol., 32*:529, 1970.

Gowland, G.: The mechanism of immunological tolerance and sensitivity. *In* Bittar, E. Edward, and Bittar, Neville (eds.): The Biological Basis of Medicine. New York, Academic Press, Inc., 1969. Vol. 4, p. 305.

Harboe, M.: Structure and metabolism of immunoglobulins. *In* Bittar, E. Edward, and Bittar, Neville (eds.): The Biological Basis of Medicine. New York, Academic Press, Inc., 1969, Vol. 4, p. 197.

Miller, J. F. A. P.: Immunology of tissue transplantation. *In* Bittar, E. Edward, and Bittar, Neville (eds.): The Biological Basis of Medicine. New York, Academic Press, Inc., 1969, Vol. 4, p. 265.

Mollison, P. L.: Blood Transfusion in Clinical Medicine. 4th ed., Philadelphia, F. A. Davis Co., 1967.

Wiener, A. S.: Elements of blood group nomenclature with special reference to the Rh-Hr blood types. *J.A.M.A., 199*:985, 1967.

Zmijewski, C. M.: Immunohematology. New York, Appleton-Century-Crofts, 1968.

CHAPTER 10

HEMOSTASIS AND BLOOD COAGULATION

EVENTS IN HEMOSTASIS

The term hemostasis means prevention of blood loss. Whenever a vessel is severed or ruptured, hemostasis is achieved by several different mechanisms including (1) vascular spasm, (2) formation of a platelet plug, (3) blood coagulation, and (4) growth of fibrous tissue into the blood clot to close the hole in the vessel permanently.

VASCULAR SPASM

Immediately after a blood vessel is cut or ruptured, the wall of the vessel contracts; this instantaneously reduces the flow of blood from the vessel rupture. The contraction results from both nervous reflexes and local myogenic spasm. The nervous reflexes presumably are initiated by pain impulses originating from the traumatized vessel or from nearby tissues. The reflex signals travel first to the spinal cord and then back through the sympathetic nerves to cause spasm of the vessel for several centimeters in either direction from the rupture.

However, this neurogenic spasm lasts for only a fraction of a minute to a few minutes. In its place a local myogenic vascular spasm ensues. This is initiated by direct damage to the vascular wall, which presumably causes transmission of action potentials along the vessel wall for several centimeters and results in constriction of the vessel.

The value of vascular spasm as a means of hemostasis is illustrated by the fact that persons whose legs have been severed by crushing types of trauma sometimes have such intense spasm in vessels as large as the anterior tibial artery that there is not serious loss of blood.

FORMATION OF THE PLATELET PLUG

The second event in hemostasis is an attempt by the platelets to plug the rent in the vessel. To understand this it is important that we first understand the nature of platelets themselves.

Platelets are minute round or oval discs about 2 to 5 microns in diameter. On coming in contact with wettable surfaces, they change their characteristics drastically, assuming bizarre and irregular forms with numerous irradiating processes protruding from their surfaces.

The normal concentration of platelets in the blood is between 150,000 and 300,000 per cubic mm.

Mechanism of the Platelet Plug. The mechanism of the platelet repair of vascular openings seems to be the following: At the site of any rent in a vessel, the endothelium loses its normal smoothness and also loses its normal nonwettability. As a result, platelets immediately begin to adhere to the endothelium at this point. On adhering, the platelets themselves then take on new characteristics; most important of all, they become sticky to other platelets. Therefore, a second layer of platelets adheres to the first layer, and third and fourth layers adhere until a small platelet plug actually fills the rent in the vessel. The platelet plugging mechanism is extremely important to close the many minute ruptures in very small blood vessels that occur hundreds of times daily.

CLOTTING IN THE RUPTURED VESSEL

The third mechanism for hemostasis is formation of the blood clot. The clot begins to develop in 15 to 20 seconds if the trauma of the vascular wall has been severe and in one to two minutes if the trauma has been minor. Activator substances both from the traumatized vascular wall and from platelets adhering to the traumatized vascular wall initiate the clotting process. The physical events of this process are illustrated in Figure 10–1, and the chemical events will be discussed in detail later in the chapter.

Within three to six minutes after rupture of a vessel, the entire end of the vessel is filled with clot. After 30 minutes to an hour, the clot reacts; this closes the vessel still further. Platelets play an important role in this clot retraction, as will also be discussed later in the chapter.

Once a blood clot has formed, it can follow two separate courses: it can become invaded by fibroblasts, which sub-

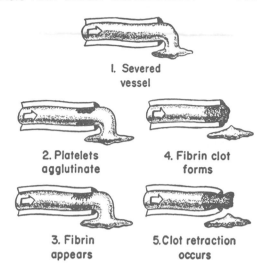

1. Severed vessel

2. Platelets agglutinate

3. Fibrin appears

4. Fibrin clot forms

5. Clot retraction occurs

FIGURE 10–1 The clotting process in the traumatized blood vessel. (Redrawn from Seegers: Hemostatic Agents. Charles C Thomas.)

sequently form connective tissue all through the clot; or it can dissolute. The usual course for a clot that forms in a small hole of a vessel wall is invasion by fibroblasts, beginning within a few hours after the clot is formed, and complete organization of the clot into fibrous tissue within approximately 7 to 10 days. On the other hand, when a large mass of blood clots, such as blood that has leaked into tissues, special substances within the tissue usually activate a mechanism for dissoluting most of the clot, which will be discussed later in the chapter.

MECHANISM OF BLOOD COAGULATION

General Mechanism. Almost all research workers in the field of blood coagulation agree that clotting takes place in three essential steps:

First, a substance called *prothrombin activator* is formed in response to rupture of the vessel or damage to the blood itself.

Second, the prothrombin activator catalyzes the conversion of prothrombin into *thrombin*.

Third, the thrombin acts as an enzyme

to convert fibrinogen into *fibrin threads* that enmesh red blood cells and plasma to form the clot itself.

CONVERSION OF PROTHROMBIN TO THROMBIN

Prothrombin. Prothrombin is a plasma protein, an alpha$_2$ globulin, having a molecular weight of 68,700. It is present in normal plasma in a concentration of about 15 mg./100 ml. It is an unstable protein that can split easily into smaller compounds, one of which is thrombin, which has a molecular weight of 33,700, almost exactly half that of prothrombin.

Prothrombin is formed continually by the liver, and it is continually being used throughout the body for blood clotting. If the liver fails to produce prothrombin, its concentration in the plasma falls too low within 24 hours to provide normal blood coagulation. Vitamin K is required by the liver for normal formation of prothrombin; therefore, either lack of vitamin K or the presence of liver disease that prevents normal prothrombin formation can often decrease the prothrombin level so low that a bleeding tendency results.

Effect of Prothrombin Activator on Prothrombin To Form Thrombin. Figure 10–2 illustrates the conversion of prothrombin to thrombin under the influence of prothrombin activator and calcium ions. The rate of formation of thrombin from prothrombin is almost directly proportional to the quantity of prothrombin activator available, which

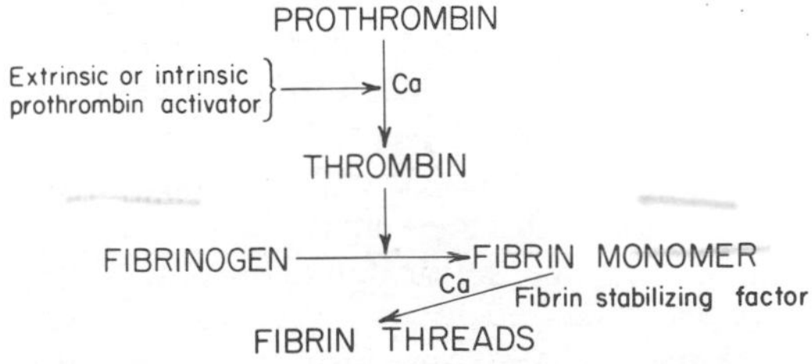

PROTHROMBIN

Extrinsic or intrinsic } ⟶ Ca
prothrombin activator }

THROMBIN

FIBRINOGEN ⟶ FIBRIN MONOMER
Ca — Fibrin stabilizing factor

FIBRIN THREADS

FIGURE 10–2 Schema for conversion of prothrombin to thrombin, and polymerization of fibrinogen to form fibrin threads.

in turn is approximately proportional to the degree of trauma to the vessel wall or to the blood. In turn, the rapidity of the clotting process is proportional to the quantity of thrombin formed.

CONVERSION OF FIBRINOGEN TO FIBRIN — FORMATION OF THE CLOT

Fibrinogen. Fibrinogen is a high molecular weight protein (340,000) occurring in the plasma in quantities of 100 to 700 mg./100 ml. Most if not all of the fibrinogen in the circulating blood is formed in the liver, and liver disease occasionally decreases the quantity of circulating fibrinogen, as it does the quantities of prothrombin, which was pointed out previously.

Action of Thrombin on Fibrinogen to Form Fibrin. Thrombin is a protein enzyme with proteolytic capabilities. It acts on fibrinogen to remove two low molecular weight peptides from each molecule of fibrinogen, forming a molecule of *activated fibrin*, also called *fibrin monomer.* Many fibrin monomer molecules polymerize rapidly into *long fibrin threads*, which form the *reticulum* of the clot. During this polymerization, *calcium ions* and another factor called *protein stabilizing factor* enhance the bonding between the fibrin monomer molecules, as well as between the polymerized chains themselves, and thereby add stability and strength to the fibrin threads.

The Blood Clot. The clot is composed of a meshwork of fibrin threads running in all directions and entrapping blood cells, platelets, and plasma. The fibrin threads adhere to damaged surfaces of blood vessels; therefore, the blood clot becomes adherent to any vascular opening and thereby prevents blood loss.

Clot Retraction — Serum. Within a few minutes after a clot is formed, it begins to contract and usually expresses most of the plasma from the clot within 30 to 60 minutes. The plasma expressed from the clot is called *serum,* for all of its fibrinogen and most of the other clotting factors have been removed. Serum obviously cannot clot because of lack of these factors.

For reasons not completely understood, large numbers of platelets are necessary for clot retraction to occur. Platelets contain tremendous quantities of the high energy compound adenosine triphosphate. It has been suggested that the energy in this compound plays a role in bonding the successive molecules in the fibrin threads or that it causes increased folding of the threads, thereby decreasing the lengths of the threads and thus expressing the serum from the clot.

As the clot retracts, the edges of the broken blood vessel are pulled together, thus contributing to the ultimate state of hemostasis.

INITIATION OF COAGULATION: FORMATION OF PROTHROMBIN ACTIVATOR

Though we know most of the final chemical events for the formation of the clot itself, the basic control system that turns the clotting process on—that is, the basic process for the formation of prothrombin activator—is still mainly shrouded in mystery. In the following few paragraphs we will attempt to relate what is known about this, even though the chemical interrelationships are still poorly understood.

The Extrinsic and Intrinsic Mechanisms for Initiating Clotting

Blood clotting can be initiated in two separate ways: (1) by the *extrinsic mechanism*, in which an extract from the damaged tissues is mixed with the blood, or (2) by the *intrinsic mechanism*, in which the blood itself is traumatized.

It is primarily the extrinsic mechanism that causes clotting when a blood vessel is damaged, since the broken wall of the blood vessel exudes a tissue extract that initiates the clotting process.

It is primarily the intrinsic mechanism that initiates clotting of blood that has been removed from the body and held in a container. If the walls of this container are completely nonwettable, such as a siliconized and very clean vessel, then clotting of the blood may be delayed as long as an hour or more. On the other hand, if the surface of the vessel is easily wettable, such as the usual glass surface, clotting usually begins within a few minutes. Contact between certain elements within the blood and the vessel wall initiates this clotting process.

The extrinsic clotting process usually occurs much more rapidly than the intrinsic one, illustrating that the clotting factors derived from tissue extracts are very powerful activators of the clotting mechanism.

Figure 10–3 depicts the basic essentials of the extrinsic and intrinsic systems for initiating clotting, illustrating that many different factors, most of which are special types of proteins in the blood plasma, supposedly enter into the reactions that initiate clotting. Note that the final products for activation of clotting are called, respectively, *extrinsic* and *intrinsic prothrombin activators*. It is possible that these two activators are the same chemical substance, but it is more likely that each of them is really a combination of several of the individual factors in the system. Furthermore, these separate factors perhaps activate separate parts of the chemical process for conversion of prothrombin into thrombin.

Characteristics of Different Clotting Factors. The clotting factors illustrated in the extrinsic and intrinsic schemata of Figure 10–3 are of three different types: calcium ions, phospholipids, and proteins. The functions of these are as follows.

Calcium ions. In the absence of calcium ions neither the extrinsic nor the intrinsic system will operate. However, in normal blood, calcium ions are

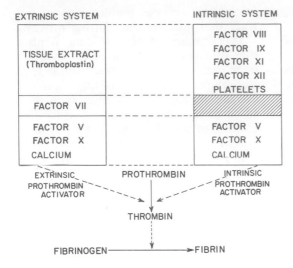

FIGURE 10-3 The extrinsic and intrinsic systems for blood clotting, illustrating the different factors that are necessary for each of these two systems. (Modified from Hougie: Fundamentals of Blood Coagulation in Clinical Medicine. McGraw-Hill Book Co.)

almost always present in sufficient excess that this is almost never a factor in determining the kinetics of the clotting reactions. Note that calcium does not actually enter into any of the reactions other than to act as a cofactor to cause the reactions to take place.

Phospholipids. In both the extrinsic and intrinsic systems, a phospholipid plays a major role in the formation of the final prothrombin activator. In the extrinsic system, this phospholipid is in thromboplastin. In the intrinsic system, it is in platelet factor 3, which is a product of the platelets. In both instances, the phospholipid responsible for the reactions is either a cephalin or a cephalin-like substance.

Protein. All the blood clotting factors from factor V through factor XII are plasma proteins; most of these are beta globulins, but in a few instances are alpha or gamma globulins. Certain of these deserve special mention.

Factor V is a labile substance easily destroyed by slight heat and therefore is often called *labile factor.*

Factors VII, IX, and X have characteristics similar to those of prothrombin and are formed in the liver along with prothrombin. Indeed, some research workers believe that these three factors actually represent variant forms of prothrombin. All these three factors as well

as prothrombin itself require vitamin K for their formation, and in the absence of vitamin K, their concentrations in the plasma diminish rapidly.

Absence of or malfunction of any one of the blood clotting factors from the plasma can cause a bleeding tendency, as is discussed later in the chapter. Factor VIII is most frequently implicated; when it fails to function the person has the disease called *hemophilia.*

TABLE 10-1 CLOTTING FACTORS IN THE BLOOD AND THEIR SYNONYMS

Clotting Factor	Synonym
Fibrinogen	Factor I
Platelets	Platelet factor 3
Prothrombin	Factor II
Factor V	Proaccelerin; labile factor; Ac-globulin; Ac-G
Factor VII	Serum prothrombin conversion accelerator; SPCA; proconvertin; autoprothrombin I
Factor VIII	Antihemophilic factor; AHF; antihemophilic globulin; AHG
Factor IX	Plasma thromboplastin component; PTC; Christmas factor; CF; autoprothrombin II
Factor X	Stuart factor; Stuart-Prower factor; Prower factor; autoprothrombin I_c
Factor XI	Plasma thromboplastin antecedent; PTA
Factor XII	Hageman factor
Prothrombin activator	Thrombokinase; complete thromboplastin

For this reason, factor VIII is frequently called *antihemophilic factor.*

Almost every investigator working in the field of blood coagulation develops new terms for different blood clotting factors. For this reason, there is tremendous confusion regarding nomenclature. The nomenclature using Roman numerals used in this chapter was adopted by an international convention on blood clotting, but synonyms in common use for the different clotting factors are given in Table 10–1.

Thromboplastin, the Initiating Factor in the Extrinsic System for Clotting. The clotting substance (or substances) in a tissue extract that initiates the extrinsic mechanism for clotting is called *thromboplastin.* This is composed primarily of lipoproteins containing one or more phospholipids. The phospholipid *cephalin* is especially active in promoting clotting.

However, thromboplastin by itself is not capable of splitting prothrombin into thrombin but instead must interact with several clotting factors of the blood itself before it becomes prothrombin activator. The following plasma factors are required in this process: *factor V, factor VII, factor X,* and *calcium ions.*

Blood Factors that Initiate the Intrinsic System for Blood Coagulation. In the intrinsic system, blood platelets play an integral role in the clotting mechanism. When platelets adhere to a broken surface of the blood vessel, many of them disintegrate and release into the blood *platelet factor 3,* which contains a phospholipid that is similar in many ways to the thromboplastin of tissue extract. However, platelet factor 3 lacks the intensity of tissue thromboplastin in initiating the clotting process, and it too must interact with several other blood factors before clotting will occur. Two of these factors are the same as those utilized for the extrinsic system, *factor V* and *factor X;* four additional factors are *factors VIII, IX, XI,* and *XII.*

In addition to the ability of platelets to disintegrate and thereby help activate the intrinsic mechanism for blood clotting, two other factors are supposedly involved in activating this system under some conditions. These are *factors XI* and *XII.* When they come in contact with a roughened or wettable surface, such as the glass wall of the test tube, they react together to form a *contact activation product;* this then acts as an enzyme to activate the other factors in the intrinsic system. When a person does not have one of these two factors, the intrinsic system for blood clotting becomes defective.

Summary

Even though the detailed mechanisms for initiating blood clotting are unknown, the physical factors that initiate clotting are well understood: namely, trauma to tissues, trauma to the endothelial surface of the vascular wall, or trauma to some of the intrinsic clotting factors of the blood itself.

PREVENTION OF BLOOD CLOTTING IN THE NORMAL VASCULAR SYSTEM — THE INTRAVASCULAR ANTICOAGULANTS

Factors that prevent intravascular clotting include:

1. Smoothness of the endothelial surface.

2. A monomolecular layer of negatively charged protein adsorbed to the surface of the endothelium, which repels the clotting factors and platelets and thereby prevents activation of prothrombin.

3. Several antithrombin factors that absorb thrombin or destroy it soon after it forms.

4. Heparin, which is discussed in the following text.

Heparin. Heparin, a powerful anticoagulant, is a conjugated polysaccharide found in the cytoplasm of many types of cells, including even the cytoplasm of unicellular animals. It is produced in especially large quantities by the basophilic *mast cells* located in the pericapillary connective tissue throughout the body. The *basophil cells* of the blood, which seem to be functionally almost identical with the mast cells, possibly also release minute quantities of heparin into the plasma.

Mechanism of heparin action. We still do not know all the different mechanisms by which heparin prevents blood coagulation. However, it is known to have at least the following four effects:

First, in low concentrations, heparin prevents the formation of intrinsic prothrombin activator, though the mechanism of this is unknown.

Second, heparin, in association with an albumin cofactor, inhibits the action of thrombin on fibrinogen and thereby prevents the conversion of fibrinogen into fibrin threads.

Third, heparin increases the rapidity with which thrombin interacts with antithrombin and therefore helps in the deactivation of thrombin.

And, fourth, heparin increases the amount of thrombin adsorbed by fibrin.

LYSIS OF BLOOD CLOTS— PLASMIN

Activation of Plasmin and Lysis of Clots. When a clot is formed, a large amount of a plasma protein called *plasminogen* is incorporated in the clot along with other plasma proteins. The tissues contain substances that can activate plasminogen to *plasmin* within a day, and the plasmin in turn acts as a digestive enzyme to dissolute the clot.

Clots that occur inside blood vessels can also be dissoluted, though this occurs less often than does dissolution of tissue clots. This illustrates that activator systems also occur within the blood itself.

CONDITIONS THAT CAUSE EXCESSIVE BLEEDING IN HUMAN BEINGS

Excessive bleeding can result from deficiency of any one of the many different blood clotting factors. Three particular types of bleeding tendencies that have been studied to the greatest extent will be discussed: (1) bleeding caused by vitamin K deficiency, (2) hemophilia, and (3) thrombocytopenia (platelet deficiency).

Decreased Prothrombin, Factor VII, Factor IX, and Factor X—Vitamin K Deficiency. Vitamin K is necessary for some of the intermediate stages in the formation of prothrombin and factors VII, IX, and X. Therefore, lack of vitamin K will lead to bleeding.

One of the most prevalent causes of vitamin K deficiency is failure of the liver to secrete bile into the gastrointestinal tract. Since lack of bile prevents adequate fat digestion and absorption, vitamin K, which is fat soluble, also fails to be absorbed. Because of this, vitamin K is injected into all patients with liver disease or obstructed bile ducts prior to performing any surgical procedure. Ordinarily, if vitamin K is given to a deficient patient four to eight hours prior to operation and the liver parenchymal cells are at least one-half normal, sufficient clotting factors will be produced to prevent excessive bleeding during the operation.

Hemophilia. The term hemophilia is loosely applied to several different hereditary deficiencies of coagulation, all of which cause bleeding tendencies hardly distinguishable from one another. The three most common causes of the hemophilic syndrome are deficiency of (1) factor VIII (classical hemophilia)—about 75 per cent of the total, (2) factor IX—about 15 per cent, and (3) factor XI—about 5 to 10 per cent. Transfusion of normal plasma into the hemophilic person in order to supply the missing clotting factor temporarily relieves his bleeding tendency for a few days.

THROMBOCYTOPENIA

Thrombocytopenia means a deficiency of platelets in the circulatory system.

Persons with thrombocytopenia have a tendency to bleed as do hemophiliacs, except that the bleeding is usually from many small capillaries rather than from large vessels as in hemophilia. As a result, small punctate hemorrhages occur throughout all the body tissues. The skin of such a person displays many small, purplish blotches, giving the disease the name *thrombocytopenic purpura*. It will be remembered that platelets are especially important for repair of minute breaks in capillaries and other small vessels. Indeed, platelets can agglutinate to fill such ruptures without actually causing clots.

Ordinarily, excessive bleeding does not occur until the number of platelets in the blood falls below a value of approximately 70,000 per cubic millimeter rather than the normal of 150,000 to 350,000.

THROMBOEMBOLIC CONDITIONS IN THE HUMAN BEING

Thrombi and Emboli. An abnormal clot that develops in a blood vessel is called a *thrombus*. Once a clot has developed, continued flow of blood past the clot is likely to break it away from its attachment, and such freely flowing clots are known as *emboli*.

The causes of thromboembolic conditions in the human being are usually twofold: First, any *roughened endothelial surface of a vessel*—as may be caused by arteriosclerosis, infection, or trauma—is likely to initiate the clotting process. Second, blood often clots *when it flows very slowly* through blood vessels, for small quantities of thrombin and other procoagulants are always being formed. If the blood is flowing too slowly, the concentrations of the procoagulants in local areas often rise high enough to initiate clotting.

Femoral Thrombosis and Massive Pulmonary Embolism. The immobility of bed patients plus the practice of propping the knees up with pillows often causes stasis of blood in one or more of the leg veins for as much as an hour at a time, and this stasis initiates the clotting process. Then the clot grows in all directions, especially in the direction of the slowly moving blood, sometimes growing the entire length of the leg veins and occasionally even into the common iliac vein and inferior vena cava. Then, about 1 time out of every 10, the clot disengages from its attachments to the vessel wall and flows freely with the venous blood into the right side of the heart and thence into the pulmonary arteries to cause *massive pulmonary embolism*. If the clot is large enough to occlude both the pulmonary arteries, immediate death ensues. If only one pulmonary artery or a smaller branch is blocked, death may not occur, or the embolism may lead to death a few hours to several days later because of further growth of the clot within the pulmonary vessels.

ANTICOAGULANTS FOR CLINICAL USE

Heparin as an Intravenous Anticoagulant. Injection of relatively small quantities of heparin causes the blood clotting time to increase from a normal of approximately 6 minutes to 30 or more minutes. Furthermore, this change in clotting time occurs instantaneously, thereby immediately preventing further development of a thromboembolic condition.

Following overtreatment with heparin, *protamine* acts specifically as an antiheparin, and the clotting mechanism can be reverted to normal by administering this substance. Protamine probably combines with heparin and inactivates it because it carries positive electrical charges, whereas heparin carries negative charges.

Dicumarol as an Anticoagulant. Dicumarol competes with vitamin K for reactive sites in the intermediate processes for formation of prothrombin and clotting factors VII, IX, and X, thereby greatly depressing the clotting process.

After administration of an effective dose of Dicumarol, the coagulant activity of the blood decreases to approximately 20 per cent of normal by the end of 24 hours.

PREVENTION OF BLOOD COAGULATION OUTSIDE THE BODY

Blood removed from the body can be prevented from clotting by:

1. Collecting the blood in *siliconized containers*, which prevent contact activation of factors XI and XII that initiates the intrinsic clotting mechanism.

2. Mixing *heparin* with the blood.

3. *Decreasing the calcium ions* in the blood. For instance, soluble *oxalate* compounds mixed in very small quantity with a sample of blood cause precipitation of calcium oxalate from the plasma and thereby block blood coagulation. Calcium deionizing agents used for preventing coagulation are *sodium, ammonium*, or *potassium citrate*, or EDTA. These substances combine with calcium in the blood to cause un-ionized calcium compounds, and the lack of ionic calcium prevents coagulation. Citrate anticoagulants have a very important advantage over the oxalate anticoagulants, for oxalate is toxic to the body, whereas small quantities of citrate can be injected intravenously. After injection, the citrate ion is removed from the blood within a few minutes by the liver and is polymerized into glucose, and then metabolized in the usual manner. Consequently, 500 ml. of blood that has been rendered incoagulable by sodium citrate can ordinarily be injected into a recipient within a few minutes without any dire consequences. Therefore, citrate is the anticoagulant used in transfusion blood.

BLOOD COAGULATION TESTS

Bleeding Time. When a sharp knife is used to pierce the tip of the finger or lobe of the ear, bleeding ordinarily lasts three to six minutes.

Clotting Time. A method widely used for determining clotting time is to collect blood in a chemically clean glass test tube and then to tip the tube back and forth approximately every 30 seconds until the blood has clotted. By this method, the normal clotting time ranges between five and eight minutes.

Prothrombin Time. The prothrombin time gives an indication of the total quantity of prothrombin in the blood. The method for determining prothrombin time is the following:

Blood removed from the patient is immediately oxalated so that none of the prothrombin can change into thrombin. At any time later, a large excess of calcium ion and thromboplastin is suddenly mixed with the oxalated blood. The calcium nullifies the effect of the oxalate, and the thromboplastin activates the prothrombin-to-thrombin reaction. The time required for coagulation to take place is known as the "prothrombin time." The normal prothrombin time is approximately 12 seconds, and clotting is often severely impaired when this increases to more than 24 seconds.

REFERENCES

Biggs, Rosemary: The nature of blood coagulation and its defects. *In* Bittar, E. Edward, and Bittar, Neville (eds.): The Biological Basis of Medicine. New York, Academic Press, Inc., 1969, Vol. 3, p. 183.

Fibrinogen and fibrin (Symposium). *Fed. Proc., 24*:783, 1965.

Kline, D. L.: Blood coagulation: reactions leading to prothrombin. *Ann. Rev. Physiol., 27*:285, 1965.

Mills, D. C. B.: Platelet aggregation. *In* Bittar, E. Edward, and Bittar, Neville (eds.): The Biological Basis of Medicine. New York, Academic Press, Inc., 1969, Vol. 3, p. 163.

Quick, A. J.: Hemorrhagic Diseases and Thrombosis. 2nd ed., Philadelphia, Lea & Febiger, 1966.

Seegers, W. H. (ed.): Blood Clotting Enzymology. New York, Academic Press, Inc., 1967.

Seegers, W. H.: Blood clotting mechanisms: three basic reactions. *Ann. Rev. Physiol., 31*:269, 1969.

PART II

THE
HEART
AND THE
CIRCULATION

THE HEART AS A PUMP, THE CARDIAC CYCLE, CARDIAC CONTRACTILITY, AND STROKE VOLUME OUTPUT

With this chapter we begin discussion of the heart and circulatory system. The heart is a pulsatile, four-chamber pump composed of two atria and two ventricles. In the present chapter we wish to explain how the heart operates as a pump: that is, to explain the function of the muscle, of the valves, and of the various chambers of the heart. Therefore, we will discuss first the basic physiology of cardiac muscle itself, especially how it differs from skeletal muscle, which was discussed in Chapter 7.

PHYSIOLOGY OF CARDIAC MUSCLE

PHYSIOLOGIC ANATOMY OF CARDIAC MUSCLE

Figure 11–1 illustrates a typical histologic section of cardiac muscle, showing arrangement of the cardiac muscle fibers in a latticework, the fibers dividing, then recombining, and spreading in all directions. One notes immediately from this figure that cardiac muscle is *striated* in the same manner as typical skeletal muscle. Furthermore, cardiac muscle has typical myofibrils that contain *actin* and *myosin filaments* almost identical to

Intercalated discs

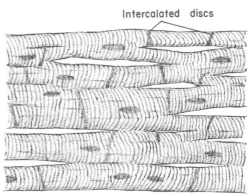

FIGURE 11–1 The "syncytial" nature of cardiac muscle.

those found in skeletal muscle, and these filaments interdigitate and slide among each other during the process of contraction in the same manner as occurs in skeletal muscle. (See Chap. 7.)

Cardiac Muscle as a "Functional Syncytium." Cardiac muscle is a *functional syncytium*, in which the cardiac muscle cells are so tightly bound that when one of these cells becomes excited, the action potential spreads to all of them, going from cell to cell and spreading laterally through the latticework interconnections.

The heart is composed of two separate functional syncytiums, the *atrial syncytium* and the *ventricular syncytium*. These are normally connected with each other only by way of specialized conductive fibers in the *A-V bundle*, which transmits impulses from the atrial muscle into the ventricular muscle. Other than for this connection, the atrial and ventricular muscles are separated from each other by the fibrous tissue surrounding the valvular rings.

All-or-Nothing Principle as Applied to the Heart. Because of the syncytial nature of cardiac muscle, stimulation of any single atrial muscle fiber causes the action potential to travel over the entire muscle mass. This is called the all-or-nothing principle; and it is precisely the same as that discussed in Chapter 7 for skeletal muscle, except that the cardiac

muscle fibers interconnect with each other, which means that the all-or-nothing principle applies to the entire functional syncytium of the heart rather than to single muscle fibers as is true in the case of skeletal muscle fibers.

ACTION POTENTIALS IN CARDIAC MUSCLE

Cardiac muscle has a peculiar type of action potential, as illustrated in Figure 11–2. After the initial *spike,* the membrane remains depolarized for 0.15 to 0.3 second, exhibiting a *plateau*, followed at the end of the plateau by abrupt repolarization. The presence of this plateau in the action potential causes the action potential to last about 100 times as long in cardiac muscle as in skeletal muscle and causes a correspondingly increased period of contraction. The cause of the plateau in the action potential is yet unknown.

Velocity of Conduction in Cardiac Muscle. The velocity of conduction of the action potential in both atrial and ventricular muscle fibers is about 0.3 to 0.4 meter per second, or about 1/300 the velocity in very large nerve fibers and about 1/10 the velocity in skeletal muscle fibers. The velocity of conduction in specialized conductive muscle fibers in the heart, which will be discussed in

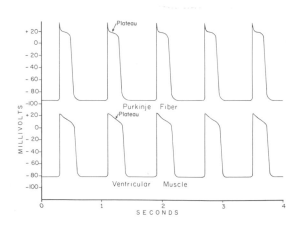

FIGURE 11-2 Rhythm action potentials from a Purkinje fiber and from a ventricular muscle fiber recorded by means of microelectrodes.

the following chapter, varies from only a small fraction of a meter to several meters per second in different parts of the conductive system.

Refractory Period of Cardiac Muscle. The refractory period of heart muscle is usually stated in terms of the *functional refractory period*, which is the interval of time during which an action potential from another part of the heart fails to re-excite an already excited area of cardiac muscle. The normal functional refractory period of the ventricle is approximately 0.25 second. There is an additional *relative refractory period* of about 0.05 second during which the muscle is more difficult than normal to excite but nevertheless can be excited.

CONTRACTION OF CARDIAC MUSCLE

The method of contraction in cardiac muscle is almost identical to that in skeletal muscle. That is, an action potential first travels over the cardiac muscle membranes, and this causes electrical current to spread by way of the T tubules to the interior of the muscle fibers. The electrical current in turn causes release of calcium ions from the longitudinal component of the sarcoplasmic reticulum. Finally, the calcium ions activate the contractile mechanism, causing the actin and myosin filaments to slide inward amongst each other, thus shortening the lengths of the cardiac muscle fibers.

A major difference between cardiac muscle contraction and skeletal muscle contraction is their duration. The very long plateau of the cardiac fiber action potential causes continued excitation of cardiac ventricular muscle for approximately 0.3 second, or 100 times as long as the excitation in skeletal muscle. Therefore, a single ventricular contraction also lasts about 0.3 second in contrast to a single skeletal muscle contraction that lasts for a much shorter period of time. The long contraction period of cardiac muscle allows sufficient time for the heart to pump blood out of its chambers prior to relaxation.

THE CARDIAC CYCLE

The period from the end of one heart contraction to the end of the next contraction is called the *cardiac cycle*. Each cycle is initiated by spontaneous generation of an action potential in the S-A node, as will be explained in detail in the following chapter. The S-A node is located in the posterior wall of the right atrium near the opening of the superior vena cava. Consequently, the atria contract ahead of the ventricles, thereby pumping blood into the ventricles prior to the very strong ventricular contraction. Thus the atria act as primer pumps for the ventricles, and the ventricles then provide the major source of power for moving blood through the vascular system.

SYSTOLE AND DIASTOLE

The cardiac cycle consists of a period of relaxation called *diastole* followed by a period of contraction called *systole*.

Figure 11-3 illustrates the different events during the cardiac cycle. The top three curves show the pressure changes in the aorta, the left ventricle, and the left atrium, respectively. The fourth curve depicts the changes in ventricular volume, the fifth the electrocardiogram, and the sixth a recording of the sounds produced by the heart as it pumps. It is especially important that the student study in detail the diagram of this figure and understand the causes of all the events illustrated. These are explained as follows:

RELATIONSHIP OF THE ELECTROCARDIOGRAM TO THE CARDIAC CYCLE

The electrocardiogram in Figure 11-3 shows the *P*, *Q*, *R*, *S*, and *T* waves,

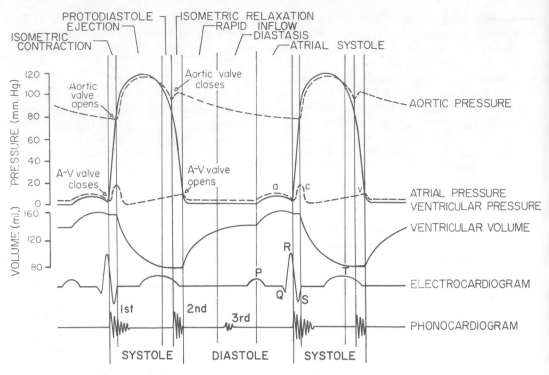

FIGURE 11-3 The events of the cardiac cycle, showing changes in left atrial pressure, left ventricular pressure, aortic pressure, ventricular volume, the electrocardiogram, and the phonocardiogram.

which will be discussed in the following chapters. The *P wave* is caused by the *spread of depolarization* through the atria. The state of depolarization, which lasts for about 0.15 second in the atria, causes the atria to contract and raise the atrial pressure.

Approximately 0.16 second after the P wave, the *QRS complex* appears as a result of depolarization of the ventricles. This initiates contraction of the ventricles and causes the ventricular pressure to begin rising, as illustrated in the figure. Therefore, the QRS complex begins approximately at the onset of ventricular systole.

Finally, one observes the *ventricular T wave* in the electrocardiogram. This represents the stage of repolarization of the ventricles, at which time the ventricular muscle fibers begin to relax. Therefore, the T wave occurs slightly prior to the end of ventricular contraction.

FUNCTION OF THE ATRIA AS PUMPS

Blood normally flows continually from the great veins into the atria, and ordinarily approximately 70 per cent of this flows directly into the ventricles even before the atria contract. Then atrial contraction causes an additional 30 per cent filling. Therefore, the primer function of the atria considerably increases the effectiveness of the ventricles as pumps.

Pressure Changes in the Atria—The a, c, and v waves. If one observes the atrial pressure curve of figure 11-3, he will note three major pressure elevations called the *a*, *c*, and *v atrial pressure waves*.

The *a wave* is caused by atrial contraction. The *c wave* occurs when the ventricles begin to contract, and it is caused primarily by bulging of the A-V valves toward the atria because of increasing pressure in the ventricles. The *v wave* occurs toward the end of ventricular contraction, and it results from slow buildup of blood in the atria because the A-V valves are closed during ventricular contraction.

FUNCTION OF THE VENTRICLES AS PUMPS

Emptying of the Ventricles During Systole. *Period of isometric contraction.* Immediately after ventricular contraction begins, the ventricular pressure abruptly rises, as shown in Figure 11–3, causing the A-V valves to close. Then an additional 0.02 to 0.03 second is required for the ventricle to build up sufficient pressure to push the semilunar (aortic and pulmonary) valves open against the pressures in the aorta and pulmonary artery. Therefore, during this period of time contraction is occurring in the ventricles but there is no emptying. This period is called the *period of isometric contraction*, meaning by this term that tension is increasing in the muscle but no shortening of the muscle fibers is occurring.

Period of ejection. When the left ventricular pressure rises slightly above 80 mm. Hg and the right ventricular arterial pressure slightly above 8 mm. Hg, the ventricular pressures now push the semilunar valves open. Immediately, blood begins to pour out of the ventricles, about half of the emptying occurring during the first quarter of systole.

Period of isometric relaxation. At the end of systole, ventricular relaxation begins suddenly, allowing the intraventricular pressures to fall rapidly. The elevated pressures in the large arteries immediately push blood back toward the ventricles, which snaps the aortic and pulmonary valves closed. For another 0.03 to 0.06 second, the ventricular muscle continues to relax, and the intraventricular pressures fall rapidly back toward their very low diastolic levels. As the ventricular pres-

sures fall below the atrial pressures, the A-V valves open to begin a new cycle of ventricular pumping.

FUNCTION OF THE VALVES

The *A-V valves* obviously prevent backflow of blood from the ventricles to the atria during systole, and the *semilunar valves* prevent backflow from the aorta and pulmonary arteries into the ventricles during diastole. All these valves, illustrated in Figure 11–4, close and open *passively.* That is, they close when a backward pressure gradient pushes blood backward, and they open when a forward pressure gradient forces blood in the forward direction. The high pressure in the arteries at the end of systole causes the semilunar valves to snap to the closed position, in comparison with a much softer closure of the A-V valves.

Function of the Papillary Muscles. Figure 11–4 also illustrates the papillary muscles attached to the vanes of the A-V valves by the *chordae tendineae.* The papillary muscles contract when the

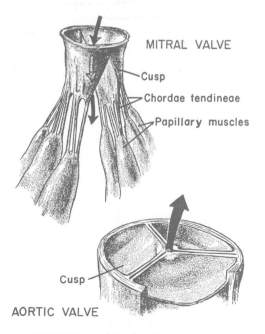

MITRAL VALVE

Cusp

Chordae tendineae

Papillary muscles

Cusp

AORTIC VALVE

FIGURE 11–4 Mitral and aortic valves.

ventricular walls contract, and, contrary to what might be expected, they *do not* help to close the valves. Instead, they pull the vanes of the valves inward toward the ventricles to prevent their bulging too far backward toward the atria during ventricular contraction. If a chorda tendinea becomes ruptured or if one of the papillary muscles becomes paralyzed as a result of ischemia or other effects, the valve bulges far backward, sometimes so far that it leaks severely and results in severe or even lethal cardiac incapacity.

THE AORTIC PRESSURE CURVE

When the left ventricle contracts, the ventricular pressure rises rapidly until the aortic valve opens. Then the pressure in the ventricle rises only slightly thereafter, as illustrated in Figure 11–3, because blood immediately flows out of the ventricle into the aorta.

The entry of blood into the arteries causes the walls of these arteries to stretch and the pressure in the arterial system to rise. Then, at the end of systole, even after the left ventricle stops ejecting blood, the extra filling of the arteries maintains high pressure in the arteries while the ventricular pressure falls. This creates a pressure differential backward across the aortic valve, and the aortic valve closes.

After the aortic valve has closed, pressure in the aorta falls slowly throughout diastole because blood stored in the distended elastic arteries flows continually through the peripheral vessels back to the veins.

Obviously, the pressure curve in the pulmonary artery is similar to that in the aorta, except that the pressures are much less, as will be discussed in Chapter 15.

THE VENTRICULAR VOLUME CURVE, AND STROKE VOLUME OUTPUT

Also shown in Figure 11–3 is a curve depicting the relative changes in ventricular volume from instant to instant. During diastole, filling of the ventricles normally increases the volume of each ventricle to about 120 to 130 ml. This volume is known as the *end-diastolic volume*. Then, as the ventricles empty during systole, the volume decreases about 70 ml., which is called the *stroke volume*. The remaining volume in each ventricle, about 50 to 60 ml., is called the *end-systolic volume*.

When the heart contracts strongly, the end-systolic volume can fall to as little as 10 to 30 ml. On the other hand, when large amounts of blood flow into the ventricles during diastole, their end-diastolic volumes can become as great as 200 to 250 ml. in the normal heart. Thus, by increasing the end-diastolic volume and decreasing the end-systolic volume, the stroke volume output can be increased to more than 150 ml. in the normal person and to as high as 225 ml. in some athletes.

RELATIONSHIP OF THE HEART SOUNDS TO HEART PUMPING

When one listens to the heart with a stethoscope, he does not hear the opening of the valves, for this is a relatively slowly developing process that makes no noise. However, when the valves close, the vanes of the valves and the surrounding fluids vibrate under the influence of the sudden pressure differentials which develop, giving off sound that travels in all directions through the chest. When the ventricles first contract, one hears a sound that is caused by closure of the A-V valves. The vibration is low in pitch and relatively long continued and is known as the *first heart sound*. When the aortic and pulmonary valves close, one hears a relatively rapid snap, for these valves close extremely rapidly, and the surroundings vibrate for only a short period of time. This sound is known as the *second heart sound*. The precise causes of these sounds will be discussed in Chapter 22 in relation to auscultation.

WORK OUTPUT OF THE HEART

The work output of the heart is the amount of energy that the heart transfers to the blood while pumping it into the arteries. Energy is transferred to the blood in two forms: First, by far the major proportion is used to move the blood from the low pressure veins to the high pressure arteries. This is *potential energy of pressure*. Second, a minor proportion of the energy is used to accelerate the blood to its velocity of ejection through the aortic and pulmonary valves. This is *kinetic energy of blood flow*.

The work performed by the left ventricle to raise the pressure of the blood during each heartbeat is equal to *stroke volume output* × (*left ventricular mean ejection pressure* minus *left atrial pressure*).

The work output of each ventricle required to create kinetic energy of blood flow is proportional to the mass of the blood ejected times the square of the velocity of ejection. However, this amount is normally only 2 to 4 per cent of the total work output of the ventricle.

ENERGY FOR CARDIAC CONTRACTION

Heart muscle, like skeletal muscle, utilizes chemical energy to provide the work of contraction. This energy is derived mainly from metabolism of glucose and fatty acids with oxygen and, to a lesser extent, from metabolism of other nutrients with oxygen. The different reactions that liberate this energy will be discussed in detail in Chapters 45 and 46.

The amount of energy expended by the heart is related to its work load in the following manner: The energy expended is *approximately proportional to the degree of tension generated by the heart musculature during contraction × the amount of time that the tension is maintained.*

Efficiency of Cardiac Contraction. During muscular contraction most of the chemical energy is converted into heat and a small portion into work output. The ratio of work output to chemical energy expenditure is called the efficiency of cardiac contraction, or simply *efficiency of the heart*. The efficiency of the normal heart, beating under normal load, is usually low, some 5 to 10 per cent. However, during maximum work output it rises as high as 15 to 20 per cent in the normal heart and perhaps 25 per cent in the heart of the well-trained athlete.

REGULATION OF CARDIAC FUNCTION

When a person is at rest, the heart must pump only 4 to 6 liters of blood each minute. However, during severe exercise it may be required to pump as much as five times this amount. The present section discusses the means by which the heart can adapt itself to such extreme increases in cardiac output.

The two basic means by which the pumping action of the heart is regulated are (1) intrinsic autoregulation in response to changes in volume of blood flowing into the heart and (2) reflex control of the heart by the autonomic nervous system.

INTRINSIC AUTOREGULATION OF CARDIAC PUMPING — THE FRANK-STARLING LAW OF THE HEART

In Chapter 21 we shall see that one of the major factors determining the amount of blood pumped by the heart each minute is the rate of blood flow into the heart from the veins, which is called *venous return*. That is, each peripheral tissue of the body controls its own blood flow, and whatever amount of blood flows through the peripheral tissues returns by way of the veins to the right atrium. The heart in turn automatically pumps this incoming blood on into the systemic arteries so that it can flow around the circuit again. Thus, the heart must adapt itself from moment to moment or even second to second to a widely varying input of blood that sometimes falls as low as 2 to 3 liters per minute and that at other times rises as high as 25 or more liters per minute.

This intrinsic ability of the heart to adapt itself to changing loads of inflowing blood is called the Frank-Starling law of the heart, in honor of Frank and Starling, two of the great physiologists of half a century ago. Basically, the Frank-Starling law states that the greater the heart is filled during diastole, the greater will be the quantity of blood pumped into the aorta. Or, expressing this law another way: *within physiologic limits, the heart pumps all the blood that comes to it without allowing excessive damming of blood in the veins.* In other words, the heart can pump either a small amount of blood or a large amount, depending on the amount that flows into it from the veins; and it automatically adapts to whatever this load might be as long as the total quantity does not rise above a physiologic limit that the heart can pump.

The primary mechanism by which the heart adapts to changing input volumes of blood is the following: When the cardiac muscle becomes stretched, as it does when extra amounts of blood enter the heart chambers, the muscle contracts with a greatly increased force, thereby automatically pumping the extra blood into the arteries. The ability of stretched muscle to contract with increased force is characteristic of all striated muscle and not simply of cardiac muscle. Referring back to Figure 7–7 in Chapter 7, one will see that stretching a skeletal muscle, within its physiological limit, also increases its force of contraction. As was also pointed out in Chapter 7, the increased force of contraction is probably caused by the fact that the actin and myosin filaments are brought to an optimal degree of interdigitation for achieving contraction.

Effect of Aortic Pressure on Pumping by the Left Ventricle. One of the most important features of the Frank-Starling law of the heart is that it makes pumping by the heart almost entirely independent of pressure changes in the aorta. For instance, the mean aortic pressure can usually increase to 180 mm. Hg, 80 per cent above the normal value, without reducing the output of the heart significantly. That is, despite marked changes in output pressure load on the heart, it is still the inflow of blood to the heart that normally controls the amount of blood pumped per minute. This is an extremely important effect because it allows the tissues of the body to control the cardiac output by simply increasing the flow through the individual tissues themselves. As the flow from the tissues returns to the heart, the heart automatically pumps the blood back into the arteries to be ready for flow once again through the tissues.

CONTROL OF THE HEART BY NERVES

The heart is well supplied with both sympathetic and parasympathetic nerves, as illustrated in Figure 11–5. These nerves affect cardiac pumping in

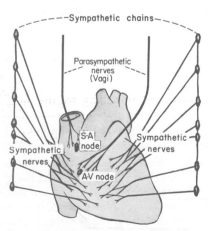

FIGURE 11–5 The cardiac nerves.

two ways: (1) by changing the heart rate and (2) by changing the strength of contraction of the heart. The effect of nerve stimulation on heart rate and rhythm will be discussed in detail in the following chapter. For the present, suffice it to say that parasympathetic stimulation decreases heart rate and sympathetic stimulation increases heart rate.

Effect of Heart Rate on Function of the Heart as a Pump. The normal heart has its peak ability to pump large quantities of blood at a heart rate of about 150 beats per minute. Below this rate, heart pumping is decreased; and above this rate, the heart fills poorly between beats.

Nervous Regulation of Contractile Strength of the Heart. The two atria are supplied with large numbers of both sympathetic and parasympathetic nerves, but the ventricles are supplied almost exclusively by sympathetic nerves with perhaps a few straggling parasympathetic fibers. Therefore, for practical purposes, the overall effect of the autonomic nerves on heart strength is mediated primarily through the sympathetic nerves. In general, sympathetic stimulation increases the strength of heart muscle contraction, whereas parasympathetic stimulation decreases the strength of contraction.

Under normal conditions the sympathetic nerve fibers to the heart continually discharge at a slow rate that maintains a strength of ventricular contraction about 20 per cent above its strength with no sympathetic stimulation at all. Therefore, one method by which the nervous system can decrease the strength of ventricular contraction is simply to slow or stop the transmission of sympathetic impulses to the heart. On the other hand, maximal sympathetic stimulation can increase the strength of ventricular contraction to approximately double normal.

Maximal parasympathetic stimulation of the heart decreases ventricular contractile strength perhaps 10 to 30 per cent. Thus, the parasympathetic effect is, by contrast with the sympathetic effect, relatively unimportant.

FUNCTION CURVES OF THE HEART

Ventricular Function Curves. One of the best ways to express the functional ability of the ventricles is by ventricular function curves, a type of which, shown in Figure 11–6, is called the *minute ventricular output curve*. The two curves of this figure represent function of the two ventricles of the human heart based on data extrapolated from lower animals. As each atrial pressure rises, the respective ventricular volume output per minute also increases.

Thus, ventricular function curves are another way of expressing the Frank-Starling law of the heart. That is, as the ventricles fill to higher atrial pressures, the strength of cardiac contraction increases, causing the heart to pump increased quantities of blood into the arteries. Ventricular function curves are exceedingly important in analyzing overall function of the circulation, for it is by such means that one can express the quantitative capabilities of the heart as a pump.

Cardiac Output Curves. Figure 11–7 illustrates a "family" of heart function curves called either *cardiac function curves* or *cardiac output curves*. These depict function of the overall heart and lung system, showing the effect of input pressure to the system (the right atrial pressure) on output from the system (cardiac output). This type of curve is

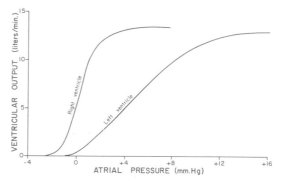

FIGURE 11–6 Approximate normal right and left ventricular output curves for the human heart as extrapolated from data obtained in dogs.

especially important in analyzing overall function of the heart.

The very dark curve of Figure 11-7 represents function of the normal heart. When the right atrial pressure is several millimeters Hg negative with respect to atmospheric pressure, there is essentially no filling of the right atrium and, consequently, no pumping by the heart. On the other hand, when the right atrial pressure rises above approximately +4 mm. Hg, the heart by then has reached its maximum pumping limit, and further increases in right atrial input pressure will not cause additional cardiac output.

Hyper- and Hypo-Effective Hearts. The three upper curves in Figure 11-7 are cardiac output curves for hyper-effective hearts — that is, hearts that have greater than normal pumping ability. Factors that can cause hypereffectivity of the heart are:

1. *Sympathetic stimulation of the heart.*

2. *Decreased parasympathetic stimulation of the heart.*

3. *Hypertrophy of the heart.*

Sympathetic stimulation can increase cardiac pumping effectiveness about 70 per cent, and hypertrophy can sometimes more than double heart pumping effectiveness.

The three lowermost curves of Figure 11-7 show the effects of decreased effectiveness of the heart as a pump. Any effect that damages the heart can obviously cause a hypoeffective heart and thereby can reduce the plateau level of the cardiac output curve. Some of the factors that cause hypoeffective hearts are:

1. *Myocardial infarction.*
2. *Valvular heart disease.*
3. *Congenital heart disease.*
4. *Disease of the myocardium.*

EFFECT OF VARIOUS IONS ON HEART FUNCTION

In Chapters 5 and 7 it was pointed out that some ions have marked effects on transmission of action potentials and also on muscle contraction. Two ions in particular have very marked effects on cardiac function; these are *potassium* and *calcium*.

Effect of Potassium Ions. Excess potassium in the extracellular fluid causes the heart to become extremely weak and dilated; also, it slows the heart rate. These effects are believed to be caused by the fact that high potassium concentration decreases the resting membrane potential. Elevation of potassium concentration to 8 to 12 mEq./liter — two to three times the normal value — will usually cause death.

Effect of Calcium Ions. Great excess of calcium ions has effects exactly opposite to those caused by potassium ions, making the heart go into spastic contraction. This probably results from the action of calcium ions to excite the cardiac contractile process, as explained in Chapter 7 in relation to muscle contraction. However, calcium ion concentration rarely changes enough, except under artificial experimental conditions, to cause significant cardiac malfunction.

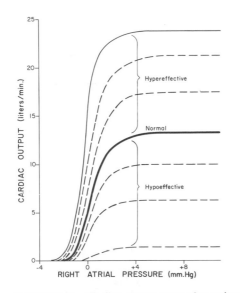

FIGURE 11-7 Cardiac output curves for various degrees of hypo- and hypereffective hearts. (From Guyton: Cardiac Output and Its Regulation.)

EFFECT OF TEMPERATURE ON THE HEART

Increased temperature causes greatly increased heart rate, and decreased temperature causes greatly decreased rate.

These effects of temperature presumably result from increased permeability of the muscle membrane to the different ions, resulting in acceleration of the self-excitation process.

REFERENCES

Brady, Allan J.: Active state in cardiac muscle. *Physiol. Rev., 48*:570, 1968.

Braunwald, E., Ross, J., and Sonnenblick, E.: Mechanism of Contractility of the Normal and Failing Heart. Boston, Little, Brown and Company, 1968.

Brecher, G. A., and Galletti, P. M.: Functional anatomy of cardiac pumping. *In* Handbook of Physiology. Baltimore, The Williams & Wilkins Co., 1963, Vol. II, Sec. II, p. 759.

Guyton, A. C.: Determination of cardiac output by equating venous return curves with cardiac response curves. *Physiol. Rev., 35*:123, 1955.

Levy, M. N., and Berne, R. M.: Heart. *Ann. Rev. Physiol., 32*:373, 1970.

Randall, W.: Nervous Control of the Heart. Baltimore, The Williams & Wilkins Co., 1965.

Sarnoff, S. J.: Myocardial contractility as described by ventricular function curves. *Physiol. Rev., 35*:107, 1955.

Starling, E. H.: The Linacre Lecture on the Law of the Heart. London, Longmans Green & Co., 1918.

CHAPTER 12

RHYTHMIC EXCITATION OF THE HEART

THE SPECIAL EXCITATORY AND CONDUCTIVE SYSTEM OF THE HEART

The adult human heart normally contracts at a rhythmic rate of about 72 beats per minute. Figure 12–1 illustrates the special excitatory and conductive system of the heart that controls these cardiac contractions. The figure shows:

(A) the *S-A node*, in which the normal rhythmic self-excitatory impulse is generated; (B) the *A-V node*, in which the impulse from the atria is delayed before passing into the ventricles; (C) the *A-V bundle*, which conducts the impulse from the atria into the ventricles; and (D) the *left* and *right bundles of Purkinje fibers*, which conduct the cardiac impulse to all parts of the ventricles.

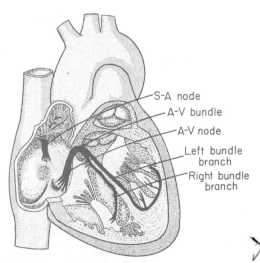

S-A node
A-V bundle
A-V node
Left bundle branch
Right bundle branch

FIGURE 12–1 The S-A node and the Purkinje system of the heart.

THE SINO-ATRIAL NODE AND ITS RHYTHMICITY

The sino-atrial (S-A) node is a small strip of specialized muscle approximately 3 mm. wide and 1 cm. long; it is located in the posterior wall of the right atrium immediately beneath and medial to the opening of the superior vena cava.

The fibers of the S-A node are self-excitatory, contracting rhythmically at a normal rate of 70 to 80 times per minute. This self-excitation process is probably caused by a high degree of permeability of the fiber membranes to sodium, an effect that re-excites the fibers soon after they repolarize.

134

Transmission of the Cardiac Impulse Through the Atria. Each time a rhythmic impulse is generated in any single fiber of the S-A node, it spreads immediately into the surrounding atrial muscle and is conducted in all directions at a velocity of about 0.3 meter per second. Though all atrial muscle fibers probably participate in conducting the cardiac impulse, a few specialized fibers seem to conduct the impulse with extra speed directly from the S-A node to the A-V node. These fibers are part of the rapid conducting system of the heart and are comparable to the very important Purkinje fibers of the ventricles, which will be discussed subsequently.

As the cardiac impulse travels through the atria, the atrial muscle contracts, forcing blood through the atrioventricular valves into the ventricles.

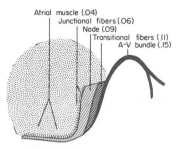

FIGURE 12-2 Organization of the A-V node. The numbers represent the interval of time from the origin of the impulse in the S-A node. The values have been extrapolated to the human being. (This figure is based on studies in lower animals discussed and illustrated in Hoffman and Cranefield: Electrophysiology of the Heart. McGraw-Hill Book Co.)

THE ATRIOVENTRICULAR (A-V) NODE AND THE PURKINJE SYSTEM

Delay in Transmission of the A-V Node. Fortunately, the conductive system is organized so that the cardiac impulse will not travel from the atria into the ventricles too rapidly; this allows time for the atria to empty their contents into the ventricles before ventricular contraction begins. It is primarily the A-V node and its associated conductive fibers that delay this transmission of the cardiac impulse from the atria into the ventricles.

Figure 12–2 shows diagrammatically the different parts of the A-V node and its connections with the atrial muscle and the A-V bundle. The figure also shows the approximate intervals of time in fractions of a second between the genesis of the cardiac impulse in the S-A node and its appearance at different points in its conductive pathway. Note that the impulse, after traveling through the atrial muscle, reaches the A-V node approximately 0.04 second after its origin in the S-A node. However, between this time and the time that the impulse emerges in the A-V bundle, another 0.11 second elapses. About one half of this time lapse occurs in the *junctional fibers*, which are very small fibers that connect the normal atrial fibers with the fibers of the node itself (illustrated in Fig. 12–2). The velocity of conduction in these fibers is 0.05 meter per second or less (about one-seventh that in normal cardiac muscle), which greatly delays entrance of the impulse into the A-V node. After entering the node, the velocity of conduction in the *nodal fibers* is still quite low, only 0.1 meter per second, about one-third to one-fourth the conduction velocity in normal cardiac muscle. Therefore, a further delay in transmission occurs as the impulse travels through the A-V node into the *transitional fibers* and finally into the *A-V bundle*.

TRANSMISSION IN THE PURKINJE SYSTEM

The *Purkinje fibers* that lead from the A-V node through the A-V bundle and into the ventricles have functional characteristics quite the opposite to those of the A-V nodal fibers; they are very large fibers, even larger than the normal ventricular muscle fibers, and they transmit impulses at a velocity of 1.5 to 2.5 meters per second, a velocity about six times that in the usual cardiac muscle and 40 times that in the junctional fibers. This allows almost immediate transmission of the cardiac impulse throughout the entire ventricular system.

Distribution of the Purkinje Fibers in the Ventricles. The Purkinje fibers, after originating in the A-V node, form the A-V bundle, which then threads between the valves of the heart and thence into the ventricular septum as shown in Figure 12–1. The A-V bundle divides almost immediately into the *left* and *right bundle branches* that lie beneath the endocardium of the respective sides of the septum. Each of these branches spreads downward toward the apex of the respective ventricle, then curves around the tip of the ventricular chamber and finally back toward the base of the heart along the lateral wall. About one-third the way up the lateral walls of the ventricles, the two bundle branches break up into many small branches of Purkinje fibers that spread in all directions beneath the endocardium of the ventricles. The terminal Purkinje fibers enter the ventricular muscle from the endocardial surface and terminate on the muscle fibers. At this termination, the cardiac impulse can pass readily from the Purkinje fiber into the cardiac muscle fiber; but, for reasons not yet understood, the impulse probably cannot pass backward from the muscle into the Purkinje fiber.

From the time that the cardiac impulse first enters the A-V bundle until it reaches the terminations of the Purkinje fibers, the total time that lapses is only 0.03 second; therefore, once a cardiac impulse enters the Purkinje system, it spreads almost immediately to the entire endocardium of the ventricles.

TRANSMISSION OF THE CARDIAC IMPULSE IN THE VENTRICULAR MUSCLE

Once the cardiac impulse has reached the ends of the Purkinje fibers, it is then transmitted through the ventricular muscle mass by the ventricular muscle fibers themselves. The velocity of transmission is now only 0.3 to 0.4 meter per second, one-sixth that in the Purkinje fibers. Because of this, transmission from the endocardial surface to the epicardial surface of the ventricle requires as much as another 0.03 second, approximately equal to the time required for transmission through the entire Purkinje system. Thus, the total time for transmission of the cardiac impulse from the origin of the Purkinje system to the last of the ventricular muscle fibers is about 0.06 second.

SUMMARY OF THE SPREAD OF THE CARDIAC IMPULSE THROUGH THE HEART

Figure 12–3 summarizes the transmission of the cardiac impulse through the human heart. The numbers on the figure represent the intervals of time in fractions of a second that lapse between the origin of the cardiac impulse in the S-A node and its appearance at each respective point in the heart. Note that the impulse spreads at moderate velocity through the atria but is delayed at least one-tenth second in the A-V nodal region before appearing in the A-V bundle.

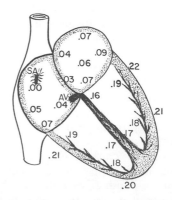

FIGURE 12–3 Transmission of the cardiac impulse through the heart, showing the time of appearance (in fractions of a second) of the impulse in different parts of the heart.

Once it has entered the bundle, it spreads rapidly through the Purkinje fibers to the entire endocardial surfaces of the ventricles. Then the impulse spreads slowly through the ventricular muscle to the epicardial surfaces.

CONTROL OF EXCITATION AND CONDUCTION IN THE HEART

THE S-A NODE AS THE PACEMAKER OF THE HEART

In the above discussion of the genesis and transmission of the cardiac impulse through the heart, it was stated that the impulse normally arises in the S-A node. However, this need not be the case under abnormal conditions, for other parts of the heart can exhibit rhythmic contraction in the same way that the fibers of the S-A node can; this is particularly true of the A-V nodal fibers, which discharge at an intrinsic rhythmic rate of 40 to 60 times per minute; and the Purkinje fibers, which discharge at a rate between 15 to 40 times per minute.

Therefore, the question that we must ask is: Why does the S-A node control the heart's rhythmicity rather than the A-V node or the Purkinje fibers? The answer to this is simply that the rate of the S-A node is considerably greater than that of either the A-V node or the Purkinje fibers. Each time the S-A node discharges, its impulse is conducted both into the A-V node and into the Purkinje fibers, discharging their excitable membranes. But the S-A node recovers much more rapidly than can either of the other two and emits still another impulse before either one of them can reach its own threshold for self-excitation. The new impulse again discharges both the A-V node and Purkinje fibers. This process continues on and on, the S-A node always exciting these other potentially self-excitatory

tissues before self-excitation can actually occur.

Thus, the S-A node controls the beat of the heart because its rate of rhythmic discharge is greater than that of any other part of the heart. Therefore, it is said that the S-A node is the normal *pacemaker* of the heart.

CONTROL OF HEART RHYTHMICITY AND CONDUCTION BY THE AUTONOMIC NERVES

The heart is supplied with both sympathetic and parasympathetic nerves, as illustrated in Figure 11–5 of the previous chapter. The parasympathetic nerves are distributed mainly to the S-A and the A-V nodes; to a lesser extent to the muscle of the two atria; but very sparsely, if at all, to the ventricular muscle. The sympathetic nerves are distributed to these same areas but with a strong representation to the ventricular muscle.

Effect of Parasympathetic (Vagal) Stimulation on Cardiac Function—Ventricular Escape. Stimulation of the parasympathetic nerves to the heart (the vagi) causes the hormone acetylcholine to be released at the vagal endings. This hormone has two major effects on the heart. First, it decreases the rate of rhythm of the S-A node, and, second, it decreases the excitability of the A-V junctional fibers between the atrial musculature and the A-V node, thereby slowing transmission of the cardiac impulse into the ventricles. Very strong stimulation of the vagi can completely stop the rhythmic contraction of the S-A node or completely block transmission of the cardiac impulse through the A-V junction. In either case, rhythmic impulses are no longer transmitted into the ventricles. The ventricles stop beating for 4 to 10 seconds, but then some point in the Purkinje fibers, usually in the A-V bundle, develops a rhythm of its own and causes ventricular contraction at a rate of 15 to 40 beats per minute. This phenomenon is called *ventricular escape.*

Mechanism of the vagal effects. The acetylcholine released at the vagal nerve endings greatly increases the permeability of the fiber membranes to potassium, which allows rapid leakage of potassium to the exterior. This causes increased negativity inside the fibers, an effect called *hyperpolarization*, which makes excitable tissue much less excitable, as was explained in Chapter 5.

Effect of Sympathetic Stimulation on Cardiac Function. Sympathetic stimulation causes essentially the opposite effects on the heart to those caused by vagal stimulation as follows: First, it increases the rate of S-A nodal discharge. Second, it increases the excitability of all portions of the heart. Third, it increases greatly the force of contraction of all the cardiac musculature, both atrial and ventricular, as discussed in the previous chapter.

In short, sympathetic stimulation increases the overall activity of the heart. Maximal stimulation can almost triple the rate of heartbeat and can increase the strength of heart contraction as much as two-fold.

Mechanism of the sympathetic effect. Stimulation of the sympathetic nerves releases the hormone norepinephrine at the sympathetic nerve endings. The precise mechanism by which this hormone acts on cardiac muscle fibers is still somewhat doubtful, but the present belief is that it increases the permeability of the fiber membrane to sodium, an effect that increases the excitability of the fibers.

ABNORMAL RHYTHMS OF THE HEART

PREMATURE CONTRACTIONS— ECTOPIC FOCI

Often, a small area of the heart becomes much more excitable than normal and causes an occasional abnormal impulse to be generated between normal impulses. A depolarization wave spreads outward from the irritable area and initiates a *premature contraction* of the heart. The focus at which the abnormal impulse is generated is called an *ectopic focus.*

The usual cause of an ectopic focus is an irritable area of cardiac muscle resulting from overuse of stimulants such as caffeine or nicotine, or resulting from lack of sleep, anxiety, or other debilitating states.

Shift of the Pacemaker to an Ectopic Focus. Sometimes an ectopic focus becomes so irritable that it establishes a rhythmic contraction of its own at a more rapid rate than that of the S-A node. When this occurs the ectopic focus becomes the pacemaker of the heart. The most common point for development of an ectopic pacemaker is the A-V node itself or the A-V bundle, as will be discussed in more detail in Chapter 13.

HEART BLOCK

Occasionally, transmission of the impulse through the heart is blocked at critical points in the conductive system. One of the most common of these points is between the atria and the ventricles; this condition is called *atrioventricular block.* Another common point is in one of the *bundle branches* of the Purkinje system. Rarely, a block also develops between the S-A node and the atrial musculature.

Atrioventricular Block. In the human being, block from the atria to the ventricles can result from localized damage or depression of the *A-V junctional* fibers or of the *A-V bundle.* The causes include different types of infectious processes, excessive stimulation by the vagus nerves (which depresses conductivity of the junctional fibers), localized destruction of the A-V bundle as a result of a coronary infarct, pressure on the A-V bundle by arteriosclerotic plaques, or depression caused by various drugs. Immediately after onset of complete heart block, the atria continue to beat at their normal rate of rhythm, while the ventricles fail to contract at all for approximately 7 seconds. Then, "ventricular escape" occurs, and a rhythmic focus in the ventricles suddenly begins to act as a ventricular pacemaker, causing ventricular contractions at a rate of 15 to 40 times per minute; these are completely dissociated from the atrial contractions.

FLUTTER AND FIBRILLATION

Frequently, either the atria or the ventricles begin to contract extremely rapidly and often incoordinately. The lower frequency and more coordinate contractions up to 200 to 300 beats per minute are generally called *flutter*; and the very high frequency and incoordinate contractions, *fibrillation*. Most instances of flutter and fibrillation result from *circus movements*, in which the impulse travels around and around through the heart muscle, never stopping.

The Circus Movement. The circus movement theory explains the course of events in most instances of atrial and ventricular fibrillation. This can be described as follows:

Figure 12–4 illustrates several small cardiac muscle strips cut in the form of circles. If such a strip is stimulated at the 12 o'clock position *so that the impulse travels in only one direction*, the impulse spreads progressively around the circle until it returns to the 12 o'clock position. If the originally stimulated muscle fibers are still in a refractory state, the impulse then dies out, for refractory muscle cannot transmit a second impulse. However, there are three different conditions that could cause this impulse to continue to travel around the circle: (1) *excess length of the pathway around the circle*, (2) *decreased velocity of conduction*,

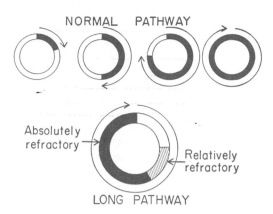

FIGURE 12–4 The circus movement, showing annihilation of the impulse in the short pathway and continued propagation of the impulse in the long pathway.

and (3) *a shortened refractory period of the muscle.*

All three of these conditions occur in different pathological states of the human heart as follows: (1) A long pathway frequently occurs in dilated hearts. (2) Decreased rate of conduction frequently results from blockage of the Purkinje system. (3) A shortened refractory period frequently occurs in response to various drugs, such as epinephrine, or following intense vagal stimulation. Thus, in many different cardiac disturbances circus movements can cause abnormal cardiac rhythmicity that completely ignores the pacesetting effects of the S-A node.

Atrial Flutter Resulting from a Circus Pathway. Figure 12–5 illustrates a circus pathway around and around the

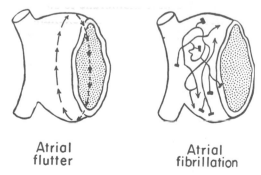

Atrial flutter

Atrial fibrillation

FIGURE 12–5 Pathways of impulses in atrial flutter and atrial fibrillation.

atria from top to bottom, the pathway lying to the left of the superior and inferior venae cavae. This causes rapid "flutter" of the atrium at a rate of 200 to 350 times per minute.

The "Chain Reaction" Mechanism of Fibrillation. Fibrillation, whether it occurs in the atria or in the ventricles, is a very different condition from flutter. One can see many separate contractile waves spreading in different directions over the cardiac muscle at the same time in either atrial or ventricular fibrillation. Obviously, then, the circus movement in fibrillation is entirely different from that in flutter. One of the best ways to explain the mechanism of fibrillation is to describe the initiation of fibrillation by stimulation with 60 cycle alternating electrical current.

Fibrillation caused by 60 cycle alternating current. At a central point in the ventricles of heart A in Figure 12–6, a 60 cycle electrical stimulus is applied through a stimulating electrode. The first cycle of the electrical stimulus causes a depolarization wave to spread in all directions, leaving all the muscle beneath the electrode in a refractory state. After about 0.25 second, this muscle begins to come out of the refractory state, some portions of the muscle coming out of refractoriness prior to other portions. This state of events is depicted in heart A by many light patches, which represent excitable

cardiac muscle, and dark patches, which represent refractory muscle. Stimuli from the electrode can now cause impulses to travel in certain directions through the heart but not in all directions. It will be observed that certain impulses travel for short distances until they reach refractory areas of the heart and then are blocked. Other impulses, however, pass between the refractory areas and continue to travel in the excitable patches of muscle. Now, several events transpire in rapid succession, all occurring simultaneously and eventuating in a state of fibrillation. These are as follows:

First, block of the impulses in some directions but successful transmission in other directions creates one of the necessary conditions for a circus movement to develop — that is, *transmission of some of the depolarization waves around the heart in only one direction.* As a result, these waves do not annihilate themselves on the opposite side of the heart but can continue around and around the ventricles.

Second, the impulses begin to divide into still many more impulses, as illustrated in heart A. That is, when a depolarization wave reaches a refractory area in the heart, it travels to both sides around the area. Thus, a single impulse becomes two impulses. Then when each of these reaches another refractory area it, too, divides to form still two more impulses. In this way many different impulses are continually being formed in the heart by a progressive *chain reaction* until, finally, there are many impulses traveling in many different directions at the same time. Furthermore, this irregular pattern of impulse travel causes a *circuitous route for the impulses to travel, greatly lengthening the conductive pathway, which is one of the conditions leading to fibrillation.*

Heart B in Figure 12–6 illustrates the final state that develops in fibrillation. Here one can see many impulses traveling in all directions, some dividing and increasing the number of impulses,

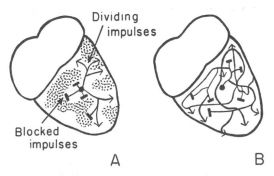

FIGURE 12–6 (*A*) Initiation of fibrillation in a heart when patches of refractory musculature are present. (*B*) Continued propagation of fibrillatory impulses in the fibrillating ventricle.

while others are blocked entirely by refractory areas. In the final state of fibrillation, the number of new impulses being formed exactly equals the number of impulses that are being blocked by refractory areas.

As is the case with many other cardiac arrhythmias, fibrillation is usually confined to either the atria or the ventricles alone and not to both syncytial masses of muscle at the same time.

Atrial Fibrillation. Atrial fibrillation occurs very frequently when the atria become greatly overdilated—in fact, many times as frequently as flutter. When flutter does occur, it usually becomes fibrillation after a few days or weeks. To the right in Figure 12–5 are illustrated the pathways of fibrillatory impulses traveling through the atria. Obviously, atrial fibrillation results in complete incoordination of atrial contraction, so that atrial pumping ceases entirely. However, even when the atria fail to act as primer pumps, the ventricles can still fill enough so that the effectiveness of the heart as a pump is reduced only 25 to 30 per cent, which is well within the "cardiac reserve" of all but severely weakened hearts. For this reason, atrial fibrillation can continue for many years without profound cardiac debility.

Irregularity of ventricular rate during atrial fibrillation. When the atria are fibrillating, impulses arrive at the A-V node very irregularly, thereby causing a very irregular ventricular rhythm. This irregularity is one of the clinical findings used to diagnose this condition.

Ventricular Fibrillation. Ventricular fibrillation is extremely important because at least one quarter of all persons die in ventricular fibrillation, especially following coronary infarction.

The likelihood of circus movements in the ventricles and, consequently, of ventricular fibrillation is greatly increased when the ventricles are dilated or when the rapidly conducting *Purkinje system is blocked* so that impulses cannot be transmitted rapidly. Also, *electrical stimuli*, as noted previously, or *ectopic foci* are common initiating causes of ventricular fibrillation.

Inability of the heart to pump blood during ventricular fibrillation. When the ventricles begin to fibrillate, the different parts of the ventricles no longer contract simultaneously. For the first few seconds the ventricular muscle undergoes rather coarse contractions which may pump a few milliliters of blood with each contraction. However, the impulses in the ventricles rapidly become divided into many much smaller impulses, and the contractions become fine rather than coarse, pumping no blood whatsoever. The ventricles dilate because of failure to pump the blood that is flowing into them, and within 60 to 90 seconds the ventricular muscle becomes too weak, because of lack of coronary blood supply, to contract strongly, even if coordinate contraction should return. Therefore, death is immediate when ventricular fibrillation begins.

Electrical defibrillation of the ventricles. A very strong electrical current passed through the ventricles for a short interval of time can stop fibrillation by simultaneously throwing all the ventricular muscle into refractoriness. This is accomplished by passing intense current through electrodes placed on two sides of the heart. The current penetrates most of the fibers of the ventricles, thus stimulating essentially all parts of the ventricles simultaneously and causing them to become refractory. All impulses stop, and the heart then remains quiescent for 3 to 5 seconds, after which it begins to beat again with the S-A node usually becoming the pacemaker.

Hand Pumping of the Heart ("Cardiac Massage") as an Aid to Defibrillation. Unless defibrillated within one minute after fibrillation begins, the heart is usually too weak to be revived. However, it is still possible to revive the heart by preliminarily pumping it by hand. In this way small quantities of blood are delivered into the aorta, and a renewed coronary blood supply develops. After a few minutes, electrical defibrillation often then becomes possible. Fibrillating hearts have been pumped by hand as long as 90 minutes before defibrillation.

REFERENCES

Brady, A. J.: Excitation and excitation-contraction coupling in cardiac muscle. *Ann. Rev. Physiol.,* *26*:341, 1964.

Guyton, A. C., and Satterfield, J.: Factors concerned in electrical defibrillation of the heart, particularly through the unopened chest. *Amer. J. Physiol., 167*:81, 1951.

Hoffman, B. F.: Physiology of atrioventricular transmission. *Circulation, 24*:506, 1961.

Hoffman, B. F., and Cranefield, P. F.: Electrophysiology of the Heart. New York, Blakiston Div., McGraw-Hill Book Company, 1960.

Levy, M. N., and Zieske, H.: Autonomic control of cardiac pacemaker activity and arterioventricular transmission. *J. Appl. Physiol., 27*:465, 1969.

Mendez, C., Mueller, W. J., and Urguiaga, X.: Propagation of impulses across the Purkinje fiber-muscle junctions in the dog heart. *Circ. Res., 26*:135, 1970.

Scher, A. M.: Excitation of the heart. *In* Handbook of Physiology. Baltimore, The Williams & Wilkins Co., 1962. Vol. I, Sec. II, 287.

CHAPTER 13

THE ELECTROCARDIOGRAM

Transmission of the impulse through the heart has been discussed in detail in Chapter 12. As the impulse passes through the heart, electrical currents spread into the tissues surrounding the heart, and a small proportion of these spreads all the way to the surface of the body. If electrodes are placed on the body surface on opposite sides of the heart, the electrical potentials generated by the heart can be recorded; the recording is known as an *electrocardiogram*. A normal electrocardiogram for two beats of the heart is illustrated in Figure 13–1.

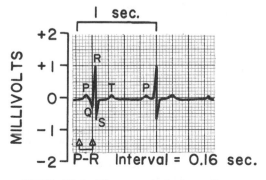

FIGURE 13–1 The normal electrocardiogram.

CHARACTERISTICS OF THE NORMAL ELECTROCARDIOGRAM

The normal electrocardiogram is composed of a P wave, a "QRS complex," and a T wave. The QRS complex is actually three separate waves, the Q wave, the R wave, and the S wave, all of which are caused by passage of the cardiac impulse through the ventricles. In the normal electrocardiogram, the Q and S waves are often much less prominent than the R wave and sometimes are actually absent, but even so, the wave is still known as the QRS complex, or simply the QRS wave.

The P wave is caused by electrical currents generated as the atria depolarize prior to atrial contraction, and the QRS complex is caused by currents generated when the ventricles depolarize prior to their contraction. Therefore, both the P wave and the components of the QRS complex are *depolarization waves*. The T wave is caused by currents generated as the ventricles recover from the state of depolarization. This process occurs in ventricular muscle about 0.25 second after depolarization,

143

and this wave is known as a *repolarization wave*. Thus, the electrocardiogram is composed of both depolarization and repolarization waves. The principles of depolarization and repolarization were discussed in Chapter 5. However, the distinction between depolarization waves and repolarization waves is so important in electrocardiography that further clarification is needed as follows:

DEPOLARIZATION WAVES
VERSUS REPOLARIZATION WAVES

Figure 13–2 illustrates a muscle fiber in four different stages of depolarization and repolarization. During the process of "depolarization" the normal negative potential inside the fiber is lost and the membrane potential actually reverses; that is, it becomes slightly positive inside and negative outside.

In Figure 13–2*A* the process of depolarization, illustrated by the positivity inside and negativity outside, is traveling from left to right, and the first half of the fiber is already depolarized while the remaining half is still polarized. This causes the meter to record positively. To the right of the muscle fiber is illustrated a record of the potential between the electrodes as recorded by a high-speed recording meter at this particular stage of depolarization.

In Figure 13–2*B* depolarization has extended over the entire muscle fiber, and the recording to the right has now returned to the zero base line because both electrodes are in areas of equal negativity. The completed wave is a *depolarization wave* because it results from spread of depolarization along the extent of the muscle fiber.

Figure 13–2*C* illustrates the repolarization process, during which the recording, as illustrated to the right, becomes negative.

Finally, in Figure 13–2*D* the muscle fiber has completely repolarized, and the recording returns once more to the zero level. This completed negative

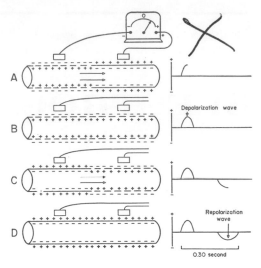

FIGURE 13–2 Recording the *depolarization wave* and the *repolarization* wave from a cardiac muscle fiber.

wave is a *repolarization wave* because it results from spread of the repolarization process over the muscle fiber.

Relationship of the Monophasic Action Potential of Cardiac Muscle to the QRS and T Waves. The monophasic action potential of ventricular muscle, which was discussed in the preceding chapter,

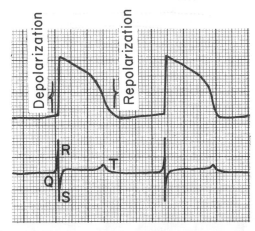

FIGURE 13–3 *Above:* Monophasic action potential from a ventricular muscle fiber during normal cardiac function, showing rapid depolarization and then repolarization occurring slowly during the plateau stage but very rapidly toward the end. *Below:* Electrocardiogram recorded simultaneously.

normally lasts between 0.25 and 0.30 second. The top part of Figure 13–3 illustrates a monophasic action potential recorded from a microelectrode inserted into the inside of a single ventricular muscle fiber. The upsweep of this action potential is caused by *depolarization*, and the return of the potential to the base line is caused by *repolarization*. Note below the simultaneous recording of the electrocardiogram from this same ventricle, which shows the QRS wave appearing at the beginning of the monophasic action potential and the T wave at the end. Note especially that *no potential at all is recorded in the electrocardiogram when the ventricular muscle is either completely polarized or completely depolarized.*

RELATIONSHIP OF ATRIAL AND VENTRICULAR CONTRACTION TO THE WAVES OF THE ELECTROCARDIOGRAM

Before contraction of muscle can occur, a depolarization wave must spread through the muscle to initiate the chemical processes of contraction. The P wave results from spread of the depolarization wave through the atria, and the QRS wave from spread of the depolarization wave through the ventricles. Therefore, the P wave occurs at the *beginning of contraction of the atria,* and the QRS wave occurs at the *beginning* of contraction of the ventricles.

An atrial repolarization wave occurs at the end of atrial contraction, but this is usually obscured by the QRS complex.

The Electrocardiogram During Ventricular Repolarization—The T Wave. Because the septum depolarizes first, it seems logical that it should repolarize first at the end of ventricular contraction, but this is not the usual case because the septum has a longer period of contraction and depolarization than do other areas of the heart. Actually, many sections of the ventricles begin to re-

polarize almost simultaneously. Yet *the greatest portion of ventricular muscle to repolarize first is that located exteriorly near the apex of the heart.* And the endocardial surfaces, on the average, repolarize last. The reason for this abnormal sequence of repolarization is reputed to be that high pressure in the ventricles during contraction greatly reduces coronary blood flow to the endocardium, thereby slowing the repolarization process on the endocardial surfaces. But whatever the cause, *the predominant direction of current flow through the heart during* repolarization *of the ventricles is from base to apex, which is the same predominant direction* of current flow during depolarization. *As a result, the T wave in the normal electrocardiogram is positive, which is also the polarity of most of the normal QRS complex.*

VOLTAGE AND TIME CALIBRATION OF THE ELECTROCARDIOGRAM

All recordings of electrocardiograms are made with appropriate calibration lines on the recording paper.

As illustrated in Figure 13–1, the horizontal calibration lines are arranged so that 10 small divisions (1 cm.) in the vertical direction in the standard electrocardiogram represents 1 millivolt.

The vertical lines on the electrocardiogram are time calibration lines. Each inch (2.5 cm.) of the standard electrocardiogram is 1 second. Each inch in turn is usually broken into five segments by dark vertical lines, the distance between which represent 0.20 second. The intervals between the dark vertical lines are broken into five smaller intervals by thin lines, and the distance between each two of the smaller lines represents 0.04 second.

The P-Q Interval. The duration of time between the beginning of the P wave and the beginning of the QRS wave is the interval between the beginning of contraction of the atria and the beginning of contraction of the ventricles. This period of time is called the P-Q interval. The normal P-Q interval is approximately 0.16 second. This interval is sometimes also called the P-R interval, because the Q wave frequently is absent.

The Q-T Interval. Contraction of the ventricle lasts essentially between the Q wave and the end of the T wave. This interval of time is called the Q-T interval and ordinarily is approximately 0.30 second.

RECORDING ELECTROCARDIOGRAMS —THE PEN RECORDER

The electrical currents generated by the cardiac muscle during each beat of the heart sometimes change potentials and polarity in less than 0.03 second. Therefore, it is essential that any apparatus for recording electrocardiograms be capable of responding rapidly to these changes in electrical potentials. The most usual type of recorder now used is the pen recorder. This instrument writes the electrocardiogram with a pen directly on a moving sheet of paper. The pen is often a thin tube connected at one end to an inkwell, with its recording end connected to a powerful electromagnet system that is capable of moving the pen back and forth at high speed. As the paper moves forward, the pen records the electrocardiogram. The movement of the pen in turn is controlled by appropriate amplifiers connected to electrocardiographic electrodes on the subject.

FLOW OF CURRENT AROUND THE HEART DURING THE CARDIAC CYCLE

Figure 13–4 illustrates the ventricular muscle mass lying within the chest. Even the lungs, though filled with air, conduct electricity to a surprising extent, and fluids of the other tissues sur-

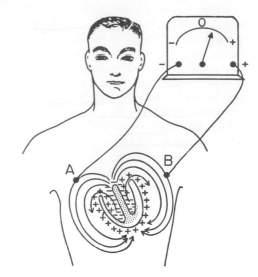

FIGURE 13–4 Flow of current in the chest around a partially depolarized heart.

rounding the heart conduct electricity even more easily. Therefore, the heart is actually suspended in a conductive medium. When one portion of the ventricles becomes electronegative with respect to the remainder, electrical current flows from the depolarized area to the polarized area in large circuitous routes, as noted in the figure.

It will be recalled from the discussion of the Purkinje system in Chapter 12 that the cardiac impulse first arrives in the ventricles in the walls of the septum and almost immediately thereafter on the endocardial surfaces of the remainder of the ventricles, as shown by the shaded areas and the negative signs in Figure 13–4. This provides electronegativity on the insides of the ventricles and electropositivity on the outer walls of the ventricles. If one algebraically averages all the lines of current flow (the elliptical lines in Figure 13–4), he finds that the average current flow is from the base of the heart toward the apex. During most of the cycle of depolarization, the current continues to flow in this direction as the impulse spreads from the endocardial surface outward through the ventricular muscle. However, immediately before the depolarization wave has completed its course through the ventricles, the direction of current flow reverses for about $1/100$ second, flowing then from the apex toward the base because the very last part of the heart to become depolarized is the outer walls of the ventricles near their base.

If a meter is connected to the surface of the body as shown in Figure 13-4, the electrode nearer the base will be negative with respect to the electrode nearer the apex, so that the meter shows a potential between the two electrodes. In making electrocardiographic recordings, various standard positions for placement of electrodes are used, and whether the polarity of the recording during each cardiac cycle is positive or negative is determined by the orientation of electrodes with respect to the current flow in the heart. Some of the conventional electrode systems, commonly called *electrocardiographic leads*, are discussed below.

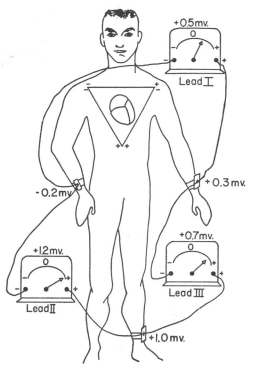

FIGURE 13-5 Conventional arrangement of electrodes for recording the standard electrocardiographic leads. Einthoven's triangle is superimposed on the chest.

ELECTROCARDIOGRAPHIC LEADS

THE THREE STANDARD LIMB LEADS

Figure 13-5 illustrates electrical connections between the limbs and the electrocardiograph for recording electrocardiograms from the so-called "standard" limb leads. The electrocardiogram in each instance is illustrated by mechanical meters in the diagram, though the actual electrocardiograph is a high-speed recording meter.

Lead I. In recording limb lead I, the negative terminal of the electrocardiograph is connected to the right arm and the positive terminal to the left arm. Therefore, when the point on the chest where the right arm connects to the chest is electronegative with respect to the point where the left arm connects, the electrocardiograph records positively—that is, above the zero voltage

line in the electrocardiogram. When the opposite is true, the electrocardiograph records below the line.

Lead II. In recording limb lead II, the negative terminal of the electrocardiograph is connected to the right arm and the positive terminal to the left leg. Thus, when the right arm is negative with respect to the left leg, the electrocardiograph records positively.

Lead III. In recording limb lead III, the negative terminal of the electrocardiograph is connected to the left arm and the positive terminal to the left leg. This means that the electrocardiograph records positively when the left arm is negative with respect to the left leg.

Normal Electrocardiograms Recorded by the Three Standard Leads. Figure 13-6 illustrates simultaneous recordings of the electrocardiogram in leads I, II,

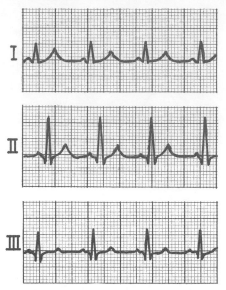

FIGURE 13-6 Normal electrocardiograms recorded from the three standard electrocardiographic leads.

and III. It is obvious from this figure that the electrocardiograms in these three standard leads are very similar to each other, for they all record positive *P* waves and positive *T* waves, and the major proportion of the QRS complex is positive in each electrocardiogram.

Because all normal electrocardiograms are very similar to each other, it does not matter greatly which electrocardiographic lead is recorded when one wishes to diagnose the different cardiac arrhythmias, for diagnosis of arrhythmias depends mainly on the time relationships between the different waves of the cardiac cycle. On the other hand, when one wishes to determine the extent and type of damage in the ventricles or in the atria, it does matter greatly which leads are recorded, for abnormalities of the cardiac muscle change the patterns of the electrocardiograms markedly in some leads and yet may not affect other leads.

Electrocardiographic interpretation of these two types of conditions—cardiac myopathies and cardiac arrhythmias—are discussed separately in later sections of this chapter.

CHEST LEADS (PRECORDIAL LEADS)

Often electrocardiograms are recorded with one electrode placed on the anterior aspect of the chest over the heart, as illustrated by the six separate points in Figure 13-7. This electrode is connected to the positive terminal of the electrocardiograph, and the negative electrode, called the *indifferent electrode*, is normally connected simultaneously through electrical resistances to the right arm, left arm, and left leg, as also shown in the figure. Usually six different standard chest leads are recorded from the anterior chest wall, the chest electrode being placed respectively at the six points illustrated in the diagram. The different leads recorded by the method illustrated in Figure 13-7 are known as leads V_1, V_2, V_3, V_4, V_5, and V_6.

Figure 13-8 illustrates the electrocardiograms of the normal heart as recorded in the six standard chest leads. Because the heart surfaces are close to the chest wall, each chest lead records mainly the electrical potential of the cardiac musculature immediately beneath the electrode. Therefore, relatively mi-

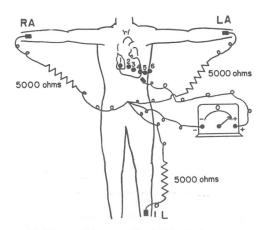

FIGURE 13-7 Connections of the body with the electrocardiograph for recording chest leads. (Modified from Burch and Winsor: A Primer of Electrocardiography. Lea & Febiger.)

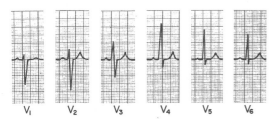

FIGURE 13-8 Normal electrocardiograms recorded from the six standard chest leads.

nute abnormalities in the ventricles, particularly in the anterior ventricular wall, frequently cause marked changes in the electrocardiograms recorded from chest leads.

ELECTROCARDIOGRAPHIC INTERPRETATION OF CARDIAC ARRHYTHMIAS

The rhythmicity of the heart and some abnormalities of rhythmicity were discussed in Chapter 12. The major purpose of the present section is to describe the electrocardiograms recorded in a few conditions known clinically as "cardiac arrhythmias."

ATRIOVENTRICULAR BLOCK

The only means by which impulses can ordinarily pass from the atria into the ventricles is through the *A-V bundle*, which is also known as the *bundle of His*. Different conditions that can either decrease the rate of conduction of the impulse through this bundle, or totally block the impulse are:

1. *Ischemia of the A-V junctional fibers.*

2. *Compression of the A-V bundle* by scar tissue or by calcified portions of the heart.

3. *Inflammation of the A-V bundle or fibers of the A-V junction.*

4. *Extreme stimulation of the heart by the vagi.*

Incomplete Heart Block. When conduction through the A-V junction is slowed until the P-R interval is 0.25 to 0.50 second, the action potentials traveling through the A-V junctional fibers are sometimes strong enough to pass on into the A-V node and at other times are not strong enough. Often the impulse passes into the ventricles on one heartbeat and fails to pass on the next one or two beats, thus alternating between conduction and nonconduction. In this instance, the atria beat at a considerably faster rate than the ventricles, and it is said that there are "dropped beats" of the ventricles. This condition is called *incomplete heart block.*

Figure 13–9 illustrates P-R intervals as long as 0.30 second, and it also illustrates one dropped beat as a result of failure of conduction from the atria to the ventricles.

At times every other beat of the ventricles is dropped so that a "2:1 rhythm" develops in the heart, with the atria beating twice for every single beat of the ventricles. Sometimes other rhythms, such as 3:2 or 3:1, also develop.

Complete Atrioventricular Block. When the condition causing poor conduction in the A-V bundle becomes extremely severe, complete block of the impulse from the atria into the ventricles occurs. In this instance the P waves become completely dissociated from the QRS-T complexes, as illustrated in Figure 13–10. Note that the rate of rhythm of the atria in this electrocardiogram is approximately 100 beats per minute, while the rate of ventricular beat is less than 40 per minute. Furthermore, there is no relationship whatsoever be-

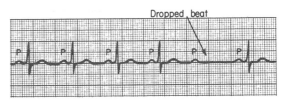

FIGURE 13-9 Partial atrioventricular block (lead V_3).

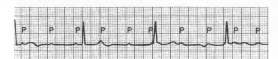

FIGURE 13-10 Complete atrioventricular block (lead II).

tween the rhythm of the atria and that of the ventricles, for the ventricles have "escaped" from control by the atria, and they are beating at their own natural rate.

Stokes-Adams syndrome — ventricular escape. In some patients with atrioventricular block, total block comes and goes — that is, all impulses are conducted from the atria into the ventricles for a period of time, and then suddenly no impulses at all are transmitted. Particularly does this condition occur in hearts with borderline ischemia.

Immediately after A-V conduction is first blocked, the ventricles stop contracting entirely for about 5 to 10 seconds. Then some part of the Purkinje system beyond the block, usually in the A-V bundle itself, begins discharging rhythmically at a rate of 15 to 40 times per minute and acting as the pacemaker of the ventricles. This is called *ventricular escape.* Because the brain cannot remain active for more than 3 to 5 seconds without blood supply, patients usually faint between block of conduction and "escape" of the ventricles. These periodic fainting spells are known as the Stokes-Adams syndrome.

PREMATURE BEATS

A premature beat is a contraction of the heart prior to the time when normal contraction would have been expected. This condition is also frequently called *extrasystole.*

Most premature beats result from *ectopic foci* in the heart, which emit abnormal impulses at odd times during the cardiac rhythm. The possible causes of ectopic foci are (1) local areas of ischemia, (2) small calcified plaques at different points in the heart, which press against the adjacent cardiac muscle so that some of the fibers are irritated, and (3) toxic irritation of the A-V node, Purkinje system, or myocardium caused by drugs, nicotine, caffeine, and so forth.

Atrial Premature Beats. Figure 13-11 illustrates an electrocardiogram showing a single atrial premature beat. The P wave of this beat is relatively normal and the QRS complex is also normal, but the interval between the preceding beat and the premature beat is shortened. Also, the interval between the premature beat and the next succeeding beat is slightly prolonged. The reason for this is that the premature beat originated in the atrium some distance from the S-A node, and the impulse of the premature beat had to travel through a considerable amount of atrial muscle before it discharged the S-A node. Consequently, the S-A node discharged very late in the cycle and made the succeeding heartbeat also late in appearing.

Ventricular Premature Beats. The electrocardiogram of Figure 13-12 illustrates a series of ventricular premature beats alternating with normal beats. Ventricular premature beats cause several special effects in the electrocardiogram, as follows: First, the QRS complex is usually considerably prolonged. The reason for this is that the impulse is conducted mainly through the muscle of the ventricle rather than through the Purkinje system.

Second, the QRS complex has a very high voltage, for the following reason: When the normal impulse passes through the heart, it passes through both ventricles approximately simultaneously; consequently, the depolarization waves of the two sides of the heart partially

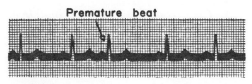

Premature beat

FIGURE 13-11 Atrial premature beat (lead I). A-V nodal premature beat (lead III).

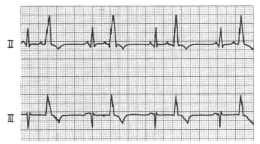

FIGURE 13–12 Ventricular premature beats (leads II and III).

neutralize each other. However, when a ventricular premature beat occurs, the impulse travels in only one direction so that there is no such neutralization effect.

Third, following almost all ventricular premature beats the T wave has a potential opposite to that of the QRS complex because the *slow conduction of the impulse* through the cardiac muscle causes the area first depolarized also to repolarize first. As a result, the direction of current flow in the heart during repolarization is opposite to that during depolarization. This is not true of the normal T wave, as was explained earlier in the chapter.

Some ventricular premature beats are benign in their origin and result from simple factors such as cigarettes, coffee, lack of sleep, various mild toxic states, and even emotional irritability. On the other hand, a large share of ventricular premature beats result from actual pathology of the heart. For instance, many ventricular premature beats occur following coronary thrombosis because of stray impulses originating around the borders of the infarcted area of the heart.

PAROXYSMAL TACHYCARDIA

Abnormalities in any portion of the heart, including the atria, the Purkinje system, and the ventricles, can sometimes cause rapid rhythmic discharge of impulses which spread in all directions throughout the heart. Because of the rapid rhythm in the irritable focus, this focus becomes the pacemaker of the heart.

Atrial Paroxysmal Tachycardia. Figure 13–13 illustrates a sudden increase in rate of heartbeat from approximately 95 beats per minute to approximately 150 beats per minute. Close analysis of the electrocardiogram shows that an inverted P wave occurs before each of the QRS-T complexes during the paroxysm of rapid heartbeat, though this P wave is partially superimposed on the normal T wave of the preceding beat. This indicates that the origin of this particular paroxysmal tachycardia is in the atrium, but, because the P wave is abnormal, the origin is not near the S-A node.

Ventricular Paroxysmal Tachycardia. Figure 13–14 illustrates a typical short paroxysm of ventricular tachycardia. The electrocardiogram of ventricular paroxysmal tachycardia has the appearance of a series of ventricular premature beats occurring one after another without any normal beats interspersed.

Ventricular paroxysmal tachycardia is usually a serious condition for two reasons. First, this type of tachycardia usually does not occur unless considerable damage is present in the ventricles. Second, ventricular tachycardia predisposes to ventricular fibrillation, which is almost invariably fatal.

ABNORMAL RHYTHMS RESULTING FROM CIRCUS MOVEMENTS

The circus movement phenomenon was discussed in detail in Chapter 12, and it was pointed out that these movements can cause atrial flutter, atrial fibrillation, and ventricular fibrillation.

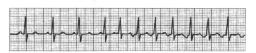

FIGURE 13–13 Atrial paroxysmal tachycardia —onset in middle of record (lead I).

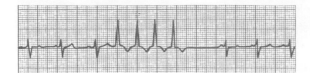

FIGURE 13-14   Ventricular paroxysmal tachycardia (lead III).

Atrial Flutter. Figure 13–15 illustrates lead II of the electrocardiogram in atrial flutter. The rate of atrial contraction (P waves) is approximately 300 times per minute, while the rate of ventricular contraction (QRS-T waves) is only 125 times per minute. From the record it will be seen that sometimes a 2:1 rhythm occurs and sometimes a 3:1 rhythm. In other words, the atria beat two or three times for every one impulse that is conducted through the A-V bundle into the ventricles.

Atrial Fibrillation. Figure 13–6 illustrates the electrocardiogram during atrial fibrillation. As discussed in Chapter 12, numerous impulses spread in all directions through the atria during atrial fibrillation. The intervals between impulses arriving at the A-V node are extremely variable. Therefore, an impulse may arrive at the A-V node immediately after the node itself is out of its refractory period from its previous discharge, or it may not arrive there for several tenths of a second. Consequently, the rhythm of the ventricles is very irregular, many of the ventricular beats falling quite close together and many far apart, the overall ventricular rate being 125 to 150 beats per minute in most instances. On the other hand, the QRS-T complexes are entirely normal unless there is some simultaneous pathology of the ventricles.

The pumping efficiency of the heart in atrial fibrillation is considerably depressed because the ventricles often do not have sufficient time to fill between beats.

Ventricular Fibrillation. In ventricular fibrillation the electrocardiogram is extremely bizarre, as shown in Figure 13–17, and ordinarily shows no tendency toward a rhythm of any type. The irregularity of this electrocardiogram is what would be expected from multiple impulses traveling in all directions through the heart, as explained in the previous chapter.

ELECTROCARDIOGRAPHIC INTERPRETATION IN CARDIAC MYOPATHIES

From the discussion in Chapter 12 of impulse transmission through the heart, it is obvious that any change in the pattern of this transmission can cause abnormal electrical currents around the heart and, consequently, can alter the shapes of the waves in the electrocardiogram. For this reason, almost all serious abnormalities of the heart muscle can be detected by analyzing the contours of the different waves in the different electrocardiographic leads. The purpose of the present section is to present several representative electrocardiograms when the muscle of the heart, especially of the ventricles, functions abnormally.

THE MEAN ELECTRICAL AXIS OF THE VENTRICLES

It was shown in Figure 13–4 that during most of the cycle of ventricular depolarization, current normally flows

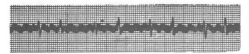

FIGURE 13-15   Atrial flutter—2:1 and 3:1 rhythm (lead II).

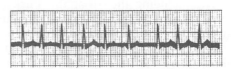

FIGURE 13-16   Atrial fibrillation (lead I).

FIGURE 13-17 Ventricular fibrillation (lead II).

from the base of the ventricle toward the apex. This preponderant direction of current flow during depolarization is called the *mean electrical axis of the ventricles*. The mean electrical axis of the normal ventricles is 59 degrees (zero degrees is toward the left side of the heart, and the axis is measured counterclockwise from this direction). However, in certain pathologi al conditions of the heart, the direction of current flow is changed markedly — sometimes even to opposite poles of the heart.

HYPERTROPHY OF ONE VENTRICLE

When one ventricle greatly hypertrophies, the principal direction of current flow through the heart during depolarization — that is, the axis of the heart — shifts toward the hypertrophied ventricle for two reasons: First, far greater quantity of muscle exists on the hypertrophied side of the heart than on the other side, and this allows excess generation of electrical currents on that side. Second, more time is required for the depolarization wave to travel through the hypertrophied ventricle than through the normal ventricle. Consequently, the normal ventricle becomes depolarized considerably in advance of the hypertrophied ventricle, and this causes strong current flow from the normal side of the heart toward the hypertrophied side. Thus the axis deviates toward the hypertrophied ventricle.

Left Axis Deviation Resulting From Hypertrophy of the Left Ventricle. Figure 13-18 illustrates the three standard leads of an electrocardiogram in which current flow is strongly positive in lead I and strongly negative in lead III. This means that the major current flow in the heart is mainly in the direction of lead I, which is from right arm toward left arm, opposite to that of lead III, which is from left arm toward left leg. That is, the axis of the heart points upward toward the left shoulder. This is called *left axis deviation.*

The electrocardiogram of Figure 13-18 is typical of that resulting from increased muscular mass of the left ventricle. In this instance the axis deviation was caused by *hypertension*, which caused the left ventricle to hypertrophy in order to pump blood against the elevated systemic arterial pressure. However, a similar picture of left axis deviation occurs when the left ventricle hypertrophies as a result of aortic valvular stenosis, aortic valvular regurgitation, or any of a number of congenital heart conditions in which the left ventricle enlarges while the right side of the heart remains relatively normal in size.

Right axis deviation occurs when the right ventricle enlarges. That is, the potential in lead I becomes negative while that in lead III becomes strongly positive.

BUNDLE BRANCH BLOCK

Ordinarily, the two lateral walls of the ventricles depolarize at almost the same time, because both the left and right bundle branches transmit the cardiac impulse to the endocardial surfaces of the two ventricular walls at almost the same instant. As a result, the currents flowing

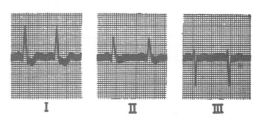

FIGURE 13-18 Left axis deviation in hypertensive heart disease. Note the slightly prolonged QRS complex.

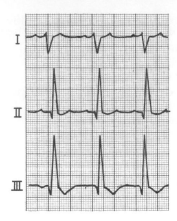

FIGURE 13-19 Right axis deviation due to right bundle branch block. Note the greatly prolonged QRS complex.

from the walls of the two ventricles almost exactly neutralize each other. However, if one of the major bundle branches is blocked, depolarization of the two ventricles does not occur even nearly simultaneously, and the depolarization currents do not neutralize each other. As a result, axis deviation occurs, as follows:

Right Bundle Branch Block. When the right bundle branch is blocked, the left ventricle depolarizes far more rapidly than the right ventricle, so that the left becomes electronegative while the right remains electropositive. Very strong current flows with its negative end toward the left ventricle and its positive end toward the right ventricle. In other words, intense right axis deviation occurs.

Right axis deviation caused by right bundle branch block is illustrated in Figure 13–19, which shows an axis of approximately 105 degrees and a prolonged QRS complex because of blocked conduction.

Left bundle branch block causes the opposite effect; namely, left axis deviation.

CURRENT OF INJURY

Many different cardiac abnormalities, especially those that damage the heart

muscle itself, often cause part of the heart to remain *depolarized all the time.* When this occurs, current flows between the pathologically depolarized and the normally polarized areas. This is called a *current of injury.* The most common cause of a current of injury is *ischemia of the muscle caused by coronary occlusion.*

Effect of Current of Injury on the QRS Complex—S-T Segment Shift. It will be recalled that in the normal heart no current flows around the heart when the heart is totally polarized during the T-P interval nor when the heart is totally depolarized during the S-T interval. Therefore, in the normal electrocardiogram, both the T-P and the S-T intervals record at the same potential level. However, when there is a current of injury, the heart cannot completely repolarize during the T-P interval. For this reason, the potential level of the T-P interval is

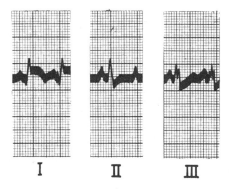

I II III

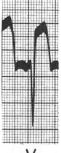

V₂

FIGURE 13-20 Current of injury in acute anterior wall infarction. Note the intense current of injury in lead V₂.

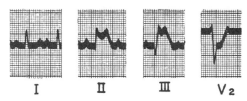

FIGURE 13–21 Current of injury in acute posterior wall, apical infarction.

different from that of the S-T interval. This effect is called the *S-T segment shift*, which unfortunately is a misnomer because the abnormality is actually in the T-P interval.

Acute Anterior Wall Infarction. Figure 13–20 illustrates the electrocardiogram in the three standard leads and in one chest lead recorded from a patient with acute anterior wall cardiac infarction caused by coronary thrombosis. The most important diagnostic feature of this electrocardiogram is the current of injury in the chest lead as denoted by the S-T segment shift.

Posterior Wall Infarction. Figure 13–21 illustrates the three standard leads and one chest lead from a patient with posterior wall infarction. The major diagnostic feature of this electrocardiogram is the S-T segment shift in the chest lead and in leads II and III.

Recovery from Coronary Thrombosis. Figure 13–22 illustrates the chest lead from a patient with posterior infarction, showing the change in this chest lead from the day of the attack to one week later, then three weeks later, and finally one year later. From this electrocardiogram it can be seen that the current of injury is strong immediately after the acute attack, but after approximately one week the current of injury has diminished considerably and after three weeks it is completely gone. After that, the electrocardiogram does not change greatly during the following year.

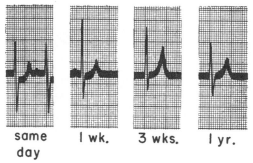

same day   **I wk.**   **3 wks.**   **I yr.**

FIGURE 13–22 Recovery of the myocardium following moderate posterior wall infarction, illustrating disappearance of the current of injury (lead V_3).

REFERENCES

Bellet, S.: Clinical Disorders of the Heart Beat. 2nd ed., Philadelphia, Lea & Febiger, 1963.

Burch, G. E., and DePasquale, N. P.: Electrocardiography in the Diagnosis of Congenital Heart Disease. Philadelphia, Lea & Febiger, 1967.

Burch, G. E., and Winsor, T.: A Primer of Electrocardiography. 5th ed., Philadelphia, Lea & Febiger, 1966.

Dimond, E. G.: Electrocardiography and Vectorcardiography. 4th ed., Boston, Little, Brown and Company, 1967.

Hurst, J. W., and Myerburg, R. J.: Introduction to Electrocardiography. New York, McGraw-Hill Book Company, 1968.

Schamroth, L.: An Introduction to Electrocardiography. 3rd ed., Philadelphia, F. A. Davis Co., 1966.

Sodi-Pallares, D., and Calder, R. M.: New Bases of Electrocardiography. St. Louis, The C. V. Mosby Co., 1956.

CHAPTER 14

PHYSICS OF BLOOD, BLOOD FLOW, AND PRESSURE: HEMODYNAMICS

Figure 14–1 illustrates the general plan of the circulation, showing the two major subdivisions, the *systemic circulation* and the *pulmonary circulation*. In the figure the arteries of each subdivision are represented by a single distensible chamber and all the veins by another even larger distensible chamber, and the arterioles and capillaries represent very small connections between the arteries and veins. Blood flows with almost no resistance in all the larger vessels of the circulation, but this is not the case in the arterioles and capillaries, where considerable resistance does occur. To cause blood to flow through these small "resistance" vessels, the heart pumps blood into the arteries under high pressure — normally at a systolic pressure of about 120 mm. Hg in the systemic circulation and about 22 mm. Hg in the pulmonary system.

As a first step toward explaining the overall function of the circulation, this chapter will discuss the physical characteristics of blood itself and then the physical principles of blood flow through the vessels, stressing the interrelation-

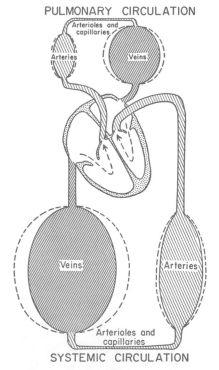

PULMONARY CIRCULATION

SYSTEMIC CIRCULATION

FIGURE 14–1 Schematic representation of the circulation, showing both the distensible and the resistive portions of the systemic and pulmonary circulations.

ships between pressure, flow, and resistance. The study of these interrelationships and other basic physical principles of blood circulation is called *hemodynamics*.

THE PHYSICAL CHARACTERISTICS OF BLOOD

Blood is a viscous fluid composed of *cells* and *plasma*. More than 99 per cent of the cells are red blood cells; this means that for practical purposes the white blood cells play almost no role in determining the physical characteristics of the blood. The fluid of the plasma is part of the extracellular fluid, and the fluid inside the cells is intracellular fluid.

THE HEMATOCRIT

The hematocrit of blood is the per cent of the blood that is cells. Thus, if one says that a person has a hematocrit of 40, he means that 40 per cent of the blood volume is cells and the remainder is plasma. The hematocrit of normal man averages about 42, whereas that of normal woman averages about 38. These values vary tremendously, depending upon whether or not the person has anemia, upon his degree of bodily activity, and on the altitude at which he resides. These effects were discussed in relation to the red blood cells and their function in Chapter 8.

Blood hematocrit is determined by centrifuging blood in a calibrated tube such as that shown in Figure 14–2. The calibration allows direct reading of the per cent of cells.

Effect of Hematocrit on Blood Viscosity. The greater the percentage of cells in the blood—that is, the greater the hematocrit—the more friction there is between successive layers of blood, and it is this friction that determines

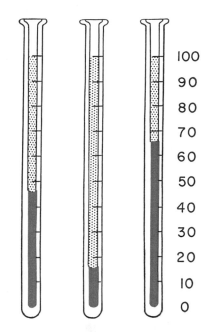

NORMAL   ANEMIA   POLYCYTHEMIA

FIGURE 14–2   Hematocrits in the normal person and in anemia and polycythemia.

viscosity. Therefore, the viscosity of blood increases drastically as the hematocrit increases, as illustrated in Figure 14–3. If we arbitrarily consider the viscosity of water to be 1, then the viscosity of whole blood is about 3 or 4; this means that 3 to 4 times as much pressure is required to force whole blood through a given tube than to force water through the same tube. Note that when the hematocrit rises to 60 or 70, which it often does in polycythemia as discussed in Chapter 8, the blood viscosity can become as great as 10 times that of water, and its flow through blood vessels becomes extremely difficult.

Another factor that affects blood viscosity is the concentration and types of proteins in the plasma, but these effects are so much less important than the effect of hematocrit that they are not significant considerations in most hemodynamic studies. The viscosity of blood

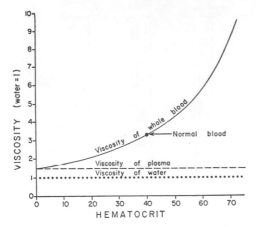

FIGURE 14-3 Effect of hematocrit on viscosity.

plasma is approximately 1.5 times that of water.

INTERRELATIONSHIPS BETWEEN PRESSURE, FLOW, AND RESISTANCE

Flow through a blood vessel is determined entirely by two factors: (1) the *pressure difference* tending to push blood through the vessel and (2) the impediment to blood flow through the vessel, which is called vascular *resistance*. Figure 14-4 illustrates these relationships quantitatively, showing a blood vessel segment located anywhere in the circulatory system.

P_1 represents the pressure at the origin of the vessel; at the other end the pressure is P_2. Since the vessel is relatively small, the blood is experiencing difficulty in flowing. This is the resistance. Stated another way, a *pressure difference* between the two ends of the vessel causes *blood* to *flow* from the

high pressure area to the low pressure area while *resistance* impedes the flow. This can be expressed mathematically as follows:

$$Q = \frac{\Delta P}{R} \tag{1}$$

in which Q is blood flow, ΔP is the pressure difference between the two ends of the vessel, and R is the resistance.

It should be noted especially that it is the *difference in pressure* between the two ends of the vessel that determines the rate of flow and not the absolute pressure in the vessel. For instance, if the pressure at both ends of the segment were 100 mm. Hg and yet no difference existed between the two ends, there would be no flow despite the presence of the 100 mm. Hg pressure.

The above formula expresses the most important of all the relationships that the student needs to understand to comprehend the hemodynamics of the circulation. Because of the extreme importance of this formula the student should also become familiar with its other two algebraic forms:

$$\Delta P = Q \times R \tag{2}$$
$$R = \frac{\Delta P}{Q} \tag{3}$$

BLOOD FLOW

Blood flow means simply the quantity of blood that passes a given point in the circulation in a given period of time. Ordinarily, blood flow is expressed in milliliters or liters per minute, but it can be expressed in milliliters per second or in any other unit of flow.

The overall blood flow in the circulation of an adult person at rest is about 5000 ml. per minute. This is called the *cardiac output* because it is the amount of blood pumped by each ventricle of the heart in a unit period of time. It is also obvious that this same amount of blood must pass through both the systemic and pulmonary circulations.

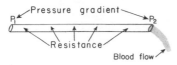

FIGURE 14-4 Relationship between pressure, resistance, and blood flow.

Measuring Blood Flow: The Electromagnetic Flowmeter. Many different mechanical or mechanoelectrical devices called *flowmeters* can be inserted in series with a blood vessel or in some instances applied to the outside of the vessel to measure flow. One of the most important of these is the electromagnetic flowmeter, the principles of which are illustrated in Figure 14–5. Figure 14–5*A* shows the generation of electromotive force in a wire that is moved rapidly through a magnetic field. This is the well-known principle for production of electricity by the electric generator. Figure 14–5*B* shows that exactly the same principle applies for generation of electromotive force in blood when it moves through a magnetic field. In this case, a blood vessel is placed between the poles of a strong magnet, and electrodes are placed on the two sides of the vessel perpendicular to the magnetic lines of force. When blood flows through the vessel, electrical voltage is generated between the two electrodes, and this is recorded using an appropriate meter or electronic apparatus. Figure 14–5*C* illustrates an actual "probe" that is placed on a large blood vessel to record its blood flow. This probe contains both the strong magnet and the electrodes.

An additional advantage of the electromagnetic flowmeter is that it can record changes in flow that occur in less than 0.01 second, allowing accurate recording of pulsatile changes in flow as well as steady state changes.

Laminar Flow of Blood in Vessels. When blood flows at a continuous rate through a long, smooth vessel, the velocity of flow in the center of the vessel is far greater than that along the outer edges. This is illustrated by the experiment shown in Figure 14–6. In vessel A are two different fluids but no flow in the vessel. Then the fluids are made to flow; and a parabolic interface

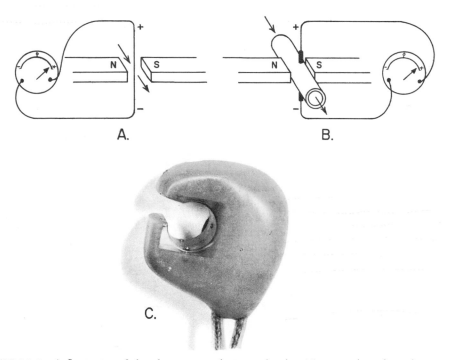

A.   B.

C.

FIGURE 14–5 A flowmeter of the electromagnetic type, showing (*A*) generation of an electromotive force in a wire as it passes through an electromagnetic field, (*B*) generation of an electromotive force in electrodes on a blood vessel when the vessel is placed in a strong magnetic field and blood flows through the vessel, and (*C*) a modern electromagnetic flowmeter "probe" for chronic implantation around blood vessels.

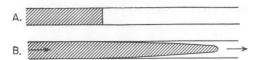

FIGURE 14–6 An experiment illustrating laminar blood flow, showing (*A*) two separate fluids before flow begins and (*B*) the same fluids 1 second after flow begins.

develops between the two fluids, as shown 1 second later in vessel B, illustrating that the portion of fluid adjacent to the walls has moved hardly at all while that in the center of the vessel has moved a long distance. This effect is called *laminar flow* or *streamline flow*.

The cause of laminar flow is the following: The fluid molecules touching the wall move essentially not at all because of adherence to the vessel wall. The next layer of molecules slips over these, the third layer over the second, the fourth layer over the third, and so forth. Therefore, the fluid in the middle of the vessel can move rapidly because many layers of molecules exist between the middle of the vessel and the vessel wall, all of these capable of slipping over each other, while those portions of fluid near the wall do not have this advantage.

Turbulent Flow of Blood Under Some Conditions. When the rate of blood flow becomes too great, when blood passes by an obstruction in a vessel, when it makes a sharp turn, or when it passes over a rough surface, the flow may then become *turbulent* rather than streamline. Turbulent flow means that the blood flows crosswise in the vessel as well as along the vessel, usually forming whorls in the blood called *eddy currents*. When eddy currents are present, blood flows with much greater resistance than when the flow is streamline because the eddies add tremendously to the overall friction of flow in the vessel.

BLOOD PRESSURE

The Standard Units of Pressure. Blood pressure is almost always measured in *millimeters of mercury* (*mm. Hg*) because the mercury manometer (shown in Fig. 14–7) has been used as the standard reference for measuring blood pressure for many years. Actually, blood pressure means the *force exerted by the blood against any unit area of the vessel wall*. When one says that the pressure in a vessel is 50 mm. Hg, this means that the force exerted would be sufficient to push a column of mercury up to a level of 50 mm. If the pressure were 100 mm. Hg, it would push the column of mercury up to 100 mm.

Occasionally, pressure is measured in *centimeters of water*. A pressure of 10 cm. of water means a pressure sufficient to raise a column of water to a height of 10 cm. *One millimeter of mercury equals 1.36 cm. of water* because the density of mercury is 13.6 times that of water, and 1 cm. is 10 times as great as 1 mm.

Measurement of Blood Pressure Using the Mercury Manometer. Figure 14–7 illustrates a standard mercury manometer for measuring blood pressure. A cannula is inserted into an artery, a vein, or even into the heart, and the pressure from the cannula is transmitted to the left-hand side of the

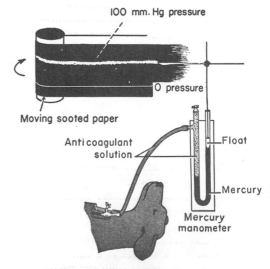

FIGURE 14–7 Recording arterial pressure with a mercury manometer.

manometer where it pushes the mercury down while raising the right-hand mercury column. The difference between the two levels of mercury is equal to the pressure in the circulation in terms of millimeters of mercury.

High-Fidelity Methods for Measuring Blood Pressure. Unfortunately, the mercury in the mercury manometer has so much *inertia* that it cannot rise and fall rapidly. For this reason the mercury manometer, though excellent for recording steady pressures, cannot respond to pressure changes that occur more rapidly than approximately one cycle every 2 to 3 seconds. Whenever it is desired to record rapidly changing pressures, some other type of pressure recorder is needed, such as an electronic pressure *transducer* recording on a high-speed electrical recorder. Such transducers employ a very thin and highly stretched metal membrane that forms one wall of a fluid chamber. The fluid chamber in turn is connected through a needle with the vessel in which the pressure is to be measured. Pressure in the chamber bulges the membrane outward slightly, which in turn activates the electronic circuit to cause the pressure recording.

Using some of these high fidelity types of recording systems, pressure cycles up to 500 cycles per second have been recorded accurately, and in common use are recorders capable of registering pressure changes as rapidly as 20 to 100 cycles per second.

RESISTANCE TO BLOOD FLOW

Units of Resistance. Resistance is the impediment to blood flow in a vessel, but it cannot be measured by any direct means. Instead, resistance must be calculated from measurements of blood flow and pressure difference in the vessel. If the pressure difference between two points in a vessel is 1 mm. Hg and the flow is 1 ml./sec., then the resistance is said to be 1 *peripheral resistance unit*, usually abbreviated *PRU*.

Expression of resistance in CGS units. Occasionally, a basic physical unit called the CGS unit is used to express resistance. This unit is *dyne seconds/centimeters*^{5}. Resistance in these units can be calculated by the following formula:

$$R \left(in \; \frac{dynes \; sec.}{cm.^5} \right) = \frac{1333 \times mm. \; Hg}{ml./sec.} \quad (4)$$

Total Peripheral Resistance and Total Pulmonary Resistance. The rate of blood flow through the circulatory system when a person is at rest is close to 100 ml./sec., and the pressure gradient from the systemic arteries to the systemic veins is about 100 mm. Hg. Therefore, in round figures the resistance of the entire systemic circulation, called the *total peripheral resistance*, is approximately 100/100 or 1 PRU. In some conditions in which the blood vessels throughout the body become strongly constricted, the total peripheral resistance rises to as high as 4 PRU, and when the vessels become greatly dilated it can fall to as little as 0.25 PRU.

In the pulmonary system the mean arterial pressure averages 13 mm. Hg and the mean left atrial pressure averages 4 mm. Hg, giving a net pressure gradient of 9 mm. Therefore, in round figures the *total pulmonary resistance* at rest calculates to be about 0.09 PRU. This can increase in disease conditions to as high as 1 PRU and can fall in certain physiological states, such as exercise, to as low as 0.03 PRU.

"Conductance" of Blood in a Vessel and Its Relationship to Resistance. Conductance is a measure of the amount of blood flow that can pass through a vessel in a given time for a given pressure gradient. This is generally expressed in terms of ml./sec./mm. Hg pressure, but it can also be expressed in terms of liters/sec./mm. Hg or in any other units of blood flow and pressure.

It is immediately evident that conductance is the reciprocal of resistance in accord with the following equation:

$$Conductance = \frac{1}{Resistance} \quad (5)$$

Effect of vascular diameter on conductance. Slight changes in diameter of a vessel cause tremendous changes in its ability to conduct blood. This is illustrated forcefully by the experiment in Figure 14–8, which shows three separate vessels with relative diameters of 1, 2, and 4 but with the same pressure difference of 100 mm. Hg between the two ends of the vessels. Though the diameters of these vessels increase only four-fold, the respective flows are 1, 16, and 256 ml./min., which is a 256-fold increase in flow. Thus, the conductance of the vessel increases in proportion to the *fourth power of the diameter*, in accord with the following formula:

$$\text{Conductance} \propto \text{Diameter}^4$$

The cause of this great increase in conductance with increase in diameter is the laminar flow effect in vessels, which was discussed earlier in the chapter. That is, the blood in the ring touching the wall of the vessel is flowing hardly at all because of its adherence to the vascular endothelium. The next ring of blood slips past the first ring and, therefore, flows at a more rapid velocity. The third, fourth, fifth, and sixth rings likewise flow at progressively increasing velocities. Thus, the blood that is very near the wall of the vessel flows extremely slowly, while that in the middle of the vessel flows extremely rapidly.

In the small vessel essentially all of the blood is very near the wall so that the extremely rapidly flowing central stream of blood simply does not exist.

Poiseuille's Law. The following equation, called Poiseuille's Law, summarizes the effects of pressure gradient (ΔP), radius (r), viscosity (η) and vessel length (l) on flow (Q):

$$Q = \frac{\pi \Delta P r^4}{8 \eta l} \qquad (7)$$

Note particularly in this equation that the rate of blood flow is directly proportional to the *fourth power of the radius* of the vessel and inversely proportional to viscosity and vessel length, which illustrates once again that the diameter of a blood vessel plays by far the greatest role of all factors in determining the rate of blood flow through the vessel.

Effect of Pressure on Vascular Resistance—Critical Closing Pressure

Since all blood vessels are distensible, increasing the pressure inside the vessels causes the vascular diameters also to increase. This in turn reduces the resistance of the vessel. Conversely, reduction in vascular pressures increases the resistance.

The middle curve of Figure 14–9 illustrates the effect on blood flow through a small tissue artery caused by changing the arterial pressure. As the arterial pressure falls from 130 mm. Hg, the flow decreases rapidly because of two factors: (1) the decreasing pressure and (2) the decreasing diameter of the vessel. Then, at 20 mm. Hg blood flow ceases entirely. This point at which the blood stops flowing is called the *critical closing pressure*, because at this point the small vessels, the arterioles in particular, close so completely that all flow through the tissue ceases.

Effect of Sympathetic Inhibition and Stimulation on Vascular Flow. Essentially all blood vessels of the body are normally excited by sympathetic impulses even under resting conditions. Different circulatory reflexes, which will be discussed in Chapter 19, can cause these impulses either to disappear, called *sympathetic inhibition*, or to increase to many times their normal rate, called *sympathetic stimulation*. As illus-

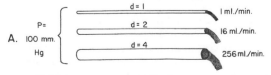

FIGURE 14–8 Demonstration of the effect of vessel diameter on blood flow.

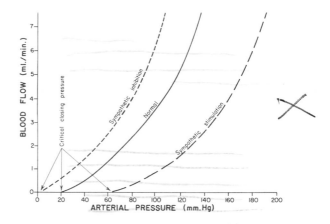

FIGURE 14–9 Effect of arterial pressure on blood flow through a blood vessel at different degrees of vascular tone, showing also the effect of vascular tone on critical closing pressure.

trated by the dashed curves of Figure 14–9, sympathetic inhibition allows far more blood to flow through a tissue for a given pressure gradient than is normally true, whereas sympathetic stimulation vastly decreases the rate of blood flow.

VASCULAR DISTENSIBILITY —PRESSURE-VOLUME CURVES

The diameter of blood vessels, unlike that of metal pipes and glass tubes, increases as the internal pressure increases, because blood vessels are *distensible*. However, the vascular distensibilities differ greatly in different segments of the circulation, and, as we shall see, this affects significantly the operation of the circulatory system under many changing physiological conditions.

Units of Vascular Distensibility. Vascular distensibility is normally expressed as the fractional increase in volume for each millimeter mercury rise in pressure in accordance with the following formula:

$$\text{Vascular distensibility} = \frac{\text{Increase in volume}}{\text{Increase in pressure} \times \text{Original volume}} \quad (8)$$

That is, if 1 mm. Hg causes a vessel originally containing 10 ml. of blood to increase its volume by 1 ml., then the distensibility would be 0.1 per mm. Hg, or 10 per cent per mm. Hg.

Difference in distensibility of the arteries and the veins. Anatomically, the walls of arteries are far stronger than those of veins. Consequently, the veins, on the average, are about 6 to 10 times as distensible as the arteries. That is, a given rise in pressure will cause about 6 to 10 times as much extra blood to fill a vein as an artery of comparable size.

In the pulmonary circulation the veins are very similar to those of the systemic veins. However, the pulmonary artery, which normally operates under pressures about one-fifth those in the systemic arterial system, has a distensibility only about one-half that of veins, rather than one-eighth, as is true of the systemic arteries.

VASCULAR COMPLIANCE

Usually in hemodynamic studies it is much more important to know the *total quantity of blood* that can be stored in a given portion of the circulation for each mm. Hg pressure rise than to know the distensibility of the individual vessels. This value is sometimes called the *overall distensibility* or *total distensibility*, or it can be expressed still more precisely by either of the terms *compliance* or *capacitance*, which are physical terms meaning the increase in volume caused by a given increase in pressure as follows:

Vascular compliance =
$$\frac{\text{Increase in volume}}{\text{Increase in pressure}} \quad (9)$$

Compliance and distensibility are quite different. A highly distensible vessel which has a very slight volume may have far less compliance than a much less distensible vessel which has a very large volume, for *compliance is equal to distensibility × volume.*

The compliance of a vein is about 24 times that of its corresponding artery because it is about 8 times as distensible and it has a volume about 3 times as great (8 × 3 = 24).

PRESSURE-VOLUME CURVES OF THE ARTERIAL AND VENOUS CIRCULATIONS

A convenient method for expressing the relationship of pressure to volume in a vessel or in a large portion of the circulation is the so-called *pressure-volume curve.* The two solid curves of Figure 14–10 represent the pressure-volume curves of, respectively, the normal arterial and venous systems, showing that the arterial system, including the larger arteries, small arteries, and arterioles, contains approximately 750 ml. of blood when the mean arterial pressure is 100 mm. Hg but only 500 ml. when the pressure has fallen to zero. The dashed curves show the effects of sympathetic stimulation or inhibition on the pressure-volume relationships.

The volume of blood normally in the entire venous tree is about 2500 ml., but only a slight change in venous pressure changes this volume tremendously.

Difference in Compliance of the Arterial and Venous Systems. Referring once again to Figure 14–10, one can see that a change of 1 mm. Hg increases the venous volume a very large amount but increases the arterial volume only slightly. That is, the *compliance of the venous system is far greater than the compliance of the arteries—about 24 times as great.*

This difference in compliance is particularly important because it means that tremendous amounts of blood can be stored in the veins with only slight changes in pressure. Therefore, the veins are frequently called the *storage areas* of the circulation.

"MEAN CIRCULATORY PRESSURE" AND PRESSURE-VOLUME CURVES OF THE ENTIRE CIRCULATORY SYSTEM

THE MEAN CIRCULATORY PRESSURE

The mean circulatory pressure is a measure of the degree of filling of the circulatory system. That is, it is the pressure that would be measured in the circulation if one could instantaneously stop all blood flow and bring all the pressures in the circulation immediately to equilibrium. The mean circulatory pressure has been measured reasonably accurately in dogs within 2 to 3 seconds

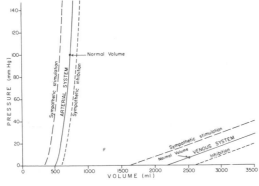

FIGURE 14–10 Pressure-volume curves of the systemic arterial and venous systems, showing also the effects of sympathetic stimulation and sympathetic inhibition.

after the heart has been stopped. To do this the heart is thrown into fibrillation by an electrical stimulus, and blood is pumped rapidly from the systemic arteries to the veins to cause equilibrium between the two major chambers of the circulation.

The mean circulatory pressure measured in the above manner is almost exactly 7 mm. Hg and almost never varies more than 1 mm. from this value in the normal resting animal. However, many different factors can change it, including especially change in the blood volume and increased or decreased sympathetic stimulation.

The mean circulatory pressure is *one of the major factors that determine the rate at which blood flows from the vascular tree into the right atrium of the heart and, therefore, that control the cardiac output itself.* This is so important that it will be explained in detail in Chapter 21; it will also be discussed in relation to blood volume regulation in Chapter 25.

PRESSURE-VOLUME CURVES OF THE ENTIRE CIRCULATION

Figure 14–11 illustrates the changes in mean circulatory pressure as the total blood volume increases (1) under normal conditions, (2) during strong sympathetic stimulation, and (3) during complete sympathetic inhibition. The point marked by the arrow is the operating point of the normal circulation: a mean circulatory pressure of 7 mm. Hg and a blood volume of 5000 ml. However, *if blood is lost from the circulatory system*, the mean circulatory pressure falls to a lower value. If increased amounts of blood are added, the mean circulatory pressure rises accordingly.

The compliance of the entire circulatory system in the human being, as estimated from experiments in dogs, is approximately 100 ml. for each 1 mm. rise in mean circulatory pressure.

Sympathetic stimulation and *inhibition* affect the pressure-volume curves

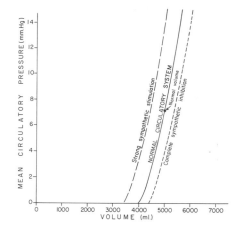

FIGURE 14–11 Pressure-volume curves of the entire circulation, illustrating the effect of strong sympathetic stimulation and complete sympathetic inhibition.

of the entire circulatory system in the same way that they affect the pressure-volume curves of the individual parts of the circulation, as illustrated by the two dashed curves of Figure 14–11. That is, for any given blood volume, the mean circulatory pressure rises two- to four-fold with strong sympathetic stimulation and falls markedly when the sympathetics are inhibited. This is an extremely important factor in the regulation of blood flow into the heart and thereby for regulating the cardiac output. For instance, during exercise, sympathetic activity increases the mean circulatory pressure and correspondingly helps to increase the cardiac output.

REFERENCES

Green, H. D.: Circulation: physical principles. *In* Glasser, O. (ed.): Medical Physics. Chicago, Year Book Medical Publishers, 1944.

Guyton, A. C.: Peripheral circulation. *Ann. Rev. Physiol., 21:*239, 1969.

Johnson, P. C.: Hemodynamics. *Ann. Rev. Physiol., 31:*331, 1969.

McDonald, D. A.: Hemodynamics. *Ann. Rev. Physiol., 30:*525, 1968.

Merrill, E. W.: Rheology of blood. *Physiol. Rev., 49:*863, 1969.

Richardson, T. Q., Stallings, J. O., and Guyton, A. C.: Pressure-volume curves in live, intact dogs. *Amer. J. Physiol., 201:*471, 1961.

Wayland, H.: Rheology and the microcirculation. *Gastroenterology, 52:*342, 1967.

CHAPTER 15

THE SYSTEMIC AND PULMONARY CIRCULATIONS

The circulation is divided into the *systemic circulation* and the *pulmonary circulation*. Though the blood vessels in each part of the body have their own special characteristics, some general principles of vascular function nevertheless apply in all parts of the circulation. It is the purpose of the present chapter to discuss these general principles.

The Functional Parts of the Circulation. Before attempting to discuss the details of function in the circulation, it is important to understand the overall role of each of its parts, as follows:

The function of the *arteries* is to transport blood under high pressure to the tissues. For this reason the arteries have strong vascular walls, and blood flows rapidly in the arteries to the tissues.

The *arterioles* are the last small branches of the arterial system, and they act as control valves through which blood is released into the capillaries. The systemic arteriole has a strong muscular wall that is capable of closing the arteriole completely or of being dilated several-fold, thereby vastly altering blood flow to the capillaries. This feature is much less developed in the pulmonary arterioles.

The function of the *capillaries* is to exchange fluid and nutrients between the blood and the interstitial spaces; and, in the lungs, to exchange gases between the alveoli and the blood. For this role, the capillary walls are very thin and permeable to small molecular substances.

The *venules* collect blood from the capillaries; they gradually coalesce into progressively larger veins.

The *veins* function as conduits for transport of blood from the tissues back to the heart. Since the pressure in the venous system is very low, the venous walls are thin. Even so, they are muscular, and this allows them to contract or expand and thereby to store either a small or large amount of blood, depending upon the needs of the body.

PHYSICAL CHARACTERISTICS OF THE SYSTEMIC CIRCULATION

Quantities of Blood in the Different Parts of the Circulation. By far the greatest amount of the blood in the circulation is contained in the systemic veins. Thus, Figure 15–1 shows that

166

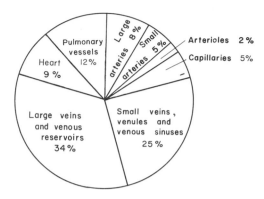

FIGURE 15–1 Percentage of the total blood volume in each portion of the circulatory system.

approximately 79 per cent of the entire blood volume of the body is in the systemic circulation, with 59 per cent in the veins, 15 per cent in the arteries, and 5 per cent in the capillaries. The heart contains 9 per cent of the blood, and the pulmonary vessels contain 12 per cent. Most surprising is the very low blood volume in the capillaries of the systemic circulation, only about 5 per cent of the total, for it is here that the most important function of the systemic circulation occurs; namely, diffusion of substances back and forth between the blood and the interstitial fluid. This function is so important that it will be discussed in detail in Chapter 16.

Pressures and Resistances in the Various Portions of the Systemic Circulation. Since the heart pumps blood continually into the aorta, the pressure in the aorta is obviously high, averaging approximately 100 mm. Hg. And, since pumping by the heart is intermittent, the arterial pressure fluctuates between a *systolic level* of 120 mm. Hg and a *diastolic level* of 80 mm. Hg, as illustrated in Figure 15–2. As the blood flows through the systemic circulation, its pressure falls progressively to approximately 0 mm. Hg by the time it reaches the right atrium.

The decrease in arterial pressure in each part of the systemic circulation is directly proportional to the vascular re-sistance. Thus, in the aorta the resistance is almost zero; therefore, the mean arterial pressure at the end of the aorta is still almost 100 mm. Hg. Likewise, the resistance in the large arteries is very slight so that the mean arterial pressure in arteries as small as 3 mm. in diameter is still 95 to 97 mm. Hg. Then the resistance begins to increase rapidly in the very small arteries, causing the pressure to drop to approximately 85 mm. Hg at the beginning of the arterioles.

The resistance of the arterioles is the greatest of any part of the systemic circulation, accounting for about half the resistance in the entire systemic circulation. Thus, the pressure decreases about 55 mm. Hg in the arterioles so that the pressure of the blood as it leaves the arterioles to enter the capillaries is only about 30 mm. Hg. Arteriolar resistance is so important to the regulation of blood flow in different tissues of the body that it is discussed in detail later in the chapter and also in Chapters 18 and 19, which consider the regulation of the systemic circulation.

The pressure at the arterial ends of the capillaries is normally about 30 mm. Hg and at the venous ends about 10 mm. Hg. Thus, the pressure decrease in the capillaries is only 20 mm. Hg, which illustrates that the capillary resistance is about two-fifths of that of the arterioles.

The pressure at the beginning of the veins is about 10 mm. Hg, and this decreases to about 0 mm. Hg at the right

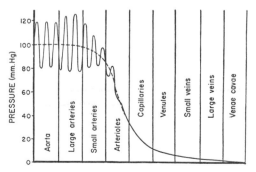

FIGURE 15–2 Blood pressures in the different portions of the systemic circulatory system.

atrium. This large decrease in pressure in the veins indicates that the veins have far more resistance than one would expect for vessels of their large sizes. This resistance is caused by compression of the veins from the outside, which keeps many of them, especially the venae cavae, collapsed a large share of the time. This effect is discussed in detail later in the chapter.

PRESSURE PULSES IN THE ARTERIES

Since the heart is a pulsatile pump, blood enters the arteries intermittently, causing *pressure pulses* in the arterial system. In the normal young adult the pressure at the height of a pulse, the *systolic pressure*, is about 120 mm. Hg and at its lowest point, the *diastolic pressure*, is about 80 mm. Hg. The difference between these two pressures, about 40 mm. Hg, is called the *pulse pressure*.

Figure 15–3 illustrates a typical *pressure pulse curve* recorded in the ascending aorta of a human being, showing a very rapid rise in arterial pressure during ventricular systole, followed by a maintained high level of pressure for 0.2 to 0.3 second. This is terminated by a sharp *incisura* at the end of systole, followed by a slow decline of pressure back to the diastolic level. The incisura occurs at the same time that the aortic valve closes, and is caused as follows: During

systole the pressure in the arteries rises to a high value. When the ventricle relaxes the intraventricular pressure begins to fall rapidly, and backflow of blood from the aorta into the ventricle allows the aortic pressure also to begin falling. However, the backflow suddenly snaps the aortic valve closed. The momentum that has built up in the backflowing blood brings still more blood into the root of the aorta, raising the pressure again and thus giving the incisura in the record.

After systole is over, the pressure in the central aorta decreases rapidly at first but progressively more and more slowly as diastole proceeds. The reason for this is that blood flows through the peripheral vessels much more rapidly when the pressure is high than when it is low.

FACTORS THAT AFFECT THE PULSE PRESSURE

Effect of Stroke Volume and Arterial Compliance. There are two major factors that affect the pulse pressure: (1) the *stroke volume output* of the heart and (2) the *compliance* (*total distensibility*) of the arterial tree.

In general, the greater the stroke volume output, the greater is the amount of blood that must be accommodated in the arterial tree with each heartbeat and, therefore, the greater is the pressure rise during systole and the pressure fall during diastole, thus causing a greater pulse pressure.

On the other hand, the greater the compliance of the arterial system the less will be the rise in pressure for a given stroke volume of blood pumped into the arteries. In effect, then, the pulse pressure is determined approximately by the *ratio of stroke volume output to compliance* of the arterial tree.

Abnormal Pressure Pulse Contours. In *arteriosclerosis* the arteries become fibrous and sometimes calcified, thereby resulting in greatly reduced arterial com-

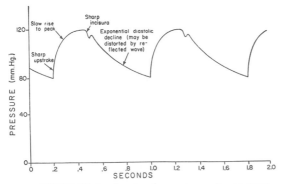

FIGURE 15–3 A normal pressure pulse contour recorded from the ascending aorta. (From Opdyke: *Fed. Proc., 11*:734, 1952.)

pliance and also resulting in greatly increased pulse pressure. The middle curve of Figure 15–4 illustrates a characteristic aortic pressure pulse contour in this condition, showing a markedly elevated pulse pressure.

In *patent ductus arteriosus* blood flows from the aorta through the open ductus into the pulmonary artery, allowing very rapid runoff of blood from the arterial tree after each heartbeat. However, this is compensated by a very large stroke volume output, giving the pressure pulse contour shown by the lowest curve in Figure 15–4. Here one finds an elevated systolic pressure, a greatly depressed diastolic pressure, and a greatly increased pulse pressure.

In *aortic regurgitation* much of the blood that is pumped into the aorta during systole flows back into the left ventricle during diastole. However, this backflow is compensated by a much greater than normal stroke volume output during systole. Thus, the condition is very similar to that of patent ductus arteriosus but not always identical, for in aortic regurgitation the valve sometimes fails entirely to close. When this is true the incisura illustrated in the lower curve of Figure 15–4 is entirely absent.

Transmission of the Pressure Pulse to the Periphery. When the heart ejects blood into the aorta during systole, only the proximal portion of the aorta becomes distended at first, and it is only

in this portion of the arterial tree that the pressure rises immediately. The cause of this is the inertia of the blood in the aorta, which prevents its sudden movement away from the central arteries into the peripheral arteries.

However, the rising pressure in the central aorta gradually overcomes the inertia of the blood, causing the pressure to rise progressively farther and farther out in the arterial tree. Figure 15–5 illustrates this *transmission of the pressure pulse* down the aorta, the aorta becoming distended as the pressure wave moves forward.

The velocity of transmission of the pressure pulse along the normal aorta is 3 to 5 meters per second, along the large arterial branches 7 to 10 meters per second, and in the smaller arteries 15 to 35 meters per second. In general, the less the compliance of each vascular segment, the faster is the velocity of transmission, which explains the slowness of transmission in the aorta versus the rapidity of transmission in the far less compliant small, distal arteries.

The *velocity of transmission of the pressure pulse is much greater than the velocity of blood flow*. Therefore, the blood ejected by the heart may have traveled only a few centimeters by the time the pressure wave has already reached the distal ends of the arteries. In the aorta, the velocity of the pressure pulse is approximately 15 times that of blood flow.

FIGURE 15–4 Pressure pulse contours in arteriosclerosis, patent ductus arteriosus, and moderate aortic regurgitation.

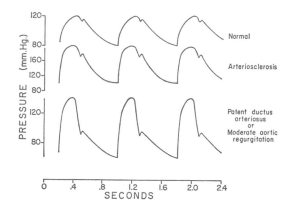

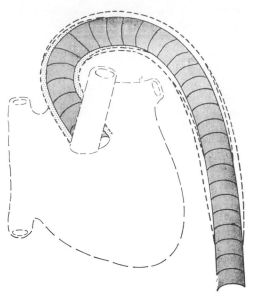

FIGURE 15–5 Progressive stages in the transmission of the pressure pulse along the aorta.

Damping of the pressure pulse in the small arteries and arterioles. The pressure pulse becomes less intense as it passes through the small arteries and arterioles, and it becomes almost absent in the capillaries. This was illustrated in Figure 15–2 and is called *damping* of the pressure pulse.

Damping of the pressure pulse is caused mainly by a combined effect of vascular distensibility and vascular resistance. That is, for a pressure wave to travel from one area of an artery to another area, a small amount of blood must flow between the two areas. The resistance in the small arteries and arterioles is great enough that this flow of blood, and consequently the transmission of pressure, is greatly impeded. At the same time, the distensibility of the small vessels is enough so that the small amount of blood that is caused to flow during a pressure pulse produces progressively less pressure rise in the more distal vessels.

Capillary Pulsation. Occasionally pressure pulses are not completely damped by the time they reach the capillaries, and abnormal capillary pul-

sations result; this can be readily demonstrated by pressing on the anterior portion of the fingernail so that the blood is pressed out of the anterior capillaries of the nail bed. This causes blanching anteriorly while the posterior half of the nail remains red because of blood still in the capillaries. If significant pulsation is occurring in the capillaries, the border between the red and white areas will shift forward as more capillaries become filled during the high pressure phase and backward during the low pressure phase.

Two major factors can cause capillary pulsation: (1) It occurs when the central pressure pulse is greatly exacerbated, as occurs when the *heart rate is very slow* or when the stroke volume output is greatly increased as a result of *aortic regurgitation, patent ductus arteriosus, or even extreme increase in venous* return. (2) Another cause of capillary pulsation is *extreme dilatation of the small arteries and arterioles*, which reduces the resistance of these vessels and thereby reduces the damping.

THE ARTERIOLES AND CAPILLARIES

Blood flow in each tissue is controlled almost entirely by the degree of contraction or dilatation of the arterioles, and it is in the capillaries where the important process of exchange between blood and the interstitial fluid occurs. These two segments of the circulation are so important that they will receive special discussion in the chapters to follow: capillary exchange phenomena in Chapters 16 and 17, and arteriolar regulation of blood flow in Chapters 18 and 19.

THE VEINS AND THEIR FUNCTIONS

For years the veins have been considered to be nothing more than passageways for flow of blood into the heart, but it is rapidly becoming apparent that the veins are capable of constricting and enlarging, of storing large quantities of blood and making this blood available when it is required by the remainder of the circulation, of actually propelling blood forward by means of the so-called venous pump that is described later, and even helping to regulate cardiac output, a function so important that it will be described in detail in Chapter 21.

CENTRAL VENOUS PRESSURE (RIGHT ATRIAL PRESSURE) AND ITS REGULATION

Blood from all the systemic veins flows into the right atrium; therefore, the pressure in the right atrium is frequently called the *central venous pressure*. The pressures in the peripheral veins are to a great extent dependent on the level of this pressure, so that anything that affects this pressure usually affects venous pressure everywhere in the body.

Right atrial pressure is regulated by a balance between, first, the *ability of the heart to pump blood* and, second, the *tendency for blood to flow from the peripheral vessels back to the heart*. If the heart is pumping strongly, the right atrial pressure tends to decrease, whereas, on the other hand, weakness of the heart tends to elevate the right atrial pressure. Likewise, any effect that causes rapid inflow of blood into the right atrium from the veins tends to elevate the right atrial pressure. Some of the factors that increase this tendency for venous return are (1) increased blood volume, (2) increase in vascular tone throughout the body with resultant increased peripheral venous pressures, and (3) dilatation of the systemic small vessels, which decreases the peripheral resistance.

The same factors that regulate right atrial pressure also enter into the regulation of cardiac output. Therefore, we will discuss the regulation of right atrial pressure with much more precision in Chapter 21 in connection with the regulation of cardiac output.

The *normal right atrial pressure* is approximately 0 mm. Hg, which is about equal to the atmospheric pressure around the body. However, it can rise to as high as 20 to 30 mm. Hg under very abnormal conditions, such as (a) in serious heart failure or (b) following massive transfusion when excessive quantities of blood are attempting to flow into the heart from the peripheral vessels.

VENOUS RESISTANCE AND PERIPHERAL VENOUS PRESSURE

As illustrated in Figure 15–6, most of the large veins entering the thorax are compressed at many points so that blood flow is impeded. For instance, the veins from the arms are compressed by their sharp angulation over the first rib. And veins coursing through the abdomen are compressed by different organs and by the intraabdominal pressure, so that often they are almost totally collapsed. For these reasons the *large veins usually offer considerable resistance to blood flow*, and because of this the pressure in the peripheral veins is usually 6 to 9 mm. Hg greater than the central venous pressure.

Note, however, that the veins inside the thorax *are not collapsed* because the *negative pressure* inside the chest distends these veins.

EFFECT OF HYDROSTATIC PRESSURE ON VENOUS PRESSURE

In a large chamber of water, the pressure at the surface of the water is equal to atmospheric pressure, but the pressure rises 1 mm. Hg for each 13.6 mm. distance below the surface of the water.

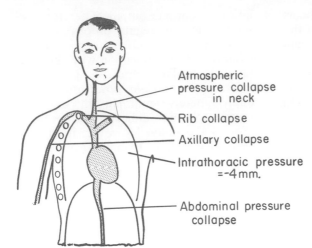

Atmospheric
pressure collapse
in neck

Rib collapse

Axillary collapse

Intrathoracic pressure
=-4mm.

Abdominal pressure
collapse

FIGURE 15–6 Factors tending to collapse the veins entering the thorax.

This pressure results from the weight of the water and therefore is called *hydrostatic pressure.*

Hydrostatic pressure also occurs in the vascular system of the human being, as is illustrated in Figure 15–7. When a person is standing, the pressure in the right atrium remains approximately 0 mm. Hg because the heart pumps into the arteries any excess blood that attempts to accumulate at this point. However, in an adult *who is standing absolutely still* the pressure in the veins of the feet is approximately +90 mm. Hg simply because of the distance from the feet to the heart. The venous pressures at other levels of the body lie proportionately between 0 and 90 mm. Hg.

The veins inside the skull are in a noncollapsible chamber, and they will not collapse. Consequently, *negative pressure can exist in the dural sinuses of the head*; in the standing position the venous pressure in the sagittal sinus is calculated to be approximately −10 mm. Hg because of the hydrostatic "suction" between the top of the skull and the base of the skull. Thus, if the sagittal sinus is entered during surgery, air can be sucked immediately into this vein; and the air may even pass downward to cause air embolism in the heart so that the heart valves do not function satisfactorily, and death ensues.

Effect of the Hydrostatic Factor on Other Pressures. The hydrostatic factor also affects the peripheral pressures in the arteries and capillaries as well as in the

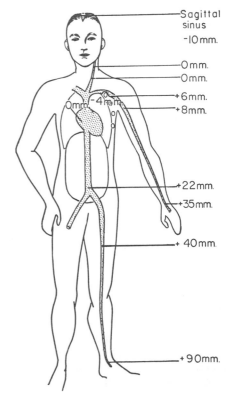

Sagittal
sinus
-10mm.

0mm.
0mm.

+6mm.
+8mm.

0mm. -4mm.

+22mm.
+35mm.

+ 40mm.

+90mm.

FIGURE 15–7 Effect of hydrostatic pressure on the venous pressures throughout the body.

veins. For instance, a standing person who has an arterial pressure of 100 mm. Hg at the level of his heart has an arterial pressure in his feet of about 190 mm. Hg. Therefore, any time one states that the arterial pressure is 100 mm. Hg, he generally means that this is the pressure at the hydrostatic level of the heart.

VENOUS VALVES AND THE "VENOUS PUMP"

Because of hydrostatic pressure, the venous pressure in the feet would always remain about +90 mm. Hg in a standing adult were it not for the valves in the veins, which are shown in Figure 15–8. Every time one moves his legs he compresses the muscles against the fascia and the fascia against the skin so that the veins of the legs are compressed, and blood tends to flow away from the areas of compression. The valves in the veins are arranged so that the direction of blood flow can be only toward the heart, as is evident in the figure. Consequently, every time a person moves his legs or even tenses his muscles, a certain amount of blood is propelled toward the heart, and the pressure in the dependent veins of the body is lowered. This pumping system is known as the "venous pump" or "muscle pump" and it is efficient enough that under ordinary circumstances the venous pressure in the feet of a walking adult remains less than 25 mm. Hg.

If the human being stands perfectly still, the venous pump does not work, and the venous pressures in the lower part of the leg can rise to the full hydrostatic value of 90 mm. Hg in about 30 seconds. Under such circumstances the pressures within the capillaries also increase greatly, and fluid leaks from the circulatory system into the tissue spaces. As a result, the legs swell, and the blood volume diminishes. Indeed, as much as 15 to 20 per cent of the blood volume is frequently lost from the circulatory system within the first 15 minutes of quiet standing.

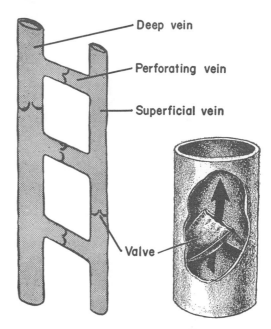

FIGURE 15–8 The venous valves of the leg.

REFERENCE POINT FOR MEASURING VENOUS AND OTHER CIRCULATORY PRESSURES

In previous discussions we have often spoken of right atrial pressure as being 0 mm. Hg and arterial pressure as being 100 mm. Hg, but we have not stated the hydrostatic level in the circulatory system to which this pressure is referred. There is one point in the circulatory system at which hydrostatic pressure factors do not affect the pressure measurements by more than 1 mm. Hg. This is at the level of the tricuspid valve, as shown by the crossed axes in Figure 15–9. Therefore, all pressure measurements discussed in this text are referred to the level of the tricuspid valve, which is called the *reference point for pressure measurement*.

The reason for lack of hydrostatic effects at the tricuspid valve is that the heart automatically prevents significant

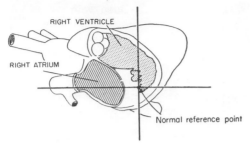

FIGURE 15–9 Location of the reference point for pressure measurement at the tricuspid valve.

hydrostatic changes in pressure at this point in the following way:

If the pressure at the tricuspid valve rises slightly above normal, then the right ventricle fills to a greater extent than usual, causing the heart to pump blood more rapidly than usual and thereby decreasing the pressure at the tricuspid valve back toward the normal mean value. On the other hand, if the pressure at this point falls, the right ventricle fails to fill adequately, its pumping decreases, and blood dams up in the venous system until the tricuspid pressure again rises to a normal value. In other words, *the heart acts as a feedback regulator of pressure* at the tricuspid valve.

MEASUREMENT OF VENOUS PRESSURE

Clinical Estimation of Venous Pressure. The venous pressure can be estimated by simply observing the distention of the peripheral veins—especially the neck veins. For instance, in the sitting position, the neck veins are never distended in the normal person. However, when the right atrial pressure becomes increased to as much as 10 mm. Hg, the lower veins of the neck begin to protrude even when one is sitting; when the right atrial pressure rises to as high as 15 mm. Hg, essentially all the veins in the neck become greatly distended, and the venous pulse becomes prominent in these veins.

Rough estimates of the venous pressure can also be made by raising or lowering an arm while observing the degree of distention of the antecubital or hand veins. As the arm is progressively raised, the veins suddenly collapse, and the level at which they collapse, when referred to the level of the heart, is a rough measure of the peripheral venous pressure.

Direct Measurement of Venous Pressure and Right Atrial Pressure. Venous pressure can be measured with ease by inserting a syringe needle connected to a water manometer directly into a vein. The venous pressure is expressed in relation to the level of the tricuspid valve, i.e., the height in centimeters of water above the level of the tricuspid valve; this can be converted to mm. Hg by dividing by a factor of 1.36.

The only means by which *right atrial pressure* can be measured accurately is by inserting a catheter through the veins into the right atrium. This catheter can then be connected to a water manometer and the pressure measured as noted previously.

BLOOD RESERVOIR FUNCTION OF THE VEINS

In discussing the general characteristics of the systemic circulation earlier in the chapter, it was pointed out that over 50 per cent of all the blood in the circulatory system is in the systemic veins. For this reason it is frequently said that the systemic veins act as a *blood reservoir* for the circulation. Also, relatively large quantities of blood are present in the veins of the lungs so that these, too, are considered to be blood reservoirs.

When blood is lost from the body to the extent that the arterial pressure begins to fall, pressure reflexes are elicited from the carotid sinuses and other pressure sensitive areas of the circulation, as will be detailed in Chapter 19; these in turn cause sympathetic constriction of the veins, which automatically takes up much of the slack in the circulatory system caused by the lost blood. Indeed, even after as much as 20 to 25 per cent of the total blood volume has been lost, the circulatory system often functions almost normally because of this variable reservoir system of the veins.

Specific Blood Reservoirs. Certain portions of the circulatory system are so extensive that they are specifically called "blood reservoirs." These include (1) the *spleen*, which can sometimes decrease in size sufficiently to release as much as 150 ml. of blood into other areas of the circulation, (2) the *liver*, the sinuses of which can release several hundred milliliters of blood into the remainder of the circulation, (3) the *large abdominal veins*, which can contribute as much as 300 ml., and (4) the *venous plexus beneath the skin*, which can probably contribute several hundred milliliters. The *heart* itself and the *lungs*, though not parts of the systemic venous reservoir system, must also be considered to be blood reservoirs. The heart, for instance, becomes reduced in size during sympathetic stimulation and in this way can contribute about 100 ml. of blood, and the lungs can contribute another 100 to 200 ml. when the pulmonary pressures fall to low values.

THE PULMONARY CIRCULATION

The quantity of blood flowing through the lungs is essentially equal to that flowing through the systemic circulation. However, there are problems related to distribution of blood flow and other hemodynamics that are special to the pulmonary circulation. Therefore, the present discussion is concerned specifically with the peculiarities of blood flow in the pulmonary circuit and the function of the right side of the heart in maintaining this flow.

The Pulmonary Vessels. The pulmonary artery extends only 4 centimeters beyond the apex of the right ventricle and then divides into the right and left main branches, which supply blood to the two respective lungs. The pulmonary artery is also a thin structure with a wall thickness approximately twice that of the venae cavae and one-third the thickness of the aorta. The pulmonary arterial branches are all very short. However, all the pulmonary arteries, even the small arteries and arterioles, have much larger diameters than their counterpart systemic arteries. This, combined with the fact that the vessels are very thin and distensible, gives the pulmonary arterial tree a compliance perhaps as great as that of the systemic arterial tree, thereby allowing the pulmonary arteries to accommodate the stroke volume output of the right ventricle.

The pulmonary veins, like the pulmonary arteries, are also short, but their distensibility characteristics are similar to those of the systemic circulation.

PRESSURES IN THE PULMONARY SYSTEM

The Pressure Pulse Curve in the Right Ventricle. The pressure pulse curves of the right ventricle and pulmonary artery are illustrated in the lower portion of Figure 15–10. These are contrasted with

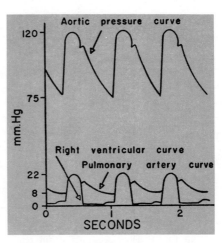

FIGURE 15–10 Pressure pulse contours in the right ventricle, pulmonary artery, and aorta. (Redrawn from Cournand: *Circulation, 2*:641, 1950.)

the much higher aortic pressure curve shown above. Approximately 0.16 second prior to ventricular systole, the right atrium pumps a small quantity of blood into the right ventricle and thereby causes about 4 mm. Hg initial rise in the right ventricular diastolic pressure even before the ventricle contracts. Immediately following this priming by the right atrium, the right ventricle contracts, and the right ventricular pressure rises rapidly until it equals the pressure in the pulmonary artery. The pulmonary valve opens, and for approximately 0.3 second blood flows from the right ventricle into the pulmonary artery. When the right ventricle relaxes, the pulmonary valve closes, and the right ventricular pressure falls to its diastolic level of about zero.

The systolic pressure in the right ventricle of the normal human being averages approximately 22 mm. Hg, and the diastolic pressure averages about 0 to 1 mm. Hg.

Pressures in the Pulmonary Artery. During systole, the pressure in the pulmonary artery is essentially equal to the pressure in the right ventricle, as shown in Figure 15–10. However, after the pulmonary valve closes at the end of systole, the ventricular pressure falls, while the pulmonary arterial pressure remains elevated and then falls gradually as blood flows through the capillaries of the lungs.

As shown in Figure 15–11, the systolic pulmonary arterial pressure averages

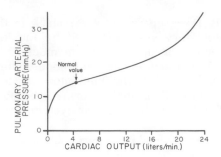

FIGURE 15–12 Effect on the pulmonary arterial pressure of increasing the cadiac output.

approximately 22 mm. Hg in the normal human being, the diastolic pulmonary arterial pressure approximately 8 mm. Hg, and the mean pulmonary arterial pressure 13 mm. Hg.

Left Atrial and Pulmonary Venous Pressure. The mean pressure in the left atrium and in the major pulmonary veins averages approximately 4 mm. Hg in the human being. However, the pressure in the left atrium varies, even among normal individuals, from as low as 1 mm. Hg to as high as 6 mm. Hg.

Effect of Increased Cardiac Output on Pulmonary Arterial Pressure. The cardiac output can increase to approximately four times normal before pulmonary arterial pressure becomes seriously elevated; this effect is illustrated in Figure 15–12. As the blood flow into the lungs increases, an initial slight rise in pressure causes the pulmonary arterioles and capillaries to expand and allows the increased blood flow to pass on through the capillary system without excessive increase in pulmonary arterial pressure. However, after cardiac output has reached three to four times normal, the expansion of the previously collapsed vessels is essentially complete, and further increase in cardiac output then causes almost proportionate increase in pulmonary arterial pressure.

This ability of the lungs to accommodate greatly increased blood flow with little increase in pulmonary arterial pressure is important because it conserves the energy of the heart. Actually, the

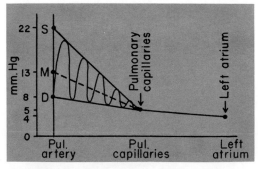

FIGURE 15–11 Pressures in the different vessels of the lungs. (Redrawn from Cournand: *Circulation,* 2:641, 1950.)

only reason for having blood flow through the lungs is to pick up oxygen and to release carbon dioxide. If the pulmonary arterial pressure should increase as a result of increased blood flow, this would cause a further unnecessary increase in load on the right ventricle.

RIGHT VENTRICULAR COMPENSATION AND FAILURE AS THE PULMONARY ARTERIAL PRESSURE RISES

When the left heart fails, the left atrial pressure sometimes rises as high as 15 to 40 mm. Hg. The mean pulmonary arterial pressure then also rises to high levels, placing a considerably greater than normal pressure load on the right ventricle. Up to 30 to 40 mm. Hg mean pulmonary arterial pressure, the right ventricle continues to pump essentially normal quantities of blood with only a slight rise in right atrial pressure. However, above this pressure the right ventricle begins to fail, so that further increases in pulmonary arterial pressure cause inordinate increases in right atrial pressure; and it is only now that the cardiac output decreases significantly.

BLOOD FLOW THROUGH THE LUNGS AND ITS DISTRIBUTION

Under most conditions, the pulmonary vessels act as passive, distensible tubes that enlarge with increasing pressure and narrow with decreasing pressure. But, for the blood to be aerated adequately, it is important for it to be distributed as evenly as possible to all the different segments of the lung. Some of the problems related to this are the following:

Effect of Low Alveolar Oxygen Pressure on Pulmonary Vascular Resistance — Autoregulation of Local Pulmonary Blood Distribution. When alveolar oxygen concentration becomes very low, the adjacent blood vessels slowly constrict during the ensuing 5 to 10 minutes, the vascular resistance increasing to as much as double normal. It should be noted specifically that this effect of oxygen lack is *opposite to the effect* normally observed in systemic vessels, and its cause is yet unexplained.

This effect of low oxygen on pulmonary vascular resistance has an important purpose: to distribute blood flow where it is most effective. That is, when some of the alveoli are poorly ventilated, the local vessels constrict, causing most of the blood to flow through other areas of the lungs that are better aerated. It is still questionable how important this autoregulation of blood flow distribution might be. Some research workers have even failed to demonstrate its existence, though most do agree that it occurs at least to a moderate extent.

Paucity of Autonomic Nervous Control of Blood Flow in the Lungs. Though autonomic nerves, both parasympathetic and sympathetic, innervate the larger pulmonary vessels, it is doubtful that these have any major function in normal control of pulmonary blood flow. Normally, stimulation of the vagal fibers to the lungs causes a slight decrease in pulmonary resistance, and stimulation of the sympathetics causes a slight increase in resistance.

THE BLOOD VOLUME OF THE LUNGS

The blood volume of the lungs, as measured by indirect methods, is approximately 12 per cent of the total blood volume of the circulatory system. In other words, in the average human being the two lungs probably contain approximately 600 ml. of blood. About 70 ml. of this is in the capillaries, and the remainder is divided about equally between the arteries and veins.

The Lungs as a Blood Reservoir. Under different physiologic and pathologic conditions, the quantity of blood in the lungs can vary from as little as 50 per cent of normal up to as high as 200 per cent of normal. For instance, when a person blows air out so hard that he builds up high pressure in his lungs — such as when blowing a trumpet — as much as 300 ml. of blood can be expelled from the pulmonary circulatory system into the systemic circulation. Also, loss of blood from the systemic circulation by hemorrhage can be partly compensated by automatic shift of blood from the lungs into the systemic vessels.

Shift of Blood Between the Pulmonary and Systemic Circulatory Systems as a Result of Cardiac Pathology. Failure of the left side of the heart or increased resistance to blood flow through the mitral valve as a result of mitral stenosis or mitral regurgitation causes blood to dam up in the pulmonary circulation, thus greatly increasing the pulmonary blood volume while decreasing the systemic volume. Concurrently, the pressures in the lung increase while the systemic pressures decrease.

On the other hand, exactly the opposite effects take place when the right side of the heart fails.

Because the volume of the systemic circulation is about seven times that of the pulmonary system, a shift of blood from one system to the other affects the pulmonary system greatly but usually has only mild systemic effects.

CAPILLARY DYNAMICS IN THE LUNGS

Pulmonary Capillary Pressure. Unfortunately, no direct measurements of pulmonary capillary pressure have been made. However, isogravimetric measurement of pulmonary capillary pressure, using the technique described in Chapter 16, has given a value of 7 mm. Hg. This is probably very nearly correct because the mean left atrial pressure is about 4 mm. Hg and the mean pulmonary arterial pressure is only 13 mm. Hg, so that the mean pulmonary capillary pressure must be between these two values.

Length of Time Blood Stays in the Capillaries. From histologic study of the total cross-sectional area of all the pulmonary capillaries, it can be calculated that when the cardiac output is normal, blood passes through the pulmonary capillaries in about 1 second. Increasing the cardiac output shortens this time sometimes to less than 0.4 second, but the shortening would not be as great as it is were it not for the fact that additional capillaries, which normally remain collapsed, open up to accommodate the increased blood flow. Thus, in less than 1 second, blood passing through the capillaries becomes oxygenated and loses its excess carbon dioxide.

CAPILLARY MEMBRANE DYNAMICS IN THE LUNGS

Negative Interstitial Fluid Pressure in the Lungs and its Significance. Because of the very low pulmonary capillary pressure, only 7 mm. Hg, the hydrostatic force tending to push fluid out the capillary pores into the interstitial spaces is very slight. Yet, the colloid osmotic pressure of the plasma, about 28 mm. Hg, is a large force tending to pull fluid into the capillaries. Therefore, there is continual osmotic tendency to dehydrate the interstitial spaces of the lungs. And it can be calculated that the normal interstitial fluid pressure of the human lungs is probably about −12 mm. Hg. That is, there is approximately 12 mm. Hg negative pressure tending to pull the alveolar epithelial membrane toward the capillary membrane, thus squeezing the pulmonary interstitial space down to almost nothing. Electron micrographic studies have demonstrated exactly this fact, the interstitial space at times being so narrow that the alveolar epithelial cells appear to be directly adherent to the pulmonary capillary endothelial cells. As a result, the distance between the air in the alveoli and the blood in the capillaries is minimal, averaging less than 0.4 micron in distance; and this obviously allows very rapid diffusion of oxygen and carbon dioxide.

Mechanism by which the alveoli remain dry. Another consequence of the negative pressure in the interstitial spaces is that it pulls fluid from the alveoli through the alveolar membrane and into the interstitial spaces, thereby tending to cause dryness in the alveoli.

Pulmonary Edema. Pulmonary edema means excessive quantities of fluid either in the pulmonary interstitial spaces or in the alveoli. The normal pulmonary interstitial fluid volume is approximately 20 per cent of the lung mass, but this can increase several-fold in serious pulmonary edema. In addition, tremendous volumes of fluid can enter the alveoli and cause *intra-alveolar edema*; intra-alveolar fluid sometimes reaches 1000 per cent of the normal interstitial fluid volume.

Safety factor against pulmonary edema. The most common cause of pulmonary edema is greatly elevated capillary pressure resulting from failure of the left heart and consequent damming of blood in the lungs. However, because of the very high dehydrating force of the colloid osmotic pressure of the blood in the lungs, pulmonary edema will rarely develop below 30 mm. Hg pulmonary capillary pressure. Thus, if the capillary pressure in the lungs is normally 7 mm. Hg and this pressure must usually rise above 30 mm. Hg before edema will occur, the lungs have a *safety factor against edema* of approximately 23 mm. Hg.

Safety factor against edema in chronic elevation of capillary pressure. Patients with chronic elevation of pulmonary capillary pressure occasionally will not develop pulmonary edema even with pulmonary capillary pressures as high as 45 mm. Hg. The reason for this is probably extremely rapid run-off of fluid from the pulmonary interstitial spaces through the lymphatics. If the pulmonary capillary pressure remains elevated for more than approximately two weeks, the pulmonary lymphatics enlarge as much as 6- to 10-fold, and lymphatic flow can increase perhaps 20- to 50-fold above the normal resting level. This extra lymphatic flow gives a safety factor perhaps 15 mm. Hg above that which one normally has against edema.

REFERENCES

Fishman, A. P.: Dynamics of the pulmonary circulation. *In* Handbook of Physiology. Baltimore, The Williams & Wilkins Co., 1963, Sec. II, Vol. II, p. 1667.

Folkow, B.: Role of the nervous system in the control of vascular tone. *Circulation, 21*:706, 1960.

Green, H. D., and Kepchar, J. H.: Control of peripheral resistance in major systemic vascular beds. *Physiol. Rev., 39*:617, 1959.

Guyton, A. C.: Peripheral circulation. *Ann. Rev. Physiol., 21*:239, 1959.

Guyton, A. C.: Introduction to Part I: Pulmonary alveolar-capillary interface and interstitium. *In* Fishman, A. P., and Hecht, H. H. (eds.): The Pulmonary Circulation and Interstitial Space. Chicago, University of Chicago Press, 1969, p. 3.

Korner, P. I.: Circulatory adaptations in hypoxia. *Physiol. Rev., 39*:687, 1959.

Mead, J., and Whittenberger, J. L.: Lung inflation and hemodynamics. *In* Handbook of Physiology. Baltimore, The Williams & Wilkins Co., 1964, Sec. III, Vol. I, p. 477.

Mellander, S.: Systemic circulation. *Ann. Rev. Physiol., 32*:313, 1970.

Sonnenschein, R. R., and White, F. N.: Systemic circulation. *Ann. Rev. Physiol., 30*:147, 1968.

West, J. B.: Ventilation/Blood Flow and Gas Exchange. Oxford, Blackwell Scientific Publications Ltd., 1965.

Wiedeman, Mary P.: Dimensions of blood vessels from distributing artery to collecting vein. *Circ. Res., 12*:375, 1963.

CHAPTER 16

CAPILLARY DYNAMICS, INTERSTITIAL FLUID, AND THE LYMPHATIC SYSTEM

It is in the capillaries that the most purposeful function of the circulation occurs, namely, interchange of nutrients and cellular excreta between the tissues and the circulating blood. About 10 billion capillaries, having a total surface area probably greater than 100 square meters, provide this function. Indeed, it is rare that any single functional cell of the body is more than 20 to 30 microns away from a capillary.

It is the purpose of this chapter to discuss the transfer of substances between the blood and interstitial fluid and especially to discuss the factors that affect the transfer of fluid volume between the circulating blood and the interstitial fluids. Return of excess fluid from the interstitial spaces to the circulation by way of the lymphatics will also be considered.

STRUCTURE OF THE CAPILLARY SYSTEM

Figure 16–1 illustrates the structure of a "unit" capillary bed, illustrating that

blood enters the capillaries through an *arteriole* and leaves by way of a *venule*. Blood from the arteriole passes into a series of *meta-arterioles*, which have a structure midway between that of arterioles and capillaries. After leaving the meta-arteriole, the blood enters the *capillaries*, some of which are large and are called the *preferential channels* and others of which are small and are the *true capillaries*. After passing through the capillaries the blood enters the venule and returns to the general circulation.

The arterioles are highly muscular,

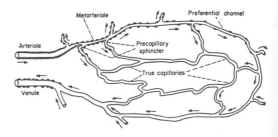

FIGURE 16–1 Overall structure of a capillary bed. (From Zweifach: Factors Regulating Blood Pressure. Josiah Macy, Jr., Foundation, 1950.)

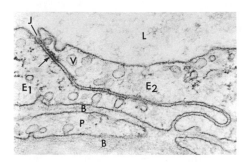

FIGURE 16-2   An electron micrograph showing a long slitlike pore that exists at the junction (J) between adjacent capillary endothelial cells (E_1 and E_2). (Approximate magnification 30,000×.) L, capillary lumen; B, basement membrane; P, pericyte; V, pinocytic vesicle. (Courtesy of Drs. Guido Majno and Ramzi Cotran.)

and their diameters can change many fold, as was discussed in Chapter 14. The meta-arterioles do not have a continuous muscular coat, but smooth muscle fibers encircle the vessel at intermediate points, as illustrated in Figure 16-1 by the large black dots to the sides of the meta-arteriole.

At the origin of the true capillaries from the meta-arterioles a smooth muscle fiber usually encircles the capillary. This is called the *precapillary sphincter.*

The venules are considerably larger than the arterioles and have a much weaker muscular coat. Yet, it must be remembered that the pressure in the venules is much less than that in the arterioles so that the venules can perhaps contract as much as can the arterioles.

Structure of the Capillary Wall.   Figure 16-2 illustrates an electron micrograph of the capillary wall. Note that the wall is primarily composed of a thin, unicellular layer of endothelial cells and is surrounded by a thin basement membrane on the outside. The diameter of the capillary is 7 to 9 microns, barely large enough for red blood cells and other blood cells to squeeze through.

Pores in the capillary membrane. On studying Figure 16-2 again, one can see a minute passageway connecting the interior of the capillary with the ex-

terior. However, such passageways are located far apart from one another, and they represent no more than 0.001 of the total surface area of the capillary. These passageways are the *pores* of the capillary membrane through which water and many dissolved substances pass between the lumen of the capillary and interstitial spaces, as is discussed later in the chapter. These pores have a diameter between 80 and 90 Ångstroms.

FLOW OF BLOOD IN THE CAPILLARIES — VASOMOTION

Blood does not flow at a continuous rate through the capillaries. Instead, it flows in intermittent spurts. The cause of this intermittency is the phenomenon called *vasomotion,* which means intermittent contraction of the meta-arterioles and precapillary sphincters. These constrict and relax in an alternating cycle 6 to 12 times per minute. The most important factor found thus far to affect the degree of opening and closing of the meta-arterioles and precapillary sphincters is the concentration of *oxygen* in the tissues. When the oxygen concentration is very low, the spurts of blood flow occur more often, and the duration of each period of flow lasts for a more prolonged interval, thereby allowing the blood to carry increased quantities of oxygen to the tissues.

EXCHANGE OF NUTRIENTS AND OTHER SUBSTANCES BETWEEN THE BLOOD AND INTERSTITIAL FLUID

DIFFUSION THROUGH THE CAPILLARY MEMBRANE

By far the most important means by which substances are transferred between the plasma and interstitial fluids is by diffusion. Figure 16-3 illustrates

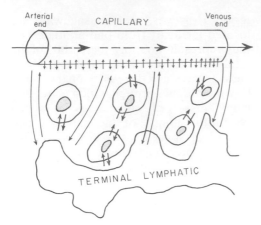

FIGURE 16–3 Diffusion of fluid and dissolved substances between the capillary and interstitial fluid spaces.

this process, showing that as the blood traverses the capillary, tremendous numbers of water molecules and dissolved particles diffuse back and forth through the capillary wall, providing continual mixing between the interstitial fluids and the plasma as was explained in Chapter 4.

Diffusion of Water-Soluble, Lipid-Insoluble Substances Through the Capillary Membrane. Despite the fact that not over $\frac{1}{1000}$ of the surface area of the capillaries is represented by pores, the rate of thermal motion in fluids is so great that even this small area of pores is sufficient to allow tremendous diffusion of water-soluble substances of small molecular weight across the membrane. To give one an idea of the extreme rapidity with which substances diffuse, *the rate at which water molecules diffuse through the capillary membrane is approximately 40 times as great as the rate at which plasma itself flows linearly along the capillary.* That is, the water of the plasma is exchanged with the water of the interstitial fluids 40 times before the plasma can go the entire distance through the capillary.

Effect of molecular size on pore permeability. The diameter of the capillary pores, 80 to 90 Ångstroms, is about 25 times the diameter of the water

molecule; and it is also so much larger than that of most other dissolved substances in the plasma, except for the plasma proteins, that there is no impediment to the diffusion of these substances. The diameters of the plasma protein molecules are slightly greater than those of the pores, and the capillary permeability for these molecules is less than $\frac{1}{10,000}$ of that for water molecules. Thus, the membrane is almost impermeable to the plasma proteins, which causes a significant concentration difference to develop between the proteins of the plasma and those of the interstitial fluid, as will become evident later in the chapter.

Diffusion of Lipid-Soluble Substances Through the Capillary Membrane. If a substance is lipid-soluble, it can diffuse directly through the cell membranes of the capillary without having to go through the pores. Lipid-soluble substances normally transported in the blood include especially oxygen and carbon dioxide. Since these substances can permeate all areas of the capillary membrane, not merely the pores, their rates of transport through the membrane are many times greater than the rates for water and other lipid-insoluble substances.

TRANSFER OF FLUID VOLUME BETWEEN THE PLASMA AND INTERSTITIAL FLUIDS

One of the most important considerations of capillary dynamics is the means by which the plasma is held in the circulation and not allowed to transude continuously through the capillary membrane into the interstitial fluid. Figure 16–4 illustrates the four primary factors that determine whether fluid will move out of the blood into the interstitial fluid or in the opposite direction; these are:

1. The *capillary pressure* (Pc), which

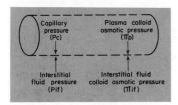

FIGURE 16–4 Forces operative at the capillary membrane tending to move fluid either outward or inward through the membrane.

tends to move fluid outward through the capillary membrane.

2. The *interstitial fluid pressure* (Pif), which tends to move fluid inward through the capillary membrane.

3. The *plasma colloid osmotic pressure* (Πp), which tends to cause osmosis of fluid inward through the membrane.

4. The *interstitial fluid colloid osmotic pressure* (Πif), which tends to cause osmosis of fluid outward through the membrane.

The regulation of fluid volumes in the blood and interstitial fluid is so important that each of these factors is discussed in turn in the following sections.

CAPILLARY PRESSURE

Unfortunately, the exact capillary pressure is not known because it has been impossible to measure capillary pressure under absolutely normal conditions. Yet two different methods have been used to estimate the capillary pressure: (1) *direct cannulation of the capillaries using a micropipet* and an appropriate manometer, which has given an average mean capillary pressure of about 25 mm. Hg, and (2) *indirect functional measurement of the capillary pressure*, which has given a capillary pressure averaging about 17 mm. Hg.

Isogravimetric Method for Measuring "Functional" Mean Capillary Pressure. Figure 16–5 illustrates an *isogravimetric* method for estimating capillary pressure. This figure shows a section of gut held by one arm of a gravimetric balance. Blood is perfused through the

gut. When the arterial pressure is decreased, the resulting decrease in capillary pressure causes osmotic absorption of fluid out of the gut wall and makes the weight of the gut decrease. This immediately causes movement of the balance arm. However, to prevent this weight decrease, the venous pressure is raised a sufficient amount to overcome the effect of decreasing the arterial pressure.

In the lower part of the figure, the changes in arterial and venous pressures that exactly balance each other in their effect on the weight of the gut are illustrated. The arterial and venous curves cross each other at a value of 17 mm. Hg. Therefore, the capillary pressure must have remained at this same level of 17 mm. Hg throughout these maneuvers. Thus, in a roundabout way, the "functional" capillary pressure is measured to be about 17 mm. Hg.

The reason the "functional" capillary pressure is lower than the pressure measured by direct cannulation is probably mainly the following: The metaarterioles and precapillary sphincters of the capillary system are normally closed during a greater part of the vaso-

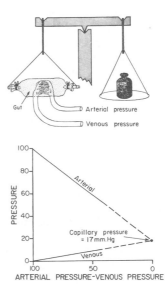

FIGURE 16–5 Isogravimetric method for measuring capillary pressure. Explained in the text.

motion cycle than they are open. When they are closed the pressure in the entire capillary system beyond the closures should be almost exactly equal to the pressure at the venous ends of the capillaries, about 10 mm. Hg. Therefore, if one considers a *weighted* average of the pressures in all capillaries, one would expect the *functional* mean capillary pressure to be much nearer to the pressure in the venous ends of the capillaries than to the pressure in the arterial ends.

Thus, there is good reason for believing *that the normal functional mean capillary pressure is about 17 mm. Hg.*

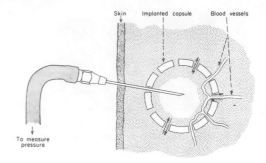

FIGURE 16–6 The perforated capsule method for interstitial fluid pressure.

INTERSTITIAL FLUID PRESSURE

The interstitial fluid pressure, like capillary pressure, has been difficult to measure, primarily because the maximum width of interstitial spaces is about 1 micron. Thus far, it has been impossible to cannulate these spaces directly to make measurements of the pressure. Therefore, several indirect methods have been used to estimate the interstitial fluid pressure; one method is the following:

Measurement of Interstitial Fluid Pressure in Implanted Perforated Spheres. Figure 16–5 illustrates an indirect method for measuring interstitial fluid pressure, which may be explained as follows: A small hollow plastic sphere perforated by several hundred small holes is implanted in the tissues, and the surgical wound is allowed to heal for approximately one month. At the end of that time, tissue will have grown inward through the holes to line the inner surface of the sphere. Furthermore, the cavity is filled with fluid that flows freely through the perforations back and forth between the fluid of the interstitial fluid spaces and the fluid of the cavity. Therefore, the pressure in the cavity should equal the pressure in the interstitial fluid spaces. A needle is inserted through the skin and through one of the perforations to the interior of the cavity, and

the pressure is measured by use of an appropriate manometer.

Interstitial fluid pressure measured by this method in normal tissues averages about −7 mm. Hg. That is, the pressure is actually *less than atmospheric pressure* or, in other words, is a semivacuum. The significance of the negativity of interstitial fluid pressure and the mechanism of its development is discussed in the following chapter in relation to interstitial fluid dynamics.

PLASMA COLLOID OSMOTIC PRESSURE

Colloid Osmotic Pressure Caused by Proteins. The proteins are the only dissolved substances of the plasma that do not diffuse readily into the interstitial fluid. Furthermore, when small quantities of protein do diffuse into the interstitial fluid, most of these are soon removed from the interstitial spaces by way of the lymph vessels. Therefore, the concentration of protein in the plasma averages about four times as much as that in the interstitial fluid, 7.3 gm./100 ml. in the plasma versus 1.8 gm./100 ml. in the interstitial fluid.

In the discussion of osmotic pressure in Chapter 4, it was pointed out that only those substances that fail to pass through the pores of a semipermeable membrane exert osmotic pressure. Therefore, the dissolved proteins of the plasma and interstitial fluids are responsible for the osmotic pressure that develops at the capillary membrane. To distinguish this osmotic pressure from that which occurs at the cell membrane, it is called *colloid osmotic pressure* or *oncotic pressure*.

Normal Value for Plasma Colloid Osmotic Pressure. The colloid osmotic pressure of normal human plasma averages approximately 28 mm. Hg.

Effect of the Different Plasma Proteins on Colloid Osmotic Pressure. The plasma proteins are a mixture of proteins that contain albumin, with an average molecular weight of 69,000; globulins, 140,000; and fibrinogen, 400,000. Thus, 1 gram of globulin contains only half as many molecules as 1 gram of albumin, and 1 gram of fibrinogen contains only one-sixth as many molecules as 1 gram of albumin. Furthermore, the average relative concentrations of these different types of proteins in the plasma are:

	gm. %
Albumin	4.5
Globulins	2.5
Fibrinogen	0.3
TOTAL	7.3

Because each gram of albumin exerts twice the osmotic pressure of a gram of globulins and because there is almost twice as much albumin in the plasma as globulins, about 70 per cent of the total colloid osmotic pressure of the plasma results from the albumin fraction and only about 30 per cent from the globulins and fibrinogens.

INTERSTITIAL FLUID COLLOID OSMOTIC PRESSURE

The total quantity of protein in the entire 12 liters of interstitial fluid of the body is almost exactly equal to the total quantity of protein in the plasma itself; but since this volume is four times the volume of plasma, the average protein *concentration* of the interstitial fluid is only one-fourth that in plasma, or approximately 1.8 grams per cent. The average colloid osmotic pressure caused by this concentration of proteins in the interstitial fluids is approximately 4.5 mm. Hg.

THE STARLING EQUILIBRIUM OF CAPILLARY EXCHANGE

E. H. Starling pointed out three-quarters of a century ago that under normal conditions a state of equilibrium exists at the capillary membrane whereby the amount of fluid leaving the circulation through the capillaries exactly equals that quantity of fluid that is returned to the circulation by reabsorption at the venous ends of the capillaries and by flow through the lymphatics. This equilibrium is caused by equilibration of the *mean* forces tending to move fluid through the capillary membranes. If we assume that the mean capillary pressure is 17 mm. Hg, the normal mean equilibrium dynamics of the capillary are the following:

Mean forces tending to move fluid outward:

	mm. Hg
Mean capillary pressure	17
Negative interstitial fluid pressure	7
Interstitial fluid colloid osmotic pressure	4.5
TOTAL OUTWARD FORCE	28.5

Mean force tending to move fluid inward:

	mm. Hg
Plasma colloid osmotic pressure	28
TOTAL INWARD FORCE	28

Summation of mean forces:

	mm. Hg
Outward	28.5
Inward	28.0
NET OUTWARD FORCE	0.5

Thus, we find a slight imbalance of forces at the capillary membranes that causes slightly more filtration of fluid into the interstitial spaces than reabsorption. This slight excess of filtration is called the *net filtration*, and it is balanced by fluid return to the circulation through the lymphatics. The normal rate of net filtration in the entire body is about 1.7 ml. per minute. This figure also represents the rate of fluid flow into the lymphatics each minute.

Filtration of Fluid at the Arterial End of the Capillary. The mean pressure at the arterial end of the capillary is 8 to 10 mm. Hg greater than that at the midpoint of the capillary. Therefore, far more fluid filters out of the capillary here than is reabsorbed; approximately 17 ml. of fluid per minute filters out at the arterial ends of all capillaries in the entire body.

Reabsorption of Fluid at the Venous End of the Capillary. The mean pressure at the venous end of the capillary is 7 to 10 mm. Hg less than that at the midpoint of the capillary. Therefore, fluid is reabsorbed. The amount reabsorbed at the venous ends of all capillaries in the body calculates to be about 15.3 ml. per minute.

Flow of Fluid Through the Tissue Spaces. Because fluid is filtered at the arterial ends and reabsorbed at the venous ends of the capillaries, there must be flow of fluid through the tissue spaces from the arterial ends to the venous ends. The total amount of this flow is 17 ml. at the arterial ends, and 15.3 ml. into the venous ends. The difference, 1.7 ml. per minute, is the amount that flows into the lymphatics, as was just explained.

CAPILLARY DYNAMICS IN THE PULMONARY CIRCULATION

Exactly the same principles as those just discussed for the systemic circulation apply also to the pulmonary circulation. However, the capillary pressure in the pulmonary circulation, as was discussed in Chapter 15, is only 7 mm. Hg instead of 17 mm. Hg. Therefore, there is far less tendency for fluid to filter out of the capillaries into the interstitial spaces of the lungs than is true in systemic tissues. As a result, the inter-

stitial fluid pressure in the lungs of human beings is probably about −12 mm. Hg, instead of the −7 mm. Hg present in peripheral tissues. For this reason there is an extreme tendency for absorption of fluid into the capillaries of the lungs.

Mechanism for Maintaining Dry Alveoli. The very negative interstitial fluid pressure in lung tissue causes continual tendency for movement of fluid inward through the epithelial membrane of the alveoli, thereby removing any excess fluid from the alveoli. Thus, this continual suction from the alveoli into the capillaries is almost certainly the mechanism by which the alveoli are maintained in a dry state rather than being continually filled with fluid.

THE LYMPHATIC SYSTEM

The lymphatic system represents an accessory route by which fluids can flow from the interstitial spaces into the blood. And, most important of all, the lymphatics can carry proteins and even large particulate matter away from the tissue spaces, neither of which can be removed by absorption directly into the blood capillary. We shall see that this removal of proteins from the interstitial spaces is an absolutely essential function, without which we would die within about 24 hours.

THE LYMPH CHANNELS OF THE BODY

All tissues of the body, with the exception of a very few, have lymphatic channels that drain excess fluid from the interstitial spaces. And even these very few tissues have minute interstitial channels through which interstitial fluid can flow into lymphatic vessels along the periphery of these tissues or, in the case of the brain, into the cerebrospinal fluid and thence directly back into the blood.

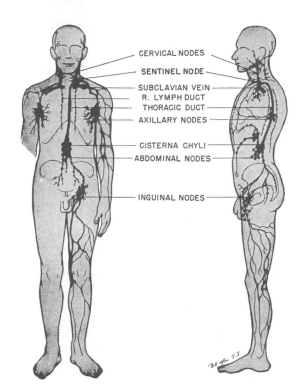

FIGURE 16–7   The lymphatic system.

Essentially all the lymph from the lower part of the body and from the upper left part of the body flows into the *thoracic duct* and empties into the venous system at the juncture of the left internal jugular vein and subclavian vein, as illustrated in Figure 16–7.

Lymph from the right side of the neck and head, from the right arm, and from parts of the right thorax enters the *right lymph duct*, which then empties into the venous system at the juncture of the right subclavian vein and right internal jugular vein.

FIGURE 16–8   Special structure of the terminal lymphatic capillaries that permits passage of substances of high molecular weight back into the circulation.

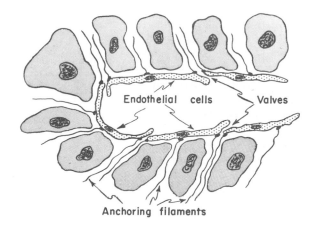

The Terminal Lymphatic Capillaries and their Permeability. Most of the fluid filtering from the arterial capillaries flows among the cells and is finally reabsorbed back into the venous capillaries; but, on the average, about one-tenth of the fluid enters the terminal lymphatic capillaries and returns to the blood through the lymphatic system rather than directly through the venous capillaries.

The minute quantity of fluid that returns to the circulation by way of the lymphatics is extremely important because substances of high molecular weight, such as proteins, cannot pass with ease through the pores of the venous capillaries, but they can enter the lymphatic capillaries almost completely unimpeded. The reason for this is a special structure of the terminal lymphatic capillaries, illustrated in Figure 16–8. This figure shows the endothelial cells of the terminal lymphatic attached by *anchoring filaments* to the connective tissue between the surrounding tissue cells. However, at the junctions between the endothelial cells there are no connections between the cells. Instead, the edge of one endothelial cell overlaps the edge of the adjacent one in such a way that the overlapping edge is free to flap. Therefore, the structure forms a minute valve that opens to the interior of the lymphatic. Interstitial fluid, along with its suspended particles, can push the valve open and flow directly into the terminal lymphatic. But this fluid cannot leave the lymphatic once it has entered because any backflow will close the flap valve. Thus, the lymphatics have valves at the very tips of the terminal lymphatics as well as valves along their entire extent up to the point where they empty into the blood vessels.

FORMATION OF LYMPH

Lymph is interstitial fluid that flows into the lymphatics. Therefore, lymph has almost the identical composition as the tissue fluid in the part of the body from which the lymph flows.

The protein concentration in the interstitial fluid averages about 1.8 gms. per cent, and this is also the protein concentration of lymph flowing from most of the peripheral tissues. On the other hand, lymph formed in the liver has a protein concentration as high as 6 gms. per cent, and lymph formed in the intestines has a protein concentration as high as 3 to 4 gms. per cent. Since approximately half of the lymph is derived from the liver and intestines, the thoracic lymph, which is a mixture of lymph from all areas of the body, usually has a protein concentration of 3 to 5 gms. per cent.

The lymphatic system is also one of the major channels for absorption of nutrients from the gastrointestinal tract, being responsible principally for the absorption of fats, as will be discussed in Chapter 44. Therefore, after a fatty meal, thoracic duct lymph sometimes contains as much as 1 to 2 per cent of fat.

Finally, even large particles, such as bacteria, can push their way between the endothelial cells of the terminal lymphatics and in this way enter the lymph. As discussed in Chapter 9, these particles are then removed by the lymph nodes, where they are destroyed.

RATE OF LYMPH FLOW

Approximately 100 ml. of lymph flow through the thoracic duct of a resting man per hour, and perhaps another 20 ml. of lymph flow into the circulation each hour through other channels, making a total estimated lymph flow of perhaps 120 ml. per hour. This rate of flow is extremely small in comparison with the rate of blood flow.

Factors that Determine the Rate of Lymph Flow

Interstitial Fluid Pressure. Elevation of interstitial fluid pressure above its normal level of -7 mm. Hg increases the flow of interstitial fluid into the terminal lymphatics and consequently also increases the rate of lymph flow. This increase in flow is approximately linear until the interstitial fluid pressure reaches 0, at which point the flow rate is 10 to 50 times normal. Therefore, any factor, besides obstruction of the lymphatic system itself, that tends to increase interstitial fluid pressure increases the rate of lymph flow. Such factors include:

1. Elevated capillary pressure.
2. Decreased plasma colloid osmotic pressure.
3. Increased interstitial fluid protein.
4. Increased permeability of the capillaries.

The Lymphatic Pump. Valves exist in all lymph channels, even down to the tips of the terminal lymphatics, as shown in Figure 16–8. In the larger lymphatics, valves exist every few millimeters, and in the smaller lymphatics the valves are much closer together than this, which illustrates the widespread existence of the valves. Every time the lymph vessel is compressed by pressure from any source, lymph tends to be squeezed in both directions, but because the valves open only in the central direction, the fluid moves unidirectionally. In order of their importance, the factors that often compress the lymphatics are:

1. Contraction of muscles.
2. Passive movement of the parts of the body.
3. Arterial pulsations.
4. Compression of the tissues from the outside.

Obviously, the lymphatic pump becomes very active during exercise but sluggish under resting conditions. During exercise the rate of lymph flow can increase to as high as 3 to 14 times normal because of the increased activity.

In summary, then, *the rate of lymph flow is determined principally by the product of tissue pressure × the activity of the lymphatic pump.*

REGULATION OF INTERSTITIAL FLUID PROTEIN BY THE LYMPHATICS

Since protein continually leaks from the capillaries into the interstitial fluid spaces, it must also be removed continually, or otherwise the tissue colloid osmotic pressure will become so high that normal capillary dynamics can no longer continue. Therefore, by far the most important of all the lymphatic functions is the maintenance of low protein concentration in the interstitial fluid. The mechanism of this is the following:

As fluid leaks from the arterial ends of the capillaries into the interstitial spaces, only small quantities of protein accompany it, but then, as the fluid is reabsorbed at the venous ends of the capillaries, most of the protein is left behind. Therefore, *protein progressively accumulates in the interstitial fluid*, and this in turn *increases the tissue colloid osmotic pressure.* This osmotic pressure then decreases reabsorption of fluid by the capillaries, thereby *promoting increased tissue fluid volume* and *increased tissue pressure.* The increased pressure then forces interstitial fluid into the lymphatic channels, and the fluid carries with it the excess protein that has accumulated. As a result, normal capillary dynamics ensue once again.

To summarize, an increase in tissue fluid protein increases the rate of lymph flow, and this washes the proteins out of the tissue spaces, automatically returning the protein concentration to its normal low level. If it were not for this continual removal of proteins, the dynamics of the capillaries would become so abnormal within only a few hours that life could no longer continue. There is certainly no other function of the lymphatics that can even approach this in importance.

MECHANISM OF NEGATIVE INTERSTITIAL FLUID PRESSURE

Until recent measurements of the interstitial fluid pressure demonstrated that the interstitial fluid pressure is negative rather than positive, as explained previously, it had been taught that the normal interstitial pressure ranges between +1 and +4 mm. Hg, and it has still been difficult to understand how negative pressures can develop in the interstitial fluid spaces. However, we can explain this negative interstitial fluid pressure by the following considerations:

First, fluid can flow into lymphatic vessels from interstitial spaces even when the mean interstitial fluid pressure is negative, because this pressure rises and falls every time a tissue is compressed, as discussed earlier. Each time the pressure rises to a value slightly above atmospheric pressure, fluid flows out of the interstitial spaces into the lymphatics. This intermittent movement of interstitial fluid into the lymphatics keeps the protein concentration of the interstitial fluid at a low value and thereby keeps the interstitial colloid osmotic pressure at a low value, usually about 4 mm. Hg. With the tissue colloid osmotic pressure so low, the plasma colloid osmotic pressure becomes greatly preponderant, causing osmosis of so much additional fluid directly into the capillaries that the average pressure in the interstitial spaces is maintained at a very negative value.

Significance of the Normally "Dry" State of the Interstitial Spaces. The normal tendency for the capillaries to absorb fluid from the interstitial spaces and thereby to create a partial vacuum causes all the minute structures of the interstitial spaces to be *compacted* together. This represents a "dry" state; that is, no *excess* fluid is present besides that required simply to fill the crevices between the tissue elements.

The "dry" state of the tissue is particularly important for optimal nutrition of the tissues, because nutrients pass from the blood to the cells by diffusion; and the rate of diffusion between two points is inversely proportional to the distance between the cells and the capillaries. Therefore, it is essential that the distances be maintained at a minimum, or otherwise nutritive damage to the cells can result.

REFERENCES

Allen, L.: Lymphatics and lymphoid tissues. *Ann. Rev. Physiol., 29*:197, 1967.

Chinard, F. P.: Starling's hypothesis in the formation of edema. *Bull. N.Y. Acad. Med., 38*:375, 1962.

Drinker, C. K.: The Lymphatic System. Stanford. Stanford University Press, 1942.

Gauer, O. H., Henry, J. P., and Behn, C.: The regulation of extracellular fluid volume. *Ann. Rev. Physiol., 32*:547, 1970.

Guyton, A. C.: Concept of negative interstitial pressure based on pressures in implanted perforated capsules. *Circ. Res., 12*:399, 1963.

Guyton, A. C.: Interstitial fluid pressure-volume relationships and their regulation. *In* Wolstenholme, G. E. W., and Knight, J. (eds.): Ciba Foundation Symposium on Circulatory and Respiratory Mass Transport. London, J. & A. Churchill Ltd., 1969, p. 4.

Guyton, A. C., and Lindsey, A. W.: Effect of elevated left atrial pressure and decreased plasma protein concentration on the development of pulmonary edema. *Circ. Res., 7*:649, 1959.

Yoffey, J. M., and Courtice, F. C.: Lymphatics, Lymph, and Lymphoid Tissue. Baltimore, The Williams & Wilkins Co., 1967.

CHAPTER 17

EDEMA; AND SPECIAL FLUID SYSTEMS OF THE BODY

EDEMA

Edema means the presence of excess interstitial fluid in the tissues, and it is one of the most important problems in clinical medicine. Obviously, any factor that increases the interstitial fluid pressure high enough can cause development of excess interstitial fluid volume and thereby cause edema to develop. However, to explain the conditions under which edema develops, we must first characterize the *pressure-volume curve* of the interstitial fluid spaces.

Pressure-Volume Curve of the Interstitial Fluid Spaces. Figure 17–1 illustrates the average relationship between pressure and volume in the interstitial fluid spaces in the human body as extrapolated from measurements in the isolated hind leg of a dog. One of the most significant features of this curve is that as long as the interstitial fluid pressure remains in the negative range there is little change in interstitial fluid volume despite marked changes in pressure. Therefore, edema will not occur so long as the interstitial fluid pressure remains

negative. However, just as soon as the interstitial fluid pressure rises to equal atmospheric pressure (zero pressure), the slope of the pressure-volume curve suddenly changes and the volume increases precipitously. An additional increase in interstitial fluid pressure of only 1 to 3 mm. Hg now causes the interstitial fluid volume to increase several hundred per cent. Finally, at the very top of the figure, the skin begins to be stretched so that the volume now increases much less rapidly.

Positive Interstitial Fluid Pressure as the Physical Basis for Edema. After studying the pressure-volume curve of Figure 17–1, one can readily see that just as soon as the interstitial fluid pressure rises above the surrounding atmospheric pressure, the tissue spaces begin to swell. Therefore, *the physical cause of edema is positive pressure in the interstitial fluid spaces.*

The Phenomenon of "Pitting" Edema. If one presses with his finger on the skin over an edematous area and then suddenly removes the finger, a small depression called a "pit" remains. Grad-

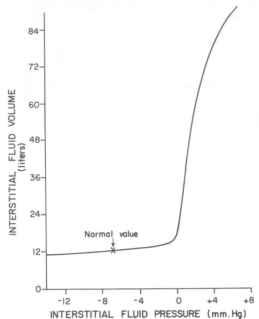

FIGURE 17-1 Pressure-volume curve of the interstitial spaces. (Extrapolated to the human being from data obtained in dogs.)

ually, within 5 to 30 seconds, the pit disappears. The cause of "pitting" is that edema fluid has been translocated away from the area beneath the pressure point. The fluid simply flows to other tissue areas. Then, when the finger is removed, 5 to 30 seconds is required for the fluid to flow back into the area from which it had been compressed.

THE CONCEPT OF A "SAFETY FACTOR" BEFORE EDEMA DEVELOPS

Safety Factor Caused by the Normal Negative Interstitial Fluid Pressure. One can also see from Figure 17–1 that the interstitial fluid pressure must rise from the normal value of −7 to above zero mm. Hg before edema begins to appear. Thus, there is a safety factor of 7 mm. Hg caused by the normal negative interstitial fluid pressure before edema appears.

Safety Factor Caused by Flow of Lymph from the Tissues. Another safety factor that helps to prevent edema is increased lymph flow. When the interstitial fluid volume rises only slightly above the normal value of −7 mm. Hg, lymph flow increases greatly, which removes a portion of the extra fluid entering the interstitial spaces. And this obviously helps to prevent development of edema.

One can estimate that maximally increased lymph flow gives approximately a 6 mm. Hg safety factor, for the extra lymph flow can carry away from the tissues approximately the extra amount of fluid that is formed by a 6 mm. Hg excess in mean capillary pressure.

Safety Factor Caused by Washout of Protein from the Interstitial Spaces. In addition to removal of fluid volume from the interstitial fluid spaces, increased lymph flow also washes out most of the proteins from the interstitial fluid spaces, decreasing the colloid osmotic pressure of the interstitial fluid from the normal value of 4.5 mm. Hg down to nearly zero mm. Hg. This provides another 4 mm. Hg safety factor.

Total Safety Factor and Its Significance. Now, let us add all the above safety factors:

	mm. Hg
Negative interstitial fluid pressure	7
Lymphatic flow	6
Lymphatic washout of proteins	4
TOTAL	17

Thus we find that a total safety factor of about 17 mm. Hg is present to prevent edema. This means that the capillary pressure must rise about 17 mm. Hg above its normal value—that is, above 34 mm. Hg—before edema begins to appear. Or the plasma colloid osmotic pressure must fall from the normal level of 28 mm. Hg to below 11 mm. Hg before edema begins to appear. This explains why the normal human being does not become edematous until severe abnormalities occur in his circulatory system.

EDEMA RESULTING FROM ABNORMAL CAPILLARY DYNAMICS

From the discussions of capillary and interstitial fluid dynamics in the preceding and present chapters, it is already evident that several different abnormalities in these dynamics can increase the tissue pressure and in turn cause extracellular fluid edema. The different causes of extracellular fluid edema are:

1. Increased capillary pressure, as occurs in venous obstruction or cardiac failure.

2. Decreased plasma proteins, as occurs in persons whose kidneys lose excess amounts of proteins or, occasionally, in persons who do not eat enough protein.

3. Increased interstitial fluid protein, as occurs when the lymphatics are obstructed or the blood capillaries have become so porous that they leak proteins too rapidly for the lymphatics to return them to the circulation.

EDEMA CAUSED BY KIDNEY RETENTION OF FLUID

When the kidney fails to excrete adequate quantities of urine, and the person continues to drink normal amounts of water and eat normal amounts of salts, the total amount of extracellular fluid in the body progressively increases. This fluid is first absorbed into the blood and elevates the capillary pressure. This in turn causes most of the fluid to pass into the interstitial fluid spaces, elevating the interstitial fluid pressure there as well. As more and more fluid is retained, both the capillary pressure and the tissue pressure rise concurrently, causing an increase especially in tissue fluid but also some increase in plasma volume. Therefore, simple retention of fluids by the kidneys can also result in extensive edema.

PULMONARY EDEMA

Almost exactly the same principles apply to the development of pulmonary edema as to the development of edema in the peripheral tissues. However, there is a significant quantitative difference because the normal pulmonary capillary pressure is about 7 mm. Hg in contrast to 17 mm. Hg capillary pressure in the peripheral tissues. Therefore, the lungs have considerably more of a safety factor against edema than do the other tissues of the body, as was discussed in Chapter 15 in relation to the pulmonary circulation.

THE PRESENCE AND IMPORTANCE OF "GROUND SUBSTANCE GEL" IN THE INTERSTITIAL SPACES

Up to this point we have talked about interstitial fluid as if it were entirely in a mobile state. But, both biochemical and physical studies indicate that when a person is not edematous, almost all of the interstitial fluid is held in a *ground substance gel* that fills the spaces between the cells. This gel contains large quantities of mucopolysaccharides, the most abundant of which is hyaluronic acid.

Most of the principles of interstitial fluid hemodynamics still hold true even though most of the fluid is in a gel state. For instance, diffusion of small molecules can occur through the gel perhaps 95 per cent as effectively as in free fluid. Therefore, nutrients can diffuse from the capillaries to the cells almost as well through the gel as through free fluid.

On the other hand, the fluid in the gel cannot be readily moved from one part of the tissue to another part. That is, *the gel structure holds the interstitial fluid in place*, except for minute rivulets of free interstitial fluid that move along the surfaces of the gel between the gel and the cells.

When edema occurs, a small portion of the edema fluid is probably imbibed by the gel, but once the extracellular volume has increased beyond approximately 20 to 30 per cent, large quantities of freely moving fluid appear in the tissues. It is this freely moving fluid that is the edema fluid.

Relationship of Interstitial Fluid Gel to the Regulation of Interstitial Fluid Volume. Since approximately 17 per cent of the average tissue is composed of interstitial fluid but almost none of this is in the mobile state, one can derive the following theory for the regulation of interstitial fluid volume. The mechanism previously described for creating a negative pressure in the tissue spaces is actually a "drying" mechanism that attempts all the time to remove any excess fluid that appears in the tissues. Thus, essentially all mobile fluid is removed, and the fluid left in the interstitial spaces is that imbibed in the gel.

SPECIAL FLUID SYSTEMS

Several special fluid systems in the body perform functions peculiar to themselves. For instance, the cerebrospinal fluid supports the brain in the cranial vault, the intraocular fluid maintains distention of the eyeballs so that the optical dimensions of the eye remain constant, and the potential spaces, such as the pleural and pericardial spaces, provide lubricated chambers in which the internal organs can move. All these fluid systems have characteristics that are similar to each other and that are also similar to those of the interstitial fluid system. However, they are also sufficiently different that they require special consideration.

THE CEREBROSPINAL FLUID SYSTEM

The entire cavity enclosing the brain and spinal cord has a volume of approximately 1650 ml., and about 140 ml. of this volume is occupied by cerebrospinal fluid. This fluid, as shown in Figure 17–2, is found in the ventricles of the brain, in the cisterns around the brain, and in the subarachnoid space around both the brain and the spinal cord. All these chambers are connected with each other, and the pressure of the fluid is regulated at a constant level.

Cushioning Function of the Cerebrospinal Fluid. A major function of the cerebrospinal fluid is to cushion the brain within its solid vault. Were it not for this fluid, any blow to the head would cause the brain to be jostled around and severely damaged. However, the brain and the cerebrospinal fluid have approximately the same specific gravity, so that the brain simply floats in the fluid. Therefore, blows on the head move the entire brain simultaneously, causing no one portion of the brain to be momentarily contorted by the blow.

Formation of Cerebrospinal Fluid. Cerebrospinal fluid is formed in several different ways; most of it is formed by the *choroid plexuses* in the ventricles, but smaller portions by the blood vessels of the meningeal and ependymal linings of the cerebrospinal fluid chambers, and still smaller portions by the blood vessels of the brain and spinal cord. The choroid plexus, which is illus-

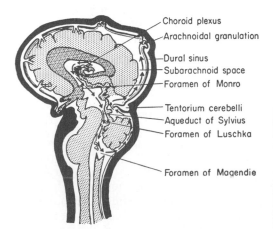

Choroid plexus
Arachnoidal granulation

Dural sinus
Subarachnoid space
Foramen of Monro

Tentorium cerebelli
Aqueduct of Sylvius
Foramen of Luschka

Foramen of Magendie

FIGURE 17–2 Pathway of cerebrospinal fluid flow. (Modified from Ranson and Clark: Anatomy of the Nervous System.)

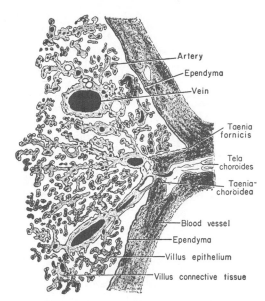

- Artery
- Ependyma
- Vein
- Taenia fornicis
- Tela choroides
- Taenia- choroidea
- Blood vessel
- Ependyma
- Villus epithelium
- Villus connective tissue

FIGURE 17-3 The choroid plexus. (Modified from Clara: Das Nervensystem des Menschen. Barth.)

trated in Figure 17–3 is a cauliflower growth of blood vessels covered by a thin coat of epithelial cells. This plexus projects into (a) the temporal horns of the lateral ventricles, (b) the posterior portions of the third ventricle, and (c) the roof of the fourth ventricle.

Cerebrospinal fluid continually exudes from the surface of the choroid plexus. This fluid is not exactly like other extra-cellular fluid but instead is a *choroid secretion.*

The probable mechanism by which the choroid plexus secretes fluid is the following: The cuboidal epithelial cells of the choroid plexus actively secrete sodium ions. The positive charges of these in turn pull negatively charged ions, particularly chloride ions, also into the cerebrospinal fluid. And all these ions cause the osmotic pressure of the ventricular fluid to be elevated to approximately 160 mm. Hg greater than that of the plasma. This osmotic force then causes large quantities of water and other dissolved substances to move through the choroidal membrane into the cerebrospinal fluid.

The rate of choroidal secretion is esti-

mated to be about 750 ml. each day, which is about five times as much as the total volume of fluid in the entire cerebrospinal cavity.

The perivascular spaces and cerebrospinal fluid. The blood vessels to the substance of the brain pass first along the surface of the brain and then penetrate inward, carrying a layer of *pia mater* with them, as shown in Figure 17–4. The pia is only loosely adherent to the vessels, so that a space, the *perivascular space*, exists between it and each vessel. Perivascular spaces follow both the arteries and the veins into the brain as far as the arterioles and venules but not to the capillaries.

The Lymphatic Function of the Perivascular Spaces. As is true elsewhere in the body, a small amount of protein leaks out of the parenchymal capillaries into the interstitial spaces of the brain, and since no true lymphatics are present in brain tissue, this protein leaves the tissues mainly through the perivascular spaces, eventually emptying into cerebrospinal fluid.

In addition to transporting fluid and proteins, the perivascular spaces also transport extraneous particulate matter from the brain into the subarachnoid space. For instance, whenever infection

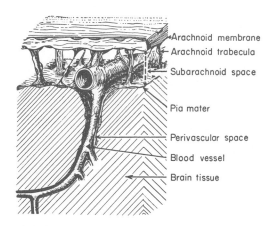

- Arachnoid membrane
- Arachnoid trabecula
- Subarachnoid space
- Pia mater
- Perivascular space
- Blood vessel
- Brain tissue

FIGURE 17-4 Drainage of the perivascular spaces into the subarachnoid space. (From Ranson and Clark: Anatomy of the Nervous System.)

occurs in the brain, dead white blood cells are carried away through the perivascular spaces.

Absorption of Cerebrospinal Fluid— The Arachnoidal Villi. Almost all the cerebrospinal fluid formed each day is reabsorbed into the blood through special structures called *arachnoidal villi* or *granulations*, which project from the subarachnoid spaces into the venous sinuses of the brain and, rarely, also into the veins of the spinal canal. The arachnoidal villi are actually arachnoidal trabeculae that protrude through the venous walls, resulting in extremely permeable areas that allow relatively free flow of cerebrospinal fluid as well as protein molecules and even small particles (less than 1 micron in size) into the blood.

Flow of Fluid in the Cerebrospinal Fluid System. The main channel of fluid flow from the choroid plexuses of the ventricles to the arachnoidal villi is illustrated in Figure 17–2. Fluid formed in the lateral ventricles passes into the third ventricle through the *foramina of Monro*, combines with that secreted in the third ventricle, and passes along the *aqueduct of Sylvius* into the fourth ventricle, where still more fluid is formed. It then passes into the *cisterna magna* through two lateral *foramina of Luschka* and a midline *foramen of Magendie*. From here it flows through the *subarachnoid spaces* upward through the subarachnoid spaces of the small *tentorial opening* around the mesencephalon. Finally, the fluid reaches the arachnoidal villi and empties into the venous sinuses.

Cerebrospinal Fluid Pressure. The normal pressure in the cerebrospinal fluid system when one is lying in a horizontal position averages 130 mm. water (10 mm. Hg), though this may be as low as 70 mm. water or as high as 180 mm. water even in the normal person. These values are considerably greater than the −7 mm. Hg pressure in the interstitial spaces elsewhere in the body.

The cerebrospinal fluid pressure is regulated by the product of, first, the *rate of fluid formation* and, second, the *resistance to absorption through the arachnoidal villi*. When either of these is increased, the pressure rises; and when either is decreased, the pressure falls.

Measurement of cerebrospinal fluid pressure. The usual procedure for measuring cerebrospinal fluid pressure is the following: First, the subject lies on his side so that the spinal fluid pressure is equal to the pressure in the cranial vault. A spinal needle is then inserted into the lumbar spinal canal below the lower end of the cord and is connected to a glass tube. The spinal fluid is allowed to rise in the tube as high as it will. If it rises to a level 100 mm. above the level of the needle, the pressure is said to be 100 mm. water pressure or, dividing this by 13.6, which is the specific gravity of mercury, about 7.5 mm. Hg pressure.

Cerebrospinal fluid pressure in pathologic conditions of the brain. Often a large *brain tumor* elevates the cerebrospinal fluid pressure to as high as 500 mm. water by decreasing the rate of absorption of fluid. And the pressure also rises considerably when *hemorrhage* or *infection* occurs in the cranial vault. In both these conditions, large numbers of cells suddenly appear in the cerebrospinal fluid, and these can almost totally block absorption through the arachnoidal villi. This sometimes elevates the cerebrospinal fluid pressure to as high as 400 to 600 mm. water.

Many babies are born with very high cerebrospinal fluid pressure which causes hydrocephalus, meaning that there is excess water in the cranial vault. Most frequently this is caused by excess formation of fluid by the choroid plexuses, but abnormal absorption or blockage of flow through one of the communicating channels is also an occasional cause.

The Blood-Cerebrospinal Fluid and Blood-Brain Barriers. Many large molecular substances hardly pass at all from the blood into the cerebrospinal fluid or into the interstitial fluids of the brain even though these same substances pass readily into the usual interstitial fluids of the body. Therefore, it is said that barriers, called the *blood-cerebrospinal* and *blood-brain barriers*, exist between the blood and the cerebrospinal fluid and brain fluids, respectively.

The cause of the low permeability of the blood-cerebrospinal barrier is the very low permeability of the secretory cells of the choroid plexus. All substances entering the cerebrospinal fluid must be transported through the secretory cells in addition to passing through the capillary membranes of the choroid plexus.

The cause of the low permeability of the blood-brain barrier is yet unknown. Some physiologists believe that many small "feet" from glial cells envelop the capillaries in the brain and prevent easy passage of ions and other substances into the brain parenchyma. However, other physiologists believe that the capillary membranes themselves are less permeable in the brain than elsewhere in the body.

Interstitial Fluid of the Brain. Research studies have suggested that the interstitial fluid in brain tissue is about 12 per cent of the tissue weight, in contrast to 17 per cent elsewhere in the body. The cause of this is lack of collagen fibers between the cells.

THE INTRAOCULAR FLUID

The eye is filled with intraocular fluid, which maintains sufficient pressure in the eyeball to keep it distended. Figure 17–5 illustrates that this fluid can be divided into two portions, the *aqueous humor*, which lies in front and to the sides of the lens, and the *vitreous humor*, which lies between the lens and the retina. The aqueous humor is a freely flowing clear fluid, whereas the vitreous humor, sometimes called the *vitreous body*, is a gelatinous mass held

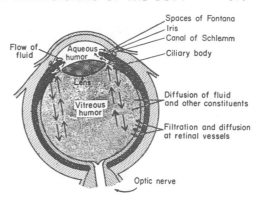

FIGURE 17–5 Formation and flow of fluid in the eye.

together by a fine fibrillar network. Substances can *diffuse* slowly in the vitreous humor, but there is little *flow* of fluid.

Aqueous humor is continually being formed and reabsorbed. The balance between formation and reabsorption of aqueous humor regulates the total volume and pressure of the intraocular fluid.

Formation of Aqueous Humor By the Ciliary Body. Aqueous humor is formed in the human eye *at an average rate of about 12 cubic millimeters each minute.* Essentially all of this is secreted by the *ciliary processes*, which are linear folds projecting from the *ciliary body* into a space called the *posterior chamber* that is located all around the eye immediately behind the lens ligaments, as seen in Figure 17–5.

Aqueous humor is formed by the ciliary processes in much the same manner that cerebrospinal fluid is formed by the choroid plexus. That is, the ciliary epithelium actively secretes sodium ions into the aqueous humor, and chloride and bicarbonate ions are pulled along with the sodium by the positive charge of the sodium. These substances increase the osmolar activity of the aqueous humor, which in turn causes osmosis of large amounts of water through the ciliary epithelium to the inside of the eye.

Outflow of Aqueous Humor From the Eye. After aqueous humor is formed by the ciliary processes, it flows, as shown in Figure 17–5, *between the ligaments of the lens*, then *through the pupil*, and finally *into the anterior chamber of the eye.* Here, the fluid flows into the *angle between the cornea and the iris* and thence through a latticework of *trabeculae*, finally into the *canal of Schlemm.* Figure 17–6 illustrates the anatomical structures at the irido-corneal angle, showing that the spaces between the trabeculae extend all the way from the anterior chamber to the canal of Schlemm. The canal of Schlemm in turn is a thin-walled vein that extends circumferentially all the way around the eye. Its endothelial membrane is so porous that even large protein molecules, as well as small particulate matter, can pass from the anterior chamber into the canal of Schlemm.

Intraocular Pressure. The average normal intraocular pressure, when measured through a small needle placed directly in the anterior chamber, is approximately 20 mm. Hg, with a range from 15 to 25 mm. Hg. The intraocular pressure of the normal eye remains almost exactly constant throughout life, illustrating that the pressure regulating mechanism is very effective. It is believed that the pressure is regulated mainly by the outflow resistance from the anterior chamber into the canal of Schlemm in the following way: A rise in pressure above normal supposedly distends the spaces between the trabeculae leading to the canal of Schlemm, thus allowing greatly increased outflow and consequently a return of the pressure to normal. On the other hand, a decrease in pressure below normal causes the spaces to narrow, causing retention of fluid until the pressure rises back to normal. One of the reasons for believing that intraocular pressure is regulated by some mechanism at the irido-corneal angle is that pathologic changes at this point almost always elevate the intraocular pressure.

Glaucoma. Glaucoma is a disease of the eye in which the intraocular pressure becomes pathologically high, sometimes rising to as high as 80 to 90 mm. Hg. Pressures rising above as little as 30 to 40 mm. Hg can cause severe pain and even blindness. As the pressure rises, the retinal artery, which enters the eyeball at the optic disc, is compressed, thus reducing the nutrition to the retina. This often results in permanent atrophy of the retina and optic nerve with consequent blindness.

Glaucoma is one of the most common causes of blindness. Very high pressures lasting only a few days can at times cause total and permanent blindness, but, in cases with only mildly elevated pressures, the blindness may develop progressively over a period of many years.

In essentially all cases of glaucoma the abnormally high pressure results from increased resistance to fluid outflow at the irido-corneal junction. In most patients, the cause of this is unknown, but in others it results from infection or trauma to the eye. In these persons large quantities of blood, white blood cells, proteins, and other matter collect in the aqueous humor and then flow into the trabecular spaces of the irido-corneal angle where they block the outflow of fluid, thereby greatly increasing the intraocular pressure.

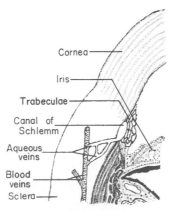

Cornea

Iris

Trabeculae

Canal of Schlemm

Aqueous veins

Blood veins

Sclera

FIGURE 17–6 Anatomy of the irido-corneal angle, showing the system for outflow of aqueous humor into the conjunctival veins.

FLUID CIRCULATION IN THE POTENTIAL SPACES OF THE BODY

Among the potential spaces of the body are the *pleural cavity*, the *pericardial cavity*, the *peritoneal cavity*, the cavity of the *tunica vaginalis* surrounding the testis, and the *synovial cavities* including both the joint cavities and the bursae. Normally, these are all empty except for a few milliliters of fluid that lubricates movement of the surfaces against each other. But in certain abnormal conditions the spaces can swell to contain tremendous quantities of fluid, which is the reason they are called "potential" spaces.

Fluid Exchange Between the Capillaries and the Potential Spaces. The membrane of a potential space does not offer significant resistance to the passage of fluid and electrolytes back and forth between the space and the surrounding interstitial fluid. Consequently, fluid leaving a capillary adjacent to the membrane, as shown in Figure 17–7, diffuses not only into the interstitial fluid but also into the potential space. Likewise, fluid can diffuse back out of the space into the interstitial fluid and thence into the capillary.

Altered capillary dynamics as a cause of increased pressure and fluid in the potential spaces. Any abnormal changes that can occur in capillary dynamics to cause extracellular edema of the tissues, as described in the preceding chapter, can also cause increased pressure and fluid in the potential spaces. Thus, *increased capillary permeability, increased capillary pressure, decreased plasma colloid osmotic pressure*, and *blockage of the lymphatics* from a potential space can all cause swelling of the space. The fluid that collects is usually called a *transudate* if it is not infected and an *exudate* if it is infected. Excessive fluid in the peritoneal space, one of the spaces most prone to develop extra fluid, is called *ascites*.

One of the most common causes of swelling in a potential space is *infection*. White blood cells and other debris caused by the infection block the lymphatics, resulting in (1) buildup of protein in the space, (2) increased colloid osmotic pressure, and (3) consequent failure of fluid reabsorption.

The Pleural Cavity. Figure 17–7 shows specifically the diffusion of fluid into and out of the pleural cavity at the parietal and the visceral pleural surfaces. This occurs in precisely the same manner as in the usual tissue spaces, except that a very porous mesenchymal *serous membrane*, the *pleura*, is interspersed between the capillaries and the pleural cavity.

Large numbers of lymphatics drain from the mediastinal and lateral surfaces of the parietal pleura, and with each expiration the intrapleural pressure rises, forcing small amounts of fluid into the lymphatics; also the respiratory movements alternately compress the lymphatic vessels, promoting continuous flow along the lymphatic channels.

Maintenance of negative pressure in the pleural cavity. The visceral pleura of the lungs continually absorbs fluid with considerable "absorptive force." This is caused by the low capillary pressure—about 5 to 10 mm. Hg—in the pulmonary system. In contrast to this low pressure, the plasma proteins exert about 28 mm. Hg colloid osmotic pressure, causing rapid osmosis of fluid

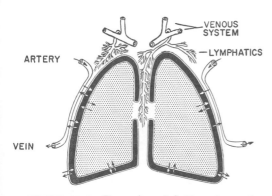

FIGURE 17–7 Dynamics of fluid exchange in the intrapleural spaces.

into the pulmonary capillaries. As a result, the pressure of the *fluid* in the intrapleural space remains negative at all times, averaging −10 to −12 mm. Hg. This negative pressure is much greater than the elastic force of the lungs (−4 mm. Hg) that tends to collapse the lungs. Therefore, it keeps the lungs expanded.

Essentially the same principles of fluid dynamics and pressures as those that pertain to the pleural cavity also apply to the *pericardial space*, the *peritoneal space*, and the *synovial cavities*.

REFERENCES

Davson, H.: Intracranial and intraocular fluids. *In* Handbook of Physiology. Baltimore, The Williams & Wilkins Co., 1960, Sec. I, Vol. III, p. 1761.

Davson, H.: The Physiology of the Cerebrospinal Fluid. Boston, Little, Brown and Company, 1967.

Hamerman, D., Barland, P., and Janis, R.: The structure and chemistry of the synovial membrane in health and disease. *In* Bittar, E., Edward, and Bittar, Neville (eds.): The Biological Basis of Medicine. New York, Academic Press, Inc., 1969, Vol. 3, p. 269.

Heisey, S. R., Held, D., and Pappenheimer, J. R.: Bulk flow and diffusion in the cerebrospinal fluid system of the goat. *Amer. J. Physiol., 203*:775, 1962.

Langham, M. E.: Aqueous humor and control of intra-ocular pressure. *Physiol. Rev., 38*:215, 1958.

Millen, J. W., and Woollam, D. H. M.: The Anatomy of the Cerebrospinal Fluid. New York, Oxford University Press, 1962.

CHAPTER 18

INTRINSIC REGULATIONS OF THE CIRCULATION

Many students are surprised to learn that the circulatory system does not require continual control by the nervous system. However, three basic intrinsic controls of the circulation are responsible for day in, day out control of the circulation: (1) The ability of the heart to respond automatically to increased input of blood; this effect is called the Frank-Starling mechanism. (2) The ability of local tissue blood vessels to adjust their blood flow in response to the needs of the surrounding tissues; this phenomenon is called *local blood flow regulation*. (3) The ability of the circulation, operating in conjunction with the kidneys and interstitial fluid system, to regulate extracellular fluid volume and blood volume, which in turn play major roles in controlling blood input to the heart and, therefore, in controlling cardiac output and arterial pressure.

The first of these three intrinsic controls, the Frank-Starling mechanism of the heart, was discussed in detail in Chapter 11; the other two mechanisms are discussed here.

ACUTE LOCAL BLOOD FLOW REGULATION IN RESPONSE TO TISSUE NEED FOR FLOW

In the majority of the tissues, the mechanism of local blood flow regulation maintains flow through each tissue almost exactly at that level required to supply the tissue's needs — no more, no less.

Table 18–1 gives relative blood flows through different organs of the body. Note the tremendous flows through the brain, liver, and kidneys despite the fact that these organs represent only a small fraction of the total body mass. Yet, in the case of the brain and liver, the metabolic rates are extremely high. Thus, the blood flow through these two organs is proportioned approximately according to their metabolic needs. In the case of the kidneys, the need is not a metabolic one, but instead it is a need to supply substances required for excretion in the urine, which is discussed later in the chapter.

TABLE 18-1   BLOOD FLOW TO DIFFERENT ORGANS
AND TISSUES UNDER BASAL CONDITIONS*

	Per cent	ml./min.
Brain	14	700
Heart	3	150
Bronchial	3	150
Kidneys	22	1100
Liver	27	1350
Portal	(21)	(1050)
Arterial	(6)	(300)
Muscle (inactive state)	15	750
Bone	5	250
Skin (cool weather)	6	300
Thyroid gland	1	50
Adrenal glands	0.5	25
Other tissues	3.5	175
Total	100.0	5000

*Based mainly on data compiled by Dr. L. A. Sapirstein.

The skeletal muscle of the body represents 35 to 40 per cent of the total body mass, and yet, in the inactive state, the blood flow through all the skeletal muscle is only 15 to 20 per cent of the total cardiac output. This accords with the fact that inactive muscle has a very low metabolic rate. Yet, when the muscles become active, their metabolic rate sometimes increases as much as 50-fold, and the blood flow in individual muscles can increase as much as 20-fold, illustrating a marked increase in blood flow in response to the increased need of the muscle for nutrients.

Local Blood Flow Regulation when Oxygen Availability Changes.   One of the most necessary of the nutrients is oxygen. Whenever the availability of oxygen to the tissues decreases, such as at high altitude, in pneumonia, in carbon monoxide poisoning, or in cyanide poisoning, the blood flow through the tissues increases markedly. Thus, Figure 18-1 shows that as the arterial oxygen saturation falls to about 25 per cent of normal, the blood flow through an isolated leg increases about three-fold; that is, the blood flow increases almost enough, but not quite, to make up for the decreased amount of oxygen in the blood, thus automatically maintaining an almost constant supply of oxygen to the tissues.

Mechanism by which oxygen concentration could regulate blood flow. Figure 18-2 illustrates what might be called a tissue unit, showing blood flowing into a tissue through a minute vessel and then entering the capillary; at the origin of the vessel is a precapillary sphincter. When one observes a local tissue under the microscope, he sees that the precapillary sphincters are normally either completely open or completely closed. And the number of precapillary sphincters that are open at any

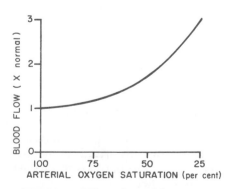

FIGURE 18-1   Effect of arterial oxygen saturation on blood flow through an isolated dog leg.

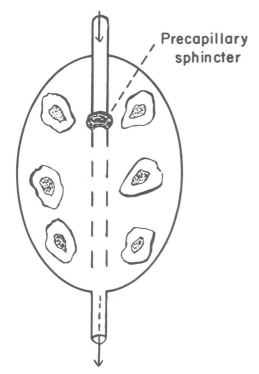

FIGURE 18-2 Diagram of a tissue unit area for explanation of local feedback control of blood flow.

given time is approximately proportional to the requirements of the tissue for nutrition, especially the requirements for oxygen. In some tissues the precapillary sphincters open and close cyclically several times per minute, the duration of the open phases being approximately proportional to the metabolic needs of the tissues. This cyclic opening and closure of the sphincter is called *vasomotion*.

Now, let us explain how oxygen concentration in the local tissue *could* regulate blood flow through the area. It is well known that smooth muscle requires oxygen to remain contracted. Therefore, we might assume that the strength of contraction of the precapillary sphincter would increase with an increase in oxygen concentration. Consequently, when the oxygen concentration in the tissue rose above a certain level, the precapillary sphincter

presumably would close and remain closed until the tissue cells consumed the excess oxygen. Then, when the oxygen concentration fell low enough, the sphincter would open once more. There is experimental evidence that this mechanism actually does exist: When a micro-"oxygen" electrode, which measures oxygen concentration, is inserted into a tissue adjacent to a precapillary sphincter, the recorded oxygen concentration goes up and down as the sphincter opens and closes, and the phase relationships are in accord with the theory just explained.

Thus, the precapillary sphincter mechanism illustrated in Figure 18-2 could easily regulate local blood flow in such a way that the tissue oxygen concentration would remain almost exactly constant, whatever the requirements of the tissue for oxygen.

The Vasodilator Theory for Local Blood Flow Regulation. Another theory to explain regulation of local blood flow is the vasodilator theory. According to this theory, the greater the rate of metabolism or the less the availability of nutrients to a tissue, the greater becomes the rate of formation of a *vasodilator substance*. The vasodilator substance then supposedly diffuses back to the precapillary sphincters, meta-arterioles, and arterioles to cause dilatation. Some of the different vasodilator substances that have been suggested are carbon dioxide, lactic acid, adenosine, adenosine phosphate compounds, histamine, potassium ions, and hydrogen ions. Thus far, none of these vasodilator substances has been proved to be formed in large enough quantities to cause local regulation, though many physiologists place greatest faith in *adenosine* as a possible dilator substance.

Some of the vasodilator theories assume that the vasodilator substance is released from the tissue in response to oxygen lack. For instance, it has been demonstrated that oxygen lack can cause both lactic acid and adenosine to be released from the tissues, and these are substances that can cause vasodilatation and therefore could be responsible, or partially responsible, for the local blood flow regulation mechanism.

*SPECIAL EXAMPLES OF ACUTE
LOCAL BLOOD FLOW
REGULATION*

Some special examples of acute local blood flow regulation are the following:

1. When the arterial pressure falls, the resultant decrease in blood flow causes reduced delivery of nutrients to the tissues; this in turn dilates the blood vessels. As a consequence, the blood flow returns essentially to normal despite the low arterial pressure. This effect is called *autoregulation of blood flow.*

2. When a blood vessel is completely occluded for a few seconds to a few minutes, upon release of the vessel the blood flow increases to as much as four to five times normal; the excess flow lasts long enough to make up for the tissue nutritional deficit that has occurred during the occlusion. This effect is called *reactive hyperemia.*

3. Local blood flow regulation also occurs in the kidneys but not in response to changes in oxygen. Instead, an increase in electrolytes or in end-products of protein metabolism in the blood causes increased blood flow through the

kidneys; the kidneys in turn eliminate these excess substances. This is called *renal autoregulation*, and it will be discussed in Chapter 25.

4. Local blood flow regulation occurs in the brain in response to carbon dioxide in the tissue fluids; that is, excess carbon dioxide produced in the tissues causes local vasodilatation.

It is clear, therefore, that not all local blood flow regulation occurs in response to oxygen changes in the tissues, for other factors such as electrolytes acting in the kidneys and carbon dioxide acting in the brain are also local factors that control blood flow in accord with the local tissue's respective function.

LONG-TERM LOCAL BLOOD FLOW REGULATION

Over a period of hours, days, and weeks a long-term type of local blood flow regulation develops in addition to the acute regulation, and this long-term regulation gives far more complete regulation than does the acute mechanism. For instance, Figure 18–3 compares the acute and long-term mechanisms for local blood flow regulation. This shows that acute changes in arterial pressure between pressure levels of 75 and 180 mm. Hg cause changes in blood flow about half as great as would occur

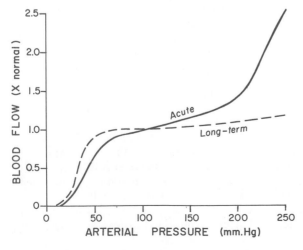

FIGURE 18–3 Effect on blood flow through a muscle of increasing arterial pressure. The solid curve shows the effect if the arterial pressure is raised over a period of a few minutes. The dashed curve shows the effect if the arterial pressure is raised extremely slowly over a period of many weeks.

if no local blood flow regulation at all occurred. On the other hand, if the blood vessels are given weeks or months to adapt to the changes in pressure, the blood flow returns almost exactly back to its normal level despite pressure levels as low as 60 mm. Hg or as high as 250 mm. Hg; this effect is illustrated by the dashed curve in the figure. Thus, one can see that over a long period of time, the long-term mechanism is extremely valuable in adjusting blood flow in each tissue to the needs of that tissue.

Change in Tissue Vascularity as the Mechanism of Long-Term Regulation. The mechanism of long-term regulation is almost certainly a change in the degree of vascularity of the tissues. That is, if the arterial pressure falls to 60 mm. Hg and remains at this level for many weeks, the number and size of vessels in the tissue increase; if the pressure then rises to a very high level, the number and size of vessels decrease. Likewise, if the metabolism in a given tissue becomes elevated for a prolonged period of time, vascularity increases; or if the metabolism is decreased, vascularity decreases.

Thus, there is continual day-by-day reconstruction of the tissue vasculature to meet the needs of the tissues. This reconstruction occurs very rapidly in extremely young animals. It also occurs rapidly in new growth of tissue, such as scar tissue or new growth of glandular tissue; but on the other hand, it occurs very slowly in old, well established tissues. Therefore, the time required for long-term regulation to take place may be only a few days in the newborn or as long as months or even years in the elderly person. Furthermore, the final degree of response is much greater in younger tissues than in older, so that in the newborn the vascularity will adjust to match almost exactly the needs of the tissue for blood flow; whereas in older tissues, vascularity frequently lags far behind the needs of the tissues.

Role of oxygen in long-term regulation. The probable stimulus for in-creased or decreased vascularity in many if not most instances is need of the tissue for oxygen. The reason for believing this is that hypoxia is known to cause increased vascularity, and hyperoxia to cause decreased vascularity. This effect is demonstrated in newborn infants who are put into an oxygen tent for therapeutic purposes. The excess oxygen causes almost immediate cessation of new vascular growth in the retina of the eye and even degeneration of some of the capillaries that have already formed. Then when the baby is taken out of the oxygen, there is an explosive overgrowth of new vessels to make up for the sudden decrease in available oxygen; indeed, there is so much overgrowth that the vessels grow into the vitreous humor and eventually cause blindness.

Significance of Long-Term Local Regulation — The Metabolic Mass to Tissue Vascularity Proportionality. From these discussions it should already be apparent to the student that there is a built-in mechanism in most tissues to keep the degree of vascularity of the tissue almost exactly that required to supply the metabolic needs of the tissue. Thus, one can state as a general rule that the vascularity of most tissues of the body is directly proportional to the local metabolism times the tissue mass. If ever this proportionality constant becomes abnormal, the long-term local regulatory mechanism automatically readjusts the degree of vascularity over a period of weeks or months. In young persons this degree of readjustment is usually very exact; in old persons it is only partial.

ROLES OF LOCAL BLOOD FLOW REGULATION AND OF THE HEART IN INTRINSIC CONTROL OF CARDIAC OUTPUT

It was pointed out in Chapter 11 that the heart is capable of pumping blood indefinitely without any external control. That is, the rhythmical beating process of the heart will continue indefinitely, and because of the Frank-Starling mechanism of the heart, the heart will pump whatever amount of blood flows into its atria up to its physiologic limit. In other words, the heart is an automatic pump, adjusting itself to its input load. As we shall see in the following chapter, this intrinsic ability of the heart can be enhanced by nervous or humoral controls; but in most resting conditions of bodily function these controls are not necessary, since the ability of the heart to adapt to increasing input load is far more than enough to supply the needs of the circulation. On the other hand, during conditions of severe circulatory stress, such as during heavy exercise, the extrinsic controls are also needed.

Therefore, in essence, cardiac output is controlled primarily by the sum of all the local regulatory processes in all the peripheral vasculature of the body. To state this in still another way, cardiac output is controlled by the tissues themselves in proportion to their needs for flow.

The more one studies the circulation, the more he recognizes the beauty of this mechanism, because it adjusts the blood flow through the tissues to that amount required for proper servicing of the tissues — no more, no less; at the same time it also adjusts the cardiac output to that amount required by the whole body, no more, no less.

INTRINSIC EFFECTS OF BLOOD AND EXTRACELLULAR FLUID VOLUMES ON CIRCULATORY CONTROL

Though the intrinsic mechanism just discussed for control of cardiac output based on local tissue regulation of blood flow and intrinsic adaptation of the heart to increasing input loads is a very beautiful one, unfortunately it will not function properly without two other types of controls: (1) control of the blood volume so that there is always blood to be pumped and (2) control of the arterial pressure so that there is a pressure head to force blood through the peripheral tissues when the local regulatory mechanisms open or close the vessels. These two controls are provided primarily by the following basic negative feedback mechanism involving the circulatory system and the kidneys:

1. A person ingests daily both water and salt, and these tend continually to increase his extracellular fluid volume and blood volume.

2. The increase in blood volume increases the availability of blood to be pumped by the heart, thus increasing the cardiac output and arterial pressure.

3. The increase in arterial pressure causes the kidneys to excrete water and salt, thus returning the fluid volumes and arterial pressures back toward their normal levels.

4. When the arterial pressure has become high enough to make the kidneys excrete water and salt at the same rate that water and salt is ingested each day, extracellular fluid volume, blood volume, and arterial pressure all become stabilized. The stabilized level of blood volume is that amount of blood needed to keep the circulatory system properly filled, and the pressure level set in this manner is that required to cause adequate blood flow through the peripheral tissues.

Effect of Arterial Pressure on Urinary Output of Extracellular Fluid. One of the most important steps in the fluid volume feedback mechanism is the effect of arterial pressure on urinary output. This effect is illustrated in Figure 18–4, which shows that at normal arterial pressure of 100 mm. Hg, the two kidneys of the normal human being excrete approximately 1 ml. of extracellular fluid each minute. This fluid contains both water and salt, the principal constituents of extracellular fluid. Figure 18–4 also illustrates that increasing the arterial pressure up to 200 mm. Hg increases the excretion of extracellular fluid (both

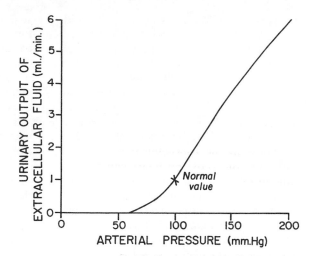

FIGURE 18-4 Effect of arterial pressure (AP) on urinary output of extracellular fluid [dE/dt$_{(o)}$].

water and salt) to about six times normal, whereas decreasing the arterial pressure below its normal value greatly diminishes urinary output, urinary output becoming zero when the arterial pressure falls to about 60 mm. Hg. The details of these effects of arterial pressure on urinary output will be discussed in Chapter 24.

Thus, the rate at which the body loses extracellular fluid is highly dependent upon the arterial pressure, the rate of loss increasing with pressure. Therefore, one can already begin to understand how the kidneys can regulate both the arterial pressure and the fluid volumes.

Effect of Extracellular Fluid Volume and Blood Volume on Cardiac Output. If the kidneys excrete excess fluid from the body, the extracellular fluid volume decreases; some of this extracellular fluid comes from the plasma portion of the blood, thus also decreasing the blood volume. One can intuitively understand that a decrease in blood volume will decrease the degree of "filling" of the circulatory system and, therefore, will also decrease the ease with which blood flows from the peripheral vessels back to the heart. Therefore, decreased extracellular fluid and blood volumes decrease cardiac output by "decreasing venous return."

Conversely, if something happens to the kidneys to cause retention of water and salt in the body, the extracellular fluid volume increases, and so does the blood volume up to a certain level, at which fluid begins to leak out of the capillaries. These increased volumes now cause exactly the opposite effect, markedly increasing venous return from the peripheral vessels and, thereby, also increasing cardiac output.

Effect of Cardiac Output on Arterial Pressure. The arterial pressure is equal to cardiac output times total peripheral resistance. Therefore, if the total peripheral resistance remains constant, an increase in cardiac output will cause a proportional increase in arterial pressure.

The Fluid Volume–Cardiac Output– Arterial Pressure Feedback Mechanism. The student will immediately recognize that in the preceding few paragraphs we have described a feedback mechanism that automatically regulates many aspects of circulatory function. For instance, if the arterial pressure suddenly rises too high, the kidneys will excrete extra quantities of fluid, the blood volume will decrease, the cardiac output will decrease, and the arterial pressure will return to its normal level. Or, conversely, a decrease in arterial pressure will stop the output of fluid by the kidneys, and the arterial pressure will eventually rise back to its normal level.

Another example of function of this mechanism is the following: If several liters of saline solution are injected into the circulation, the extracellular fluid volume and blood volume will become far too great. This increases the cardiac output and arterial pressure, causing the kidneys to excrete large volumes of fluid. Within a few hours, the fluid volumes, the cardiac output, and the arterial pressure all return to their normal steady state values.

EFFECT OF LOCAL BLOOD FLOW REGULATION ON THE BASIC FLUID PRESSURE CONTROL SYSTEM

There are times when the mechanisms for local regulation of blood flow through isolated tissues conflict with the basic fluid pressure control system just described. For example, if any factor causes an increase in blood volume and therefore also an increase in cardiac output, too much blood will then flow through the tissues to supply the nutrition required. According to the local flow regulation concept, the local tissue vessels should then constrict progressively over a period of days, weeks, or months until finally the blood flow through the tissues would return near to normal. Therefore, eventually, despite the elevated blood volume, the cardiac output would be returned almost to normal. But, in the meantime, the total peripheral resistance would increase markedly, and this in turn would keep the arterial pressure at its elevated level even though the cardiac output returns essentially to its normal values.

Dissociation of Cardiac Output Regulation from Arterial Pressure Regulation in the Chronic State. Under acute conditions, cardiac output and arterial pressure tend to increase and decrease simultaneously. For instance, infusion of excess blood into an animal increases both the cardiac output and the arterial pressure, or, conversely, loss of blood causes the opposite effects on both of these.

However, under long-term conditions, regulation of cardiac output and arterial pressure are mainly dissociated from each other. The cause of this difference is the long-term effect of local blood flow regulation as just explained. That is, despite an increase in arterial pressure caused by abnormal retention of fluid in the body, the local blood flow regulating mechanism slowly constricts the local blood vessels until blood flow throughout the body returns to normal. Therefore, the cardiac output also returns to normal even though the arterial pressure still remains elevated.

This dissociation of cardiac output regulation from arterial pressure regulation is an extremely important effect because (1) it allows the blood flow to the tissues to be adjusted to that amount required to supply the nutritional needs of the tissues, with neither excess flow nor too little flow, and (2) it allows the arterial pressure to be regulated to that level required by the kidneys to excrete the necessary amount of water and salt from the body; that is, to excrete the same amount of water and salt each day as is ingested.

To state this another way, it is the local blood flow regulation mechanism that in the long run sets the cardiac output; and it is the balance between intake and output of fluid, operating in conjunction with the kidneys, that sets the arterial pressure. Furthermore, the local blood flow regulation mechanism allows eventual dissociation between cardiac output regulation and arterial pressure regulation, the cardiac output being adjusted to that value required to supply the needs of the tissues and the arterial pressure being adjusted to that value required to give adequate function of the kidneys as well as to maintain an adequate pressure head in the arteries to keep blood flowing through the system.

SUMMARY

The purpose of this chapter has been to point out that most of the really important features of circulatory control are vested in intrinsic functions of the circulatory elements themselves and that the circulatory system can function quite well without superimposed nervous or humoral controls. This does not mean that these other superimposed controls are not important, for they are; it is merely that the intrinsic controls are capable of performing most of the basic day in and day out controls of the circulation.

The intrinsic controls of the circulation that have been presented in this chapter are the following:

1. The phenomenon of local blood flow regulation controls blood flow in individual tissues in accordance with the needs of the tissues for flow. Acute local regulation gives a small degree of control that takes place in a matter of minutes. Long-term regulation gives an intense degree of control, but it is slow to develop fully, requiring days, weeks, or months.

2. The intrinsic controls of the heart, coupled with the local blood flow regulation mechanism, are basic to the control of cardiac output. That is, the heart automatically pumps the blood that comes to it from the tissues, and the amount of blood that comes from the tissues is automatically set by the sum of local regulatory mechanisms in the tissues. Therefore, except when the heart becomes overloaded, long-term regulation of cardiac output is controlled primarily by the local blood flow regulatory mechanism of the tissues.

3. The fluid volume control system allows control of the extracellular fluid volume, blood volume, and arterial pressure. Control of these factors is essential to the operation of the local blood flow regulatory mechanism, for without blood volume no blood at all can circulate, and without pressure blood will not flow through the tissues when local tissue vasodilatation occurs.

4. The local blood flow regulation mechanism allows eventual dissociation between arterial pressure regulation and cardiac output regulation because the long-term local regulation mechanism, once it has had time to take effect, sets the level of cardiac output; while on the other hand, the fluid feedback system and the kidneys set the arterial pressure at that level required for adequate urinary output of water and salt.

5. The fluid feedback control system and the long-term local blood flow regulation mechanism are both slow to come to equilibrium, requiring days in the case of the fluid mechanism and days, weeks, or months in the case of the long-term local flow regulation mechanism. It is because of the slowness of action of these two mechanisms that the circulatory system needs additional rapidly acting controls — the nervous and humoral controls — to handle acute emergencies. The next chapter will deal with these rapidly acting controls.

6. The three factors that play the dominant role in the overall intrinsic regulation of the circulation, therefore, are: (a) the intrinsic ability of the heart to pump blood automatically into the arteries whenever blood appears at the input side of the heart (the Frank-Starling mechanism), (b) the long-term local blood flow regulation mechanism, and (c) the effect of arterial pressure on output of extracellular fluid from the kidneys into the urine.

If the student is not by now thoroughly familiar with the above concepts, he should restudy each of them in detail before going to subsequent chapters, because the future discussions are all highly dependent upon the basic principles presented in this chapter.

REFERENCES

Donald, D. E., Rowlands, D. J., and Ferguson, D. A.: Similarity of blood flow in the normal and the sympathectomized dog hind limb during graded exercise. *Circ. Res.,* 26:185, 1970.

Duling, B. R., and Berne, R. M.: Propagated vasodilation in the microcirculation of the hamster cheek pouch. *Circ. Res.,* 26:163, 1970.

Guyton, A. C., and Coleman, T. G.: Long-term regulation of the circulation interrelationships with body fluid volumes. *In* Reeve, B., and Guyton, A. C. (eds.): Physical Bases of Circulatory Transport: Regulation and Exchange. Philadelphia, W. B. Saunders Company, 1967, p. 179.

Guyton, A. C., Ross, J. M., Carrier, O., Jr., and Walker, J. R.: Evidence for tissue oxygen demand as the major factor causing autoregulation. *Circ. Res.,* 14:60, 1964.

Harrison, R. G.: Functional aspects of the vascularization of tissues. *Scient. Basis Med. Ann. Rev., 76,* 1963.

Herd, J. A.: Overall regulation of the circulation. *Ann. Rev. Physiol.,* 32:289, 1970.

Lamport, H., and Baez, S.: Physical properties of small arterial vessels. *Physiol. Rev., 42*(Suppl. 5):328, 1962.

Sagawa, K.: Overall circulatory regulation. *Ann. Rev. Physiol., 31*:295, 1969.

CHAPTER 19

NERVOUS AND HUMORAL REGULATION OF THE CIRCULATION— THE CARDIOVASCULAR REFLEXES

Superimposed onto the intrinsic regulations of the circulation discussed in the previous chapter are two additional types of regulation: (1) *nervous* and (2) *humoral*. These regulations are not necessary for basic function of the circulation, but they do provide greatly increased effectiveness of control.

NERVOUS REGULATION OF THE CIRCULATION

There are two very important features of nervous regulation of the circulation: First, nervous regulation can function extremely rapidly, some of the nervous reflexes beginning to take effect within two to three seconds and reaching full development within 20 to 30 seconds. Second, the nervous system provides a means for controlling large parts of the circulation simultaneously, irrespective of the needs of the individual tissues. For instance, when it is important to raise the arterial pressure temporarily, the nervous system can arbitrarily cut off, or at least greatly decrease, blood flow to major segments of the circulation despite the fact that the local blood flow regulatory mechanisms oppose this.

On the other hand, nervous control of the circulation usually cannot last for long periods of time because of "adaptation" of parts of the nervous control system and also because the local blood flow regulatory mechanisms eventually become powerfully activated and either override or partially override the nervous controls. Nevertheless, for minutes and sometimes even for hours the nervous system can dominate circulatory control.

AUTONOMIC REGULATION OF THE CIRCULATION

The autonomic nervous system will be discussed in detail in Chapter 40. However, it is so important to the regulation of the circulation that its specific anatomical and functional characteristics relating to the circulation deserve special attention.

By far the most important part of the autonomic nervous system for regulation of the circulation is the *sympathetic system*. The parasympathetic nervous system is important only in its regulation of heart function, as we shall see later in the chapter.

The Sympathetic Nervous System. Figure 19–1 illustrates the anatomy of sympathetic nervous control of the circulation. Sympathetic vasomotor nerve fibers leave the spinal cord through all the thoracic and the first one to two lumbar spinal nerves. These pass into the sympathetic chain and thence by two routes to the blood vessels throughout the body: (1) through the *peripheral sympathetic nerves* and (2) through the *spinal nerves*. Almost all vessels of the body except the capillaries are supplied with sympathetic nerve fibers.

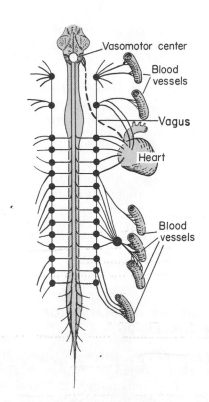

FIGUE 19–1 The vasomotor center and its control of the circulatory system through the sympathetic and vagus nerves.

In addition to fibers supplying the blood vessels, sympathetic fibers spread also to the heart from the upper four thoracic sympathetic ganglia, from the stellate ganglion, and from the cervical ganglia. This innervation was discussed in Chapter 11; and it will be recalled that sympathetic stimulation markedly increases the activity of the heart, increasing the heart rate and enhancing its strength of pumping.

Parasympathetic Control of the Heart. The parasympathetic nervous system plays only a minor role in regulation of the circulation. Its only important effect is control of heart rate, though it probably also has a slight but relatively unimportant influence on the control of cardiac contractility. Parasympathetic nerves pass to the heart in the vagus nerve, as illustrated in Figure 19–1. Principally, parasympathetic stimulation causes a marked decrease in heart rate, and inhibition of the parasympathetics causes an increase in heart rate.

The Sympathetic Vasoconstrictor System and Its Control by the Central Nervous System

The sympathetic nerves carry both vasoconstrictor and vasodilator fibers, but by far the most important of these are the *sympathetic vasoconstrictor* fibers. Sympathetic vasoconstrictor fibers are distributed to essentially all segments of the circulation. However, this distribution is greater in some tissues than in others. It is rather poor in both skeletal and cardiac muscle and in the brain, while it is powerful in the kidneys, the gut, the spleen, and the skin.

The Vasomotor Center and Its Control of the Vasoconstrictor System. Located bilaterally in the reticular substance of the lower third of the pons and upper two thirds of the medulla, as illustrated in Figure 19–2, is an area called the *vasomotor center*. The lateral portions of the vasomotor center stimulate the vasoconstrictor fibers throughout the body. These fibers have an inherent

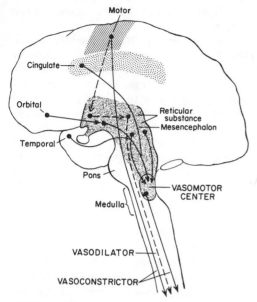

FIGURE 19-2 Areas of the brain that play important roles in the nervous regulation of the circulation.

tendency to transmit nerve impulses all the time, thereby maintaining even normally a slow rate of firing in essentially all vasoconstrictor nerve fibers of the body at a rate of about one-half to two impulses per second. This continual firing is called *sympathetic vasoconstrictor tone.* These impulses maintain a partial state of contraction in the blood vessels, a state called *vasomotor tone.*

The medial portion of the vasomotor center does not participate in excitation of the vasoconstrictor fibers. Instead, stimulation of this medial area transmits *inhibitory* impulses into the lateral parts of the vasomotor center, thereby *decreasing the degree of sympathetic vasoconstrictor tone* and consequently allowing dilatation of the blood vessels. Thus, the vasomotor center is composed of two parts, a bilateral *excitatory part* that can excite the vasoconstrictor fibers and cause vascular constriction and a medial *inhibitory part* that can inhibit vasoconstriction, thus allowing vasodilatation.

Control of heart activity by the vasomotor center. At the same time that the vasomotor center is controlling the degree of vascular constriction, it also regulates heart activity. The lateral portions of the vasomotor center transmit excitatory impulses through the sympathetic nerve fibers to the heart, while the medial portion of the vasomotor center, which lies in immediate apposition to the *dorsal motor nucleus of the vagus nerve,* transmits impulses through the parasympathetic fibers of the vagus nerve to the heart to decrease heart rate. Therefore, the activity of the heart ordinarily increases at the same time that vasoconstriction occurs throughout the body, and ordinarily decreases at the same time that vasoconstriction is inhibited. However, these interrelationships are not invariable, because some nerve impulses passing down the vagus nerves to the heart bypass the vasomotor center.

Control of the vasomotor center by higher nervous centers. Large numbers of areas throughout the *reticular substance of the pons, the mesencephalon, the hypothalamus,* and *the cerebral cortex* can either excite or inhibit the vasomotor center. Some of these areas are shown in Figure 19-2, and some of their functions will be discussed later.

The Sympathetic Vasodilator System and Its Control by the Central Nervous System

The sympathetic nerves to skeletal muscles carry sympathetic vasodilator fibers as well as constrictor fibers. These fibers release *acetylcholine* at their endings, which acts on the smooth muscle of the blood vessels to cause vasodilatation in contrast to the vasoconstrictor effect of norepinephrine.

The pathways for central nervous system control of the vasodilator system are illustrated by the dashed lines in Figure 19-2. The principal area of the brain controlling this system is the *anterior hypothalamus,* which transmits impulses to a relay station in the *subcollicular region* of the mesencephalon. From here impulses are transmitted down the cord to the *sympathetic preganglionic neurons* in the lateral horns of the cord. These then transmit vasodilator impulses to the blood vessels of the muscles.

Note also in Figure 19–2 that stimulation of the *motor cortex excites* the *sympathetic vasodilator system* by transmitting impulses downward into the hypothalamus.

Importance of the Sympathetic Vasodilator System. It is doubtful that the sympathetic vasodilator system plays a very important role in the control of the circulation. Yet, it is possible, if not probable, that at the onset of exercise the sympathetic vasodilator system causes initial vasodilatation in the skeletal muscles to allow an *anticipatory increase in blood flow* even before the muscles require increased nutrients.

"PATTERNS" OF CIRCULATORY RESPONSES ELICITED BY DIFFERENT CENTRAL NERVOUS SYSTEM CENTERS

Stimulation of the Vasomotor Center— The Mass Action Effect. Diffuse stimulation of the lateral portions of the vasomotor center causes widespread activation of the vasoconstrictor fibers throughout the body, while stimulation of the medial portion of the vasomotor center causes widespread *inhibition of vasoconstriction*. In many conditions the entire vasomotor center acts as a unit, stimulating all vasoconstrictors throughout the body, stimulating the heart, and stimulating the adrenal medullae to secrete norepinephrine and epinephrine that circulate in the blood to excite the circulation still further. The results of this "mass action" are three-fold: First, the peripheral resistance increases in all parts of the circulation, thereby elevating the arterial pressure. Second, the capacity vessels, particularly the veins, are excited at the same time, greatly decreasing their capacity; this forces increased quantities of blood into the heart, thereby increasing the cardiac output. Third, the heart is simultaneously stimulated so

that it can handle the increased cardiac output.

Stimulation of the Hypothalamus— The "Alarm" Pattern. Diffuse stimulation of the hypothalamus activates the vasodilator system to the muscles, thereby increasing the blood flow through the muscles; at the same time it causes intense vasoconstriction throughout the remainder of the body and an intense increase in heart activity. The arterial pressure rises, the cardiac output increases, the heart rate increases, and the circulation is ready to supply nutrients to the muscles if there be need. Also, impulses are transmitted simultaneously throughout the central nervous system to cause a state of generalized excitement and attentiveness, these often increasing to such a pitch that the overall pattern of the reaction is that of *alarm*. This pattern seems to have the purposeful effect of preparing the animal or person to perform on a second's notice whatever activity is required.

The "Motor" Pattern of Circulatory Stimulation. When the motor cortex transmits signals to the skeletal muscle to cause motor activities, it also sends signals to the circulatory system to cause the following effects: First, it excites the alarm system just described, and this generally activates the heart, causes muscular vasodilatation, and increases arterial pressure, making the circulatory system ready to supply increased blood flow to the muscles. Second, impulses pass directly from the motor cortex to the sympathetic neurons of the spinal cord. These enhance the vasoconstriction in the nonmuscular parts of the body, thereby raising the arterial pressure still more and helping to increase blood flow.

Emotional fainting. A particularly interesting circulatory reaction occurs in persons who faint because of intense emotional experiences. In this condition, the muscle cholinergic vasodilator system becomes powerfully activated, so that blood flow through the muscles increases several-fold. This decreases the arterial pressure, which in turn reduces the blood flow to the brain and causes the person to lose consciousness. It is probable, therefore, that emotional fainting results from powerful stimulation of the anterior hypothalamic vasodilator center.

EFFECT OF ISCHEMIA AND CARBON DIOXIDE ON THE ACTIVITY OF THE VASOMOTOR CENTER — THE CNS ISCHEMIC RESPONSE

Ischemia means lack of sufficient blood flow to maintain normal metabolic function of a tissue. When ischemia of the vasomotor center occurs as a result of occluding the arteries to the brain, systemic arterial pressure rises markedly. It is believed that this is caused mainly by failure of the slowly flowing blood to carry carbon dioxide away from the vasomotor center; the local concentration of carbon dioxide supposedly increases greatly and elevates the arterial pressure. This arterial pressure elevation in response to cerebral ischemia is known as the *central nervous system ischemic response* or simply CNS ischemic response.

The magnitude of the ischemic effect on vasomotor activity is almost identical to that caused by excess carbon dioxide in the blood; both elevate the arterial pressure to as high as 270 mm. Hg. *The degree of sympathetic vasoconstriction caused by intense cerebral ischemia is often so great that some of the peripheral vessels become totally or almost totally occluded.* The kidneys, for instance, will entirely cease their production of urine because of arteriolar constriction in response to the sympathetic discharge. Therefore, *the CNS ischemic response is one of the most powerful of all the activators of the sympathetic vasoconstrictor system.*

Importance of the CNS Ischemic Response as a Regulator of Arterial Pressure. The CNS ischemic response does not

become very active until the arterial pressure falls far below normal, down to levels of 20 to 40 mm. Hg. Therefore, it is not one of the mechanisms for regulating normal arterial pressure. Instead, it operates principally as an *emergency arterial pressure control system that acts rapidly and extremely powerfully to prevent further decrease in arterial pressure whenever blood flow to the brain decreases dangerously close to the lethal level.*

THE ARTERIAL BARORECEPTOR CONTROL SYSTEM — BARORECEPTOR REFLEXES

The nervous system also controls many circulatory functions by means of circulatory reflexes. By far the most important and best known of the circulatory reflexes is the *baroreceptor reflex*. Basically, this reflex is initiated by pressure receptors, called either *baroreceptors* or *pressoreceptors*, located in the walls of the large systemic arteries. A rise in pressure causes the baroreceptors to transmit signals into the central nervous system, and reflex signals are sent back to the circulation to reduce arterial pressure toward the normal level.

Physiologic Anatomy of the Baroreceptors. Baroreceptors are spray-type nerve endings that are stimulated when stretched. As illustrated in Figure 19–3, they are extremely abundant, in (1) the walls of the internal carotid arteries slightly above the carotid bifurcations, areas known as the *carotid sinuses*, and (2) the walls of the aortic arch.

Figure 19–3 also shows that impulses are transmitted from each carotid sinus through the very small *Hering's nerve* to the glossopharyngeal nerve and thence to the medulla. Impulses from the arch of the aorta are transmitted through the vagus nerves to the medulla.

Response of the Baroreceptors to Pressure. The baroreceptors are not stimulated at all by pressures between 0 and 50 mm. Hg, but above 50 mm. Hg they

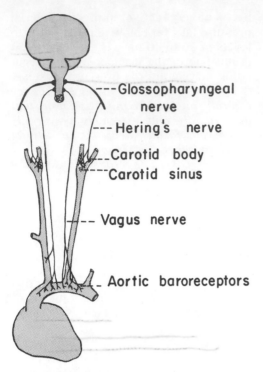

FIGURE 19-3 The baroreceptor system.

respond more and more rapidly and reach a maximum at about 200 mm. Hg.

It is especially important that the increase in number of impulses for each unit change in arterial pressure is greatest at about the level of the normal mean arterial pressure. This means that the baroreceptors respond most markedly to changes in arterial pressure just at the level where the response needs to be most marked; that is, in the normal operating range of arterial pressure even a slight change in pressure at this level causes strong sympathetic reflexes to readjust the arterial pressure when it becomes even slightly abnormal.

The Reflex Initiated by the Baroreceptors. The baroreceptor impulses *inhibit the vasomotor center* of the medulla and *excite* the vagal center. The net effects are (1) *vasodilatation* throughout the peripheral circulatory system and (2) *decreased cardiac rate* and *strength of contraction*. Therefore, excitation of the baroreceptors by pressure in the arteries reflexly *causes the arterial pres-*

sure to decrease. Conversely, low pressure has opposite effects, reflexly causing the pressure to rise back toward normal.

Figure 19–4 illustrates a typical reflex change in arterial pressure caused by clamping the common carotid arteries. This procedure reduces the carotid sinus pressure; as a result, the baroreceptors become inactive and lose their inhibitory effect on the vasomotor center. The vasomotor center then becomes much more active than usual, causing the arterial pressure to rise and to remain elevated as long as the carotids are clamped. Removal of the clamps allows the pressure to fall immediately to slightly below normal as a momentary overcompensation and then to return to normal in another minute or so.

It is obvious that the baroreceptor system opposes increases and decreases in arterial pressure. For this reason it is often called a *buffer system*, and the nerves from the baroreceptors are called *buffer nerves*.

The baroreceptor control system is probably of no importance in long-term regulation of arterial pressure for a very simple reason: The baroreceptors themselves eventually adapt to whatever

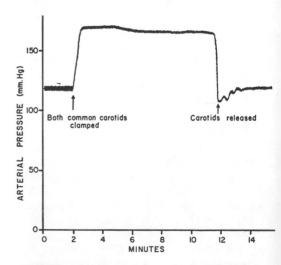

FIGURE 19–4 Typical carotid sinus reflex effect on arterial pressure caused by clamping both common carotids.

pressure level they are exposed to. That is, if the pressure rises from the normal value of 100 mm. Hg up to 200 mm. Hg, extreme numbers of baroreceptor impulses are at first transmitted. During the next few seconds, the rate of firing diminishes considerably; then it diminishes very slowly during the next two to four days, at the end of which time the rate will have returned essentially to the normal level despite the fact that the arterial pressure remains 200 mm. Hg.

This adaptation of the baroreceptors obviously prevents the baroreceptor reflex from continuing to function as a control system for longer than a few days at a time. Therefore, prolonged regulation of arterial pressure requires other control systems, principally the intrinsic control systems discussed in the previous chapter.

LESS IMPORTANT CIRCULATORY REFLEXES

Many other circulatory reflexes have been described, though none of them seems to be nearly as powerful or as important as the baroreceptor reflex. Some of these reflexes are the following:

Chemoreceptor Reflexes — Effect of Oxygen Lack on Arterial Pressure. Located in the bifurcations of the carotid arteries and also along the arch of the aorta are several small bodies, 1 to 2 mm. in size, called respectively *carotid* and *aortic bodies*. These contain specialized sensory *receptors sensitive to oxygen lack*, called *chemoreceptors*. Whenever the oxygen concentration in the arterial blood falls too low, the chemoreceptors become excited, and impulses are transmitted into the vasomotor center to *excite* the vasomotor center, thus reflexly elevating the arterial pressure. However, this reflex is not powerful at all in the normal arterial pressure range. It does exert reasonable feedback effects on arterial pressure when the pressure is in the range of 40 to 80 mm. Hg.

The chemoreceptors will be discussed in much more detail in Chapter 29 in relation to respiratory control.

Atrial Reflexes. Increased pressure in the atria can cause reflex increases in heart rate, arterial pressure, and urinary output. The afferent pathway for the reflexes is through the vagus nerves, whereas the efferent pathways are the usual vasomotor pathways. The effect of elevated right atrial pressure to cause increased heart rate is called the *Bainbridge reflex*, but it is still questionable how important this reflex is, since an increase in right atrial pressure also has a direct effect — because of stretch of the S-A node — to increase the heart rate as much as 10 to 15 per cent, as was discussed in Chapter 12.

The increase in urinary output in response to increased atrial pressure has been called a *volume reflex*, and the atrial receptors responsible for this are frequently called *volume receptors* because this effect helps to control the blood volume.

These atrial reflexes, like the baroreceptor reflex, are known to be transient in nature, lasting no longer than a few days at most. Therefore, sustained elevation of atrial pressures, as occurs in heart failure, does not cause long-term continuation of the reflex effects.

PARTICIPATION OF THE VEINS IN NERVOUS REGULATION OF THE CIRCULATION

The veins offer relatively little resistance to blood flow in comparison with the arterioles and arteries. Therefore, sympathetic constriction of the veins does not change significantly the overall total peripheral resistance. Instead, the important effect of sympathetic stimulation of the veins is a *decrease in their capacity*. This means that the veins then hold less blood at any given venous pressure, which increases the translocation of blood out of the systemic veins into the heart, lungs, and systemic arteries. The distention of the heart in turn causes the heart to pump with increasing effectiveness, in accordance with the Frank-Starling law of the heart, as discussed in Chapter 11. Therefore, the net effect of sympathetic stimulation of the veins is to increase the cardiac output.

HUMORAL REGULATION OF THE CIRCULATION

The term "humoral regulation" means regulation by substances such as hormones and ions in the body fluids. Among the most important of these factors are the following:

Effect of Aldosterone in Circulatory Regulation. Probably the most important of all humoral factors affecting circulatory regulation is the hormone *aldosterone* secreted by the adrenal cortex. This hormone helps to regulate salt and water in the extracellular fluid, thereby also helping to regulate blood volume. Since this hormone will be discussed in detail in Chapter 25 in relation to renal function and in Chapter 50 in relation to the hormones, only its function in overall circulatory regulation will be presented briefly here.

Whenever the blood volume, the sodium concentration in the body fluids, or the cardiac output decreases, the rate of secretion of aldosterone automatically increases. The precise cause or causes of the increase in aldosterone secretion in all these instances has not yet been determined. Yet, the fact remains that almost any type of circulatory debility causes increased output of aldosterone.

The aldosterone in turn acts on the kidney to cause increased reabsorption of salt and water, thereby increasing the extracellular fluid volume and blood volume. Obviously, these effects are additive to the effects of the intrinsic fluid and pressure control system that was discussed in the previous chapter, making the intrinsic system perhaps two or three times as effective as it would be without the aldosterone feedback mechanism. Therefore, one can readily understand the extreme importance of this humoral factor in the control of the circulation.

Effects of Epinephrine and Norepinephrine. When the sympathetic nervous system is stimulated to cause direct effects on the blood vessels, it also causes the adrenal medullae to secrete the two hormones epinephrine and norepinephrine, which then circulate everywhere in the body fluids and act on all vasculature. Norepinephrine has vasoconstrictor effects in almost all vascular beds of the body, and epinephrine has similar effects in some but not all beds. For instance, epinephrine causes vasodilatation in both skeletal and cardiac muscle. The actions of these hormones obviously support the direct actions of sympathetic nervous stimulation on the circulation. Epinephrine and norepinephrine will be treated in more detail in Chapter 40 in the discussion of the autonomic nervous system.

REFERENCES

Cushing, H.: Concerning a definite regulatory mechanism of the vasomotor center which controls blood pressure during cerebral compression. *Bull. Hopkins Hosp., 12*:290, 1901.

Guyton, A. C., Batson, H. M., Smith, C. M., and Armstrong, G. G.: Method for studying competence of the body's blood pressure regulatory mechanisms and effect of pressoreceptor denervation. *Amer. J. Physiol., 164*:360, 1951.

Hilton, S. M.: Hypothalamic regulation of the cardiovascular system. *Brit. Med. Bull., 22*: 243, 1966.

Korner, P. I.: Circulatory adaptations in hypoxia. *Physiol. Rev., 39*:687, 1959.

Mellander, S.: Systemic circulation. *Ann. Rev. Physiol., 32*:313, 1970.

Mellander, S., and Johansson, B.: Control of resistance, exchange, and capacitance functions in the peripheral circulation. *Pharmacol. Rev., 20*:117, 1968.

Sagawa, K., Ross, J. M., and Guyton, A. C.: Quantitation of cerebral ischemic pressor response in dogs. *Amer. J. Physiol., 200*:1164, 1961.

Schachter, M.: Kallikreins and kinins. *Physiol. Rev., 49*:509, 1969.

CHAPTER 20

REGULATION OF MEAN ARTERIAL PRESSURE; AND HYPERTENSION

Most of the basic principles of arterial pressure regulation have been presented in the previous two chapters, Chapter 18 dealing with the intrinsic mechanisms of pressure regulation and Chapter 19 dealing with the reflex and hormonal controls of pressure. It will be the purpose of the present chapter to see how these two types of regulatory mechanisms work together to maintain normal mean arterial pressure. We will also discuss the clinical problem of *hypertension*, or "high blood pressure," which is caused by abnormalities of arterial pressure regulation.

NORMAL ARTERIAL PRESSURES

The systolic pressure of a normal young adult averages about 120 mm. Hg and his diastolic pressure about 80 mm. Hg—that is, his arterial pressure is said to be 120/80.

The Mean Arterial Pressure. The mean arterial pressure is the average pressure throughout the pressure pulse cycle. Off hand one might suspect that it would be equal to the average of systolic and diastolic pressures, but this is not true; the arterial pressure usually remains nearer to diastolic level than systolic level during a greater portion of the pulse cycle. Therefore, the mean arterial pressure is usually slightly less than the average of systolic and diastolic pressures.

The mean arterial pressure of the normal young adult averages about 96 mm. Hg, which is slightly less than the average of his systolic and diastolic pressures, 120 and 80 mm. Hg, respectively. However, for purposes of discussion, the mean arterial pressure is usually considered to be 100 mm. Hg because this value is easy to remember.

The Auscultatory Method for Measuring Systolic and Diastolic Pressure. Figure 20–1 illustrates the usual clinical method, called the "auscultatory method," for determining systolic and diastolic arterial pressures. A stethoscope is placed over the antecubital artery while a blood pressure cuff is inflated around the upper arm. As long as the cuff presses against the arm with so little pressure that the artery remains

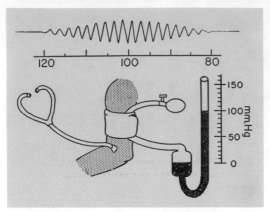

FIGURE 20-1 The auscultatory method for measuring systolic and diastolic pressures.

to equal diastolic pressure. Below this pressure the sounds suddenly change to muffled thumping and usually disappear entirely after another 5 to 10 mm. drop in cuff pressure. One notes the manometer pressure when the Korotkow sound quality changes from tapping to thumping, and this pressure is approximately equal to the diastolic pressure.

The auscultatory method for determining systolic and diastolic pressures is not entirely accurate, but it usually gives values within 10 per cent of those determined by direct measurements from the arteries.

distended with blood, no sounds whatsoever are heard by the stethoscope despite the fact that the blood pressure within the artery is pulsating. But when the cuff pressure is great enough to close the artery during part of the arterial pressure cycle, a sound is heard in the stethoscope with each pulsation. These sounds are called *Korotkow sounds*. *Korotkow sounds* are believed to be caused by blood jetting through the compressed vessel when it is just opening. The jet causes turbulence that sets up the vibrations heard through the stethoscope.

In determining blood pressure by the auscultatory method, the pressure in the cuff is first elevated well above arterial systolic pressure. Then the cuff pressure is gradually reduced. As long as this pressure is higher than systolic pressure, the brachial artery remains collapsed and no Korotkow sounds are heard in the lower artery. Just as soon as the pressure in the cuff falls below systolic pressure, blood slips through the artery beneath the cuff during the peak of systolic pressure, and one begins to hear *tapping* sounds in the antecubital artery in synchrony with the heart beat. As soon as these sounds are heard, the pressure in the cuff is approximately equal to the systolic pressure.

As the pressure in the cuff is lowered still more, the Korotkow sounds continue, until the pressure in the cuff falls

REGULATION OF ARTERIAL PRESSURE — THE OVERALL SYSTEM: NERVOUS, INTRINSIC, HUMORAL

Before discussing the overall regulation of arterial pressure, it is good to remember the basic relationship between arterial pressure, cardiac output, and total peripheral resistance, which was discussed in detail in Chapter 14, as follows:

$$\text{Pressure} = \text{Cardiac Output} \times \text{Total Peripheral Resistance}$$

Thus, it is obvious from this formula that any factor that increases either the cardiac output or total peripheral resistance will cause an increase in mean arterial pressure. We have seen in the previous two chapters that both of these factors are manipulated in the control of arterial pressure, but except under temporary conditions, it is usually total peripheral resistance that is manipulated. We shall have more to say about this later in the chapter.

RAPIDLY ACTING ARTERIAL PRESSURE REGULATION — NERVOUS REGULATION

It is primarily the circulatory reflexes that keep the arterial pressure from rising extremely high or falling extremely low from minute to minute. For instance, when a person first stands up there is a hydrostatic tendency for arterial pressure in his head to fall very low, but the nervous reflexes take over immediately and cause vasoconstriction throughout the body, thus maintaining an almost normal pressure in the head. Likewise, when a person bleeds severely and his pressure tends to fall, the nervous reflexes play an immediate role in bringing the arterial pressure back toward normal.

There are three circulatory reflexes that play major roles in arterial pressure regulation: (1) the baroreceptor reflex, (2) the chemoreceptor reflex, and (3) the central nervous system (CNS) ischemic response. However, each of these reflexes operates in a different pressure range, the baroreceptor mechanism operating from 60 to 200 mm. Hg, the chemoreceptor mechanism from 40 to 100 mm. Hg, and the CNS ischemic mechanism from 15 to 50 mm. Hg. All these were discussed in the previous chapter.

Temporary Nature of the Nervous Regulatory Mechanisms. Unfortunately, these rapidly acting nervous regulatory mechanisms can act maximally for only a few hours or, at most, a few days, because they all adapt either completely or almost completely. Therefore, it is doubtful that the rapidly acting nervous arterial pressure regulatory mechanisms are significant in long-term regulation of arterial pressure.

INTERMEDIATELY ACTING ARTERIAL PRESSURE REGULATION — THE CAPILLARY FLUID SHIFT MECHANISM

When the arterial pressure changes, this is often also associated with a similar change in capillary pressure. One can readily understand that this will cause fluid to begin moving across the capillary membrane between the blood and the interstitial fluid compartment. Within a few minutes to an hour a new state of equilibrium usually will be achieved, but in the meantime this shift of fluid has played a very beneficial role in the control of arterial pressure. For instance, if the arterial pressure rises too high, loss of fluid through the capillaries into the interstitial spaces causes the blood volume to fall and thereby causes return of arterial pressure back toward normal.

LONG-TERM ARTERIAL PRESSURE REGULATION — ROLE OF THE KIDNEY

Now we come to the truly important mechanism for regulating arterial pressure, the long-term regulation of pressure day in and day out and month in and month out, without which we could not live. This regulation seems to be vested mainly, if not almost entirely, in the kidneys, and the kidneys seem to function mainly by controlling the fluid volumes of the body, as was discussed in relation to the intrinsic fluid volume mechanism for overall circulatory regulation in Chapter 18. This basic mechanism is so important that it will be described briefly once more at this point, but it would also be wise to study its details in Chapter 18 if these are not already well understood.

The Renal–Fluid Volume Mechanism for Arterial Pressure Regulation. The output of the kidneys is highly responsive to the arterial pressure level. For instance, an increase in arterial pressure from 100 to 200 mm. Hg increases the urinary output of both water and salt approximately six-fold. Thus, the kidney is geared to act as a feedback regulator of body fluid volumes and, at the same time, arterial pressure. That is, if the arterial pressure falls too low, the extracellular fluid volume continues to rise until the arterial pressure returns to normal. Conversely, if the arterial pressure rises too high, excess quantities of fluid are lost, causing decreased extracellular fluid volume, blood volume, mean systemic pressure, venous return, and arterial pressure. Thus, the arterial pressure once again returns to normal. The phenomenon of long-term local tissue blood flow regulation (autoregulation) plays an especially important role in keeping the cardiac output relatively normal even though the renal-fluid volume mechanism alters the arterial pressure markedly. The details of this dissociation of arterial pressure and cardiac output regulation were presented in Chapter 18.

Importance of Aldosterone in the Renal-Fluid Volume–Arterial Pressure Regulating System. In Chapter 19 it was pointed out that aldosterone greatly helps to control extracellular fluid salt and water and, also, extracellular fluid volume. We can now see how aldosterone plays an important role in the overall renal-fluid volume–arterial pressure regulating system.

For reasons discussed in Chapter 19, the rate of aldosterone secretion is in some way inversely related to the adequacy of circulation throughout the body. Therefore, when arterial pressure increases, there is an indirect effect operating through the aldosterone mechanism to increase urinary output; that is, the increased pressure decreases the aldosterone output, and the decrease in aldosterone allows excessive loss of salt and water from the kidneys. Therefore, at each given increase in arterial pressure, the output of water and salt is much greater from the kidneys if the aldosterone feedback mechanism is operative than is true if it is not operative.

Conversely, at low pressures the decrease in arterial pressure decreases urinary output indirectly through the aldosterone mechanism; that is, decreased pressure causes excessive production of aldosterone that in turn causes the kidneys to retain salt and water.

The net effect, therefore, of the aldosterone mechanism is to increase greatly the effectiveness of the renal-fluid volume mechanism for regulating arterial pressure.

Time Required for the Renal–Fluid Volume Mechanism to Act. In acute situations, such as following severe loss of blood, the fluid volume mechanism begins to return arterial pressure back toward normal within hours. But often, almost a day will lapse before significant response can be seen. Therefore, one can understand the extreme importance of the rapidly acting and intermediately acting pressure control mechanisms discussed earlier, for these tide the circulation over until the long-term regulatory mechanism can take effect.

HYPERTENSION

Hypertension means, simply, high arterial pressure. This subject is extremely important because about 12 per cent of all persons die as a direct result of hypertension, and about 20 per cent of all people can expect to have high blood pressure at some time during their lives.

Relative Importance of Total Peripheral Resistance and Cardiac Output in Causing Hypertension. Since arterial pressure is the product of cardiac output times total peripheral resistance, it is obvious that an increase in either of these two factors can cause hypertension. However, in most types of hypertension, one finds that the total peripheral resistance is greatly increased while the cardiac output is very near to normal. This has usually been interpreted as meaning that hypertension is caused by some primary factor that increases the resistance. As a result, many research workers have searched for vasoconstrictor substances circulating in the blood, hoping to prove that one of these constricts the peripheral vessels and thereby causes hypertension. To date, there is no proof that such a vasoconstrictor substance is the usual cause of hypertension, though several substances have been postulated, as will be discussed later in the chapter.

Can a primary increase in cardiac output cause increased total peripheral resistance? If there should be no vasoconstrictor substance to increase the total peripheral resistance in hypertension, how could the measured increase in resistance come about? One answer to this would be a primary increase in cardiac output followed by activation of the local blood flow regulation mechanism discussed in Chapter 18. That is, if the blood volume increases and in turn causes an increase in cardiac output, there would be too much blood flow through the tissues; in response, the local flow regulating mechanism in the tissues would gradually cause the total peripheral resistance to increase. It was pointed out in Chapter 18 that the long-term local flow regulatory mechanism is probably very powerful. It takes days to weeks to react fully, but when it does do so, a primary increase in cardiac output is followed by an increase in total peripheral resistance and return of the cardiac output essentially to its normal value.

Therefore, the concept that most hypertension is caused by a primary increase in total peripheral resistance could easily be a false assumption, for it is obviously possible for a primary increase in cardiac output to cause a secondary rise in total peripheral resistance.

HYPERTENSION INVOLVING THE KIDNEY

In general, anything that decreases the overall function of the kidney causes extracellular fluid retention and hypertension. Consequently, hypertension can be caused by many different conditions involving the kidneys. These include the following:

Removal of Both Kidneys — "Renoprival Hypertension." With the advent of artificial kidneys, more and more human beings are being kept alive with no functional kidney mass at all. When the person without kidneys is properly dialyzed, his arterial pressure can almost always be kept in a normal range. But, on the other hand, if the amount of water and salt in the patient's body is allowed to build up, so also does his arterial pressure increase. The increase in pressure does not occur instantaneously with the increase in extracellular fluid volume; instead, it occurs a few days later, preceded by a transient increase in cardiac output. This is what would be predicted from the renal-fluid volume-arterial pressure regulatory mechanism discussed in Chapter 18 in relation to intrinsic circulatory control. That is, increase in fluid volume causes a transient increase in cardiac output, and the excess blood flow through the tissues during the transient increase in cardiac output causes progressive increase in total peripheral resistance because of the ability of the tissues to readjust their flows to normal by increasing the resistance.

Hypertension Caused by Renal Insufficiency Plus Water and Salt Loading. Renal hypertension occurs very commonly in patients with severely damaged kidneys that cannot excrete adequate quantities of salt. Retention of salt in the body causes secondary retention of water, as will be explained in Chapter 25, thereby building up the extracellular fluid volume. Concomitantly, the patient develops hypertension.

This type of hypertension can be created experimentally in animals, a typical example of which is illustrated in Figure 20–2. This figure illustrates average results from four dogs in which 70 per cent of the renal tissue was removed during the first few weeks of the experiment. The arterial pressure remained normal until the animals were required to drink 0.9 per cent saline solution instead of the usual drinking water. Within a few days the arterial pressure had increased almost 50 per cent. Two weeks later the animals were

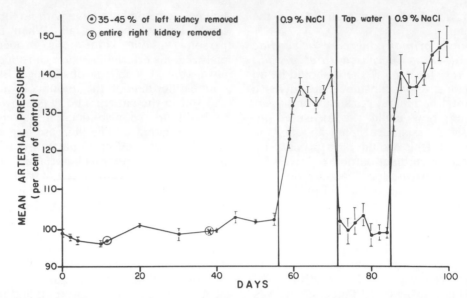

FIGURE 20–2 Effect on arterial pressure of requiring four dogs that had had 70 per cent of their renal tissue removed to drink 0.9 per cent saline solution instead of water. (Courtesy Dr. Jimmy B. Langston.)

given pure water to drink once again, and the arterial pressure fell back to normal within 24 hours. A few weeks later the animals were again required to drink 0.9 per cent saline solution, and hypertension reappeared immediately. Measurements of cardiac output and total peripheral resistance in these experiments showed that the cardiac output rises about 35 per cent immediately after the animals begin drinking the saline solution. Then, the total peripheral resistance begins to rise three days later, increasing approximately 40 per cent while the cardiac output returns almost all the way back to normal. It has been experiments such as these that have demonstrated the importance of fluid volume retention and the local tissue blood flow regulatory mechanism in the causation of many types of renal hypertension.

Goldblatt Hypertension. Almost 40 years ago Goldblatt demonstrated that hypertension develops regularly in animals whose renal arteries are constricted. Furthermore, the degree of hypertension can be altered by the degree of renal artery constriction. Therefore, this preparation has proved to be

extremely valuable in studying renal hypertension, for which reason the hypertension caused in this manner is called "Goldblatt" hypertension.

Figure 20–3 illustrates a typical arterial pressure record in a dog whose renal arteries are constricted. Note that immediately after constriction the renal arterial pressure falls to a very low value, but gradually over a period of one to two weeks the systemic arterial pressure rises to a very high level while the renal arterial pressure returns either to normal or almost to normal. Thus, once again, the principle that the kidney regulates its own arterial pressure to a normal mean value is demonstrated.

Goldblatt hypertension is caused at least partially by an increase in extracellular fluid volume that is caused by failure of the kidneys to excrete fluid when the clamps are first applied. The volume increases about 35 per cent and remains increased until the arterial pressure rises to a high value, at which time the volume decreases back almost to the normal level.

However, there is also reason to believe that Goldblatt kidneys secrete renin, which in turn causes vascular

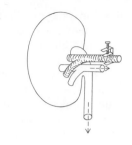

FIGURE 20-3 Changes in systemic arterial pressure and renal arterial pressure distal to the clamp following renal artery constriction.

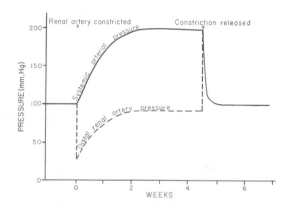

constriction throughout the body, as subsequently described. Therefore, part of the increase in pressure in Goldblatt hypertension might also be ascribed to this vasoconstrictor mechanism.

The Renin-Angiotensin Mechanism. We have several times suggested that some types of hypertension might be caused or partially caused by a vasoconstrictor substance secreted by the kidneys. The substance *renin* is secreted into the bloodstream from the kidneys when blood flow through the kidneys is so poor that it causes renal ischemia. It is postulated that the renin then elevates the arterial pressure and thereby corrects the poor renal blood flow. Figure 20-4 illustrates the schema of events that occurs in this mechanism, as follows:

FIGURE 20-4 The renin-angiotensin theory of arterial pressure regulation.

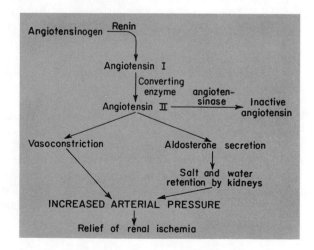

Renin itself is an enzyme that catalyzes the conversion of one of the plasma proteins, a glycoprotein called *angiotensinogen*, into a substance called *angiotensin I*. This reaction continues for about 20 minutes, and within another minute or so, angiotensin I is converted into *angiotensin II* by an enzyme called *converting enzyme* that is present in the lung tissue. Angiotensin II persists in the blood for 30 seconds to 2 minutes, but it is gradually inactivated by a number of different blood and tissue enzymes collectively called *angiotensinase*.

During its persistence in the blood, angiotensin II has two effects which can elevate arterial pressure: (1) It causes *vasoconstriction of the systemic arterioles* throughout the body; the arteriolar constriction in turn increases the total peripheral resistance and thereby increases the arterial pressure. (2) It causes *increased production of aldosterone* by the adrenal cortex, and the aldosterone in turn acts on the kidneys to cause salt and water retention, thereby elevating the arterial pressure as a result of increased extracellular fluid volume.

Quantitative Importance of the Renin-Angiotensin Mechanism in Renal Hypertension. In most types of hypertension, chemical measurements of the concentrations of renin or angiotensin in the circulating blood have shown that the quantities of these are probably much too little to be the cause of the hypertension. Therefore, most renal hypertension is probably caused by water and salt retention or other factors besides the renin-angiotensin mechanism. However, there are times when large amounts of renin are produced, and this mechanism might then be important, particularly when the kidneys are severely ischemic or diseased.

Hypertension in Coarctation of the Aorta. One of the most interesting examples of the Goldblatt type of hypertension occurs in persons who have *coarctation of the aorta*; in this condition, the aorta is greatly constricted or totally occluded, usually in its thoracic portion. Therefore, blood cannot flow through the aorta to the lower part of the body but instead must flow through collateral vessels, such as the internal mammary arteries and arteries along the lateral chest wall and ribs. The resistance to blood flow from the upper part of the body to the lower part through these accessory channels is great. As a result, the arterial pressure in the upper part of the body is usually 50 to 100 mm. Hg greater than that in the lower part. Yet the pressure below the aortic constriction is essentially normal because the kidneys maintain their pressure at a normal value, while pressure in the upper aorta is always greatly elevated.

It is even more interesting that the blood flow to the tissues is approximately normal in both the upper and lower body. That is, the vascular resistance is very high in the upper body but normal in the lower body. The normal resistance in the lower body demonstrates that there probably is little if any generalized vasoconstrictor substance acting on the circulation.

HYPERTENSION CAUSED BY A PHEOCHROMOCYTOMA TUMOR

Hypertension is occasionally caused by a small tumor, called a pheochromocytoma, that develops in one of the adrenal medullae or, rarely, in other chromaffin tissue of the sympathoadrenal system. This tumor is composed of neuron-like cells similar to the normal secreting cells of the adrenal medulla. It secretes epinephrine and norepinephrine. However, it usually secretes these only in response to sympathetic stimuli in the same manner that the normal adrenal medullae secrete these two hormones.

The pheochromocytoma secretes tremendously more epinephrine and norepinephrine than do the normal adrenal medullae. Therefore, when a person who has a pheochromocytoma becomes excited or for any other reason increases his level of sympathetic activity, these two hormones pour into his circulation, and his arterial pressure goes sky-high. The person can frequently be treated by giving a drug that blocks the activity of the sympathetic nervous system, but obviously the more effective procedure is to remove the tumor.

"ESSENTIAL" HYPERTENSION

The term essential hypertension means hypertension without known cause. Any person having an arterial pressure greater than 140/90 and who also has no obvious cause for the hypertension is considered to have essential hypertension. About 90 per cent of all persons who have hypertension have this essential variety.

Role of the Kidneys in Essential Hypertension. Because of the dominant role of the kidneys in so many known types of hypertension, most research workers have the feeling that the kidneys are in some way involved in essential hypertension. Indeed, one important functional abnormality of the kidneys in this disease is known to occur as follows: The kidneys require a far higher than normal arterial pressure to make them excrete a normal amount of urine. For instance, if the mean arterial pressure of an essential hypertensive person is dropped from his usual level of 150 mm. Hg down to 100 mm. Hg, which is the proper level for a normal person, his kidneys excrete only one-quarter to one-sixth the normal amount of urine. Thus, the person with essential hypertension definitely requires a far higher than normal arterial pressure to make his kidneys operate normally.

Treatment of Essential Hypertension by Decreasing Extracellular Fluid Volume. The kidneys of the essential hypertensive person can be made to excrete extra quantities of extracellular fluid by giving a drug called a *natriuretic*, which promotes the loss of salt and water into the urine. Therefore, the extracellular fluid volume becomes decreased, and the pressure of many essential hypertensive patients can in this way be lowered to normal.

The fact that essential hypertension can usually be effectively treated by reducing extracellular fluid volume is further reason to believe that essential hypertension is caused basically by functional retention of water and salt by the kidneys.

REFERENCES

Blalock, A.: Experimental hypertension. *Physiol. Rev., 20*:159, 1940.

Freis, E. D.: Hemodynamics of hypertension. *Physiol. Rev., 40*:27, 1960.

Granger, H. J., and Guyton, A. C.: Autoregulation of the total systemic circulation following destruction of the central nervous system in the dog. *Circ. Res., 25*:379, 1969.

Guyton, A. C., and Coleman, T. G.: Quantitative analysis of the pathophysiology of hypertension. *Circ. Res., 24*:I–1, 1969.

Kezdi, P.: Baroreceptors and Hypertension. New York, Pergamon Press, Inc., 1968.

Langston, J. B., Guyton, A. C., Douglas, B. H., and Dorsett, P. E.: Effect of changes in salt intake on arterial pressure and renal function in partially nephrectomized dogs. *Circ. Res., 12*:508, 1963.

McCubbin, J., and Page, I.: Renal Hypertension. Chicago, Year Book Medical Publishers, Inc., 1968.

Peart, W. S.: The renin-angiotensin system. *Pharm. Rev., 17*:143, 1965.

Vander, A. J.: Control of renin release. *Physiol. Rev., 47(3)*:359, 1967.

CHAPTER 21

CARDIAC OUTPUT; AND CIRCULATORY SHOCK

CARDIAC OUTPUT

Cardiac output is perhaps the single most important factor that we have to consider in relation to the circulation, for it is cardiac output that is responsible for transport of blood with its contained nutrients.

Cardiac output is the quantity of blood pumped by the left ventricle into the aorta each minute, and *venous return* is the quantity of blood flowing from the veins into the right atrium each minute. Obviously, over any prolonged period of time, venous return must equal cardiac output. However, for short intervals, venous return and cardiac output need not be the same, since blood can be temporarily stored in or lost from the central circulation.

NORMAL VALUES FOR CARDIAC OUTPUT

The normal cardiac output for the young healthy male adult averages approximately 5.6 liters per minute. How-

ever, if we consider all adults, including older people and females, the average cardiac output is very close to 5 liters per minute. In general, the cardiac output of females is about 10 per cent less than that of males of the same body size.

Cardiac Index. The cardiac output changes markedly with body size. Therefore, the cardiac output is frequently stated in terms of the cardiac index, which is the *cardiac output per square meter of body surface area*. The normal human being weighing 70 kg. has a body surface area of approximately 1.7 square meters, which means that the normal average cardiac index for adults is approximately 3.0 liters per minute.

Effect of Age. Figure 21–1 illustrates the change in cardiac index with age. Rising rapidly to a level greater than 4 liters per minute per square meter at 10 years of age, the cardiac index declines to only slightly above 2 liters per minute at the age of 80.

Effect of Metabolism and Exercise. The cardiac output usually remains almost proportional to the overall metabolism of the body. This relationship is illus-

228

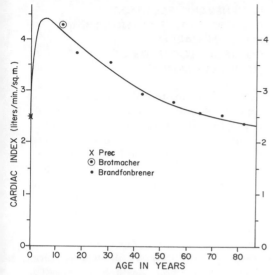

FIGURE 21–1 Cardiac index at different ages. (From Guyton: Cardiac Output and Its Regulation.)

trated in Figure 21–2, which shows that as the work output during exercise increases, the cardiac output also increases in almost linear proportion. Note that in very intense exercise the cardiac output can rise to as high as 30 to 35 liters per minute in the young, well-trained athlete, which is about five to six times the normal control value.

Figure 21–2 also demonstrates that

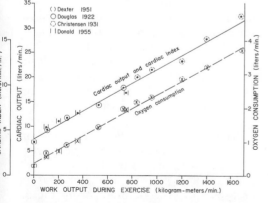

FIGURE 21–2 Relationship between cardiac output and work output (solid curve) and between oxygen consumption and work output (dashed curve) during different levels of exercise. (From Guyton: Cardiac Output and Its Regulation.)

oxygen consumption increases in almost direct proportion to work output during exercise. We shall see later in the chapter that the increase in cardiac output probably results primarily from increased oxygen consumption.

REGULATION OF CARDIAC OUTPUT

PRIMARY ROLE OF PERIPHERAL CIRCULATION AND PERMISSIVE ROLE OF HEART IN CARDIAC OUTPUT REGULATION

The peripheral circulatory system is primarily responsible for controlling cardiac output in the normal person. This seems strange, because at first one would think that the heart would be the controller of cardiac output. Instead, the heart normally plays what may be described as a *permissive role* in cardiac output regulation. That is, it is capable of pumping a certain amount of blood each minute and, therefore, will *permit* the cardiac output to be regulated at any value below this given permitted level. The normal human heart, under resting conditions, permits a maximum heart pumping of about 13 to 15 liters per minute, but the actual cardiac output under resting conditions is only approximately 5 liters per minute. It is the peripheral circulatory system, not the heart, that sets this level at 5 liters per minute. That is, under normal resting conditions the amount of blood that normally flows into the right atrium from the peripheral circulation is about 5 liters; and since this 5 liters is within the permissive range of heart pumping, it is pumped on into the aorta. Therefore, the peripheral circulatory system controls the cardiac output whenever the permissive level of heart pumping is greater than the required cardiac output.

Increase in Permissive Level for Heart Pumping in Hypertrophied Hearts. Heavy athletic training causes the heart to enlarge sometimes as much as 50 per cent. Coincident with this enlargement is an increase in the permissive level to which the heart can pump. Thus, even under resting conditions, the permissive level for a well-trained athlete might be greater than 20 liters per minute, rather than the normal value of about 13 to 15 liters per minute.

Increase in Permissive Level of Heart Pumping by Autonomic Stimulation of the Heart. There are times when the cardiac output must rise temporarily to levels greater than the normal permissive level of the heart. For instance, in heavy exercise by well-trained athletes, cardiac outputs as high as 35 liters per minute have been measured. Obviously, the *resting* heart, even of the well-trained athlete, would not be able to pump this amount of blood. On the other hand, stimulation of the heart by the sympathetic nervous system increases the permissive level of heart pumping to approximately 70 per cent above normal. This effect comes about by enhancement of both heart rate and strength of heart contraction. Furthermore, the increase in permissive level of heart pumping occurs within a few seconds after exercise begins.

Reduction in Permissive Level for Heart Pumping in Heart Disease. Though the normal permissive level for heart pumping is usually much higher than the cardiac output, this is not necessarily true when the heart is diseased. Such conditions as myocardial infarction, valvular heart disease, myocarditis, congenital heart abnormalities, and so forth, can reduce the pumping effectiveness of the heart. In these instances, the permissive level for heart pumping may fall as low as 5 liters per minute or, for a few hours, even as low as 2 to 3 liters per minute. When this happens, the heart becomes unable to cope with the amount of blood that is attempting to flow into the right atrium from the peripheral circulation. Therefore, the heart is said to *fail*, meaning simply that it fails to pump the amount of blood that is demanded of it.

PERIPHERAL CIRCULATORY FACTORS THAT PLAY SIGNIFICANT ROLES IN CARDIAC OUTPUT CONTROL — LOCAL TISSUE BLOOD FLOW REGULATION

We have discussed in previous chapters, especially in Chapter 18, the concept of local control of blood flow in the tissues themselves. In general, blood flow through each individual tissue is locally adjusted to the changing needs of the tissues for oxygen or other nutrients; or in some tissues, blood flow is controlled by other factors such as the blood carbon dioxide concentration in the brain or blood electrolyte concentrations in the kidneys. Therefore, blood flow into the heart, which is called *venous return*, is controlled mainly by the tissues themselves.

However, local blood flow regulation is not the only peripheral circulatory factor that plays a significant role in cardiac output regulation. For instance, skin blood flow is controlled almost entirely by nerve signals from the brain, which are activated by the heat control centers of the body, as discussed in Chapter 47. Thus, when a person is very hot, skin blood flow may be as much as 2 liters per minute or more; when one is very cold, skin blood flow may be as little as 100 milliliters per minute. Likewise, pathologic conditions frequently cause blood to flow directly from arteries to veins without having any relation to the nutrition of the tissues. This occurs in A-V fistulae, in dilated arteriovenous anastomoses, and through some very large capillaries that are concerned more with direct transmission of blood from arteries to veins than with nutrition. Such direct flow of blood from arteries to veins also represents part of the venous return to the heart.

Thus, any peripheral factor that increases the flow of blood from the arteries to the veins increases venous return and thereby increases the cardiac output.

Cardiac Output Regulation as the Sum of Local Tissue Blood Flow Regulations. It is obvious from this discussion that under normal conditions the amount of blood that flows into the heart is determined by the sum of the blood flows through all the local tissues throughout the body. Therefore, it follows also that normal cardiac output regulation is achieved by the sum of the local peripheral blood flow regulations. This does not hold true when the pumping ability of the heart is so greatly curtailed that the heart is unable to pump all the blood required by the tissues, as we shall see subsequently.

Effect of Mean Systemic Pressure on Cardiac Output Regulation. Up to this point we have discussed changes in peripheral resistance to blood flow as the primary factor regulating cardiac output. However, another circulatory factor that can have important effects on cardiac output regulation is the mean systemic pressure. Basically, the mean systemic pressure is a measure of the degree of filling of the systemic circulation with blood. It is measured by stopping the circulation and pumping blood from the arteries to the veins within a few seconds so that all pressures in the systemic circulation come to an equilibrium level. This equilibrium level of pressure is the mean systemic pressure, and its normal value is about 7 mm. Hg.

Obviously, if the systemic circulation contains a small amount of blood, there will be essentially zero cardiac output. Conversely, if the circulatory system is greatly overloaded with blood, cardiac output can be increased to levels as great as three to four times normal. Therefore, *blood volume*, because of its effect on the mean systemic pressure, is another extremely important factor in the regulation of cardiac output.

In addition to the effect of blood volume on mean systemic pressure, the mean systemic pressure can also be altered by sympathetic stimulation or by tightening the skeletal muscles around the peripheral blood vessels. For instance, maximal *sympathetic stimulation* increases the mean systolic pressure as much as two and one-half–fold, and *skeletal muscle contraction* can increase it as much as four-fold. In each instance the blood in the peripheral vessels is squeezed toward the heart, thereby increasing cardiac output.

REGULATION OF CARDIAC OUTPUT IN IMPORTANT CONDITIONS

Effect of Arteriovenous Fistulae on Cardiac Output. The effect of an A-V fistula on cardiac output is discussed here because of the extremely important lessons that can be learned from this condition. An A-V fistula is a direct opening from an artery into a vein and allows rapid flow of blood directly from the artery to the vein. In experimental animals, such a fistula can be opened and closed artificially, and one can study the changes in fistula flow, cardiac output, and arterial pressure. A typical experiment of this type in the dog is illustrated in Figure 21–3, showing an initial cardiac output of 1300 ml. per minute. After 15 seconds of recording normal conditions, a fistula is opened through which 1200 ml. of blood flows per minute. Note that the cardiac output increases instantaneously about 1000 ml. per minute, or about 84 per cent as much increase in cardiac output as there is fistula flow. Then, when the fistula is closed, the cardiac output returns instantaneously back to its original level.

Now for the lesson to be learned from this study: When the fistula is opened, the pumping ability of the heart is not changed at all; there is only a change in the peripheral circulation—opening of the A-V fistula. Nevertheless, within seconds after the fistula is opened, the cardiac output increases almost as much as there is increase in blood flow through the fistula. In other words, the fistula

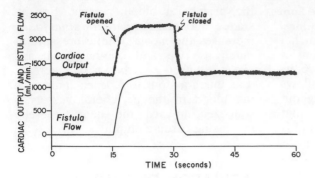

FIGURE 21-3 Effect of suddenly opening and suddenly closing an A-V fistula, showing changes in fistula flow and cardiac output.

flow is added to the flow through the remainder of the body, and the two of these summate to determine the cardiac output. This type of experiment demonstrates the two cardinal features of cardiac output regulation: (1) almost all normal cardiac output regulation is determined by peripheral circulatory factors, and (2) the normal heart has a permissive level of pumping that is normally considerably above the actual cardiac output, so that the heart does not have to be stimulated for an increase in cardiac output to come about.

Regulation of Cardiac Output in Heavy Exercise. Heavy exercise is one of the most stressful situations to which the body is ever subjected. The tissues can require as much as 20 times the normal amounts of oxygen and other nutrients, so that simply to transport enough oxygen from the lungs to the tissues sometimes demands a minimal cardiac output increase of five to six-fold. To insure this result, almost every factor that is known to increase cardiac output is pulled into play. These may be listed as follows:

1. Even before exercise begins, the thought of exercise stimulates the autonomic nervous system, which increases the heart rate and strength of the heart contraction, thus immediately increasing the permissive level of heart pumping from the normal level of 13 to 15 liters per minute up to as high as 25 to 30 liters per minute. Simultaneously, the sympathetic stimulation constricts the veins throughout the body, and this can increase the mean systemic pressure to as high as two and one-half times normal, thus pushing extra quantities of blood from the peripheral circulation toward the heart. Together, these effects increase the cardiac output instantaneously, even before exercise begins, as much as 50 per cent.

2. At the very onset of exercise, the motor cortex transmits signals through the sympathetic cholinergic nerve fibers (described in Chapter 19) to cause vasodilatation of the muscle blood vessels. This causes a further instantaneous increase in cardiac output, perhaps to as high as double normal.

3. Also, at the onset of exercise, the motor cortex transmits signals directly into the sympathetic nervous system to intensify further the degree of sympathetic activity, further enhancing heart activity, mean systemic pressure, and arterial pressure.

4. Still another effect at the very onset of exercise is tightening of the skeletal muscles and of the abdominal wall around the peripheral blood vessels. This additionally increases the mean systemic pressure, now to as high as four times normal. However, compression of blood vessels in the muscles during muscle contraction has a mechanical effect to increase their resistance, and this partially offsets the value of the additional increase in mean systemic pressure.

5. Finally, the most important effect

of all is the direct effect of increased metabolism in the muscle itself. This increased metabolism causes a tremendously increased usage of oxygen and other nutrients, as well as formation of vasodilator substances, all of which cause marked local vasodilatation, thereby greatly increasing local blood flow. This local vasodilatation requires 5 to 20 seconds to reach full development after a person begins to exercise strongly; but, once it does reach full development, it by all means is the most important of all the factors that increase cardiac output during exercise.

In summary, there is tremendous setting of the background conditions of the circulatory system so that it can supply the required blood flow to the muscles during heavy exercise. These conditions include increased mean systemic pressure, increased arterial pressure, and greatly increased activity of the heart. But it is the local vasodilatation in the muscles themselves, occurring as a direct consequence of the muscle activity, that finally sets the level to which the cardiac output rises.

Increased Cardiac Output Caused by Decreased Total Peripheral Resistance. Figure 21–4 illustrates the cardiac output in different pathological conditions,

showing excess cardiac output in some conditions and decreased cardiac output in others.

In almost all conditions in which there is excess cardiac output, one finds decreased total peripheral resistance. For instance, in *beriberi* the total peripheral resistance is decreased by dilatation of the peripheral vessels caused by thiamine deficiency. In *A-V fistula*, decreased total peripheral resistance is caused by direct flow of blood from the arteries to the veins. In *hyperthyroidism* it is caused by increased peripheral metabolism, which results in local vasodilatation. In *hypoxia* (anoxia) it is caused by local vasodilatation resulting from lack of oxygen. In *pregnancy* it is caused by excess blood flow through the placenta. And in *Paget's disease* it is caused by large numbers of minute A-V anastomoses in the bones. Obviously, all of these conditions can increase return of blood to the heart and thereby cause excess cardiac output.

Decrease in Cardiac Output Caused by Decreased Blood Volume. The only peripheral factor that usually decreases cardiac output is decreased mean systemic pressure, most often caused by decreased blood volume. This obviously reduces the available peripheral pressures to return blood to the heart. Ordinarily, rapid bleeding of a person of approximately 30 per cent of his blood volume will reduce the cardiac output to zero. This is such an important problem in the condition known as "circulatory shock" that it will be discussed in detail later in this chapter.

Decrease in Cardiac Output Caused by Cardiac Debility. Most students will be surprised to learn that an increase in cardiac activity caused by direct electrical stimulation of the heart or caused by sympathetic stimulation of the heart (without simultaneous sympathetic stimulation elsewhere in the body) will hardly affect cardiac output. What autonomic stimulation does is simply to increase the permissive level of the heart

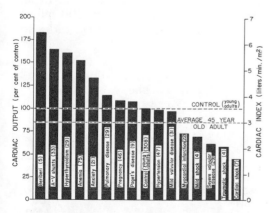

FIGURE 21–4 Cardiac output in different pathological conditions. (From Guyton: Cardiac Output and Its Regulation.)

pumping rather than to increase the actual level.

On the other hand, whenever the heart becomes severely damaged, its permissive level of pumping falls below that needed for adequate blood flow to tissues. Therefore, the heart can be the cause of *decreased* cardiac output, even though it ordinarily cannot by itself be the cause of increased cardiac output. Thus, note in Figure 21–4 that cardiac output is decreased in *myocardial infarction*, in severe *valvular heart disease*, and in *cardiac shock*; in all these conditions the heart simply is not capable of pumping an adequate cardiac output. These conditions will be discussed more fully in Chapter 22 in relation to cardiac failure.

RIGHT VENTRICULAR OUTPUT VERSUS LEFT VENTRICULAR OUTPUT— BALANCE BETWEEN THE VENTRICLES

Obviously, the output of one ventricle must remain almost exactly the same as that of the other. Therefore, an intrinsic mechanism must be available for automatically adjusting the outputs of the two ventricles to each other. This operates as follows:

Let us assume that the strength of the left ventricle decreases suddenly so that left ventricular output falls below right ventricular output. This immediately causes more blood to be pumped into the lungs than is pumped into the systemic circulation. Consequently, the mean pulmonary pressure and the left atrial pressure rise, causing the left ventricular output to increase, while the mean systemic pressure and right atrial pressure fall, causing the right ventricular output to decrease. This process continues until the output of the left ventricle rises to equal the falling output of the right ventricle. Thus, the outputs of the two ventricles become rebalanced

with each other within a few beats of the heart. The same effects occur in the opposite direction when the strength of the right heart diminishes.

These problems of balance between the two ventricles are especially important when myocardial failure occurs in one ventricle or when valvular lesions cause poor pumping by one side of the heart. These conditions will be discussed in more detail in Chapter 22 in relation to cardiac failure.

METHODS FOR MEASURING CARDIAC OUTPUT

MEASUREMENT OF CARDIAC OUTPUT BY THE OXYGEN FICK METHOD

The Fick procedure for measuring cardiac output is best explained by Figure 21–5, which shows absorption of 200 ml. of oxygen from the lungs into the pulmonary blood each minute and also illustrates that the blood entering the right side of the heart has an oxygen concentration of approximately 16 volumes per cent (16 ml. of oxygen per 100 ml. of blood) while that leaving the left side has an oxygen concentration of approximately 20 volumes per cent. From these data we see that each 100 ml. of blood passing through the lungs picks up

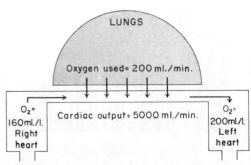

FIGURE 21–5 The Fick principle for determining cardiac output.

4 ml. of oxygen. And, since the total quantity of oxygen absorbed into the blood from the lungs each minute is 200 ml., a total of 50 100-ml. portions of blood must pass through the pulmonary circulation each minute to absorb this amount of oxygen. Therefore, the quantity of blood flowing through the lungs each minute is 5000 ml., which is also a measure of the cardiac output. Thus, the cardiac output can be calculated by the following formula:

Cardiac output (ml./min.) =

$$\frac{\text{Oxygen absorbed per minute by the lungs (ml./min.)}}{\text{Arteriovenous oxygen difference (ml./ml. of blood)}}$$

In applying the Fick procedure, mixed venous blood is obtained via catheter from the right ventricle; or preferably from the pulmonary artery, for blood in the right atrium often is not mixed satisfactorily.

Blood used for determining the oxygen saturation in arterial blood can be obtained from any artery in the body, because all arterial blood is thoroughly mixed before it leaves the heart and therefore has the same oxygen concentration.

The rate of oxygen absorption by the lungs is usually measured by a "respirometer," which will be described in Chapter 47. In essence, this device is a floating chamber that sinks in the water as oxygen is removed from it, thus measuring the use of oxygen by the rate at which it falls.

THE INDICATOR DILUTION METHOD

In measuring the cardiac output by the indicator dilution method a small amount of indicator, such as a dye, is injected into a large vein or preferably into the right side of the heart itself. This then passes rapidly through the right heart, the lungs, the left heart, and finally into the arterial system. If one records the concentration of the dye as it passes through one of the peripheral arteries, a curve such as one of those illustrated in Figure 21–6 will be obtained. In each of these instances 5 mg. of T-1824 dye was injected at zero time. In the top recording none of the dye passed into the arterial tree until approximately 3 seconds after the injection, but then the arterial concentration of the dye rose rapidly to a maximum in approximately 6 to 7 seconds. After that, the concentration fell rapidly. However, before the concentration reached the zero point, some of the dye had already circulated all the way through some of the peripheral vessels and returned through the heart for a second time. Consequently, the dye concentration in the artery began to rise again. For the purpose of calculation, however, it is necessary to extrapolate the early downslope of the curve to the zero point, as shown by the dashed portion of the curve. In this way, the *time-concentration curve* of the dye in an artery can be measured in its first portion and estimated reasonably accurately in its latter portion.

Once the time-concentration curve has been determined, one can then calculate the mean concentration of dye in the arterial blood for the duration of the curve. In Figure 21–6, this was done by measuring the area under the entire curve, and then averaging the concentration of dye for the duration of the curve; one can see from the shaded rectangle straddling the upper curve of the figure that the average concentration of dye was approximately 0.25 mg./100 ml. blood and that the duration of the curve was 12 seconds. However, a total of 5 mg. of dye was injected at the beginning of the experiment. In order for blood carrying only 0.25 mg. of dye in each 100 ml. to carry the entire 5 mg. of dye through the heart and lungs in 12 seconds, it would be necessary for a total of 20 100-ml. portions of blood to pass through the heart during this time, which would be the same as a cardiac output of 2 liters per 12 seconds, or 10 liters per minute.

In the bottom curve of Figure 21–6

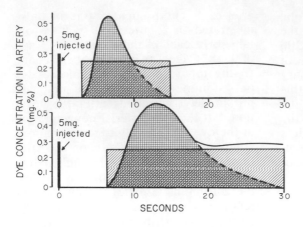

FIGURE 21-6 Dye concentration curves used to calculate the cardiac output by the dilution method. (The rectangular areas are calculated average concentrations of dye in the arterial blood for the durations of the respective curves.)

the blood flow through the heart was considerably slower, and the dye did not appear in the arterial system until approximately 6 seconds after it had been injected. It reached a maximum height in 12 to 13 seconds and was extrapolated to 0 at approximately 30 seconds. Averaging the dye concentrations over the 24-second duration of the curve one finds again an average concentration of 0.25 mg. of dye in each 100 ml. of blood, but this time for a 24-second time interval instead of 12 seconds. To transport the total 5 mg. of dye, 20 100-ml. portions of blood would have had to pass through the heart during the 24-second time interval. Therefore, the cardiac output was 2 liters per 24 seconds, or 5 liters per minute.

To summarize, the cardiac output can be determined from the following formula:

Cardiac output (ml./min.) =

$$\frac{\text{Milligrams of dye injected} \times 60}{\left(\begin{array}{l}\text{Average concentration of} \\ \text{dye in each milliliter of} \\ \text{blood for the duration of} \\ \text{the curve}\end{array}\right) \times \left(\begin{array}{l}\text{Duration of the} \\ \text{curve in seconds}\end{array}\right)}$$

Substances That Can Be Injected for Determining Cardiac Output by the Indicator Dilution Method. Almost any substance that can be analyzed satisfactorily in the arterial blood can be injected when making use of the indicator dilution method for determining cardiac output. However, for optimum accuracy it is necessary that the injected substance not be lost into the tissues of the lungs during its passage to the sampling site. The most widely used substance is *Cardio-Green*, a dye that combines with the plasma proteins and, therefore, is not lost from the blood.

CIRCULATORY SHOCK

Circulatory shock means *reduced cardiac output* to the extent that the tissues of the body are damaged for lack of adequate tissue blood flow. Even the cardiovascular system itself—the heart musculature, the walls of the blood vessels, the vasomotor center, and other parts—begins to weaken so that the shock itself often becomes progressively worse.

At times a person can be in severe shock and still have a normal arterial pressure because of nervous reflexes that keep the pressure from falling. At other times the arterial pressure can fall to as low as one-half normal but the person can still have a normal cardiac output and therefore not be in shock. However, in many types of shock, especially that caused by severe blood loss, the arterial pressure does usually fall at the same time that the cardiac output decreases. Therefore, measurements of arterial pressure are often of value in assessing the degree of shock.

SHOCK CAUSED BY HYPOVOLEMIA— HEMORRHAGIC SHOCK

Hypovolemia means diminished blood volume, and hemorrhage is perhaps the most common cause of hypovolemic shock. Hemorrhage *decreases the mean systemic pressure* and as a consequence decreases venous return. As a result, the cardiac output falls below normal, and shock ensues.

Relationship of Bleeding Volume to Cardiac Output and Arterial Pressure. Figure 21–7 illustrates the effect on both cardiac output and arterial pressure of removing blood from the circulatory system over a period of about half an hour. Approximately 10 per cent of the total blood volume can be removed with no significant effect on arterial pressure or cardiac output, but greater blood loss usually diminishes the cardiac output first and later the pressure, both of these falling to zero when about 35 to 45 per cent of the total blood volume has been removed.

Sympathetic reflex compensation in shock. Fortunately, the decrease in arterial pressure caused by blood loss initiates powerful sympathetic reflexes that stimulate the sympathetic vasoconstrictor system throughout the body, resulting in three important effects: (1) The arterioles constrict in most parts of the body, thereby greatly increasing the total peripheral resistance. (2) The veins and venous reservoirs constrict, thereby helping to maintain adequate venous return despite diminished blood volume. And (3) heart activity increases markedly, sometimes increasing the heart rate from the normal value of 72 beats per minute to as much as 200 beats per minute.

Value of the reflexes. In the absence of the sympathetic reflexes, only 15 to 20 per cent of the blood volume can be removed over a period of half an hour before a person will die; this is in contrast to 30 to 35 per cent when the reflexes are intact.

Especially interesting are the two plateaus in the arterial pressure curve of Figure 21–7. The failure of the arterial pressure to decrease when one first begins to remove blood from the circulation is caused mainly by strong stimulation of the *baroreceptor reflexes*. Then, as the arterial pressure falls to about 30 mm. Hg, a second plateau results from activation of the *CNS ischemic response*, which was discussed in Chapter 19. This effect of the CNS ischemic response can be called the "last-ditch stand" of the sympathetic reflexes in their attempt to keep the arterial pressure from falling too low.

Protection of coronary and cerebral blood flow by the reflexes. A special value of the maintenance of normal arterial pressure even in the face of decreasing cardiac output is protection of blood flow through the coronary and cerebral circulatory systems. Sympathetic stimulation does not cause significant constriction of either the cerebral or cardiac vessels—if anything, it causes slight vasodilatation of the coronary vessels. Also, in both these vascular beds local blood flow regulation is excellent, which prevents moderate changes in arterial pressure from significantly affecting their blood flows. Therefore, blood flow through the heart and brain is maintained essentially at normal levels as long as the arterial pres-

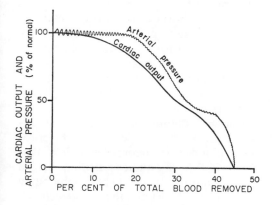

FIGURE 21–7 Effect of hemorrhage on cardiac output and arterial pressure.

sure does not fall below about 70 mm. Hg, despite the fact that blood flow in many other areas of the body might be decreased almost to zero because of vasospasm.

NONPROGRESSIVE SHOCK— COMPENSATED SHOCK

If shock is not severe enough to cause its own progression, the person eventually recovers. Therefore, shock of this lesser degree can be called *nonprogressive shock*. It is also frequently called *compensated shock*, meaning that the sympathetic reflexes and other factors have compensated enough to prevent deterioration of the circulation.

The factors that cause a person to recover from moderate degrees of shock are the negative feedback control mechanisms of the circulation that attempt to return cardiac output and arterial pressure to normal levels. These include (1) the *baroreceptor reflexes*, which elicit powerful sympathetic stimulation of the circulation; (2) the *central nervous system ischemic response*, which elicits even more powerful sympathetic stimulation throughout the body but is not activated until the arterial pressure falls below 50 mm. Hg; (3) *reverse stress relaxation of the circulatory system*, which causes the blood vessels to contract down around the diminished blood volume so that the blood volume that is available will more adequately fill the circulation; and (4) *compensatory mechanisms that return the blood volume back toward normal*, including absorption of large quantities of fluid from the intestinal tract, absorption of fluid from the interstitial spaces of the body, and increased thirst and increased appetite for salt which make the person drink water and eat salty foods if he is able.

The sympathetic reflexes provide immediate help in bringing about recovery, for they become maximally activated within 30 seconds after hemorrhage. The reverse stress relaxation that causes contraction of the blood vessels and venous reservoirs around the blood requires some 10 minutes to an hour to occur completely, but, nevertheless, this aids greatly in increasing the mean systemic pressure and thereby increasing the return of blood to the heart. Finally, the readjustment of blood volume by absorption of fluid from the interstitial spaces and from the intestinal tract, as well as the ingestion and absorption of additional quantities of fluid and salt, may require from 1 to 48 hours, but eventually recovery takes place provided the shock does not become severe enough to enter the progressive stage.

PROGRESSIVE SHOCK—THE VICIOUS CYCLE OF CARDIOVASCULAR DETERIORATION

Once shock has become severe enough, the structures of the circulatory system themselves begin to deteriorate, and various types of positive feedback develop that can cause a vicious cycle of progressively decreasing cardiac output. Figure 21–8 illustrates some of these different types of positive feedback that further depress the cardiac output in shock. These include:

1. Cardiac depression.
2. Failure of the vasomotor center of the brain.
3. Thrombosis of minute vessels.
4. Increased capillary permeability with resultant loss of blood volume.
5. Release of toxins from ischemic tissues and from the gut, especially release of *endotoxin* from "gram-negative" bacteria in the gut.

Cardiac Depression. When the arterial pressure falls low enough, coronary blood flow decreases below that required for adequate nutrition of the myocardium itself. This obviously weakens the heart and thereby decreases the cardiac output still more. As a consequence, the arterial pressure falls still

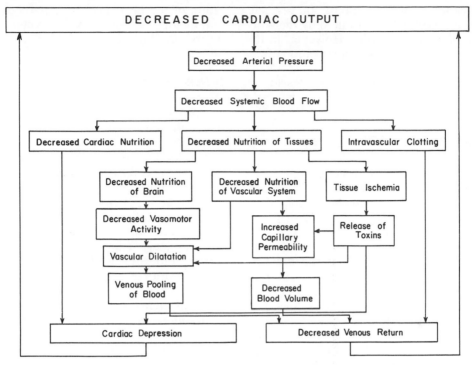

FIGURE 21-8 Different types of feedback that can lead to progression of shock.

further, and the coronary blood flow decreases some more, making the heart still weaker. Thus, a positive feedback cycle has developed whereby the shock becomes more and more severe.

Figure 21-9 illustrates cardiac function curves from experiments in dogs, showing progressive deterioration of the heart at different times following the onset of shock. The dog was bled until the arterial pressure fell to 30 mm. Hg, and the pressure was held at this level by further bleeding or retransfusion of blood as required. Note that there was little deterioration of the heart during the first two hours, but by four hours the heart had deteriorated about 40 per cent; and, finally, during the last hour of the experiment the heart deteriorated almost completely.

Thus, one of the important features of progressive shock, whether it be hemorrhagic shock or any other type of shock, is eventual progressive deterioration of the heart. In the early stages of shock, this plays very little role in the condition of the person, partly because deterioration of the heart itself is not very severe during the first hour or so of shock but mainly because the heart has a tremendous reserve that makes it normally

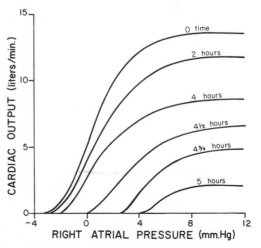

FIGURE 21-9 Function curves of the heart at different times after hemorrhagic shock begins. (These curves are extrapolated to the human heart from data obtained in dog experiments by Dr. J. W. Crowell.)

capable of pumping 300 to 400 per cent more blood than is required by the body for adequate nutrition. However, in the late stages of shock, deterioration of the heart is probably by far the most important factor in the further progression of the shock.

IRREVERSIBLE SHOCK

After shock has progressed to a certain stage, transfusion or any other type of therapy becomes incapable of saving the life of the person. Therefore, the person is then said to be in the *irreversible stage of shock*. Paradoxically, therapy often returns the arterial pressure and even the cardiac output to normal, but the circulatory system continues to deteriorate and death ensues in another few minutes to hours.

Relationship of Oxygen Deficit to the Irreversible Stage of Shock. The diminished cardiac output during shock obviously causes diminished transport of oxygen to the tissues. Therefore, the tissues have a *deficit* of oxygen utilization during shock. Recently, Crowell has demonstrated that shock becomes irreversible in a dog when his oxygen deficit has reached an average of 120 ml. of oxygen per kg. Once the dogs have accumulated the critical oxygen deficit, regardless of how long a period of time this requires, the shock becomes irreversible to normal types of therapy, and the animals die.

This experiment demonstrates the importance of oxygen in preventing death from hemorrhagic shock and probably also in preventing death from other types of shock. Indeed, animals administered oxygen in a pressure tank in which oxygen pressure can be increased to extremely high levels can survive longer periods of hypotension than can animals exposed only to air. Therefore, it is probable that the most detrimental factor in shock is diminished transport of oxygen to the tissues and that it is mainly lack of oxygen that leads to the severe deterioration of the heart and other parts of the circulatory system, finally eventuating in irreversible shock and death.

OTHER CAUSES OF SHOCK

Plasma Loss Shock. Loss of plasma from the circulatory system can sometimes be severe enough to reduce the total blood volume markedly, in this way causing typical hypovolemic shock similar in almost all details to that caused by hemorrhage. Severe plasma loss occurs in the following conditions:

1. *Intestinal obstruction*, in which plasma leaks from the intestinal capillaries into the intestinal walls and intestinal lumen.

2. *Severe burns*, in which plasma is lost through the denuded body surface.

Neurogenic Shock — Increased Vascular Capacity. Occasionally shock results without any loss of blood volume whatsoever. Instead, the *vascular capacity* increases so much that even the normal amount of blood becomes incapable of adequately filling the circulatory system. One of the major causes of this is *loss of vasomotor tone* throughout the body, and the resulting condition is then known as *neurogenic shock*.

The relationship of vascular capacity to blood volume was discussed in Chapter 14, where it was pointed out that either an increase in vascular capacity or a decrease in blood volume *reduces the mean systemic pressure*, which in turn reduces the venous return to the heart. This effect is often called "venous pooling" of blood.

Some of the different factors that can cause sudden loss of vasomotor tone include:

1. *Deep general anesthesia.*

2. *Spinal anesthesia*, especially when this extends all the way up the spinal cord and blocks the sympathetic outflow.

3. *Brain damage.*

Anaphylactic Shock. "Anaphylaxis" is an allergic condition in which the cardiac output and arterial pressure often fall drastically. It results primarily from an antigen-antibody reaction that takes place all through the body immediately after an antigen to which the person is sensitive has entered the circulatory system. Such a reaction is detrimental to the circulatory system in several important ways. First, if the antigen-antibody reaction takes place in direct contact with the vascular walls or cardiac musculature, damage to these tissues presumably can result directly. Second, cells damaged anywhere in the body by the antigen-antibody reaction release several highly toxic substances into the blood. Among these is *histamine* or a *histamine-like substance* that has a strong vasodilator effect. The histamine in turn causes (1) an increase in vascular capacity because of venous dilatation and (2) dilatation of the arterioles with resultant greatly reduced arterial pressure. The effect is great reduction in venous return and often such serious shock that the person dies within minutes.

Septic Shock. The condition that was formerly known by the popular name of "blood poisoning" is now called *septic shock* by most clinicians. This means simply widely disseminated infection in many areas of the body, the infection often being borne through the blood from one tissue to another and causing extensive damage.

Septic shock is extremely important to the clinician because it is this type of shock that, more frequently than any other kind of shock besides cardiac shock, causes death in the modern hospital. Some of the typical causes of septic shock include:

1. Peritonitis caused by spread of infection from the uterus and fallopian tubes, frequently resulting from instrumental abortion.

2. Peritonitis resulting from rupture of the gut, sometimes caused by intestinal disease and sometimes by wounds.

3. Generalized infection resulting from spread of a simple skin infection such as streptococcal or staphylococcal infection.

4. Generalized gangrenous infection resulting specifically from gas gangrene bacilli, spreading first through the tissues themselves and finally by way of the blood to the internal organs, especially to the liver.

In the early stages of septic shock the patient usually does not have signs of circulatory collapse but, instead, simply signs of the bacterial infection itself. However, as the infection becomes more severe, the circulatory system usually becomes involved either directly or as a secondary result of toxins from the bacteria, and there finally comes a point at which deterioration of the circulation becomes progressive in the same way that progression occurs in all other types of shock.

IMPAIRED RENAL FUNCTION IN SHOCK

The very low blood flow during shock greatly diminishes urine output because glomerular pressure falls below the critical value required for filtration of fluid into Bowman's capsule, as explained in Chapter 24. Also, the kidney has such a high rate of metabolism and requires such large amounts of nutrients that the reduced blood flow often causes *tubular necrosis*, which means death of the tubular epithelial cells, with subsequent sloughing and blockage of the tubules, causing total loss of function of the respective nephrons. This is often a serious after-effect of shock that occurs during major surgical operations; the patient sometimes survives the shock associated with the surgical procedure and then dies a week or so later of uremia.

BLOOD REPLACEMENT THERAPY IN SHOCK

If a person is in shock caused by hemorrhage, the best possible therapy is usually transfusion of whole blood. If the shock is caused by plasma loss, the best therapy is administration of plasma. However, at times neither blood nor plasma is available, in which case a *plasma substitute* is often used. One of these is *dextran solution*. Dextrans of appropriate molecular size do not pass through the capillary pores and, therefore can replace plasma proteins as colloid osmotic agents. Consequently, the solution will stay in the blood and expand the blood volume.

CIRCULATORY ARREST

A condition closely allied to circulatory shock is circulatory arrest, in which all blood flow completely stops. This occurs frequently on the surgical operating table as a result of *cardiac arrest* or of *ventricular fibrillation*.

Ventricular fibrillation can usually be stopped by electrical defibrillation, a process described in Chapter 12.

Cardiac arrest usually results from too little oxygen in the anesthetic gaseous mixture or from a depressant effect of the anesthesia itself. A normal cardiac rhythm can usually be restored by removing the anesthetic and then pumping the heart by hand for a few minutes while supplying the patient's lungs with adequate quantities of ventilatory oxygen.

Effect of Circulatory Arrest on the Brain. The real problem in circulatory arrest is usually not to restore cardiac function but instead to prevent detrimental effects in the brain as a result of the circulatory arrest. In general, four to five minutes of circulatory arrest causes permanent brain damage in over half the patients, and circulatory arrest for as long as 10 minutes almost universally destroys most, if not all, of the mental powers.

For many years it has been taught that these detrimental effects on the brain were caused by the cerebral hypoxia that occurs during circulatory arrest. However, recent studies have shown that dogs can almost universally stand 30 minutes of circulatory arrest without brain damage if the blood is removed from the circulation prior to the arrest. On the basis of these studies, it is postulated that the circulatory arrest causes vascular *clots* to develop throughout the brain and that these cause permanent or semipermanent ischemia of brain areas.

REFERENCES

Brecher, G. A.: Venous Return. New York, Grune & Stratton, Inc., 1956.

Crowell, J. W.: Cardiac deterioration as the cause of irreversibility in shock. *In* Mills, Lewis J., and Moyer, John H. (eds.): Shock and Hypotension: Pathogenesis and Treatment. Grune & Stratton, Inc., 1965, p. 605.

Crowell, J. W., Jones, C. E., and Smith, E. E.: Effect of allopurinol on hemorrhagic shock. *Amer. J. of Physiol., 216*:744, 1969.

Grodins, F. S.: Integrative cardiovascular physiology: a mathematical synthesis of cardiac and blood vessel hemodynamics. *Quart. Rev. Biol., 34*:93, 1959.

Guyton, A. C.: Circulatory Physiology: Cardiac Output and Its Regulation. Philadelphia, W. B. Saunders Company, 1963.

Guyton, A. C.: Determination of cardiac output by equating venous return curves with cardiac response curves. *Physiol, Rev., 35*:123, 1955.

Guyton, A. C.: Venous return. *In* Handbook of Physiology. Baltimore, The Williams & Wilkins Co., 1963, Sec. II, Vol. II, p. 1099.

Hamilton, W. F.: Measurement of the cardiac output. *In* Handbook of Physiology. Baltimore, The Williams & Wilkins Co., 1962, Sec. II, Vol. I, p. 551.

Jones, C. E., Crowell, J. W., and Smith, E. E.: A cause-effect relationship between oxygen deficit and irreversible hemorrhagic shock. *Surgery, 127*:93, 1968.

Stone, H. L., Bishop, V. S., and Dong, E., Jr.: Ventricular function in cardiac-denervated and cardiac-sympathectomized conscious dogs. *Circ. Res., 20*:587, 1967.

Weil, M., and Shubin, H.: Shock. Baltimore, The Williams & Wilkins Co., 1967.

CHAPTER 22

CARDIAC FAILURE; HEART SOUNDS; AND DYNAMICS OF VALVULAR AND CONGENITAL HEART DEFECTS

Perhaps the most important ailment that must be treated by the physician is cardiac failure, which can result from any heart condition that reduces the ability of the heart to pump blood. Usually the cause is decreased contractility of the myocardium caused by diminished coronary blood flow, but failure to pump adequate quantities of blood can also be caused by damage to the heart valves, external pressure around the heart, vitamin deficiency, or any other abnormality that makes the heart a hypoeffective pump.

Definition of Cardiac Failure. The term cardiac failure means simply *failure of the heart to pump blood adequately*. This does not mean that the cardiac output in all instances of failure is less than normal, for the output can be normal or sometimes above normal provided the tendency for venous return is high enough to offset the diminished strength of the heart. Therefore, cardiac failure may be manifest in either of two ways: (1) by a decrease in cardiac output or (2) by an increase in either left or right atrial pressure even though the cardiac output is normal or above normal.

Unilateral versus Bilateral Cardiac Failure. Since the left and right sides of the heart are two separate pumping systems, it is possible for one of these to fail independently of the other. For instance, unilateral failure can result from coronary thrombosis in one or the other of the ventricles. Because debilitating thrombosis occurs approximately 30 times as often in the left ventricle as in the right ventricle, there is a tendency among clinicians to view failure following myocardial infarction as almost always primarily left-sided. Occasionally, however, right-sided failure does occur with no left-sided failure at all; this happens most frequently in persons with pulmonary stenosis or some other congenital disease affecting primarily the right heart.

Usually, though, when one side of the heart becomes weakened, this causes a sequence of events that makes the opposite side of the heart also fail. For instance, in left-sided failure the left atrial pressure increases greatly, resulting in considerable back pressure in the pulmonary system and a rise in pulmonary arterial pressure sometimes to two to three times normal. This loads the right ventricle, causing combined failure of both ventricles, even though the initiating cause was in the left side of the heart only.

DYNAMICS OF THE CIRCULATION IN CARDIAC FAILURE

DYNAMICS OF MODERATE CARDIAC FAILURE

Acute Effects. If a heart suddenly becomes severely damaged in any way, such as by myocardial infarction, the pumping ability of the heart is immediately depressed. As a result, two essential effects occur: (a) reduced cardiac output and (b) increased systemic venous pressure.

Compensation for Acute Cardiac Failure by Sympathetic Reflexes. When the cardiac output falls precariously low, many of the different circulatory reflexes discussed in Chapter 19 immediately become active. The best known of these reflexes is the baroreceptor reflex, which is activated by diminished arterial pressure. As a result, the sympathetics become strongly stimulated within a few seconds, and the parasympathetics become reciprocally inhibited at the same time.

Strong sympathetic stimulation has two major effects on the circulation: first, on the heart itself, and, second, on the peripheral vasculature. Even a damaged myocardium usually responds with increased force of contraction following sympathetic stimulation. Therefore, *the heart becomes a stronger pump, often as much as 100 per cent stronger, under the influence of the sympathetic impulses.*

Sympathetic stimulation also increases the tendency for venous return, for it increases the tone of most of the blood vessels of the circulation, *raising the mean systemic pressure* to 12 to 14 mm. Hg, almost 100 per cent above normal. As will be recalled from the discussion in Chapter 21, this greatly increases the tendency for blood to flow back to the heart. Therefore, the damaged heart becomes primed with more inflowing blood than usual; and, if the heart has not been too greatly damaged, this increased priming helps the heart to pump larger quantities of blood.

As a result, the person who has a sudden moderate heart attack might experience nothing more than a few seconds of fainting. Shortly thereafter, with the aid of the sympathetic compensation, his circulation may be completely adequate as long as he remains quiet.

The Chronic Stage of Failure—Compensation by Renal Retention of Fluid. A low cardiac output has a profound effect on renal function, sometimes causing anuria when the cardiac output falls to as low as one-half to two-thirds normal. This is caused by (a) *decreased glomerular filtration* and (b) *increased aldosterone secretion* with resultant increased salt and water reabsorption, as will be discussed in Chapter 25.

Effect of fluid retention on cardiovascular dynamics. Fluid retention does not have any significant effect on the pumping ability of the heart; but it does have an extreme effect on the tendency for venous return. The fluid retained by the kidneys causes a progressive increase in both extracellular fluid volume and blood volume. For two different reasons this increases the tendency for venous return to the heart: First, both the increased extracellular fluid and blood volume increase the mean systemic pressure, which *increases the pressure gradient for flow of blood toward the heart.* Second, *reduced venous resistance* caused by distention of the veins allows increased flow of blood toward the heart. Therefore, if the heart is not too severely damaged, this fluid retention can be a valuable aid in helping to compensate for reduced cardiac function.

DYNAMICS OF SEVERE CARDIAC FAILURE — DECOMPENSATED HEART FAILURE

If the heart becomes severely damaged, then no amount of compensation, either by sympathetic nervous reflexes or by fluid retention, can make this weakened heart pump a normal cardiac output. As a consequence, the cardiac output cannot rise to a high enough value to bring about return of normal renal function. Fluid continues to be retained, the person develops progressively more and more edema, and this state of events, eventually leads to his death. This is called decompensated heart failure. The basis of decompensated heart failure is *failure of the heart to pump sufficient blood to make the kidneys function adequately.*

Treatment of Decompensation. Two ways in which the decompensation process can often be stopped are as follows.

1. Treatment by strengthening the heart. A weakened heart can frequently be strengthened by administering digitalis or a similar cardiotonic drug. Recent experimental evidence indicates that these drugs strengthen the heart by increasing the rate of diffusion of calcium through the membrane to the interior of the muscle fibers; the calcium then promotes increased force of contraction by the actomyosin complex, as explained in Chapter 7. The chronically weakened heart can often be strengthened by as much as 50 per cent by these drugs.

Another means by which the heart often becomes strengthened is simply prolonged bed rest, for this protects the heart against excessive overloading and allows maximum recovery from any myocardial disease.

2. Treatment of decompensation by dietary restriction of water and salt and administration of diuretics. The second means by which the process of decompensation can be stopped is to administer a diuretic that makes the kidneys put out far greater quantities of urine than normally or to restrict the person's intake of salt and water. By these two measures the output of fluid can be increased to a greater value than the intake, thereby preventing further retention of fluid even though the cardiac output is considerably less than normal.

EDEMA IN CARDIAC FAILURE

On first thought, it seems that cardiac edema is a simple problem — that increased right atrial pressure causes increased capillary pressure, which causes the edema. However, generalized edema does not develop until a day or more after a heart attack, and then it occurs *because of fluid retention by the kidneys.* The retention of fluid increases the mean systemic pressure, resulting in increased tendency for blood to return to the heart. This now elevates the right atrial pressure to a still higher value, and at the same time it keeps the cardiac output and arterial pressure from decreasing very much despite the failing myocardium. Therefore, the rising right atrial pressure now *causes the capillary pressure to rise markedly,* thus causing loss of fluid into the tissues and development of severe edema.

Edema of cardiac failure can occur either in the systemic circulation or in the lungs, or in both, depending on which side or sides of the heart might be failing. Pulmonary edema is further discussed below in relation to unilateral left heart failure.

CARDIAC SHOCK

In the description of acute heart failure, it was pointed out that the cardiac output can fall very low immediately after heart damage occurs. This obviously leads to greatly diminished blood flow throughout the body and can lead to a typical picture of circulatory shock as described in the previous chapter. This type of shock is called *cardiac shock*. Cardiac shock is extremely important to the clinician because approximately one-tenth of all patients who have severe acute myocardial infarction will die of an ensuing shock syndrome before the physiologic compensatory measures can come into play to save life.

Vicious Cycle of Cardiac Deterioration in Cardiac Shock. The discussion of circulatory shock in the previous chapter emphasized the tendency for the heart itself to become progressively damaged when its coronary blood supply is reduced during the course of shock. That is, shock reduces the coronary supply, which makes the heart still weaker, which makes the shock still worse. In cardiac shock caused by myocardial infarction, this problem is further compounded by the already existing coronary thrombosis. For instance, in a normal heart, the arterial pressure usually must be reduced below about 45 mm. Hg before cardiac deterioration sets in. However, in a heart that already has a major coronary vessel blocked, a similar degree of deterioration will set in when the arterial pressure falls as low as 80 to 90 mm. Hg. In other words, even the minutest amount of fall in arterial pressure can then set off a vicious cycle of cardiac deterioration. For this reason, in treating myocardial infarction it is extremely important to prevent even short periods of hypotension.

Treatment of cardiac shock is one of the most important problems in the management of acute heart attacks. Immediate digitalization of the heart is often employed for strengthening the heart, but more frequently intravenous infusion of whole blood, plasma, or a blood pressure raising drug is used to sustain the arterial pressure. If this is done, the coronary blood flow can often be elevated to a high enough value to prevent the vicious cycle of deterioration until appropriate compensatory mechanisms in the body have corrected the shock.

UNILATERAL LEFT HEART FAILURE

When the left side of the heart fails without concomitant failure of the right side, blood continues to be pumped into the lungs by the normal right heart, while it is not pumped adequately out of the lungs by the left heart. As a result, the *mean pulmonary pressure* rises while the *mean systemic pressure* falls because of shift of large volumes of blood from the systemic circulation into the pulmonary circulation.

As the volume of blood in the lungs increases, the pulmonary vessels enlarge, and, if the pulmonary capillary pressure rises above 28 mm. Hg, that is, above the colloid osmotic pressure of the plasma, fluid begins to filter out of the capillaries into the interstitial spaces and alveoli, resulting in pulmonary edema.

Thus, among the most important problems of left heart failure are *pulmonary vascular congestion* and *pulmonary edema*, which are discussed in detail in Chapters 15 and 17 in relation to the pulmonary circulation and capillary dynamics. As long as the pulmonary capillary pressure remains less than the normal colloid osmotic pressure of the blood, about 28 mm. Hg, the lungs remain "dry." But even a millimeter rise in capillary pressure above this critical level causes progressive transudation of fluid into the interstitial spaces and alveoli, leading rapidly to death. Pulmonary edema can occur so rapidly that it

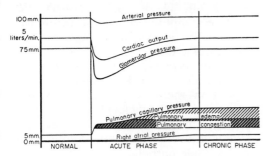

FIGURE 22–2 The overall effects, acute and chronic, of left-sided heart failure.

can cause death after only 20 to 30 minutes of severe acute left heart failure.

Course of Events in Chronic Left Heart Failure. In chronic left heart failure, one additional feature must be added to the acute picture. This is retention of fluid resulting from reduced renal function. In moderate acute left heart failure the pulmonary capillary pressure sometimes rises only to 15 to 20 mm. Hg, not enough to cause pulmonary edema. Yet following retention of fluid for the next few days, the blood volume increases, and more blood is pumped into the lungs by the right ventricle. Then, the pulmonary capillary pressure often rises above the colloid osmotic pressure, resulting in severe pulmonary edema, as shown in Figure 22–1. Indeed, this is a common occurrence, the patient suddenly developing severe pulmonary edema a week or more after the acute attack and dying a respiratory death—not a death resulting from diminished cardiac output.

UNILATERAL RIGHT HEART FAILURE

In unilateral right heart failure, blood is pumped normally by the left ventricle from the lungs into the systemic circulation, but it is not pumped adequately from the systemic circulation into the lungs. Therefore, blood shifts from the lungs into the systemic circulation, causing systemic congestion. However, in acute right heart failure this conges-

tion is hardly noticeable for the following reason: The total amount of blood in the lungs is only about one-seventh that in the systemic circulation. Therefore, even in severe right heart failure, the systemic blood volume increases only a few per cent, and this is not sufficient to cause significant systemic congestion.

Low Cardiac Output in Acute Right Heart Failure. On the other hand, acute right heart failure causes far greater depression of the cardiac output than acute left heart failure of the same degree. This again stems from the far greater compliance of the systemic circulation than of the pulmonary circulation. Not enough blood can transfer from the lungs into the systemic vessels to raise the systemic pressures to a very high level, and these pressures are not enough to make the weakened right ventricle pump adequate quantities of blood. Therefore, the cardiac output falls greatly in acute right heart failure, often resulting in cardiac shock.

The Chronic Stage of Unilateral Right Heart Failure. The chronic stage of unilateral right heart failure is much the same as that discussed earlier for the entire heart. The depressed cardiac output results in progressively more and more retention of fluid by the kidneys until the cardiac output either rises back nearly to normal or until the person goes into decompensation and dies. Figure 22–2 illustrates the progressive changes that occur in chronic right-sided heart

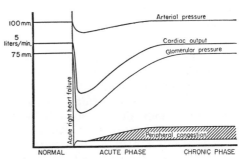

FIGURE 22–2 The overall effects, acute and chronic, of right-sided heart failure.

failure, showing a gradual return of cardiac output to normal or near normal, a return along with increased urinary output, and, finally, progressive development of peripheral congestion and edema.

PHYSIOLOGICAL CLASSIFICATION OF CARDIAC FAILURE

From the above discussions, it is apparent that the symptoms of cardiac failure fall into the following three physiological classifications:

1. Low cardiac output.
2. Pulmonary congestion.
3. Systemic congestion.

Low cardiac output usually occurs immediately after a heart attack. If the attack is mainly right-sided, a slight amount of systemic congestion may occur along with the low cardiac output, but in the early stages of acute attacks of any type the degree of systemic congestion is slight until fluids are retained by the kidneys. If the acute heart attack is mainly left-sided, concurrent pulmonary congestion usually occurs along with the low cardiac output, but here again the pulmonary congestion may be mild until after considerable fluid has been retained by the kidneys. Thus, low cardiac output may be the only significant clinical effect observed in many persons who have sudden heart attacks. This results in the following symptoms:

1. Generalized weakness.
2. Fainting.
3. Symptoms of increased sympathetic activity such as high heart rate, thready pulse, cold skin, sweating, and so forth.

Pulmonary congestion may be the only effect in patients with pure left-sided heart failure, for, after the fluid that shifts from the systemic circulation into the lungs during acute left-sided heart failure has been replenished by renal retention of fluid, the heart sometimes pumps normal quantities of blood, and the right side of the heart may not fail at all. Therefore, pulmonary congestive symptoms alone can occur with essentially no systemic congestion nor low cardiac output.

Systemic congestion alone can occur in pure right-sided heart failure. In this condition there is no pulmonary congestion, and, if sufficient fluids have been retained in the blood to prime the heart sufficiently, the heart may pump a normal cardiac output.

Obviously, all the above classes of heart failure can occur together or in any combination.

THE HEART SOUNDS

The function of the heart valves was discussed in Chapter 11, and it was pointed out that closure of the valves is associated with an audible sound, though no sound whatsoever occurs when the valves open. The purpose of the present section is to discuss the causes of sounds in the heart, under both normal and abnormal conditions.

NORMAL HEART SOUNDS

When one listens with a stethoscope to a normal heart, he hears a sound usually described as "lub, dub, lub, dub ----." The "lub" is associated with closure of the A-V valves at the beginning of systole and the "dub" with closure of the semilunar valves at the beginning of diastole. The "lub" sound is called the *first heart sound* and the "dub" the *second heart sound* because the normal cycle of the heart is considered to start with the beginning of systole and to end with the end of diastole. The first record in Figure 22–4 illustrates graphically the timing of these sounds.

Causes of the First and Second Heart Sounds. The earliest suggestion for the cause of the heart sounds was that slapping together of the vanes of the valves sets up vibrations, but this has now been shown to cause little if any of the sound because of the cushioning effect of the blood. Instead, the cause is *vibration of the walls of the heart and major vessels around the heart*. For instance, contraction of the ventricles causes sudden backflow of blood against the A-V valves, causing the valves to close and to bulge toward the atria. The elasticity of the valves then causes the back-surging blood to bounce backward into each respective ventricle. This effect sets the walls of the ventricles into vibration, and the vibrations travel away from the valves. When the vibrations reach the chest wall, sound waves are created that can be heard by the stethoscope.

The second heart sound results from vibration of the walls of the pulmonary artery, the aorta, and, to a much less extent, the ventricles. When the semilunar valves close, they bulge backward toward the ventricles, and then the elastic stretch recoils the blood in the bulge back into the arteries, which causes a short period of reverberation of blood back and forth between the walls of the arteries and the valves. The vibrations set up in the arterial walls are then transmitted along the arteries at the velocity of the pulse wave. When these vibrations come into contact with a "sounding board," such as the chest wall or the lung, they create sound that can be heard.

Duration and Frequencies of the First and Second Heart Sounds. The duration of each of the heart sounds is slightly more than 0.1 second; the first sound lasts about 0.14 second and the second about 0.11 second. And both of them are described as very low-pitch sounds, the first lower than the second.

Figure 22–3 illustrates by the shaded area the amplitudes of the different frequencies in the heart sounds and murmurs, illustrating that these are composed of frequencies ranging all the way from a few cycles per second to more than 1000 cycles per second, with the maximum amplitude of vibration occur-

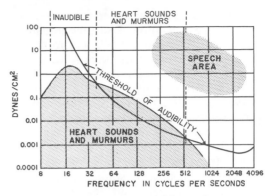

FIGURE 22–3 Amplitude of different frequency vibrations in the heart sounds and heart murmurs in relation to the threshold of audibility, showing that the range of sounds that can be heard is between about 40 and 500 cycles per second. (Modified from Butterworth, Chassin, and McGrath: Cardiac Auscultation. Grune & Stratton.)

ring at a frequency of about 24 cycles per second.

Also shown in Figure 22–3 is another curve called the "threshold of audibility," which depicts the capability of the ear to hear sounds of different amplitudes. Note that in the very low frequency range the heart vibrations have a high degree of amplitude, but the threshold of audibility is so high that ordinarily the heart vibrations below approximately 30 to 50 cycles per second are not heard by the ears. Then above about 500 cycles per second, the heart sounds are so weak that despite a low threshold of audibility no frequencies in this range are heard. For practical purposes, then, we can consider that all the *audible* heart sounds lie in the range of approximately 40 to 500 cycles per second despite the fact that the maximum amplitude of vibration occurs at the very low frequency of 24 cycles per second.

The Third Heart Sound. Occasionally a third heart sound is heard at the beginning of the middle third of diastole. The most logical explanation of this sound is oscillation of blood back and forth between the walls of the ventricles initiated by inrushing blood from the atria. This is analogous to running water from a faucet into a sack, the inrushing water reverberating back and forth between the walls of the sack to cause vibrations in the walls.

The Atrial Heart Sound (Fourth Heart Sound). An atrial heart sound can be recorded in many persons in the phonocardiogram, but it can almost never be heard with a stethoscope because of its low frequency—usually 20 cycles per second or less. This sound occurs when the atria contract, and presumably it is caused by an inrush of blood into the ventricles, which initiates vibrations similar to those of the third heart sound.

THE PHONOCARDIOGRAM

If a microphone specially designed to detect low-frequency sound waves is placed on the chest, the heart sounds can be amplified and recorded by a special high-speed recording apparatus, such as an oscilloscope or a high-speed pen recorder, which are described in Chapters 5 and 13. The recording is called a *phonocardiogram*, and the heart sounds appear as waves in the record, as illustrated in Figure 22–4. Record A illustrates a recording of normal heart sounds, showing the vibrations of the first, second, and third heart sounds and even the atrial sound. Note

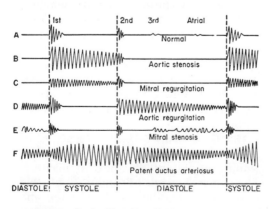

FIGURE 22–4 Phonocardiograms from normal and abnormal hearts.

specifically that the third and atrial heart sounds are each a very low rumble. The third heart sound can be recorded in only one third to one half of all persons, and the atrial heart sound can be recorded in perhaps one fourth of all persons.

ABNORMAL HEART SOUNDS

Rheumatic Valvular Lesions. By far the greatest number of valvular lesions result from rheumatic fever. Rheumatic fever is an autoimmune disease in which the heart valves are likely to be damaged or destroyed. It is initiated by some abnormal immunological or allergic reaction initiated by streptococcal toxin. Large hemorrhagic, bulbous lesions grow along the inflamed edges of the heart valves. Because the mitral valve receives more trauma during valvular action than do any of the other valves, this valve is the one most often affected, and the aortic valve is second most frequently involved. The tricuspid and pulmonary valves are rarely involved because the stresses acting on these valves are slight compared to those in the left ventricle.

Scarring of the valves. The bulbous, hemorrhagic lesions of acute rheumatic fever frequently adhere to adjacent valve leaflets simultaneously so that the edges of the leaflets become stuck together. Then, in the late stages of rheumatic fever, the lesions become scar tissue, permanently fusing portions of the leaflets. Also, the free edges of the leaflets, which are normally filmy and free-flapping, become solid, scarred masses.

A valve in which the leaflets adhere to each other so extensively that blood cannot pass satisfactorily is said to be *stenosed.* On the other hand, when the valve edges are so destroyed by scar tissue that they cannot close together, *regurgitation*, or backflow, of blood occurs when the valve should be closed. Stenosis usually does not occur without coexistence of at least some degree of regurgitation, and vice versa. Therefore, when a person is said to have stenosis or regurgitation, it is usually meant that one merely predominates over the other.

Abnormal Heart Sounds Caused by Valvular Lesions. As illustrated by the phonocardiograms of Figure 22–4, many abnormal heart sounds, known as "murmurs," occur in abnormalities of the valves.

In *aortic stenosis*, blood is ejected from the left ventricle through only a small opening at the aortic valve. This causes severe *turbulence* in the root of the aorta. The turbulent blood impinging against the aortic walls causes intense vibration, and a loud murmur is transmitted throughout the upper aorta and even into the larger vessels of the neck.

In *aortic regurgitation* no sound is heard during systole, but *during diastole* blood flows backward from the aorta into the left ventricle, causing a "blowing" murmur of relatively high pitch heard maximally over the left ventricle.

In *mitral regurgitation* blood flows backward through the mitral valve *during systole*. This also causes a high frequency "blowing" sound, which is transmitted most strongly into the left atrium, but the left atrium is so deep within the chest that it is difficult to hear this sound directly over the atrium. As a result, the sound of mitral regurgitation is transmitted to the chest wall mainly through the left ventricle, and it is usually heard best at the apex of the heart.

In *mitral stenosis*, blood passes with difficulty from the left atrium into the left ventricle and, because the pressure in the left atrium never rises to a very high value, a great pressure differential forcing blood from the left atrium into the left ventricle never develops. Consequently, the abnormal sounds heard in mitral stenosis are usually extremely weak. Also, the murmur is of such low pitch that it is difficult to hear; but, with the aid of a proper stethoscope (the "bell" type), one can usually discern very low frequency rumbles of 30 to 50 cycles per second.

Phonocardiograms in valvular murmurs. Phonocardiograms *B, C, D,* and *E* of Figure 22–4 illustrate, respectively, idealized records obtained from patients with aortic stenosis, mitral regurgitation, aortic regurgitation, and mitral stenosis. Note especially that the murmurs of aortic stenosis and mitral regurgitation occur only during systole, whereas the murmurs of aortic regurgitation and mitral stenosis occur only during diastole — if the student does not understand this timing, he should pause a moment until he does understand it.

ABNORMAL CIRCULATORY DYNAMICS IN VALVULAR HEART DISEASE

Aortic stenosis means a constricted aortic valve with a tubular opening too small for easy ejection of blood from the left ventricle. *Aortic regurgitation* means failure of the aortic valve to close completely and, therefore, failure to prevent backflow of blood from the aorta into the left ventricle during diastole. In aortic stenosis the left ventricle fails to empty adequately, whereas in aortic regurgitation blood returns to the ventricle after it has been emptied. Therefore, in either case, the net stroke volume output of the heart is reduced, and this in turn tends to reduce the cardiac output, resulting eventually in typical left heart failure. The principal results are (1) tremendously increased work load on the left ventricle, resulting in left ventricular hypertrophy; and (2) back pressure in the lungs when the left heart finally fails, often leading to death from pulmonary edema.

In *mitral stenosis* blood flow from the left atrium into the left ventricle is impeded, and in *mitral regurgitation* much of the blood that has flowed into the left ventricle is pumped back into the left atrium during systole rather than being pumped into the aorta. Therefore, the effect is reduced net movement of blood from the left atrium into the left ventricle. Obviously, the buildup of blood in the left atrium causes progressive increase in left atrial pressure, and this can result eventually in the development of lethal pulmonary edema. The high left atrial pressure also causes progressive enlargement of the left atrium, which increases the distance that the cardiac impulse must travel through the atrial walls. Eventually, this pathway becomes so long that it predisposes to development of circus movements. Therefore, in late stages of mitral valvular disease, atrial fibrillation is an almost invariable result. This state further reduces the pumping effectiveness of the heart and, therefore, causes still further cardiac debility.

Blood Volume Compensation in Valvular Disease. In most patients with severe valvular disease and in many types of congenital heart disease, the blood volume increases. This tends to increase venous return to the heart, thereby helping to overcome the cardiac debility. Therefore, cardiac output does not fall greatly until the late stages of valvular disease.

ABNORMAL CIRCULATORY DYNAMICS IN CONGENITAL HEART DEFECTS

Occasionally, the heart or its associated blood vessels are malformed during fetal life; the defect is called a congenital anomaly. Two common and representative types of congenital anomalies are *patent ductus arteriosus* and *tetralogy of Fallot.*

Patent Ductus Arteriosus. During fetal life the lungs are collapsed, and the elastic factors that keep the alveoli collapsed also keep the blood vessels collapsed, causing almost all the pulmonary arterial blood to flow through the ductus arteriosus into the aorta rather than through the lungs. As soon as the baby is born, his lungs inflate; and not only do the alveoli fill, but the resistance to blood flow through the pulmonary vascular tree decreases tremendously, allowing pulmonary arterial pressure to fall. Simultaneously, the aortic pressure rises because of sudden cessation of blood flow through the placenta. Thus, the pressure in the pulmonary artery falls, while that in the aorta rises. As a result, forward blood flow through the ductus ceases suddenly at birth, and blood even flows backward from the aorta to the pulmonary artery. This new state of blood flow causes the ductus arteriosus to become occluded within a few hours to a few days in most babies so that blood flow through the ductus does not persist. The possible causes of ductus closure will be discussed in Chapter 56. In many instances it takes several months for the ductus to close completely, and in about 1 out of every 5500 babies the ductus never closes, causing the condition known as *patent ductus arteriosus*, which is illustrated in Figure 22–5.

Dynamics of persistent patent ductus. During the early months of an infant's life a patent ductus usually does not cause severely abnormal dynamics because the blood pressure of the aorta then is not much higher than the pressure in the pulmonary artery, and only a small amount of blood flows backward into the pulmonary system. However, as the child grows older, the differential between the pressure in the aorta and that in the pulmonary artery progressively increases with corresponding increase in the backward flow of blood from the aorta to the pulmonary artery. Soon, as much as half to three-fourths of the aortic blood flows into the pulmonary artery, then through the lungs, into the left atrium, and finally back into the left ventricle, passing through this circuit two or more times for every one time that it passes through the systemic circulation.

The major effect of patent ductus arteriosus on the patient is low cardiac

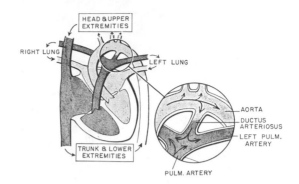

FIGURE 22-5 Patent ductus arteriosus, illustrating the degree of blood oxygenation in the different parts of the circulation.

and respiratory reserve. The left ventricle is already pumping approximately two or more times the normal cardiac output, and the maximum that it can possibly pump is about four to six times normal. Therefore, during exercise the cardiac output can be increased much less than usual. Also, the high pressures in the pulmonary vessels soon lead to pulmonary congestion.

The entire heart usually hypertrophies greatly in patent ductus arteriosus. The left ventricle hypertrophies because of the excessive work load that it must perform in pumping a far greater than normal cardiac output, while the right ventricle hypertrophies because of increased pulmonary arterial pressure resulting from, first, increased flow of blood through the lungs caused by the extra blood from the patent ductus and, second, increased resistance to blood flow through the lungs caused by progressive sclerosing of the vessels as they are exposed year in and year out to excessive pulmonary blood flow.

As a result of the increased load on the heart and because of the pulmonary congestion, most patients with patent ductus die between the ages of 20 and 40 unless the defect is corrected by surgery.

The machinery murmur. In the infant with patent ductus arteriosus, occasionally no abnormal heart sounds are heard because the quantity of reversed blood flow may be insignificant. As the baby grows older, reaching the age of one to three years, a harsh, blowing murmur begins to be heard in the pulmonic area of the chest. This sound is much more intense during systole when the aortic pressure is high and much less intense during diastole, so that it waxes and wanes with each beat of the heart, creating the so-called "machinery murmur." The idealized phonocardiogram of this murmur is shown in Figure 22-4E.

Surgical treatment. Surgical treatment of patent ductus arteriosus is extremely simple, for all one needs to do is to ligate the patent ductus.

Tetralogy of Fallot. Tetralogy of Fallot is illustrated in Figure 22-6, from which it will be noted that four different abnormalities of the heart occur simultaneously.

First, the aorta originates from the right ventricle rather than the left, or it overrides the septum as shown in the figure.

Second, the pulmonary artery is stenosed so that a much less than normal amount of blood passes from the right side of the heart into the lungs; instead the blood passes into the aorta.

Third, blood from the left ventricle flows through a ventricular septal defect into the right ventricle and then into the aorta or directly into the overriding aorta.

Fourth, because the right side of the heart must pump large quantities of blood against the high pressure in the

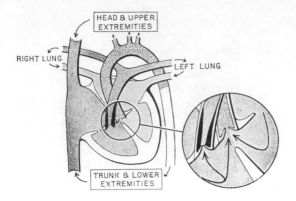

FIGURE 22–6 Tetralogy of Fallot, illustrating the degree of blood oxygenation in the different parts of the circulation.

aorta, its musculature is highly developed, causing an enlarged right ventricle.

Abnormal dynamics. It is readily apparent that the major physiological difficulty caused by tetralogy of Fallot is shunt of blood past the lungs without its becoming oxygenated. As much as 75 per cent of the venous blood returning to the heart may pass directly from the right ventricle into the aorta without becoming oxygenated. Tetralogy of Fallot is the major cause of cyanosis in babies ("blue babies").

Surgical treatment. In recent years tetralogy of Fallot has been treated relatively successfully by surgery. One type of treatment is to create an artificial ductus arteriosus by making a small opening between the aorta and pulmonary artery, thereby correcting the cyanosis and increasing the life expectancy of the patient with tetralogy of Fallot from approximately 1 to 10 years to an age of perhaps 50 or more years. Another operation is to open the pulmonary stenosis and close the septal defect in those cases in which this is possible.

USE OF EXTRACORPOREAL CIRCULATION IN CARDIAC SURGERY

It is almost impossible to repair intracardiac defects while the heart is still pumping. Therefore, many different types of artificial *heart-lung machines* have been developed to take the place of the heart and lungs during the course of operation. Such a system is called an *extracorporeal circulation.* The system consists principally of (1) a pump and (2) an oxygenating device. Almost any type of pump that does not cause hemolysis of the blood seems to be suitable.

The different principles that have been used for oxygenating blood are (1) bubbling oxygen through the blood and then removing the bubbles from the blood before passing it back into the patient, (2) dripping the blood downward over the surfaces of large areas of screen wire, (3) passing the blood over the surfaces of rotating discs, and (4) passing the blood between thin membranes that are porous to oxygen and carbon dioxide.

The different principles that have been fraught with many difficulties, including hemolysis of the blood, development of small clots in the blood, likelihood of small bubbles of oxygen or small emboli of antifoaming agent passing into the arteries of the patient, necessity for large quantities of blood to prime the entire system, failure to exchange adequate quantities of oxygen, and the necessity to use heparin in the system to prevent blood coagulation, the heparin also preventing adequate hemostasis during the surgical procedure. Yet, despite these difficulties, in the hands of experts patients can be kept on artificial heart-lung machines for several hours while operations are performed on the inside of the heart.

HYPERTROPHY OF THE HEART IN VALVULAR AND CONGENITAL HEART DISEASE

Hypertrophy of cardiac muscle is one of the most important mechanisms by which the heart adapts to increased work loads, whether these loads be caused by increased pressure against which the heart muscle must contract or by increased volume that must be pumped. Some investigators believe that it is the increased work load itself that causes the hypertrophy; others believe the increased metabolic rate of the muscle to be the primary stimulus and the work load simply to be the cause of the increase in the metabolic rate. Regardless of which of these is correct, one can calculate approximately how much hypertrophy will ocur in each chamber of the heart by multiplying ventricular output times the pressure against which the ventricle must work. In *aortic stenosis* and *aortic regurgitation*, the left ventricular musculature hypertrophies tremendously, sometimes to as much as four to five times normal, so that the weight of the heart on occasion may be as great as 1000 grams instead of the normal 300 grams.

REFERENCES

Braunwald, E., Ross, J., and Sonnenblick, E.: Mechanism of Contractility of the Normal and Failing Heart. Boston, Little, Brown and Company, 1968.

Butterworth, J. S., Chassin, M. R., McGrath, R., and Reppert, E. H.: Cardiac Auscultation— Including Audio-Visual Principles. 2nd ed., New York, Grune & Stratton, Inc., 1960.

Cooley, D. A., and Hallman, G. L.: Surgical Treatment of Congenital Heart Disease. Philadelphia, Lea & Febiger, 1966.

Friedberg, C. K.: Diseases of the Heart. 3rd ed., Philadelhia, W. B. Saunders Company, 1966.

Guyton, A. C.: Cardiac output and venous return in heart failure. *In* Luisada, A. A. (ed): Cardiology. New York, McGraw-Hill Book Co., 1959, Vol. IV, p. 18–8.

Harrison, T. R., and Reeves, T. J.: Principles and Problems of Ischemic Heart Disease. Chicago, Year Book Medical Publishers, Inc., 1968.

Myerson, R. M., and Pastor, B. H.: Congestive Heart Failure. St. Louis, The C. V. Mosby Co., 1967.

Ravin, A.: Auscultation of the Heart. Baltimore, The Williams & Wilkins Co., 1967.

Taussig, H.: Congenital Malformations of the Heart. 2nd ed., Cambridge, Mass., Harvard University Press, 1960.

CHAPTER 23

BLOOD FLOW IN SPECIAL AREAS OF THE BODY: CORONARY, CEREBRAL, SPLANCHNIC, MUSCLE, AND SKIN

THE CORONARY CIRCULATION

Approximately one-third of all deaths result from coronary artery disease, and almost all elderly persons have at least some impairment of coronary artery circulation. For this reason, the purpose of this chapter is to present in detail the subject of coronary circulation, emphasizing especially the physiology of recovery from coronary thrombosis.

NORMAL CORONARY BLOOD FLOW AND ITS VARIATIONS

Physiologic Anatomy of the Coronary Blood Supply. In the human being, the left coronary artery supplies a large portion of the left ventricle, and the right coronary artery usually supplies the right ventricle plus a large part of the

left ventricle as well. In about one-half of all human beings, more blood flows through the right coronary artery than through the left, whereas the left artery predominates in only 20 per cent.

Most of the venous blood flow from the left ventricle leaves by way of the *coronary sinus* — about 70 per cent of the total coronary blood flow — and most of the venous blood from the right ventricle flows through the small *anterior cardiac veins*, which empty directly into the right atrium and are not connected with the coronary sinus.

Normal Coronary Flow. The resting coronary blood flow in the human being averages approximately 225 ml. per minute, which is about 0.8 ml. per gram of heart muscle, or 4 to 5 per cent of the total cardiac output.

In strenuous exercise, the coronary blood flow increases four- to five-fold to supply the extra nutrients needed by the heart.

256

LOCAL BLOOD FLOW REGULATION AS THE PRIMARY CONTROLLER OF CORONARY FLOW

Blood flow through the coronary system is regulated almost entirely by local blood flow regulation in the heart tissue itself in response to the needs of the cardiac musculature for nutrition. This mechanism works equally well when the nerves to the heart are intact or are removed. Whenever the vigor of contraction is increased, regardless of cause, the rate of coronary blood flow simultaneously increases, and, conversely, decreased activity is accompanied by decreased coronary flow. It is immediately obvious that this local regulation of blood flow is almost identical with that which occurs in many other tissues, especially in the skeletal muscles of all the body.

Oxygen Demand as the Mechanism of Local Blood Flow Regulation. Blood flow in the coronaries is regulated almost exactly in proportion to the need of the cardiac musculature for oxygen. Even in the normal resting state, about 65 per cent of the oxygen in the arterial blood is removed as the blood passes through the heart; and, because not much oxygen is left in the blood, little additional oxygen can be removed from the blood unless the flow increases. Fortunately, the blood flow does increase, and almost directly in proportion to the need of the heart for oxygen. Therefore, it is believed that *oxygen lack dilates the coronary arterioles.*

Yet the means by which oxygen lack causes coronary dilatation has not been determined. The two principal possibilities that have been suggested are: (1) Decreased oxygen tension in the cardiac tissues reduces the amount of oxygen available to the coronary vessels themselves and this causes the coronaries to become weakened and, therefore, to dilate automatically. (2) Oxygen lack causes vasodilator substances, such as adenosine compounds, to be released by the tissues. (Small amounts of adenosine are released during oxygen lack, but it has not been shown that it is released in sufficient quantity to cause the extreme degree of vasodilatation that results.) Therefore, for the present we can simply say that oxygen lack is followed by coronary vasodilatation and in this way regulates blood flow to the cardiac musculature in proportion to the metabolic need for oxygen by the muscle fibers.

Tension X Time as the Primary Determinant of Oxygen Consumption. Perhaps the best relationship that has yet been found between cardiac function and oxygen consumption is: *oxygen consumption is proportional to myocardial muscle tension × time of contraction.* Thus, when the arterial pressure rises, the muscle tension increases and oxygen consumption also increases. Likewise, when the heart dilates, which makes it necessary (because of the law of Laplace) for the muscle to generate increased tension to pump against even a normal arterial pressure, oxygen consumption also increases even though the work output of the heart does not increase.

Reactive Hyperemia in the Coronary System. If the coronary flow to the

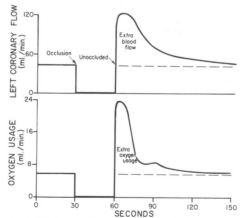

FIGURE 23-1 Reactive hyperemia in the coronary system caused by a 30 second period of coronary occlusion. Note the extra blood flow called "reactive hyperemia" and the extra oxygen usage after the period of occlusion was over. (Reconstructed from data in Gregg and Fisher: Handbook of Physiology, Section II, Vol. I. The Williams & Wilkins Co.)

heart is completely occluded for a few seconds to a few minutes and then suddenly unoccluded, the blood flow increases to as high as three to six times normal, as shown in Figure 23–1. It remains high for a few seconds to several minutes, depending on the period of occlusion. This extra flow of blood is called reactive hyperemia, which was explained in Chapter 18. During the period of excess flow, the heart removes a large amount of extra oxygen from the blood, as shown by the bottom curve of the figure, to make up for the deficiency of oxygen during the period of occlusion. Reactive hyperemia is simply another manifestation of the ability of the coronary system to adjust its flow to the need of the heart for oxygen.

NERVOUS CONTROL OF CORONARY BLOOD FLOW

Stimulation of the autonomic nerves to the heart can affect coronary blood flow in two ways—directly and indirectly. The direct effects result from direct action of the nervous transmitter substances, acetylcholine and norepinephrine, on the coronary vessels themselves. The indirect effects result from secondary changes in coronary blood flow caused by increased or decreased activity of the heart.

The indirect effects probably play the more important role in control of coronary blood flow. Thus, sympathetic stimulation increases both heart rate and heart contractility as well as its rate of metabolism. In turn, the increased activity of the heart sets off the local flow regulatory mechanism to increase blood flow approximately in proportion to the metabolic needs of the heart muscle. However, sympathetic stimulation also has a weak direct effect that dilates the coronaries, thus further increasing the coronary blood flow.

The distribution of parasympathetic (vagal) nerve fibers to the ventricular coronary system is so slight that para-sympathetic stimulation has an almost negligible direct effect, but the effect of vagal stimulation to slow heart rate can at times indirectly decrease coronary flow.

In summary, the direct nervous influences on coronary blood flow are unimportant in comparison with the local ability of the blood vessels to adjust blood flow according to the metabolic needs of the cardiac musculature. But it is possible that sympathetic stimulation of the coronaries does on some occasions moderately enhance or diminish blood flow to the heart muscle.

CORONARY OCCLUSION

The coronary blood vessels sometimes become occluded rapidly—within a few minutes or a few hours—as a result of blood clots or other abnormalities that can plug the lumina of the vessels. At other times the vessels are slowly and progressively occluded over a period of years, in which case a collateral blood supply can usually develop to take over the function of the primary coronary blood supply. Unfortunately, the collateral blood supply usually does not become well developed until after the coronaries are occluded. Therefore, whether or not the heart is greatly damaged by coronary occlusion depends mainly on the rapidity with which the occlusion occurs.

Collateral Circulation in the Heart. Relatively few communications exist among the larger coronary arteries, but there are many anastomoses between the small arteries approximately 20 to 350 microns in diameter, as shown in Figure 23–2.

When a sudden occlusion occurs in one of the larger coronary vessels, the minute anastomoses increase to their maximum physical diameters within a few seconds. The blood flow through these minute collaterals is less than one-half that needed to keep alive the cardiac muscle that they supply; and unfor-

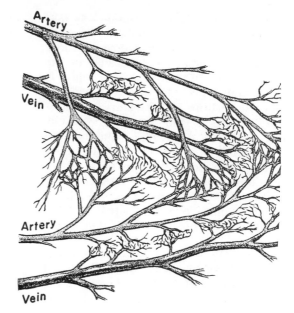

FIGURE 23–2 Minute anastomoses of the coronary arterial system.

tunately the diameters of the collateral vessels do not enlarge further for the next 8 to 24 hours. But then collateral flow begins to increase, doubling in the next one to two days, often reaching normal or almost normal coronary supply within about one month. It is because of these developing collateral channels that a patient recovers from the various types of coronary occlusion.

Atherosclerosis and Coronary Thrombosis. The most frequent cause of coronary occlusion is *thrombosis* resulting from atherosclerosis. The atherosclerotic process will be discussed in connection with lipid metabolism in Chapter 46, but, to summarize this process, in various diseases affecting lipid metabolism and also in old age, lipids containing mainly cholesterol and cholesterol salts are deposited beneath the intima of the major arteries. These deposits become calcified over a period of years, and considerable fibrous tissue also invades the walls of the degenerating arteries. Furthermore, "atherosclerotic plaques" occasionally break through the intima of the blood vessel and protrude into the lumen. The presence of such a rough

surface inside a vessel initiates the clotting process. When a small clot has developed, platelets become entrapped and cause more clot to develop until the vessel is plugged; or the clot breaks away and plugs smaller vessels farther downstream. This mechanism causes most coronary occlusions.

If the blood flow to cardiac muscle is diminished beyond a critical level, the muscle not only becomes nonfunctional but actually begins to die. Death of the muscle fibers begins within about one hour in total ischemia and is complete in four to five hours.

CAUSES OF DEATH IN SEVERE CORONARY ISCHEMIA

The four major causes of death following coronary occlusion are (1) decreased cardiac output; (2) damming of blood in the pulmonary or systemic veins with death resulting from edema, especially pulmonary edema; (3) fibrillation of the heart; and, (4) occasionally, rupture of the heart.

Decreased Cardiac Output—Systolic Stretch. The overall pumping strength of the heart is often decreased more than one might expect because of the phenomenon of *systolic stretch*, which is illustrated in Figure 23–3. When the normal portions of the ventricular muscle contract, the ischemic muscle, whether this be dead or simply nonfunctional, instead of contracting is actually forced outward by the pressure that develops inside the ventricle. Owing to this effect, much of the pumping force of the ventricle is dissipated, and the cardiac output is often diminished severely.

Damming of Blood in the Venous System. When the heart fails to pump, it dams blood in the venous systems of the lungs or the systemic circulation. In the early stages of coronary thrombosis this is not usually as serious as it is later, for during the first few days after a coronary occlusion progressively more fluid collects in the body because of renal shutdown, adding progressively to the venous congestive symptoms. When the congestion becomes intense, death frequently results from pulmonary edema or, rarely, from systemic congestive symptoms.

Rupture of the Ischemic Area. During the first day of an acute coronary infarct there is little danger of rupture of the ischemic portion of the heart, but a few days after a large infarct occurs, the dead muscle fibers begin to degenerate, and the dead heart musculature is likely to become very thin and to tear.

When a ventricle does rupture, loss of blood into the pericardial space causes rapid development of *cardiac tamponade*—that is, compression of the heart from the outside by blood collecting in the pericardial cavity. Because the heart is then compressed, blood cannot flow into the right atrium with ease, and the patient dies of decreased cardiac output.

Fibrillation of the Ventricles Following Coronary Occlusion. Many persons who die of coronary occlusion die because of ventricular fibrillation. The tendency to develop fibrillation is especially great following a large occlusion, but fibrillation occasionally occurs even following a small occlusion.

There are two especially dangerous periods during which fibrillation is likely to occur. The first of these is during the first 10 minutes after the occlusion occurs. Then there is a period of relative safety, followed by a secondary period of cardiac irritability beginning three to five hours after the occlusion and lasting for another five to six hours.

At least three different factors enter into the tendency for the heart to fibrillate:

First, acute loss of blood supply to the cardiac muscle causes rapid depletion of potassium from the ischemic musculature. This increases the potassium concentration in the extracellular fluids surrounding the cardiac muscle fibers, which in itself increases the irritability of the cardiac musculature.

Second, ischemia of the muscle causes an "injury current," which was described in Chapter 16 in relation to electrocardiograms in patients with acute coronary occlusion. The ischemic

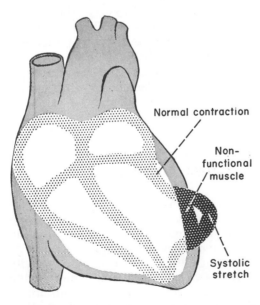

Normal contraction

Non-functional / muscle

Systolic stretch

FIGURE 23–3 Systolic stretch in an area of ischemic cardiac muscle.

musculature cannot repolarize its membranes so that this muscle remains negative with respect to the normal polarized cardiac muscle membrane. Therefore, electrical current flows from this ischemic area of the heart to the normal area and can elicit abnormal impulses which can cause fibrillation.

Third, powerful sympathetic reflexes develop following massive coronary occlusion, principally because the heart does not pump an adequate volume of blood into the arterial tree. The sympathetic stimulation also increases the irritability of the cardiac muscle and thereby predisposes to fibrillation.

Shortly after a coronary occlusion, muscle fibers begin to die in the very center of the ischemic area, and during the ensuing days this area of dead fibers grows, because many of the marginal fibers finally succumb to the prolonged ischemia. At the same time, owing to the enlargement of collateral arterial channels, the nonfunctional (but not dead) area of the heart muscle becomes smaller and smaller. After approximately two to three weeks most of the nonfunctional area of muscle becomes functional again. In the meantime, fibrous tissue begins developing among the dead fibers, for ischemia stimulates growth of fibroblasts and promotes development of greater than normal quantities of fibrous tissue. Therefore, the dead muscular tissue is gradually replaced by fibrous tissue. Then, because it is a general property of fibrous tissue to undergo progressive elastomeric contraction and dissolution, the size of the fibrous scar becomes smaller and smaller over a period of several months to a year.

During progressive recovery of the infarcted area of the heart, the development of a strong, fibrous scar which becomes smaller and smaller finally stops the original systolic stretch, and the functional musculature once again becomes capable of exerting its entire force for pumping blood. Furthermore, the normal areas of the heart gradually hypertrophy to compensate at least partially for the lost cardiac musculature. By these means the heart recovers.

PAIN IN CORONARY DISEASE — ANGINA PECTORIS

Normally, a person cannot "feel" his heart, but ischemic cardiac muscle does exhibit pain sensation. Exactly what causes this pain is not known, but it is believed that ischemia causes the muscle to release acidic or other pain-promoting products such as histamine or kinins that are not removed rapidly enough by the slowly moving blood. The high concentrations of these abnormal products then stimulate the pain endings in the cardiac muscles, and pain impulses are conducted through the sympathetic afferent nerve fibers into the central nervous system. The pain is usually referred to surface areas of the body, mainly the left arm and left shoulder; this will be discussed in detail in Chapter 33.

Ischemic pain that occurs intermittently from the heart is known as *angina pectoris*, and it occurs whenever the load on the heart becomes too great in relation to the blood flow through the coronary vessels. In general, most persons who have chronic angina pectoris feel the pain only when they exercise or when they experience emotions that increase the metabolism of the heart. However, some patients have such severe and lasting cardiac ischemia that pain is present all the time.

CEREBRAL CIRCULATION

NORMAL RATE OF CEREBRAL BLOOD FLOW

The normal blood flow through brain tissue averages 50 to 55 ml. per 100 grams of brain per minute. For the entire brain of the average adult, this is approximately 750 ml. per minute, or 15 per cent of the total resting cardiac output. This blood flow through the brain, even under extreme conditions, usually does not vary greatly from the normal value because the control systems are especially geared to maintain constant cerebral blood flow. Some exceptions to this occur when the brain is subjected to excess carbon dioxide or extreme lack of oxygen, as is discussed subsequently.

REGULATION OF CEREBRAL CIRCULATION

Regulation in Response to Excess Carbon Dioxide. The rate of blood flow through the brain is regulated principally by the concentration of carbon dioxide in the cerebral tissues. Excess carbon dioxide causes marked vasodilatation. Therefore, any time the cerebral blood flow becomes sluggish, causing a buildup of carbon dioxide in the cerebral blood, the blood flow automatically increases, thereby correcting this condition. Here again, one finds an important local regulatory mechanism for maintenance of blood flow to the brain in proportion to its metabolism.

Oxygen Deficiency as a Regulator of Cerebral Blood Flow. The rate of metabolism and also the rate of oxygen utilization by the brain are extremely constant under many different conditions, varying little during intense mental activity, during muscular exercise, or even during sleep. Indeed, the rate of oxygen utilization remains within a few per cent of 3.5 ml. of oxygen per 100 gm. of brain tissue per minute. If ever the blood flow to the brain becomes insufficient to supply this amount of oxygen, the oxygen deficiency mechanism for vasodilatation, present in almost all tissues of the body, immediately causes vasodilatation, allowing the transport of oxygen to the cerebral tissues to return toward normal. Thus, this autoregulatory mechanism is much the same as that which exists in the coronary and muscle circulations and in many other circulatory areas of the body. The need for constant excitability of the neurons — never depressed and never overexcited — makes this mechanism of special importance in the cerebral circulation.

Effect of Cerebral Activity on Blood Flow. Only rarely does neuronal activity increase sufficiently to increase the overall rate of metabolism in the brain. And, in those few instances in which this does occur, the cerebral blood flow increases only a moderate amount. For instance, a convulsive attack in an animal, which causes extreme activity throughout the entire brain, can result in an overall increase in cerebral blood flow of as much as 50 per cent. On the other hand, administration of anesthetics sometimes reduces brain metabolism and also cerebral blood flow as much as 30 to 40 per cent.

Blood flow in localized portions of the brain can increase as much as 40 to 50 per cent as a result of intense localized neuronal activity even though the total cerebral flow is hardly affected. This effect is illustrated in Figure 23–4, which shows the effect of shining an intense light into one eye of a cat; this results in increased blood flow in the occipital cortex that becomes excited.

THE SPLANCHNIC CIRCULATION

A large share of the cardiac output flows through the vessels of the intestines and through the spleen, finally coursing into the portal venous system

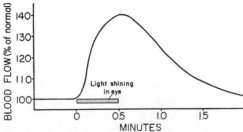

FIGURE 23–4 Increase in blood flow to the occipital regions of the brain when a light is flashed in the eyes of an animal.

and then through the liver, as illustrated in Figure 23–5. This is called the portal circulatory system, and it, plus the arterial blood flow into the liver, is called the splanchnic circulation.

Blood Flow Through the Liver. About 1100 ml. of portal blood enter the liver each minute. This flows through the *hepatic sinuses* in close contact with the cords of liver parenchymal cells. Then it enters the *central veins* of the liver and from there flows into the vena cava.

In addition to the portal blood flow, approximately 350 ml. of blood flows into the liver each minute through the hepatic artery, making a total hepatic flow of almost 1500 ml. per minute, or an average of 29 per cent of the total cardiac output. This arterial blood flow maintains nutrition of the connective tissue and especially of the walls of the bile ducts. Therefore, loss of hepatic arterial blood flow can be lethal because this often causes necrosis of the basic liver structures. The blood from the hepatic artery, after it supplies the structural elements of the liver, empties into the hepatic sinuses to mix with the portal blood.

Reservoir Function of the Liver. Because the liver is an expandable and compressible organ, large quantities of blood can be stored in its blood vessels. Its normal blood volume, including both that in the hepatic veins and hepatic sinuses, is about 650 ml., or 13 per cent of the total blood volume. However, when high pressure in the right atrium causes back pressure in the liver, the liver expands, and as much as 1 liter of extra blood occasionally is thereby stored in the hepatic veins and sinuses.

Permeability of the Hepatic Sinuses. The hepatic sinuses are lined with an endothelium similar to that of the capillaries, but its permeability is extreme in comparison with that of usual capillaries — so much so that even the proteins of the blood diffuse into the extravascular spaces of the liver almost as easily as fluids.

This extreme permeability of the liver

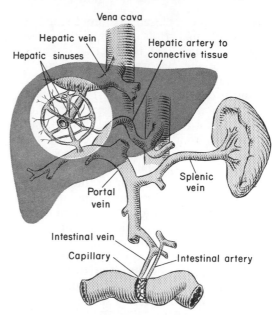

FIGURE 23–5 The portal and hepatic circulations.

sinuses brings the fluids of the hepatic blood into extremely close contact with the liver parenchymal cells, thus facilitating rapid exchange of nutrient materials between the blood and the liver cells.

The Blood Cleansing Function of the Liver. Special high-speed motion pictures of the action of Kupffer cells, the large phagocytic cells that line the hepatic sinuses, have demonstrated that these cells can cleanse blood extremely efficiently as it passes through the sinuses; when a bacterium comes into momentary contact with a Kupffer cell, in less than 0.01 second the bacterium passes inward through the wall of the Kupffer cell to become permanently lodged therein until it is digested. Probably not over 1 per cent of the bacteria entering the portal blood from the intestines succeeds in passing through the liver into the systemic circulation.

The Metabolic "Buffering" Function of the Liver. The function of blood flow through the intestines is to absorb into the body fluids the products of digestion. These products, after entering the portal system, pass through the venous sinuses of the liver in close proximity to the liver parenchymal cells. Before the blood leaves the sinuses, the liver cells remove most of the glucose absorbed from the intestinal tract, most of the amino acids, and probably most of the fat that occasionally is absorbed in special forms into the portal blood. Removal of these substances from the portal blood prevents excessive increase in their concentrations in the systemic blood immediately after a meal. Then, during the following 8 to 24 hours or more, the removed products of digestion are released by the liver into the systemic circulatory system so that they flow continually at well-controlled blood concentrations to the remainder of the body. Also, the liver changes some of these products of digestion into new forms so that they can be used to better advantage by the remainder of the body, as will be discussed later in the chapters on metabolism.

BLOOD FLOW THROUGH THE INTESTINAL VESSELS

About four-fifths of the portal blood flow originates in the intestines and stomach (about 850 ml. per minute), and the remaining one-fifth originates in the spleen and pancreas.

Control of Gastrointestinal Blood Flow. Blood flow in the gastrointestinal tract seems to be controlled in almost exactly the same way as in most other areas of the body: that is, mainly by local regulatory mechanisms. Furthermore, blood flow to the mucosa and submucosa, where the glands are located and where absorption occurs, is controlled separately from blood flow to the musculature. When glandular secretion increases, so does mucosal and submucosal blood flow. Likewise, when motor activity of the gut increases, blood flow in the muscle layers increases.

However, the precise mechanisms by which alterations in gastrointestinal activity alter the blood flow are not completely understood. It is known that decreased availability of oxygen to the gut increases local blood flow in the same way that this occurs elsewhere in the body, so that local regulation of blood flow in the gut might occur entirely secondarily to changes in metabolic rate.

Nervous control of gastrointestinal blood flow. Sympathetic stimulation, on the other hand, has a direct effect on essentially all blood vessels of the gastrointestinal tract to cause intense vasoconstriction, sometimes reducing intestinal blood flow almost to zero. A major value of sympathetic vasoconstriction in the gut is that it allows shutting off of blood flow when increased flow is needed by the muscles and heart during extreme exercise.

PORTAL VENOUS PRESSURE

The liver offers a moderate amount of resistance to the blood flow from the portal system to the vena cava. As a result, the pressure in the portal vein averages 10 mm. Hg, which is considerably higher than the almost zero pressure in the vena cava. Because of this high portal venous pressure, the pressures in the portal venules and capillaries have a much greater tendency to become abnormally high than is true elsewhere in the body. This occurs especially in *cirrhosis of the liver*, in which much of the liver tissue has been replaced by fibrous scar tissue.

Ascites as a Result of Portal Obstruction. Ascites is free fluid in the peritoneal cavity. It results from exudation of fluid either from the surface of the liver or from the surfaces of the gut and its mesentery. Ascites usually will develop only in case outflow of blood from the liver into the inferior vena cava is blocked. This causes extremely high pressure in the liver sinusoids, which in turn causes fluid to weep from the surfaces of the liver. The weeping fluid is almost pure plasma, containing tremendous quantities of protein. The protein, because it causes a high colloid osmotic pressure in the abdominal fluid, then pulls (by osmosis) additional fluid from the surfaces of the gut and mesentery.

On the other hand, obstruction of the portal vein, without directly involving the liver, rarely causes ascites. If obstruction occurs acutely as a result of a portal vein blood clot, the person is likely to die of shock within hours because of plasma loss into the gut; and if it occurs slowly, collateral vessels can usually develop rapidly enough to prevent ascites.

THE SPLENIC CIRCULATION

The Spleen as a Reservoir. The capsule of the spleen in many lower animals contains large amounts of smooth muscle; and sympathetic stimulation causes intense contraction of this capsule, whereas sympathetic inhibition results in considerable splenic expansion with consequent storage of blood.

In man the splenic capsule is nonmuscular, but even so, constriction or dilatation of vessels within the spleen can still cause the spleen to store several hundred milliliters of blood at times, and then, under the influence of sympathetic stimulation, to express most of this blood into the general circulation. Unfortunately, these effects in man are poorly understood.

As illustrated in Figure 23–6, two areas exist in the spleen for the storage of blood; these are the venous sinuses and the pulp. Small vessels flow directly into the venous sinuses, and when the spleen distends, the venous sinuses swell, storing large quantities of blood.

In the splenic pulp, the capillaries are very permeable, so that much of the blood passes first into the pulp and then oozes through this before entering the venous sinuses. As the spleen enlarges, many cells become stored in the pulp. Therefore, the net quantity of red blood cells in the general circulation decreases slightly when the spleen enlarges. The spleen can store enough cells that splenic contraction can cause the hema-

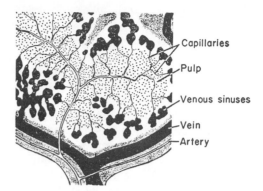

FIGURE 23–6 The functional structures of the spleen. (Modified from Bloom and Fawcett: Textbook of Histology.)

tocrit of the systemic blood to increase as much as 3 to 4 per cent.

This ability of the spleen to store and release blood is important (at least in lower animals) in times of stress such as during strenuous exercise, because the circulatory system then needs an increased volume of blood, as well as an increased concentration of red cells.

The Blood Cleansing Function of the Spleen. The pulp of the spleen contains many large phagocytic reticuloendothelial cells, and the venous sinuses are lined with similar cells. These cells act as a cleansing system for the blood, similar to that in the venous sinuses of the liver.

Much of the spleen is filled with *white pulp*, which is in reality a large quantity of lymphocytes and plasma cells. These function in exactly the same way in the spleen as in the lymph glands to cause either humoral or lymphocytic immunity against toxins, bacteria, and so forth, as described in Chapter 9.

BLOOD FLOW THROUGH THE SKELETAL MUSCLES

RATE OF BLOOD FLOW THROUGH THE MUSCLES

During rest, blood flow through skeletal muscle averages 4 to 7 ml. per 100 grams of muscle. However, during extreme exercise this rate can increase as much as 15- to 20-fold, rising to more than 100 ml. per 100 grams of muscle.

Intermittent Flow During Muscle Contraction. Figure 23-7 illustrates a study of blood flow changes in the calf muscles of the human leg during strong rhythmic contraction. Note that the flow increases and decreases with each muscle contraction, decreasing during the contraction phase and increasing between contractions. At the end of the rhythmic contractions, the blood flow remains very high for a minute or so and then gradually fades toward normal.

The cause of the decreased flow during sustained muscle contraction is compression of the blood vessels by the contracted muscle. During strong *tetanic* contraction, blood flow can be almost totally stopped.

Opening of Muscle Capillaries During Exercise. During rest, only about 10 per cent of the muscle capillaries are open. But during strenuous exercise all the capillaries open up, which can be demonstrated by studying histologic specimens removed from active muscles. It is this opening up of dormant capillaries that allows most of the increased blood flow. It also diminishes the dis-

tance that oxygen and other nutrients must diffuse from the capillaries to the muscle fibers.

CONTROL OF BLOOD FLOW THROUGH THE SKELETAL MUSCLES

Local Regulation. The tremendous increase in blood flow that occurs during skeletal muscle contraction is caused primarily by local effects in the muscles acting directly on the arterioles to cause vasodilatation. These effects occur even after all sympathetic nerves to the muscle have been cut. Furthermore, the degree of dilatation is as great (except at the very beginning of muscle activity, as will be noted later) without these nerves as it is when they are present.

The local increase in blood flow during muscle contraction is probably caused by several different factors all operating at the same time. One of the most important of these is reduction of dissolved oxygen in the muscle tissues. That is, during muscle activity the muscle utilizes oxygen very rapidly, thereby decreasing its concentration in the tissue fluids. This in turn causes vasodilatation either because the vessel walls cannot maintain contraction in the absence of oxygen or because oxygen deficiency causes release of some vasodilator material. The vasodilator material that has been suggested most widely in recent years has been adenosine.

Other vasodilator substances released during muscle contraction include potassium ions, acetylcholine, adenosine triphosphate, lactic acid, and carbon dioxide. Unfortunately, we still do not know quantitatively how much role each of these plays in increasing muscle blood flow during muscle activity.

Finally, an additional factor that increases blood flow during muscle contraction is elevation of arterial pressure, an effect that usually accompanies mus-

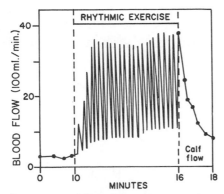

FIGURE 23-7 Effect of muscle exercise on blood flow in the calf of a leg during strong rhythmic contraction. The blood flow was much less during contraction than between contractions. (From Barcroft and Dornhorst: *J. Physiol., 109*: 402, 1949.)

cle exercise. Because even slight elevation in pressure stretches the arterioles considerably, as little as a 20 to 30 mm. Hg rise in pressure can often double the blood flow.

Nervous Control of Muscular Blood Flow. In addition to the basic autoregulatory mechanism, the skeletal muscles are also provided with both sympathetic vasoconstrictor fibers and sympathetic vasodilator fibers.

Sympathetic vasoconstrictor fibers. The sympathetic vasoconstrictor fibers secrete norepinephrine and when maximally stimulated can decrease blood flow through the muscles to about one-fourth normal. This represents rather poor vasoconstriction in comparison with that caused by sympathetic nerves in some other areas of the body in which blood flow can be almost completely blocked. Yet under some conditions even this degree of vasoconstriction may be of physiological importance, such as during shock and other periods of stress when it would be desirable to reduce blood flow through the many muscles of the body.

Sympathetic vasodilator fibers. The sympathetic vasodilator fibers to the skeletal muscles secrete acetylcholine and on maximal stimulation can increase blood flow by 400 per cent. These vasodilator fibers are activated by a special nervous pathway beginning in the cerebral cortex in close association with the motor areas for control of muscular activity and passing downward through the hypothalamus and brain stem into the spinal cord, as was explained in Chapter 19. When the motor cortex initiates muscle activity it simultaneously excites the vasodilator fibers to the active muscles, and vasodilatation occurs immediately, several seconds before the local regulatory vasodilatation can take place. Thus, it seems that this vasodilator system has the important *function of initiating extra blood flow through the muscles at the onset of muscular activity.*

CIRCULATION IN THE SKIN

PHYSIOLOGIC ANATOMY OF THE CUTANEOUS CIRCULATION

Circulation through the skin subserves two major functions: first, *nutrition of the skin tissues* and, second, *conduction of heat* from the internal structures of the body to the skin so that the heat can be removed from the body. To perform these two functions the circulatory apparatus of the skin is characterized by two major types of vessels, illustrated diagrammatically in Figure 23–8: (1) the usual nutritive arteries, capillaries, and veins and (2) vascular structures concerned with heating the skin, consisting principally of (a) an extensive *subcutaneous venous plexus*, which holds large quantities of blood that can heat the surface of the skin, and (b) in some skin areas, *arteriovenous anastomoses*, which are large vascular communications directly between the arteries and the venous plexuses. The walls of these anastomoses have strong muscular coats innervated by sympathetic vasoconstrictor nerve fibers that secrete norepinephrine. When constricted, they reduce the flow of blood into the venous plexuses to almost nothing; or when maximally dilated, they allow extremely rapid flow of warm blood into the plexuses. The arterio-

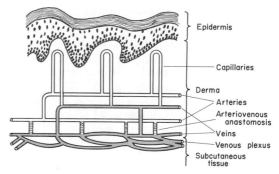

FIGURE 23–8 Diagrammatic representation of the skin circulation.

venous anastomoses are found principally in the volar surfaces of the hands and feet, the lips, the nose, and the ears, which are areas of the body most often exposed to maximal cooling.

Rate of Blood Flow Through the Skin. The rate of blood flow through the skin is among the most variable of those in any part of the body, because the flow required to regulate body temperature changes markedly in response to, first, the rate of metabolic activity of the body, and, second, the temperature of the surroundings. This will be discussed in detail in Chapter 47. The blood flow required for nutrition is slight, so that this plays almost no role in controlling normal skin blood flow. At ordinary skin temperatures, the amount of blood flowing through the skin vessels to subserve heat regulation is about 10 times as much as that needed to supply the nutritive needs of the tissues. But, when the skin is exposed to extreme cold, the blood flow may become so slight that nutrition begins to suffer—even to the extent, for instance, that the fingernails grow considerably more slowly in arctic climates than in temperate climates.

Under ordinary cool conditions the blood flow to the skin is about 0.25 liter/sq. meter of body surface area, or a total of about 400 ml. per minute, in the average adult. On the other hand, when the skin is heated until maximal vasodilatation has resulted, the blood flow can be as much as 7 times this value, or a total of about 2.8 liters per minute, thus illustrating both the extreme variability of skin blood flow and the great drain on cardiac output that can occur under hot conditions. Indeed, many persons with borderline cardiac failure become decompensated in hot weather because of the extra load on the heart and recompensated in cool weather.

REGULATION OF BLOOD FLOW IN THE SKIN

Nervous Control of Cutaneous Blood Flow. Since the principal function of blood flow through the skin is to control body temperature, and since this function in turn is regulated by the nervous system, the blood flow through the skin is principally regulated by nervous mechanisms rather than by local regulation, which is opposite to the regulation in most parts of the body. The temperature control center of the hypothalamus controls the nerve impulse traffic to the skin blood vessels, modulating the skin blood flow in accord with the body's need to conserve or eliminate heat. This mechanism of body heat control will be discussed in detail in Chapter 47.

REFERENCES

Barcroft, H.: Circulation in skeletal muscle. *In* Handbook of Physiology. Baltimore, The Williams & Wilkins Co., 1963, Sec. II, Vol. II, p. 1353.

Berne, R. M.: Regulation of coronary blood flow. *Physiol. Rev., 44*:1, 1964.

Bevegard, B. S., and Shepherd, J. T.: Regulation of the circulation during exercise in man. *Physiol. Rev., 47*:178, 1967.

Blumgart, H.: Symposium on Coronary Heart Disease. 2nd ed., American Heart Association, 1967.

Bradley, S. E.: The hepatic circulation. *In* Handbook of Physiology. Baltimore, The Williams & Wilkins Co., 1963, Sec. II, Vol. II, p. 1387.

Cerebral blood flow (Symposium). *Acta Physiol. Scand., 66*, Suppl. 258, 1966.

Green, H. D., and Kepchar, J. H.: Control of peripheral resistance in major systemic vascular beds. *Physiol. Rev., 39*:617, 1959.

Gregg, D. E.: Coronary Circulation in Health and Disease. Philadelphia, Lea & Febiger, 1950.

Jacobson, E. D.: The gastrointestinal circulation. *Ann. Rev. Physiol., 30*:133, 1968.

Shepherd, J. T.: Physiology of Circulation in the Limbs. Philadelphia, W. B. Saunders Company, 1963.

PART III

THE
BODY FLUIDS
AND
KIDNEYS

CHAPTER 24

FORMATION OF URINE BY THE KIDNEY

The kidneys perform two major functions: first, they excrete most of the end-products of bodily metabolism, and, second, they control the concentrations of most of the constituents of the body fluids. The purpose of the present chapter is to discuss the principles of urine formation and especially the mechanisms by which the kidneys excrete the end-products of metabolism.

PHYSIOLOGIC ANATOMY OF THE KIDNEY

Each kidney contains about 1,000,000 nephrons, every one of which is capable of forming urine by itself. The nephron is composed basically of (1) a *glomerulus* through which fluid is filtered out of the blood and (2) a long coiled *tubule* in which the filtered fluid is converted into urine on its way to the *pelvis* of the kidney. Figure 24–1 shows the general organizational plan of the kidney, illustrating especially the distinction between the *cortex* of the kidney and the *medulla*. Figure 24–2 illustrates the basic anatomy of the nephron, which may be described as follows: Blood enters the glomerulus through the *afferent arteriole* and then leaves through the *efferent arteriole*. The glomerulus is a network of up to 50 parallel capillaries encased in *Bowman's capsule*.

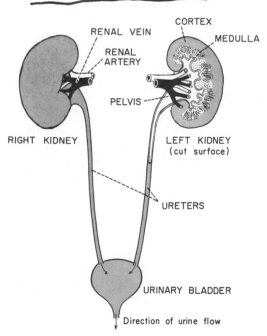

FIGURE 24–1 The general organizational plan of the kidney.

271

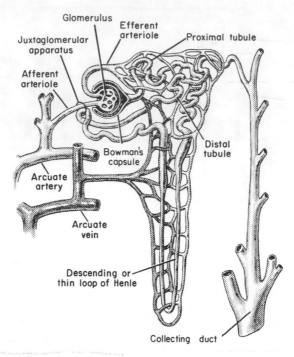

FIGURE 24-2 The nephron. (Redrawn from Smith: The Kidney. Oxford University Press.)

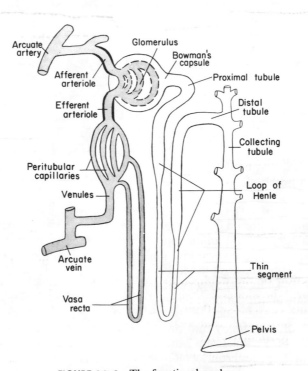

FIGURE 24-3 The functional nephron.

Pressure of the blood in the glomerulus causes fluid to filter into Bowman's capsule, from which it flows first into the *proximal tubule* that lies in the cortex of the kidney along with the glomerulus. From there the fluid passes into the *loop of Henle*, which sometimes passes deep into the *medulla*. From the loop of Henle the fluid flows next into the *distal tubule*, which again is in the renal cortex. Finally, the fluid flows into a *collecting duct* that collects fluid from several nephrons. The collecting tubule passes from the cortex back downward through the medulla, paralleling the loops of Henle. Then it empties into the pelvis of the kidney.

As the glomerular filtrate flows through the tubules, most of its water and varying amounts of its solutes are reabsorbed into the tubular capillaries. The water and solutes that are not reabsorbed become urine.

After blood passes into the efferent arteriole from the glomerulus, most of it flows through the *peritubular capillary network* that surrounds the tubules. From this network, straight capillary loops, called *vasa recta*, extend downward into the medulla to envelop the lower parts of the thin segments before looping back upward to empty into the cortical veins.

Functional Diagram of the Nephron. Figure 24–3 illustrates a simplified diagram of the "physiologic nephron." This diagram contains most of the nephron's functional structures, and it is used in the present discussion to explain many aspects of renal function.

BASIC THEORY OF NEPHRON FUNCTION

The basic function of the nephron is to clean, or "clear," the blood plasma of unwanted substances as it passes through the nephron. The substances that must be cleared include particularly the end-products of metabolism, such as urea, creatinine, uric acid, sulfates and phe-

nols. In addition, many nonmetabolic substances, such as sodium ions, potassium ions, and chloride ions, frequently accumulate in the body in excess quantities; it is the function of the nephron also to clear the plasma of these excesses.

The mechanism by which the nephron clears the plasma of unwanted substances is as follows: (1) It filters a large proportion of the plasma, usually about one-fifth of it, through the glomerular membrane into the tubules of the nephron. (2) Then, as this filtered fluid flows through the tubules, the unwanted substances fail to be reabsorbed while the wanted substances, especially the water and many of the electrolytes, are reabsorbed back into the plasma of the peritubular capillaries. In other words, the tubules separate the unwanted portions of the tubular fluid from the wanted portions, and the wanted portions are returned to the blood, while the unwanted portions pass into the urine.

RENAL BLOOD FLOW AND PRESSURES

BLOOD FLOW THROUGH THE KIDNEYS

The rate of blood flow through both kidneys of a 70 kg. man is about 1200 ml./minute. The per cent of the total cardiac output that passes through the kidneys is called the *renal fraction*, and this is normally 21 per cent.

Special Aspects of Blood Flow and Pressures in the Nephron. Note in Figure 24–3 that there are two capillary beds supplying the nephron: (1) the *glomerulus* and (2) the *peritubular capillaries*. These two capillary beds are separated from each other by the *efferent arteriole*, which offers considerable resistance to blood flow. As a result, the glomerular capillary bed is a *high pressure bed*, having an average pressure of about 70 mm. Hg; while the peritubular capillary bed is a *low pressure bed*, having a mean pressure of only 13 mm. Hg. Because of the high pressure in the glomerulus, it functions in much the same way as the usual arterial ends of the tissue capillaries, with fluid filtering continually out of the glomerulus into Bowman's capsule. On the other hand, the low pressure in the peritubular capillary system causes it to function in much the same way as the venous ends of the tissue capillaries, with fluid being continually absorbed from the tubules into the capillaries.

The vasa recta. A special portion of the peritubular capillary system is the vasa recta, which are a network of capillaries that descend around the lower portions of the loops of Henle. These capillaries form loops into the medulla of the kidney and then return to the cortex before emptying into the veins. The vasa recta play a special role in the formation of concentrated urine, as is discussed later in the chapter. Only a small proportion of the total renal blood flow, about 1 to 2 per cent, flows through the vasa recta.

PRESSURES IN THE RENAL CIRCULATION

In the afferent arteriole the pressure falls from 100 mm. Hg at its arterial end to an estimated mean pressure of about 70 mm. Hg in the glomerulus. As the blood flows through the efferent arterioles from the glomerulus to the peritubular capillary system, the pressure falls another 57 mm. Hg to a mean peritubular capillary pressure of about 13 mm. Hg.

GLOMERULAR FILTRATION AND THE GLOMERULAR FILTRATE

The Glomerular Membrane. The fluid that filters through the glomerular membrane into Bowman's capsule is called *glomerular filtrate*, and the membrane of the glomerular capillaries is called the *glomerular membrane*. Though, in general, this membrane is similar to that of other capillaries throughout the body, it has a permeability about 25 times as great as that of the usual capillary. It is extremely permeable to almost all substances having molecular sizes less than those of plasma proteins (molecular weight less than 70,000); almost no plasma proteins or other substances of larger molecular size are able to filter into Bowman's capsule.

Composition of the Glomerular Filtrate. The glomerular filtrate has almost exactly the same composition as the fluid that filters from the arterial ends of the capillaries into the interstitial fluids. It contains no red blood cells and less than 0.03 per cent protein, or less than $1/200$ the protein in the plasma. Therefore, for all practical purposes, glomerular filtrate is the same as plasma except that it has no significant amount of proteins.

THE GLOMERULAR FILTRATION RATE

The quantity of glomerular filtrate formed each minute in all nephrons of both kidneys is called the *glomerular filtration rate*. In the normal person this averages approximately 125 ml./min. To express this differently, the total quantity of glomerular filtrate formed each day averages about 180 liters, or more than two times the total weight of the body. Over 99 per cent of the filtrate is usually reabsorbed in the tubules. The remaining fraction of a per cent passes into the urine.

The Filtration Fraction. The filtration fraction is the fraction of the renal plasma flow that becomes glomerular filtrate. Since the normal plasma flow through both kidneys is 650 ml./min. and the normal glomerular filtration rate in both kidneys is 125 ml., *the average filtration fraction is approximately* $125/650$, *or 19 per cent.*

DYNAMICS OF GLOMERULAR FILTRATION

Glomerular filtration occurs in almost exactly the same manner that fluid filters out of any high pressure capillary in the body. That is, *pressure inside the glomerular capillaries* causes filtration of fluid through the capillary membrane into Bowman's capsule. On the other hand, *colloid osmotic pressure in the blood and pressure in Bowman's capsule* oppose the filtration. These pressures are normally:

	mm. Hg
Glomerular pressure	70
Colloid osmotic pressure	
(This is high because of rapid	
filtration of fluid.)	32
Bowman's capsule pressure	14

Filtration Pressure. The filtration pressure is the net pressure forcing fluid through the glomerular membrane, and this is *equal to the glomerular pressure minus the sum of glomerular colloid osmotic pressure and capsular pressure.* Using the values just given, we can calculate *the normal filtration pressure to be about 24 mm. Hg.*

REGULATION OF GLOMERULAR FILTRATION RATE

The rate at which glomerular filtrate is formed is directly proportional to the filtration pressure. Therefore, any factor that changes the filtration pressure also changes the filtration rate.

Effect of Afferent Arteriolar Constriction. Afferent arteriolar constriction decreases the rate of blood flow into the glomerulus and thereby decreases the glomerular pressure and the filtration rate.

Effect of Efferent Arteriolar Constriction. Constriction of the efferent arteriole increases the resistance to outflow from the glomeruli. This obviously increases the glomerular pressure and usually increases the glomerular filtration rate as well. However, if the degree of efferent arteriolar constriction is severe, blood flow through the glomerulus may become sluggish. Then, because the plasma remains for a long period of time in the glomerulus and extra large portions of plasma are lost into the capsule, the plasma colloid osmotic pressure rises to excessive levels, causing glomerular filtration paradoxically to fall to a low value despite the elevated glomerular pressure.

Effect of Sympathetic Stimulation. In mild sympathetic stimulation of the kidneys, the afferent and efferent arterioles constrict approximately proportionately to each other so that the glomerular filtration rate neither rises nor falls. This allows blood flow to be shunted from the kidneys to other parts of the body during emergency states without significantly altering renal function.

With strong sympathetic stimulation, however, the effects are different, because all the arterioles then become so greatly constricted that glomerular blood flow is reduced almost to zero, thereby greatly reducing the filtration rate. Thus, strong sympathetic stimulation reduces renal output almost to zero.

REABSORPTION AND SECRETION IN THE TUBULES

The glomerular filtrate entering the tubules of the nephron flows (1) through the *proximal tubule*, (2) through the *loop of Henle*, (3) through the *distal tubule*, and (4) through the *collecting duct* into the pelvis of the kidney. Along this course, substances are selectively reabsorbed or secreted by the tubular epithelium, and the resultant fluid entering the pelvis is *urine*.

BASIC MECHANISMS OF ABSORPTION AND SECRETION IN THE TUBULES

The basic mechanisms for transport through the tubular membrane can be

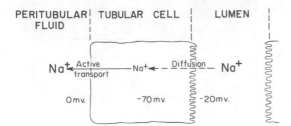

FIGURE 24-4 Mechanism for active transport of sodium from the tubular lumen into the peritubular fluid, illustrating active transport at the base of the epithelial cell and diffusion through the luminal border of the cell.

divided into *active transport* and *passive transport*. The essentials of these mechanisms are described here, but for details refer to Chapter 4.

Active Transport. Figure 24-4 illustrates the mechanism for active transport of sodium from the lumen of the proximal tubule into the peritubular fluid. Active transport of sodium occurs through only one side of the epithelial cell membrane, the side adjacent to the peritubular fluid; on the other hand, the "brush" border of the cell is reasonably permeable to the diffusion of sodium. The active transport of sodium out of the cell and into the peritubular fluid diminishes the concentration of sodium inside the cell and also causes marked *electronegativity* in the cell. Then, because of the low concentration and negativity inside the cell, sodium ions diffuse from the lumen of the tubule to the interior of the cell; once there, the active transport process carries the sodium the rest of the way into the peritubular fluid.

In brief, the mechanism for active transport of sodium through the cell membrane itself at the cell's posterior border is the following: The sodium combines with a *carrier* in the substance of the membrane and in this form diffuses to the opposite wall of the membrane where it is released. Metabolic processes supply the energy for the chemical reactions that cause sodium to be transported through the membrane. Other substances beside sodium that are actively absorbed through the tubular epithelial cells include *glucose, amino acids, calcium ions, potassium ions, phosphate ions,* and *urate ions.* The basic principles of their absorption are also discussed in Chapter 4.

In addition, some substances, especially *hydrogen ions*, are *actively secreted* into certain portions of the tubules. Active secretion occurs in the same way as active absorption except that the cell membrane transports the secreted substance in the opposite direction.

Diffusion Through the Tubular Membrane. Diffusion can occur through the tubular membrane all the way from the lumen of the tubule to the peritubular fluid or in the opposite direction as the result of either a concentration difference or an electrical difference. Some of the substances that normally diffuse through the tubular membrane include water, urea, and potassium, each of which will be discussed in subsequent sections of this chapter.

Diffusion as a result of a concentration difference—osmosis of water. To give an example of diffusion as a result of a concentration difference, let us consider the transport of water through the tubular membrane. When sodium and other solutes are transported from the lumen of the tubule, this decreases the concentration of solutes in the tubular fluid and thereby *increases* the tubular concentration of water. Conversely, the increase in solute concentration in the peritubular fluid *decreases* the concentration of water there. As a result, a concentration difference of water develops between the tubular lumen and the peritubular fluid and causes net diffusion of water from the lumen into the peritubular fluid. This process is called *osmosis*. The basic principles of diffusion and osmosis are described in detail in Chapter 4.

REABSORPTION AND SECRETION OF INDIVIDUAL SUBSTANCES IN DIFFERENT SEGMENTS OF THE TUBULES

Transport of Water and Flow of Fluid at Different Points in the Tubular System. Water transport occurs entirely by osmotic diffusion. That is, whenever some solute in the glomerular filtrate is absorbed either by active absorption or by diffusion caused by an electrochemical gradient, the resulting decreased concentration of solute in the tubular fluid and increased concentration in the peritubular fluid causes osmosis of water out of the tubules.

Figure 24–5 depicts aggregate rate of water flow axially along the lumens of each segment of the tubular system. In both kidneys of man, the total volumes of water flowing into each segment of the tubular system each minute (under normal resting conditions) are the following:

	ml./min.
Glomerular filtrate	125
Flowing into the loops of Henle	25
Flowing into the distal tubules	18
Flowing into the collecting ducts	6
Flowing into the urine	1

From this chart one can also deduce the per cent of the glomerular filtrate water that is reabsorbed in each segment of the tubules, as follows:

	Per Cent
Proximal tubules	80
Loop of Henle	6
Distal tubules	9
Collecting ducts	4
Passing into the urine	1

We shall see later in the chapter that these values vary greatly under different operational conditions of the kidney, particularly when the kidney is forming very dilute or very concentrated urine.

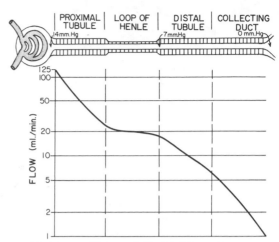

FIGURE 24–5 Volume flow of fluid in each segment of the tubular system per minute. Note that the flow is plotted on a semi-logarithmic scale, illustrating the tremendous difference in flow between the earlier and later segments of the tubules.

Transport of Substances of Nutritional Value to the Body — Glucose, Proteins, Amino Acids, Acetoacetate Ions, and Vitamins. Five different substances in the glomerular filtrate of particular importance to bodily nutrition are glucose, proteins, amino acids, acetoacetate ions, and the vitamins. Normally all of these are completely or almost completely reabsorbed by active processes in the *proximal tubules* of the kidney, as shown in Figure 24–6 for glucose, protein, and amino acids. Therefore, almost none of these substances remain in the tubular fluid entering the loop of Henle.

Special mechanism for absorption of protein. As much as 30 grams of protein filters into the glomerular filtrate each day. This would be a great metabolic drain on the body if the protein were not returned to the body fluids. Because the protein molecule is much too large to be transported by the usual transport processes, protein is absorbed through the brush border of the proximal tubular epithelium by pinocytosis, which means simply that the protein attaches itself to the membrane and this portion of the membrane then invaginates to the interior of the cell. Once

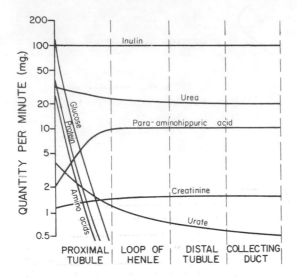

FIGURE 24–6 Reabsorption of the nutritionally important substances in the proximal tubules, and poor reabsorption of the metabolic end-products in all segments of the tubules. Note the total absence of reabsorption of inulin and the secretion of para-aminohippuric acid into the proximal tubules.

inside the cell the protein is probably digested into its constituent amino acids which are then actively absorbed through the base of the cell into the peritubular fluids. Details of the pinocytosis mechanism are discussed in Chapter 4.

Poor Transport of the Metabolic End-Products Urea, Creatinine, Urates, and Others. Figure 24–6 illustrates the rates of flow of three major metabolic end-products in the different segments of the tubular system—urea, creatinine, and urate ions. Note, especially, that only small quantities of *urea*—about 40 per cent of the total—are reabsorbed during the entire course through the tubular system.

Creatinine is not reabsorbed in the tubules at all; and, indeed, small quantities of creatinine are actually secreted into the tubules by the proximal tubules so that the total quantity of creatinine increases about 20 per cent.

The *urate ion* is absorbed much more than urea—about 86 per cent reabsorption. But even so, large quantities of urate remain in the fluid that finally issues into the urine.

Several other end-products, such as *sulfates*, *phosphates*, and *nitrates*, are transported in much the same way as urate ions. All of these are normally reabsorbed to a far less extent than water, so that their concentrations become greatly increased as they flow along the tubules. Yet, each is actively reabsorbed under some conditions, which keeps their concentrations in the extracellular fluid from ever falling too low.

Transport of Inulin and Para-aminohippuric Acid by the Tubules. Note also in Figure 24–6 that the rate of flow of *inulin*, a large polysaccharide, remains exactly the same throughout the entire tubular system. The cause of this is simply that inulin is neither reabsorbed nor secreted in any segment of the tubules.

Finally, Figure 24–6 shows that the rate of flow of *para-aminohippuric acid* (PAH) increases about five-fold as the tubular fluid passes through the proximal tubules; then its rate of flow remains constant in the other tubules. This is because large quantities of PAH are *secreted* into the tubular fluid by the proximal tubular epithelial cells, and it is not reabsorbed in any segment of the tubular system.

These two substances play an important role in experimental studies of tubular function, as will be discussed later in the chapter.

Transport of Ions by the Tubular Epithelium. The rates of flow of different ions decrease markedly as the tubular fluid progresses from glomerular filtrate to urine. Positive ions are generally transported through the tubular epithelium by active transport processes (except for potassium secretion, as noted subsequently), whereas negative ions are usually transported passively as a result of electrical differences developed across the membrane when the positive ions are transported. For instance, when sodium ions are transported out of the tubular fluid, the resulting electronegativity that develops in the tubular fluid causes chloride ions to follow in the wake of the sodium ions.

Transport of other ions will be discussed later in this chapter and in the following chapters in relation to acid-base balance in the body fluids.

CONCENTRATION OF DIFFERENT SUBSTANCES AT DIFFERENT POINTS IN THE TUBULES

Whether or not a substance becomes concentrated in the tubular fluid as it moves along the tubules is determined by the *relative reabsorption of the substance versus the reabsorption of water.* If a greater percentage of water is reabsorbed, the substance becomes more and more concentrated. Conversely, if a greater percentage of the substance is reabsorbed, it becomes more and more dilute. Figure 24–7 illustrates this effect, showing three different classes of substances as follows:

First, the nutritionally important substances — glucose, protein, and amino acids — are reabsorbed so much more rapidly than water that their concentrations fall extremely rapidly in the proximal tubules and remain essentially zero throughout the remainder of the tubular system as well as in the urine.

Second, the concentration of the meta-

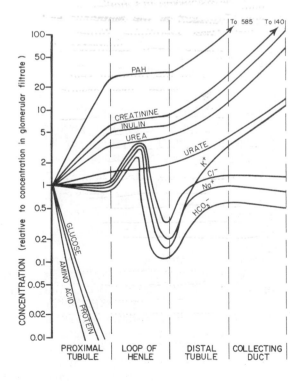

FIGURE 24–7 Composite figure, showing average changes in concentrations of different substances at different points in the tubular system.

TABLE 24–1 RELATIVE CONCENTRATIONS OF SUBSTANCES IN THE
GLOMERULAR FILTRATE AND IN THE URINE

	Glomerular Filtrate (125 ml./min.)		Urine (1 ml./min.)		Conc. Urine/ Conc. Plasma (Plasma Clearance per Minute)
	QUANTITY/MIN.	CONCENTRATION	QUANTITY/MIN.	CONCENTRATION	
Na+	17.7 mEq.	142 mEq./l.	0.128 mEq.	128 mEq./l.	0.9
K+	0.63	5	0.06	60	12
Ca++	0.5	4	0.0048	4.8	1.2
Mg++	0.38	3	0.015	15	5.0
Cl⁻	12.9	103	0.134	134	1.3
HCO₃⁻	3.5	28	0.014	14	0.5
H₂PO₄⁻ HPO₄⁻⁻	0.25	2	0.05	50	25
SO₄⁻⁻	0.09	0.7	0.033	33	47
Glucose	125 mg.	100 mg.%	0 mg.	0 mg.%	0.0
Urea	33	26	18.2	1820	70
Uric acid	3.8	3	0.42	42	14
Creatinine	1.4	1.1	1.96	196	140
Inulin	...	...	...	...	125
Diodrast	...	...	...	...	560
PAH	...	...	...	...	585

bolic end-products as well as of inulin and PAH becomes progressively greater throughout the tubular system, for all these substances are reabsorbed to a far less extent than is water. Note that potassium ions also fall into this category because a far greater proportion of potassium ions than water is normally removed each day from the extracellular fluid.

Third, many other ions are normally excreted into the urine in concentrations not greatly different from those in the glomerular filtrate and extracellular fluid. That is, water, sodium, chloride, and bicarbonate ions, on the average, are normally reabsorbed from the tubules in not too dissimilar proportions.

Table 24–1 summarizes the average concentrating ability of the tubular system for different substances. It also gives the actual quantities of the different substances normally handled by the tubules each minute.

CONCENTRATING AND DILUTING MECHANISM OF THE KIDNEY—THE COUNTER-CURRENT MECHANISM

The process for concentrating urine is called the *counter-current mechanism.* This depends on a special anatomical arrangement of the loops of Henle and vasa recta. The *loops of Henle* of approximately one-fifth of the nephrons, called the *juxtamedullary nephrons*, dip deep into the medulla and then return to the cortex, some dipping all the way to the tips of the papillae that project into the renal pelvis. Also, the straight peritubular capillaries called the *vasa recta* loop down into the medulla and then back out of the medulla to the cortex. These arrangements of the different parts of the juxtamedullary nephron and of the vasa recta are illustrated diagrammatically in Figure 24–8.

Hyperosmolality of the Medullary Fluid. The normal osmolality of the glomerular filtrate as it enters the proximal tubules is about 300 milliosmols/liter. However, as shown by the numbers in Figure 24–8, the osmolality of the interstitial fluid in the medulla of the kidney becomes progressively greater the more deeply one goes into the medulla, increasing from 300 milliosmols/liter in the cortex to 1200 milliosmols/liter at the pelvic tip of the medulla.

The cause of this greatly increased osmolality of the medullary interstitial fluid is active transport of sodium out of the loop of Henle into the medullary interstitial fluid. The very dark arrows through the wall of the ascending limb of the loop of Henle illustrate this active absorption of sodium. In addition, a slight amount of sodium is actively absorbed through the walls of the collecting duct into the interstitial fluid, and it is likely that sodium is actively absorbed from the descending limb of the loop of Henle. As the sodium is actively absorbed, an electrical difference develops across the walls of the tubules which pulls chloride ions and some other negative ions into the interstitial fluid as well. Thus, the concentrations of these ions increase markedly in the interstitial fluid of the medulla.

The question that must now be asked is: Why does the osmolar concentration of solutes rise so high in the medulla but not in the fluid of the cortex where equally large quantities of sodium are reabsorbed from the proximal and distal

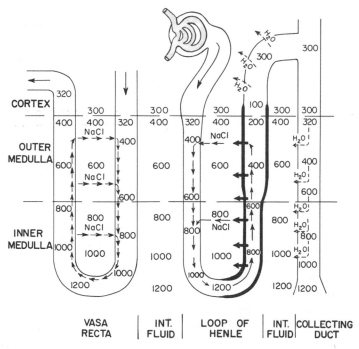

FIGURE 24–8 The "counter-current" mechanism for concentrating the urine.

tubules? The answer to this is two-fold: First, the blood flow through the vasa recta is extremely sluggish. Only about 1 to 2 per cent of the total renal blood flow passes through these vessels in comparison with 90 per cent through the cortical peritubular capillaries. Obviously, the sluggish flow in the medulla would help to prevent removal of the large quantities of sodium chloride being continually absorbed into the medullary interstitial fluid.

The second cause of the high concentration of osmotically active substances in the medulla is the presence of counter-current mechanisms operating in both the loop of Henle and the vasa recta. These may be explained as follows:

Counter-Current "Multiplier" of the Loop of Henle. A counter-current fluid mechanism is one in which fluid flows through a long U tube, the fluid flowing in one arm of the U and then out the opposite arm. When the fluids in the two parallel streams of flow interact appropriately with each other, tremendous concentrations of solute can be built up at the tip of the loop. As an example, let us first observe the loop of Henle. The descending limb of this loop is highly permeable to sodium chloride, while the ascending limb has a strong transport mechanism to remove sodium chloride from the tubule into the interstitial fluid. Therefore, each time sodium chloride is transported out of the ascending limb, it almost immediately diffuses into the descending limb, thereby increasing the concentration of sodium chloride in the tubular fluid that flows downward toward the tip of the loop. As illustrated by the arrows in the diagram, this sodium chloride flows on around the loop and is then actively transported again out of the ascending limb. In addition, new sodium chloride is entering the tubular system in the glomerular filtrate and is also transported out of the ascending limb. Thus, by this retransport again and again of sodium chloride, plus constant addition of more and more sodium chlo-

ride, one can readily see why the concentration of sodium chloride in the medullary fluid rises very high. This mechanism of the loop of Henle is called a counter-current multiplier.

Counter-Current Mechanism in the Vasa Recta. Another factor that helps explain the very high concentration of sodium chloride in the medulla is the *counter-current flow of blood in the vasa recta*. This is illustrated to the left in Figure 24–8. As blood flows down the descending limbs of the vasa recta, sodium chloride diffuses into the blood from the interstitial fluid, causing the osmolar concentration to rise progressively higher, to a maximum concentration of 1200 milliosmols/liter at the tips of the vasa recta. Then as the blood flows back up the loop, the extreme diffusibility of sodium chloride through the capillary membrane allows essentially all of the extra sodium chloride to be lost once again back into the interstitial fluid. Therefore, by the time the blood finally leaves the medulla, its osmolar concentration is only slightly greater than that of the blood that had initially entered the vasa recta. As a result, blood flowing through the vasa recta carries only a minute amount of sodium chloride away from the medulla.

Dilution of the Tubular Fluid in the Ascending Limb of the Loop of Henle. The ascending limb of the loop of Henle is highly impermeable to water, and yet tremendous quantities of sodium chloride are transported out of the ascending limb into the interstitial fluid. Therefore, as illustrated in Figure 24–8, the fluid in the ascending limb of the loop of Henle becomes progressively more dilute as it ascends toward the cortex, decreasing to an osmolality of about 100 milliosmols/liter before leaving the loop of Henle.

Mechanism for Excreting a Dilute Urine. Now that some of the basic aspects of the counter-current mechanism of the kidney have been described, we can explain easily how the kidney secretes a dilute urine. Since the tubular fluid has a concentration of only 100 milliosmols/ liter as it leaves the ascending limb of the loop of Henle, the kidney can excrete a dilute urine by simply allowing this fluid to empty directly into the pelvis of the kidney. And the kidney has a mechanism for doing this. In the absence of *antidiuretic hormone*, which is secreted by a hypothalamic-neurohypophyseal mechanism, the pores of the epithelial cells in the distal tubules and collecting ducts become very small, too small for absorption of water or almost any other substance that is normally absorbed by diffusion. As a result, in the absence of this hormone almost no water is absorbed from the distal tubules and collecting ducts, and the dilute tubular fluid flushes rapidly on through these tubules into the urine. However, a slight amount of sodium and a few other substances are actively absorbed, which reduces the osmolality sometimes to as low as 70 milliosmols/liter by the time the fluid enters the urine; this is about one-fourth the osmolality of the plasma and glomerular filtrate.

Mechanism for Excreting a Concentrated Urine. The mechanism for excreting a concentrated urine is exactly opposite to that for excreting a dilute urine. In this case, large quantities of *antidiuretic hormone* are secreted into the body fluids, and this in turn increases the sizes of the pores in the distal tubules and collecting ducts until water molecules can be absorbed with ease. The very dilute fluid leaving the ascending limb of the loop of Henle rapidly loses water in the distal tubule by osmosis, and its osmolality rises to equilibrium with that in the cortical interstitial fluids, rising to 300 milliosmols/liter as illustrated in Figure 24–8. Now, the tubular fluid passes back through the medulla a second time, this time downward through the collecting duct. Here it is exposed to the hyperosmolality of the medullary interstitial fluid, for the collecting duct passes directly through the medulla on its way to the pelvis of the kidney. Therefore, as illustrated in Figure 24–8, large quantities of water are absorbed by osmosis into the medullary interstitial fluid from the collecting duct. Therefore, the concentration of the fluid in the duct rises to approach the osmolality of the fluid in the interstitial fluid, about 1200 milliosmols per liter, so that the fluid leaving the collecting duct to enter the pelvis of the kidney is very concentrated urine. Thus, the maximum concentration of the urine that can usually be achieved is about four times the osmolality of the plasma and glomerular filtrate.

Thus, all degrees of concentration of the urine can occur between the two extremes of 70 and 1200 milliosmols/liter. This, of course, depends mainly on the quantity of antidiuretic hormone present in the body fluid at any given time. The importance of the antidiuretic hormone and its relationship to control of osmolality of the body fluids is discussed in the following chapter.

AUTOREGULATION OF RENAL BLOOD FLOW AND OF GLOMERULAR FILTRATION RATE

Blood flow through the kidneys and also the glomerular filtration rate remain nearly constant despite widely varying arterial pressure. For instance, the arterial pressure can change from as low as 70 mm. Hg to as high as 200 or more mm. Hg, and yet the renal blood flow and glomerular filtration rate change only a few per cent. This effect is illustrated in Figure 24–9. It is called *autoregulation of renal blood flow and autoregulation of glomerular filtration rate*. The exact cause of this phenomenon of autoregulation has not been discovered. However, an exciting concept for which there is at least some experimental support is that it results from a feedback mechanism operating from the distal tubule to control blood flow into the glomerulus, as follows:

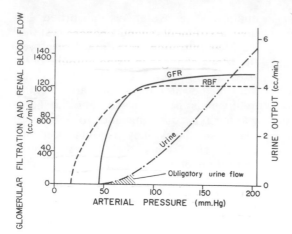

FIGURE 24-9 Autoregulation of renal blood flow (*RBF*) and glomerular filtration rate (*GFR*), and lack of autoregulation of urine flow.

The Juxtaglomerular Apparatus and its Possible Feedback Role in the Nephron. Figure 24–10 illustrates a peculiar characteristic of each of the two million nephrons in the two kidneys. Immediately beyond the point where the ascending limb of the loop of Henle empties into the distal tubule, the distal tubule returns to the glomerulus and passes between the afferent and efferent arterioles. The epithelial cells of the distal tubule where it lies in apposition to the afferent arteriole become condensed into small, tightly packed cells that are collectively called the *macula densa*. And the smooth muscle cells of the afferent arteriole at this point of contact are swollen and are filled with granules of the substance *renin*, which was discussed in relation to hypertension in Chapter 20.

It is almost certain that the intimate relationship between the distal tubule and the afferent arteriole is the basis for some type of feedback mechanism whereby conditions in the distal tubule control blood flow through the afferent arteriole.

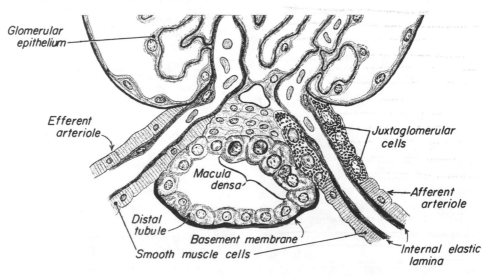

FIGURE 24-10 Structure of the juxtaglomerular apparatus, illustrating its possible feedback role in the control of nephron function. (Modified from Ham: Histology. J. B. Lippincott Co.)

Two types of feedback have received special attention. One of these is *osmotic feedback* and the other is *feedback based on the concentration of sodium ions* in the distal tubule. Although the exact nature of the feedback is yet to be discovered, such a mechanism would allow exactly the amount of glomerular filtrate to enter the tubules that can be processed each minute, no more, no less.

RELATIONSHIP OF GLOMERULAR FILTRATION RATE TO URINE OUTPUT—THE OVERFLOW PHENOMENON

Note in Figure 24–9 that, even though the glomerular filtration rate *is* autoregulated, *the rate of urine output is not autoregulated.* Instead, the urinary output rises greatly as the arterial pressure increases. The cause of this difference is the *overflow phenomenon,* which can be explained as follows:

As the glomerular filtration rate rises, the amount of substances that enters the tubules each minute eventually exceeds the amount that can be reabsorbed. Beyond this glomerular filtration rate, the substances begin to spill into the urine, and proportionate quantities of water also fail to be reabsorbed by the tubules, thus increasing the quantity of water in the urine. In other words, when the absorptive ability of the tubules for each substance is exceeded, the phenomenon of *overflow* takes place.

Now, let us return to Figure 24–9 and observe the relationship between glomerular filtration rate and urine flow. Between the limits of 100 and 200 mm. Hg arterial pressure, the glomerular filtration rate increases only about 15 per cent. Yet, because of the overflow phenomenon, the urine flow increases 450 per cent, which is 30 times as great an increase as the glomerular filtration rate increase. Thus, one can readily see that extremely slight changes in glomerular filtration rate can cause tremendous changes in urine output.

THE CONCEPT OF "PLASMA CLEARANCE"

The term "plasma clearance" is used to express the ability of the kidneys to remove various substances from the plasma. If the plasma passing through the kidneys contains 0.1 gram of a substance in each 100 ml., and 0.1 gram of this substance passes into the urine each minute, then 100 ml. of the plasma are cleaned or "cleared" of the substance per minute.

Referring to Table 24–1, note that the normal concentration of urea in each 100 ml. of plasma and glomerular filtrate is 26 mg., and the quantity of urea that passes into the urine each minute is approximately 18.2 mg. Therefore, the quantity of plasma that completely loses its entire content of urea each minute may be calculated by dividing the quantity of urea entering the urine each minute by the quantity of urea in each milliliter of plasma. Thus, $18.2 \div 0.26 = 70$; that is, 70 ml. of plasma, by filtering through the glomerulus and then being reabsorbed by the tubules, *is cleaned* or *cleared* of urea each minute. This is known as the *plasma clearance* of urea.

Plasma clearance for any substance can be calculated by using the formula at the bottom of the page.

The concept of plasma clearance is important because it is an excellent measure of kidney function, and the clearance of different substances can be determined by simply analyzing the concentrations of the substances simultaneously in the plasma and in the urine.

The plasma clearances of the usual

$$\text{Plasma clearance (ml./min.)} = \frac{\text{Quantity of urine (ml./min.)} \times \text{Concentration in urine}}{\text{Concentration in plasma}}$$

constituents of urine are shown in the last column of Table 24–1.

INULIN CLEARANCE AS A MEASURE OF GLOMERULAR FILTRATION RATE

Inulin is a polysaccharide that has the specific attributes of not being reabsorbed to any extent by the tubules of the nephron and yet being of small enough molecular weight (about 5200) that it passes through the glomerular membrane as freely as the crystalloids and water of the plasma. Therefore, *all the glomerular filtrate formed is cleared of inulin, and this is equal to the amount of plasma that is simultaneously "cleared."* Therefore, the plasma clearance per minute of inulin is also equal to the glomerular filtration rate.

PARA-AMINOHIPPURIC ACID (PAH) CLEARANCE AS A MEASURE OF PLASMA FLOW AND BLOOD FLOW THROUGH THE KIDNEYS

PAH, like inulin, passes through the glomerular membrane with perfect ease along with the remainder of the glomerular filtrate. However, it is different from inulin in that almost all the PAH remaining in the plasma after the glomerular filtrate is formed is secreted into the tubules by the tubular epithelium (if the plasma load of the PAH is considerably less than the transport maximum for PAH). Indeed, less than one-tenth of the original PAH remains in the plasma when the blood leaves the kidneys.

Therefore, one can use the clearance of PAH for estimating the *flow of plasma* through the kidneys. That is, however much plasma is cleared of PAH, at least this much plasma must have passed through the kidneys in this same period of time. Usually the plasma flow is about 10 per cent greater than the PAH clearance, because the clearance averages only 91 per cent of complete clearance.

EFFECT OF "TUBULAR LOAD" AND "TRANSPORT MAXIMUM" ON URINE CONSTITUENTS

Plasma Load and Tubular Load. The *plasma load* of a substance means the total amount of the substance in the plasma that passes through the kidney each minute. For instance, if the concentration of glucose in the plasma is 100 mg./100 ml., and 600 ml. of plasma passes through both kidneys each minute, the plasma *load of glucose* is 600 mg./min.

Normally a fraction of the plasma load filters into the glomeruli, and this portion is called the *tubular load.* In the example just given, if 125 ml. of glomerular filtrate is formed each minute with a glucose concentration of 100 mg. per cent, the tubular load of glucose is 100 mg. × 1.25, or 125 mg. of glucose per minute. Similarly, the load of sodium that enters the tubules each minute is approximately 18 mEq./min., the load of chloride ion is about 13 mEq./min., and the load of urea is approximately 33 mg./min., etc.

Maximum Rate of Transport of Actively Reabsorbed or Secreted Substances — The "Transport Maximum" (Tm). Since each substance that is actively reabsorbed or secreted requires a specific transport system in the tubular epithelial cells, the maximum amount that can be reabsorbed often depends on the maximum rate at which the transport system itself can operate, and this in turn depends on the total amounts of carrier and specific enzymes available. Consequently, for almost every actively reabsorbed or secreted substance, there is a maximum rate at which it can be transported; this is called the *transport maximum* for the substance, and the abbreviation is *Tm*. For instance, the Tm for glucose averages 320 mg./min. for the adult human being, and if the tubular load becomes greater than 320 mg./min., the excess above this amount is not reabsorbed but instead passes on into the urine. Ordinarily, though, the tubular load of glucose is only 125 mg./min., so that for practical purposes, all of it is reabsorbed.

Transport maximums of important substances absorbed from the tubules. Some of the important transport maximums for substances absorbed from the tubules are the following:

Glucose	320	mg./min.
Phosphate	0.1	mM./min.
Sulfate	0.06	mM./min.
Amino acids	1.5	mM./min.
Vitamin C	1.77	mg./min.
Urate	15	mg./min.
Plasma protein	30	mg./min.
Hemoglobin	1	mg./min.
Lactate	75	mg./min.
Acetoacetate	variable (about 30 mg./min.)	

Transport maximums for secretion. Many substances secreted by the tubules also exhibit transport maximums as follows:

	mg./min.
Creatinine	16
PAH	80
Diodrast	57 (of iodine)
Phenol red	56

REFERENCES

De Wardener, H. E.: The Kidney. 3rd ed., Boston, Little, Brown and Company, 1968.

Forster, R. P.: Kidney, water, and electrolytes. *Ann. Rev. Physiol., 27*:183, 1965.

Giebisch, G.: Kidney, water and electrolyte metabolism. *Ann. Rev. Physiol., 24*:357, 1962.

Pitts, R.: Physiology of the Kidney and Body Fluids. 2nd ed., Chicago, Year Book Medical Publishers, Inc., 1968.

Thurau, K., Valtin, H., and Schnermann, J.: Kidney. *Ann. Rev. Physiol., 30*:441, 1968.

Wesson, L. G., Jr.: Physiology of the Human Kidney. New York, Grune & Stratton, Inc., 1963.

Windhager, E. E.: Kidney, water, and electrolytes. *Ann. Rev. Physiol., 31*:117, 1969.

CHAPTER 25

REGULATION OF EXTRACELLULAR FLUID COMPOSITION; OSMOTIC EQUILIBRIA BETWEEN EXTRACELLULAR AND INTRACELLULAR FLUIDS; AND REGULATION OF BLOOD VOLUME

In the discussion of homeostasis in Chapter 1 it was pointed out that the cells of the body can continue to live and function normally as long as the composition of the extracellular fluid remains within reasonably normal limits. Though most organs of the body are at least partially concerned with maintenance of constancy in the extracellular fluids, the kidneys are more directly responsible than any other organ for regulation of most of the constituents.

REGULATION OF ION CONCENTRATIONS IN THE EXTRACELLULAR FLUIDS

REGULATION OF SODIUM ION CONCENTRATION — THE ALDOSTERONE FEEDBACK MECHANISM

Sodium ions represent approximately 90 per cent of all the extracellular positive ions. Therefore, sodium is the single most important ion that needs to be regulated. A specific mechanism for regulating its concentration is vested in the kidneys and the adrenal cortex. In the preceding chapter it was noted that sodium is actively reabsorbed by the tubules and that this active reabsorption of sodium seems to occur in all segments of the tubules. The sodium may be almost completely reabsorbed, with essentially none of it passing into the urine, or, on the other hand, large portions of the sodium may fail to be reabsorbed and are lost in the urine. In this way, the rate of sodium loss from the extracellular fluid can be altered from as little as a fraction of a gram a day to as great as 30 grams per day. The precise amount of sodium reabsorbed is regulated mainly by the concentration of *aldosterone*, a hormone secreted by the adrenal cortices, in the body fluids. The mechanism of this regulation is explained in the following paragraphs.

Aldosterone and its Effect on Sodium Reabsorption in the Tubules. Aldosterone is one of several steroid hormones secreted by the adrenal cortex; these will be discussed in detail in Chapter 50 in relation to the overall function of the adrenal gland.

Aldosterone increases the rate of sodium reabsorption in all segments of the tubular system; but its function in the ascending limbs of the loops of Henle, the distal tubules, and the collecting ducts is more important than in the proximal tubules and thin segments of the loops of Henle. In the absence of aldosterone, almost no sodium is reabsorbed by the distal segments of the tubular system. On the other hand, in the presence of excess aldosterone almost all sodium that reaches these segments is reabsorbed. Therefore, the presence or absence of aldosterone determines whether or not significant quantities of sodium will be lost in the urine.

Regulation of Aldosterone Secretion. The rate of secretion of aldosterone can be increased greatly by any one or a combination of the following stimuli: (1) reduced sodium concentration in the extracellular fluids, (2) high extracellular concentration of potassium, (3) reduced blood volume or cardiac output, and (4) physical stress caused by trauma, burns, and so forth. The means by which these factors control aldosterone secretion will be discussed in Chapter 50. From the point of view of our present discussion, almost any effect that compromises the circulation, such as decreasing blood flow through the body, decreasing body fluid volumes, or decreasing sodium in the fluid, will lead to aldosterone secretion.

Regulation of Sodium Concentration. One of the most important stimuli that increase the rate of aldosterone secretion is low sodium concentration in the extracellular fluid. A 5 per cent decrease in sodium concentration continued over a period of several days approximately doubles the rate of aldosterone secretion. Conversely, a similar increase in sodium concentration reduces aldosterone secretion to levels far below normal.

Obviously, this effect of sodium concentration on aldosterone secretion provides a feedback system for maintaining a constant concentration of sodium ions in the extracellular fluid. When their concentration falls below normal, increased aldosterone secretion causes increased reabsorption of sodium by the renal tubules, thus conserving sodium in the body while the daily intake of sodium returns the sodium concentration to normal. Conversely, elevated sodium concentration reduces aldosterone secretion, which allows rapid loss of sodium into the urine with consequent return of the sodium concentration to normal.

REGULATION OF OTHER POSITIVE IONS

Even though the concentrations of the other positive ions in the extracellular fluids are also regulated exactly, the mechanisms for their regulation are less well known than those for sodium. Nevertheless, we know that the kidneys play key roles in the regulation of at least potassium, calcium, and magnesium.

Regulation of Potassium Ion Concentration. Potassium ion concentration is regulated in at least two different ways: (1) by an aldosterone feedback method that functions exactly oppositely to that for sodium regulation and (2) by direct secretion of potassium into the distal tubules and collecting ducts when the extracellular fluid potassium concentration rises too high.

Aldosterone feedback regulation of potassium. The aldosterone feedback mechanism for control of potassium ion concentration is the following: When aldosterone causes sodium to be reabsorbed in the distal tubules and collecting ducts, the sodium ions carry with them large quantities of positive charges from the tubules into the peritubular fluids. As a result, strong electronegativity develops in the tubules, −40 to −120 millivolts, and this pulls positive potassium ions from the tubular cells into the tubules, thereby causing passive secretion of potassium into the urine. Thus, in effect, exchange of potassium ions for sodium ions occurs in the distal tubules. This exchange increases manyfold under the influence of aldosterone.

To complete the feedback loop, high potassium concentration in the extracellular fluid increases the rate of aldosterone secretion, as was discussed earlier. The aldosterone then simultaneously increases sodium reabsorption and potassium excretion, decreasing the potassium back to a normal level. Thus, in effect, the aldosterone feedback mechanism controls the *sodium to potassium ratio* in the extracellular fluids.

Excretion of potassium as a direct response to increased extracellular fluid potassium. In addition to the aldosterone mechanism for control of extracellular fluid potassium ion concentration, an increase in potassium concentration also has a direct effect on the distal tubules and collecting ducts to cause potassium secretion into the tubular fluid. This has been postulated to result from the following mechanism: Increased potassium concentration in the extracellular fluids causes increased potassium ions in the epithelial cells of the distal tubules and collecting ducts. This in turn provides increased quantities of potassium to diffuse into the tubules. Under the influence of the electronegativity in the tubules, the potassium is passively secreted.

The combined effects of the aldosterone feedback mechanism and the direct mechanism for secretion of potassium cause the potassium ion concentration in the extracellular fluids to remain almost exactly 4 to 5 milliequivalents per liter, rarely rising or falling more than ±10 per cent.

Regulation of Calcium Ion Concentration. Though calcium ion concentration is controlled mainly by the parathyroid glands and their effects on bone, which will be discussed in Chapter 53, calcium is also regulated to some extent by the kidney tubules. Large amounts of calcium are known to be lost into the urine when its extracellular concentration is high, while very little is lost when the concentration is low. Unfortunately, the total mechanism for this regulation is unknown, though at least part of it results from a direct effect of parathyroid hormone on the renal tubules to increase calcium reabsorption (see Chap. 53) when the extracellular concentration of calcium falls below normal.

Regulation of Magnesium Concentration. Almost nothing is understood about the regulation of magnesium concentration other than that decreased magnesium concentration in the extracellular fluids promotes increased reabsorption of magnesium from the tubules, while increased concentration promotes decreased reabsorption. The few available experiments indicate that the kidneys regulate extracellular magnesium concentration in much the same way that they regulate potassium concentration. This is logical since both these ions function principally as intracellular rather than as extracellular ions.

REGULATION OF NEGATIVE ION CONCENTRATIONS IN THE EXTRACELLULAR FLUIDS

Regulation of Total Negative Ion Concentration. In general, the total negative ion concentration is regulated secondarily to the regulation of the positive ions for the following reasons: Each time a positive ion is absorbed from the tubules, a state of electronegativity is created in the lumens of the tubules. Such a potential immediately causes negative ions, primarily chloride ions, to diffuse through the tubular wall to provide electrical neutrality.

Therefore, the same regulatory mechanisms that increase the reabsorption of positive ions also promote reabsorption of negative ions at the same time. For instance, secretion of aldosterone by the adrenal glands directly increases the reabsorption of sodium but indirectly promotes absorption of negative ions at the same time; and since about three-fourths of the negative ions in the glomerular filtrate are chloride ions, it frequently is said that aldosterone promotes chloride reabsorption.

Regulation of Chloride to Bicarbonate Ratio—Relation to Acid-Base Balance. Much more important than the absolute concentrations of chloride and bicarbonate in the extracellular fluid is the *ratio of chloride to bicarbonate*. When the extracellular fluids become exceedingly acidic, the renal tubules reabsorb large quantities of bicarbonate ions, and chloride ion reabsorption becomes greatly diminished. The excessive reabsorption of bicarbonate ions shifts the pH of the buffer systems in the extracellular fluid toward a normal pH. These relationships between chloride and bicarbonate ions are discussed in detail in the following chapter.

Regulation of Phosphate Concentration. Phosphate concentration is regulated primarily by an *overflow* mechanism, which can be explained as follows: Phosphate ion spills into the urine when its concentration in the plasma is above the threshold value of approximately 0.8 millimol/liter. Any time the concentration falls below this value, all the phosphate is conserved in the plasma, and the daily ingested phosphate accumulates in the extracellular fluid until its concentration rises above the threshold. Since most people ingest large quantities of phosphate day in and day out, either in milk or in meat, the concentration of phosphate is usually maintained at a level of about 1.0 millimol/liter, a level at which there is continual overflow of excess phosphate into the urine.

OSMOTIC EQUILIBRIA AND FLUID SHIFTS BETWEEN THE EXTRACELLULAR AND INTRACELLULAR FLUIDS

One of the most troublesome of all problems in clinical medicine is maintenance of adequate body fluids and proper balance between the extracellular and intracellular fluid volumes in seriously ill patients. The purpose of the following discussion, therefore, is to explain the interrelationships between extracellular and intracellular fluid volumes and the osmotic factors that cause shifts of fluid between the extracellular and intracellular compartments. However, let us first recall what constitutes extracellular and intracellular fluids.

The Extracellular and Intracellular Fluid Compartments. All the fluids outside the cells compose the extracellular fluid and are said to be in the *extracellular fluid compartment*. These fluids include (1) the *interstitial fluid*, (2) the *plasma*, (3) the *cerebrospinal fluid*, (4) the *intraocular fluid*, (5) the *fluid in the gastrointestinal* tract, and (6) the *fluid in the potential spaces of the body*.

On the other hand, all the fluid inside the cells of the body is collectively called the intracellular fluid and is said

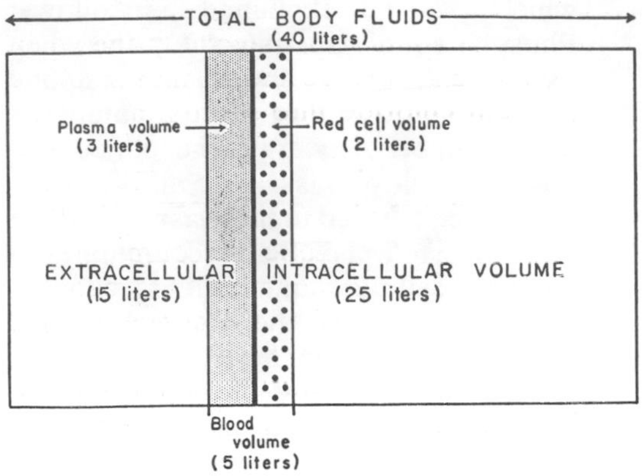

FIGURE 25-1 Diagrammatic representation of the body fluids, showing the extracellular fluid volume, intracellular fluid volume, blood volume, and total body fluids.

to be in the *intracellular fluid compartment*.

The distribution of fluid in the average man (70 kg.) is shown in Figure 25–1, with an extracellular fluid volume of 15 liters, an intracellular fluid volume of 25 liters, and a total fluid volume of 40 liters (or 57 per cent of the total body).

BASIC PRINCIPLES OF OSMOSIS AND OSMOTIC PRESSURE

The basic principles of osmosis and osmotic pressure were presented in Chapter 4. However, these principles are so important to the following discussion that they are reviewed briefly here.

Whenever a membrane between two fluid compartments is permeable to water but not to some of the dissolved solutes (this is called a *semipermeable membrane*) and the concentration of nondiffusible substances is greater on one side of the membrane than on the other, water passes through the membrane toward the side with the greater concentration of nondiffusible substances. This phenomenon is called *osmosis*.

Osmosis results from the kinetic motion of the molecules in the solutions on the two sides of the membrane and can be explained in the following manner: The individual molecules on both sides of the membrane are equally active because the temperature, which is a measure of the kinetic activity of the molecules, is the same on both sides. However, the nondiffusible solute on one side of the membrane displaces some of the water molecules, thereby reducing the concentration of water molecules. As a result, the so-called *chemical activity* of water molecules on this side is less than on the other side, so that fewer water molecules strike each pore of the membrane each second on the solute side of the pore than on the pure water side, resulting in net diffusion of water molecules from the water side to the solute side. This net rate of diffusion is the *rate of osmosis*.

Osmosis of water molecules can be opposed by applying a pressure across the semipermeable membrane in the direction opposite to that of the osmosis. The amount of pressure required exactly to oppose the osmosis is called the *osmotic pressure*.

Relationship of the Molecular Concentration of a Solution to Its Osmotic Pressure. Each nondiffusible molecule dissolved in water dilutes the "activity" of the water molecules by a given amount. Consequently, the tendency for the water in the solution to diffuse through a membrane is reduced in direct proportion to the numbers of nondiffusible molecules. And, as a corollary, the osmotic pressure of the solution is also proportional to the numbers of non-diffusible molecules in the solution. This relationship holds true for all nondiffusible molecules almost regardless of their molecular weight. For instance, one molecule of albumin with a molecular weight of 70,000 has the same osmotic effect as a molecule of glucose with a molecular weight of 180.

Osmotic Effect of Ions. Nondiffusible ions cause osmosis and osmotic pressure in exactly the same manner as do nondiffusible molecules. Furthermore, when a molecule dissociates into two or more ions, each of the ions then exerts osmotic pressure individually. Therefore, to determine the osmotic effect, all the nondiffusible ions must be added to all the nondiffusible molecules; but note that a bivalent ion, such as calcium, exerts no more osmotic pressure than does a univalent ion, such as sodium.

Osmols. The ability of solutes to cause osmosis and osmotic pressure is measured in terms of "osmols," and the osmol is a measure of solute concentration in terms of the number of dissolved particles. *One gram mol of dissolved nondiffusible and nonionizable substance is equal to 1 osmol.* On the other hand, if a substance ionizes into two ions (sodium chloride into sodium and chloride ions, for instance), then 0.5 gram mol of the substance would equal 1 osmol. The obvious reason for using the osmol is that osmotic pressure is proportional to the concentration in number of particles instead of the concentration in terms of dissolved mass.

In general, the osmol is too large a unit for satisfactory use in expressing osmotic activity of solutions in the body. Therefore, the term *milliosmol, which equals* $1/1000$ osmol, is commonly used. The osmotic pressure of a solution *at body temperature* can be approximately determined from the following formula:

Osmotic pressure (mm. Hg)
$= 19.3 \times$ Concentration in milliosmols

OSMOLALITY OF THE BODY FLUIDS

Table 25–1 lists the osmotically active substances in plasma, interstitial fluid,

TABLE 25–1   OSMOLAR SUBSTANCES IN EXTRACELLULAR AND INTRACELLULAR FLUIDS

	Plasma (mOsmols./L. of H_2O)	Interstitial (mOsmols./L. of H_2O)	Intracellular (mOsmols./L. of H_2O)
Na^+	144	137	10
K^+	5	4.7	141
Ca^{++}	2.5	2.4	0
Mg^{++}	1.5	1.4	31
Cl^-	107	112.7	4
HCO_3^-	27	28.3	10
HPO_4^{--} $H_2PO_4^-$	2	2	11
SO_4	0.5	0.5	1
Phosphocreatine			45
Carnosine			14
Amino acids	2	2	8
Creatine	0.2	0.2	9
Lactate	1.2	1.2	1.5
Adenosine triphosphate			5
Hexose monophosphate			3.7
Glucose	5.6	5.6	
Protein	1.2	0.2	4
Urea	4	4	4
TOTAL mOsmols.	303.7	302.2	302.2
Total osmotic pressure at 37° C. (mm. Hg)	5454	5430	5430

and intracellular fluid. The milliosmols of each of these per liter of water is given. Note especially that approximately four-fifths of the total osmolality of the interstitial fluid and plasma is caused by sodium and chloride ions, while approximately half of the intracellular osmolality is caused by potassium ions, the remainder being divided among the many other intracellular substances.

As noted at the bottom of Table 25–1, the total osmolality of each of the three compartments is almost exactly 300 milliosmols, with that of the plasma 1.3 milliosmols greater than that of the interstitial and intracellular fluids. This slight difference between plasma and interstitial fluid is caused by the osmotic effect of the plasma proteins, which maintains about 24 mm. Hg greater pressure in the capillaries than in the surrounding interstitial fluid spaces, as was explained in Chapter 16.

Total Osmotic Pressure Exerted by the Body Fluids. At the bottom of Table 25–1 is shown the total osmotic pressure in mm. Hg that would be exerted by each of the different fluids if it were placed on one side of a cell membrane with pure water on the other side. Note that this total pressure averages about 5450 mm. Hg and also that the osmotic pressure of plasma is 24 mm. Hg greater than that of the interstitial fluids, this difference equaling the approximate hydrostatic pressure difference between the pressure of the blood inside the capillaries and the pressure in the interstitial fluid outside the capillaries.

MAINTENANCE OF OSMOTIC EQUILIBRIUM BETWEEN EXTRACELLULAR AND INTRACELLULAR FLUIDS

The tremendous osmotic pressure that can develop across the cell membrane when one side is exposed to pure water —about 5400 mm. Hg—illustrates how much force could become available to

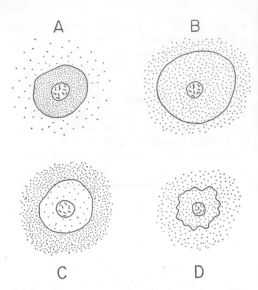

FIGURE 25–2 Establishment of osmotic equilibrium when cells are placed in a hypo- or hypertonic solution.

push water molecules through the membrane should the solutions on the two sides of the membrane not be in osmotic equilibrium. For instance, in Figure 25–2, a cell is placed in a solution that has an osmolality far less than that of the intracellular fluid. As a result, osmosis of water begins immediately from the extracellular fluid to the intracellular fluid, causing the cell to swell and diluting the intracellular fluid, while concentrating the extracellular fluid. When the fluid inside the cell becomes diluted sufficiently to equal the concentration of the fluid on the outside, further osmosis then ceases. This condition is shown in Figure 25–2B. In Figure 25–2C, a cell is placed in a solution having a much higher concentration outside the cell than inside. This time, water passes by osmosis to the exterior, diluting the extracellular fluid while concentrating the intracellular fluid. In this process the cell shrinks until the two concentrations become equal, as shown in Figure 25–2D.

Rapidity of Attaining Extracellular and Intracellular Osmotic Equilibrium. The transfer of water through the cell membrane by osmosis occurs so rapidly that any lack of osmotic equilibrium between the two fluid compartments is usually corrected within a few seconds and at most within a minute. Because of this rapidity of equilibration, it is generally considered that a state of osmotic equilibrium between the two compartments is maintained constantly.

Isotonicity, Hypotonicity, and Hypertonicity. A fluid into which normal body cells can be placed without causing either swelling or shrinkage of the cells is said to be *isotonic* with the cells. A 0.9 per cent saline solution is approximately isotonic.

A solution that will cause the cells to swell is said to be *hypotonic*; any solution of sodium chloride with less than 0.9 per cent concentration is hypotonic.

A solution that will cause the cells to shrink is said to be *hypertonic*; sodium chloride solutions of greater than 0.9 per cent concentration are all hypertonic.

CHANGES IN THE VOLUMES AND OSMOLALITIES OF THE EXTRACELLULAR AND INTRACELLULAR FLUID COMPARTMENTS IN ABNORMAL STATES

Effect of Adding Water to the Extracellular Fluid. Water can be added to the extracellular fluid by injection into the bloodstream, injection beneath the skin, or by ingesting water, which is followed by absorption from the gastrointestinal tract into the blood. The water dilutes the extracellular fluid, causing it to become hypotonic with respect to the intracellular fluids. Osmosis begins immediately at the cell membranes, with large amounts of water passing to the interior of the cells. Within a few minutes the water becomes distributed almost evenly among all the extracellular and intracellular fluid compartments.

Effect of Dehydration. Water can be removed from the body by evaporation from the skin, evaporation from the lungs, or excretion of a very dilute urine. In all these conditions the water leaves the extracellular fluid compartment, but on doing so some of the intracellular water passes immediately into the extracellular compartment by osmosis, thus keeping the osmolalities of the extracellular and intracellular fluids equal to each other. The overall effect is called *dehydration*.

Effect of Adding Saline Solutions to the Extracellular Fluid. If an *isotonic* saline solution is added to the extracellular fluid compartment, the osmolality of the extracellular fluid does not change, and no osmosis results. The only effect is an increase in extracellular fluid volume.

However, if a *hypertonic* solution is added to the extracellular fluid, the osmolality increases and causes osmosis of water out of the cells into the extracellular compartment.

Finally, if a *hypotonic* solution is added, the osmolality of the extracellular fluid decreases, and some of the extracellular fluids pass into the cells.

Effect of Infusing Hypertonic Glucose, Mannitol, or Sucrose Solutions. Very concentrated glucose, mannitol, or sucrose solutions are often injected into patients to cause immediate decrease in intracellular fluid volume. For instance, often in severe cerebral edema the patient dies because of too much pressure in the cranial vault, which obstructs the flow of blood to the brain. The condition can be relieved, however, in a few minutes by injecting a hypertonic solution of a substance that will not enter the intracellular compartment. The dynamics of the resulting changes are shown in Figure 25–3, illustrating that the intracellular fluid volume can be decreased by several liters in a few minutes.

But glucose, mannitol, and sucrose are all excreted rapidly by the kidneys, and glucose is also metabolized by the cells for energy. Therefore, within two to four hours the osmotic effects of these substances are lost so that large quantities of water can then rediffuse into the intra-

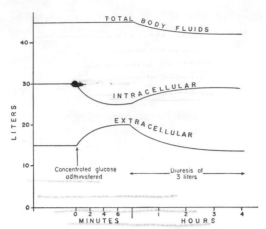

FIGURE 25-3 Time course of the body fluid changes following infusion of very concentrated glucose into the extracellular fluids.

cellular compartment. Thus, this procedure is temporarily beneficial, lasting only a few hours. Nevertheless, from the standpoint of saving a life of a patient it is often very valuable.

Glucose and Other Solutions Administered for Nutritive Purposes. Many different types of solutions are often administered intravenously to provide nutrition to patients who cannot otherwise take adequate amounts of food. Especially used are glucose solutions, to a lesser extent amino acid solutions, and, rarely, homogenized fat solutions. In administering all these, their concentrations are adjusted nearly to isotonicity, or they are given slowly enough that they do not upset the osmotic equilibria of the body fluids. However, after the glucose or other nutrient is metabolized, an excess of water often remains. Ordinarily, the kidneys excrete this in the form of a very dilute urine. Thus, the net result is only addition of the nutrient to the body.

But occasionally the kidneys are functioning poorly, such as often occurs after a surgical operation, and the body becomes greatly overhydrated, resulting sometimes in "water intoxication,"

which is characterized by mental irritability and even convulsions.

REGULATION OF BODY WATER

INTAKE VERSUS OUTPUT OF WATER

Daily Intake of Water. Most of our daily intake of water enters by the oral route. Approximately two-thirds is in the form of pure water or some other beverage, and the remainder is in the food that is eaten. A small amount is also synthesized in the body as the result of oxidation of hydrogen in the food; this quantity ranges between 150 and 250 ml. per day, depending on the rate of metabolism. The normal intake of fluid, including that synthesized in the body, averages about 2400 ml. per day.

Daily Loss of Body Water. Table 25-2 shows the routes by which water is lost from the body under different conditions. Normally, at an atmospheric temperature of about 68 degrees, approximately 1400 ml. of the 2400 ml. of water intake is lost in the *urine*, 100 ml. is lost in the *sweat*, and 200 ml. in the *feces*. The remaining 700 ml. is lost by *evaporation through the lungs* or by *diffusion through the skin*, which is called *insensible water loss*.

TABLE 25-2   DAILY LOSS OF WATER (IN MILLILITERS)

	Normal Temperature	Hot Weather	Prolonged Heavy Exercise
Insensible Loss:			
Skin	350	350	350
Lungs	350	250	650
Urine	1400	1200	500
Sweat	100	1400	5000
Feces	200	200	200
Total	2400	3400	6700

REGULATION OF WATER LOSS IN THE URINE — THE OSMORECEPTOR-ANTIDIURETIC HORMONE SYSTEM

In addition to regulation of the different ions in the extracellular fluids, regulation of the total water is equally important, for this controls the osmotic concentration in the extracellular fluid. This regulation is vested in the so-called *osmoreceptor-antidiuretic hormone system* involving the *hypothalamus*, the *posterior pituitary gland, antidiuretic hormone*, and the *renal tubules*. This system, when activated, promotes water conservation in a complex manner, as described in the following paragraphs.

Antidiuretic Hormone and Its Effect on the Kidney Tubules. *Antidiuretic hormone* (ADH) is secreted into the blood by the supraoptico-hypophyseal axis of of the hypothalamus and posterior pituitary gland. This hormone promotes increased water reabsorption from both the distal tubules and the collecting ducts of the kidneys, as was discussed in the preceding chapter. When no ADH is secreted, the amount of water passing into the urine each day is 5 to 15 times normal. And, because of this, the extracellular fluids become more concen-trated. On the other hand, when large quantities of ADH are secreted, water is reabsorbed to an extreme degree, so that the volume of urine formed each day may be as little as 400 to 500 ml., or one-third the normal volume. As a result, water is conserved, and the extracellular fluids become more dilute.

Function of the Osmoreceptors in Regulating Osmolality. Antidiuretic hormone would not be of any value to the body unless there existed a concomitant system for regulating the secretion of antidiuretic hormone in proportion to its need. To provide this, special neurons of the anterior hypothalamus, located, as shown in Figure 25–4, mainly in the supraoptic nuclei and called *osmorecep-tors*, respond specifically to changes in extracellular fluid osmolality. When the osmolality becomes very low, osmosis of water into the osmoreceptors causes them to swell, which decreases their rate of impulse discharge. Conversely, increased osmolality pulls water out of the osmoreceptors, causing them to shrink and thereby increasing their rate of discharge.

The impulses from the osmoreceptors are transmitted from the supraoptic nucleus through the pituitary stalk into the posterior pituitary gland where they promote the release of antidiuretic hor-

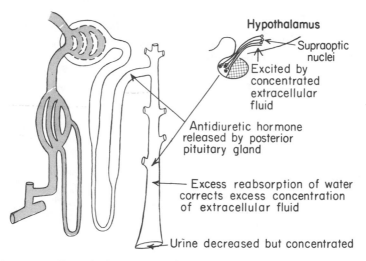

Hypothalamus

Supraoptic nuclei

Excited by concentrated extracellular fluid

Antidiuretic hormone released by posterior pituitary gland

Excess reabsorption of water corrects excess concentration of extracellular fluid

Urine decreased but concentrated

FIGURE 25–4 Control of extracellular fluid osmolality by the osmoreceptor system.

mone (ADH), the details of which will be discussed in Chapter 49 in relation to the endocrinology of the pituitary gland.

Thus, ADH is secreted in relation to the osmolality of the extracellular fluids — the greater the osmolality, the greater the rate of ADH secretion; and the less the osmolality, the less the rate of ADH secretion.

Summary of Feedback Control of Extracellular Fluid Osmolality. Now to synthesize the whole mechanism: An increase in osmolality of the extracellular fluids excites the osmoreceptors; this promotes ADH secretion and causes marked reabsorption of water by the renal tubules while solutes continue to be lost into the urine. As a consequence, the extracellular fluid becomes diluted and its concentrations of ions and other solutes return toward normal. On the other hand, low osmolality of the extracellular fluids decreases the activity of the osmoreceptors, thus decreasing ADH secretion, and large amounts of water are lost into the urine until the extracellular fluid osmolality again returns to normal.

The osmoreceptor system is sufficiently active that overconcentration of the extracellular fluid constituents of only 1 to 2 per cent causes marked retention of water, or a similar decrease causes rapid loss of water.

Diabetes Insipidus. Destruction of the supraoptic nuclei or high level destruction of the nerve tract from the supraoptic nuclei to the posterior pituitary gland causes antidiuretic hormone production to cease or to become greatly reduced. When this happens the person thereafter secretes nothing but a dilute urine, and his daily urine volume may be anywhere from 5 to 15 liters per day, a condition called diabetes insipidus.

THIRST

The phenomenon of thirst is just as important for regulating body water as are the renal mechanisms discussed above, for the amount of water in the body is determined by the balance between both *intake* and *output* of water each day. Thirst is the primary regulator of the intake. A definition of thirst is the *conscious desire for water.*

Neural Integration of Thirst — The "Thirst" or "Drinking" Center. An area called the *drinking center* has been localized in each side of the hypothalamus, lateral and caudal to the supraoptic nuclei. This center, illustrated in Figure 25–5, overlaps with the area of the hypothalamus that causes the release of ADH, thus explaining why factors that cause increased intake of water often cause increased conservation of water by the kidneys at the same time.

Electrical stimulation of the thirst center by implanted electrodes causes an animal to begin drinking within seconds and to continue drinking until the electrical stimulus is stopped. Also, injection of hypertonic salt solutions into the area, which causes osmosis of water out of the neuronal cells and results in intracellular dehydration, also causes drinking.

Basic Stimulus for Exciting the Thirst Center — Intracellular Dehydration. Among the many factors affecting thirst the most important are *extracellular dehydration, intracellular dehydration,* and *circulatory failure.* All three of these stimuli seem to affect the neurons of the thirst center directly, and it is

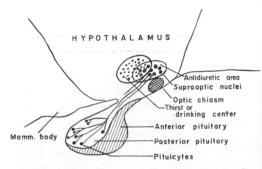

FIGURE 25–5 Location of the thirst center in the hypothalamus in relation to the supraoptico-pituitary antidiuretic system.

possible to explain all their effects on the basis of one single stimulus; namely, *intracellular dehydration*. For instance, extracellular dehydration is known to promote loss of fluid from inside cells. Thus, extracellular dehydration could well cause sufficient dehydration of the neurons in the thirst center to promote thirst. In the case of circulatory failure it is also known that an inadequate supply of nutrients to the cells depresses the active transport mechanisms that are responsible for maintaining normal intracellular electrolyte compositions, and this, too, under some conditions can promote loss of intracellular electrolytes with resultant intracellular dehydration.

This attempt to unify the different stimuli that excite thirst remains at present mainly speculation. But, nevertheless, almost any type of dehydration or failure of the circulation can promote thirst.

Temporary Relief of Thirst Caused by the Act of Drinking. A thirsty person receives relief from his thirst immediately after drinking water even before the water has been absorbed from the gastrointestinal tract. In fact, in persons who have esophageal fistulae (a condition in which the water never goes into the gastrointestinal tract) relief of thirst still occurs following the act of drinking, but this relief is only temporary and thirst returns after 15 minutes or more. If the water does enter the gut, distention of the upper gastrointestinal tract, particularly the stomach, provides still further temporary relief of thirst. For instance, simple inflation of a balloon in the stomach can relieve thirst for 5 to 30 minutes.

One might wonder what the value of this temporary relief of thirst could be, but there is a good reason for its occurrence as follows: After a person has drunk water, as long as a half to one hour may be required for the water to be absorbed into the extracellular fluids. Were his thirst sensations not temporarily relieved following the drinking of water, he would continue to drink more

and more. When all this water should finally become absorbed, his body water would be far greater than normal, and he would have created a condition opposite to that which he was attempting to correct. It is well known that a thirsty animal almost never drinks more than the amount of water needed to relieve his state of dehydration but ordinarily drinks almost exactly the right amount.

REGULATION OF EXTRACELLULAR FLUID VOLUME AND BLOOD VOLUME

Many aspects of extracellular fluid volume and blood volume have already been discussed—especially the regulation of interstitial fluid volume in the discussion of capillary dynamics in Chapter 16, and the regulation of blood volume in the discussion of overall regulation of the circulation in Chapter 18. Therefore, the purpose of the present section will be to put together in a concise form the overall picture of these volume regulations.

Basically, there are three different mechanisms for extracellular fluid volume and blood volume regulation: (1) an intrinsic mechanical system, which includes the dynamics of fluid exchange at the capillary membranes and the effects of hemodynamics on fluid excretion by the kidneys; (2) hormonal factors that affect fluid volume regulation, especially those that affect the ability of the kidneys to excrete extracellular fluid; and (3) nervous reflexes that also affect primarily the rate of excretion of fluid by the kidneys.

THE INTRINSIC MECHANICAL SYSTEM FOR EXTRACELLULAR FLUID VOLUME AND BLOOD VOLUME REGULATION

Regulation of Interstitial Fluid Volume. As discussed in Chapter 16, the interstitial fluid is composed of two separate components: (1) mobile fluid that is capable of moving freely through the interstitial fluid spaces and (2) immobile fluid that is bound in the ground substance of the interstitial spaces. It was also pointed out in Chapter 16 that the quantity of mobile fluid in the interstitial spaces is normally almost zero because two basic mechanisms cause mobile fluid to be absorbed almost as fast as it is formed. These mechanisms are (a) pumping of fluid from the spaces by the lymphatic system and (b) osmotic absorption of fluid directly into the capillaries.

Therefore, the normal interstitial fluid is almost entirely bound fluid, and its volume is controlled by the amount of mucopolysaccharides in the interstitial spaces. For instance, in young children the quantity of mucopolysaccharides is far greater than in older people, and this causes a much higher percentage of interstitial fluid in the tissues of a baby than in those of an adult. On the other hand, abnormal retention of fluid in the body, such as occurs in kidney disease or as a result of abnormal capillary dynamics, can cause excess quantities of *mobile* fluid in the tissues, giving rise to the condition called *edema* (see Chap. 16).

Regulation of Blood Volume. An overall schema for the regulation of extracellular fluid volume, blood volume, cardiac output, arterial pressure, and total peripheral resistance was presented in Chapter 18 to show the interrelationships between all these separate factors. It is important at the present time that the student review this schema to understand the intrinsic mechanical system for regulation of blood volume.

Basically, this system is the following: When the extracellular fluid volume becomes too high, the blood volume also increases. This increases venous return and cardiac output. The rise in cardiac output increases arterial pressure, which causes the kidneys to excrete the excess extracellular fluid volume. This reduces the extracellular fluid volume back toward normal and thereby also returns the plasma volume and blood volume back toward normal.

This description of the intrinsic mechanical feedback mechanism for regulation of extracellular fluid volume and blood volume is only a skeleton of the factors involved. If the student will return to Chapter 18 and study all of these factors in detail he will understand the beautiful integration of the circulatory and fluid volume regulatory systems.

HORMONAL FACTORS THAT AFFECT VOLUME REGULATION

Two hormonal systems play major roles in fluid volume regulation. These are (1) the aldosterone system for regulating many of the ions in the extracellular fluids, and (2) the ADH system for regulating water excretion. These have already been discussed earlier in the chapter, but it is important also to understand how they affect the overall regulation of extracellular fluid volume and blood volume.

Effect of the Aldosterone System. As discussed earlier, aldosterone has a direct effect on the kidneys to increase the retention of both sodium chloride and water, thereby increasing the extracellular fluid volume.

In the absence of aldosterone secretion, the kidneys lose tremendous quantities of sodium and chloride in the urine; consequently, large amounts of water are also lost, thereby reducing extracellular fluid volume and blood volume below normal. If this loss continues for more than 4 to 5 days the effect can actually cause death because of development of circulatory shock. This condition is called *Addison's disease*, which is discussed in more detail in Chapter 50.

The ADH System. The ADH system can alter by a few per cent the total quantity of water in the body fluids, excess ADH increasing body water as much as 3 to 5 per cent and lack of ADH decreasing body water perhaps 3 to 5 per cent.

The ADH system, as well as the aldosterone system, is to a great extent opposed by the intrinsic mechanical system for regulation of extracellular fluid volume and blood volume. Therefore, the ADH system is not nearly so important for regulation of fluid volumes as it is for regulation of body fluid osmolality.

NERVOUS REFLEXES THAT AFFECT VOLUME REGULATION

Almost all the circulatory reflexes play at least some role in the regulation of blood volume. For instance, the *baroreceptor reflex* in response to low arterial pressure causes afferent arteriolar constriction in the kidney and thereby causes retention of extracellular fluid. The buildup in extracellular fluid eventually helps to increase the arterial pressure back toward normal. Likewise, the *CNS ischemic response* has a tremendous effect in decreasing urinary output, sometimes reducing urinary output to absolute zero.

However, a few circulatory reflexes seem to play more specific roles in helping to control blood volume; these are mainly reflexes that have sensors that are excited by excess filling of the atria and great veins. Located in the walls of these structures, particularly in the walls of the left atrium, are specific stretch receptors called *volume receptors*, which elicit strong renal reflexes to cause increased excretion of urine. Also, signals are transmitted to the hypothalamus to reduce the output of ADH, which also causes the kidneys to lose extra quantities of water. Thus, whenever the atria and great veins become overfilled with blood, these volume receptors set into play almost instantaneous effects to increase the loss of fluid from the body, thereby helping to readjust both extracellular fluid volume and blood volume

back toward normal. Unfortunately, these reflexes seem to adapt within a few days, so that they are of only transient benefit.

MEASUREMENT OF BODY FLUID VOLUMES

THE DILUTION PRINCIPLE FOR MEASURING FLUID VOLUMES

The volume of any fluid compartment of the body can be measured by placing a substance in the compartment, allowing it to disperse evenly throughout the fluid, and then measuring the extent to which the substance becomes diluted. Figure 25–6 illustrates this "dilution" principle for measuring the volume of any fluid compartment of the body. A small quantity of dye or other foreign substance is placed in fluid chamber A, and the substance is allowed to disperse throughout the chamber until it becomes mixed in equal concentrations in all areas, as shown in chamber B. Then a sample of the dispersed fluid is removed and the concentration of the substance is analyzed chemically, photoelectrically, or by any other means. The volume of the chamber can then be determined from the following formula:

$$\text{Volume in ml.} = \frac{\text{Quantity of test substance instilled}}{\text{Concentration per ml. of dispersed fluid}}$$

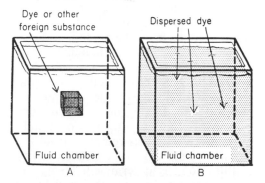

FIGURE 25–6 Principles of the dilution method for measuring fluid volumes (explained in text).

Note that all one needs to know is (1) the *total quantity of the test substance* put into the chamber and (2) the *concentration in the fluid after dispersement.*

DETERMINATION OF BLOOD VOLUME

Substances Used in Determining Blood Volume. A substance used for measuring blood volume must be capable of dispersing throughout the blood with ease, and it must remain in the circulatory system long enough for measurements to be made. The two major groups of substances that satisfy these conditions for measurement of blood volume are substances that combine with the red blood cells or substances that combine with the plasma proteins, for both the red blood cells and the plasma proteins remain reasonably well in the circulatory system, and any foreign substance that combines with either of them likewise remains in the blood stream.

Substances that combine with red blood cells and that are used for determining blood volume are *radioactive iron, radioactive chromium*, and *radioactive phosphate*. Substances that combine with plasma proteins are the *vital dyes* and *radioactive iodine*.

Radioactive red blood cells. The method most often used to make red blood cells radioactive is to tag the red blood cells with radioactive chromium (Cr^{51}). A small quantity of Cr^{51} is mixed with a few milliliters of blood that have been removed from the person, and this is incubated at 36° C. for half an hour or more. After this time, Cr^{51} will have entered the red blood cells. Then the radioactive cells are reinjected into the person. After mixing in the circulatory system has continued for approximately 10 minutes, blood is removed from the circulatory system, and the radioactivity in this blood is determined. Using the above dilution formula, the total blood volume is calculated.

Vital dyes for measurement of plasma volume. A number of dyes, generally known as "vital dyes," have the ability to combine with proteins. When such a dye is injected into the blood, it immediately forms a slowly dissociable union with the plasma proteins. Thereafter, the dye travels where the proteins travel.

The dye almost universally used for measuring plasma volume is *T-1824*, also called *Evans blue*. In making determinations of plasma volume, a known quantity of the dye is injected, and it immediately combines with the proteins and disperses throughout the circulatory system within approximately 10 minutes. A sample of the blood is then taken, and the red blood cells are removed from the plasma by centrifugation. Then, by spectrophotometric analysis of the plasma, one can determine the exact quantity of dye in the sample of plasma. From the determined quantity of dye in each milliliter of plasma and the known quantity of dye injected, the *plasma volume* is calculated using the dilution formula noted above.

Note that neither T-1824 nor any other vital dye enters the red blood cells. Therefore, this method *does not measure the total blood volume*. However, the blood volume can be calculated from the plasma volume, provided that the hematocrit is known by using the following formula:

$$\text{Blood volume} = \text{Plasma volume} \times \frac{100}{100 - 0.87\ \text{Hematocrit}}$$

(The factor 0.87 is used for two reasons. First, when blood is centrifuged to determine hematocrit, some of the plasma remains trapped between the red cells. Second, there is about 20 per cent more plasma in the blood of the small blood vessels than in the large vessels from which blood samples are taken.)

Radioactive protein. If a sample of plasma is allowed to incubate with radioactive iodine (I^{131}) for 30 minutes or more, some of the protein combines with the iodine, and the iodinated protein can be separated from the remaining iodine by dialysis. The radioactive protein is then injected into the subject, and plasma and blood volumes are determined in the same manner as that discussed for the vital dyes.

MEASUREMENT OF THE EXTRACELLULAR FLUID VOLUME

To use the dilution principle for measuring the volume of the extracellular fluid one injects into the blood stream a substance that can diffuse readily throughout the entire extracellular fluid chamber, passing easily through the capillary membranes but as little as possible through the cell membranes into the cells. After half an hour or more of mixing, a sample of extracellular fluid is obtained by removing blood and separating the plasma from the cells by centrifugation. The plasma, which is actually a part of the extracellular fluid, is then analyzed for the injected substance.

Substances Used in Measuring Extracellular Fluid Volume—The Concept of "Fluid Space." Substances that have been used for measuring extracellular fluid volume are *radioactive sodium, radioactive chloride, radioactive bromide, thiosulfate ion, thiocyanate ion, inulin*, and *sucrose*. Some of these, sucrose and inulin especially, do not diffuse readily into all the out-of-the-way places of the extracellular fluid compartment. Therefore, the volume of extracellular fluid measured with these is likely to be lower than the actual volume of the compartment. On the other hand, others of these substances—radioactive chloride, radioactive bromide, radioactive sodium, and thiocyanate ion, for instance—are likely to penetrate into the cells to a slight extent and, therefore, are likely to measure a space somewhat in excess of the extracellular fluid volume.

Because there is no single substance that measures the exact extracellular fluid volume, one usually speaks of the *sodium space*, the *thiocyanate space*, the *inulin space*, and so forth, rather than the extracellular volume. Measurements for the normal 70 kg. adult, when different ones of the above substances have been used, have ranged from 8 liters to 22 liters; but the average measurement has been about 15 liters.

MEASUREMENT OF TOTAL BODY WATER

The total body water can be measured in exactly the same way as the extracellular fluid volume except that a substance must be used that will diffuse into the cells as well as throughout the extracellular compartment. The substance that gives best results is water containing radioactive hydrogen (tritium).

Measurements of total body water in the 70 kg. adult have ranged from as low as 30 liters up to as high as 50 liters, with a reasonable average of approximately 40 liters, or 57 per cent of the total body mass.

CALCULATION OF INTERSTITIAL FLUID VOLUME

Since any substance that passes into the interstitial fluid also passes into almost all other portions of the extracellular fluid, there is no direct method for measuring interstitial fluid volume separately from the entire extracellular fluid volume. However, if the extracellular fluid volume and plasma volume have both been measured, the interstitial fluid volume can be approximated by *subtracting the plasma volume from the total extracellular fluid volume*. This calculation gives a normal interstitial fluid volume of 12 liters in the 70 kg. adult.

REFERENCES

Adolph, E. F.: Regulation of water intake in relation to body water content. *In* Handbook of Physiology. Baltimore, The Williams & Wilkins Co., 1967, Sec. VI, Vol. I, p. 163.

Black, D.: Essentials of fluid balance. 4th ed., Philadelphia, F. A. Davis Co., 1968.

Fitzsimons, J. T.: The hypothalamus and drinking. *Brit. Med. Bull., 22*:232, 1966.

Frazier, H. S.: Renal regulation of sodium balance. *New Eng. J. Med., 279*:868, 1968.

Gottschalk, C. W., and Mylle, M.: Micropuncture study of the mammalian urinary concentrating mechanism: evidence for the countercurrent hypothesis. *Amer. J. Physiol., 196*:927, 1959.

Guyton, A. C., Langston, J. B., and Navar, G.: Theory for renal autoregulation by feedback at the juxtaglomerular apparatus. *Circ. Res., 14*:187, 1964.

Moore, F. D., Olesen, K. H., McMurrey, J. D., Parker, H. V., Ball, M. R., and Boyden, C. M.: The Body Cell Mass and Its Supporting Environment. Philadelphia, W. B. Saunders Company, 1963.

Potts, W. T. W.: Osmotic and ionic regulation. *Ann. Rev. Physiol., 30*:73, 1968.

Windhager, E. E.: Glomerulo-tubular balance of salt and water. *Physiologist, 11*:103, 1968.

Wolf, A. V.: Thirst: Physiology of the Urge to Drink and Problems of Water Lack. Springfield, Ill., Charles C Thomas, Publishers, 1958.

Wolf, A. V., and Crowder, N. A.: Introduction to Body Fluid Metabolism. Baltimore, The Williams & Wilkins Co., 1964.

CHAPTER 26

ACID-BASE BALANCE; MICTURITION; AND RENAL DISEASE

ACID-BASE BALANCE

When one speaks of the regulation of acid-base balance he actually means regulation of hydrogen ion concentration in the body fluids. The hydrogen ion concentration in different solutions can vary from less than 10^{-14} equivalents per liter to higher than 10^1, which means a total variation of more than a quadrillion-fold. Only slight changes in hydrogen ion concentration from the normal value can cause marked alterations in the rates of chemical reactions in the cells, some being depressed and others accelerated.

Normal Hydrogen Ion Concentration and Normal pH of the Body Fluids — Acidosis and Alkalosis. The hydrogen ion concentration in the extracellular fluid is normally regulated at a constant value of approximately 4×10^{-8} Eq./liter; this value can vary from as low as 1.6×10^{-8} to as high as 1.2×10^{-7}.

From these values, it is already apparent that expressing hydrogen ion concentration in terms of actual concentrations is a cumbersome procedure. Therefore, the symbol *pH* has come into usage for expressing the concentration, and pH is related to the actual hydrogen ion concentration by the following formula (when H^+ conc. is expressed in equivalents per liter):

$$pH = \log \frac{1}{H^+ \text{ conc.}} = -\log H^+ \text{ conc.} \quad (1)$$

Note from this formula that a low pH corresponds to a high hydrogen ion concentration, which is called *acidosis*; and, conversely, a high pH corresponds to a low hydrogen ion concentration, which is called *alkalosis*.

The normal pH of arterial blood is 7.4, while the pH of venous blood and of interstitial fluids is about 7.35 because of extra quantities of carbon dioxide which form carbonic acid in these fluids.

Since the normal pH of the arterial blood is 7.4, a person is considered to have acidosis whenever the pH is below this value and to have alkalosis when it rises above 7.4. The lower limit at which a person can live more than a few minutes is about 7.0, and the upper limit approximately 7.8.

DEFENSE AGAINST CHANGES IN HYDROGEN ION CONCENTRATION

To prevent acidosis or alkalosis, several special control systems are available: (1) All the body fluids are supplied with acid-base *buffer systems* that immediately combine with any acid or alkali and thereby prevent excessive changes in hydrogen ion concentration. (2) If the hydrogen ion concentration does change measurably, the *respiratory center is immediately stimulated* to alter the rate of pulmonary ventilation. As a result, the rate of carbon dioxide removal from the body fluids is automatically changed, and, for reasons that will be presented later, this causes the hydrogen ion concentration to return toward normal. (3) When the hydrogen ion concentration changes from normal, *the kidneys excrete either an acid or alkaline urine*, thereby also helping to readjust the hydrogen ion concentration of the body fluids toward normal.

FUNCTION OF ACID-BASE BUFFERS

An acid-base buffer is a solution of two or more chemical compounds that prevents marked changes in hydrogen ion concentration when either an acid or a base is added to the solution. As an example, if only a few drops of concentrated hydrochloric acid are added to a beaker of pure water, the pH of the water might immediately fall to as low as 1.0. However, if a satisfactory buffer system is present, the hydrochloric acid combines instantaneously with the buffer, and the pH falls only slightly. Perhaps the best way to explain the action of an acid-base buffer is to consider an actual simple buffer system, such as the bicarbonate buffer, which is extremely important in regulation of acid-base balance in the body.

THE BICARBONATE BUFFER SYSTEM

A typical bicarbonate buffer system consists of a mixture of carbonic acid (H_2CO_3) and sodium bicarbonate ($NaHCO_3$) in the same solution. It must first be noted that carbonic acid is a very weak acid because its degree of dissociation into hydrogen ions and bicarbonate ions is poor in comparison with that of many other acids.

When an acid such as hydrochloric acid is added to a solution containing bicarbonate buffer, the following reaction takes place:

$$HCl + NaHCO_3 \rightarrow H_2CO_3 + NaCl \quad (2)$$

From this equation it can be seen that the strong hydrochloric acid is converted into the very weak carbonic acid. Therefore, the HCl lowers the pH of the solution only slightly.

Though this bicarbonate buffer system has been illustrated in the above reactions as a mixture of carbonic acid and sodium bicarbonate, any other salt of bicarbonate besides sodium bicarbonate has identically the same function.

The Henderson-Hasselbalch Equation. The relationship between pH and the concentration of HCO_3^- and dissolved CO_2 (which combines with water to give carbonic acid) is shown by the following formula:

$$pH = 6.1 + \log \frac{HCO_3^-}{CO_2} \quad (3)$$

This is called the Henderson-Hasselbalch equation, and by using it one can calculate the pH of a solution with reasonable accuracy after measuring the concentrations of bicarbonate ion and dissolved carbon dioxide, respectively. If the bicarbonate concentration is equal to the dissolved carbon dioxide concentration, the second member of the right-hand portion of the equation becomes log of 1, which is equal to zero. Therefore, under these conditions the pH of the solution is equal to 6.1.

From the Henderson-Hasselbalch equation one can readily see that an increase in bicarbonate ion concentration causes the pH to rise, or, in other words, shifts the acid-base balance toward the alkaline side. On the other hand, an increase in the concentration of dissolved carbon dioxide decreases the pH, or shifts the acid-base balance toward the acid side. These effects are illustrated in Figure 26-1.

THE BUFFER SYSTEMS OF THE BODY FLUIDS

The three major buffer systems of the body fluids are the *bicarbonate buffer*, which has been described above, the *phosphate buffer*, and the *protein buffer*. Each of these performs major buffering functions under different conditions.

Importance of the Bicarbonate Buffer System. The bicarbonate buffer system is probably equally as important as all the others in the body *because the concentrations of each of the two elements of the bicarbonate system can be regulated*, carbon dioxide by the respiratory system and bicarbonate ion by the kidneys. As a result, the pH of the blood can be shifted up or down by the respiratory and renal regulatory systems.

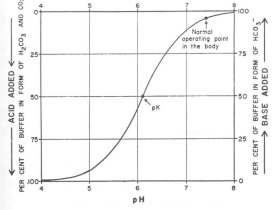

FIGURE 26-1 Reaction curve for the bicarbonate buffer system.

The Phosphate Buffer System. The phosphate buffer system acts in almost identically the same manner as the bicarbonate buffer system, but it is composed of the following two elements: NaH_2PO_4 and Na_2HPO_4. When a strong acid, such as hydrochloric acid, is added to a mixture of these two substances, the following reaction occurs:

$$HCl + Na_2HPO_4 \rightarrow NaH_2PO_4 + NaCl \quad (4)$$

The net result of this reaction is that the hydrochloric acid is removed, and in its place an additional quantity of NaH_2PO_4 is formed. Since NaH_2PO_4 is only weakly acidic, a strong acid has been traded for a very weak acid, and the pH changes relatively slightly.

The phosphate buffer is especially important in the tubular fluids of the kidneys because phosphate usually becomes greatly concentrated in the tubules, thereby increasing the buffering power of the phosphate system. The phosphate buffer is also extremely important in intracellular fluids because the concentration of phosphate in these fluids is many times that in the extracellular fluids.

The Protein Buffer System. By far the most plentiful buffer of the body is the proteins of the cells and plasma. There is a slight amount of diffusion of hydrogen ions through the cell membrane, and even more important, carbon dioxide can diffuse readily through cell membranes and bicarbonate ions can diffuse to some extent. The diffusion of the two elements of the bicarbonate buffer system causes the pH in the intracellular fluids to change approximately in proportion to the changes in pH in the extracellular fluids. Thus, all the buffer systems inside the cells help to buffer the extracellular fluids as well. These include the extremely large amounts of proteins inside the cells. Indeed, experimental studies have shown that at least three quarters of all the *chemical* buffering power of the body fluids is inside the cells, and most of this results from the intracellular proteins.

The method by which the protein buffer system operates is precisely the same as that of the bicarbonate buffer system. It will be recalled that a protein is composed of amino acids bound together by peptide linkages, but some of the different amino acids have free acidic radicals in the form of $-COOH$, and these can dissociate into $-COO^-$ and H^+. Also, some have free basic radicals in the form of $-NH_3OH$, which can dissociate into $-NH_3^+$ and OH^-. The OH^- in turn can react with hydrogen ions to form water, in this way reducing the hydrogen ion concentration. Thus, proteins can operate in both *acidic* and *basic* buffering systems, the basic systems operating oppositely but similarly to the acidic systems.

THE ISOHYDRIC PRINCIPLE

Each of the above buffer systems has been discussed as if it could operate individually in the body fluids. However, they all actually work together, for the hydrogen ion is common to the chemical reactions of all the systems. Therefore, whenever any condition causes the hydrogen ion concentration to change, it causes the balance of all the buffer systems to change at the same time. This phenomenon is called the *isohydric principle*. The important feature of this principle is that any condition that changes the balance of any one of the buffer systems also changes the balance of all the others, for *the buffer systems actually buffer each other.*

RESPIRATORY REGULATION OF ACID-BASE BALANCE

In the discussion of the Henderson-Hasselbalch equation, it was noted that an increase in carbon dioxide concentration in the body fluids lowers the pH toward the acidic side, whereas a decrease in carbon dioxide raises the pH toward the alkaline side. It is on the basis of this effect that the respiratory system is capable of altering the pH either up or down.

Effect of Increasing or Decreasing the Alveolar Ventilation on pH of the Extracellular Fluids. If we assume that the rate of metabolic formation of carbon dioxide remains constant, then the only factor that affects the carbon dioxide concentration in the body fluids is the rate of alveolar ventilation as expressed by the following formula:

$$CO_2 \propto \frac{1}{\text{Alveolar ventilation}} \qquad (5)$$

And, since an increase in carbon dioxide decreases the pH, changes in alveolar ventilation also change the hydrogen ion concentration.

Figure 26–2 illustrates the approximate change in pH in the blood that can be effected by increasing or decreasing the rate of alveolar ventilation. Note that an increase in alveolar ventilation to two times normal raises the pH of the extracellular fluids by about 0.23 pH unit. This means that if the pH of the body fluids has been 7.4 with normal alveolar ventilation, doubling the ventilation raises the pH to 7.63. Conversely, a decrease in alveolar ventilation to one-quarter normal reduces the pH 0.4 pH

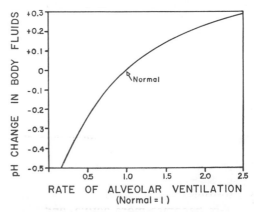

FIGURE 26–2 Approximate change in body fluid pH caused by increased or decreased rate of alveolar ventilation.

unit. That is, if at normal alveolar ventilation the pH had been 7.4, reducing the ventilation to one-quarter normal reduces the pH 0.4 unit to 7.0. Since alveolar ventilation can be reduced to zero ventilation or increased to about 15 times normal, one can readily understand how much the pH of the body fluids can be changed by alterations in the activity of the respiratory system.

EFFECT OF HYDROGEN ION CONCENTRATION ON ALVEOLAR VENTILATION

Not only does the rate of alveolar ventilation affect the hydrogen ion concentration of the body fluids, but, in turn, the hydrogen ion concentration can affect the rate of alveolar ventilation. This results from a *direct action of hydrogen ions on the respiratory center in the medulla oblongata* that controls breathing.

Figure 26–3 illustrates the changes in alveolar ventilation caused by changing the pH of arterial blood from 7.0 to 7.6. From this graph it is evident that a decrease in pH from the normal value of 7.4 to the strongly acidic range can increase the rate of alveolar ventilation to as much as four to five times normal, while an increase in pH into the alkaline range can decrease the rate of alveolar ventilation to as little as 50 to 75 per cent of normal.

Feedback Regulation of Hydrogen Ion Concentration by the Respiratory System. Because of the ability of the respiratory center to respond to hydrogen ion concentration, and because changes in alveolar ventilation in turn alter the hydrogen ion concentration in the body fluids, the respiratory system acts as a typical feedback regulatory system for controlling hydrogen ion concentration. That is, any time the hydrogen ion concentration becomes high, the respiratory system becomes more active, and alveolar ventilation increases. As a result, the carbon dioxide concentration in the

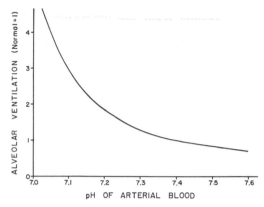

FIGURE 26–3 Effect of blood pH on the rate of alveolar ventilation. (Constructed from data obtained by Gray: Pulmonary Ventilation and Its Regulation. Charles C Thomas.)

extracellular fluids decreases, thus reducing the hydrogen ion concentration back toward a normal value. Conversely, if the hydrogen ion concentration falls too low, the respiratory center becomes depressed, alveolar ventilation also decreases, and the hydrogen ion concentration rises back toward normal.

Buffering power of the respiratory system. In effect, respiratory regulation of acid-base balance is a *physiological type of buffer* system having almost identically the same importance as the chemical buffering systems of the body discussed earlier in the chapter. The overall "buffering power" of the respiratory system is one to two times as great as that of all the chemical buffers combined. That is, one to two times as much acid or base can normally be buffered by this mechanism as by the chemical buffers.

RENAL REGULATION OF HYDROGEN ION CONCENTRATION

In the discussion of the Henderson-Hasselbalch equation it was pointed out that the kidneys can regulate hydrogen ion concentration principally by increasing or decreasing the bicarbonate ion concentration in the body fluid. To do this, a complex series of reactions occurs in the renal tubules, including reactions for hydrogen ion secretion, sodium ion reabsorption, bicarbonate ion excretion into the urine, and ammonia secretion into the tubules. The following sections describe the interplay of these different tubular mechanisms and the manner in which they help to regulate the hydrogen ion concentration of the body fluids.

TUBULAR SECRETION OF HYDROGEN IONS

The epithelial cells of the proximal tubules, distal tubules, and collecting ducts all secrete hydrogen ions into the tubular fluid, the chemical reactions of which are illustrated in Figure 26–4. Carbon dioxide, under the influence of an enzyme, *carbonic anhydrase*, combines with water to form *carbonic acid.* This then dissociates into *bicarbonate ion* and *hydrogen ion*, and the hydrogen ion is secreted through the cell membrane into the tubule. In the collecting ducts this process can continue until the concentration of hydrogen ions in the tubules becomes as much as 900 times that in the extracellular fluid or, in other words, until the pH of the tubular fluids falls to about 4.5. This represents a limit to the ability of the tubular epithelium to secrete hydrogen ions.

Regulation of Hydrogen Ion Secretion by the Carbon Dioxide Concentration in the Extracellular Fluid. Since the chemical reactions for secretion of hydrogen ions begin with carbon dioxide, the greater the carbon dioxide concentration in the extracellular fluid, the more rapidly the reactions proceed, and the greater becomes the rate of hydrogen ion secretion. Therefore, any factor that increases the carbon dioxide concentration in the extracellular fluids also increases the rate of hydrogen ion secretion.

INTERACTION OF BICARBONATE IONS WITH HYDROGEN IONS IN THE TUBULES — "REABSORPTION" OF BICARBONATE IONS

Bicarbonate ions are continually being filtered through the glomerular membrane into the glomerular filtrate, and the hydrogen ions secreted by the tubular epithelium combine with the bicarbonate ions to form carbonic acid, as illustrated in Figure 26–4. This carbonic acid in turn dissociates into carbon dioxide and water; essentially all the carbon dioxide then diffuses through the epithelial cells into the peritubular fluids while the water passes into the urine. Thus, *if plenty of hydrogen ions are available, the bicarbonate ions are almost completely removed from the tubules*, so that for practical purposes *none* remains to pass into the urine. If we now

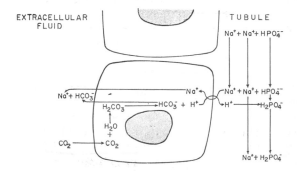

FIGURE 26–4 Chemical reactions for (1) hydrogen ion secretion, (2) sodium ion absorption in exchange for a hydrogen ion, and (3) combination of hydrogen ions with bicarbonate ions in the tubules.

note in Figure 26–4 the chemical reactions that are responsible for formation of hydrogen ions in the epithelial cells, we will see that a bicarbonate ion is formed inside these cells each time a hydrogen ion is secreted. The bicarbonate ion then diffuses into the peritubular fluid in combination with a sodium ion that has been absorbed from the tubule.

The net effect of all these reactions is a mechanism for "reabsorption" of bicarbonate ions from the tubules, though the bicarbonate ions that enter the peritubular fluid are not the same bicarbonate ions that had been in the tubular fluid.

RENAL CORRECTION OF ALKALOSIS — DECREASE IN BICARBONATE IONS IN THE EXTRACELLULAR FLUID

Now that we have described the mechanisms by which the renal tubules secrete hydrogen ions and reabsorb bicarbonate ions, we can explain the manner in which the kidneys readjust the pH of the extracellular fluids when it becomes abnormal.

The initial step in this explanation is to understand what happens to the concentrations of carbon dioxide and bicarbonate ions in the extracellular fluids in alkalosis and acidosis. First, let us consider *alkalosis*. Referring again to Equation 3, the Henderson-Hasselbalch equation, we see that the *ratio* of bicarbonate ions to dissolved carbon dioxide molecules increases when the pH rises into the alkalosis range above 7.4. The effect of this in the tubules is to increase the *ratio* of bicarbonate ions filtered into the tubules to hydrogen ions secreted. Therefore, the fine balance that normally exists between these two no longer occurs. Instead, far greater quantities of bicarbonate ions now enter the tubules than do hydrogen ions, and all the excess bicarbonate ions pass into the urine and carry with them sodium ions or other positive ions. Thus, in effect, sodium bicarbonate is removed from the extracellular fluid.

Loss of sodium bicarbonate from the extracellular fluid decreases the bicarbonate ion portion of the bicarbonate buffer system, and, in accordance with the Henderson-Hasselbalch equation, this shifts the pH of the body fluids back in the acid direction. Thus, the alkalosis is corrected.

RENAL CORRECTION OF ACIDOSIS

In acidosis, the *ratio* of carbon dioxide to bicarbonate ions in the extracellular fluid increases, which is exactly opposite to the effect in alkalosis. Therefore, in acidosis, the *rate of hydrogen ion secretion* rises to a level far greater than the *rate of bicarbonate ion filtration* into the tubules. As a result, excess hydrogen ions are secreted into the tubules and combine with the buffers in the tubular fluid, as is explained in the following paragraphs, and are excreted into the urine. Thus, *the net effect of secreting excess hydrogen ions into the tubules is to remove hydrogen ions from the body fluids, thereby correcting the acidosis.*

TRANSPORT OF EXCESS HYDROGEN IONS FROM THE TUBULES INTO THE URINE

When excess hydrogen ions are secreted into the tubules, they combine with other substances in the tubules and are transported into the urine. Two distinct mechanisms exist by which the excess hydrogen ions are transported from the tubules: first, by combination with the buffers of the tubular fluid, especially phosphate buffers as discussed earlier; and, second, by combination with ammonia secreted by the tubular epithelium.

Ammonia Secretion by the Tubular Epithelium. The epithelial cells of the proximal tubules, distal tubules, and collecting ducts all continually synthesize ammonia, and this diffuses into the tubules. The ammonia then reacts with hydrogen ions, as illustrated in Figure 26–5, forming ammonium ions that are excreted into the urine in combination with chloride or other tubular anions. Note in the figure that the net effect of these reactions is again to remove hydrogen ions from the body fluids.

This mechanism for removal of hydrogen ions from the tubules is especially important because most of the negative ions of the tubular fluid are chloride. Only a few hydrogen ions could be transported into the urine in direct combination with chloride, because hydrochloric acid is a very strong acid and the tubular pH would fall rapidly to the critical value of 4.5, below which further hydrogen ion secretion would cease. However, when hydrogen ions combine with ammonia and the resulting ammonium ions then combine with chloride, the pH does not fall significantly because ammonium chloride is a neutral salt instead of an acid.

If the tubular fluids remain highly acidic for long periods of time, the formation of ammonia steadily increases during the first two to three days, rising as much as 10-fold. For instance, immediately after acidosis begins, as little as 30 millimols of ammonia might be secreted each day, but after two days as much as 200 to 300 millimols can be secreted, illustrating that the ammonia-secreting mechanism can adapt readily to handle greatly increased loads of acid elimination.

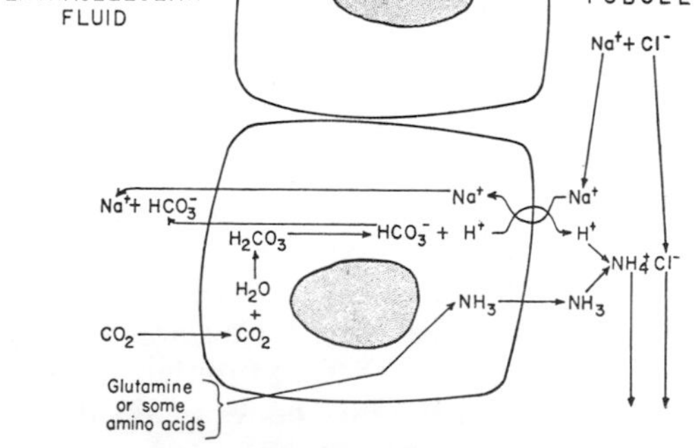

RAPIDITY OF ACID-BASE REGULATION BY THE KIDNEYS

The total amount of buffers in the entire body is approximately 1000 millimols. If all these should be suddenly shifted to the alkaline or acidic side by injecting an alkali or an acid, the kidneys would be able to return the pH of the body fluids back almost to normal in 10 to 20 hours. Though this mechanism is slow to act, it continues acting until the pH returns almost exactly to normal rather than a certain percentage of the way. Therefore, the real value of the renal mechanism for regulating hydrogen ion concentration is not rapidity of action but instead its ability in the end to neutralize completely any excess acid or alkali that enters the body fluids.

Range of Urinary pH. In the process of adjusting the hydrogen ion concentration of the extracellular fluids, the kidneys often excrete urine at pH's as low as 4.5 or as high as 8.0.

FIGURE 26–5 Secretion of ammonia by the tubular epithelial cells, and reaction of the ammonia with hydrogen ions in the tubules.

CLINICAL ABNORMALITIES OF ACID-BASE BALANCE

RESPIRATORY ACIDOSIS AND ALKALOSIS

From the discussions earlier in the chapter it is obvious that any factor that decreases the rate of pulmonary ventilation increases the concentration of dissolved carbon dioxide in the extracellular fluid, which in turn leads to increased carbonic acid and hydrogen ions, thus resulting in acidosis. Because this type of acidosis is caused by an abnormality of respiration, it is called *respiratory acidosis*.

On the other hand, excessive pulmonary ventilation reverses the process, resulting in *respiratory alkalosis*.

A person can cause respiratory acidosis in himself by simply holding his breath, which he can do until the pH of the body fluids falls to as low as perhaps 7.1. On the other hand, he can voluntarily overbreathe and cause alkalosis to a pH of about 7.8.

METABOLIC ACIDOSIS AND ALKALOSIS

The terms "metabolic acidosis" and "metabolic alkalosis" refer to all other abnormalities of acid-base balance besides those caused by excess or insufficient carbon dioxide in the body fluids. Use of the word "metabolic" in this instance is unfortunate, because carbon dioxide is also a metabolic product. Yet, by convention, carbonic acid resulting from dissolved carbon dioxide is called a *respiratory acid* while any other acid in the body, whether it be formed by metabolism or simply ingested by the person, is called a *metabolic acid*.

Causes of Metabolic Acidosis. Metabolic acidosis can result from (1) the formation of metabolic acids in the body, (2) intravenous administration of metabolic acids, or (3) addition of metabolic acids by way of the gastrointestinal tract; or metabolic acidosis can result also from (4) loss of alkali from the body fluids. Two of the specific conditions that cause metabolic acidosis are the following:

Severe diarrhea is one of the most frequent causes of metabolic acidosis for the following reasons: The gastrointestinal secretions normally contain large amounts of sodium bicarbonate. Therefore, excessive loss of these secretions during a bout of diarrhea is exactly the same as excretion of large amounts of sodium bicarbonate into the urine. In accordance with the Henderson-Hasselbalch equation, this results in a shift of the bicarbonate buffer system toward the acid side and results in metabolic acidosis. In fact, acidosis resulting from severe diarrhea can be so serious that it is one of the most common causes of death in young children.

Diabetes mellitus is an extremely important cause of metabolic acidosis. In this condition, lack of insulin secretion by the pancreas prevents normal use of glucose for metabolism. Instead, the stored fats are split into acetoacetic acid, and this in turn is metabolized by the tissues for energy in place of glucose. Simultaneously, the concentration of acetoacetic acid in the extracellular fluids often rises very high, causing extreme acidosis.

Causes of Metabolic Alkalosis. Metabolic alkalosis does not occur nearly so often as metabolic acidosis. It most frequently follows *excessive ingestion of alkaline drugs*, such as sodium bicarbonate, for the treatment of gastritis or peptic ulcer. However, metabolic alkalosis occasionally results from *excessive vomiting of gastric contents* without vomiting of lower gastrointestinal contents, which causes excessive loss of the hydrochloric acid secreted by the stomach mucosa. The net result is loss of acid from the extracellular fluids and development of metabolic alkalosis.

EFFECTS OF ACIDOSIS AND ALKALOSIS ON THE BODY

Acidosis. The major effect of acidosis is depression of the central nervous system. When the pH of the blood falls below 7.0, the nervous system becomes so depressed that the person first becomes disoriented and, later, comatose. Therefore, patients dying of diabetic acidosis, uremic acidosis, and other types of acidosis usually die in a state of coma.

Alkalosis. The major effect of alkalosis on the body is *overexcitability of the nervous system.* This effect occurs both in the central nervous system and in the peripheral nerves, but usually the peripheral nerves are affected before the central nervous system. The nerves become so excitable that they automatically and repetitively fire even when they are not stimulated by normal stimuli. As a result, the muscles go into a state of *tetany*, which means a state of tonic spasm. Only occasionally does an alkalotic person develop severe symptoms of central nervous system overexcitability. The symptoms may manifest themselves as extreme nervousness or, in susceptible persons, as convulsions. For instance, in persons who are predisposed to epileptic fits, simply overbreathing often results in an attack.

PHYSIOLOGY OF TREATMENT IN ACIDOSIS OR ALKALOSIS

Obviously, the best treatment for acidosis or alkalosis is to remove the condition causing the abnormality, but, if this cannot be effected, different drugs can be used to neutralize the excess acid or alkali.

To neutralize excess acid, large amounts of sodium bicarbonate can be ingested by mouth. This is absorbed into the blood stream and increases the bicarbonate ion portion of the bicarbonate buffer, thereby shifting the pH to the alkaline side. Sodium bicarbonate is occasionally used also for intravenous therapy, but this has such strong and often dangerous physiological effects that other substances are more often used, such as sodium lactate, sodium gluconate, or other organic compounds of sodium. The lactate and gluconate portions of the molecules are metabolized in the body, leaving the sodium in the extracellular fluids in the form of sodium bicarbonate, and thereby shifting the pH of the fluids in the alkaline direction.

For treatment of alkalosis, ammonium chloride is usually administered by mouth. When this is absorbed into the blood, the ammonia portion of the ammonium chloride is converted by the liver into urea; this reaction liberates hydrochloric acid, which immediately reacts with the buffers of the body fluids to shift the hydrogen ion concentration in the acid direction.

Carbon Dioxide Combining Power of the Plasma. To determine the degree of metabolic acidosis or alkalosis in a patient, the usual clinical laboratory measures the so-called *carbon dioxide combining power of the plasma*, which is also called the *alkali reserve*. This measurement is made as follows: Blood is removed from the patient, it is rendered incoagulable with potassium oxalate, and the plasma is separated from the cells. No special precautions are taken about exposure to air. Instead, in the laboratory the plasma is re-exposed to 7 per cent carbon dioxide until it equilibrates. Then the total carbon dioxide in the plasma is determined chemically, including all the dissolved carbon dioxide as well as that which can be liberated from the bicarbonate in the plasma by the addition of acid. Since about $19/20$ is in the form of bicarbonate, this is principally a measure of plasma bicarbonate, and it is called the *carbon dioxide combining power of the plasma*.

The normal carbon dioxide combining power is 28 mM./liter, but in severe metabolic acidosis it may be reduced to as low as 5 to 10 mM./liter. Conversely, in severe metabolic alkalosis, it may rise to as high as 40 to 50 mM./liter.

MICTURITION

Micturition is the process by which the urinary bladder empties itself when it becomes filled. Basically the bladder (1) progressively fills until the tension in its walls rises above a threshold value, at which time (2) a nervous reflex called the "micturition reflex" is elicited that (3) greatly exacerbates the pressure in the bladder and simultaneously causes a conscious desire to urinate. The micturition reflex also initiates appropriate signals from the nervous system to relax the external sphincter of the bladder, thereby allowing micturition.

Physiologic Anatomy of the Bladder and its Nervous Connections. The urinary bladder, which is illustrated in Figure 26–6, is mainly a smooth muscle vesicle composed of two principal parts: (a) the *body*, which is comprised mainly of the *detrusor muscle*, and (b) the *trigone*, a small triangular area near the mouth of the bladder through which both the *ureters* and the *urethra* pass. During bladder expansion the body of the bladder stretches, and during the micturition reflex the detrusor muscle contracts to empty the bladder.

The trigonal muscle is interlaced around the opening of the urethra and maintains tonic closure of the urethral opening until the pressure in the bladder rises high enough to overcome the tone of the trigonal muscle. The trigonal muscle, therefore, is called the *internal sphincter of the bladder*. About 2 cm. beyond the bladder, the urethra passes through the *urogenital diaphragm*, the muscle of which constitutes the *external sphincter* of the bladder. This muscle is a voluntary skeletal muscle, in contrast to the other muscle of the bladder, which is entirely smooth muscle. Normally, this muscle remains tonically contracted,

which prevents constant dribbling of urine, but it can be reflexly or voluntarily relaxed at the time of micturition.

Figure 26–6 also illustrates the basic nervous connections with the spinal cord for bladder control.

Transport of Urine Through the Ureters. The ureters are small smooth muscle tubes that originate in the pelves of the two kidneys and pass downward to enter the bladder. Each ureter is innervated by both sympathetic and parasympathetic nerves, and each also has an intramural plexus of neurons and nerve fibers that extends along its entire length.

As urine collects in the pelvis, the pressure in the pelvis increases and initiates a peristaltic contraction beginning in the pelvis and spreading down along the ureter to force urine toward the bladder. A peristaltic wave occurs from once every 10 seconds to once every two to three minutes. Parasympathetic stimulation increases, and sympathetic stimulation decreases, the frequency. Transmission of the peristaltic wave is probably caused mainly by nerve impulses passing along the intramural plexus in the same manner that the intramural plexus functions in the gut.

At its lower end, the ureter penetrates the bladder obliquely through the *trigone*, as illustrated in Figure 26–6. The ureter courses for several centimeters under the bladder epithelium so that pressure in the bladder compresses the

FIGURE 26–6 The urinary bladder and its innervation.

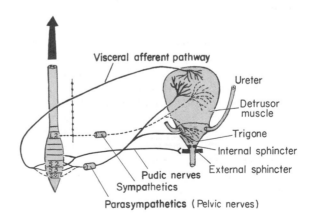

Visceral afferent pathway

Ureter

Detrusor muscle

Trigone

Internal sphincter

External sphincter

Pudic nerves

Sympathetics

Parasympathetics (Pelvic nerves)

ureter, thereby preventing backflow of urine.

Tonus of the Bladder. The solid curve of Figure 26–7 is called the *cystometrogram* of the bladder. When no urine at all is in the bladder, the intravesical pressure is approximately zero, but, by the time 100 ml. of urine has collected, the pressure will have risen to about 10 cm. water. Additional urine up to 300 to 400 ml. can collect without significant further rise in pressure; the pressure remains only 10 cm. water. Beyond this point, collection of more urine causes the pressure to rise very rapidly.

Superimposed on the tonic pressure changes during filling of the bladder are periodic acute increases in pressure which last from a few seconds to more than a minute. The pressure can rise only a few centimeters of water or it can rise to over 100 cm. water. These are *micturition waves* in the cystometrogram caused by the micturition reflex.

The Micturition Reflex. Referring once again to Figure 26–7, it is evident that as the bladder fills to the upper end of the cystometrogram, many superimposed *micturition contractions* begin to appear. These are the result of a stretch reflex initiated by stretch receptors in the bladder wall. Sensory signals are conducted to the sacral segments of the cord through the pelvic nerves and then back again to the bladder through the parasympathetic fibers in these same nerves.

Once a micturition reflex begins, it is "self-regenerative." That is, initial contraction of the bladder further stretches the receptors to cause still further increase in afferent impulses from the bladder, which causes further increase in reflex contraction of the bladder, the cycle thus repeating itself again and again until the bladder has reached a strong degree of contraction. Then, after a few seconds to more than a minute, the reflex begins to fatigue, and the regenerative cycle of the micturition reflex ceases, allowing rapid reduction in bladder contraction. In other words, the micturition reflex is a single, complete cycle of (a) progressive and rapid increase in pressure, (b) a period of sustained pressure, and (c) return of the pressure back to the basal tonic pressure of the bladder. Once a micturition reflex has occurred, the nervous elements of this reflex usually remain in an inhibited state for at least a few minutes or sometimes as long as an hour or more before another micturition reflex occurs. However, as the bladder becomes more and more filled, micturition reflexes occur more and more often and more and more powerfully.

Control of Micturition by the Brain. The micturition reflex is a completely

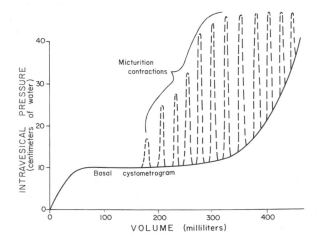

FIGURE 26–7 A normal cystometrogram showing also acute pressure waves (the dashed curves) caused by micturition reflexes.

automatic cord reflex, but it can be inhibited or facilitated by centers in the brain. The higher centers normally exert final control of micturition by the following means:

1. The higher centers keep the micturition reflex partially inhibited all the time except when it is desired to micturate.

2. The higher centers prevent micturition by continual tonic contraction of the external urinary sphincter until a convenient time presents itself.

3. When the time to urinate arrives, the cortical centers can (a) facilitate the sacral micturition centers to initiate a micturition reflex and (b) inhibit the external urinary sphincter so that urination can occur.

RENAL DISEASE

ACUTE RENAL SHUTDOWN

Renal Shutdown Caused by Acute Glomerular Nephritis. Acute glomerular nephritis is a disease in which the glomeruli become markedly inflamed. Large numbers of white blood cells collect in the inflamed glomeruli, and the endothelial cells on the vascular side of the glomerular membrane, as well as the epithelial cells on the Bowman's capsule side of the membrane, proliferate, sometimes completely filling the glomeruli and the capsule. These inflammatory reactions cause total or partial blockage of large numbers of glomeruli, and those glomeruli that are not blocked develop greatly increased permeability of the glomerular membrane, allowing large amounts of protein to leak into the glomerular filtrate. Also, rupture of the membrane in severe cases often allows large numbers of red blood cells to pass into the glomerular filtrate. Therefore, all degrees of glomerular malfunction can occur, including total renal shutdown.

The inflammation of acute glomerular nephritis occurs one to three weeks following, elsewhere in the body, an infection caused by group A beta streptococci, such as a streptococcal sore throat, streptococcal tonsillitis, scarlet fever, or even streptococcal infection of the skin. The infection itself does not cause damage to the kidneys, but when antibodies develop following the infection, the glomeruli are damaged. Some clinicians believe that this effect results from precipitation of streptococcal antigen with its specific antibody and then subsequent entrapment of the precipitate in the glomerular membrane. But others believe that the streptococcal antigen causes antibodies to be formed that in turn attack the proteins of the glomerular membrane, thus resulting in a type of "autoimmunity." Regardless of the specific cause, it is clear that the damage to the glomeruli results from antibodies that develop following a streptococcal infection and not from the streptococcal toxins themselves.

The acute inflammation of the glomeruli most often subsides in 10 days to two weeks, but it frequently becomes *chronic glomerulonephritis*, in which mild episodes of recurring glomerulonephritis return periodically, and with each attack progressively more nephrons are destroyed.

Tubular Necrosis. Another common cause of acute renal shutdown is *tubular necrosis*, which means destruction of epithelial cells in the tubules. Some causes of tubular necrosis include (1) *various poisons* that destroy the tubular epithelial cells, (2) *severe acute ischemia* of the kidneys, and (3) transfusion reactions.

CHRONIC RENAL INSUFFICIENCY — DECREASED NUMBERS OF NEPHRONS

Even though the nephrons in a human kidney may be completely normal or almost normal, their total number may be greatly reduced. This leads to insufficient quantity of renal tissue to perform the normal functions of the kidneys. The different causes of this include *chronic glomerulonephritis, traumatic loss of a kidney, congenital absence of a kidney, congenital polycystic disease* (in which large cysts develop in the kidneys and destroy surrounding nephrons by compression), *urinary tract obstruction* resulting from renal stones, *pyelonephritis* (kidney infection) and *arteriosclerosis.*

Inability of Insufficient Kidneys to Excrete Excess "Loads" of Excretory Products. The most important effect of renal insufficiency is the inability of the kidneys to cope with large "loads" of electrolytes or other substances that must be excreted. Normally, one-fourth of the nephrons can eliminate essentially all the normal load of waste products from the body. However, man sometimes engages in voracious ingestion of salts, meats, and so forth, which is the reason for the excess renal functional capacity.

Function of the Remaining Nephrons in Renal Insufficiency. In renal insufficiency, the still functioning nephrons usually become greatly overloaded in several different ways. First, for reasons not completely understood, the blood flow through the glomerulus and the amount of glomerular filtrate formed each minute by each nephron often increase to as much as double normal. Second, large amounts of excretory substances, such as urea, phosphates, sulfates, uric acid, and creatinine, accumulate in the extracellular fluid. On entering the glomerular filtrate, these constitute markedly increased tubular loads of substances that are poorly reabsorbed. These increasing loads are partially compensated by as much as 50 per cent increase in reabsorptive power of each tubule, but even this increase is often far too little to keep pace with the loads, which may be increased 1000 per cent. Therefore, only a small fraction of the solutes are reabsorbed, and this also prevents reabsorption of water, resulting in rapid flushing of tubular fluid through the tubules. Consequently, the volume of urine formation by each nephron can rise to as much as 20 times normal, and a person will occasionally have *as much as three times normal urine output* despite the fact that he has significant renal insufficiency. This paradoxical situation is caused by a greater increase in urine volume output per nephron than reduction in numbers of nephrons.

Hypertension in Renal Insufficiency. Either acute renal shutdown or chronic renal insufficiency can cause hypertension, but it more frequently occurs in the chronic condition. The most likely cause of the hypertension is retention of large quantities of water and salt, which increases the extracellular fluid volume, thereby causing a hyperactive circulation. This theory and others were discussed in detail in Chapter 20.

Effects of Renal Insufficiency on the Body Fluids — Uremia

The effect of acute renal shutdown, or of severe renal insufficiency caused in any other way, on the body fluids depends to a great extent on the water and food intake of the person after shutdown has occurred. Assuming that the person continues to ingest small amounts of water and food, the concentration changes of different substances in the extracellular fluid are approximately those shown in Figure 26–8. The most important effects are (1) *generalized edema* resulting from water retention, (2) *acidosis* resulting from failure of the kidneys to rid the body of normal acidic end-products of metabolism, (3) *high potassium concentration* resulting from failure of potassium excretion, and (4) *high concentrations of the nonprotein nitrogens*, especially *urea*, resulting from failure of the body to excrete the metabolic end-products. This condition is called *uremia* because of the high concentrations of normal urinary excretory products that collect in the body fluids.

Uremic Coma. After a week or more of renal shutdown the sensorium of the patient becomes clouded, and he soon

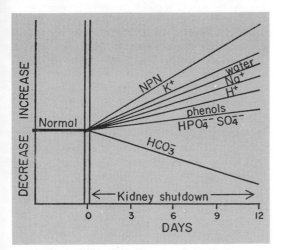

FIGURE 26–8 Effect of kidney shutdown on extracellular fluid constituents.

progresses into a state of coma. The acidosis is believed to be the principal factor responsible for the coma because acidosis caused by other conditions, such as severe diabetes mellitus, also causes coma. However, many other abnormalities could also be contributory — the generalized edema, the high potassium concentration, and possibly even the high nonprotein nitrogen concentration.

The respiration usually is deep and rapid in coma, which is a respiratory attempt to compensate for the metabolic acidosis. In addition to this, during the last day or so before death, the arterial pressure falls progressively, then rapidly in the last few hours. Death occurs usually when the pH of the blood falls to about 7.0.

Dialysis of Uremic Patients with the Artificial Kidney

Artificial kidneys have now been used for about 25 years to treat patients with severe renal insufficiency. In certain types of acute renal shutdown, the artificial kidney is used simply to tide the patient over for a few weeks until the renal damage heals so that the kidneys can resume function. Also, the artificial kidney has now been developed to the point that several thousand persons with permanent renal insufficiency or even total kidney removal are being maintained in health for years at a time, their lives depending entirely on the artificial kidney.

The basic principle of the artificial kidney is to pass blood through very minute channels between thin cello-

FIGURE 26–9 Schematic diagram of the Skeggs-Leonards artificial kidney. (Redrawn from Bluemle et al.: *Trans. Stud. Coll. Physcns. Philad.*, 20:157, 1953.)

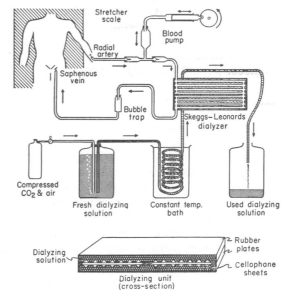

phane membranes. On the other sides of the membranes is a *dialyzing fluid* into which unwanted substances in the blood pass by diffusion.

Figure 26–9 illustrates an artificial kidney in which blood is pumped continually between two thin sheets of cellophane; on the outside of the sheets is the dialyzing fluid. The cellophane is porous enough to allow all constituents of the plasma except the plasma proteins to diffuse freely in both directions — from plasma into the dialyzing fluid and from the dialyzing fluid back into the plasma. If the concentration of a substance is greater in the plasma than in the dialyzing fluid, there will be net transfer of the substance from the plasma into the dialyzing fluid.

The Dialyzing Fluid. Sodium, potassium, and chloride concentrations in the dialyzing fluid are usually adjusted so that they are the same as those in normal plasma.

On the other hand, there is no phosphate, urea, urate, or creatinine in the dialyzing fluid. Therefore, when the uremic patient is dialyzed, these substances are lost in large quantities into the dialyzing fluid, thereby removing major proportions of them from the plasma.

Thus, the constituents of the dialyzing fluid are chosen so that those substances in excess in the extracellular fluid in uremia can be removed at rapid rates, while the normal electrolytes remain.

THE NEPHROTIC SYNDROME — INCREASED GLOMERULAR PERMEABILITY

Large numbers of patients with renal disease develop a so-called *nephrotic syndrome*, which is characterized especially by *loss of large quantities of plasma proteins into the urine*. In some instances this occurs without evidence of any other abnormality of renal function, but more often it is associated with some degree of renal insufficiency.

The cause of the protein loss in the urine is increased permeability of the glomerular membrane. Therefore, any disease condition that can increase the permeability of this membrane can cause the nephrotic syndrome. Such diseases include the subacute phase of *chronic glomerulonephritis* (in the previous discussion, it was noted that this disease primarily affects the glomeruli and causes a greatly increased permeability of the glomerular membrane); *amyloidosis*, which results from the deposition of an abnormal proteinous substance in the walls of blood vessels and thereby seriously damages the basement membrane; *syphilis*, which often results in fibrotic reactions around the blood vessels; and *disseminated lupus*, which is an abnormality distantly related to glomerulonephritis.

Protein Loss. In the nephrotic syndrome, as much as 30 to 40 grams of plasma proteins can be lost into the urine each day. Though the resulting low plasma protein concentration stimulates the liver to produce far more plasma proteins than usual, nevertheless, the liver often cannot keep up with the loss. Therefore, in severe nephrosis the colloid osmotic pressure sometimes falls extremely low, often from the normal level of 28 mm. Hg to as low as 6 to 8 mm. Hg.

Edema. The low colloid osmotic pressure in turn allows large amounts of fluid to filter into the interstitial spaces and also into the potential spaces of the body, thus causing serious *edema*. The nephrotic person has been known on occasion to develop as much as 40 liters of excess extracellular fluid, and as much as 15 liters of this has been *ascites* in the abdomen. Also, the joints swell, and the pleural cavity and the pericardium become partially filled with fluid.

SPECIFIC TUBULAR DISORDERS

In the discussion of active reabsorption and secretion by the tubules in

Chapter 24, it was pointed out that the active transport processes are carried out by various carriers and enzymes in the tubular epithelial cells. Furthermore, there are a number of different carrier mechanisms for the different individual substances. In Chapter 3 it was also pointed out that each cellular enzyme and probably also each carrier substance are formed in response to respective genes in the nucleus. If any of the respective genes happens to be absent or abnormal, the tubules might be deficient in one of the appropriate enzymes or carriers. For this reason many different specific tubular disorders are known to occur in the transport of individual or special groups of substances through the tubular membrane. Essentially all these are hereditary disorders. Some of the more important ones are:

1. *Renal glycosuria*, in which glucose is lost into the urine because the tubules fail to reabsorb glucose.

2. *Nephrogenic diabetes insipidus*, in which excess water is lost into the urine because the tubules do not respond to antidiuretic hormone.

3. *Renal hypophosphatemia*, in which the blood phosphates become greatly reduced because the tubules fail to reabsorb phosphate.

RENAL FUNCTION TESTS

Renal Clearance Tests. Any of the renal clearance tests, including clearance of para-aminohippuric acid, Diodrast, inulin, mannitol, or other substances, as described in Chapter 25, can be used as a renal function test. Indeed, if all these are run, one can determine the glomerular filtration rate, the effective blood flow through the kidney per minute, the filtration fraction, and many other characteristics of renal function. However, most of these clearance tests are not satisfactory for routine clinical use because of their complexities. Instead, other types of renal clearance tests are normally used. These include phenolsulfonphthalein clearance and x-ray measurements of Diodrast clearance into the renal pelves.

Phenolsulfonphthalein clearance. Phenolsulfonphthalein, an alkaline dye, is injected either intravenously or subcutaneously, and successive samples of urine are collected. This dye is cleared into the urine both by the glomeruli and by tubular secretion. When injected intravenously, at least 60 per cent, and usually much more, of the dye is returned to the urine during the first hour if the kidneys are normal. The quantity of dye in the urine is determined by alkalinizing the urine with sodium hydroxide and checking it against a color standard. In severe cases of renal insufficiency the amount cleared into the urine during the first hour is often depressed to as low as 5 to 10 per cent of the quantity originally injected.

Intravenous pyelography. Several substances containing large quantities of iodine in their molecules — Diodrast, Hippuran, and Iopax — are excreted into the urine either by glomerular filtration or by active tubular secretion. Consequently, the concentration in the urine becomes very high within a few minutes after intravenous injection. Also, the iodine in the compounds makes them relatively opaque to x-rays. Therefore, x-ray pictures can be made showing shadows of the renal pelves, of the ureters, and even of the urinary bladder. Ordinarily, a sufficient quantity is excreted within five minutes after injection to give good shadows of the kidney pelves. Failure to show a distinct shadow within this time indicates depressed renal function.

Blood Analyses as Tests of Renal Function. One can also estimate how well the kidneys are performing their functions by measuring the concentrations of various substances in the blood. For instance, the normal concentration of urea in the blood is 26 mg./100 ml., but in severe cases of renal insufficiency this can rise to as much as 200 mg. per cent. The normal concentration of creatinine in the blood is 1.3 mg. per cent, but this, too, can increase as much as 10-fold. To determine the degree of metabolic acidosis resulting from renal dysfunction, one can measure the carbon dioxide combining power as discussed in the preceding chapter. Though these different tests are not so satisfactory as clearance tests for determining the functional capabilities of the kidneys, they are easy to perform, and they do tell the physician how seriously the internal environment has been disturbed.

REFERENCES

Brest, A. M., and Moyer, J. H.: Renal Failure. Philadelphia, J. B. Lippincott Co., 1968.

Christensen, H. N.: Body Fluids and the Acid-Base Balance. 2nd ed., Philadelphia, W. B. Saunders Company, 1964.

Davenport, H. W.: The ABC of Acid-Base Chemistry. 4th ed., Chicago, University of Chicago Press, 1958.

Goldberger, E.: A Primer of Water, Electrolyte, and Acid-Base Syndromes. 3rd ed., Philadelphia, Lea & Febiger, 1965.

Heptinstall, R. H.: Pathology of the Kidney. Boston, Little, Brown and Company, 1966.

Kuru, M.: Nervous control of micturition. *Physiol. Rev., 45*:425, 1965.

Merrill, J. P.: The Treatment of Renal Failure. 2nd ed., New York, Grune & Stratton, Inc., 1965.

Robinson, J. R.: Fundamentals of Acid-Base Regulation. 3rd ed., Philadelphia, F. A. Davis Co., 1967.

Waddell, W. J., and Bates, R. G.: Intracellular pH. *Physiol. Rev., 49*:285, 1969.

PART IV

RESPIRATION, AND ENVIRONMENTAL PHYSIOLOGY

CHAPTER 27

PULMONARY VENTILATION; AND PHYSICS OF RESPIRATORY GASES

Respiration means the transport of oxygen from the atmosphere to the cells and, in turn, the transport of carbon dioxide from the cells back to the atmosphere. This process begins with pulmonary ventilation, which means the actual inflow and outflow of air between the atmosphere and the alveoli.

MECHANICS OF RESPIRATION

BASIC MECHANISMS OF LUNG EXPANSION AND CONTRACTION

The lungs can be expanded and contracted by (1) downward or upward movement of the diaphragm to lengthen or shorten the chest cavity and (2) elevation or depression of the ribs to increase or decrease the anteroposterior diameter of the chest cavity. Figure 27–1 illustrates these two methods.

It is readily evident that contraction of the diaphragm pulls the lower boundary of the chest cavity downward and therefore increases its longitudinal length. Upward movement of the diaphragm can be caused either by simple relaxation of the diaphragm or by active contraction of the abdominal muscles, which forces the abdominal contents upward against the bottom of the diaphragm.

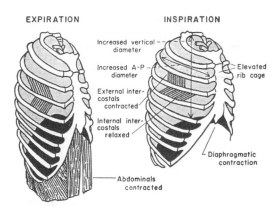

FIGURE 27-1 Expansion and contraction of the thoracic cage during expiration and inspiration, illustrating especially diaphragmatic contraction, elevation of the rib cage, and function of the intercostals.

325

Elevation of the anterior portion of the chest cage causes the anteroposterior dimension of the chest cavity to increase by the following mechanism: During expiration the ribs are pulled downward mainly by the anterior abdominal muscles, and during inspiration they are pulled upward by the anterior neck muscles. The anteroposterior diameter of the chest is about 20 per cent greater during inspiration than during expiration. Therefore, muscles that elevate the chest cage can be classified as muscles of inspiration, and muscles that depress the chest cage as muscles of expiration. Normal quiet breathing is accomplished almost entirely by movement of the diaphragm; but during maximal breathing, increase in chest thickness might account for more than half of the lung enlargement.

RESPIRATORY PRESSURES

Intra-alveolar Pressure. The respiratory muscles cause pulmonary ventilation by alternatively compressing and distending the lungs, which in turn causes the pressure in the alveoli to rise and fall. During inspiration the intra-alveolar pressure becomes slightly negative *with respect to atmospheric pressure*, normally about −3 mm. Hg, and this causes air to flow inward through the respiratory passageways. During normal expiration, on the other hand, the intra-alveolar pressure rises to approximately +3 mm. Hg, which causes air to flow outward through the respiratory passageways.

During maximum expiratory effort the intra-alveolar pressure can usually be increased to well over 100 mm. Hg, and during maximum inspiratory effort it can usually be reduced to as low as −80 mm. Hg.

Recoil Tendency of the Lungs and the Intrapleural Pressure. The lungs have a continual tendency to collapse and therefore to recoil away from the chest wall. This tendency is caused by two different factors. First, throughout the lungs are many *elastic fibers* that are constantly stretched and are attempting to shorten. Second, and even more important, the *surface tension* of the fluid lining the alveoli causes a continual tendency for the alveoli to collapse. This effect is caused by intermolecular attraction between the surface molecules of the fluid that tends continually to reduce the surface areas of the individual alveoli; all these minute forces added together tend to collapse the whole lung and therefore to cause its recoil away from the chest wall.

Ordinarily, the elastic fibers in the lungs account for about one-third of the recoil tendency, and the surface tension phenomenon accounts for about two-thirds.

The total recoil tendency of the lungs can be measured by the amount of negative pressure in the intrapleural spaces required to prevent collapse of the lungs, and this pressure is called the *intrapleural pressure* or, occasionally, the *recoil pressure*. It is normally about −4 mm. Hg. When the lungs are stretched to very large size, such as at the end of deep inspiration, the intrapleural pressure required then to expand the lungs may be as great as −9 to −12 mm. Hg.

"Surfactant" in the alveoli, and its effect on the collapse tendency. A lipoprotein mixture called "surfactant" is secreted by the alveolar epithelium into the alveoli and respiratory passages. This mixture, containing especially the phospholipid *dipalmityl lecithin*, acts in much the same way that a detergent acts. That is, it decreases the surface tension of fluids lining the alveoli and respiratory passages. In the absence of surfactant, lung expansion is extremely difficult, which illustrates that the presence of surfactant is exceedingly important to minimize the effect of surface tension in causing collapse of the lungs.

Surfactant acts by forming a monomolecular layer at the interface between the fluids lining the alveoli and the air in the alveoli. This prevents the development of a water-air interface, which has 7 to 14 times as much surface tension as does the surfactant-air interface.

EXPANSIBILITY OF THE LUNGS AND THORAX: "COMPLIANCE"

The expansibility of the lungs and thorax is called compliance. This is expressed as the *volume increase in the lungs for each unit increase in intra-alveolar pressure*. The compliance of the normal lungs and thorax combined is 0.13 liter per centimeter of water pressure. That is, every time the alveolar pressure is increased by 1 cm. water, the lungs expand 130 ml.

Compliance of the Lungs Alone. The lungs alone, when removed from the chest, are almost twice as distensible as the lungs and thorax together, because the thoracic cage itself must also be stretched when the lungs are expanded in situ. Thus, the compliance of the normal lungs when removed from the thorax is about 0.22 liter per cm. water. This illustrates that the muscles of inspiration must expend energy not only to expand the lungs but also to expand the thoracic cage around the lungs.

Measurement of Lung Compliance. Compliance of the lungs is measured in the following way: The glottis must first be completely open and remain so. Then air is inspired in steps of approximately 50 to 100 ml. at a time, and pressure measurements are made from an intra-esophageal balloon (which measures almost exactly the intrapleural pressure) at the end of each step until the total volume of air in the lungs is equal to the normal tidal volume of the person. Then the air is expired also in steps until the lung volume returns to the expiratory resting level. The relationship of lung volume to pressure is then plotted as illustrated in Figure 27–2. This graph shows that the plot during inspiration is a different curve from the return during expiration. A diagonal line is drawn between the two ends of the curves, and the compliance of the lungs is determined from this diagonal line. Thus, in the figure the lung volume increases 600 ml. for a change in trans-lung pressure (atmospheric pressure in the alveoli minus intra-esophageal pressure) of 2.0 cm. water. The compliance in this instance is 0.3 liter per cm. water.

Slight modifications of this procedure can be used to measure the compliance of the lungs and thoracic cage together.

FIGURE 27–2 Compliance diagram in a large subject. This diagram shows compliance of the lungs alone.

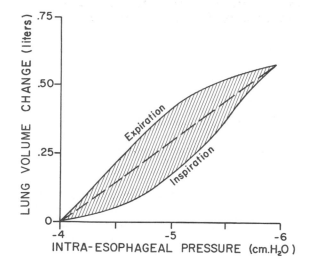

Factors that Cause Abnormal Compliance. Any condition that destroys lung tissue, causes it to become fibrotic or edematous, blocks the alveoli, or in any other way impedes lung expansion and contraction causes decreased lung compliance. When considering the compliance of both the lung and thorax together, we must also include any abnormality that reduces the expansibility of the thoracic cage. Thus, deformities of the chest cage, such as kyphosis, severe scoliosis, and other restraining conditions, such as fibrotic pleurisy or paralyzed and fibrotic muscles, can all reduce the expansibility of the lungs and thereby reduce the total pulmonary compliance.

The "Work" of Breathing. When the lungs are stretched, energy is expended by the respiratory muscles to cause the stretching. This is part of the "work" of breathing. But, in addition to the work required simply to expand the lungs, work is needed to overcome two other factors that impede the expansion and contraction of the lungs. These are (1) *viscosity* of the pulmonary tissues, which is also called *nonelastic tissue resistance*, and (2) *airway resistance*. In normal quiet breathing, the energy required for breathing is 5 to 10 per cent of the total body energy expenditure, but when the airway becomes obstructed as a result of asthma, obstructive emphysema, diphtheria, or other diseases,

the airway resistance may become so greatly increased that the energy for breathing becomes 30 per cent or more of the total body energy expenditure.

THE PULMONARY VOLUMES AND CAPACITIES

Figure 27–3 gives a graphical representation of changes in lung volume under different conditions of breathing. For ease in describing the events of pulmonary ventilation, the air in the lungs has been subdivided at different points on this diagram into four different *volumes* and four different *capacities*, which are as follows:

THE PULMONARY "VOLUMES"

To the left in Figure 27–3 are listed four different pulmonary lung "volumes" which, when added together, equal the maximum volume to which the lungs can be expanded. The significance of each of these volumes is the following:

1. The *tidal volume* is the volume of air inspired and expired with each normal breath, and it amounts to about 500 ml. in the normal young male adult.

2. The *inspiratory reserve volume* is the extra volume of air that can be inspired over and beyond the normal tidal

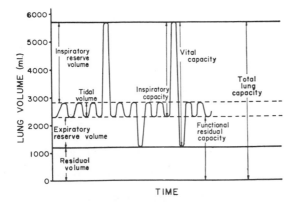

FIGURE 27–3 Diagram showing respiratory excursions during normal breathing and during maximal inspiration and maximal expiration.

volume, and it is usually equal to approximately 3000 ml. in the young male adult.

3. The *expiratory reserve volume* is the amount of air that can still be expired by forceful expiration after the end of a normal tidal expiration; this normally amounts to about 1100 ml. in the young male adult.

4. The *residual volume* is the volume of air remaining in the lungs even after the most forceful expiration. This volume averages about 1200 ml. in the young male adult.

THE PULMONARY "CAPACITIES"

In describing events in the pulmonary cycle, it is sometimes desirable to consider two or more of the above volumes together. Such combinations are called pulmonary capacities. To the right in Figure 27–3 are listed the different pulmonary capacities, which can be described as follows:

1. The *inspiratory capacity* equals the *tidal volume* plus the *inspiratory reserve volume*. This is the amount of air (about 3500 ml.) that a person can breathe beginning at the normal expiratory level and distending his lungs to the maximum amount.

2. The *functional residual capacity* equals the *expiratory reserve volume* plus the *residual volume*. This is the amount of air remaining in the lungs at the end of normal expiration (about 2300 ml.).

3. The *vital capacity* equals the *inspiratory reserve volume* plus the *tidal volume* plus the *expiratory reserve volume*. This is the maximum amount of air that a person can expel from his lungs after first filling his lungs to their maximum extent and then expiring to the maximum extent (about 4600 ml.).

4. The *total lung capacity* is the maximum volume to which the lungs can be expanded with the greatest possible inspiratory effort (about 5800 ml.).

All pulmonary volumes and capacities

are about 20 to 25 per cent less in the female than in the male, and they obviously are greater in large and athletic persons than in small and asthenic persons.

SIGNIFICANCE OF THE RESIDUAL VOLUME

The residual volume represents the air that cannot be removed from the lungs even by forceful expiration. This is important because it provides air in the alveoli to aerate the blood even between breaths. Were it not for the residual air, the concentrations of oxygen and carbon dioxide in the blood would rise and fall markedly with each respiration, which would certainly be disadvantageous to the respiratory process.

SIGNIFICANCE OF THE VITAL CAPACITY

Other than the anatomical build of a person, the major factors that affect vital capacity are (1) the position of the person during the vital capacity measurement, (2) the strength of the respiratory muscles, and (3) the distensibility of the lungs and chest cage, which is called "pulmonary compliance."

The average vital capacity in the young adult male is about 4.6 liters, and in the young adult female about 3.1 liters, though these values are much greater in some persons of the same weight than in others. A tall, thin person usually has a higher vital capacity than an obese person, and a well-developed athlete may have a vital capacity as great as 30 to 40 per cent above normal —that is, 6 to 7 liters.

MEASUREMENT OF PULMONARY VOLUMES AND CAPACITIES— SPIROMETRY

A simple method by which most of the pulmonary volumes and capacities can

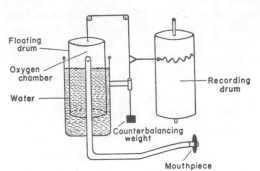

FIGURE 27–4 A spirometer.

be measured is spirometry. A typical spirometer is illustrated in Figure 27–4. This consists of a drum inverted over a chamber of water, the drum counterbalanced by a weight. In the drum is a breathing mixture of gases, usually air or oxygen, and a tube connects the mouth with this gas chamber. On breathing in and out of the chamber the drum rises and falls, and an appropriate *spirogram* is recorded, such as that already discussed in Figure 27–3. From this, most of the lung volumes and capacities can be measured.

THE MINUTE RESPIRATORY VOLUME — RESPIRATORY RATE AND TIDAL VOLUME

The *minute respiratory volume* is the total amount of new air moved into the respiratory passages each minute, and this is equal to the *tidal volume* × the *respiratory rate*. The normal tidal volume of a young adult male, as pointed out above, is 500 ml., and the normal respiratory rate is approximately 12 breaths per minute. Therefore, the *minute respiratory volume averages about 6 liters per minute*. A person can occasionally live for short periods of time with a minute respiratory volume as low as 1.5 liters per minute and with a respiratory rate as low as two to four breaths per minute. Obviously, a greatly increased tidal volume can compensate for a markedly reduced respiratory rate.

The respiratory rate occasionally rises to as high as 40 to 50 per minute, and the tidal volume can become as great as the vital capacity, averaging about 4600 ml. in the young adult male. However, at very rapid breathing rates, a person usually cannot sustain a tidal volume greater than about one-half the vital capacity.

Maximum Breathing Capacity. A young male adult forcing himself to breathe as much volume of air as possible can usually breathe 150 to 170 liters per minute for about 15 seconds. This is called the maximum breathing capacity. On the average, this same person can maintain for long periods of strenuous exercise a minute respiratory volume as high as 100 to 120 liters per minute.

VENTILATION OF THE ALVEOLI

The truly important factor of the entire pulmonary ventilatory process is the rate at which the alveolar air is renewed each minute by atmospheric air; this is called *alveolar ventilation*. One can readily understand that alveolar ventilation per minute is not equal to the minute respiratory volume because a large portion of the inspired air goes to fill the respiratory passageways, the membranes of which are not capable of significant gaseous exchange with the blood.

THE DEAD SPACE

Effect of Dead Space on Alveolar Ventilation. The air that goes to fill the respiratory passages with each breath, which amounts to about 150 ml., is called *dead space air*. On inspiration, much of the new air must first fill the different dead space areas — the nasal passageways, the pharynx, the trachea, and the bronchi — before any reaches the alveoli. Then, on expiration, all the air in the dead space is expired first before any of the air from the alveoli reaches the atmosphere. *The volume of air that enters the alveoli with each breath, therefore, is equal to the tidal volume minus the dead space volume.*

Anatomical versus Physiological Dead Space. The volume of all the respiratory passageways is called the *anatomical dead space*. On occasion, however, some of the alveoli are not functional and therefore must be considered along with the respiratory passages to be dead space because there is no blood flow through the adjacent pulmonary vessels. Also, at other times the *ratio of pulmonary blood flow to ventilation* in certain alveoli is so low that these alveoli are only partially functional, so that they, too, can be considered to be partially dead space. When the alveolar dead space is included in the total measurement of dead space this is then called *physiological dead space* in contradistinction to the anatomical dead space. In the normal person, the anatomical and the physiological dead spaces are essentially equal because all alveoli are functional in the normal lung, but in persons with nonfunctional alveoli or with abnormal ratios of blood flow to alveolar ventilation in some parts of the lungs, the physiological dead space is sometimes as much as 10 times the anatomical dead space, or as much as 1 to 2 liters.

RATE OF ALVEOLAR VENTILATION

Alveolar ventilation per minute is the total volume of new air entering the alveoli each minute. It is equal to the respiratory rate times the amount of new air that enters the alveoli with each breath:

Alveolar ventilation per minute = Respiratory
    rate × (Tidal volume − Dead space volume)

Thus, with a normal tidal volume of 500 ml., a normal dead space of 150 ml., and a respiratory rate of 12 times per minute, alveolar ventilation equals 12 × (500 − 150), or 4200 ml. per minute.

Alveolar ventilation is one of the major factors determining the concentrations of oxygen and carbon dioxide in the alveoli. Therefore, almost all discussions of gaseous exchange problems in the following chapters emphasize alveolar ventilation. *The respiratory rate, the tidal volume, and the minute respiratory volume are of importance only in so far as they affect alveolar ventilation.*

FUNCTIONS OF THE RESPIRATORY PASSAGEWAYS

FUNCTIONS OF THE NOSE

As air passes through the nose, three distinct functions are performed by the nasal cavities: First, the *air is warmed* by the extensive surfaces of the turbinates and septum, which are illustrated in Figure 27–5. Second, the *air is moistened* to a considerable extent even before it passes beyond the nose. Third, the *air is filtered*. All these functions together are called the *air conditioning function* of the upper respiratory passageways. Ordinarily, the air rises to within 2 to 3 per cent of body temperature and within 2 to 3 per cent of full saturation with water vapor before it reaches the lower trachea. When a person breathes air through a tube directly into his trachea (as through a tracheostomy), the cooling and, especially, the drying effect in the lower lung can lead to lung infection.

Filtration Function of the Nose. The hairs at the entrance to the nostrils are important because they remove large particles. Much more important, though, is the removal of particles by *turbulent precipitation*. That is, the air passing through the nasal passageways hits many obstructing vanes, such as the turbinates, the septum, and the pharyngeal

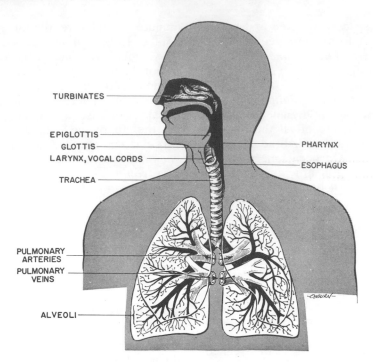

FIGURE 27–5 The respiratory passages.

wall. Each time air hits one of these obstructions it must change its direction of movement; and the particles suspended in the air, having far more mass and momentum than air, cannot change their direction of travel as rapidly as can the air. Therefore, they continue forward, striking the surfaces of the obstructions. This nasal turbulence mechanism for removing particles from air is so effective that almost no particles larger than 4 to 6 microns in diameter enter the lungs through the nose. This size is smaller than the size of red blood cells.

THE COUGH REFLEX

The cough reflex is almost essential to life, for the cough is the means by which the passageways of the lungs are maintained free of foreign matter.

The bronchi and the trachea are so sensitive that any foreign matter or other cause of irritation initiates the cough reflex. The larynx and carina (the point where the trachea divides into the bronchi) are especially sensitive, and the terminal bronchioles and alveoli are especially sensitive to corrosive chemical stimuli, such as sulfur dioxide gas and chlorine. Afferent impulses pass from the respiratory passages mainly through the vagus nerve to the medulla. There, an automatic sequence of events is triggered by the neuronal circuits of the medulla, causing the following effects:

First, about 2.5 liters of air is inspired. Second, the epiglottis closes, and the vocal cords shut tightly to entrap the air within the lungs. Third, the abdominal muscles contract forcefully, pushing against the diaphragm while other expiratory muscles, such as the internal intercostals, also contract forcefully. Consequently, the pressure in the lungs rises usually to 100 or more mm. Hg. Fourth, the vocal cords and the epiglottis suddenly open widely so that the air under pressure in the lungs *explodes* outward. Indeed, this air is sometimes expelled at velocities as high as 75 to 100 miles an hour. Furthermore, and very important, the strong compression of the lungs also collapses the bronchi and trachea (the noncartilaginous part of the trachea invaginating inward) so that the exploding air actually passes through bronchial and tracheal slits. The rapidly moving air usually carries with it any foreign matter that is present in the bronchi or trachea.

ACTION OF THE CILIA TO CLEAR RESPIRATORY PASSAGEWAYS

In addition to the cough mechanism, the respiratory passageways of the trachea, lungs, and nose are lined with a ciliated, mucus-coated epithelium that aids in clearing the passages, for the cilia beat toward the pharynx. Thus, small foreign particles and mucus are mobilized at a velocity of as much as a centimeter per minute along the surface of the trachea toward the pharynx.

THE SNEEZE REFLEX

The sneeze reflex is very much like the cough reflex except that it applies to the nasal passageways instead of to the lower respiratory passages. The initiating stimulus of the sneeze reflex is irritation in the nasal passageways, the afferent impulses passing in the fifth nerve to the medulla where the reflex is triggered. A series of reactions similar to those for the cough reflex takes place; however, the uvula is depressed so that large amounts of air pass rapidly through the nose, as well as through the mouth, thus helping to clear the nasal passages of foreign matter.

VOCALIZATION

Speech involves the respiratory system particularly, but it also involves (1) specific speech control centers in the cerebral cortex, which will be discussed in Chapter 39, (2) respiratory centers of the brain stem, and (3) the articulation and resonance structures of the mouth and nasal cavities. Basically, speech is composed of two separate mechanical functions: (1) *phonation*, which is achieved by the larynx, and (2) *articulation*, which is achieved by the structures of the mouth.

Phonation. The larynx is specially adapted to act as a vibrator. The vibrating element is the *vocal cords*, which are folds along the lateral walls of the larynx that are stretched and positioned by several specific muscles within the confines of the larynx itself. Figure 27–6*A* illustrates the basic structure of the larynx, showing that each vocal cord is stretched between the *thyroid cartilage* and an *arytenoid cartilage*.

Vibration of the vocal cords. One might suspect that the vocal cords would vibrate in the direction of the flowing air. However, this is not the case. Instead, they vibrate laterally. The cause of the vibration is the following: When the vocal cords are approximated and

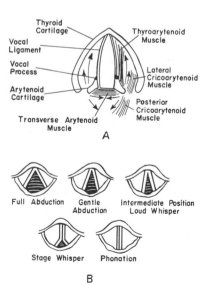

FIGURE 27–6 Laryngeal function in phonation. (Modified from Greene: The Voice and Its Disorders. Pitman Medical Publishing Co.)

air is expired, pressure of the air from below first pushes the vocal cords apart, which allows rapid flow of air between their margins. The rapid flow of air then immediately creates a partial vacuum between the vocal cords, which pulls them once again toward each other. This stops the flow of air, pressure builds up behind the cords, and the cords open once more, thus continuing in a vibratory pattern.

Frequency of vibration. The pitch of the sound emitted by the larynx can be changed in two different ways: first, by *stretching or relaxing the vocal cords*, and, second, by *changing the shape and mass of the vocal cord edges*. When very high frequency sounds are emitted, different slips of the thyroarytenoid muscles contract in such a way that the edges of the vocal cords are sharpened and thinned, whereas when bass frequencies are emitted, the muscles contract in a different pattern so that broad edges with a large mass are approximated. Figure 27–6*B* shows some of the positions and shapes of the vocal cords during different types of phonation.

Articulation and Resonance. The three major organs of articulation are the *lips*, the *tongue*, and the *soft palate*. These need not be discussed in detail because all of us are familiar with their movements during speech and other vocalizations.

The resonators include the *mouth*, the *nose and associated nasal sinuses*, the *pharynx*, and even the *chest cavity* itself. Here again we are all familiar with the resonating qualities of these different structures. For instance, the function of the nasal resonators is illustrated by the change in quality of the voice when a person has a severe cold.

PHYSICS OF RESPIRATORY GASES

In the following chapters we will discuss principally the transport of oxygen and carbon dioxide between the alveoli and the cells. However, to understand the mechanisms of this transport, its regulation, and its abnormalities, it is first necessary to understand some of the basic physics related to respiratory gases.

Vapor Pressure of Water. All gases in the body are in direct contact with water. Therefore, all gaseous mixtures in the body are saturated with water vapor, and this must always be considered when the dynamics of gaseous exchange are discussed.

The vapor pressure of water depends entirely on the temperature of the water. The greater the temperature, the greater is the activity of the molecules in the water, and the greater is the likelihood these molecules will escape from the surface of the water into the gaseous phase. When dry air is suddenly mixed with water, water molecules immediately begin escaping from the surface of the water into the air. As the air becomes progressively more and more humidified, an equilibrium vapor pressure is approached at which the rate of condensation of water becomes equal to the rate of water vaporization. At body temperature, this vapor pressure is 47 mm. Hg; this value will appear in most of our subsequent discussions.

The Solution of Gases in Water. When a gas remains in contact with water for long periods of time, much of it becomes dissolved in the water. Two factors determine the quantity of gas that will become dissolved. These factors are (1) the pressure of the gas and (2) the solubility coefficient of the gas in water at the temperature of the water. These relationships are expressed by the following formula, which is Henry's law:

Volume = Pressure × Solubility coefficient

When volume is expressed in volumes of gas dissolved in each volume of water at 0° C. and pressure is expressed in atmospheres, the solubility coefficients for important respiratory gases at body temperatures are the following:

Oxygen	0.024
Carbon dioxide	0.57
Nitrogen	0.012

Pressure of Dissolved Gases. The molecules of a gas dissolved in water, because of their molecular motion, are always attempting to escape from the water, and this creates a pressure of gas even in the dissolved state. This pressure is written as P_{O_2}, P_{CO_2}, P_{N_2}, etc., for each of the gases dissolved in the liquid.

The pressure of a dissolved gas is proportional to the quantity of gas dissolved in the fluid divided by the solubility coefficient of the gas in that particular fluid. Molecules of gases that are highly soluble are attracted by the fluid molecules so that, as they strike the surface of the fluid, this attraction prevents many of these highly soluble molecules from leaving the surface. Consequently, the total number of molecules that must be dissolved to exert a given pressure is considerably greater for very soluble gases than for gases not so soluble.

Partial Pressures. An understanding of partial pressures is an important preliminary to understanding gaseous diffusion from the alveoli to the pulmonary blood, for it is the partial pressure of a gas that determines the force it exerts in attempting to diffuse through the pulmonary membrane.

From the kinetic theory of gases we know that pressure against any surface is determined by the number of molecules striking a unit area in any given instant times the average kinetic energy of the molecules. Therefore, the partial pressure of a gas in a mixture is the sum of the force of impact of the molecules of that particular gas against the unit area of surface. For instance, at sea level, the nitrogen in air exerts enough force to elevate the mercury in a manometer 600 mm., while the oxygen in the air will raise the mercury another 160 mm., making a total of 760 mm. Therefore, the partial pressures of nitrogen and oxygen are, respectively, 600 and 160 mm. Hg. The partial pressures of the gases in a mixture are designated by the same terms used to designate pressures in liquids; i.e., P_{O_2}, P_{CO_2}, P_{N_2}, and so forth.

Pressures of dissolved gases in equilibrium with a mixture of gases. Because each gas in a mixture exerts its own partial pressure in proportion to the concentration of its molecules, when the gases of a mixture dissolve in a liquid and come to equilibrium with the gaseous phase of the mixture, the pressure of each dissolved gas is equal to the partial pressure of the same gas in the gaseous mixture. In other words, *each gas is independent of the others in its ability to dissolve in a liquid*; obviously, this principle also applies to the solution of gases in blood.

Diffusion of Gases Through Liquids. Figure 27–7 illustrates a chamber filled with water. The water at one end of this chamber contains a relatively large amount of dissolved oxygen. Because the dissolved gaseous molecules are constantly undergoing molecular motion, the oxygen molecules are bouncing in all directions, and this process is called *diffusion*. Furthermore, it is obvious that more of them will bounce from the area of high concentration toward the area of low concentration. However, some molecules do bounce from the area of low pressure toward the area of high pressure. Therefore, the *net diffusion* of gas from the area of high pressure to the area of low pressure is

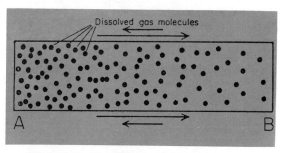

FIGURE 27–7 Net diffusion of oxygen from one end of a chamber to the other.

equal to the number of molecules bouncing in this direction minus the number bouncing in the opposite direction, and this in turn is proportional to the gas pressure difference between the two areas.

This principle of diffusion from an area of high pressure to an area of low pressure holds true for diffusion of gases in a gaseous mixture, diffusion of dissolved gases in a solution, and even diffusion of gases from the gaseous phase into the dissolved state in liquids. That is, *there is always net diffusion from areas of high pressure to areas of low pressure.*

Rate of diffusion. In addition to the pressure difference, several other factors affect the rate of gas diffusion in a fluid. These are (1) the solubility of the gas in the fluid, (2) cross-sectional area of the fluid, (3) the distance through which the gas must diffuse, (4) the molecular weight of the gas, (5) the viscosity of the fluid, and (6) the temperature of the fluid. In the body, the last two of these factors remain reasonably constant and usually need not be considered.

These factors can be expressed in a single formula, as follows:

$$DR \propto \frac{PD \times A \times S}{D \times MW}$$

in which DR is the diffusion rate, PD is the pressure difference between the two areas, A is the cross-sectional area, S is the solubility of the gas, D is the distance of diffusion, and MW is the molecular weight of the gas.

It is obvious from this formula that the characteristics of the gas itself determine two factors of the formula: solubility and molecular weight. There-fore, the *diffusion coefficient* — that is, the rate of diffusion through a given area for a given distance — for any given gas — is proportional to S/MW. Considering the diffusion coefficient for oxygen to be 1, the diffusion coefficients for different gases of respiratory importance in the body fluids are:

Oxygen	1.0
Carbon dioxide	20.3
Nitrogen	0.53

REFERENCES

Agostoni, E.: Thickness and pressure of the pleural liquid. *In* Fishman, A. P., and Hecht, H. H. (eds.): The Pulmonary Circulation and Interstitial Space. Chicago, University of Chicago Press, 1969, p. 65.

Bernstein, L.: Respiration. *Ann. Rev. Physiol.,* 29:133, 1967.

Campbell, E. J. M.: Respiration. *Ann. Rev. Physiol., 30:*105, 1968.

Clements, J. A., and Tierney, D. F.: Alveolar instability associated with altered surface tension. *In* Handbook of Physiology. Baltimore, The Williams & Wilkins Co., 1965, Sec. III, Vol. II, p. 1565.

Comroe, J. H., Jr.: The Physiology of Respiration. Chicago, Year Book Medical Publishers, Inc., 1965.

Comroe, J. H., Jr., et al.: The Lung: Clinical Physiology and Pulmonary Function Tests. 2nd ed., Chicago, Year Book Medical Publishers, Inc., 1963.

Luchsinger, R., and Arnold, G.: Voice-Speech-Language: Clinical Communicology — Its Physiology and Pathology. Wadsworth Publishing Co., Inc., 1965.

Mead, J., and Agostoni, E.: Dynamics of breathing. *In* Handbook of Physiology. Baltimore, The Williams & Wilkins Co., 1964, Sec. III, Vol. I, p. 411.

Otis, A. B.: The work of breathing. *In* Handbook of Physiology. Baltimore, The Williams & Wilkins Co., 1964, Sec. III, Vol. I, p. 463.

Permutt, S.: Respiration. *Ann. Rev. Physiol., 28:* 177, 1966.

CHAPTER 28

TRANSPORT OF OXYGEN AND CARBON DIOXIDE BETWEEN THE ALVEOLI AND THE TISSUE CELLS

After the alveoli are ventilated with fresh air, the next step in the respiratory process is *diffusion* of oxygen from the alveoli into the pulmonary blood and diffusion of carbon dioxide in the opposite direction—from the pulmonary blood into the alveoli. The process of diffusion is simple, involving merely random molecular motion of molecules, these intertwining their ways back and forth through the respiratory membrane. However, in respiratory physiology we are not only concerned with the basic mechanism by which diffusion occurs but also with the *rate* at which it occurs.

The major problems discussed in this chapter, therefore, are, first, the physical factors that determine the alveolar concentrations of gases, particularly of oxygen and carbon dioxide; second, the factors that affect the rate at which these gases can diffuse through the respiratory membrane; and, third, the transport of oxygen and carbon dioxide in the blood and body fluids.

COMPOSITION OF ALVEOLAR AIR

Humidification of the Air as It Enters the Respiratory Passages. Column 1 of Table 28-1 shows that atmospheric air is composed almost entirely of nitrogen and oxygen; it normally contains almost no carbon dioxide and little water vapor. However, as soon as the atmospheric air enters the respiratory passages, it is exposed to the fluids covering the respiratory surfaces. Even before the air enters the alveoli, it becomes totally humidified.

The partial pressure of water vapor at normal body temperature of 37° C. is 47 mm. Hg, which, therefore, is the partial pressure of water in the alveolar air. Since the total pressure in the alveoli cannot rise to more than the atmospheric pressure, this water vapor simply expands the volume of the air and thereby *dilutes* all the other gases in the inspired air. In column 2 of Table 28-1 it can be seen that humidification of the

TABLE 28-1 PARTIAL PRESSURES OF RESPIRATORY GASES AS THEY ENTER AND LEAVE THE LUNGS (AT SEA LEVEL)—PER CENT CONCENTRATIONS ARE GIVEN IN PARENTHESES

	Atmospheric Air* (mm. Hg)	Humidified Air (mm. Hg)	Alveolar Air (mm. Hg)	Expired Air (mm. Hg)
N_2	597.0 (78.62%)	563.4 (74.09%)	569.0 (74.9%)	566.0 (74.5%)
O_2	159.0 (20.84%)	149.3 (19.67%)	104.0 (13.6%)	120.0 (15.7%)
CO_2	0.3 (0.04%)	0.3 (0.04%)	40.0 (5.3%)	27.0 (3.6%)
H_2O	3.7 (0.50%)	47.0 (6.20%)	47.0 (6.2%)	47.0 (6.2%)
TOTAL	760.0 (100.0%)	760.0 (100.0%)	760.0 (100.0%)	760.0 (100.0%)

*On an average cool, clear day.

air reduces the oxygen partial pressure at sea level from an average of 159 mm. Hg in atmospheric air to 149 mm. Hg, and it reduces the nitrogen partial pressure from 597 to 563 mm. Hg.

Rate at which Alveolar Air Is Renewed by Atmospheric Air. In the preceding chapter it was pointed out that the *functional residual capacity* of the lungs, which is the amount of air remaining in the lungs at the end of normal expiration, measures approximately 2300 ml. Furthermore, only 350 ml. of new air is brought into the alveoli with each normal respiration. Therefore, the amount of alveolar air replaced by new atmospheric air with each breath is only one-seventh of the total, so that many breaths are required to exchange most of the alveolar air. With normal alveolar ventilation, approximately half of the alveolar gases is renewed each 17 seconds. When a person's rate of alveolar ventilation is only half normal, half of the gases is renewed every 34 seconds.

Value of the slow replacement of alveolar air. This slow replacement of alveolar air is of particular importance in preventing sudden changes in gaseous concentrations in the blood. This makes the respiratory control mechanism much more stable than it would otherwise be and helps to prevent excessive increases and decreases in tissue oxygenation, tissue carbon dioxide concentration, and tissue pH when the respiration is temporarily interrupted.

OXYGEN CONCENTRATION IN THE ALVEOLI

Oxygen is continually being absorbed into the blood of the lungs, and new oxygen is continually entering the alveoli from the atmosphere. The more rapidly oxygen is absorbed, the lower becomes its concentration in the alveoli; on the other hand, the more rapidly new oxygen is brought into the alveoli from the atmosphere, the higher becomes its concentration. Therefore, oxygen concentration in the alveoli is controlled by (1) the rate of absorption of oxygen into the blood and (2) the rate of entry of new oxygen into the lungs.

The fourth column of Table 28-1 shows the alveolar P_{O_2} to be only 104 mm. Hg, a value far below that in the air that is breathed. Obviously slow breathing or rapid use of oxygen by the body will decrease this even more.

CO_2 CONCENTRATION IN THE ALVEOLI

Carbon dioxide is continually being formed in the body, then discharged into the alveoli; and it is continually being removed from the alveoli by the process of ventilation. Therefore, the two factors that determine carbon dioxide partial pressure (P_{CO_2}) in the lungs are (1) the rate of excretion of carbon dioxide from the blood into the alveoli and (2) the rate at which carbon dioxide is removed from the alveoli by alveolar ventilation.

The third column of Table 28–1 shows the P_{CO_2} of alveolar air normally to be 40 mm. Hg. With decreased respiration or increased excretion of CO_2 from the blood, this can increase to as much as 80 to 100 mm. Hg. Or with very rapid breathing, it can decrease to as little as 5 to 10 mm. Hg.

EXPIRED AIR

Expired air is a combination of dead space air and alveolar air. Normal expired air therefore has gaseous concentrations approximately as shown in column 4 of Table 28–1—that is, concentrations somewhere between those of humidified atmospheric air and alveolar air; under normal resting conditions expired air is about two-thirds alveolar air and one-third dead space air.

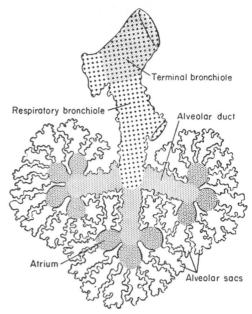

FIGURE 28–1 The respiratory lobule. (From Miller: The Lung. Charles C Thomas.)

DIFFUSION OF GASES THROUGH THE RESPIRATORY MEMBRANE

The Respiratory Unit. Figure 28–1 illustrates the respiratory unit, which is comprised of a *respiratory bronchiole, alveolar ducts, atria,* and *alveolar sacs* or *alveoli* (of which there are about 250 million in the two lungs, each alveolus having an average diameter of about 0.1 mm.). The epithelium of these structures is a very thin membrane, and the alveolar gases are in close proximity to the blood of the capillaries. Consequently, gaseous exchange between the alveolar air and the pulmonary blood occurs through the membranes of all these terminal portions of the lungs. These membranes are collectively known as the *respiratory membrane,* also called the *pulmonary membrane.*

The Respiratory Membrane. Figure 28–2 illustrates the ultrastructure of the respiratory membrane. It shows also the diffusion of oxygen from the alveolus

into the red blood cell and diffusion of carbon dioxide in the opposite direction. Note the following different layers of the respiratory membrane:

1. A monomolecular layer of surfactant, a mixture of phospholipids and perhaps other substances, that spreads out over the surface of the fluid lining the alveolus.

2. A very thin layer of fluid lining the alveolus.

3. The alveolar epithelium comprised of very thin epithelial cells.

4. A very thin interstitial space between the alveolar epithelium and capillary membrane.

5. A capillary basement membrane.

6. The capillary endothelial membrane.

Despite the large number of layers, the overall thickness of the respiratory membrane in some areas is less than 0.1 micron and in essentially all areas is less than 1 micron.

From histologic studies it has been estimated that the total surface area of the respiratory membrane is approxi-

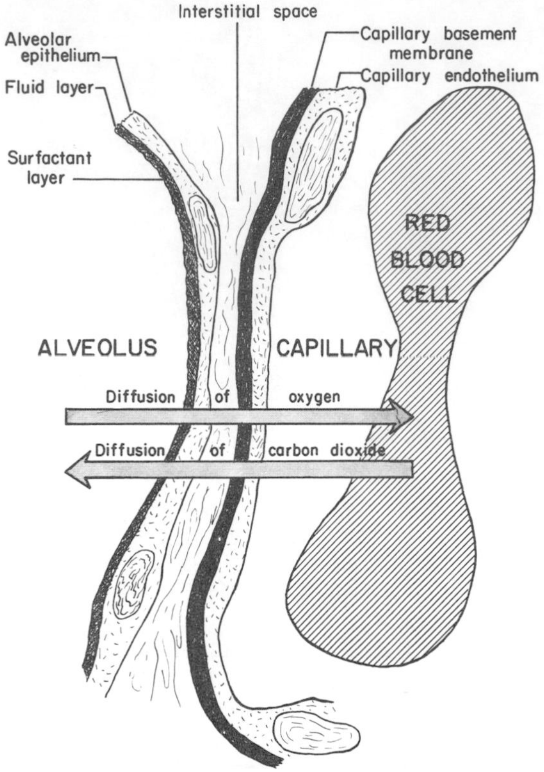

FIGURE 28-2 Ultrastructure of the respiratory membrane.

mately 70 square meters in the normal adult. This is equivalent to the floor area of a room approximately 30 feet long by 25 feet wide. The total quantity of blood in the capillaries of the lung at any given instant is about 60 to 100 ml. If this small amount of blood were spread over the entire surface of a 25 by 30 foot floor, one could readily understand how respiratory exchange of gases occurs as rapidly as it does.

The average diameter of the pulmonary capillaries is only 7 microns, which means that red blood cells must actually squeeze through them. Therefore, the red blood cell membrane usually touches the capillary wall so that oxygen and carbon dioxide need not pass through plasma as they diffuse between the red cell and alveolus. Obviously, this aids the rapidity of diffusion.

Permeability of the respiratory membrane. The limiting factor in the permeability of the respiratory membrane to gases is the rate at which the gases can diffuse through the water in the membrane. All gases of respiratory importance are highly soluble in the lipid substances of the cell membranes and for this reason can diffuse through these with great ease. Earlier in the chapter it was pointed out that the rate of carbon dioxide diffusion in water is about 20 times as rapid as the rate of oxygen diffusion. Therefore, it follows that the rate of diffusion of carbon dioxide through the respiratory membrane is also about 20 times as rapid as the diffusion of oxygen, as would be predicted.

FACTORS AFFECTING GASEOUS DIFFUSION THROUGH THE RESPIRATORY MEMBRANE

Referring to the earlier discussion of diffusion through water, one can apply the same principles and same formula to diffusion of gases through the respiratory membrane. Thus, (1) the *thickness of the membrane*, (2) the *surface area of the membrane*, (3) the *diffusion coefficient* of the gas in the substance of the membrane—that is, in water, and (4) the *pressure difference* between the two sides of the membrane, all combined, determine how rapidly a gas will pass through the membrane.

The *thickness of the respiratory membrane* occasionally increases markedly, usually as a result of the presence of edema fluid in the interstitial space of the membrane. Also, fluid may collect in the alveoli, so that the respiratory gases must diffuse not only through the membrane but also through this fluid. Finally, some pulmonary diseases cause fibrosis of the lungs, which can increase the thickness of some portions of the respiratory membrane. Because the rate of diffusion through the membrane is inversely proportional to the thickness of the membrane, any factor that increases the thickness more than two to three times above normal can interfere markedly with normal respiratory exchange of gases.

The *surface area of the respiratory membrane* may be greatly decreased by many different conditions. For instance, removal of an entire lung decreases the surface area to half normal. Also, in the condition known as *emphysema* many of the alveoli coalesce, with dissolution of the septa between the alveoli. Therefore, the new chambers are much larger than the original alveoli, but the total surface area of the respiratory membrane is considerably decreased because of loss of the alveolar septa. When the total surface area is decreased to approximately a third to one-fourth normal, exchange of gases through the membrane is impeded to a significant degree even under resting conditions. And, during heavy exercise, even the slightest decrease in surface area of the lungs can be a detriment to respiratory exchange of gases.

The *pressure difference* across the respiratory membrane is the difference between the partial pressure of the gas in the alveoli and the pressure of the gas in the blood. The partial pressure represents a measure of the total number of molecules of a particular gas striking a unit area of the alveolar surface of the membrane, and the pressure of the same gas in the bood represents the number of molecules striking the same area of the membrane from the opposite side. Therefore, the differences between these two pressures, the pressure difference, is a measure of th *net tendency* for the particular gas to pass through the membrane. Obviously, when the partial pressure of a gas in the alveoli is greater than the pressure of the gas in the blood, as is true of oxygen, net diffusion from the alveoli into the blood occurs, but, when the pressure of the gas in the blood is greater than the partial pressure in the alveoli, as is true of carbon dioxide, net diffusion from the blood into the alveoli occurs.

DIFFUSING CAPACITY OF THE RESPIRATORY MEMBRANE

The overall ability of the respiratory membrane to exchange a gas between the alveoli and the pulmonary blood can be expressed in terms of its *diffusing capacity*, which is defined as the *volume of a gas that diffuses through the membrane each minute for a pressure difference of 1 mm. Hg*.

The Diffusing Capacity for Oxygen. In the average young male adult the *diffusing capacity for oxygen* under resting conditions averages 21 ml. per minute. Since the oxygen pressure difference across the respiratory membrane during quiet breathing averages 11 mm. Hg, this gives a total of about 250 ml. of oxygen diffusing through the respiratory membrane each minute.

Change in oxygen-diffusing capacity during exercise. During strenuous exercise, or during other conditions that greatly increase pulmonary activity, the diffusing capacity for oxygen increases in young male adults to a maximum of about 65 ml. per minute, which is three times the diffusing capacity under resting conditions. This increase is caused by three different factors: (1) opening up of a number of previously dormant pulmonary capillaries, thereby increasing the surface area of the blood into which the oxygen can diffuse, (2) dilatation of all the pulmonary capillaries that were already open, thereby further increasing the surface area, and (3) stretching of the alveolar membranes, increasing their surface area and decreasing their thickness. Therefore, during exercise, the oxygenation of the blood is increased not only by increased alveolar ventilation but also by a greater capacity of the respiratory membrane for transmitting oxygen into the blood.

Diffusing Capacity for Carbon Dioxide. The diffusing capacity for carbon dioxide has never been measured because of the following technical difficulty: Carbon dioxide diffuses through the respiratory membrane so rapidly that the average P_{CO_2} in the pulmonary blood is not far different from the P_{CO_2} in the alveoli — the difference is less than 1 mm. Hg — and with the available techniques, this difference is too small to be measured accurately.

Nevertheless, measurements of diffusion of other gases have shown that the diffusing capacity varies directly with the diffusion coefficient of the particular gas. Since the diffusion coefficient of carbon dioxide is 20 times that of oxygen, one would expect a diffusing capacity for carbon dioxide under resting conditions of about 400 to 450 ml. and during exercise of about 1200 to 1300 ml. per minute.

The importance of these high diffusing capacities for carbon dioxide is this: When the respiratory membrane becomes progressively damaged, its decreased capacity for transporting oxygen into the blood is always a much more serious problem than is the decreased capacity for transporting carbon dioxide.

TRANSPORT OF OXYGEN AND CARBON DIOXIDE

Once oxygen has diffused from the alveoli into the pulmonary blood, it is transported to the tissue capillaries where it is released for use by the cells. In the cells, oxygen reacts with various foodstuffs to form large quantities of carbon dioxide. This in turn enters the tissue capillaries and is transported by the blood back to the lungs. The purpose of the remainder of this chapter, therefore, is to present both qualitatively and quantitatively the physical and chemical principles of oxygen and carbon dioxide transport in the blood.

TRANSPORT OF OXYGEN TO THE TISSUES

Uptake of Oxygen by the Pulmonary Blood. The top part of Figure 28–3 illustrates a pulmonary alveolus adjacent to a pulmonary capillary, showing the diffusion of oxygen molecules between the alveolar air and the pulmonary blood. However, the P_{O_2} of the venous blood entering the capillary is only 40 mm. Hg because a large amount of oxygen was removed from this blood when it passed through the tissue capillaries. The P_{O_2} in the alveolus is 104 mm. Hg, giving an initial pressure difference for diffusion of oxygen into the pulmonary capillary of 64 mm. Hg. Therefore, far more oxygen diffuses into the pulmonary capillary than in the opposite direction. The graph below the capillary in Figure 28–3 shows the progressive rise in blood P_{O_2} as the blood passes through the capillary. This illustrates that the P_{O_2} rises essentially to equal that of the alveolar air before reaching the midpoint of the capillary, becoming approximately 104 mm. Hg.

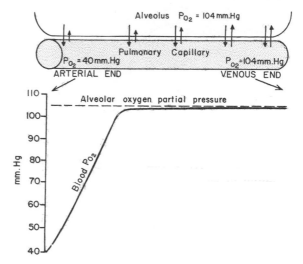

FIGURE 28-3 Uptake of oxygen by the pulmonary capillary blood. (The curve in this figure was constructed from data in Milhorn and Pulley: *Biophys. J., 8:*337, 1968.)

Diffusion of Oxygen from the Capillaries to the Cells. In the tissue capillaries, oxygen diffuses into the tissues by a process essentially the reverse of that which takes place in the lungs. That is, the P_{O_2} in the interstitial fluid immediately outside a capillary is low and very variable but averaging about 40 mm. Hg, while that in the arterial blood is high, about 95 mm. Hg. Therefore, at the arterial end of the capillary, a pressure difference of 55 mm. Hg exists for diffusion of oxygen. But, as illustrated in Figure 28–4, before the blood has gone more than a short distance into this capillary a large portion of the oxygen has diffused into the tissues, and the capillary P_{O_2} has approached the 40 mm. Hg oxygen pressure in the tissue fluids. Consequently, the venous blood leaving the tissue capillaries contains oxygen at essentially the same pressure as that immediately outside the tissue capillaries, 40 mm. Hg.

Diffusion of Oxygen from the Interstitial Fluids into the Cells. Since oxygen is always being used by the cells, the intracellular P_{O_2} remains lower than the interstitial fluid P_{O_2}. Therefore, oxygen diffuses into the cells to replenish that being used for metabolism.

The intracellular P_{O_2} ranges from as low as zero to as high as 40 mm. Hg, averaging (by direct measurement in the dog) 6 mm. Hg, which is the value given for the cell in Figure 28–4. Since only 1 to 5 mm. Hg oxygen pressure is required for full support of the metabolic processes of the cell, one can see that even this low cellular P_{O_2} is adequate.

TRANSPORT OF CARBON DIOXIDE TO THE LUNGS

Diffusion of Carbon Dioxide from the Cells to the Tissue Capillaries. Because of the large quantities of carbon dioxide

FIGURE 28-4 Diffusion of oxygen from a tissue capillary to the cells.

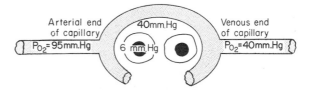

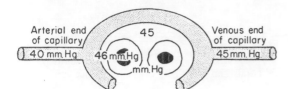

FIGURE 28–5 Uptake of carbon dioxide by the blood in the capillaries.

formed in the cells, the intracellular P_{CO_2} tends to rise. However, carbon dioxide diffuses about 20 times as easily as oxygen, diffusing from the cells extremely rapidly into the interstitial fluids and thence into the capillary blood. Thus, in Figure 28–5 the intracellular P_{CO_2} is shown to be 46 mm. Hg, while that in the interstitial fluids immediately adjacent to the capillaries is about 45 mm. Hg, a pressure differential of only 1 mm. Hg.

Arterial blood entering the tissue capillaries contains carbon dioxide at a pressure of approximately 40 mm. Hg. As the blood passes through the capillaries, the blood P_{CO_2} rises to approach the 45 mm. Hg P_{CO_2} of the interstitial fluid. Because of the very large diffusion coefficient for carbon dioxide, the P_{CO_2} of venous blood is also about 45 mm. Hg, within a fraction of a millimeter of reaching complete equilibrium with the P_{CO_2} of the interstitial fluid.

Removal of Carbon Dioxide From the Pulmonary Blood. On arriving at the lungs, the P_{CO_2} of the venous blood is about 45 mm. Hg while that in the alveoli is 40 mm. Hg. Therefore, as illustrated in Figure 28–6, the initial pressure difference for diffusion is only 5 mm. Hg, which is far less than that for diffusion of oxygen across the membrane. Yet, even so, because of the 20 times greater diffusion coefficient for carbon dioxide than for oxygen, the excess carbon dioxide in the blood is rapidly transferred into the alveoli. Indeed, the figure shows that the P_{CO_2} of the pulmonary capillary blood becomes almost equal to that of the alveoli within the first four-tenths of the blood's transit through the pulmonary capillary.

OXYGEN TRANSPORT IN THE BLOOD

Normally, about 97 per cent of the oxygen transported from the lungs to the tissues is carried in chemical combination with hemoglobin in the red blood cells, and the remaining 3 per cent is carried in the dissolved state in the water of the plasma and cells. However, when a person breathes oxygen at very high pressures, as much oxygen can sometimes be transported in the dissolved state as in chemical combination with hemoglobin. Therefore, the present

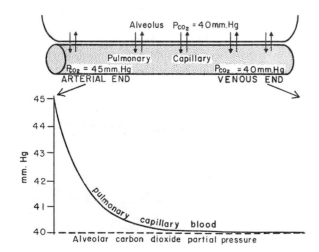

FIGURE 28–6 Diffusion of carbon dioxide from the pulmonary blood into the alveolus. (This curve was constructed from data in Milhorn and Pulley: *Biophys. J.,* 8:337, 1968.)

discussion takes up the transport of oxygen first in combination with hemoglobin and then in the dissolved state.

THE REVERSIBLE COMBINATION OF OXYGEN WITH HEMOGLOBIN

The chemistry of hemoglobin was presented in Chapter 8, where it was pointed out that the oxygen molecule combines loosely and reversibly with the heme portion of the hemoglobin. When the P_{O_2} is high, as in the pulmonary capillaries, oxygen binds with the hemoglobin, but, when the P_{O_2} is low, as in the tissue capillaries, oxygen is released from the hemoglobin. This is the basis for oxygen transport from the lungs to the tissues.

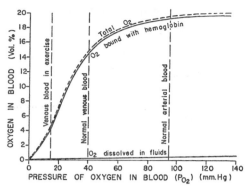

FIGURE 28–7 The oxygen-hemoglobin dissociation curve, showing the quantity of oxygen in each 100 ml. of normal blood bound with hemoglobin. Also shown is the oxygen dissolved in the water of the blood (between the two curves) and the total oxygen in both forms.

The Oxygen-Hemoglobin Dissociation Curve. Figure 28–7 illustrates the oxygen-hemoglobin dissociation curve, which shows the progressive increase in volumes per cent of oxygen bound with hemoglobin as the P_{O_2} increases. The blood leaving the lungs usually has a P_{O_2} of about 100 mm. Hg, and the amount of the hemoglobin that is bound with oxygen at this P_{O_2}, called the *hemoglobin saturation, is about 97 per cent.* On the other hand, in normal venous blood the P_{O_2} is about 40 mm. Hg, and *the per cent saturation of the hemoglobin is about 75 per cent.*

The blood of a normal person contains approximately 15 grams of hemoglobin in each 100 ml. of blood, and each gram of hemoglobin can bind with a maximum of about 1.34 ml. of oxygen. Therefore, on the average, the hemoglobin in 100 ml. of blood can combine with a total of about 20 ml. of oxygen. This is usually expressed as 20 *volumes per cent.*

Amount of Oxygen Released from the Hemoglobin in the Tissues. Note in Figure 28–7 that the total quantity of oxygen bound with hemoglobin in the normal arterial blood is approximately 19.4 ml. On passing through the tissue capillaries, this amount is reduced to 14.4 ml. (P_{O_2} of 40 mm. Hg, 75 per cent saturated), or a total loss of about 5 ml.

of oxygen from each 100 ml. of blood. Then, when the blood returns to the lungs, the same quantity of oxygen diffuses from the alveoli to the hemoglobin, and this too is carried to the tissues. Thus, *under normal conditions about 4 ml. of oxygen is transported by each 100 ml. of blood during each cycle through the tissues.*

Transport of Oxygen During Strenuous Exercise. In heavy exercise the muscle cells utilize oxygen at a rapid rate, which causes the interstitial fluid P_{O_2} to fall as low as 15 mm. Hg. At this pressure only 4.4 ml. of oxygen remains bound with the hemoglobin in each 100 ml. of blood, as shown in Figure 28–7. Thus, 19.4 — 4.4, or 15 ml., is the total quantity of oxygen transported by each 100 ml. of blood in each cycle through the tissues. This, obviously, is three times as much as that normally transported by the same amount of blood, illustrating that simply an increase in rate of oxygen utilization by the tissues causes an automatic increase in the rate of oxygen release from the hemoglobin.

The Utilization Coefficient. The fraction of the hemoglobin that gives up its oxygen as it passes through the tissue capillaries is called the *utilization coefficient.* Normally, this is approximately 25 per cent. During strenuous exercise, it can increase to over 75 per cent.

THE OXYGEN BUFFER FUNCTION OF HEMOGLOBIN

Though hemoglobin is necessary for transport of oxygen to the tissues, it performs still another major function essential to life. This is the function of hemoglobin as an "oxygen buffer" system. That is, the hemoglobin in the blood is mainly responsible for controlling the oxygen pressure in the tissues, as follows:

The tissues normally require 5 to 15 ml. oxygen from every 100 ml. of blood passing through the tissue capillaries. Referring back to the oxygen-hemoglobin dissociation curve in Figure 28–7, one can see that for 5 ml. of oxygen to be released from each 100 ml. of blood, the P_{O_2} must fall to about 40 mm. Hg, and for 15 ml. of oxygen to be released the P_{O_2} must fall to about 20 mm. Hg. It can be seen, then, that hemoglobin automatically delivers oxygen to the tissues at a pressure between approximately 20 and 40 mm. Hg.

Value of Hemoglobin for Automatically Adjusting the Tissue P_{O_2} to Different Concentrations of Atmospheric Oxygen. The normal P_{O_2} in the alveoli is approximately 104 mm. Hg, but, as one ascends a mountain or goes high in an airplane, the P_{O_2} falls considerably; or, when one enters areas of compressed air, such as deep below the sea or in tunnels, the P_{O_2} may rise to very high values. It will be seen from the oxygen-hemoglobin dissociation curve of Figure 28–7 that, when the P_{O_2} is decreased to as low as 60 mm. Hg, the hemoglobin is still 89 per cent saturated, only 8 per cent below the normal saturation of 97 per cent. Furthermore, when the P_{O_2} rises to as high as 500 mm. Hg, the hemoglobin saturation rises only 3 per cent. Therefore, a change in atmospheric P_{O_2} from 60 to 500 mm. Hg hardly affects the tissue P_{O_2}.

METABOLIC USE OF OXYGEN BY THE CELLS

Relationship Between Intracellular P_{O_2} and Rate of Oxygen Usage. Only minute oxygen pressure is required in the cells for normal intracellular chemical reactions to take place. The reason for this is that the respiratory enzyme systems of the cell (discussed in Chap. 45) are geared so that when the cellular P_{O_2} is more than 5 mm. Hg, oxygen availability is no longer a limiting factor in the rates of the chemical reactions. Instead, the main limiting factor then is the *concentration of adenosine diphosphate* (ADP) in the cells, as was explained in Chapter 3. It will be recalled that when adenosine triphosphate (ATP) is utilized in the cells to provide energy, it is converted into ADP. The increasing concentration of ADP then increases the metabolic usage of both oxygen and the various nutrients that combine with oxygen to release energy. Therefore, under normal operating conditions the rate of oxygen utilization by the cells is controlled by the rate of energy expenditure within the cells and not by the rate of delivery of oxygen to the cells.

COMBINATION OF HEMOGLOBIN WITH CARBON MONOXIDE

Carbon monoxide combines with hemoglobin at the same point on the hemoglobin molecule as does oxygen. Furthermore, it binds with approximately 210 times as much tenacity as oxygen. Therefore, a carbon monoxide pressure of only 0.5 mm. Hg in the alveoli, 210 times less than that of the alveolar oxygen, allows the carbon monoxide to compete equally with the oxygen for combination with the hemoglobin and causes half the hemoglobin in the blood to become bound with carbon monoxide instead of with oxygen. A carbon monoxide pressure of 0.7 mm. Hg (a concentration of about 0.1 per cent) is lethal.

CARBON DIOXIDE TRANSPORT IN THE BLOOD

Transport of carbon dioxide is not nearly so great a problem as transport of oxygen, because even in the most abnormal conditions carbon dioxide can usually be transported by the blood in far greater quantities than can oxygen. Under normal resting conditions, *an average of 4 ml. of carbon dioxide is transported from the tissues to the lungs in each 100 ml. of blood.*

CHEMICAL FORMS IN WHICH CARBON DIOXIDE IS TRANSPORTED

To begin the process of carbon dioxide transport, carbon dioxide diffuses out of the tissue cells. On entering the capillary, the chemical reactions illustrated in Figure 28–8 occur immediately; the quantitative aspects of these can be described as follows:

Transport of Carbon Dioxide in the Dissolved State. A small portion of the carbon dioxide is transported in the dissolved state to the lungs; this is about 7 per cent of all the carbon dioxide transported.

Transport of Carbon Dioxide in the Form of Bicarbonate Ion. By far the most important means for transporting carbon dioxide is in the form of bicarbonate ion. The dissolved carbon dioxide in the blood reacts with water to form carbonic acid. Inside the red blood cells the enzyme *carbonic anhydrase* catalyzes the reaction, accelerating its rate many thousand-fold. Consequently, the reaction occurs so rapidly in the red blood cells that it reaches almost complete equilibrium within a fraction of a second. This allows tremendous amounts of carbon dioxide to react with the red cell water even before the blood leaves the tissue capillaries. In another small fraction of a second the carbonic acid dissociates into hydrogen and bicarbonate ions. Most of the hydrogen ions formed react rapidly with hemoglobin, which is a powerful acid-base buffer. Then, many of the bicarbonate ions diffuse into the plasma while chloride ions diffuse into the red cells. Thus, the chloride content of venous red blood cells is greater than that of arterial cells, a phenomenon called the *chloride shift*.

The reversible combination of carbon dioxide with water under the influence of carbonic anhydrase in the red blood cells accounts for 60 to 90 per cent of all the carbon dioxide transported from the tissues to the lungs.

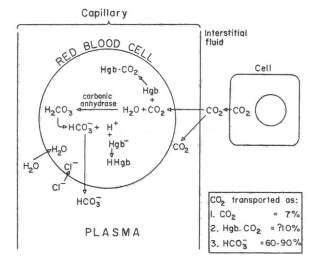

FIGURE 28–8 Transport of carbon dioxide in the blood.

Transport of Carbon Dioxide in Combination with Hemoglobin and Plasma Protein—Carbaminohemoglobin. In addition to reacting with water, carbon dioxide also reacts directly with hemoglobin and also with plasma protein. The combination of carbon dioxide with hemoglobin is a reversible reaction that occurs with a very loose bond. The compound formed by this reaction is known as "carbaminohemoglobin." A small amount of carbon dioxide reacts in this same way with the plasma proteins, but this is much less significant because of the one-fourth as great quantity of these proteins in the blood.

The *maximum theoretical* quantity of carbon dioxide that can be carried from the tissues to the lungs in combination with hemoglobin and plasma proteins is approximately 30 per cent of the total quantity transported. However, this reaction is much slower than the reaction of carbon dioxide with water inside the red blood cells. Therefore, it is doubtful that this mechanism provides transport of more than a few per cent of the total quantity of carbon dioxide.

THE CARBON DIOXIDE DISSOCIATION CURVE

It is now apparent that carbon dioxide can exist in blood in many different forms, (1) as free carbon dioxide and (2) in chemical combination with water, hemoglobin, and plasma protein. The curve shown in Figure 28–9 depicts the dependence of total blood CO_2 in all its forms on P_{CO_2}; this curve is called the *carbon dioxide dissociation curve.*

Note that the normal resting P_{CO_2} ranges between the limits of 40 and 45 mm. Hg, which is a very narrow range. Note also that the normal concentration of carbon dioxide in the blood is about 50 volumes per cent but that only 4 volumes per cent of this is actually exchanged in the process of transporting carbon dioxide from the tissues to the lungs. That is, the concentration rises to about 52 volumes per cent as the blood passes through the tissues, and falls to about 48 volumes per cent as it passes through the lungs.

THE RESPIRATORY EXCHANGE RATIO

The discerning student will have noted that normal transport of oxygen from the lungs to the tissues by each 100 ml. of blood is about 5 ml., while normal transport of carbon dioxide from the tissues to the lungs is about 4 ml. Thus, under normal resting conditions only about 80 per cent as much carbon dioxide is expired from the lungs as there is oxygen uptake by the lungs. The ratio of carbon dioxide output to oxygen uptake is

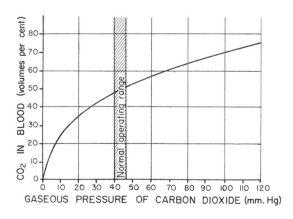

FIGURE 28–9 The carbon dioxide dissociation curve.

called the respiratory exchange ratio (R). That is,

$$R = \frac{\text{Rate of carbon dioxide output}}{\text{Rate of oxygen uptake}}$$

The value for R changes under different metabolic conditions. When a person is utilizing entirely carbohydrates for body metabolism, R rises to 1.00. On the other hand, when the person is utilizing fats almost entirely for his metabolic energy, the level falls to as low as 0.7. The reason for this difference is that when oxygen is metabolized with carbohydrates one molecule of carbon dioxide is formed for each molecule of oxygen consumed, while when oxygen reacts with fats a large share of the oxygen combines with hydrogen atoms to form water instead of carbon dioxide. Therefore, the quantity of carbon dioxide released from these reactions is far less than the quantity of oxygen consumed. In other words, the *respiratory quotient of the chemical reactions* in the tissues is now about 0.70 instead of 1.00 as is the case when carbohydrates are being utilized. The tissue respiratory quotient will be discussed in Chapter 47.

For a person on a normal diet consuming average amounts of carbohydrates, fats, and proteins, the average value for R is considered to be 0.825.

REFERENCES

Antonini, E.: Interrelationship between structure and function in hemoglobin and myoglobin. *Physiol. Rev., 45*:123, 1965.

Chance, B.: Regulation of intracellular oxygen. *Proc. Int. Union of Physiol. Sciences, 6*:13, 1968.

Crowell, J. W., and Smith, E. E.: Determinants of the optimal hematocrit. *J. Appl. Physiol., 22*:501, 1967.

Forster, R. E.: Diffusion of gases. *In* Handbook of Physiology. Baltimore, The Williams & Wilkins Co., 1964, Sec. III, Vol. I, p. 839.

Haldane, J. S., and Priestley, J. G.: Respiration. New Haven, Yale University Press, 1935.

Rahn, H., and Farhi, L. E.: Ventilation, perfusion, and gas exchange—the Va/Q concept. *In* Handbook of Physiology. Baltimore, The Williams & Wilkins Co., 1964, Sec. III, Vol. I, p. 735.

Riggs, A.: Functional properties of hemoglobins. *Physiol. Rev., 45*:619, 1965.

West, J.: Ventilation: Blood Flow and Gas Exchange. Philadelphia, F. A. Davis Co., 1965.

CHAPTER 29

REGULATION OF RESPIRATION; AND PHYSIOLOGY OF RESPIRATORY INSUFFICIENCY

During exercise and other physiological states that increase the metabolic activity of the body, the respiratory system is called upon to supply increased quantities of oxygen to the tissues and to remove increased quantities of carbon dioxide. The *respiratory center* in the brain stem adjusts the rate of alveolar ventilation almost exactly to the demands of the body, so that, as a result, the arterial oxygen pressure (Po_2) and carbon dioxide pressure (Pco_2) are hardly altered even during strenuous exercise or other types of respiratory stress.

The present chapter describes the operation of this neurogenic system for regulation of respiration.

THE RESPIRATORY CENTER

The so-called "respiratory center" is a widely dispersed group of neurons located bilaterally in the reticular sub-stance of the medulla oblongata and pons, as illustrated in Figure 29–1*A*. It is divided into three major areas: (1) the medullary rhythmicity area, (2) the apneustic area, and (3) the pneumotaxic area. The medullary rhythmicity area seems to be by far the most important. Therefore, it deserves particular attention.

The Medullary Rhythmicity Area. The medullary rhythmicity area, which is also frequently referred to simply as the *medullary respiratory center*, is located beneath the lower part of the floor of the fourth ventricle in the medial half of the medulla.

Microelectrodes inserted into this center detect some neurons that discharge during inspiration and others that discharge during expiration. There have been some attempts to show that part of this center is primarily related to inspiration and part primarily related to expiration; however, most efforts in this direction have not been really success-

350

A.

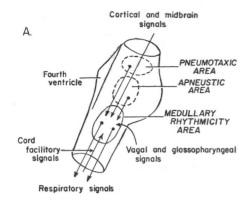

B.

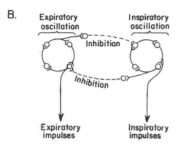

FIGURE 29-1 (*A*) The respiratory center located bilaterally in the lateral reticular substance of the medulla and lower pons. (*B*) Theoretical mechanism for the rhythmicity of the respiratory center.

ful. Thus, in general, *inspiratory neurons* and *expiratory neurons* seem to intermingle in the medullary rhythmicity center.

Basic rhythmicity in the medullary rhythmicity area. It is in the medullary rhythmicity area that the basic rhythm of respiration is established. In the normal resting person, inspiration usually lasts for about two seconds and expiration for about three seconds. Both of these are correspondingly shortened during increased respiration and lengthened during decreased respiration.

The medullary rhythmicity area by itself is not capable of giving a normal smooth pattern of respiration. If the medulla is transected immediately above the rhythmicity area but with this area still connected to the spinal cord, respiration occurs in gasps rather than in normal smooth inspirations and expira-

tions. Furthermore, the rhythmical activity of the medullary rhythmicity center is very weak when afferent signals do not reach it from other sources. Thus, in Figure 29-1*A*, signals are shown entering the medullary rhythmicity area from the spinal cord, from the cerebral cortex and midbrain, from the pneumotaxic area in the upper pons, and from the apneustic area in the lower pons. All of these signals modify the rhythm of respiration and contribute to the normal smooth pattern of respiration.

Function of the Apneustic and Pneumotaxic Areas. The apneustic and pneumotaxic areas are located in the reticular substance of the pons. They are not necessary for maintenance of the basic rhythm of respiration. However, when the apneustic area is still connected to the medullary rhythmicity area but the pons has been transected between the apneustic and pneumotaxic areas, the animal breathes with a pattern of prolonged inspiration and very short expiration, which is exactly opposite to the pattern that occurs when breathing is accomplished by the medullary rhythmicity area alone. This apneustic pattern becomes especially marked when the afferent nerve fibers of the vagus and of the glossopharyngeal nerves have been transected.

If the pneumotaxic area is also connected to the rhythmicity center, the pattern of respiration becomes essentially normal, having a reasonable balance between inspiration and expiration. Also, stimulation of the pneumotaxic area can change the rate of respiration, which is the reason for its name, pneumotaxic area.

Basic Mechanism of the Rhythmicity Observed in the Medullary Rhythmicity Area. Though the cause of respiratory rhythmicity is mainly unknown, the following is a recent theory that is at least partially substantiated by experimental studies:

Figure 29-1*B* shows four expiratory neurons to the left and four inspiratory neurons to the right. Each of these neuronal groups is arranged in an oscillating network that is believed to function in the following manner: Let us first consider the inspiratory neurons. If one of the neurons becomes excited, it excites the next one, which in turn excites the third one, and so forth. This excitation continues around and around the "inspiratory" circuit, and inspiratory impulses are transmitted to the inspiratory muscles. This process continues until the oscillation stops after about 2 seconds. The factor that stops the oscillation is perhaps fatigue of the neurons themselves, for we know that neurons cannot transmit large numbers of impulses in rapid succession without decreasing their excitability.

In addition to transmitting impulses to the inspiratory muscles, the inspiratory neurons also transmit *inhibitory* impulses into the expiratory network. This keeps the expiratory network from oscillating while the inspiratory network is oscillating. But just as soon as inspiratory oscillation ceases, there is no longer inhibition of the expiratory network. In the absence of this inhibition, the natural excitability of the expiratory neurons causes the expiratory network to begin oscillating. It then oscillates for about 3 seconds until its neurons likewise fatigue. During this oscillation, expiratory impulses are transmitted to the expiratory muscles, and *inhibitory* impulses are transmitted into the inspiratory network to keep it from oscillating. However, when the expiratory network has fatigued and stopped oscillating, the inspiratory network begins to oscillate once again.

In summary, two oscillating circuits are postulated: one for inspiration and one for expiration. However, the two circuits cannot oscillate simultaneously because they inhibit each other. Thus, when the inspiratory neurons are active, the expiratory neurons are inactive; and the reverse effects occur when the expiratory neurons become active. There-

fore, alternation occurs back and forth between inspiratory signals and expiratory signals, this process continuing indefinitely and causing the act of respiration.

THE HERING-BREUER REFLEXES — THE INFLATION AND DEFLATION REFLEXES

Many receptors in the lungs detect stretch or compression of the lungs. Some of these are located in the visceral pleura, and others in the bronchi, bronchioles, and even the alveoli. When the lungs become stretched, the stretch receptors, especially those of the bronchioles, transmit impulses through the vagus nerves into the *tractus solitarius* of the brain stem and thence into the respiratory center where they inhibit inspiration and thereby prevent further inflation. The effect is called the *Hering-Breuer inflation reflex*. This reflex prevents overdistention of the lungs.

A *Hering-Breuer deflation reflex* also occurs during expiration: that is, as the stretch receptors become unstretched the impulses from these receptors cease, which allows inspiration to begin again. Also, it is likely that compression receptors (not well defined as yet but possibly in the walls of the alveoli) transmit impulses that inhibit expiration in the same manner that the stretch receptors inhibit inspiration. Regardless of the precise mechanism, the degree of lung deflation is reduced by this reflex. However, this deflation reflex is normally much less active than the inflation reflex.

Importance of the Hering-Breuer Reflexes in Maintaining Respiratory Rhythmicity. In addition to limiting lung inflation and deflation, the Hering-Breuer receptors feed into the respiratory center and help to inhibit inspiration as the lungs fill, thereby shifting activity from the inspiratory network to the expiratory network. Then, during expiration, the deflation reflex works exactly oppositely.

Therefore, when the Hering-Breuer reflexes are blocked by cutting the vagus nerves, it is considerably easier to cause respiratory arrest than it is when the Hering-Breuer reflexes are intact.

HUMORAL REGULATION OF RESPIRATION

The ultimate goal of respiration is to maintain proper concentrations of oxygen, carbon dioxide and hydrogen ions in the body fluids. It is fortunate, therefore, that respiratory activity is highly responsive to even slight changes in any one of these in the fluids. Therefore, when one speaks of *humoral* regulation of respiration, he is referring primarily to the regulation of respiratory activity by changes in concentrations of oxygen, carbon dioxide, or hydrogen ions in the body fluids.

Carbon dioxide and hydrogen ions exert their effects primarily on the respiratory center in the brain, whereas oxygen has its effect almost entirely on the peripheral chemoreceptors, which in turn affect the respiratory center.

EFFECTS OF CARBON DIOXIDE AND HYDROGEN ION CONCENTRATIONS ON RESPIRATORY CENTER ACTIVITY

The respiratory center, meaning by this all of the respiratory areas of the brain stem, is greatly stimulated by an increase in either carbon dioxide or hydrogen ion concentration in the fluids of the respiratory center. These effects are illustrated by two of the curves in Figure 29–2, which show, respectively, (1) that an increase in carbon dioxide concentration (while all other factors remain constant) can increase the rate of alveolar ventilation as much as six- to seven-fold and (2) that a decrease in pH, which means an increase in hydrogen ion concentration, can also increase the rate of

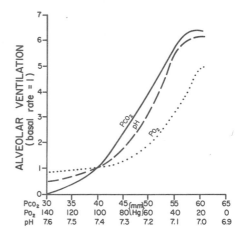

FIGURE 29–2   Approximate effects on alveolar ventilation of changing the concentrations of carbon dioxide, hydrogen ions, and oxygen in the arterial blood when only one of the humoral factors is changed at a time and the other two are maintained at absolutely normal levels.

alveolar ventilation almost exactly as much as can an excess of carbon dioxide (provided that all other factors are held constant).

Stimulation of Respiration by Carbon Dioxide and Hydrogen Ions in Cerebrospinal Fluid—The Brain Stem Chemosensitive Areas.   Recent experiments have shown the existence of a very small chemosensitive respiratory area on the surface of each side of the medulla oblongata at approximately the entry points of the eighth, ninth, and tenth cranial nerves; this area is especially sensitive to changes in hydrogen ion concentration in the cerebrospinal fluid. An increase in hydrogen ions greatly increases respiratory activity.

Even though these chemosensitive areas are primarily responsive to hydrogen ions, it is mainly carbon dioxide in the blood that causes the hydrogen ions in the cerebrospinal fluids to increase. The reason for this is that carbon dioxide diffuses from the blood into the cerebrospinal fluid extremely rapidly, but hydrogen ions do not. The carbon dioxide immediately combines with water of the cerebrospinal fluid to form carbonic acid. This in turn dissociates to increase the hydrogen ion concentration. Therefore, the cerebrospinal fluid mechanism for exciting respiration is primarily responsive to changes in carbon dioxide concentration in the blood.

Unfortunately, we still do not know how important the chemosensitive areas on the surface of the medulla are to overall respiratory regulation. Some respiratory physiologists believe that they account for as much as one-half of the respiratory drive caused by carbon dioxide, whereas others believe that the respiratory drive caused by this mechanism is of almost no importance.

A Unified Concept of Humoral Stimulation of Respiratory Neurons—Role of the Hydrogen Ion. When carbon dioxide diffuses to the interior of a cell, it reacts with water inside the cell to form carbonic acid, and this in turn dissociates rapidly to increase the hydrogen ion concentration. Therefore, either an increase in hydrogen ion concentration in the extracellular fluids or an increase in extracellular carbon dioxide concentration increases the hydrogen ion concentration inside the cell. For this reason, it has been postulated that the hydrogen ion concentration inside the respiratory neurons is the common factor that actually determines their levels of activity. Though this is only a theory, it could easily explain how carbon dioxide and hydrogen ions have almost exactly the same effect in stimulating the respiratory center.

THE CHEMORECEPTOR SYSTEM FOR CONTROL OF RESPIRATORY ACTIVITY—ROLE OF OXYGEN IN RESPIRATORY CONTROL

Aside from the direct sensitivity of the respiratory center itself to humoral factors, special chemical receptors located outside the central nervous system and called *chemoreceptors* are responsive to changes in oxygen, carbon dioxide, and hydrogen ion concentrations. These transmit signals to the respiratory center to help regulate respiratory activity. These chemoreceptors are located primarily in association with the large arteries of the thorax and neck; most of them are in the *carotid* and *aortic bodies*, which are illustrated in Figure 29–3 along with their afferent nerve connections to the respiratory center. The *carotid bodies* are located bilaterally in the bifurcations of the common carotid arteries, and their afferent nerve fibers pass through Hering's nerves to the glossopharyngeal nerves and thence to the medulla. The *aortic bodies* are located along the arch of the aorta; their afferent nerve fibers pass to the medulla through the vagi. Each of these bodies receives a special blood supply through a minute artery directly from the adjacent arterial trunk.

Stimulation of the Chemoreceptors by Decreased Arterial Oxygen. Changes in arterial oxygen concentration have almost no direct effect on the respiratory center itself, but when the oxygen concentration in the arterial blood falls below normal, the chemoreceptors become strongly stimulated.

If care is taken to prevent simultaneous changes in either pH or P_{CO_2}, changes in arterial P_{O_2} have the effect

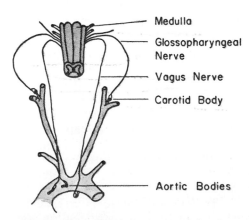

Medulla

Glossopharyngeal Nerve

Vagus Nerve

Carotid Body

Aortic Bodies

FIGURE 29–3 Respiratory control by the carotid and aortic bodies.

illustrated by the dotted curve of Figure 29–2 on the rate of alveolar ventilation. This shows that so long as arterial P_{O_2} is above 100 mm. Hg there is relatively little effect on alveolar ventilation, but as the alveolar P_{O_2} falls progressively below this normal level, alveolar ventilation can increase to *as much as five times normal.* That is, a decrease in arterial P_{O_2} stimulates the chemoreceptors, which then stimulate the respiratory center to increase alveolar ventilation. Thus, the chemoreceptor system provides a very valuable feedback mechanism to help maintain normal arterial oxygen concentration. However, for reasons that are discussed later in the chapter, this mechanism does not operate very powerfully under normal respiratory conditions, but only under certain abnormal conditions.

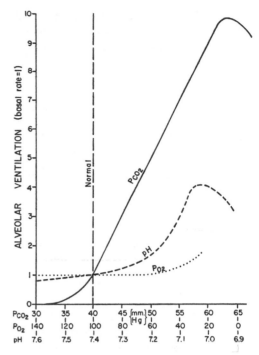

FIGURE 29–4 Effects of increased arterial P_{CO_2}, decreased arterial P_{O_2}, and decreased arterial pH on the rate of alveolar ventilation.

QUANTITATIVE ASPECTS OF HUMORAL CONTROL OF ALVEOLAR VENTILATION

The effects of the three individual humoral factors, P_{CO_2}, P_{O_2}, and pH, on alveolar ventilation *when the other two respective factors are kept artificially at absolutely normal levels* were illustrated in Figure 29–2. This figure showed that the effects of the individual factors are all very powerful and that each of them can be the basis for an important feedback regulatory system to maintain proper concentration of itself in the body fluids. However, Figure 29–4 illustrates the effect of changing each factor when the other two respective factors *are not* artificially maintained at constant normal values. Under these conditions, one sees that the effect of P_{CO_2} on alveolar ventilation is tremendous, the effect of pH is intermediate, and the effect of P_{O_2} is slight. The purpose of this section, therefore, is to explain these differences.

CARBON DIOXIDE, THE MAJOR CHEMICAL FACTOR REGULATING ALVEOLAR VENTILATION

Note in Figure 29–4 that an increase in alveolar P_{CO_2} from the normal level of 40 mm. Hg to 63 mm. Hg causes a 10-fold increase in alveolar ventilation; this is about one and a half times as great as that which occurs when pH and P_{O_2} are controlled. The cause of this difference is that an increase in P_{CO_2} causes an almost proportional increase in hydrogen ion concentration in all the body fluids, as was explained in Chapter 26. Therefore, an increase in P_{CO_2} stimulates alveolar ventilation directly, and also indirectly through its effect on hydrogen ion concentration.

Value of Carbon Dioxide as a Regulator of Alveolar Ventilation. Since carbon dioxide is one of the end-products of metabolism, its concentration in the body fluids greatly affects the chemical reactions of the cells. For this reason, the tissue fluid P_{CO_2} must be regulated exactly. In the preceding chapter, it was pointed out that blood and interstitial fluid P_{CO_2} are determined to a great extent by the rate of alveolar ventilation. Therefore, stimulation of the respiratory center by carbon dioxide provides an important feedback mechanism for regulation of the concentration of carbon dioxide throughout the body. That is, (1) an increase in P_{CO_2} stimulates the respiratory center; (2) this increases alveolar ventilation and reduces the alveolar carbon dioxide; (3) as a result, the tissue P_{CO_2} returns to normal. In this way, the respiratory center maintains the P_{CO_2} of the tissue fluids at a relatively constant level and, therefore, might well be called a "carbon dioxide pressostat."

CONTROL OF ALVEOLAR VENTILATION BY THE HYDROGEN ION CONCENTRATION OF THE EXTRACELLULAR FLUIDS

Figure 29–4 illustrates that changes in pH have considerably less effect on alveolar ventilation when the other humoral factors are not controlled than when they are controlled. The reason for this is that an increase in hydrogen ion concentration causes the person to breathe more rapidly than usual, which decreases the carbon dioxide concentration in the body fluids and at the same time increases the oxygen concentration. The decreased carbon dioxide *inhibits* the respiratory center and the increased oxygen *inhibits* the chemoreceptors. Therefore, opposing effects are caused by these other two factors, thus causing a "braking" effect on the hydrogen ion stimulation of respiration.

Nevertheless, despite the "braking" effect of the P_{CO_2} and P_{O_2} changes, a decrease in pH to approximately 7.1 can increase alveolar ventilation about four-fold, and an increase in pH to 7.6 can decrease ventilation to about 80 per cent of normal.

CONTROL OF ALVEOLAR VENTILATION BY ARTERIAL OXYGEN SATURATION

Unimportance of Oxygen Regulation of Respiration Under Normal Conditions — The "Braking" Effect of the P_{CO_2} and pH Regulatory Mechanisms. Figure 29–4 illustrates that changes in alveolar P_{O_2} have extremely little effect on alveolar ventilation when P_{CO_2} and pH are not controlled. This is in marked contrast to the effect when these other two factors are controlled, as was illustrated in Figure 29–2. This lack of importance of oxygen in normal regulation of respiration is caused mainly by a "braking" effect of both the carbon dioxide and hydrogen ion control mechanisms. That is, when low oxygen begins to increase the ventilation, this blows off carbon dioxide from the blood and therefore decreases the P_{CO_2}; at the same time it also decreases the hydrogen ion concentration. Therefore, two powerful respiratory inhibitory effects are caused by (a) diminished carbon dioxide and (b) diminished hydrogen ions. These two exert an inhibitory braking effect that opposes the excitatory effect of the diminished oxygen. As a result, they keep the decreased oxygen from causing marked increase in ventilation. Therefore, the maximum effect of decreased alveolar oxygen on alveolar ventilation is normally only a 66 per cent increase. This is in contrast to about a 400 per cent increase caused by decreased pH and about a 1000 per cent increase caused by increased carbon dioxide.

Yet, despite its weak response to diminished oxygen, the chemoreceptor mechanism is very important, for it is the only mechanism by which low oxygen concentration in the arterial blood can increase alveolar ventilation, and even this slight increase can sometimes be a life saver.

Reason oxygen regulation of respiration is not normally needed. On first thought, it seems strange that oxygen should play so little role in the normal regulation of respiration, particularly since one of the primary functions of the respiratory center is to provide adequate intake of oxygen. However, usually it does not matter whether alveolar ventilation is normal or 10 times normal, the blood will still be essentially fully saturated. Also, alveolar ventilation can decrease to as low as one-half normal, and the blood still remains within 10 per cent of complete saturation. Therefore, one can see that alveolar ventilation can change tremendously without significantly affecting oxygen transport to the tissues. Instead, the factors that play the major role in controlling Po_2 in the tissues under normal conditions are the hemoglobin-oxygen buffer system and the blood flow to the tissues, which were discussed in the preceding chapter. On the other hand, carbon dioxide concentration in the tissues is controlled almost entirely by alveolar ventilation itself, as was explained earlier in the chapter.

Effect of the Oxygen Lack Mechanism in Pneumonia or at High Altitudes. When a person develops pneumonia, or ascends to high altitudes, or in any other way is exposed to prolonged oxygen deficiency, the respiratory center gradually becomes "adapted," for reasons yet unknown, to the diminished carbon dioxide and hydrogen ion concentrations in the blood; therefore, these no longer depress the respiratory center significantly. As a result the "braking" effect of the oxygen control is gradually lost, and alveolar ventilation then rises to as high as seven times normal. This is part of the acclimatization that occurs as a person slowly ascends a mountain, thus allowing the person to adjust his respiration gradually to a level fitted for the higher altitudes.

REGULATION OF RESPIRATION DURING EXERCISE

In strenuous exercise, oxygen utilization and carbon dioxide formation can increase as much as 20-fold (see Fig. 29–5). Yet, measurements of arterial Pco_2, pH, and Po_2 show that none of these changes sufficiently to account for more than a small portion of the increase in ventilation. Therefore, the question must be asked: What is it during exercise that causes the intense respiration? This question has not been answered completely, but at least two different effects, illustrated in Figure 29–6, seem to be predominantly concerned, as follows:

1. The motor cortex, on transmitting impulses to the contracting muscles, is believed to transmit collateral impulses into the reticular substance of the brain stem to excite the respiratory center. This is analogous to the stimulatory effect that causes the arterial pressure to rise during exercise when similar collateral impulses pass to the vasomotor center.

2. During exercise, the body movements, especially of the limbs, are believed to increase pulmonary ventilation by exciting joint proprioceptors that

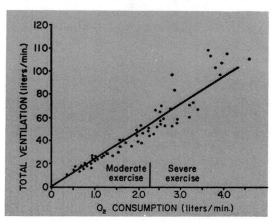

FIGURE 29–5 Effect of exercise on oxygen consumption and ventilatory rate. (From Gray: Pulmonary Ventilation and Its Physiological Regulation. Charles C Thomas.)

then transmit excitatory impulses to the respiratory center. The reason for believing this is that even passive movements of the limbs often increase pulmonary ventilation several-fold.

Most times when a person exercises, the nervous factors seem to stimulate the respiratory center almost exactly the proper amount to supply the extra oxygen requirements for the exercise and to blow off the extra carbon dioxide. But, occasionally, the nervous signals are either too strong or too weak in their stimulation of the respiratory center. Then, the humoral factors play a very significant role in bringing about the final adjustment in respiration required to keep the carbon dioxide, oxygen, and hydrogen ion concentrations of the body fluids as nearly normal as possible, as is illustrated in Figure 29–6. Thus, the neurogenic factors initiate the increase in alveolar ventilation during exercise and the humoral factors make additional adjustments, either upward or downward, to balance the rate of alveolar ventilation with the rate of metabolism in the body.

PERIODIC BREATHING — AN ABNORMALITY OF RESPIRATORY REGULATION

An abnormality of respiration called periodic breathing occurs in a number of different disease conditions. The person breathes deeply for a short interval of time and then breathes slightly or not at all for an additional interval, the cycle repeating itself over and over again.

The most common type of periodic breathing, *Cheyne-Stokes breathing*, is characterized by slow waxing and waning of respiration, occurring over and over again every 45 seconds to 3 minutes.

Basic Mechanism of Cheyne-Stokes Breathing. Let us assume that the respiration becomes much more rapid and deeper than usual. This causes the P_{CO_2} in the pulmonary blood to decrease. A few seconds later the pulmonary blood reaches the brain, and the decreased P_{CO_2} inhibits respiration. As a result, the pulmonary blood P_{CO_2} gradually increases. After another few seconds the blood carrying the increased CO_2 arrives at the respiratory center and stimulates respiration again, thus making the person overbreathe once again and initiating a new cycle that continues on and on, causing Cheyne-Stokes periodic breathing.

Fortunately, this oscillatory process is usually blocked by slowness of the control system to react to changing respiratory stimuli, but in pathological conditions the respiratory drive can become so extreme that it overcomes this natural "damping" of the system. Such occurs when blood flow becomes very sluggish in heart failure — this sometimes causing Cheyne-Stokes breathing for weeks at a time — or when the respiratory center reacts abnormally to the humoral respiratory stimuli after brain damage

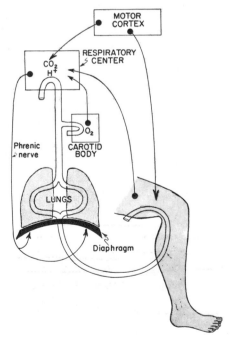

FIGURE 29–6 The different factors that enter into regulation of respiration during exercise.

from head injury, brain tumor, or other causes.

HYPOXIA

In some abnormalities of the lungs, oxygen therapy is of great value; in others it is of moderate value, while in still others it is of almost no value. Therefore, it is important to classify the different types of hypoxia; then from this classification we can readily discuss the physiological principles of therapy. The following is a descriptive classification of the different causes of hypoxia:
1. Inadequate oxygenation of the lungs because of extrinsic reasons.
   a. Deficiency of oxygen in atmosphere.
   b. Hypoventilation (neuromuscular disorders).
2. Pulmonary disease.
   a. Hypoventilation due to airway or pulmonary disease.
   b. Uneven alveolar ventilation-perfusion ratio.
   c. Diminished diffusing capacity.
3. Venous-to-arterial shunts (intrapulmonary or intracardiac), the blood thus by-passing the lungs.
4. Inadequate transport and delivery of oxygen.
   a. Anemia, abnormal hemoglobin.
   b. General circulatory deficiency.
   c. Localized circulatory deficiency (peripheral, cerebral, coronary vessels).
5. Inadequate tissue oxygenation or oxygen use.
   a. Tissue edema.
   b. Abnormal tissue demand.
   c. Poisoning of cellular enzymes.

This classification of the different types of hypoxia is mainly self-evident from previous discussions. Only one of the types of hypoxia needs further elaboration; this is inadequate tissue oxygenation.

Inadequate Tissue Oxygenation. The most classic type of inadequate tissue oxygenation is caused by cyanide poisoning in which the action of cytochrome oxidases in the cells is completely blocked—to such an extent that the tissues simply cannot utilize the oxygen even though plenty is available. This type of hypoxia is frequently also called *histotoxic hypoxia.*

A common type of tissue hypoxia also occurs in tissue edema, which causes increased distances through which the oxygen must diffuse before it can reach the cells. This type of hypoxia can become so severe that the tissues in edematous areas actually die, as is often illustrated by serious ulcers in edematous skin.

Finally, tissues can become hypoxic when the cells themselves demand more oxygen than can be made available to them by the normal respiratory and oxygen transport systems. For instance, in strenuous exercise one of the major limiting factors to the degree of exercise that can be performed is the tissue hypoxia that develops.

Effects of Hypoxia on the Body. Hypoxia, if severe enough, can actually cause death of the cells, but in less severe degrees it results principally in (1) depressed mental activity, sometimes culminating in coma, and (2) reduced work capacity of the muscles. These effects are discussed in the following chapter in relation to aviation medicine and, therefore, are only mentioned here.

CYANOSIS

The term "cyanosis" means blueness of the skin, and its cause is excessive amounts of deoxygenated hemoglobin in the skin blood vessels, especially in the capillaries. This deoxygenated hemoglobin has an intense dark blue color that is transmitted through the skin.

One of the most important factors determining the degree of cyanosis is the *quantity of deoxygenated hemoglobin in the arterial blood.* It is not the percentage deoxygenation of the hemoglobin that causes the bluish hue of the skin, but principally the *concentration of deoxygenated hemoglobin without regard to the concentration of oxygenated hemoglobin*. The reason for this is that the red color of oxygenated blood is weak in comparison with the dark blue color of deoxygenated blood. Therefore, when the two are mixed together, the oxygenated blood has relatively little coloring effect in comparison with that of the deoxygenated blood. In general, definite cyanosis appears whenever the arterial blood contains more than 5 grams per cent of deoxygenated hemoglobin.

Another important factor that affects the degree of cyanosis is the *rate of blood flow through the skin*, because this determines the amount of deoxygenation that occurs as the blood passes through the capillaries. Ordinarily, the metabolism of the skin is relatively low so that little deoxygenation occurs as the blood passes through the skin capillaries. However, if the blood flow becomes extremely sluggish, even a low metabolism can cause marked desaturation of the blood and therefore can cause cyanosis. This explains the cyanosis that appears in very cold weather, particularly in children who have thin skins.

A final factor that affects the blueness of the skin is *skin thickness*. For instance, in newborn babies, who have very thin skin, cyanosis occurs readily, particularly in highly vascular portions of the body, such as the heels. Also, in adults the lips and fingernails often appear cyanotic before the remainder of the body shows any blueness.

HYPERCAPNIA

Hypercapnia means excess carbon dioxide in the body fluids and especially refers to excess carbon dioxide at the cellular level.

One might suspect at first thought that any respiratory condition that causes hypoxia also causes hypercapnia. However, hypercapnia usually occurs in association with hypoxia only when the hypoxia is caused by *hypoventilation* or by *circulatory deficiency*. The reasons for this are the following:

Obviously, hypoxia caused by *too little oxygen in the air*, by *too little hemoglobin*, or by *poisoning of the oxidative enzymes* has to do only with the availability of oxygen or use of oxygen by the tissues. Therefore, it is readily understandable that hypercapnia is *not* a concomitant of these types of hypoxia.

Also, in hypoxia resulting from poor diffusion through the pulmonary membrane or through the tissues, serious hypercapnia usually does not occur because carbon dioxide diffuses 20 times as rapidly as oxygen. Therefore, even when oxygen diffusion is greatly depressed, carbon dioxide diffusion usually is still sufficient for adequate carbon dioxide transfer.

However, in hypoxia caused by hypoventilation, including hypoventilation of only some of the alveoli (abnormal ventilation-perfusion ratio), carbon dioxide transfer between the alveoli and the atmosphere is affected as much as is oxygen transfer. Then hypercapnia always results along with hypoxia. And in circulatory deficiency, diminished flow of blood decreases the removal of carbon dioxide from the tissues, resulting in tissue hypercapnia. However, the transport capacity of the blood for carbon dioxide is about three times that for oxygen, so that even here the hypercapnia is much less than the hypoxia.

EFFECTS OF HYPERCAPNIA ON THE BODY

When the alveolar P_{CO_2} rises above approximately 60 to 75 mm. Hg, dyspnea usually becomes intolerable, and as the P_{CO_2} rises to 70 to 80 mm. Hg, the person becomes lethargic and sometimes even semi-comatose. Total anesthesia and death result when the P_{CO_2} rises to 100 to 150 mm. Hg.

DYSPNEA

Dyspnea means primarily a desire for air or mental anguish associated with the effort to ventilate enough to satisfy the air demand. A common synonym is "air hunger."

At least three different factors often enter into the development of the sensation of dyspnea. These are: (1) abnormality of the respiratory gases in the body fluids, especially hypercapnia and to a much less extent hypoxia, (2) the amount of work that must be performed by the respiratory muscles to provide adequate ventilation, and (3) the state of the mind itself.

At times, the levels of both carbon dioxide and oxygen in the body fluids are completely normal, but to attain this the person has to breathe forcefully. In these instances the forceful activity of the respiratory muscles themselves gives the person a sensation of air hunger. Indeed, the dyspnea can be so intense, despite normal gaseous concentrations in the body fluids, that clinicians often overemphasize this cause of dyspnea and forget that dyspnea can also result purely and simply from abnormal gaseous concentrations.

Also, the person's respiratory functions may be completely normal, and still he experiences dyspnea because of an abnormal state of mind. This is called *neurogenic dyspnea* or, sometimes, *emotional dyspnea*. For instance, almost anyone momentarily thinking about his act of breathing will suddenly find himself taking breaths a little more deeply than ordinarily because of a feeling of mild dyspnea. This feeling is greatly enhanced in persons who have a psychic fear of not being able to receive a sufficient quantity of air. For example, many persons on entering small or crowded rooms immediately experience emotional dyspnea, and patients with "cardiac neurosis" who have heard that dyspnea is associated with heart failure frequently experience severe psychic dyspnea even though the blood gases are completely normal. Neurogenic dyspnea has been known to be so intense that the person over-respires and causes alkalotic tetany.

OXYGEN THERAPY IN THE DIFFERENT TYPES OF HYPOXIA

Oxygen therapy can be administered by (1) placing the patient's head in a "tent" which contains air fortified with oxygen, (2) allowing the patient to breathe either pure oxygen or high concentrations of oxygen from a mask, or (3) administering oxygen through an intranasal tube.

Oxygen therapy is of great value in certain types of hypoxia but of almost no value at all in other types. However, if one will simply recall the basic physiological principles of the different types of hypoxia, he can readily decide when oxygen therapy is of value and, if so, how valuable. For instance, oxygen therapy is very beneficial in:

1. *Atmospheric hypoxia.*
2. *Hypoventilation hypoxia.*
3. *Hypoxia caused by decreased diffusing capacity.*

Therapy is of moderate benefit in:

1. *Anemia.*
2. *Carbon monoxide poisoning.*

It is of slight benefit in *hypoxia caused by circulatory failure.*

It is of almost no value in *hypoxia caused by inability of the tissues to utilize oxygen,* such as in cyanide poisoning.

DANGER OF HYPERCAPNIA DURING OXYGEN THERAPY

Much of the stimulus that helps to maintain ventilation in hypoxia results from hypoxic stimulation of the aortic and carotid chemoreceptors. Earlier in this chapter it was noted that in chronic hypoxia, oxygen lack becomes a far more powerful stimulus to respiration than usual, sometimes increasing the ventilation as much as 5- to 7-fold. Therefore, during oxygen therapy, relief of the hypoxia occasionally causes pulmonary ventilation to decrease so low that lethal levels of hypercapnia develop. For this reason, oxygen therapy in hypoxia is sometimes contraindicated.

PHYSIOLOGY OF SPECIFIC PULMONARY DISEASES

TUBERCULOSIS

In tuberculosis the tubercle bacilli cause a peculiar tissue reaction in the lungs including, first, invasion of the infected region by macrophages and, second, walling off of the lesion by fibrous tissue to form the so-called "tubercle." This walling-off process helps to limit further transmission of the tubercle bacilli in the lungs and, therefore, is part of the protective process against the infection. However, in approximately 3 per cent of all persons who contract tuberculosis (perhaps half the population becomes infected at some time or other), the walling-off process

fails and tubercle bacilli spread throughout the lungs, causing fibrotic tubercles in many areas. In the late stages of tuberculosis, secondary infection by other bacteria causes extensive destruction of the lungs. Thus, tuberculosis causes many areas of fibrosis throughout the lungs, and it reduces the total amount of functional lung tissue. These effects cause (1) increased effort on the part of the respiratory muscles to cause pulmonary ventilation and therefore *reduced vital capacity and maximum breathing capacity*; (2) *reduced total respiratory membrane surface* and *increased thickness of the respiratory membrane*, these causing progressively diminished pulmonary diffusing capacity; and (3) *abnormal ventilation-perfusion ratio* in the lungs, further reducing the pulmonary diffusing capacity—that is, some areas of the lungs have too little blood flow even though other areas might be overperfused, and when alveoli are overperfused, there is not enough oxygen to oxygenate the blood.

PNEUMONIA

The term pneumonia describes any lung condition in which the alveoli become filled with fluid or blood cells, or both, as shown in Figure 29–7. Thus the filling of the alveoli with fluid in severe pulmonary edema is a type of pneumo-

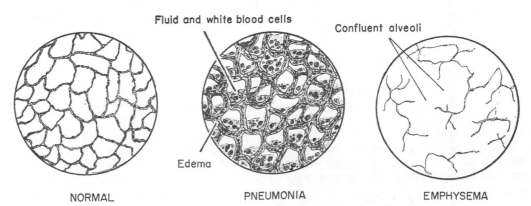

Figure **29–7** Pulmonary changes in pneumonia and emphysema.

nia, but another even more common type of pneumonia is *bacterial pneumonia* caused most frequently by pneumococci. This disease begins with infection in the alveoli of one part of the lungs; the alveolar membrane becomes edematous and highly porous so that fluid and often even red and white blood cells pass out of the blood into the alveoli. Thus, the infected alveoli become progressively filled with fluid and cells, and the infection spreads by extension of bacteria from alveolus to alveolus. Eventually, large areas of the lungs, sometimes whole lobes or even a whole lung, become "consolidated," which means that they are filled with fluids and cellular debris.

The pulmonary function of the lungs during pneumonia changes in different stages of the disease. In the early stages, the pneumonia process might well be localized to only one lung, and alveolar ventilation may be reduced even though blood flow through the lung continues normally. This results in two major pulmonary abnormalities: (1) reduction in the total available surface area of the respiratory membrane and (2) abnormal ventilation-perfusion ratio — that is, some areas of the lungs do not receive an adequate amount of ventilatory air to aerate the blood. Both these effects cause reduced diffusing capacity, which results in hypoxemia. Figure 29–8 illustrates the effect of an abnormal ventilation-perfusion ratio in pneumonia, showing that the blood passing through the aerated lung becomes 97 per cent saturated while that passing through the unaerated lung remains only 60 per cent saturated, causing the mean saturation of the aortic blood to be about 78 per cent, which is far below normal.

In other stages of pneumonia the blood flow through the diseased areas decreases concurrently with the decrease in ventilation. This gives much less debility than that resulting when the blood flow through the unventilated lung is normal.

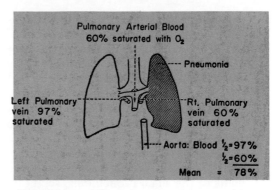

FIGURE 29–8 Effect of pneumonia on arterial blood oxygen saturation.

ATELECTASIS

Atelectasis means collapse of the alveoli. It can occur in a localized area of a lung, in an entire lobe, or in an entire lung. It has three causes: (1) obstruction of the airway followed by absorption of the air in the blocked alveoli, (2) external compression of the lung, or (3) lack of surfactant in the fluids lining the alveoli, thus causing such high surface tension of the fluids that they collapse the alveoli.

Atelectasis not only occludes the alveoli but also increases the resistance to blood flow through the pulmonary vessels as much as five-fold. Most of this resistance increase occurs instantaneously with the collapse because the collapse compresses and folds the vessels as the volume of the lung decreases. Hypoxia in the collapsed alveoli might also cause additional vasoconstriction, but this occurs slowly, as was explained in Chapter 15.

Because of the vascular constriction, blood flow through the atelectatic lung becomes slight. Essentially all the blood is routed through the ventilated lung and therefore becomes well aerated. For instance, following massive collapse of an entire lung (atelectasis), as much as five-sixths of the blood might pass through the aerated lung and only one-sixth through the unaerated lung. As a result, the aortic blood becomes almost fully saturated with oxygen despite the collapse of the entire lung.

BRONCHIAL ASTHMA

Bronchial asthma is usually caused by allergic hypersensitivity of the person to foreign substances in the air—especially to plant pollens. The allergic reaction causes localized edema in the walls of the small bronchioles, secretion of thick mucus into their lumens, and spasm of their smooth muscle walls. These effects greatly increase the airway resistance. The bronchiolar diameter becomes even further reduced during expiration, because expiratory compression of the lungs compresses the bronchioles as well as the alveoli. This causes the functional residual capacity and the residual volume of the lung to become greatly increased during the asthmatic attack because of the difficulty in expiring air from the lungs. Over a long period of time the chest cage becomes permanently enlarged, causing a "barrel chest," and the functional residual capacity and residual volume also become permanently increased.

CHRONIC EMPHYSEMA

A disease called chronic emphysema is becoming highly prevalent, mainly because of the effects of smoking. It resu ts from two major pathophysiological changes in the lungs. First, air flow through many of the terminal bronchioles is obstructed. Second, much of the lung parenchyma is destroyed.

Many clinicians believe that chronic emphysema begins with chronic infection in the lung that causes *bronchiolitis*, which means inflammation of the small air passages of the lungs. At any rate, many bronchioles become irreparably obstructed, and the total surface of the respiratory membrane becomes greatly decreased as illustrated in Figure 29–8, sometimes to less than one-quarter

normal. Among the physiological effects of chronic emphysema are:

1. *Increased airway resistance* with greatly increased work of breathing.

2. *Greatly decreased diffusing capacity* of the lung, which reduces the ability of the lungs to oxygenate the blood.

3. *Pulmonary hypertension* because of decreased number of pulmonary capillaries through which blood can pass; this in turn overloads the right heart and frequently causes right-heart failure.

Chronic emphysema usually progresses slowly over many years. The person develops hypoxia and hypercapnia because of hypoventilation of the alveoli and because of loss of lung parenchyma. The net result of all these effects is prolonged, severe air hunger that can last for years until the hypoxia and hypercapnia cause death.

REFERENCES

American Physiological Society: Handbook of Physiology. Sec. III, Respiration. Vol. I, 1964; Vol. II, 1965. Baltimore, The Williams & Wilkins Co.

Bendixen, H., et al.: Respiratory Care. 2nd ed., St. Louis, The C. V. Mosby Co., 1969.

Burns, B. D.: The central control of respiratory movements. *Brit. Med. Bull., 19*:7, 1963.

Comroe, J. H., Jr.: The peripheral chemoreceptors. *In* Handbook of Physiology. Baltimore, The Williams & Wilkins Co., 1964, Sec. III, Vol. I, p. 557.

Comroe, J. H., Jr., Forster, R. E., II, Dubois, A. B., Briscoe, W. A., and Carlsen, E.: The Lung: Clinical Physiology and Pulmonary Function Tests. 2nd ed., Chicago, Year Book Medical Publishers, Inc., 1962.

Cotes, J. E.: Lung Function: Assessment and Application in Medicine. Philadelphia, F. A. Davis Co., 1965.

Cunningham, D. J. C., and Lloyd, B. B. (eds.): The Regulation of Human Respiration. Philadelphia, F. A. Davis Co., 1963.

Filley, G.: Pulmonary Insufficiency and Respiratory Failure. Philadelphia, Lea & Febiger, 1967.

Milhorn, H. T., Jr., Benton, R., Ross, R., and Guyton, A. C.: A mathematical model of the human respiratory control system. *Biophys. J., 5*:27, 1965.

Mitchell, R. A.: Respiration. *Ann. Rev. Physiol., 32*:415, 1970.

CHAPTER 30

AVIATION, HIGH ALTITUDE, SPACE, AND UNDERSEA PHYSIOLOGY

As man has ascended to higher and higher altitudes in aviation, in mountain climbing, and in space vehicles, it has become progressively more important to understand the effects of altitude and low environmental gas pressures on the human body. In the early days of aviation only two factors were of concern: (1) the effects of hypoxia on the body and (2) the effects of physical factors of high altitude, such as temperature and ultraviolet radiation. When airplanes were made to go still higher and to make sharper turns, the problems of hypoxia became compounded, and, in addition, it was soon learned that airplanes could be built to withstand acceleratory forces far greater than the human body can stand. Now, with the space age at hand, all these problems have become multiplied to the point that the physical conditions in the spacecraft must be created artificially.

The present chapter deals with all these problems: first, the hypoxia at high altitudes, second, the other physical factors affecting the body at high altitudes, and, third, the tremendous acceleratory forces that occur in both aviation and space physiology.

EFFECTS OF LOW OXYGEN PRESSURE ON THE BODY

Alveolar Po_2 at Different Elevations. Obviously, when the Po_2 in the atmosphere decreases at higher elevations, a decrease in alveolar Po_2 is also to be expected. At low altitudes the alveolar Po_2 does not decrease quite so much as the atmospheric Po_2 because increased pulmonary ventilation helps to compensate for the diminished atmospheric oxygen. But at higher altitudes the alveolar Po_2 decreases even more than atmospheric Po_2 for peculiar reasons that are explained as follows:

Effect of carbon dioxide and water vapor on alveolar oxygen. Even at high altitudes carbon dioxide is continually excreted from the pulmonary blood into the alveoli. Also, water vaporizes into the alveolar space from the respiratory surfaces. Therefore, these two gases dilute the oxygen and nitrogen already in the alveoli, reducing the amount of space available in the alveoli for oxygen.

The presence of carbon dioxide and water vapor in the alveoli becomes exceedingly important at high altitudes because the total barometric pressure falls to low levels while the pressures of carbon dioxide and water vapor do not fall comparably. For instance, water vapor pressure remains 47 mm. Hg as long as the body temperature is normal, regardless of altitude; and the pressure of carbon dioxide falls from about 40 mm. Hg at sea level to about 24 mm. Hg at extremely high altitudes because of increased respiration.

Now let us see how the pressures of these two gases affect the available space for oxygen. Let us assume that the total barometric pressure falls to 100 mm. Hg; 47 mm. Hg of this must be water vapor, leaving only 53 mm. Hg for all the other gases. Under acute exposure to high altitude, 24 mm. Hg of the 53 mm. Hg must be carbon dioxide, leaving a remaining space of only 29 mm. Hg. If there were no use of oxygen by the body, one-fifth of this 29 mm. Hg would be

oxygen, and four-fifths would be nitrogen; or, the P_{O_2} in the alveoli would be 6 mm. Hg. However, by this time the person's tissues would be almost totally anoxic, for which reason even this last vestige of oxygen would be absorbed into the blood, leaving not more than 1 mm. Hg oxygen pressure in the alveoli. Therefore, at a barometric pressure of 100 mm. Hg, the person could not possibly survive. This effect is somewhat different if the person is breathing pure oxygen, as we shall see in the following discussions.

A simple formula for calculating alveolar P_{O_2} is the following:

$$\text{Alveolar } P_{O_2} = \frac{P_B - P_{CO_2} - 47}{5} - P_{O_2} \text{ LOSS}$$

In this formula P_B is barometric pressure, the value 47 is the vapor pressure of water, and P_{O_2} LOSS is the oxygen pressure decrease caused by oxygen uptake into the blood.

Table 30–1 shows the P_{O_2} in the alveoli at different elevations when one is breathing air and when one is breathing pure oxygen. When breathing air, the alveolar P_{O_2} is 104 mm. Hg at sea level; it falls to approximately 67 mm. Hg at 10,000 feet and to only 1 mm. Hg at 50,000 feet.

Saturation of Hemoglobin with Oxygen at Different Altitudes. Figure 30–1 illustrates arterial oxygen saturation at different altitudes when breathing air and when breathing oxygen, and the

TABLE 30–1   EFFECTS OF LOW ATMOSPHERIC PRESSURES ON ALVEOLAR GAS CONCENTRATIONS AND ON ARTERIAL OXYGEN SATURATION

Altitude (ft.)	Barometric Pressure (mm. Hg)	P_{O_2} in Air (mm. Hg)	Breathing Air			Breathing Pure Oxygen		
			P_{CO_2} in Alveoli (mm. Hg)	P_{O_2} in Alveoli (mm. Hg)	Arterial Oxygen Saturation (%)	P_{CO_2} in Alveoli (mm. Hg)	P_{O_2} in Alveoli (mm. Hg)	Arterial Oxygen Saturation (%)
0	760	159	40	104	97	40	673	100
10,000	523	110	36	67	90	40	436	100
20,000	349	73	24	40	70	40	262	100
30,000	226	47	24	21	20	40	139	99
40,000	141	29	24	8	5	36	58	87
50,000	87	18	24	1	1	24	16	15

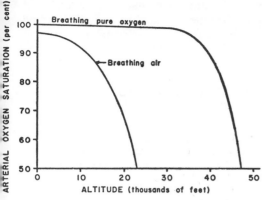

FIGURE 30-1 Effect of low atmospheric pressure on arterial oxygen saturation when breathing air and when breathing pure oxygen.

actual per cent saturations at certain altitudes are given in Table 30–1. Up to an elevation of approximately 10,000 feet, even when breathing air, the arterial oxygen saturation remains at least as high as 90 per cent. However, above 10,000 feet the arterial oxygen saturation falls progressively, as illustrated by the left-hand curve of the figure, until it is only 70 per cent at 20,000 feet elevation and still less at higher altitudes.

Effect of Breathing Pure Oxygen on the Alveolar Po₂ at Different Altitudes. If an aviator breathes pure oxygen instead of air, most of the space in the alveoli formerly occupied by nitrogen now becomes occupied by oxygen instead. For instance, at 30,000 feet the aviator could have an alveolar Po_2 of 139 mm. Hg instead of the 21 mm. Hg that he has when he breathes air.

The second curve of Figure 30–1 illustrates the arterial oxygen saturation at different altitudes when one is breathing pure oxygen. Note that the saturation remains above 90 per cent until the aviator ascends to approximately 39,000 feet; then it falls rapidly to approximately 50 per cent at about 47,000 feet.

The "Ceiling" when Breathing Air and when Breathing Oxygen in an Unpressurized Airplane. Comparing the two arterial oxygen saturation curves in Figure 30–1, one notes that an aviator breathing oxygen can ascend to far higher altitudes than one not breathing oxygen. For instance, the arterial saturation at 47,000 feet when breathing oxygen is about 50 per cent and is equivalent to the arterial oxygen saturation at 23,000 feet when breathing air. And, because a person ordinarily can remain conscious until the arterial oxygen saturation falls to 40 to 50 per cent, the ceiling for an unacclimatized aviator in an unpressurized airplane when breathing air is approximately 23,000 feet and when breathing pure oxygen about 47,000 feet, provided the oxygen-supplying equipment operates perfectly.

Effects of Hypoxia. Beginning at an elevation of approximately 12,000 feet, a person often develops symptoms of drowsiness, lassitude, mental fatigue, sometimes headache, occasionally nausea, and sometimes euphoria. Most of these symptoms increase in intensity at still higher altitudes, the headache often becoming especially prominent and the cerebral symptoms sometimes progressing to the stage of twitchings or convulsions and ending, above 23,000 feet, in coma.

One of the most important effects of hypoxia is decreased judgment. For instance, if an aviator stays at 15,000 feet for one hour without supplementary oxygen, his mental proficiency ordinarily will have fallen to approximately 50 per cent below normal; and after 18 hours at this level, to approximately 80 per cent below normal.

ACCLIMATIZATION TO LOW Po₂

If a person remains at high altitudes for days, weeks, or years, he gradually becomes acclimatized to the low Po_2, so that it causes fewer and fewer deleterious effects to his body and also so that it becomes possible for him to work harder or to ascend to still higher altitudes. The five principal means by which acclimatization comes about are: (1) increased pulmonary ventilation, (2) increased hemoglobin and increased blood volume, (3) increased diffusing capacity of the lungs because of more blood in the capillaries, (4) increased vascularity of the tissues, and (5) increased ability of the cells to utilize oxygen despite the low P_{O_2}.

Cause of the Increased Pulmonary Ventilation. On immediate exposure to low P_{O_2}, the hypoxic stimulation of the chemoreceptors increases alveolar ventilation to a maximum of about 65 per cent. Then, if he remains at a very high altitude for several days, his ventilation gradually increases to as much as five to seven times normal. The basic cause of this gradual increase is the following:

The immediate 65 per cent increase in pulmonary ventilation on rising to a high altitude blows off large quantities of carbon dioxide, reducing the P_{CO_2} and increasing the pH of the body fluids. Both of these changes *inhibit* the respiratory center and thereby *oppose the stimulation by the hypoxia.* However, during the ensuing three to five days, this inhibition fades away, allowing the respiratory center now to respond with full force to the chemoreceptor stimuli resulting from hypoxia, and the ventilation increases to about five to seven times normal.

EFFECTS OF ACCELERATORY FORCES ON THE BODY IN AVIATION AND SPACE PHYSIOLOGY

Because of rapid changes in velocity and direction of motion in airplanes and spacecraft, several types of acceleratory forces often affect the body during flight. At the beginning of flight, simple linear acceleration occurs; at the end of flight, deceleration; and, every time the airplane turns, angular acceleration occurs. In aviation physiology it is usually angular acceleration that demands greatest consideration, because the structure of the airplane is capable of withstanding much greater angular acceleration than is the human body.

ANGULAR ACCELERATORY FORCES

When an airplane makes a turn, the force of angular acceleration is determined by the following relationship:

$$f = \frac{mv^2}{r}$$

in which f is the angular acceleratory force, m is the mass of the object, v is the velocity of travel, and r is the radius of curvature of the turn. From this formula it is obvious that as the velocity increases, the force of angular acceleration increases in proportion to the square of the velocity. It is also obvious that the force of acceleration is directly proportional to the sharpness of the turn.

Measurement of Angular Acceleratory Force — "G." When a subject is simply sitting in his seat, the force with which he is pressing against the seat results from the pull of gravity, and it is equal to his weight. The intensity of this force is 1 "G" because it is equal to the pull of gravity. If the force with which he presses against his seat becomes five times his normal weight during a pull-out from a dive, the force acting upon the seat is 5 G.

If the airplane goes through an outside loop so that the pilot is held down by his seat belt, *negative G* is applied to his body, and, if the force with which he is thrown against his belt is equal to the weight of his body, the negative force is −1 G.

Effects of Angular Acceleratory Force on the Body. *Effects on the circulatory system.* The most important effect of angular acceleration is on the circulatory system because blood is mobile and can be translocated by angular forces. Angular forces also tend to displace the tissues, but, because of their more solid structure, they only sag—ordinarily not enough to cause abnormal function.

When the aviator is subjected to *positive G* his blood is centrifuged toward the lower part of his body. Thus, if the angular acceleratory force is 5 G and the subject is in a standing position, the hydrostatic pressure in the veins of the feet is five times normal, or approximately 450 mm. Hg, and even in the sitting position this pressure is nearly 300 mm. Hg. As the pressure in the vessels of the lower part of the body increases, the vessels passively dilate, and a major proportion of the blood from the upper part of the body is translocated into these lower vessels. Because the heart cannot pump unless blood returns to it, the greater the quantity of blood "pooled" in the lower body, the less becomes the cardiac output.

Positive G diminishes blood flow to the brain. Acceleration greater than 4 to 6 G ordinarily causes "blackout" of vision within a few seconds and then unconsciousness shortly thereafter.

Transverse G. The human body can withstand tremendous transverse acceleratory forces (forces applied along the anteroposterior axis of the body—lying down in the airplane, for instance). If the transverse acceleratory forces are applied rather uniformly over large areas of the body, as much as 100 G can be withstood for a fraction of a second, and as much as 15 to 25 G can be withstood for many seconds without serious effects other than occasional collapse of a lung, which is not a lethal effect. Therefore, when very large acceleratory forces are involved, the aviator or astronaut flies in the semi-reclining or lying position.

Protection of the Body Against Angular Acceleratory Forces. Specific procedures and apparatus have been developed to protect aviators against the circulatory collapse that occurs during positive G. First, if the aviator tightens his abdominal muscles to an extreme degree and leans forward to compress his abdomen, he can prevent some of the pooling of blood in the large vessels of the abdomen, thereby delaying the onset of blackout. Also, special "anti-G" suits have been devised to prevent pooling of blood in the lower abdomen and legs. The simplest of these applies positive pressure to the legs and abdomen by inflating compression bags as the G increases.

EFFECTS OF LINEAR ACCELERATORY FORCES ON THE BODY

Acceleratory Forces in Space Travel. In contrast to aviation physiology, the spacecraft cannot make rapid turns; therefore, angular acceleration is of little importance except when the spacecraft goes into abnormal gyrations. On the other hand, takeoff acceleration and landing deceleration might be tremendous; both of these are types of linear acceleration.

Figure 30–2 illustrates a typical profile of the acceleration during take-off in a three-stage spacecraft, showing that the first stage booster causes acceleration as high as 9 G and the second stage booster as high as 8 G. Also, similar degrees of deceleratory forces occur when the spacecraft re-enters the atmos-

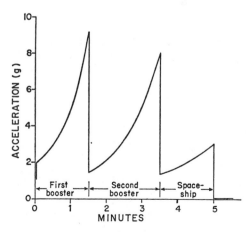

FIGURE 30–2 Acceleratory forces during the take-off of a spacecraft.

phere. In the standing position the human body could not withstand this much acceleration, but in a semi-reclining position *transverse to the axis of acceleration*, this amount of acceleration can be withstood with ease despite the fact that the acceleratory forces continue for as long as five minutes at a time. Therefore, we see the reason for the reclining seats used by the astronauts.

Deceleratory Forces Associated with Parachute Jumps. When the parachuting aviator leaves the airplane, his velocity of fall is exactly 0 feet per second at first. However, because of the acceleratory force of gravity, within 1 second his velocity of fall is 32 feet per second; in two seconds it is 64 feet per second, etc. However, as his rate of fall increases, the air resistance tending to slow his fall also increases. Finally, the air resistance exactly opposes the acceleratory force of gravity, so that by the time he has fallen for approximately 12 seconds and a distance of 1400 feet, he will be falling at a "terminal velocity" of 109 to 119 miles per hour (175 feet per second).

If the parachutist has already reached the terminal velocity of fall before he opens his parachute, an "opening shock load" of approximately 1200 pounds occurs on the parachute strands.

The usual size parachute slows the fall of the parachutist to approximately one-ninth the terminal velocity. In other words, the speed of landing is approximately 20 feet per second, and the force of impact against the earth is approximately $1/81$ the force of impact without a parachute. Even so, the force of impact is still approximately the same as that which would be experienced from jumping from a height of about 7 feet. If the parachutist is not careful, his senses will allow him to strike the earth with his legs still extended, and this will result in tremendous deceleratory forces along the skeletal axis of his body, resulting in fracture of the pelvis, of a vertebra, or a leg.

"ARTIFICIAL CLIMATE" IN THE SEALED SPACECRAFT

Since there is no atmosphere in outer space, an artificial atmosphere and other artificial conditions of climate must be provided. The ability of a person to survive in this artificial climate depends entirely on appropriate engineering design.

Most important of all, the oxygen concentration must remain high enough and the carbon dioxide concentration low enough for healthy survival. The gas mixture most suitable for spacecraft travel is one of almost pure oxygen but also having just enough nitrogen to prevent fire hazards. The total pressure is maintained as low as 400 mm. Hg. This allows more than adequate oxygenation of the astronauts, and yet the low pressure minimizes the hazard of explosive decompression of the spacecraft.

For space travel lasting more than several weeks, it will be impractical to carry along an adequate oxygen supply and enough carbon dioxide absorbent. For this reason, "recycling techniques" have been developed for continual re-use of oxygen. These techniques also frequently include re-use of the same food and water. Basically, they involve (1) a method for removing oxygen from carbon dioxide, (2) a method for removing water from the human excreta, and (3) use of the human excreta for resynthesizing or regrowing an adequate food supply.

Large amounts of energy are required for these processes, and the real problem at present is to derive enough energy from the sun's radiation to energize the necessary chemical reactions. Some recycling processes depend on purely physical procedures such as distillation, electrolysis of water, and capture of the sun's energy by solar batteries, whereas others depend on biologic methods, such as use of algae with its large store of chlorophyll to generate foodstuffs by photosynthesis. Unfortunately, a completely practical system for recycling is yet to be achieved. The problem is the weight of the equipment that must be carried.

WEIGHTLESSNESS IN SPACE

A person in an orbiting satellite or in any nonpropelled spacecraft experiences a feeling of weightlessness. That is, he is not drawn toward the bottom, sides, or top of the spacecraft but simply floats inside its chambers. The cause of this is not failure of gravity to pull on the body, because gravity from any nearby heavenly body is still active. However, the gravity acts on both the spacecraft and the person at the same time, and since there is no resistance to movement in space, both are pulled with exactly the same acceleratory forces and in the same direction. For this reason, the person simply is not attracted toward any wall of the spacecraft.

Weightlessness fortunately has not proved to be a physiological problem; rather it is an engineering problem to provide special techniques for eating and drinking (since food and water will not stay in open plates or glasses), special waste disposal systems, and adequate hand holds or other means for stabilizing the person in the spacecraft so that he can adequately control the ship.

RADIATION HAZARDS IN SPACE PHYSIOLOGY

Lethal radiations occur at altitudes between 300 and 20,000 miles from the earth and therefore must be considered in planning space voyages. The cause of this problem is that large quantities of cosmic particles continually bombard the earth, some originating from the sun and some from outer space. The magnetic field of the earth traps many of these cosmic particles in two major belts around the earth called *Van Allen radiation belts*, illustrated in Figure 30–3. The inner belt begins at an altitude

FIGURE 30–3 The Van Allen belts of radiation around the earth. (From Newell: *Science 131*:385, 1960.)

of about 300 miles and extends to about 3000 miles. The outer belt begins at about 6000 miles and extends to 20,000 miles. As also illustrated in Figure 30–3, the inner belt extends only 30 degrees on each side of the equatorial plane, and the outer belt 70 to 75 degrees.

The types of radiation in the two Van Allen belts are almost entirely high energy electrons and protons, the outer belt being comprised almost entirely of electrons and the inner belt of both protons and electrons. The energy level in the inner belt is extremely high, many of the particles having energies as high as 40 Mev., which makes it almost impossible to shield a spacecraft adequately against this radiation. Even with best possible shielding, a person traversing these two belts in an interplanetary space trip would be expected to receive up to 10 roentgens of radiation, which is about one-fortieth the lethal dose; and a person in a spacecraft orbiting the earth within one of these two belts could receive enough radiation in only a few hours to cause death. During solar flares, the intensity of the outer Van Allen belt increases tremendously, which could make space travel through this belt at this time extremely dangerous.

Thus, it is important to orbit spacecraft below an altitude of 200 to 300 miles, an altitude at which the radiation hazard is slight. Also, it is possible to minimize the radiation hazard during interplanetary space travel by leaving the earth or returning to earth near one of the poles rather than near the equator.

PHYSIOLOGY OF DEEP SEA DIVING AND OTHER HIGH PRESSURE OPERATIONS

When a person descends beneath the sea, the pressure around him increases tremendously. To keep his lungs from collapsing, air must be supplied to him also under high pressure, which exposes the blood in his lungs to extremely high alveolar gas pressures. Beyond certain limits these high pressures can cause tremendous alterations in the physiology of the body, often resulting in serious problems as follows:

Nitrogen Narcosis. Approximately four-fifths of the air is nitrogen. At sea level pressure this has no known effect on bodily function, but at high pressures it can cause varying degrees of narcosis. At 150 to 200 feet depths, the diver becomes drowsy. At 200 to 250 feet, his strength wanes considerably, and he often becomes too clumsy to perform the work required of him. Beyond 300 feet (10 atmospheres pressure), the diver usually becomes almost useless as a result of nitrogen narcosis, and he becomes unconscious at 350 to 400 feet.

Nitrogen narcosis has characteristics very similar to those of alcohol intoxication, and for this reason it has frequently been called "raptures of the depths."

The mechanism of the narcotic effect is believed to be the same as that of essentially all the gas anesthetics. That is, nitrogen dissolves freely in the fats of the body, and it is presumed that it, like most other anesthetic gases, dissolves in the membranes or other lipid structures of the neurons and thereby reduces their excitability.

Oxygen Toxicity at High Pressures. Breathing oxygen under very high partial pressures can be detrimental to the central nervous system, sometimes resulting in epileptic convulsions followed by coma. Indeed, exposure to 3 atmospheres pressure of oxygen or 15 atmospheres of air ($Po_2 = 2280$ mm. Hg) will cause convulsions and coma in most persons after about one hour. These convulsions often occur without any warning, and they obviously are likely to be lethal to a diver submerged beneath the sea.

Decompression of the Diver After Exposure to High Pressures. When a person breathes air under high pressure for a long time, the amount of nitrogen dissolved in his body fluids becomes great. The reason for this is the followng: The blood flowing through the pulmonary capillaries becomes saturated with nitrogen to the same pressure as that in the breathing mixture. Over several hours, nitrogen is carried to all the tissues of the body to saturate them also with dissolved nitrogen. After the diver has become totally saturated with nitrogen the *sea level volume of nitrogen* dissolved in his body fluids at the different depths is:

feet	liters
33	2
100	4
200	7
300	10

Decompression sickness (synonyms: compressed air sickness, bends, caisson disease, diver's paralysis, dysbarism). If a diver has been beneath the sea long enough so that large amounts of nitrogen dissolve in his body, and then he suddenly comes back to the surface of the sea, significant quantities of nitrogen bubbles can develop in his body fluids either intracellularly or extracellularly, and these can cause minor or serious damage in almost any area of the body, depending on the amount of bubbles formed.

The principles underlying this effect are shown in Figure 30–4. To the left,

the diver's tissues have become equilibrated to a very high nitrogen pressure. However, as long as the diver remains deep beneath the sea, the pressure against the outside of his body compresses all the body tissues sufficiently to keep the dissolved gases in solution. Then, when the diver suddenly rises to sea level, the pressure on the outside of his body becomes only 1 atmosphere (760 mm. Hg), while the pressure inside the body fluids is the sum of the pressures of water vapor, carbon dioxide, oxygen, and nitrogen, or a total of 4045 mm. Hg, which is far greater than the pressure on the outside of the body. Therefore, the gases can escape from the dissolved state and form actual bubbles inside the tissues.

In serious cases of decompression sickness, bubbles frequently disrupt important nerve pathways in the brain or spinal cord, and bubbles in the peripheral nerves can cause severe pain. Formation of large bubbles in the central nervous system occasionally leads to permanent paralysis or permanent mental disturbances.

Fortunately, if a diver is brought to the surface slowly, the dissolved nitrogen is eliminated through his lungs rapidly enough to prevent decompression sickness. After long periods at deep levels beneath the sea (200 feet deep for 2 hours, for instance), the time required to bring the diver to the surface or to decompress him safely in a decompression tank is often six hours or more.

In deep dives, helium instead of nitrogen is usually mixed with oxygen because of several properties of helium: (1) it has no narcotic effect; (2) only 40 per cent as much helium as nitrogen dissolves in the fluids; and (3) because of its small atomic size, helium diffuses through the tissues at a velocity about $2\frac{1}{2}$ times that of nitrogen.

Effect of Rapid Descent — "The Squeeze." On rapid descent, the volumes of all gases in the body become greatly reduced because of increasing pressure applied to the outside of the body. If

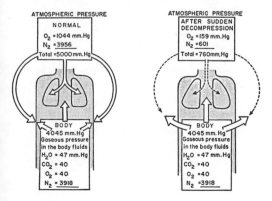

FIGURE 30–4 Gaseous pressures responsible for bubble formation in the body tissues.

additional quantities of air are supplied to the gas cavities during descent — especially to the lungs — no harm is done, but, if the person continues to descend without addition of gas to his cavities, the volume becomes greatly reduced, and serious physical damage results; this is called "the squeeze."

The most damaging effects of the squeeze occur in the lungs, for the smallest volume that the lungs can normally achieve is approximately 1.5 liters. Even if the diver inspires a maximal breath prior to descending, he can go down no farther than 100 feet before his chest begins to cave in. Therefore, to prevent lung squeeze, the diver must inspire additional air as he descends.

electrolyte composition of the plasma, plus anoxia, cause the heart to fibrillate within one to three minutes after the person first inhales the water, causing death much earlier than would have occurred had the water not been inhaled.

Effect of Inhaling Salt Water. If a drowning person inhales salt water instead of fresh water, osmosis of water occurs in the opposite direction through the pulmonary membrane because the total osmotic pressure of salt water is several times that of blood. Loss of water out of the blood causes marked hemoconcentration but not hemolysis or cardiac fibrillation. Instead, the person dies of asphyxia in five to eight minutes.

DROWNING

Asphyxia in Drowning. Approximately 30 per cent of all persons who drown do not inhale water, and even after recovery of the body from the water the lungs are still dry. These persons simply die from asphyxiation. The reason they do not inhale water is that water, in attempting to enter the trachea, elicits a powerful laryngeal reflex that causes spastic closure of the vocal cords.

Inhalation of Fresh Water — Cardiac Fibrillation. When a drowning person inhales fresh water, water is absorbed through the alveolar membrane into the blood extremely rapidly by osmosis because of the total osmotic pressure of the pulmonary capillary blood (which gives an absorption gradient of about 5400 mm. Hg). Several liters of water are absorbed into the blood within one to three minutes. This greatly dilutes the electrolytes of the blood and even causes hemolysis of many red cells, with spillage of large quantities of potassium into the plasma. The resulting changes in

REFERENCES

Andersen, H. T.: Physiological adaptations in diving vertebrates. *Physiol. Rev., 46*:212, 1966.

Armstrong, H. G.: Aerospace Medicine. Baltimore, The Williams & Wilkins Co., 1961.

Behnke, A. R., Jr., and Lanphier, E. H.: Underwater physiology. *In* Handbook of Physiology. Baltimore, The Williams & Wilkins Co., 1965, Sec. III, Vol. II, p. 1159.

Clamann, H. G.: Space physiology. *In* Handbook of Physiology. Baltimore, The Williams & Wilkins Co., 1965, Sec. III, Vol. II, p. 1147.

Haugaard, N.: Cellular mechanisms of oxygen toxicity. *Physiol. Rev., 48*:311, 1968.

Hurtado, A.: Animals in high altitudes: resident man. *In* Handbook of Physiology. Baltimore, The Williams & Wilkins Co., 1964, Sec. IV, p. 843.

Kellogg, R. H.: Altitude acclimatization, a historical introduction emphasizing the regulation of breathing. *Physiologist, 11*:37, 1968.

Luft, U. C.: Aviation physiology — the effects of altitude. *In* Handbook of Physiology. Baltimore, The Williams & Wilkins Co., 1965, Sec. III, Vol. II, p. 1099.

McCally, M.: Hypodynamics and Hypogravics, The Physiology of Inactivity and Weightlessness. New York, Academic Press, Inc., 1969.

Miles, S.: Underwater Medicine. 2nd ed., Philadelphia, J. B. Lippincott Co., 1966.

PART V

THE
NERVOUS SYSTEM
AND
SPECIAL SENSES

CHAPTER 31

BASIC ORGANIZATION OF THE NERVOUS SYSTEM; AND PROCESSING OF SIGNALS

The nervous system, along with the endocrine system, provides the control functions for the body. In general, the nervous system controls the rapid activities of the body such as muscular contractions, rapidly changing visceral events, and even the rates of secretion of some endocrine glands. The endocrine system regulates principally the metabolic functions of the body.

The purpose of this chapter is to present a general outline of the basic mechanisms by which the nervous system performs its functions, and succeeding chapters analyze in detail the functions of the individual parts of the nervous system. Before beginning this discussion, however, the reader should refer to Chapters 5 and 6, which present, respectively, the principles of membrane potentials and transmission of impulses through synapses.

GENERAL DESIGN OF THE NERVOUS SYSTEM

THE SENSORY DIVISION— SENSORY RECEPTORS

Most activities of the nervous system are originated by sensory experience emanating from *sensory receptors*, whether these be visual receptors, auditory receptors, tactile receptors on the surface of the body, or other kinds of receptors. This sensory experience can cause an immediate reaction, or its memory can be stored in the brain for minutes, weeks, or years and then can help to determine the bodily reactions at some future date.

Figure 31–1 illustrates a minute portion of the *somatic* sensory system, which transmits sensory information from the receptors of the entire surface of the body and deep structures. This information enters the nervous system through the spinal nerves and is conducted into (a) the spinal cord at all levels, (b) the reticular substance of the medulla, pons, and mesencephalon, (c) the cerebellum, (d) the thalamus, and (e) the somesthetic areas of the cerebral cortex. But in addition to these "primary sensory" areas, signals are then relayed to essentially all other segments of the nervous system.

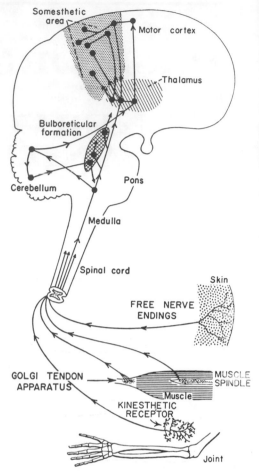

FIGURE 31–1 The somatic sensory axis of the nervous system.

THE MOTOR DIVISION — THE EFFECTORS

The most important ultimate role of the nervous system is control of bodily activities. This is achieved by controlling (a) contraction of skeletal muscles throughout the body, (b) contraction of smooth muscle in the internal organs, and (c) secretion by both exocrine and endocrine glands in many parts of the body. These activities are collectively called *motor functions* of the nervous system, and the muscles and glands are called *effectors* because they perform the functions dictated by the nerve signals. That portion of the nervous system directly concerned with transmitting signals to the muscles and glands is called the motor division of the nervous system.

Figure 31–2 illustrates the *motor axis* of the nervous system for controlling skeletal muscle contraction. Operating parallel with this axis is another similar system for control of the smooth muscles and glands; it is called the *autonomic nervous system*. Note in Figure 31–2 that the skeletal muscles can be controlled from many different levels of the nervous system, including (a) the spinal cord, (b) the substance of the medulla, pons, and mesencephalon, (c) the basal ganglia, and (d) the motor cortex. Each of these different areas plays its own specific role in the control of body move-

ments, the lower regions being concerned primarily with automatic, instantaneous responses of the body to sensory stimuli and the higher regions with deliberate movements controlled by the thought processes of the cerebrum.

PROCESSING OF INFORMATION

The nervous system would not be at all effective in controlling bodily functions if each bit of sensory information caused some motor reaction. Therefore, one of the major functions of the nervous system is to process incoming information in such a way that *appropriate* motor responses occur. Indeed, more than 99 per cent of all sensory information is continually discarded by the brain as unimportant. For instance, one is ordinarily totally unaware of the parts of his body that are in contact with his clothes and is also unaware of the pressure on his seat when he is sitting. Likewise, his attention is drawn only to an occasional object in his field of vision, and even the perpetual noise of his surroundings is usually relegated to the background.

After the important sensory information has been selected, it must be channeled into proper motor regions of the brain to cause the desired response. Thus, if a person places his hand on a hot stove, the desired response is to lift the hand, plus other associated responses, such as moving the entire body away from the stove and perhaps even

shouting with pain. Yet even these responses represent activity by only a small fraction of the total motor system of the body.

Function of the Synapse in Processing Information. Transmission of impulses from one neuron to another through the synapse was discussed in Chapter 6. It will be recalled that different synapses have widely varying properties: Some transmit signals from one neuron to the next with ease while others transmit signals only with difficulty. Some postsynaptic neurons respond with large numbers of impulses while others respond with only a few. Thus, the synapses perform a selective action, often blocking the weak signals while allowing the strong signals to pass, often selecting and amplifying certain weak signals, and often channeling the signal in many different directions rather than simply in one direction. The basic principles of this processing of information by the synapses are so important that they are discussed in detail in the following chapter.

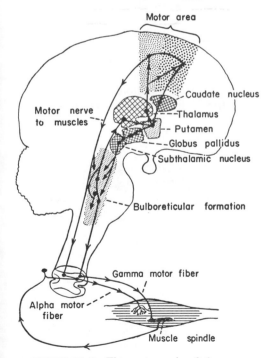

FIGURE 31–2 The motor axis of the nervous system.

STORAGE OF INFORMATION — MEMORY

Only a small fraction of the important sensory information causes an immediate motor response. The remainder is stored for future control of motor activities and for use in the thinking processes. Most of this storage of information occurs in the cerebral cortex, but not all, for even the basal regions of the brain and perhaps even the spinal cord can store small amounts of information.

The storage of information is the process we call *memory*, and this too is a function of the synapses. That is, each time a particular sensory signal passes through a sequence of synapses, the respective synapses become more capable of transmitting the same signal the next time, which process is called *facilitation*. After the sensory signal has passed through the synapses a large number of times, the synapses become so facilitated that signals from other parts of the brain can also cause transmission of impulses through the same sequence of synapses even though the sensory input has not been excited. This gives the person a feeling of experiencing the original sensation, though in effect it is only a memory of the sensation.

Unfortunately, we do not know the precise mechanism by which facilitation of synapses occurs in the memory process, but what is known about this and other details of the memory process will be discussed in Chapter 39.

Once memories have been stored in the nervous system, they become part of the processing mechanism. The thought processes of the brain compare new sensory experiences with the stored memories; the memories help to select the important new sensory information and to channel this into appropriate storage areas for future use or into motor areas to cause bodily responses.

THE THREE MAJOR LEVELS OF NERVOUS SYSTEM FUNCTION

The human nervous system has inherited specific characteristics from each stage of evolutionary development. From this heritage, there remain three major levels of the nervous system that have special functional significance: (1) the spinal cord level, (2) the lower brain level, and (3) the higher brain or cortical level.

THE SPINAL CORD LEVEL

The spinal cord of the human being still retains many functions of the multisegmental animal. Sensory signals are transmitted through the spinal nerves into each *segment* of the spinal cord, and these signals can cause localized motor responses either in the segment of the body from which the sensory information is received or in adjacent segments. Essentially all the spinal cord motor responses are automatic and occur almost instantaneously in response to the sensory signal. In addition, they occur in specific patterns of response called *reflexes*.

Figure 31–3 illustrates two of the simpler cord reflexes. To the left is the neural control of the *muscle stretch reflex*. If a muscle suddenly becomes stretched, a sensory nerve receptor in the muscle called the *muscle spindle* becomes stimulated and transmits nerve impulses through a sensory nerve fiber into the spinal cord. This fiber synapses directly with a motoneuron in the anterior horn of the cord gray matter, and the motoneuron in turn transmits impulses back to the muscle to cause the muscle, the effector, to contract. Thus, this reflex acts as a *feedback* mechanism, operating from a receptor to an effector, to prevent sudden changes in length of the muscle. This allows a person to maintain his limbs and other parts of his body in desired positions despite sudden outside forces that tend to move the parts out of position.

To the right in Figure 31–3 is illus-

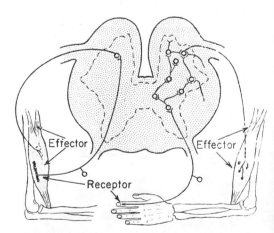

FIGURE 31–3 *Left:* The simple stretch reflex. *Right:* A withdrawal reflex.

trated the neural control of another reflex called the *withdrawal reflex*. This is a protective reflex that causes withdrawal of any part of the body from an object that is causing pain. For instance, let us assume that the hand is placed on a sharp object. Pain signals are transmitted into the gray matter of the spinal cord, and, after appropriate selection of information by the synapses, signals are channeled to the appropriate motoneurons to cause flexion of the biceps muscle. This obviously lifts the hand away from the sharp object.

We see, then, that the withdrawal reflex is much more complex than the stretch reflex, for it involves many neurons in the gray matter of the cord and signals are transmitted to many adjacent segments of the cord to cause contraction of the appropriate muscles.

Cord Functions After the Brain Is Removed. The many reflexes of the spinal cord will be discussed in Chapter 37; however, the following list of a few of the important cord reflex functions that occur even after the brain is removed illustrates the many capabilities of the spinal cord:

1. The animal can under certain conditions be made to stand up. This is caused primarily by reflexes initiated from the pads of the feet. Sensory signals from the pads cause the extensor muscles of the limbs to tighten, which in turn allows the limbs to support the animal's body.

2. A spinal animal held in a sling so that his feet hang downward often begins walking or galloping movements involving one, two, or all his legs. This illustrates that the basic patterns for causing the limb movements of locomotion are present in the spinal cord.

3. A flea crawling on the skin of a spinal animal causes reflex to-and-fro scratching by the paw, and the paw can actually localize the flea on the surface of the body.

4. Cord reflexes exist to cause emptying of the urinary bladder or of the rectum.

5. Segmental temperature reflexes are present throughout the body. Local cooling of the skin causes vasoconstriction, which helps to conserve heat in the body. Conversely, local heating in the skin causes vasodilatation, resulting in loss of heat from the body.

This list of some of the segmental and multisegmental reflexes of the spinal cord demonstrates that many of our day-by-day and moment-by-moment activities are controlled locally by the respective segmental levels of the spinal cord, and the brain plays only a modifying role in these local controls.

THE LOWER BRAIN LEVEL

Many if not most of what we call subconscious activities of the body are controlled in the lower areas of the brain — in the medulla, pons, mesencephalon, hypothalamus, thalamus, and basal ganglia. Subconscious control of arterial blood pressure and respiration is achieved primarily in the reticular substance of the medulla and pons. Control of equilibrium is a combined function of the older portions of the cerebellum and the reticular substance of the medulla, pons, and mesencephalon. The coordinated turning movements of the head, of the entire body, and of the eyes are controlled by specific centers located in the mesencephalon, paleocerebellum, and lower basal ganglia. Feeding reflexes, such as salivation in response to taste of food and licking of the lips, are controlled by areas in the medulla, pons, mesencephalon, amygdala, and hypothalamus. And many emotional patterns, such as anger, excitement, sexual activities, reactions to pain, or reactions of pleasure, can occur in animals without a cerebral cortex.

In short, the subconscious but coordinate functions of the body, as well as many of the life processes themselves — arterial pressure and respiration, for instance — are controlled by the lower regions of the brain, regions that usually, but not always, operate below the conscious level.

THE HIGHER BRAIN OR CORTICAL LEVEL

We have seen from the discussion so far that many of the intrinsic life processes of the body are controlled by subcortical regions of the brain or by the spinal cord. What then is the function of the cerebral cortex? The cerebral cortex is primarily a vast information storage area. Approximately three-quarters of all the neuronal cell bodies of the entire nervous system are located in the cerebral cortex. It is here that most of the memories of past experiences are stored, and it is here that many of the patterns of motor responses are stored, which information can be called forth at will to control motor functions of the body.

Relation of the Cortex to the Thalamus and Other Lower Centers. The cerebral cortex is actually an outgrowth of the lower regions of the brain, particularly of the thalamus. For each area of the cerebral cortex there is a corresponding and connecting area of the thalamus, and activation of a minute portion of the thalamus activates the corresponding and much larger portion of the cerebral cortex. It is presumed that in this way the thalamus can call forth cortical activities at will. Also, activation of regions in the mesencephalon transmits diffuse signals to the cerebral cortex, partially through the thalamus and partially directly, to activate the entire cortex. This is the process that we call *wakefulness*. On the other hand, when these areas of the mesencephalon become inactive, the thalamic and cortical regions also become inactive, which is the process we call *sleep*.

Some areas of the cerebral cortex are not directly concerned with either sensory or motor functions of the nervous system—for example, the prefrontal lobe and large portions of the temporal and parietal lobes. These areas are set aside for the more abstract processes of thought, but even they also have direct nerve connections with the lower regions of the brain.

Large areas of the cerebral cortex can be destroyed without blocking the subconscious, and even some of the involuntary conscious, activities of the body. For instance, destruction of the somesthetic cortex does not destroy one's ability to feel objects touching his skin, but it does destroy his ability to distinguish the shapes of objects, their character, and the precise points on the skin where the objects are touching. Thus, the cortex is not required for perception of sensation, but it does add immeasurably to its depth of meaning. Likewise, destruction of the prefrontal lobe does not destroy one's ability to think, but it does destroy his ability to think in abstract terms. In other words, each time a portion of the cerebral cortex is destroyed, a vast amount of information is lost to the thinking process and some of the mechanisms for processing this information are also lost. The result is a vegetative type of existence rather than a "living" existence.

TRANSMISSION AND PROCESSING OF INFORMATION IN THE NERVOUS SYSTEM

INFORMATION, SIGNALS, AND IMPULSES

The term *information*, as it applies to the nervous system, means a variety of different things. For instance, pain from a pin prick is information, pressure on the bottom of the feet is information, degree of angulation of the joints is information, and a stored memory in the brain is information.

The primary function of the brain is to transmit information from one point to another, to store it, and to process it so that it can be used advantageously or so that its meaning can become clear to the mind. However, information cannot be transmitted in its original form but only in the form of nerve impulses. A part of the body that is subjected to pain must first convert this information into *nerve impulses*, or areas of the brain must convert abstract thoughts into nerve impulses. These are then transmitted either elsewhere in the brain or into peripheral nerves to motor effectors throughout the body.

Signals. In the transmission of information, it is frequently not desirable to speak of the individual impulses but instead of the overall pattern of impulses; this pattern is called a *signal*. As an example: When pressure is applied to a large area of skin, impulses are transmitted by large numbers of parallel nerve fibers, and the total pattern of impulses transmitted by all these fibers is a signal. Thus, we can speak of visual signals, auditory signals, somesthetic sensory signals, motor signals, and so forth.

SPATIAL ORIENTATION OF SIGNALS IN FIBER TRACTS

How does the brain detect the precise position on the body that is receiving a sensory stimulus, and how does the brain transmit impulses precisely to individual skeletal muscle bundles? The answer to this is by transmitting their signals in a precise spatial pattern through the nerve tracts. All the different nerve tracts, both in the peripheral nerves and in the fiber tracts of the central nervous system, are spatially organized. For instance, in the dorsal columns of the spinal cord the sensory fibers from the feet lie toward the midline, while those fibers entering the dorsal columns at higher levels of the body lie progressively more toward the lateral sides of the dorsal columns. This spatial organization is maintained with precision

FIGURE 31-4 Spatial pattern of nerve fiber stimulation in a nerve trunk following stimulation of the skin by three separate but simultaneous pinpricks.

throughout the sensory pathway all the way to the somesthetic cortex. Likewise, the fiber tracts within the brain and those extending into motor nerves are spatially oriented in the same way.

As an example, Figure 31-4 illustrates to the left three separate pins stimulating the skin; to the right the spatial orientation of the stimulated fibers in the nerve are shown. Each pin in this example stimulates a single fiber strongly and adjacent fibers less strongly. This spatial orientation of the three separate bundles of fibers is maintained all the way to the cerebral cortex and obviously allows the brain to localize the three different pinpricks to their points of stimulation.

TRANSMISSION AND PROCESSING OF SIGNALS IN NEURONAL POOLS

The central nervous system is made up of literally hundreds of separate neuronal pools, some of which are extremely small and some very large. For instance, the entire cerebral cortex could be considered to be a single large neuronal pool. If all the surface area of the cerebral cortex were flattened out, including the surfaces of the penetrating folds, the total area of this large flat pool would be several square feet. It has many separate fiber tracts coming to it (afferent fibers) and others leaving it (efferent fibers). Furthermore it maintains the same quality of spatial orientation as that found in the nerve bundles, individual points of the cortex connecting with specific points elsewhere in the nervous system or connecting through the peripheral nerves with specific points in the body. However, within this pool of neurons are large numbers of short nerve fibers whereby signals spread horizontally from neuron to neuron within the pool itself.

Other neuronal pools include the different basal ganglia, and the specific nuclei in the cerebellum, mesencephalon, pons, and medulla. Also, the entire dorsal gray matter of the spinal cord could be considered to be one long pool of neurons, and the entire anterior gray matter another long neuronal pool. Each pool has its own special characteristics of organization which cause it to process signals in its own special way. It is these special characteristics of the different pools that allow the multitude of functions of the nervous system. Yet, despite their differences in function, the pools also have many similarities of function which are described in the following pages.

Organization of Neurons in the Neuronal Pools. Figure 31–5 is a schematic diagram of the organization of neurons

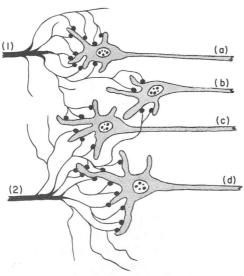

FIGURE 31–5 Basic organization of a neuronal pool.

in a neuronal pool, showing "input" fibers to the left and "output" fibers to the right. Each input fiber divides as many as several hundred times and provides an average of several hundred terminal fibrils that spread over a large area in the pool and that finally synapse with the dendrites or cell bodies of the neurons in the pool. The area into which the endings of each incoming nerve fiber spread is called its *excitatory field*. Note that each input fiber arborizes, so that large numbers of its presynaptic terminals lie on the centermost neurons in its "field," but progressively fewer terminals lie on the neurons farther from the center of the field.

Threshold and Subthreshold Stimuli— Facilitation. Going back to the discussion of synaptic function in Chapter 6, it will be recalled that stimulation of a single excitatory presynaptic terminal almost never stimulates the postsynaptic neuron. Instead, large numbers of presynaptic terminals must discharge on the same neuron either simultaneously or in rapid succession to cause excitation. For instance, let us assume that six separate presynaptic terminals must discharge simultaneously or in rapid succession to excite any one of the neurons in Figure 31–5. By counting terminals, one can readily see that an incoming impulse in fiber 1 will cause neuron *a* to discharge, and an impulse in fiber 2 will excite neuron *d*.

Though fiber 1 fails to discharge neurons *b* and *c*, discharge of the presynaptic terminals changes the membrane potentials of these neurons so that they can be more easily excited by other incoming signals. Thus, there are two types of stimuli that enter a neuronal pool, *threshold stimuli* and *subthreshold stimuli*. A threshold stimulus causes actual excitation. A subthreshold stimulus fails to excite the neuron but does make the neuron more excitable to impulses from other sources. The neuron that is made more excitable but does not discharge is said to be *facilitated*.

"Convergence" of subthreshold stimuli to cause excitation. Let us now assume that both fibers 1 and 2 in Figure 31–5 transmit concurrent impulses into the neuronal pool. In this case, neuron *c* is stimulated by six presynaptic terminals simultaneously, which is the required number to cause excitation. Thus, subthreshold stimuli can *converge* from several sources and *summate* at the neuron to cause a threshold stimulus.

The field of terminals. It must be recognized that Figure 31–5 represents a highly condensed version of the neuronal pool, for each input nerve fiber gives off terminals that spread to perhaps 100 or more separate neurons, and the surface of each neuron is covered by many hundred synapses. In the center of the *field* of terminals, almost all the neurons are stimulated by the incoming fiber; whereas farther toward the periphery of the field, the neurons are facilitated but do not discharge. Figure 31–6 illustrates this effect, showing the field of a single input nerve fiber. The area in the neuronal pool in which all the neurons discharge is called the *discharge* or *threshold* or *liminal zone*, and the area to either side in which the neurons are facilitated but do not discharge is called the *facilitated* or *subthreshold* or *subliminal zone*.

Inhibition in a Neuronal Pool — The Inhibitory Circuit. It will be recalled from the discussions in Chapter 6 that some neurons of the central nervous system secrete an inhibitory transmitter substance instead of an excitatory trans-

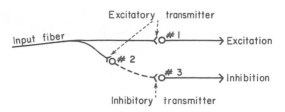

FIGURE 31–7 Inhibitory circuit. Neuron #2 is an inhibitory neuron.

mitter substance. Figure 31–7 illustrates the so-called inhibitory circuit that can change an excitatory signal into an inhibitory signal. In this figure the input fiber divides and secretes excitatory transmitter at both its endings. This causes excitation of both neurons 1 and 2. However, neuron 2 is an inhibitory neuron that secretes inhibitory transmitter at its terminal nerve endings. Excitation of this neuron therefore inhibits neuron 3.

In short, one way to cause inhibition is *to transmit a signal through an inhibitory neuron*, which then secretes the inhibitory transmitter substance.

Convergence. The term "convergence" means simultaneous control of a single neuron by two or more separate input nerve fibers. One type of convergence was illustrated in Figure 31–5 in which two excitatory input nerve fibers from the same source converged upon several separate neurons to stimulate them. This type of convergence from a single source is illustrated again in Figure 31–8*A*.

However, convergence can also result from input signals (excitatory or inhibitory) from several different sources, which is illustrated in Figure 31–8*B*. For instance, the internuncial cells of the spinal cord receive converging signals from (a) peripheral nerve fibers entering the cord, (b) propriospinal fibers passing from one segment of the cord to another, (c) corticospinal fibers from the cerebral cortex, and (d) probably several other long pathways descending from the brain into the spinal cord.

Such convergence allows summation

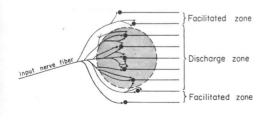

FIGURE 31–6 "Discharge" and "facilitated" zones of a neuronal pool.

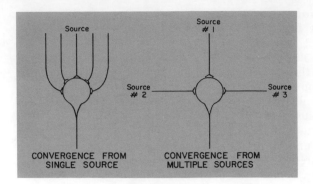

FIGURE 31–8 "Convergence" of multiple input fibers on a single neuron: (A) Input fibers from a single source. (B) Input fibers from multiple sources.

of information from several different sources, and the resulting response is a summated effect of all these different types of information. Obviously, therefore, convergence is one of the important means by which the central nervous system correlates, summates, and sorts different types of information.

Divergence. Divergence means that excitation of a single input nerve fiber stimulates multiple output fibers from the neuronal pool. The two major types of divergence are illustrated in Figure 31–9 and may be described as follows:

An *amplifying* type of divergence often occurs, illustrated in Figure 31–9A. This means simply that an input signal spreads to an increasing number of neurons as it passes through successive pools of a nervous pathway. This type of divergence is characteristic of the corticospinal pathway in its control of skeletal muscles as follows: Stimulation of a single large pyramidal cell in

the motor cortex transmits a single impulse into the spinal cord. Yet, under appropriate conditions, this impulse can stimulate perhaps 15 to 20 internuncial cells, each of which in turn stimulates perhaps as many as several hundred anterior motoneurons. Each of these then stimulates as many as 100 to 300 muscle fibers. Thus, there is a total divergence, or amplification, of as much as 50,000-fold.

The second type of divergence, illustrated in Figure 31–9B, is *divergence into multiple tracts*. In this case, the signal is transmitted in two separate directions from the pool. This allows the same information to be transmitted to several different parts of the nervous system where it is needed. For instance, information transmitted in the dorsal columns of the spinal cord takes two courses in the lower part of the brain, (1) into the cerebellum and (2) on through the lower regions of the brain

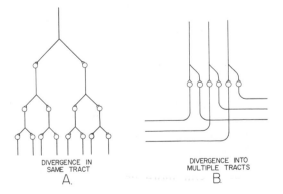

FIGURE 31–9 "Divergence" in neuronal pathways: (A) Divergence within a pathway to cause "amplification" of the signal. (B) Divergence into multiple tracts to transmit the signal to separate areas.

to the thalamus and cerebral cortex. Likewise, in the thalamus almost all sensory information is relayed both into deep structures of the thalamus and to discrete regions of the cerebral cortex.

PROLONGATION OF A SIGNAL BY A NEURONAL POOL— "AFTER-DISCHARGE"

Thus far, we have considered signals that are transmitted instantaneously through a neuronal pool. However, in many instances, a signal entering a pool causes a prolonged output discharge, called after-discharge, even after the incoming signal is over. The three basic mechanisms by which after-discharge occurs are as follows:

Synaptic After-Discharge. When presynaptic terminals discharge on the surfaces of dendrites or the soma of a neuron, a postsynaptic potential develops in the neuron and lasts for many milliseconds—in the anterior motoneuron for about 15 milliseconds, though perhaps much longer in other neurons. As long as this potential lasts it can excite the neuron, causing it to transmit output impulses as was explained in Chapter 6. Thus, as a result of this synaptic after-discharge mechanism alone, it is possible for a single instantaneous input to cause a sustained signal output lasting as long as 15 milliseconds.

The Parallel Circuit Type of After-Discharge. Figure 31–10 illustrates a second type of neuronal circuit that can cause short periods of after-discharge. In this case, the input signal spreads through a series of neurons in the neuronal pool, and from many of these neurons impulses keep converging on an output neuron. It will be recalled that a signal is delayed at each synapse for about 0.5 millisecond, which is called the *synaptic delay*. Therefore, signals that pass through a succession of intermediate neurons reach the output neuron one by one after varying periods of de-

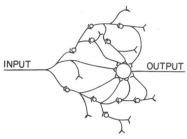

FIGURE 31–10 The parallel after-discharge circuit.

lay. Therefore, the output neuron continues to be stimulated for many milliseconds.

It is doubtful that more than a few dozen successive neurons ordinarily enter into a parallel after-discharge circuit. Therefore, one would suspect that this type of after-discharge circuit could cause after-discharges that last for no more than perhaps 25 to 50 milliseconds. Yet, this circuit does represent a means by which a single input signal, lasting less than 1 millisecond, can be converted into a sustained output signal lasting many milliseconds.

The Reverberating (Oscillatory) Circuit as a Cause of After-Discharge. One of the most important of all circuits in the entire nervous system is the reverberating, or oscillatory, circuit, several different varieties of which are illustrated in Figure 31–11. The simplest theoretical reverberating circuit, even though such may not actually exist in the nervous system, is that illustrated in Figure 31–11*A*, which involves a single neuron. In this case, the output neuron simply sends a collateral nerve fiber back to its own dendrites or soma to restimulate itself; therefore, once the neuron should discharge, the feedback stimuli could theoretically keep the neuron discharging for a long time thereafter.

Figure 31–11*B* illustrates a few additional neurons in the feedback circuit, which would give a longer period of time between the initial discharge and the feedback signal. Figure 31–11*C* illustrates a still more complex system in

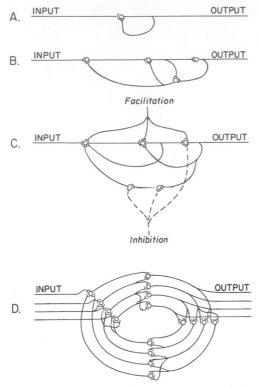

FIGURE 31-11 Reverberatory circuits of increasing complexity.

which both facilitatory and inhibitory fibers impinge on the reverberating pool. A facilitatory signal increases the ease with which reverberation takes place, while an inhibitory signal decreases the ease with which reverberation takes place.

Figure 31-11D illustrates that most reverberating pathways are constituted of many parallel fibers, and at each cell station the terminal fibrils diffuse widely. In such a system the total reverberating signal can be either weak or strong, depending on how many parallel nerve fibers are momentarily involved in the reverberation.

Finally, reverberation need not occur only in a single neuronal pool, for it can occur through a circuit of successive pools.

Characteristics of after-discharge from a reverberating circuit. Figure 31-12 illustrates postulated output sig-

nals from a reverberating after-discharge circuit. The input stimulus need last only 1 millisecond or so, and yet the output can last for many milliseconds or even minutes. The figure demonstrates that the intensity of the output signal increases to a reasonably high value early in the reverberation, then decreases to a critical point, and suddenly ceases entirely. Furthermore, the duration of the after-discharge is determined by the degree of inhibition or facilitation of the neuronal pool. In this way, signals from other parts of the brain can control the reaction of the pool to the input stimulus.

Almost these exact patterns of output signals can be recorded from the motor nerves exciting a muscle involved in the flexor reflex (discussed in Chap. 37), which is believed to be caused by a reverberating type of after-discharge following stimulation of a pain fiber.

Importance of synaptic fatigue in determining the duration of reverberation. It was pointed out in Chapter 6 that synapses fatigue if stimulated for prolonged periods of time. One of the most important factors that determine the duration of the reverberatory type of after-discharge is probably the rapidity with which the involved synapses fatigue. Rapid fatigue would obviously tend to shorten the period of after-discharge and slow fatigue to lengthen it.

Furthermore, the greater the number of neurons in the reverberatory pathway and the greater the number of collateral

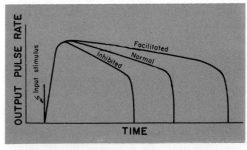

FIGURE 31-12 Typical pattern of the output signal from a reverberatory circuit following a single input stimulus, showing the effects of facilitation and inhibition.

feedback fibrils, the easier it would be to keep the reverberation going. Therefore, it is to be expected that longer reverberating pathways would in general sustain after-discharges for longer periods of time.

Duration of reverberation. Typical after-discharge patterns of different reverberatory circuits have durations from as short as 10 milliseconds to as long as several minutes, or perhaps even hours. Indeed, as will be explained in Chapter 41, wakefulness is perhaps an example of reverberation of neuronal circuits in the basal region of the brain. In this case "arousal impulses" are postulated to set off the wakefulness reverberation at the beginning of each day and thereby to cause sustained excitability of the brain, this excitability lasting 14 or more hours.

RHYTHMIC SIGNAL OUTPUT

Many neuronal circuits emit rhythmic output signals—for instance, the rhythmic respiratory signal originating in the reticular substance of the medulla and pons. This repetitive rhythmic signal continues throughout life, while other rhythmic signals, such as those that cause scratching movements by the hind leg of a dog or the walking movements in an animal, require input stimuli into the respective circuits to initiate the signals.

Rhythmic signals probably result from reverberating pathways that are self-excitatory. After a period of reverberation, fatigue or some other inhibitory factor stops the reverberation until the self-excitatory process begins again.

REFERENCES

Anokhin, P. K.: Convergence of excitations on a neuron as the basis of integrative brain activity. *Proc. Int. Union of Physiol. Sciences,* 6:3, 1968.

Ashby, W. R.: Mathematical models and computer analysis of the function of the central nervous system. *Ann. Rev. Physiol.,*

Bullock, T. H.: Comparative physiology of sensory systems: problems of meaning and coding. *Proc. Int. Union of Physiol. Sciences,* 6:55, 1968.

Bunge, R. P.: Glial cells and the central myelin sheath. *Physiol. Rev.,* 48:197, 1968.

Corning, W. C., and Balaban, M. W. (eds.): The Mind: Biological Approaches to Its Functions. New York, John Wiley & Sons, Inc., 1968.

Eldred, E., and Buchwald, J.: Central nervous system. *Ann. Rev. Physiol.,* 29:573, 1967.

Harmon, L. D., and Lewis, E. R.: Neural modeling. *Physiol. Rev.,* 46:513, 1966.

Konorski, J.: Integrative Activity of the Brain. Chicago, University of Chicago Press, 1967.

Moore, G. P., Perkel, D. H., and Segundo, J. P.: Statistical analysis and functional interpretation of neuronal spike data. *Ann. Rev. Physiol.,* 28:493, 1966.

CHAPTER 32

SOMATIC SENSATIONS: I. SENSORY RECEPTORS AND THE MECHANORECEPTIVE SENSATIONS

Input to the nervous system is provided by the sensory receptors that detect such sensory stimuli as touch, sound, light, cold, and warmth. The purpose of the present chapter is to begin discussing the basic mechanisms by which these receptors change sensory stimuli into nerve signals and the transmission of these signals into the central nervous system.

TYPES OF SENSORY RECEPTORS AND THE SENSORY STIMULI THEY DETECT

Table 32–1 gives a list and classification of most of the body's sensory receptors. This table shows that there are basically five different types of sensory receptors: (1) *mechanoreceptors*, which detect mechanical deformation of the

receptor or of cells adjacent to the receptor; (2) *thermoreceptors*, which detect changes in temperature, some receptors detecting cold and others detecting warmth; (3) *nociceptors* (pain receptors), which detect damage in the tissues, whether it be physical damage or chemical damage; (4) *electromagnetic receptors*, which detect light on the retina of the eye; and (5) *chemoreceptors*, which detect taste in the mouth, smell in the nose, oxygen level in the arterial blood, osmolality of the body fluids, carbon dioxide concentration, and perhaps other factors that make up the chemistry of the body.

This chapter will discuss functions of some of these receptors, primarily the mechanoreceptors, to illustrate the basic principles by which receptors in general operate. Other receptors will be discussed in relation to the sensory systems that they subserve, which will be presented mainly in the next few chap-

390

TABLE 32-1 CLASSIFICATION OF SENSORY RECEPTORS

Mechanoreceptors	Thermoreceptors
	Cold
Skin tactile sensibilities (epidermis and dermis)	Probably free nerve endings
Free nerve endings	Warmth
Expanded tip endings	Probably free nerve endings
Ruffini's endings	
Merkel's discs	**Nociceptors**
Hederiform terminations	
Plus several other variants	Pain
Encapsulated endings	Free nerve endings
Meissner's corpuscles	
Krause's corpuscles	**Electromagnetic Receptors**
Pacinian corpuscles	
Hair end-organs	Vision
Deep tissue sensibilities	Rods
Free nerve endings	Cones
Expanded tip endings	
Ruffini's endings	**Chemoreceptors**
Plus a few other variants	
Encapsulated endings	Taste
Pacinian corpuscles	Receptors of taste buds
Plus a few other variants	Smell
Specialized endings	Receptors of olfactory epithelium
Muscle spindles	Arterial oxygen
Golgi tendon receptors	Receptors of aortic and carotid bodies
Hearing	Osmolality
Sound receptors of cochlea	Probably neurons of supraoptic nuclei
Equilibrium	Blood CO_2
Vestibular receptors	Receptors in or on surface of medulla and in
Arterial pressure	aortic and carotid bodies
Baroreceptors of carotid sinuses and aorta	Perhaps blood glucose, amino acids, fatty acids
	Receptors in hypothalamus

ters. Figure 32-1 illustrates some of the different types of mechanoreceptors found in the skin or in the deep structures of the body.

MODALITY OF SENSATION — LAW OF SPECIFIC NERVE ENERGIES

Each of the principal types of sensation that we can experience—pain, touch, sight, sound, and so forth—is called a *modality* of sensation. Yet, despite the fact that we experience these different modalities of sensation, nerve fibers transmit only impulses. Therefore, how is it that different nerve fibers transmit different modalities of sensation?

The answer to this is that each nerve tract terminates at a specific point in the central nervous system, and the type of sensation felt when a nerve fiber is stimulated is determined by the specific area in the nervous system to which the fiber leads. For instance, if a pain fiber is stimulated, the person perceives pain regardless of what type of stimulus excites the fiber. This stimulus could be electricity, heat, crushing, or stimulation of the pain nerve ending by damage to the tissue cells. Yet, whatever the means of stimulation, the person still perceives pain. Likewise, if a touch fiber is stimulated by exciting a touch receptor or in any other way, the person perceives touch because touch fibers lead to specific touch areas in the brain.

This specificity of nerve fibers for transmitting only one modality of sen-

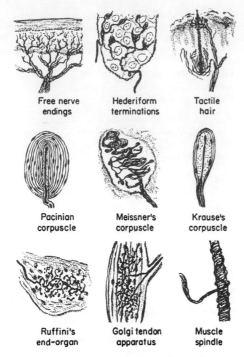

Free nerve endings	Hederiform terminations	Tactile hair
Pacinian corpuscle	Meissner's corpuscle	Krause's corpuscle
Ruffini's end-organ	Golgi tendon apparatus	Muscle spindle

FIGURE 32–1 Several types of somatic sensory nerve endings. (Modified from Ramon y Cajal: Histology. William Wood and Co.)

sation is called the *law of specific nerve energies*.

TRANSDUCTION OF SENSORY STIMULI INTO NERVE IMPULSES

LOCAL CURRENTS AT NERVE ENDINGS — GENERATOR POTENTIALS AND RECEPTOR POTENTIALS

All sensory receptors studied thus far have one feature in common. Whatever the type of stimulus that excites the ending, it first causes a *local flow of current* in the neighborhood of the nerve ending, and it is this current that in turn excites action potentials in the nerve fiber.

There are two different ways in which local current can be elicited. One of these is to deform or chemically alter the terminal nerve ending itself. This causes ions to diffuse through the nerve membrane, thereby setting up a local current flow. Electrical voltages caused in this manner are called *generator potentials*, because the nerve fiber itself generates the current and the voltage.

The second method for causing local current involves specialized receptor cells that lie adjacent to the nerve endings. For instance, when sound enters the cochlea of the ear, the hair cells that lie on the basilar membrane emit electrical currents that in turn stimulate the terminal nerve fibrils entwining the hair cells. Potentials resulting from these currents are called *receptor potentials* because they are caused by specialized, non-neuronal "receptor" cells. However, in some instances it is difficult to tell whether the potential comes from the terminal nerve fibril itself or from a specialized receptor cell. Therefore, many physiologists use the terms generator potential and receptor potential interchangeably, meaning simply the local potential created around a terminal nerve fibril in response to a specific sensory stimulus.

The Generator Potential of the Pacinian Corpuscle. The pacinian corpuscle is a very large and easily dissected sensory receptor. For this reason, one can study in detail the mechanism by which tactile stimuli excite it and by which it causes action potentials in the sensory fiber leading from it. Note in Figure 32–1 that the pacinian corpuscle has a long central nonmyelinated nerve fiber extending through its core. Surrounding this fiber are several capsule layers so that compression on the outside of the corpuscle tends to elongate or shorten or otherwise deform the central core of the fiber, depending on how the compression is applied. The deformation causes a sudden change in membrane potential, as illustrated in Figure 32–2. This perhaps results from stretching the nerve fiber membrane, thus increasing its permeability and allowing positively charged sodium ions to leak to the interior of the

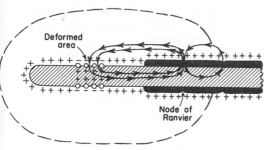

FIGURE 32-2 Excitation of a sensory nerve fiber by a generator potential produced in a pacinian corpuscle.

fiber. This change in local potential causes local current flow that spreads along the nerve fiber to its myelinated portion. At the first node of Ranvier, which itself lies inside the capsule of the pacinian corpuscle, the local circuit of current flow initiates action potentials in the nerve fiber.

Relationship of Generator Potential to Stimulus Strength. Figure 32-3 illustrates the effect on the amplitude of the generator potential caused by progressively stronger stimuli applied to the central core of the pacinian corpuscle. Note that the amplitude increases rapidly at first but then progressively less rapidly at high stimulus strengths. The maximum amplitude that can be achieved by generator potentials is

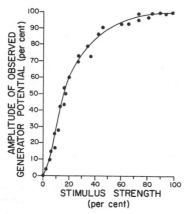

FIGURE 32-3 Relationship of amplitude of generator potential to strength of a stimulus applied to a pacinian corpuscle. (From Loewenstein: *Ann N.Y. Acad. Sci., 94:*510, 1961.)

around 100 millivolts. That is, a generator potential can have almost as high a voltage as an action potential.

ADAPTATION OF RECEPTORS

A special characteristic of all sensory receptors is that they *adapt* either partially or completely to their stimuli after a period of time. That is, when a continuous sensory stimulus is applied, the receptors respond at a very high impulse rate at first, then progressively less rapidly, until finally many of them no longer respond at all.

Figure 32-4 illustrates typical adaptation of certain types of receptors. Note that the pacinian corpuscle and the receptor at the hair bases adapt extremely rapidly, while joint capsule and muscle spindle receptors adapt very slowly.

Furthermore, some sensory receptors adapt to a far greater extent than others. For example, the pacinian corpuscles adapt to "extinction" within a fraction of a second, and the hair base receptors adapt to extinction within a second or more.

Mechanisms by which Receptors Adapt. Adaptation of receptors seems to be an individual property of each type of receptor. For instance, in the eye, the rods and cones adapt by changing their chemical compositions (which will be discussed in Chap. 34). In the case of the pacinian corpuscle, the corpuscular structure itself very rapidly adjusts to the deformation of the tissue. Also, the ability of the central core of the pacinian corpuscle to emit local current also becomes depressed, so that the rate of firing of the nerve fiber becomes decreased. Presumably these same two principles apply to other types of mechanoreceptors; that is, part of the adaptation is often in the structure of the receptor itself, and part is in the terminal nerve fibril of the receptor.

Function of the Poorly Adapting Receptors—The "Static" Receptors. The poorly adapting receptors, those that adapt very slowly and do not adapt to extinction,

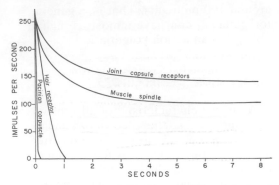

FIGURE 32-4 Adaptation of different types of receptors, showing rapid adaptation of some receptors and slow adaptation of others.

continue to transmit impulses to the brain as long as the stimulus is present (or at least for many minutes or hours). Therefore, they keep the brain constantly apprised of the status of the body and its relation to its surroundings. For instance, impulses from the joint capsule receptors allow the person to "know" the positions of the different parts of his body at all times, while impulses from the baroreceptors allow the central nervous system to know the pressure in the arteries.

Other types of poorly adapting receptors include the receptors of the macula in the vestibular apparatus, the sound receptors of the ear, the pain receptors, and the chemoreceptors of the carotid and aortic bodies.

Because the poorly adapting receptors can continue to transmit information for many hours, they are called *static* receptors. It is probable that many of these poorly adapting receptors would adapt to extinction if the intensity of the stimulus should remain absolutely constant over many days. Fortunately, because of our continually changing bodily state, the static receptors almost never reach a state of complete adaptation.

Function of the Rapidly Adapting Receptors — The Rate Receptors or Movement Receptors or Phasic Receptors. Obviously, receptors that adapt rapidly cannot be used to transmit a continuous sig-

nal, because these receptors are stimulated only when the stimulus strength changes. Yet they react strongly *while a change is actually taking place*. Furthermore, the number of impulses transmitted is directly related to the *rate at which the change takes place*. Therefore, these receptors are called *rate* receptors, *movement* receptors, or *phasic* receptors. Thus, in the case of the pacinian corpuscle, sudden pressure applied to the skin excites this receptor for a few milliseconds, and then its excitation is over even though the pressure continues. In other words, the pacinian corpuscle is exceedingly important in transmitting information about rapid changes in tissue stress, but it is useless in transmitting information about constant stress.

Importance of the rate receptors — their predictive function. If one knows the rate at which some change in his bodily status is taking place, he can predict ahead to the state of the body a few seconds or even a few minutes later. For instance, the receptors of the semicircular canals in the vestibular apparatus of the ear detect the rate at which the head begins to turn when one runs around a curve. Using this information, a person can predict that he will turn 10, 30, or some other number of degrees within the next 10 seconds, and he can adjust the motion of his limbs *ahead of time* to keep from losing his balance. Likewise, pacinian corpuscles located in the joint capsules help to detect the rates of movement of the different parts of the body. Therefore, when one is running, information from these receptors allows the nervous system to predict ahead of time where the feet will be during any precise fraction of a second, and appropriate motor signals can be transmitted to the muscles of the legs to make any necessary anticipatory corrections in limb position so that the person will not fall. Loss of this predictive function makes it impossible for the person to run.

JUDGMENT OF STIMULUS STRENGTH

Physiopsychologists have evolved numerous methods for testing one's judgment of sensory stimulus strength, but only rarely do the results from the different methods agree with each other. Two principles are widely discussed: the *Weber-Fechner principle* and the *power law*.

History and Significance of the Weber-Fechner Principle. In the mid eighteen hundreds, Weber first and Fechner later proposed the principle that gradations of stimulus strength are discriminated approximately in proportion to the logarithm of stimulus strength. This law is based primarily on one's ability to judge minimal changes in stimulus strength that can be detected; that is, a person can barely detect a 5 gram increase in weight when holding 50 grams, or a 50 gram increase when holding 500 grams. Thus, the *ratio* of the change in stimulus strength required for perception of a change remains essentially constant, which is what the logarithmic principle means.

Because the Weber-Fechner principle offers a ready explanation for the tremendous range of stimulus strength that our nervous system can discern (ranges as great as 100,000 to 1 for somatic sensations and one trillion to one for auditory sensations), it unfortunately became widely accepted for all types of sensory experience and for all levels of background sensory intensity. More recently it has become evident that this principle applies mainly to higher levels of visual, auditory, and cutaneous sensory experience and that it applies only poorly to most other types of sensory experience.

Yet, the Weber-Fechner principle is still a good one to remember because it emphasizes that the greater the background sensory stimulus, the greater also must be the additional change in stimulus strength in order for the psyche to detect the change.

The Power Law. Another attempt by physiopsychologists to find a good mathematical relationship between actual stimulus strength and interpretation of stimulus strength is the following formula, known as the power law:

Interpreted strength = K (Stimulus strength)^x

In this formula K is a constant, and x is the power to which the stimulus strength is raised. The exponent x and the constant K are different for each type of sensation.

Unfortunately, the power law is not of great philosophical importance for a very simple mathematical reason. Almost any curve (even the curve of a hill) that has a progressive change in slope in the same direction can be fitted over most of its range to the power law equation, provided one simply finds proper values for the exponent x and constant K. Therefore, use of the power law is more an exercise in mathematical curve fitting than it is a valuable tool in understanding sensory experience.

PHYSIOLOGICAL CLASSIFICATION OF NERVE FIBERS

A general classification of nerve fibers is given in Table 32–2. The fibers are divided into types A, B, and C; and the type A fibers are further subdivided into α, β, γ, and δ fibers.

Type A fibers are the typical myelinated fibers of spinal nerves. The type B fibers differ from very small type A fibers only in the fact that they do not display a negative after-potential following stimulation. However, they also are myelinated like type A fibers. They are the preganglionic autonomic nerve fibers.

Type C fibers are the very small unmyelinated nerve fibers that conduct impulses at low velocities. These constitute more than half the number of sensory fibers and also all of the postganglionic autonomic fibers.

TABLE 32–2 PROPERTIES OF DIFFERENT MAMMALIAN NERVE FIBERS*

Type of Fiber	Diameter of Fiber (μ)	Velocity of Conduction (meters/sec.)	Duration of Spike (msec.)	Duration of Negative After-potential (msec.)	Duration of Positive After-potential (msec.)	Function
A (α)	13–22	70–120	0.4–0.5	12–20	40–60	Motor, muscle proprioceptors
A (β)	8–13	40–70	0.4–0.6	(?)	(?)	Touch, kinesthesia
A (γ)	4–8	15–40	0.5–0.7	(?)	(?)	Touch, excitation of muscle spindles, pressure
A (δ)	1–4	5–15	0.6–1.0	(?)	(?)	Pain, heat, cold, pressure
	1–3	3–14	1.2	None	100–300	Preganglionic autonomic
C	0.2–1.0	0.2–2	2.0	50–80	300–1000	Pain, heat(?), cold(?), pressure(?), postganglionic autonomic, smell

*Compiled from various sources but mainly from Grundfest.

The sizes, velocities of conduction, and functions of the different nerve fibers are given in the table. Note that the very large fibers can transmit impulses as rapidly as 120 meters per second, a distance longer than a football field. On the other hand, the smallest fibers transmit impulses as slowly as 0.5 meter per second, which is the distance from the foot to the knee.

Over two-thirds of all the nerve fibers in peripheral nerves are type C fibers. Because of their great number, these can transmit tremendous amounts of information from the surface of the body, even though their velocities of transmission are very slow. Utilization of type C fibers for transmitting this great mass of information represents an important economy of space in the nerves, for use of type A fibers would require peripheral nerves the sizes of large ropes and a spinal cord almost as large as the body itself.

SOMATIC SENSATIONS

The *somatic senses* are the nervous mechanisms that collect sensory information from the body. These senses are in contradistinction to the *special senses*, which mean specifically sight, hearing, smell, taste, and equilibrium.

The somatic senses can be classified into three different physiological types: (1) The *mechanoreceptive somatic senses*, stimulated by mechanical displacement of some tissue of the body, (2) the *thermoreceptive senses*, which detect heat and cold, and (3) the *pain sense*, which is activated by any factor that damages the tissues. The present section deals with the mechanoreceptive somatic senses; the following chapter deals with the thermoreceptive and pain senses.

The mechanoreceptive senses include *touch*, *pressure*, and *vibration* senses (which are collectively called the *tactile senses*) and the *kinesthetic sense*, which determines the relative positions and rates of movement of the different parts of the body.

DETECTION AND TRANSMISSION OF TACTILE SENSATIONS

Interrelationship Between the Tactile Sensations of Touch, Pressure, and Vibration. Though touch, pressure, and vibration are frequently classified as separate sensations, they are all detected by the same types of receptors. The only differences among these three types of sensations are (1) touch sensation generally results from stimulation of tactile receptors in the skin or in tissues immediately beneath the skin, (2) pressure sensation generally results from deformation of deeper tissues, and (3) vibration sensation results from rapidly repetitive sensory signals, but the same types of receptors as those for both touch and pressure are utilized.

The Tactile Receptors. At least six entirely different types of receptors are known, but many more similar to these probably also exist. Some of these receptors were illustrated in Figure 32–1, and their special characteristics are the following:

First, some *free nerve endings*, which are found everywhere in the skin and in many other tissues, can detect touch and pressure. For instance, even light contact with the cornea of the eye, which contains no other type of nerve ending besides free nerve endings, can nevertheless elicit touch and pressure sensations.

Second, a touch receptor of special sensitivity is *Meissner's corpuscle*, an encapsulated nerve ending that excites a large myelinated sensory nerve fiber. These receptors are particularly abundant in the fingertips, lips, and other areas of the skin where one's ability to discern spatial characteristics of touch sensations is highly developed. Meissner's corpuscles probably adapt within a second or more after they are stimulated, which means that they are particularly sensitive to movement of very light objects over the surface of the skin.

Third, the fingertips and other areas that contain large numbers of Meissner's corpuscles also contain *expanded tip tactile receptors*. These receptors differ from Meissner's corpuscles in that they transmit long-continuing signals rather than adapting rapidly.

Fourth, slight movement of any hair on the body stimulates the nerve fiber entwining its base. Thus, each hair and its basal nerve fiber, called the *hair end-organ*, is also a type of touch receptor.

Fifth, located in the deeper layers of the skin and also in deeper tissues of the body are many *Ruffini's end-organs*, which are multibranched endings. These endings do not adapt rapidly and, therefore, are important for signaling continuous states of deformation of the deeper tissues, such as heavy and continuous touch signals and pressure signals.

Sixth, many *pacinian corpuscles*, which were discussed earlier in the chapter, lie both immediately beneath the skin and also deep in the tissues of the body. These are stimulated only by very rapid movement of the tissues because these receptors adapt in a small fraction of a second. Therefore, they are particularly important for detecting tissue vibration or other extremely rapid changes in the mechanical state of the tissues.

Transmission of Tactile Sensations in Peripheral Nerve Fibers. The specialized sensory receptors, such as Meissner's corpuscles, expanded tip endings, pacinian corpuscles, and Ruffini's endings, all transmit their signals in beta type A nerve fibers that have transmission velocities of 30 to 60 meters per second. On the other hand, free nerve ending receptors and probably some of the hair end-organs transmit signals via very small delta type A nerve fibers and perhaps via some type C fibers that conduct at velocities of 1 to 8 meters per second. Thus, the more critical types of sensory signals—those that help to determine precise localization on the skin, minute gradations of intensity, or rapid changes in sensory signal intensity—are all transmitted in the rapidly conducting types of sensory nerve fibers. On the other hand, the cruder types of signals, such as crude deep pressure and poorly localized touch, are transmitted via much slower nerve fibers, fibers that also require much less space in the nerves.

KINESTHETIC SENSATIONS

The term "kinesthesia" means conscious recognition of the orientation of the different parts of the body with respect to each other as well as of the rates of movement of the different parts of the body. These functions are subserved principally by extensive sensory endings in the joint capsules and ligaments.

The Kinesthetic Receptors. By far the most abundant type of nerve ending in the joint capsules and ligaments is a spray type of ending similar to *Ruffini's end-organ*, which was illustrated in Figure 32–1. These endings are stimulated strongly when the joint is suddenly moved; they adapt slightly at first but then transmit a steady signal thereafter.

Stimulation of the joint receptors. Figure 32–5 illustrates the excitation of seven different nerve fibers leading from separate joint receptors in the capsule of a cat's knee joint. Note that at 180 degrees angulation one of the receptors is stimulated; then at 150 degrees another is stimulated; at 140 degrees two are stimulated, and so forth. The information from these joint receptors continually apprises the central nervous system of the momentary angulation of the joint. That is, the angulation determines *which* receptor is stimulated, and from this the brain can know how far the joint is angulated.

Detection of rate of movement at the joint. Rate of movement at the joint is probably detected mainly in the following way: The Ruffini endings in the joint tissues are stimulated very strongly at first by the process of joint movement, but within a fraction of a second this strong level of stimulation fades to a lower, steady state rate of firing. Nevertheless, this early overshoot in receptor stimulation is directly proportional to the rate of joint movement and is believed

to be the signal used by the brain to discern the rate of movement. However, it is likely that a few pacinian corpuscles in the joint ligaments also play at least some role in this process.

Transmission of Kinesthetic Signals in the Peripheral Nerves. Kinesthetic signals, like those from the critical tactile sensory receptors, are transmitted almost entirely in the beta type A sensory nerve fibers, which carry signals very rapidly to the cord and, thence, to the brain. This rapid transmission of kinesthetic signals is particularly important when parts of the body are being moved rapidly, because it is essential for the central nervous system to "know" at each small fraction of a second the exact locations of the different parts of the body; otherwise one would not be capable of controlling further movements.

THE DUAL SYSTEM FOR TRANSMISSION OF MECHANORECEPTIVE SOMATIC SENSATIONS IN THE CENTRAL NERVOUS SYSTEM

All sensory information from the somatic segments of the body enters the spinal cord through the posterior roots. On entering the cord, the fibers separate into medial and lateral divisions. The medial fibers immediately enter the *dorsal columns* of the cord and ascend the entire length of the cord, while the lateral fibers soon synapse in the spinal cord with neurons that form the *spinothalamic tracts*. These ascend the cord in the ventral and lateral columns of the cord.

Comparison of the Dorsal Column and Spinothalamic Systems. The dorsal column system is composed of large, heavily myelinated nerve fibers and transmits signals to the brain at velocities of 35 to 70 meters per second. Also, there is a high degree of spatial orientation of the nerve fibers with respect to their origin on the surface of the body. On the other

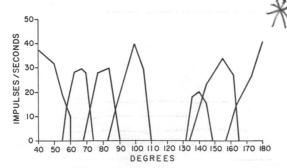

FIGURE 32–5 Responses of seven different nerve fibers from knee joint receptors in a cat (Modified from Skogland: *Acta Physiol. Scand.*, Suppl. 124, *36*:1, 1956.)

hand, the spinothalamic system is composed mainly of small fibers, some of which are not myelinated at all, or are poorly myelinated, and which transmit impulses at low velocities. These fibers are spatially oriented but only poorly so.

With this differentiation in mind we can now list the types of sensations transmitted in the two systems.

The Dorsal Column System
1. Touch sensations requiring a high degree of localization of the stimulus and requiring the transmission of fine gradations of intensity.
2. Phasic sensations, such as vibratory sensations.
3. Kinesthetic sensations.
4. Pressure sensations having to do with fine degrees of judgment of pressure intensity.

The Spinothalamic System
1. Pain.
2. Thermal sensations, including both warm and cold sensations.
3. Crude touch sensations capable of much less localizing ability on the surface of the body and perhaps requiring slightly increased intensity of stimulus to elicit.
4. Pressure sensations of a somewhat cruder nature than those transmitted by the dorsal column system.
5. Tickle and itch sensations.
6. Sexual sensations.

TRANSMISSION IN THE DORSAL COLUMN SYSTEM

ANATOMY OF THE DORSAL COLUMN PATHWAY

The dorsal column system is illustrated in Figure 32–6. The nerve fibers entering the *dorsal columns* pass all the way up these columns to the medulla, where they synapse in the *dorsal column nuclei* (the *cuneate* and *gracile nuclei*). From here, *second order neurons* decussate immediately to the opposite side and then pass upward to the thalamus through bilateral pathways called the *medial lemnisci*. Each medial lemniscus

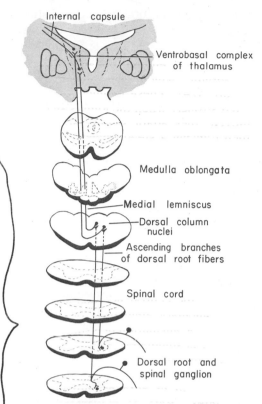

FIGURE 32–6 The dorsal column pathway for transmitting critical types of mechanoreceptive signals. (Modified from Ransom and Clark: Anatomy of the Nervous System.)

terminates in a *ventrobasal complex of nuclei* located posteriorly and ventrally in each side of the *thalamus*. In its pathway through the hindbrain, the medial lemniscus is joined by additional fibers from the *main sensory nucleus of the trigeminal nerve and from the upper portion of its descending nuclei*; these fibers subserve the same sensory functions for the head that the dorsal column fibers subserve for the body.

From the ventrobasal complex, *third order neurons* project mainly to the *postcentral gyrus* of the *cerebral cortex*. But, in addition, neurons also project to closely associated regions of the cortex behind and in front of the postcentral gyrus.

Spatial Orientation of the Nerve Fibers in the Dorsal Column Pathway. All the way from the origin of the dorsal columns to the cerebral cortex, a distinct spatial orientation of the fibers from individual parts of the body is maintained. The fibers from the lower parts of the body lie toward the center of the dorsal columns, while those that enter the dorsal columns at progressively higher levels form successive layers on the lateral sides of the dorsal columns.

In the thalamus, the distinct spatial orientation is maintained, with the tail end of the body represented by the most lateral portions of the ventrobasal complex and the head and face represented in the medial component of the complex.

THE SOMESTHETIC CORTEX

Two distinct cortical areas are known to receive direct afferent nerve fibers from the relay nuclei of the thalamus; these, called *somatic sensory area I* and *somatic sensory area II*, are illustrated in Figure 32–7.

Though the term "somesthetic cortex" actually means all the areas of the cerebral cortex that receive primary sensory information from the body, somatic sensory area I is so much more important to the sensory functions of the body than is somatic sensory area II that in popular usage "somesthetic cortex" almost always designates area I exclusive of area II.

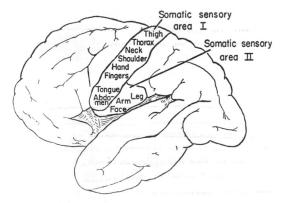

FIGURE 32-7 The two somesthetic cortical areas, somatic sensory areas I and II.

Projection of the Body in Somatic Sensory Area I. Somatic sensory area I lies in the postcentral gyrus of the human cerebral cortex. Figure 32–8 illustrates a cross-section through the brain at the level of the postcentral gyrus, showing the representations of the different parts of the body in separate regions of somatic sensory area I. Note, however, that each side of the cortex receives sensory information almost exclusively from the opposite side of the body (except for a small amount of sensory information from the same side of the face).

Some areas of the body are represented by large areas in the somatic cortex—the lips by far the greatest area of all, followed by the face and thumb—while the trunk and lower part of the body are represented by relatively small areas. The sizes of these areas are directly proportional to the number of specialized sensory receptors in each respective peripheral area of the body. For instance, a great number of specialized nerve endings are found in the lip and thumb, while only a few are present in the skin of the trunk.

Note also that the head is represented in the lower or lateral portion of the postcentral gyrus, while the lower part of the body is represented in the medial or upper portion of the postcentral gyrus.

Functions of Somatic Sensory Area I. Widespread excision of somatic sensory area I causes loss of the following types of sensory judgment:

1. The person is unable to localize discretely the different sensations in the different parts of the body. However, he can localize these sensations crudely, which indicates that the thalamus or parts of the cerebral cortex not normally considered to be concerned with somatic sensations can perform some degree of localization.

2. He is unable to judge critical degrees of pressure against his body.

3. He is unable to judge exactly the weights of objects.

4. He is unable to judge shapes or

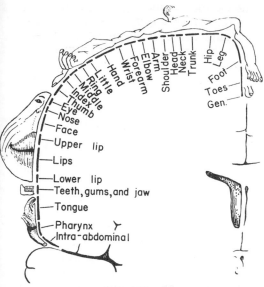

FIGURE 32–8 Representation of the different areas of the body in the somatic sensory area I of the cortex. (From Penfield and Rasmussen: The Cerebral Cortex in Man. The Macmillan Co.)

forms of objects. This is called *astereognosis*.

5. He is unable to judge texture of materials, for this type of judgment depends on highly critical sensations caused by movement of the skin over the surface to be judged.

6. He is unable to judge fine gradations in temperature.

7. He is unable to recognize the relative orientation of the different parts of his body with respect to each other.

Somatic Sensory Area II. The second cortical area to which somatic afferent fibers project, somatic sensory area II, lies posterior and inferior to the lower end of the postcentral gyrus and on the upper wall of the lateral fissure, as shown in Figure 32–7. The degree of localization of the different parts of the body is far less acute in this area than in somatic sensory area I. The face is represented anteriorly, the arms centrally, and the legs posteriorly.

So little is known about the function of somatic sensory area II that it cannot be discussed intelligently. It has been sug-

gested that somatic sensory area II might be a cortical terminus of pain information, though this is questionable.

CHARACTERISTICS OF TRANSMISSION IN THE DORSAL COLUMN SYSTEM

Cortical Control of Sensory Sensitivity. "Centrifugal" impulses are transmitted from the cortex to the still lower relay stations in the sensory pathways to facilitate or inhibit transmission. For instance, centrifugal pathways control the sensitivity of the synapses in the dorsal column nuclei and also in the posterior horn relay station of the spinothalamic system. And, similar systems are known for the visual, auditory, and olfactory senses, which are discussed in later chapters. Each centrifugal pathway begins in the cortex where the sensory pathway that it controls terminates. Thus, a feedback loop exists for each sensory pathway.

Centrifugal control of sensory input could allow the brain to determine the threshold for different sensory signals. Also, it might help the brain focus its attention on specific types of information, which is an important and necessary quality of nervous system function.

Basic Neuronal Circuit and Discharge Pattern in the Dorsal Column System. The lower part of Figure 32–9 illustrates the basic organization of the neuronal circuit of the dorsal column system, showing that at each synaptic stage a moderate degree of divergence occurs. However, the upper part of Figure 32–9 shows that a single receptor stimulus on the skin does not cause all the cortical neurons with which that receptor connects to discharge at the same rate. Instead, the cortical neurons that discharge to the greatest extent are those in a central part of the cortical "field" for each respective receptor. Thus, a weak stimulus causes only central neurons to fire. A moderate stimulus causes still more neurons to fire, but those in the

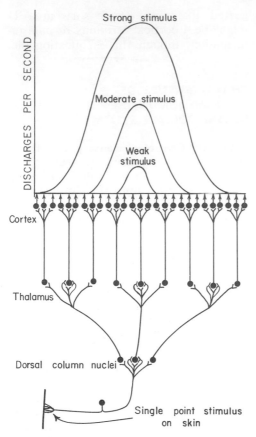

FIGURE 32–9 Transmission of a pinpoint stimulus signal to the cortex.

one can detect two separate points. The reason for this is that there are many specialized tactile receptors in the tips of the fingers in comparison with a small number in the skin of the back. Referring back to Figure 32–8, we can see also that the portions of the body that have a high degree of two-point discrimination have a correspondingly large cortical representation in somatic sensory area I.

Figure 32–10 illustrates the probable mechanism by which the dorsal column pathway transmits two-point discriminatory information. This shows two adjacent points on the skin that are strongly stimulated, and it shows the small area of the somesthetic cortex (greatly enlarged) that is excited by signals from the two stimulated points in the skin. The two dashed curves show the individually excited cortical fields, and the solid curve shows the resultant cortical excitation when both the skin points are stimulated simultaneously. Note that the resultant zone of excitation has two separate peaks. It is believed to be these two peaks, separated by a valley, that allow the sensory cortex to detect the presence of two stimulatory points rather than a single stimulatory point.

center discharge at a considerably more rapid rate than do those farther away from the center. Finally, a strong stimulus causes widespread discharge in the cortex but again a much more rapid discharge of the central neurons in comparison with the peripheral neurons.

Two-Point Discrimination. A method frequently used to test a person's tactile capabilities is to determine his so-called "two-point discriminatory ability." In this test, two needles are pressed against the skin, and the subject determines whether he feels two points of stimulus or one point. On the tips of the fingers a person can distinguish two separate points even when the needles are as close together as 1 to 2 mm. However, on the back, the needles must usually be as far apart as 10 to 15 mm. before

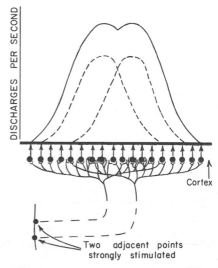

FIGURE 32–10 Transmission of signals to the cortex from two adjacent pinpoint stimuli.

Increase in Contrast in the Spatial Pattern Caused by the Phenomenon of "Lateral Inhibition." Contrast in sensory patterns is increased by inhibitory signals transmitted laterally in the sensory pathway. For instance, in the case of the dorsal column system, an excited sensory receptor in the skin transmits not only excitatory signals to the somesthetic cortex but also inhibitory signals laterally to adjacent fibers. These inhibitory signals help to block lateral spread of the excitatory signal. As a result, the peak of excitation stands out, and much of the surrounding diffuse stimulation is blocked. Obviously, this mechanism accentuates the contrast between the areas of peak stimulation and the surrounding areas, thus greatly increasing the contrast or sharpness of the perceived spatial pattern.

Transmission of Rapidly Changing and Repetitive Sensations. The dorsal column system is of particular value for apprising the sensorium of rapidly changing peripheral conditions. This system can "follow" changing stimuli up to at least 100 cycles per second and can detect vibratory sensations up to 700 cycles per second.

TRANSMISSION IN THE SPINOTHALAMIC SYSTEM

The spinothalamic system transmits sensory signals that do not require rapid transmission or highly discrete localization in the body. These include the pain, heat, cold, crude touch, crude pressure, and sexual sensations. In the following chapter, pain and temperature sensations are discussed specifically; the present section is concerned principally with transmission of crude touch and pressure in the spinothalamic system.

THE SPINOTHALAMIC PATHWAY

Figure 32–11 illustrates the spinothalamic pathway, which is divided into

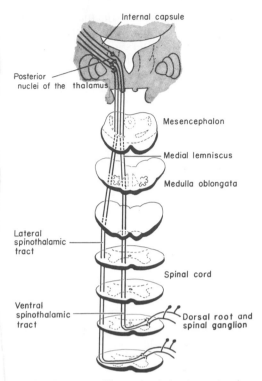

FIGURE 32–11 The spinothalamic pathways. (Modified from Ranson and Clark: Anatomy of the Nervous System.)

two separate tracts: the ventral spinothalamic tract and the lateral spinothalamic tract.

Ventral Spinothalamic Tract. The ventral spinothalamic tract transmits crude touch and pressure signals. The fibers of this tract first enter the dorsal columns along with the fibers of the dorsal column system, but they soon terminate on second order neurons in the dorsal horns. Then, fibers from these neurons cross through the anterior commissure to the opposite anterior column and form the *ventral spinothalamic tract* that passes all the way to the thalamus.

The ventral spinothalamic tract terminates, along with the terminations of the dorsal column system, mainly in the ventrobasal nuclear complex of the thalamus, though there are a few terminations in the bulbar and mesencephalic reticular areas. The fibers of the ventral spinothalamic tract are mainly larger size delta type A nerve fibers, in contrast to smaller fibers found in the lateral tract. Furthermore, there is crude localization of touch and pressure sensations transmitted through this tract, indicating a moderate degree of *spatial orientation* of the fibers in the tract.

Lateral Spinothalamic Tract. The lateral spinothalamic tract transmits pain and temperature signals. On entering the cord through the posterior roots, the fibers travel upward for one to three segments and then terminate on *second order neurons* in the gray matter of the dorsal horns. The fibers from these neurons then cross through the anterior commissure of the cord gray matter, pass to the opposite lateral column of the spinal cord where they form the *lateral spinothalamic tract*, and eventually terminate mainly in the intralaminar nuclei of the thalamus.

Many second order neurons of the lateral spinothalamic tract terminate in the reticular substance of the bulbar and mesencephalic areas or send collaterals into these areas. These fibers, though often considered to be part of the spinothalamic system, are also frequently designated as the *spinotectal* or the *spinobulbar tract*. The fibers of the lateral spinothalamic and spinotectal tracts are mainly very small delta type A or type C fibers.

Projection of Spinothalamic Signals from the Thalamus to the Cortex. Third order neurons from both the ventrobasal complex of the thalamus and the intralaminar nuclei probably project primarily to somatic area I. However, on the basis of inadequate information, it has been suggested that the lateral spinothalamic system, which transmits pain and temperature sensations, might project also to somatic area II.

Characteristics of Transmission in the Spinothalamic Tracts. In general, the same principles apply to transmission in the spinothalamic tracts as apply in the dorsal column system, except for the following differences: (a) the velocities of transmission in the spinothalamic tracts are only one-fifth to a half those in the dorsal column system; (b) the spatial localization of signals transmitted in the spinothalamic tracts is far less acute than in the dorsal column system, especially in the lateral spinothalamic tract that transmits pain and temperature signals; (c) the gradations of intensities are also far less acute, most of the sensations being recognized in 10 to 20 gradations of strength rather than the 100 or more gradations for the dorsal column system; and (d) the ability to transmit rapidly repetitive sensations is almost nil in the spinothalamic system.

Thus, it is evident that the spinothalamic tracts are by far a cruder type of transmission system than is the dorsal column system. However, certain types of sensations are transmitted only by the spinothalamic tracts, including pain, thermal sensations, and sexual sensations. Only the mechanoreceptive sensations are transmitted by both systems, the dorsal column system having to do with the critical types of mechanoreceptive sensations, and the ventral spinothalamic tract having to do with the cruder types of mechanoreceptive sensations.

REFERENCES

Adrian, E. D.: The Physical Background of Perception. Oxford, Clarendon Press, 1947.
Bishop, P. O.: Central nervous system: afferent mechanisms and perception. *Ann. Rev. Physiol., 29:*427, 1967.
Darian-Smith, I.: Somatic sensation. *Ann. Rev. Physiol., 31:*417, 1969.
Goldberg, J. M., and Lavine, R. A.: Nervous system: afferent mechanisms. *Ann. Rev. Physiol., 30:*319, 1968.
Granit, R.: Receptors and Sensory Perception. New Haven, Yale University Press, 1955.
Loewenstein, W. R.: Biological transducers. *Sci. Amer., 203(2):*98, 1960.
Sinclair, D. C.: Cutaneous Sensation. New York, Oxford University Press, 1967.
Zotterman, Y.: Sensory Mechanisms. Progress in Brain Research. New York, American Elsevier Publishing Co., Inc., 1967, Vol. 23.

CHAPTER 33

SOMATIC SENSATIONS: II. PAIN, VISCERAL PAIN, HEADACHE, AND THERMAL SENSATIONS

Pain is a protective mechanism for the body; it occurs whenever tissues are being damaged, and induces the individual to react reflexly to remove the cause of the pain. Even such simple activities as sitting for a long time on the ischia can cause tissue destruction because of lack of blood flow to the skin. When the skin becomes painful because of the ischemia, the person unconsciously shifts his weight. A person who has lost his pain sense, such as after spinal cord injury, fails to shift his weight, and his skin soon ulcerates.

Pain has been classified into three different major types: pricking, burning, and aching pain. It is not necessary to describe these different qualities of pain in detail because they are well known to all persons.

Uniformity of Pain Threshold in Different People. One method for testing the minimum stimulus that will cause pain is to determine the minimal skin temperature that will cause pain. Figure 33–1 shows graphically the skin temperature at which pain is first perceived by different persons. By far the greatest number of subjects perceive pain when the skin temperature reaches almost exactly 45° C., and almost everyone perceives pain before the temperature reaches 47° C. In other words, it is almost never true that some persons are unusually sensitive or insensitive to pain. Indeed, measurements in people as widely different as Eskimos, Indians, and whites have shown no significant differences in their *thresholds for pain*. However, different people do *react* very differently to pain, as is discussed below.

THE PAIN RECEPTORS AND THEIR STIMULATION

Free Nerve Endings as Pain Receptors. The pain receptors in the skin and other tissues are all free nerve endings. They are widespread in the superficial layers

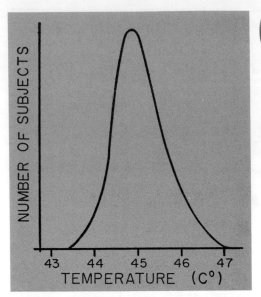

FIGURE 33–1 Distribution curve obtained from a large number of subjects of the minimal skin temperature that causes pain. (Modified from Hardy: *J. Chronic Dis., 4:22*, 1956.)

of the *skin* and also in certain internal tissues, such as the *periosteum*, the *arterial walls*, the *joint surfaces*, and the *falx* and *tentorium* of the cranial vault. Most of the other deep tissues are not extensively supplied with pain endings but are diffusely supplied; nevertheless, any widespread tissue damage can still summate to cause the aching type of pain in these areas.

Rate of Tissue Damage as the Cause of Pain. The average critical value of 45° C. at which a person first begins to perceive pain is also the temperature at which tissues begin to be damaged by heat; indeed, tissues are eventually completely destroyed if the temperature remains at this level indefinitely. Therefore, it is immediately apparent that pain resulting from heat is closely correlated with the ability of heat to damage the tissues.

Furthermore, in studying soldiers who had been severely wounded in World War II, it was found that the majority of them felt little or no pain except for a short time after the severe wound had been sustained. This, too, indicates that pain generally is not felt after damage has been done but only *while damage is being done*, or for a short time thereafter.

Bradykinin and Histamine as Possible Stimulators of Pain Endings. The precise mechanism by which tissue damage stimulates pain endings is not known. However, there are many reasons to believe that the substance *bradykinin* or some similar product might be the principal substance that stimulates pain endings. For instance, when this substance is injected in extremely minute quantities underneath the skin, severe pain is felt. Furthermore, cell damage releases proteolytic enzymes that almost immediately split bradykinin and other similar substances from the globulins in the interstitial fluid. And, finally, bradykinin and similar substances appear in the skin when painful stimuli are applied.

Thus, the postulated mechanism for eliciting pain is as follows: Damage to cells releases proteolytic enzymes that split bradykinin and associated substances from globulin, and these in turn stimulate the nerve endings.

Another substance possibly involved in at least some types of pain is histamine, because damaged cells also release this substance and because almost infinitesimal amounts of it, too, can cause very severe pain upon injection beneath the skin.

Tissue Ischemia and Muscle Spasm as Causes of Pain. When blood flow to a tissue is blocked, the tissue becomes very painful within a few minutes. And the greater the rate of metabolism of the tissue, the more rapidly the pain appears. Also, muscle spasm is a frequent cause of pain. The reason for this is probably two-fold. First, the contracting muscle compresses the intramuscular blood vessels and either reduces or cuts off the blood flow. Second, muscle contraction increases the rate of metabolism of the muscle. Therefore, muscle spasm probably causes relative muscle ischemia so that typical ischemic pain results.

The cause of pain in ischemia is yet unknown; however, it is relieved by supplying oxygen to the ischemic tissue. Flow of unoxygenated blood to the tissue will not relieve the pain.

One of the suggested causes of pain in ischemia is accumulation of large amounts of lactic acid in the tissues, formed as a consequence of the anaerobic metabolism (metabolism without oxygen) that occurs during ischemia. However, it is also possible that other chemical agents, such as bradykinin and histamine, are formed in the tissues because of muscle cell damage and that these, rather than lactic acid, stimulate the pain nerve endings.

Tickling and Itch. The phenomenon of tickling and itch has often been stated to be caused by very mild stimulation of pain nerve endings, since whenever pain is blocked by anesthesia of a nerve or by compressing the nerve, the phenomenon of tickling and itch also disappears. However, recent neurophysiologic studies have demonstrated the existence of a few very sensitive free nerve endings that respond to extremely light touch. These are suspected to carry the sensation of tickle and itch, and their signals are transmitted over primitive type C fibers similar to those that transmit the burning type of pain.

TRANSMISSION OF PAIN SIGNALS INTO THE CENTRAL NERVOUS SYSTEM

"Fast" Pain Fibers and "Slow" Pain Fibers. Pricking pain signals are transmitted by small delta type A fibers at velocities of between 3 and 10 meters per second, while burning and aching pain signals are transmitted by type C fibers at velocities between 0.5 and 2 meters per second. Therefore, a sudden painful stimulus gives a "double" pain sensation: a fast pricking pain sensation followed a second or so later by a slow burning pain sensation. The pricking pain presumably apprises the person very rapidly of a damaging influence and, therefore, plays an important role in making the person react immediately to remove himself from the stimulus. On the other hand, the slow burning sensation tends to become more and more painful over a period of time. It is this sensation that gives one the intolerable feeling toward pain.

Transmission in the Spinothalamic and Spinotectal Tracts. The type C fibers of the spinothalamic and spinotectal tracts that transmit burning and aching pain terminate in reticular areas of the medulla, pons, and mesencephalon and also in the intralaminar nuclei of the thalamus. However, the small myelinated type A delta fibers that transmit pricking pain terminate in the posterior nuclear group. It is in this same area that the ventral spinothalamic fibers for transmission of touch and pressure terminate.

Pain Pathways in the Brain. Unfortunately, very little is known about the transmission of pain signals in the bulbar, diencephalic, and cortical regions of the brain. It is believed that the pricking pain signals that enter the posterior nuclear group are thence conducted by way of third order neurons to the somesthetic areas of the cortex along much the same pathways that touch and pressure signals are transmitted.

On the other hand, the burning and aching types of pain signals seem to spread through the reticular regions of the brain stem along pathways illustrated in Figure 33–2 and, finally, pass into the intralaminar nuclei of the thalamus. What happens beyond here is very unclear.

The pain signals that enter the reticular substance of the hindbrain probably are of special importance in causing the suffering quality of pain, because stimulation in these regions apparently causes very severe suffering in animals.

Function of the Thalamus and Cerebral Cortex in the Appreciation of Pain. Complete removal of the sensory areas of the cerebral cortex does not destroy one's ability to perceive pain. Therefore, it is believed that pain impulses entering only the thalamus and even lower cen-

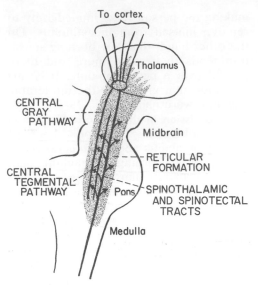

FIGURE 33–2 Transmission of pain signals into the hindbrain, thalamus, and cortex through multiple diffuse pathways.

ters cause at least some conscious perception of pain. However, this does not mean that the cerebral cortex has nothing to do with normal pain appreciation; indeed, electrical stimulation of points in the somesthetic cortical areas causes a person to perceive mild pain in approximately 3 per cent of the stimulations. Furthermore, lesions in these areas, particularly in somatic sensory area II, at times give rise to severe pain. Thus, there is reason to believe that somatic sensory area II may be more concerned with pain sensation than is somatic sensory area I.

Localization of Pain in the Body. Most localization of pain probably results from simultaneous stimulation of tactile receptors along with the pain stimulation. However, the pricking type of pain, transmitted through delta type A fibers, can be localized at least crudely.

THE REACTION TO PAIN

Even though the threshold for recognition of pain remains approximately equal from one person to another, the degree to which each one reacts to pain varies tremendously. Stoical persons, such as members of the American Indian race, react to pain far less intensely than do more emotional persons. In the preceding chapter, it was pointed out that conditioning impulses entering the sensory areas of the central nervous system from various portions of the central and peripheral nervous systems can determine whether incoming sensory impulses will be transmitted extensively or weakly to other areas of the brain. It is probably some such mechanism as this that determines how much one reacts to pain.

Pain causes both reflex motor reactions and psychic reactions. Some of the reflex actions occur directly from the spinal cord, for pain impulses entering the gray matter of the cord can directly initiate "withdrawal reflexes" that remove the body or a portion of the body from the noxious stimulus, as will be discussed in Chapter 37. These primitive spinal cord reflexes, though important in lower animals, are mainly suppressed in the human being by the higher centers of the central nervous system. Yet, in their place, much more complicated and more effective reflexes from the motor cortex are initiated by the pain stimuli to eliminate the painful stimulus.

The psychic reactions to pain are likely to be far more subtle; they include all the well-known aspects of pain such as anguish, anxiety, crying, depression, nausea, and excess muscular excitability throughout the body. These reactions vary tremendously from one person to another following comparable degrees of pain stimuli.

VISCERAL PAIN

In clinical diagnosis, pain from the different viscera of the abdomen and chest is one of the few criteria that can be used for diagnosing visceral inflammation, disease, and other ailments. In general, the viscera have sensory receptors for no other modalities of sensation besides pain, and visceral pain differs from surface pain in important aspects.

One of the most important differences between surface pain and visceral pain is that highly localized types of damage to the viscera rarely cause severe pain. For instance, a surgeon can cut the gut entirely in two in a patient who is awake without causing significant pain. On the other hand, any stimulus that causes *diffuse stimulation of pain nerve endings throughout a viscus causes pain* that can be extremely severe. For instance, occluding the blood supply to a large area of gut stimulates many diffuse pain fibers at the same time and can result in extreme pain. Also, chemical damage to the surfaces of the viscera, spasm of the smooth muscle in a hollow viscus, distention of a hollow viscus, or stretching of the ligaments can all cause severe visceral pain.

The Visceral Pathway for Transmission of Pain. Most of the internal organs of the body are supplied by pain fibers that pass along the visceral sympathetic nerves into the spinal cord and thence up the lateral spinothalamic tract along with the pain fibers from the body's surface. A few visceral pain fibers—those from the distal portions of the colon, from the rectum, and from the bladder—enter the spinal cord through the sacral parasympathetic nerves, and some enter the central nervous system through various cranial nerves. These include fibers in the glossopharyngeal and vagus nerves, which transmit pain from the pharynx, trachea, and upper esophagus. And fibers from the surfaces of the diaphragm as well as from the lower esophagus are carried in the phrenic nerves.

"Referred Pain" Transmitted by the Visceral Pathways. The position in the cord to which visceral afferent fibers pass from each organ depends on the segment of the body from which the organ developed embryologically. For instance, the heart originated in the neck

and upper thorax. Consequently, the heart's visceral pain fibers enter the cord in the neck and upper five segments of the thoracic cord. The stomach had its origin approximately from the seventh to the ninth thoracic segments of the embryo, and consequently the visceral afferents from the stomach enter the spinal cord between these levels. The gallbladder had its origin almost entirely in the ninth thoracic segment, so that the visceral afferents from the gallbladder enter the spinal cord at T-9.

Because the visceral afferent pain fibers are responsible for transmitting pain from the viscera, the location of the pain on the surface of the body is in the dermatome of the segment from which the visceral organ was originally derived in the embryo. Since this surface area usually is not directly over the visceral organ, it is called *referred pain*. Some of the areas of referred pain on the surface of the body are shown in Figure 33–3.

Mechanism of referred pain. Figure 33–4 illustrates the most generally accepted mechanism by which most pain is referred. In the figure, branches of visceral pain fibers are shown to synapse in the spinal cord with some of the same

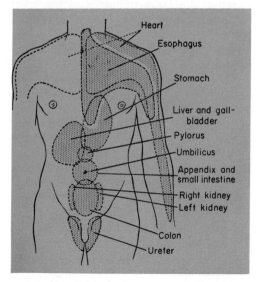

FIGURE 33–3 Surface areas of referred pain from different visceral organs.

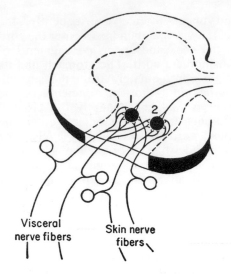

FIGURE 33–4 Mechanism of referred pain and referred hyperalgesia.

second order neurons that receive pain fibers from the skin. When the visceral pain fibers are stimulated intensely, pain sensations from the viscera spread into some of the neurons that normally conduct pain sensations only from the skin, and the person has the feeling that the sensations actually originate in the skin itself. It is also possible that some referred pain results from convergence of visceral and skin impulses at the level of the thalamus rather than in the spinal cord.

The Parietal Pathway for Transmission of Abdominal and Thoracic Pain. Where skeletal nerves overlie the abdomen or thorax, pain fibers penetrate inward to innervate the parietal peritoneum, parietal pleura, and parietal pericardium. Also, retroperitoneal visceral organs and portions of the mesentery are innervated to some extent by parietal pain fibers. The kidney, for instance, is supplied by both visceral and parietal fibers.

Pain from the viscera is frequently localized in two surface areas of the body because of the dual pathways for transmission of pain. Figure 33–5 illustrates dual transmission of pain from an inflamed appendix. Impulses pass from the appendix through the sympathetic visceral pain fibers into the sympathetic chain and then into the thoracic spinal cord at approximately T-10 or T-11; this pain is referred to an area around the umbilicus. On the other hand, pain impulses also often originate in the parietal peritoneum where the inflamed appendix touches the abdominal wall, and these impulses pass directly through the skeletal nerves into the lumbar spinal cord at a level of approximately L-1 or L-2. This pain is localized directly over the irritated peritoneum in the right lower quadrant of the abdomen.

SOME CLINICAL ABNORMALITIES OF PAIN AND OTHER SENSATIONS

Paresthesia. Sometimes a sensory nerve tract becomes irritated by a lesion in a peripheral nerve or in the central nervous system. Spontaneous impulses are then transmitted from the irritated nerve fiber all the way to the brain, and a sensation is localized to a peripheral

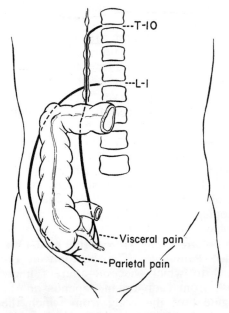

FIGURE 33–5 Visceral and parietal transmission of pain from the appendix.

area of the body which has not been directly stimulated. This sensation is called a paresthesia. Obviously, the irritative lesion can cause any one of the different modalities of sensation, depending on which type of fiber is involved.

Hyperesthesia. Hyperesthesia is very similar to paresthesia except that there is no spontaneous sensation. Instead, the sensory fibers are merely facilitated so that they are excessively sensitive to sensory stimuli.

The basic causes of hyperesthesia are, first, excessive sensitivity of the receptors themselves, which is called *primary hyperesthesia*, or, second, facilitation of sensory transmission, which is called *secondary hyperesthesia*.

Hyperesthesia that involves only pain sensations is called *hyperalgesia*. An example of *primary hyperalgesia* is the extreme sensitivity of sunburned skin. An example of *secondary hyperalgesia* is extreme pain sensitivity caused by lesions in some areas of the hindbrain or thalamus.

Tic Douloureux. Lancinating pains occur in some persons over one side of the face in part of the sensory distribution area of the fifth or ninth nerve; this phenomenon is called tic douloureux. The pains feel like sudden electric shocks, and they may appear for only a few seconds at a time or they may be almost continuous. Often, they are set off by exceedingly sensitive "trigger areas" on the surface of the face, in the mouth, or in the nose. For instance, when the patient swallows a bolus of food, as the food touches a tonsil it might set off a severe lancinating pain in the mandibular portion of the fifth nerve.

The pain of tic douloureux can usually be blocked by cutting the peripheral nerve from the hypersensitive area. The fifth nerve is often sectioned immediately inside the cranium, where the motor and sensory roots of the fifth nerve can be separated so that the motor portions, which are needed for many of the jaw movements, are spared while the sensory elements are destroyed. Obviously, this operation leaves the side of the face anesthetic, which in itself may be annoying. Therefore, still another operation is often performed in the medulla for relief of this pain, as follows:

The sensory nuclei of the fifth nerve divide into the main sensory nucleus, which is located in the pons, and the spinal sensory nucleus, which descends into the upper part of the spinal cord. It is the spinal nucleus that subserves the function of pain. Therefore, the spinal tract of the fifth nerve can be cut in the medulla as it passes to the spinal nucleus. This blocks pain sensations from the side of the face but does not block the sensations of touch and pressure.

HEADACHE

Headaches are actually referred pain to the surface of the head from the deep structures. Most headaches result from pain stimuli arising inside the cranium, but at least some result from pain arising outside the cranium.

HEADACHE OF INTRACRANIAL ORIGIN

Pain-Sensitive Areas in the Cranial Vault. The brain itself is almost totally insensitive to pain. Even cutting or electrically stimulating the somesthetic centers of the cortex only occasionally causes pain; instead, it causes tactile paresthesias on the area of the body represented by the portion of the somesthetic cortex stimulated.

On the other hand, *tugging on the venous sinuses, damaging the tentorium*, or *stretching the dura at the base of the brain* can all cause intense pain that is recognized as headache. Also, almost any type of traumatizing, crushing, or stretching stimulus to the *blood vessels of the dura* can cause headache.

Areas of the Head to which Intracranial Headache Is Referred. Stimulation of pain receptors in the intracranial vault above the tentorium, including the upper surface of the tentorium itself, initiates impulses in the fifth nerve and, therefore, causes referred headache in the area supplied by the fifth cranial nerve. This area includes the upper part of the head anterior to the ear, as outlined by the dark shaded area of Figure 33–6. Thus, pain arising above the tentorium causes what is called "frontal headache."

On the other hand, pain impulses from beneath the tentorium enter the central nervous system mainly through the second cervical nerve, which also supplies the scalp behind the ear. Therefore, subtentorial pain stimuli cause "occipital headache" referred to the posterior part of the head, as also shown in Figure 33–6.

Types of Intracranial Headache. One of the most severe headaches of all is that resulting from *meningitis*, which causes inflammation of all the meninges, including the sensitive areas of the dura and the sensitive areas around the venous sinuses.

Another type of meningeal irritation that almost invariably causes headache

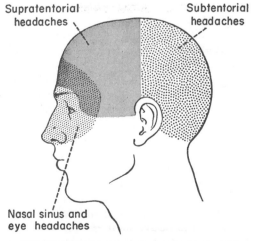

Supratentorial headaches

Subtentorial headaches

Nasal sinus and eye headaches

FIGURE 33–6 Areas of headache resulting from different causes.

is that resulting from brain tumor. Since any tumor above the tentorium refers its pain to the frontal areas and any tumor below the tentorium refers its pain to the occipital region of the skull, the general location of an intracranial tumor can often be predicted from the area of the headache.

Migraine headache. Migraine headache is a special type of headache that is thought to result from abnormal vascular phenomena, though the exact mechanism is unknown.

Migraine headaches often begin with various prodromal sensations, such as nausea, loss of vision in part of the fields of vision, visual aura, or other types of sensory hallucinations. Ordinarily, the prodromal symptoms begin half an hour to an hour prior to the beginning of the headache itself. Therefore, any theory that explains migraine headache must also explain these prodromal symptoms.

One of the theories of the cause of migraine headaches is that prolonged emotion or tension causes reflex vasospasm of some of the arteries of the head, including arteries that supply the brain itself. The vasospasm theoretically produces ischemia of portions of the brain, and this is responsible for the prodromal symptoms. Then, as a result of the intense ischemia, something happens to the vascular wall to allow it to become flaccid and incapable of maintaining vascular tone for 24 to 48 hours. The blood pressure in the vessels causes them to dilate and pulsate intensely, and it is supposedly the excessive stretching of the walls of the arteries that causes the actual pain of migraine headaches. However, it is possible that diffuse after-effects of ischemia in the brain itself are responsible for this type of headache.

Alcoholic headache. As many people have experienced, a headache usually follows an alcoholic binge. It is most likely that alcohol, because it is toxic to tissues, directly irritates the meninges and causes the cerebral pain.

EXTRACRANIAL TYPES OF HEADACHE

Headache Resulting from Muscular Spasm. Emotional tension often causes many of the muscles of the head, including especially those muscles attached to the scalp and the neck muscles attached to the occiput, to become moderately spastic, and it is postulated that this causes headache. The pain of the spastic head muscles supposedly is referred to the overlying areas of the head and gives one the same type of headache as do intracranial lesions.

Headache Caused by Irritation of the Nasal and Accessory Nasal Structures. The mucous membranes of the nose and also of all the nasal sinuses are sensitive to pain, but not intensely so. Nevertheless, infection or other irritative processes in widespread areas of the nasal structures usually cause headache that is referred behind the eyes or, in the case of frontal sinus infection, to the frontal surfaces of the forehead and scalp, as illustrated in Figure 33–6. Also, pain from the lower sinuses—such as the maxillary sinuses—can be felt in the face.

Headache Caused by Eye Disorders. Difficulty in focusing one's eyes clearly may cause excessive contraction of the ciliary muscles in an attempt to gain clear vision. Even though these muscles are extremely small, tonic contraction of them can be the cause of retroorbital headache.

THERMAL SENSATIONS

THERMAL RECEPTORS AND THEIR EXCITATION

The human being can perceive different gradations of cold and heat, progressing from *cold* to *cool* to *indifferent* to *warm* to *hot*; some persons perceive even freezing cold and burning hot.

Both heat and cold can be perceived from areas of the body that contain only free nerve endings, such as the cornea. Therefore, it is known that free nerve endings can subserve these sensory functions. Thus far, no complex specialized receptors for either cold or warmth have been discovered.

Stimulation of Thermal Receptors — Sensations of Cold, Warmth, and Hot. Figure 33–7 illustrates the responses to different temperatures of three different types of nerve fibers: (1) a fiber from a *cold receptor*, (2) a fiber from a *warm receptor*, and (3) a pain fiber. Note especially that these fibers respond differently at different levels of temperature. For instance, in the *very* cold region only the pain fibers are stimulated (if the skin becomes even colder so that it nearly freezes or actually does freeze, even these fibers cannot be stimulated). As the temperature rises to 10 to 15° C., pain impulses cease, but the cold receptors begin to be stimulated. Then, above about 25° C. the warm receptors become stimulated, while the cold receptors fade out at about 35° C. Finally, at around 45° C. the warm endings become nonresponsive once more; but the cold endings begin to respond again, and the pain endings also begin to be stimulated.

One can understand from Figure 33–7, therefore, that a person determines the different gradations of thermal sensations by the relative degrees of stimulation of the different types of endings. One can understand also from this figure why extreme degrees of cold or heat can be painful and why both these sensations, when intense enough, may give almost exactly the same quality of sensation—freezing cold and burning hot sensations feel almost alike to the person.

It is particularly interesting that a few areas of the body, such as the tip of the penis, do not contain any warmth receptors; but these areas can experience the sensations of cold with ease and of "hot," which depends on stimulating

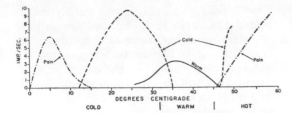

FIGURE 33-7 Frequencies of discharge of a cold receptor, a warm receptor, and a pain nerve fiber at different temperatures. (The responses of the cold and warm receptors are modified from Zotterman: *Ann Rev. Physiol.*, 15:357, 1953.)

cold and pain endings, but never the sensation of warmth.

Stimulating Effects of Rising and Falling Temperature — Adaptation of Thermal Receptors. When a cold receptor is suddenly subjected to an abrupt fall in temperature, it becomes strongly stimulated at first, but this stimulation fades rapidly during the first minute and progressively more slowly during the next half hour or more. In other words, the receptor adapts to a great extent; this is illustrated in Figure 33-8, which shows that the frequency of discharge of a cold receptor rose approximately four-fold when the temperature fell suddenly from 32° to 30° C., but in less than a minute the frequency fell about five sixths of the way back to the original control value. Later, the temperature was suddenly raised from 30° to 32° C. At this point the cold receptor stopped firing entirely for a

short time, but after adaptation, the firing returned to its original control level. The same type of response occurs in warm receptors.

Thus, it is evident that thermal receptors respond markedly to *changes in temperature* in addition to being able to respond to steady states of temperature. This explains the extreme degree of heat that one feels when he first enters a hot tub of water and the extreme degree of cold he feels when he goes from a heated room to the out-of-doors on a cold day.

Mechanism of Stimulation of the Thermal Receptors. It is believed that the thermal receptors are stimulated by changes in their metabolic rates, these changes resulting from the fact that temperature alters the rates of intracellular chemical reactions about 2.3 times for each 10° C. change. In other words, thermal detection probably results not from direct physical stimulation but instead from chemical stimulation of the endings as modified by the temperature.

Spatial Summation of Thermal Sensations. The number of cold or warmth endings in any small surface area of the body is very slight, so that it is difficult to judge gradations of temperature when small areas are stimulated. However, when a large area of the body is stimulated, the thermal signals from the entire area summate. Indeed, one reaches his maximum ability to discern minute temperature variations when his entire body is subjected to a temperature change all at once. For instance, rapid changes in temperature as little as 0.01° C. can be detected if this change affects the entire surface of the body simultaneously. On

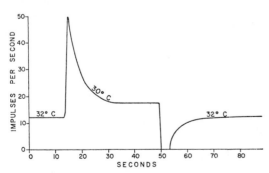

FIGURE 33-8 Response of a nerve fiber from a cold receptor following, first, instantaneous change in skin temperature from 32° to 30° C. and, second, instantaneous change back to 32° C. Note the adaptation of the receptor and also the higher steady state level of discharge at 30° than at 32°.

the other hand, temperature changes 100 times this great might not be detected when the skin surface affected is only a square centimeter or so in size.

TRANSMISSION OF THERMAL SIGNALS IN THE NERVOUS SYSTEM

Thermal signals are transmitted by the smallest myelinated fibers of the peripheral nerves, the delta group, with diameters of only 2 to 3 microns.

On entering the central nervous system, the thermal fibers travel in the lateral spinothalamic tract. The exact point in the thalamus at which the impulses synapse is not known, but since other myelinated spinothalamic fibers terminate mainly in the posterior group nuclei of the thalamus, it is believed that the thermal tracts do too. Then, third order neurons travel to the somesthetic cortex, but the portion of the cerebral cortex to which the major share of thermal fibers radiate is also unknown. Occasionally, a neuron in somatic sensory area I has been found by microelectrode studies to be directly respon-

sive to either cold or warm stimuli in specific areas of the skin. Furthermore, it is known that removal of the postcentral gyrus in the human being reduces his ability to distinguish different gradations of temperature. Therefore, at present it is believed that somatic sensory area I, rather than area II, is the major somesthetic area concerned with thermal sensations.

REFERENCES

Ciba Foundation Symposium: Touch, Heat and Pain. Boston, Little, Brown and Company, 1965.

Hardy, J. D.: The nature of pain. *J. Chronic Dis.,* 4:22, 1956.

Henry Ford Hospital Symposium: Pain. Boston, Little, Brown and Company, 1966.

Lim, R. K. S.: Pain. *Ann. Rev. Physiol.,* 32:269, 1970.

Sweet, W. H.: Pain. *In* Handbook of Physiology. Baltimore, The Williams & Wilkins Co., 1959, p. 459, Sec. I, Vol. 1.

Way, E. L.: New Concepts in Pain and Its Clinical Management. Philadelphia, F. A. Davis Co., 1967.

Wolff, H. G.: Headache and Other Head Pain. 2nd ed., New York, Oxford University Press, 1963.

Zotterman, Y.: Thermal sensations. *In* Handbook of Physiology. Baltimore, The Williams & Wilkins Co., 1959, p. 431, Sec. I, Vol. 1.

CHAPTER 34

THE EYE: I. OPTICS OF VISION; AND RECEPTOR FUNCTIONS OF THE RETINA

PHYSICAL PRINCIPLES OF LENSES

Before it is possible to understand the optical systems of the eye, the student must be familiar with a few of the basic physical principles of lenses. Therefore, in our present study of the optics of the eye, a brief review of these principles is first presented, and then the optics of the eye are discussed.

The Convex Lens. Figure 34–1 shows parallel light rays entering a convex lens. The light rays passing through the center of the lens strike the lens exactly perpendicular to the lens surfaces and therefore pass through the lens without being refracted at all. Toward either edge of the lens, however, the light rays strike a progressively more angulated interface. Therefore, the outer rays bend more and more toward the center. Thus, parallel light rays entering an appropriately formed convex lens come to a single

point focus at some distance beyond the lens.

The Concave Lens. Figure 34–2 shows the effect of a concave lens on parallel light rays. The rays that enter the very center of the lens strike an interface that is absolutely perpendicular to the beam and, therefore, do not refract at all. The rays at the edge of the lens enter the lens ahead of the rays toward the center. This is opposite to the effect in the convex lens, and it causes the peripheral light rays to *diverge* away from the

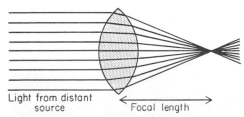

Light from distant source   Focal length

FIGURE 34–1 Bending of light rays at each surface of a convex spherical lens, showing that parallel light rays are focused to a point focus.

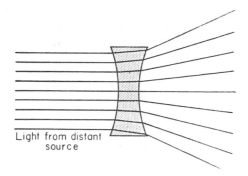

FIGURE 34-2 Bending of light rays at each surface of a concave spherical lens, illustrating that parallel light rays are diverged by a concave lens.

light rays that pass through the center of the lens.

Thus, the concave lens *diverges* light rays, whereas the convex lens *converges* light rays.

Spherical versus Cylindrical Lenses. Figure 34-3 illustrates both a convex *spherical* lens and a convex *cylindrical* lens. Note that the convex cylindrical lens bends light rays from the two sides of the lens but not from either the top or the bottom. Therefore, parallel light rays are bent to a focal *line*. On the other hand, the light rays that pass through the spherical lens are refracted at all edges of the lens toward the central ray, and all the rays come to a focal *point*.

FORMATION OF AN IMAGE BY A CONVEX LENS

The upper drawing of Figure 34-4 illustrates a convex lens with two point sources of light to the left. Because light rays from any point source pass through the center of a convex lens without being refracted in either direction, the light rays from both point sources of light are shown to pass straight through the lens center. Furthermore, the other light rays from each point source of light, whether they pass through the upper edge of the lens, the sides of the lens, or the lower edge, all come to the same point focus behind the lens *directly in*

line with the point source of light and the center of the lens.

Any object in front of the lens is in reality a mosaic of point sources of light. Some of these points are very bright and some are very weak, and they vary in color. The light rays from each point source of light are focused behind the lens in line with the rays that pass through the center. If all portions of the object are the same distance in front of the lens, all the focal points behind the lens will fall in a common plane a certain distance behind the lens. If a white piece of paper is placed at this distance, one can see an image of the object, as is illustrated in the lower portion of Figure 34-4. However, this image is upside

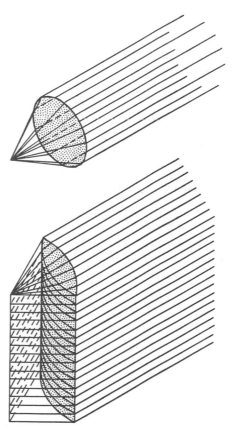

FIGURE 34-3 *Top:* Point focus of parallel light rays by a spherical convex lens. *Bottom:* Line focus of parallel light rays by a cylindrical convex lens.

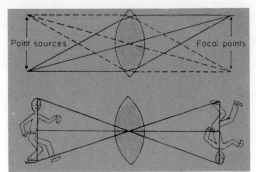

FIGURE 34–4 The top drawing illustrates two point sources of light focused at two separate points on the opposite side of the lens. The lower drawing illustrates formation of an image by a convex spherical lens.

down with respect to the original object, and the two lateral sides of the image are reversed with respect to the original object. This is the method by which the lens of a camera focuses light rays on the camera film.

Measurement of the Refractive Power of a Lens — The Diopter. The more a lens bends light rays, the greater is its "refractive power." This refractive power is measured in terms of *diopters*. The refractive power of a convex lens is equal to 1 meter divided by its focal length. Thus a lens has a refractive power of +1 diopter when it is capable of converging parallel light rays to a focal point 1 meter beyond the lens, as illustrated in Figure 34–5. Likewise, if a concave lens diverges light rays the same amount that a

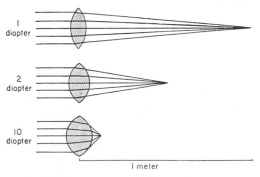

FIGURE 34–5 Effect of lens strength on the focal distance.

1 diopter convex lens converges them, the concave lens is said to have a dioptric strength of −1.

THE EYE AS A CAMERA

The eye, as illustrated in Figure 34–6 is optically equivalent to the usual photographic camera, for it has a lens system, a variable aperture system, and a retina that corresponds to the film. The lens system of the eye is composed of (1) the interface between air and the anterior surface of the cornea, (2) the interface between the posterior surface of the cornea and the aqueous humor, (3) the interface between the aqueous humor and the anterior surface of the lens, and (4) the interface between the posterior surface of the lens and the vitreous humor. At each of these interfaces the light rays are bent, and the net result is a convex lens system capable of forming images on the retina.

The Reduced Eye. If all the refractive surfaces of the eye are algebraically added together and then considered to be one single lens, the optics of the normal eye may be simplified and represented schematically as a "reduced eye." This is useful in simple calculations. In the reduced eye, a single lens is considered to exist with its central point 17 mm. in front of the retina and have a total refractive power of approximately 59 diopters when the lens is accommodated for distant vision.

FOCUSING OF THE EYE — THE MECHANISM OF ACCOMMODATION

The refractive power of the crystalline lens of the eye makes up 18 diopters of the total 59 diopters of the eye lens system; the remaining 41 diopters results almost entirely from the anterior interface of the cornea with air. However, the strength of the lens can be voluntarily increased from 18 diopters to approximately 32 diopters in young

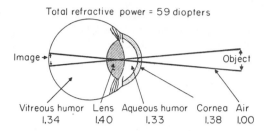

Total refractive power = 59 diopters

Image → | Object

Vitreous humor | Lens | Aqueous humor | Cornea | Air
1.34 | 1.40 | 1.33 | 1.38 | 1.00

FIGURE 34–6 The eye as a camera. The numbers are the refractive indices.

children; this is a total "accommodation" of 14 diopters. The mechanism of this is the following:

Normally, the lens is composed of a strong elastic capsule filled with viscous proteinaceous, but transparent, fibers. When the lens is in a relaxed state, with no tension on its capsule, it assumes a spherical shape, owing entirely to the elasticity of the lens capsule. However, as illustrated in Figure 34–7, approximately 70 ligaments attach radially around the lens, pulling the lens edges toward the ciliary body. These ligaments are constantly tensed by the elastic pull of their attachments at the ciliary body, and the tension on the ligaments causes the lens to remain relatively flat under normal resting conditions of the eye. At the insertions of the tendons in the

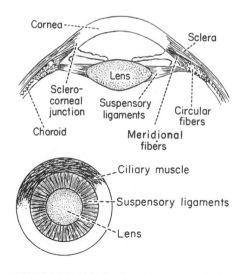

Cornea
Sclera
Sclero-corneal junction
Lens
Suspensory ligaments
Circular fibers
Choroid
Meridional fibers

Ciliary muscle
Suspensory ligaments
Lens

FIGURE 34–7 Mechanism of accommodation.

ciliary body is the ciliary muscle. When this muscle contracts, the insertions of the ligaments are pulled forward, thereby releasing the tension on the crystalline lens, and the lens assumes a more spherical shape, like that of a balloon, because of elasticity of its capsule. When the ciliary muscle is completely relaxed, the dioptric strength of the lens is as weak as it can become, about 18 diopters. On the other hand, when the ciliary muscle contracts as strongly as possible, the dioptric strength of the lens becomes maximal.

The ciliary muscle is controlled by the parasympathetic nervous system, which will be discussed later in the chapter.

Presbyopia. As a person grows older, his lens loses its elastic nature and becomes a relatively solid mass, probably because of progressive denaturation of the proteins. Therefore, the ability of the lens to assume a spherical shape progressively decreases, and the power of accommodation decreases from approximately 14 diopters shortly after birth to approximately 2 diopters at the age of 45 to 50. Thereafter, the lens of the eye may be considered to be almost totally nonaccommodating, which condition is known as "presbyopia."

Once a person has reached the state of presbyopia, each eye remains focused permanently at an almost constant distance; this distance depends on the physical characteristics of each individual's eyes. Obviously, the eyes can no longer accommodate for both near and far vision.

THE PUPILLARY APERTURE

A major function of the iris is to increase the amount of light that enters the eye during darkness and to decrease the light that enters the eye in bright light. The reflexes for controlling this mechanism will be considered later in the chapter. The amount of light that enters the eye through the pupil is proportional to the area of the pupil or to the *square of the diameter* of the pupil. The pupil of the human eye can become as small as approximately 1.5 mm. and can become almost as large as 8 mm. in diameter. Therefore, the quantity of light entering the eye may vary approximately 30 times as a result of changes in pupillary aperture size.

Depth of Focus of the Lens System of the Eye. Figure 34–8 illustrates two separate eyes that are exactly alike except that the diameters of the pupillary apertures are different. In the upper eye the pupillary aperture is small, and in the lower eye the aperture is large. In front of each of these two eyes are two small point sources of light, and light from each passes through the pupillary apertures and focuses on the retina. Consequently, in both eyes the retina sees two spots of light in perfect focus. It is evident from the diagrams, however, that if the retina is moved forward or backward, the size of each spot will not change much in the upper eye, but in the lower eye the size of the spot will increase greatly, and it becomes a "blur circle." In other words, the lens system with the small aperture has far greater *depth of focus* than the lens system with the large aperture. When a lens system has great depth of focus, the retina can be considerably displaced from the focal plane and still discern the various points of an image distinctly; whereas, when a

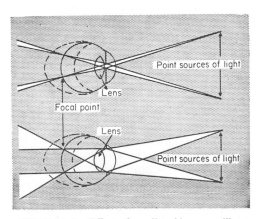

FIGURE 34–8 Effect of small and large pupillary apertures on the depth of focus.

lens system has shallow depth of focus, having the retina only slightly away from the focal plane causes extreme blurring of the image.

ERRORS OF REFRACTION

Emmetropia. As shown in Figure 34–9, the eye is considered to be normal or "emmetropic" if, when the ciliary muscle is completely relaxed, parallel light rays from distant objects are in sharp focus on the retina. This means that the emmetropic eye can, with its ciliary muscle completely relaxed, see all distant objects clearly, but to focus objects at close range it must contract its ciliary muscle and thereby provide various degrees of accommodation.

Hypermetropia (Hyperopia). Hypermetropia, which is also known as "far-sightedness," is due either to an eyeball that is too short or to a lens system that is too weak when the ciliary muscle is completely relaxed. In this condition, parallel light rays are not bent sufficiently by the lens system to come to a focus by the time they reach the retina.

Myopia. In myopia, or "near-sightedness," even when the ciliary muscle is completely relaxed, the strength of the lens is still so great that light rays coming from distant objects are focused in front of the retina. This is usually due to too long an eyeball, but it can rarely result from too much power of the lens system of the eye.

Correction of Myopia and Hypermetropia by Use of Lenses. It will be recalled that light rays passing through a concave lens diverge. Therefore, if the refractive surfaces of the eye have too much refractive power, as in myopia, some of this excessive refractive power can be neutralized by placing in front of the eye a concave spherical lens, which will diverge the rays. On the other hand, in a person who has hypermetropia—that is, one who has too weak a lens for the distance of the retina away from the lens—the abnormal vision can be cor-

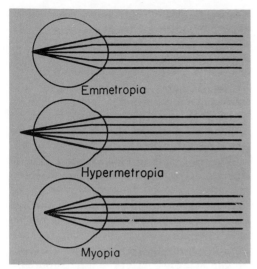

FIGURE 34–9 Parallel light rays focus on the retina in emmetropia, behind the retina in hypermetropia, and in front of the retina in myopia.

rected by adding refractive power with a convex lens in front of the eye. One usually determines the strength of the concave or convex lens needed for clear vision by "trial and error"—that is, by trying first a strong lens and then a stronger or weaker lens until the one that gives the best visual acuity is found.

Astigmatism. Astigmatism is a refractive error of the lens system of the eye caused usually by an oblong shape of the cornea or, rarely, by an oblong shape of the lens. A lens surface like the side of a football lying edgewise to the incoming light, for instance, would be an example of an astigmatic lens. The degree of curvature in a plane through the long axis of the football is not nearly so great as the degree of curvature in a plane through the short axis. The same is true of an astigmatic lens. Because the curvature of the astigmatic lens along one plane is less than the curvature along the other plane, light rays striking the peripheral portions of the lens in one plane are not bent nearly so much as are rays striking the peripheral portions of the other plane.

This is illustrated in Figure 34–10,

which shows what happens to rays of light emanating from a point source and passing through an astigmatic lens. The light rays in the vertical plane, which is indicated by plane BD, are refracted greatly by the astigmatic lens because of the greater curvature in the vertical direction than in the horizontal direction. However, the light rays in the horizontal plane, indicated by plane AC, are bent not nearly so much as the light rays in the vertical plane. It is obvious, therefore, that the light rays passing through an astigmatic lens do not all come to a common focal point because the light rays passing through one plane of the lens focus far in front of those passing through the other plane.

Correction of astigmatism with a cylindrical lens. To correct the focusing of an astigmatic eye lens system, a glass lens having equal but reciprocal astigmatism is placed in front of the eye. That is, a cylindrical lens must be found that will make the refractive power of one lens plane equal to that of the other lens plane. To do this, it is necessary to determine both the *strength* of the cylindrical lens needed to neutralize the excess cylindrical power of the eye lens and the *axis* of this abnormal cylindrical lens.

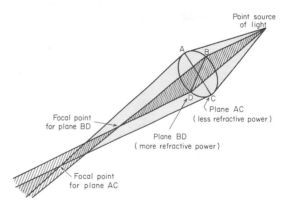

FIGURE 34–10 Astigmatism, illustrating that light rays focus at one focal distance in one focal plane and at another focal distance in the plane at right angles.

SIZE OF THE IMAGE ON THE RETINA AND VISUAL ACUITY

Theoretically, a point of light from a distant point source, when focused on the retina, would be infinitely small. However, since the lens system of the eye is not absolutely perfect, such a retinal spot ordinarily has a total diameter of about 11 microns even with maximum resolution of the optical system. However, it is brightest in its very center and shades off gradually toward the edges.

The average diameter of cones *in the fovea* of the retina is approximately 1.5 microns, which is one-seventh the diameter of the spot of light. Nevertheless, since the spot of light has a bright center point and shaded edges, a person can distinguish two separate points if their centers lie approximately 2 microns apart on the retina, which is slightly greater than the width of a foveal cone. This discrimination between points is illustrated in Figure 34–11.

The maximum visual acuity of the human eye for discriminating between point sources of light is 26 seconds. That is, when light rays from two separate points strike the eye with an angle of at least 26 seconds between them, they can usually be recognized as two points instead of one. This means that a person with maximal acuity looking at two bright pinpoint spots of light 10 meters away can barely distinguish the spots as separate entities when they are 1 millimeter apart.

The visual acuity is no more than one-tenth this good in the retina outside the foveal region, becoming poorer and poorer as the periphery is approached. This is caused by the connection of many rods and cones to the same nerve fiber, as discussed later in the chapter.

Clinical Method for Stating Visual Acuity. Usually the test chart for testing eyes is placed 20 feet away from the tested person, and if the person can see the letters of the size that he should be able to see at 20 feet, he is said to have 20/20 vision: that is, normal vision. If he can see only letters that he should be able to see at 200 feet, he is said to have 20/200 vision. On the other hand, if he can see at 20 feet letters that he should be able to see only at 15 feet, he is said to have 20/15 vision. In other words, the clinical method for expressing visual acuity is to use a mathematical fraction that expresses the ratio of two distances, which is also the ratio of one's visual acuity to that of the normal person.

DETERMINATION OF DISTANCE OF AN OBJECT FROM THE EYE —DEPTH PERCEPTION

The visual apparatus normally perceives distance, a phenomenon known as depth perception, in three ways: (1) by relative sizes of objects, (2) by moving parallax, and (3) by stereopsis.

Determination of Distance by Relative Sizes. If a person knows that a man is six feet tall and then he sees this man even with only one eye, he can determine how far away the man is simply by the size of the man's image on his retina.

Determination of Distance by Moving Parallax. When a person moves his head to one side or the other, the images of objects close to him move rapidly across his retinae while the images of distant objects remain rather stationary.

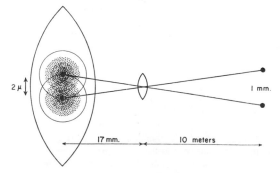

FIGURE 34–11 Maximal visual acuity for two point sources of light.

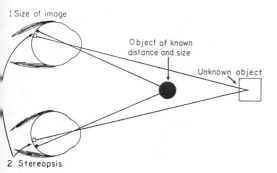

2. Stereopsis

FIGURE 34-12 Perception of distance (1) by the size of the image on the retina and (2) as a result of stereopsis.

Determination of Distance by Stereopsis.

Another method by which one perceives parallax is that of binocular vision. Because one eye is a little more than 2 inches to one side of the other eye, the images on the two retinae are different one from the other—that is, an object that is 1 inch in front of the bridge of the nose forms images on the temporal portions of the two retinas, whereas a small object 20 feet in front of the nose forms its images at closely corresponding points in the middle of the two retinas. This type of parallax is illustrated in Figure 34–12, which shows the images of a black spot and a square actually reversed on the retinae because they are at different distances in front of the eyes. This gives a type of parallax that is present all the time when both eyes are being used. It is almost entirely this binocular parallax (or stereopsis) that gives a person with two eyes far greater ability to judge relative distances *when objects are nearby* than a person who has only one eye. However, stereopsis is virtually useless for depth perception at distances beyond 200 feet.

ANATOMY AND FUNCTION OF THE STRUCTURAL ELEMENTS OF THE RETINA

The retina is the light-sensitive portion of the eye, containing the cones, which detect specific colors, and the rods, which detect light of any color besides deep red. When the rods and cones are excited, signals are transmitted through successive neurons in the retina itself and finally into the optic nerve fibers and cerebral cortex. The purpose of the remainder of this chapter is to explain specifically the mechanisms by which the rods and cones detect both white and colored light.

The Layers of the Retina. Figure 34–13 shows the functional components of the retina arranged in layers from the outside to the inside as follows: pigment layer, layer of rods and cones projecting into the pigment, outer limiting membrane, outer nuclear layer, outer plexiform layer, inner nuclear layer, inner plexiform layer, ganglion layer, layer of optic nerve fibers, and inner limiting membrane.

Light entering the retina passes from the bottom upwards in Figure 34–13. Since the rods and cones are the light-sensitive portion of the retina, the light must pass through all the other layers besides the pigment layer before reaching the excitable cells, and this obviously decreases the visual acuity.

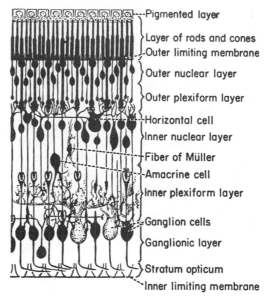

FIGURE 34-13 Plan of the retinal neurons. (From Polyak: The Retina. University of Chicago Press.)

The Rods and Cones. The nuclei of the rods and cones are located in the outer nuclear layer while the light-sensitive portions of both these receptors project through the outer limiting membrane and protrude part way into the pigment layer of the retina. The rods are relatively slender, having a diameter of about 2 microns in the more central portions of the retina and 4 to 5 microns in the more peripheral portions. The cones, on the other hand, have a diameter of about 1.5 microns in the *fovea*, the centralmost portion of the retina, and about 5 to 8 microns in the periphery.

The sensitive portions of both the rods and cones contain light-sensitive chemicals which decompose on exposure to light. The decomposition products in turn stimulate the cell membranes of the rods and cones, eliciting signals that are then transmitted into the nervous system. Indeed, the rods and cones themselves are modified neurons and can actually be considered to be part of the nervous system.

The Foveal Region of the Retina and Its Importance in Acute Vision. A minute area in the center of the retina called the *macula* and occupying a total area of about 1 square millimeter is especially capable of acute and detailed vision. This area is composed entirely of cones, but the cones are very much elongated and have a diameter of only 1.5 microns in contradistinction to the very large cones located farther peripherally in the retina. The central portion of the macula is called the *fovea*; in this region the blood vessels, the ganglion cells, the inner nuclear layer of cells, and the plexiform layers are all displaced to one side rather than resting directly on top of the cones. This allows light to pass unimpeded to the cones rather than through several layers of retina, which aids in the acuity of visual perception by this region of the retina.

The Pigment Layer of the Retina. The black pigment *melanin* in the pigment layer prevents light reflection throughout the globe of the eyeball, and this is extremely important for acute vision. This pigment performs the same function in the eye as the black paint inside the bellows of a camera.

Relationship of the Retina to the Choroid. The retina adheres to the choroid, which is a highly vascular layer of the eye underlying the sclera. The outer layers of the retina, including especially the rods and cones, are at least partially dependent on diffusion of nutrients from the capillaries of the choroid.

Retinal detachment. The neural portion of the retina occasionally detaches from the pigment epithelium, and the retina then hangs free in the eyeball separated from the choroid. Fortunately, partly because of diffusion across the detachment gap and partly because of an independent blood supply to the retina through the retinal artery, the retina can resist degeneration for many days and can become functional once again if surgically replaced in its normal relationship with the pigment epithelium.

PHOTOCHEMISTRY OF VISION

Both the rods and cones contain chemicals that decompose on exposure to light and, in the process, excite the nerve fibers leading from the eye. The chemical in the *rods* is called *rhodopsin* or *visual purple*, and the light-sensitive chemicals in the *cones* have chemical compositions only slightly different from that of rhodopsin.

THE RHODOPSIN-RETINENE VISUAL CYCLE, AND EXCITATION OF THE RODS

Rhodopsin and its Decomposition by Light Energy. The portion of the rod that projects into the pigment layer of the retina has a concentration of about 40 per cent of *rhodopsin*. This substance is a combination of the protein *scotopsin* and the carotenoid pigment 11–*cis reti-nene*. This *cis* form of the retinene is important because only this type of retinene can combine with scotopsin to form rhodopsin.

When light energy is absorbed by rhodopsin, the rhodopsin immediately begins to decompose as shown in Figure 34–14. The cause of this is an instantaneous change of the *cis* form of retinene into an all-*trans* form, which still has the same chemical structure as the *cis* form but has a different physical structure. Yet, the reactive sites of the all-*trans* retinene no longer fit with the reactive sites of the protein scotopsin, and it begins to pull away from the scotopsin. The immediate product is *lumi-rhodopsin*, which is a partially split combination of the all-*trans* retinene and scotopsin. However, lumi-rhodopsin is an extremely unstable compound and decays in a small fraction of a second to meta-rhodopsin, which is still another loose combination of the all-*trans* retinene and scotopsin. This compound, too, is unstable, and it slowly decomposes during the next few seconds into completely split products – scotopsin and all-*trans* retinene. During the process of splitting, the rods become excited, and signals are transmitted into the central nervous system.

Reformation of Rhodopsin.   The first stage in reformation of rhodopsin, as shown in Figure 34–14, is to reconvert

the all-*trans* retinene into the *cis* form of this pigment. This process is catalyzed by the enzyme *retinene isomerase*. However, this process requires active metabolism and energy transfer to form the 11–*cis* retinene; once 11–*cis* retinene is formed, it automatically recombines with the scotopsin to perform rhodopsin, an exergonic process (which means that it gives off energy). The product, rhodopsin, is a stable compound until its decomposition is again triggered by absorption of light energy.

Excitation of the Rod as Rhodopsin Decomposes.   The exact method by which rhodopsin decomposition excites the rod is still speculative. However, it is known that when the all-*trans* retinene begins to split away from the scotopsin, several ionized radicals are momentarily exposed, and this could excite the rod in one of several ways. One suggestion has been that the instantaneous ionic field of the radicals causes the membrane potential of the rod to change and that this in turn causes excitation. The physical construction of the pigment portion of the rod is compatible with this concept because the cell membrane folds inward to form tremendous numbers of discs lying one on top of the other, and each one is surfaced on the inside with aggregates of rhodopsin. This extensive relationship between rhodopsin and the cell membrane could explain why the rod is so exquisitely sensitive to light, being capable of detectable excitation following absorption of only one quantum of light energy.

Relationship Between Retinene and Vitamin A.   The lower part of the scheme in Figure 34–14 illustrates that each of the two types of retinene can be converted into a corresponding type of vitamin A, and in turn the two types of vitamin A can be reconverted into the two types of retinene. However, these processes require much longer time to approach equilibrium than for conversion of retinene and scotopsin into rhodopsin, or for conversion (under the influence of strong light energy) of rhodopsin into retinene and scotopsin.

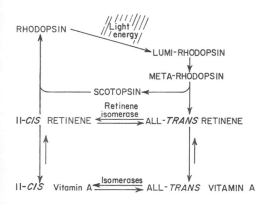

FIGURE 34–14 Photochemistry of the rhodopsin-retinene-vitamin A visual cycle.

Therefore, all the reactions of the upper part of Figure 34–14 can take place relatively rapidly in comparison with the slow interconversions between retinene and vitamin A.

Night blindness. Night blindness is associated with vitamin A deficiency, for, when the total quantity of vitamin A in the blood becomes greatly reduced, the quantities of vitamin A, retinene, and rhodopsin in the rods, as well as the color photosensitive chemicals in the cones, are all depressed, thus decreasing the sensitivities of the rods and cones. This condition is called night blindness because at night the amount of available light is far too little to permit adequate vision, whereas in daylight, sufficient light is available to excite the rods and cones despite their reduction in photochemical substances.

AUTOMATIC REGULATION OF RETINAL SENSITIVITY — LIGHT AND DARK

If a person has been in bright light for a long time, large proportions of the photochemicals in both the rods and cones are reduced to retinene and opsins (scotopsin in rods, photopsins in cones). Furthermore, much of the retinene in both the rods and cones is converted into vitamin A. Because of these two effects, the concentrations of the photochemicals are considerably reduced, and the sensitivity of the eye to light is even more reduced. This is called *light adaptation.*

On the other hand, if the person remains in the darkness for a long time, essentially all the retinene and opsins in the rods and cones become converted into light-sensitive pigments. Furthermore, large amounts of vitamin A are converted into retinene, which is then changed into additional light-sensitive pigments, the final limit being determined by the amount of opsins in the rods and cones. Because of these two effects, the visual receptors gradually become so sensitive that even the mi-

nutest amount of light causes excitation. This is called *dark adaptation.*

Figure 34–15 illustrates the course of dark adaptation when a person is exposed to total darkness after having been exposed to bright light for several hours. Note that retinal sensitivity is very low when one first enters the darkness, but within 1 minute the sensitivity has increased 10-fold—that is, the retina can respond to light of one-tenth intensity. At the end of 20 minutes the sensitivity has increased about 6000-fold, and at the end of 40 minutes it has increased about 25,000-fold.

The resulting curve of Figure 34–15 is called the *dark adaptation curve.* Note, however, the inflection in the curve. The early portion of the curve is caused by adaptation of the cones, for these adapt much more rapidly than the rods because of a basic difference in the rate at which they resynthesize their photosensitive pigments. On the other hand, the cones do not achieve anywhere near the same degree of sensitivity as the rods. Therefore, despite rapid adaptation by the cones, they cease adapting after only a few minutes, while the slowly adapting rods continue to adapt for many minutes or even hours, their sensitivity increasing tremendously. However, a large share of the greater sensi-

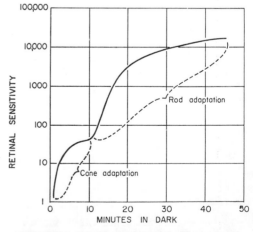

FIGURE 34–15 Dark adaptation, illustrating the relationship of cone adaptation to rod adaptation.

tivity of the rods is caused by convergence of as many as 100 rods onto a single cell in the retina; these rods summate to increase their sensitivity, as will be discussed later in the chapter.

Value of Light and Dark Adaptation in Vision. Between the limits of maximal dark adaptation and maximal light adaptation, the retina of the eye can change its sensitivity to light by as much as 500,000 to 1,000,000 times, the sensitivity automatically adjusting to changes in illumination.

Since the registration of images by the retina requires detection of both dark and light spots in the image, it is essential that the sensitivity of the retina always be adjusted so that the receptors respond to the lighter areas and not to the darker areas. An example of maladjustment of the retina occurs when a person leaves a movie theater and enters the bright sunlight, for even the dark spots in the images then seem exceedingly bright, and, as a consequence, the entire visual image is bleached, having little contrast between its different parts. Obviously, this is poor vision, and it remains poor until the retina has adapted sufficiently that the dark spots of the image no longer stimulate the receptors excessively.

Negative After-Images. If one looks steadily at a scene for a while, the areas of the retina that are stimulated by light become less sensitive while areas that are exposed only to darkness gain in sensitivity. If the person then moves his eyes away from the scene and looks at a bright white surface he sees exactly the same scene that he had been viewing, but the light areas of the scene now appear dark, and the dark areas appear light. This is known as the negative after-image, and it is a natural consequence of light and dark adaptation.

FUSION OF FLICKERING LIGHTS BY THE RETINA

A flickering light is one whose intensity alternately increases and decreases rapidly. An instantaneous flash of light excites the visual receptors for as long as $1/10$ to $1/5$ second, and because of this *persistence* of excitation, rapidly successive flashes of light become *fused* together to give the appearance of being continuous. This effect is well known when one observes motion pictures.

The critical frequency at which flicker fusion occurs varies with light intensity. At low intensity, fusion results even when the rate of flicker is as low as 5 to 6 per second. However, in bright illumination, the critical frequency for fusion rises to as great as 60 flashes per second. This difference results at least partly from the fact that the cones, which operate mainly at high levels of illumination, can detect much more rapid alterations in illumination than can the rods.

COLOR VISION

PHOTOCHEMISTRY OF COLOR VISION BY THE CONES

The only differences between the photochemicals of the cones and the rhodopsin of the rods is that the protein portions, the opsins, called *photopsin* in the cones, are different from the scotopsin of the rods. The retinene portions are exactly the same. The color-sensitive pigments of the cones, therefore, are combinations of retinene and photopsins.

Three different types of pigments are present in different cones, thus making these cones selectively sensitive to the colors red, green, and blue. The absorption characteristics of the pigments in the three types of cones show peak absorbences at, respectively, 430, 535, and 575 millimicrons. The approximate absorption curves for these three pigments are shown in Figure 34–16. The peak absorption for the rhodopsin of the rods, on the other hand, occurs at 505 millimicrons.

Spectral Sensitivities of the Three Types of Cones. On the basis of psychological tests, the spectral sensitivities of the three different types of cones in human

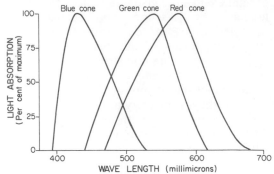

FIGURE 34–16 Light absorption by the respective pigments of the three color receptive cones of the human retina. (Drawn from curves recorded by Marks, Dobelle, and MacNichol, Jr.: *Science, 143*:1181, 1964, and by Brown and Wald: *Science, 144*:45, 1964.)

beings are essentially the same as the light absorption curves for the three types of pigment found in the respective cones, as illustrated in Figure 34–16, and they can readily explain almost all the phenomena of color vision.

INTERPRETATION OF COLOR IN THE NERVOUS SYSTEM—THE TRI-COLOR THEORY OF COLOR VISION

Referring again to Figure 34–16, one can see that a red monochromatic light with a wavelength of 610 millimicrons stimulates the red cones to a stimulus value of approximately 0.75 while it stimulates the green cones to a stimulus value of approximately 0.13 and the blue cones not at all. Thus, the ratios of stimulation of the three different types of cones in this instance are 75:13:0. The nervous system interprets this set of ratios as the sensation of red. On the other hand, a monochromatic blue light with a wavelength of 450 millimicrons stimulates the red cones to a stimulus value of 0, the green cones to a value of 0.14 and the blue cones to a value of 0.86. This set of ratios—0:14:86—is interpreted by the nervous system as blue. Likewise, ratios of 100:50:0 are

interpreted as orange-yellow and 50:85:15 as green.

This mechanism for color vision is called the tri-color theory, or the Young-Helmholtz theory, of color vision. It also shows how it is possible for a mixture of two different monochromatic colors to give the sensation of seeing still another color. For instance, a person perceives a sensation of yellow when a red light and a green light are shown into the eye at the same time, for this stimulates the red and green cones approximately equally, which gives a sensation of yellow even though no wavelength of light corresponding to yellow is present.

Perception of White Light. Approximately equal stimulation of all the red, green, and blue cones gives one the sensation of seeing white. Yet there is no wavelength of light corresponding to white; instead, white is a combination of all the wavelengths of the spectrum. Furthermore, the sensation of white can be achieved by stimulating the retina with a proper combination of only three chosen colors that stimulate the respective types of cones equally.

COLOR BLINDNESS

Red-Green Color Blindness. When a single group of color receptive cones is missing from the eyes because of a genetic deficiency, the person is unable to distinguish some colors from others. As can be observed by studying Figure 34–16, if the red cones are missing, light of 525 to 625 millimicrons wavelength can stimulate only the green-sensitive cones, so that the *ratio* of stimulation of the different cones does not change as the color changes from green all the way through the red spectrum. Therefore, within this wavelength range, all colors appear to be the same to this "color blind" person.

On the other hand, if the green-sensitive cones are missing, the colors in the range from green to red can stimulate only the red-sensitive cones, and the

FIGURE 34–17 Two Ishihara charts: *Upper:* In this chart, the normal person reads "74," whereas the red-green color blind person reads "21." *Lower:* In this chart, the red-blind person (protanope) reads "2," while the green-blind person (deuteranope) reads "4." The normal person reads "42." (From Ishihara: Tests for Colour-Blindness. Tokyo, Kanehara and Co.)

person also perceives only one color within these limits. Therefore, when a person lacks either the red or green types of cones, he is said to be "red-green" color blind.

Blue Weakness. Rarely, a person has "blue weakness," which results from diminished or absent blue receptors.

Tests for Color Blindness—Stilling and Ishihara Charts. A rapid method for determining color blindness is based on the use of spot-charts such as those illustrated in Figure 34–17. These charts are composed of a confusion of spots of several different colors. In the top chart, the normal person reads "74," while the red-green color blind person reads "21." In the bottom chart, the normal person reads "42," while the red blind "protanope" reads "2," and the green blind "deuteranope" reads "4."

If one will study these charts while at the same time observing the spectral sensitivity curves of the different cones in Figure 34–16, he can readily understand how excessive emphasis can be placed on spots of certain colors by color blind persons in comparison with normal persons.

REFERENCES

Adler, F. H.: Physiology of the Eye. 4th ed., St. Louis, The C. V. Mosby Co., 1965.

Alpern, M.: Distal mechanisms of vertebrate color vision. *Ann. Rev. Physiol., 30*:279, 1968.

Armington, J. C.: Vision. *Ann. Rev. Physiol., 27*:163, 1965.

Brindley, G. S.: Physiology of the Retina and Visual Pathway. Baltimore, The Williams & Wilkins Co., 1960.

Ciba Foundation: Colour Vision: Physiology and Experimental Psychology. Boston, Little, Brown and Company, 1965.

Davson, H.: Physiology of the Eye. 2nd ed., Boston, Little, Brown and Co., 1962.

Davson, H. (ed.): The Eye. 4 volumes. New York, Academic Press, Inc., 1962.

Duke-Elder, S.: Diseases of the Retina. *In* Systems of Ophthalmology, St. Louis, The C. V. Mosby Co., 1967, Vol. 10.

Linksz, A.: Physiology of the Eye: Vol. I. Optics. New York, Grune & Stratton, Inc., 1950.

MacNichol, E. F., Jr.: Three-pigment color vision. *Sci. Amer., 211*:48(6), 1964.

CHAPTER 35

THE EYE:
II. NEUROPHYSIOLOGY
OF VISION

THE VISUAL PATHWAY

Figure 35–1 illustrates the visual pathway from the two retinae back to the *visual cortex*. After impulses leave the retinae they pass backward through the *optic nerves*. At the *optic chiasm* all the fibers from the nasal halves of the retinae cross to the opposite side where they join the fibers from the opposite temporal retinae to form the two respective *optic tracts*. The fibers of each optic tract synapse in the *lateral geniculate body*, and from here, the *geniculocalcarine fibers* pass through the *optic radiation*, or *geniculocalcarine tract*, to the *optic* or *visual cortex* in the calcarine area of the occipital lobe.

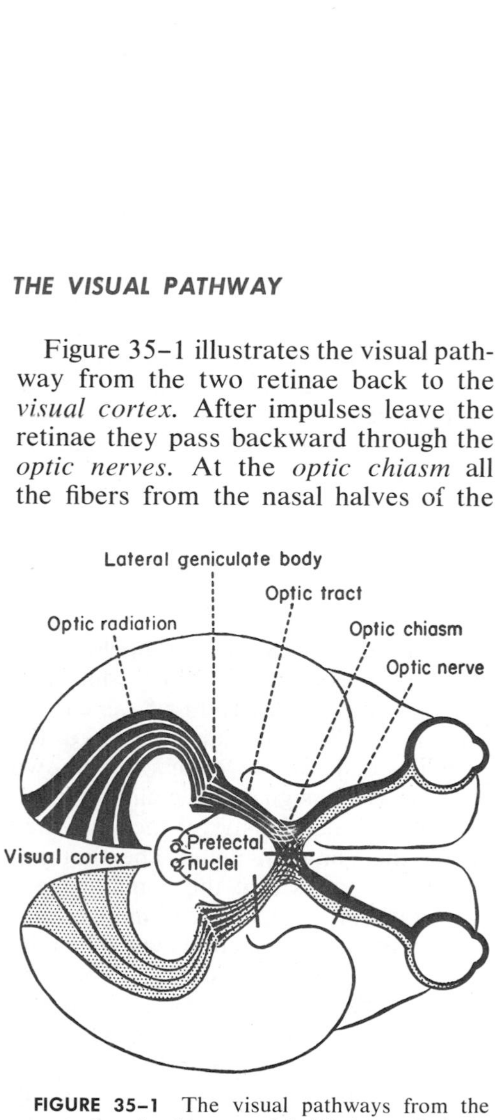

In addition, visual fibers pass to three other areas of the brain: (1) from the lateral geniculate body to the lateral thalamus, the ventral thalamus, the superior colliculi, and the pretectal nuclei; (2) from the optic tracts directly to the superior colliculi; and (3) from the optic tracts directly into the pretectal nuclei of the brain stem.

FIGURE 35–1 The visual pathways from the eyes to the visual cortex. (Modified from Polyak: The Retina. University of Chicago Press.)

NEURAL FUNCTION OF THE RETINA

NEURAL ORGANIZATION OF THE RETINA

Figure 35–2 illustrates the basic essentials of the retina's neural connections; to the left is the general organization of

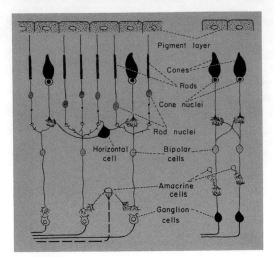

FIGURE 35–2 Neural organization of the retina: peripheral area to the left, foveal area to the right.

the neural elements in a peripheral retinal area and to the right the organization in the foveal area. Note that in the peripheral region both rods and cones converge on *bipolar cells* which in turn converge on *ganglion cells*. In the fovea, where only cones exist, there is little or no convergence; instead, the cones are represented by approximately equal numbers of bipolar and ganglion cells.

Each retina contains about 125 million rods and 5.5 million cones; yet, as counted with the light microscope, only 900,000 optic nerve fibers lead from the retina to the brain. Thus, an average of 140 rods plus 6 cones converge on each optic nerve fiber. Nearer the fovea, fewer and fewer rods and cones converge on each optic fiber, and the rods and cones both become slenderer. These two effects progressively increase the acuity of vision toward the central retina.

Another difference between the peripheral and central portions of the retina is that there is considerably greater sensitivity of the peripheral retina to weak light. This is believed to result mainly from the fact that as many as 600 rods converge and summate on the same optic nerve fiber in the most peripheral portions of the retina.

STIMULATION OF THE BIPOLAR AND HORIZONTAL CELLS

Neither the rods nor the cones generate action potentials. Instead, when light acts on these receptors, it stimulates a "receptor potential" that is approximately proportional to the logarithm of the light intensity, and this receptor potential is transmitted unchanged to the axons of the rods and cones.

In the inner nuclear layer the horizontal cells transmit signals laterally in the retina and the bipolar cells transmit signals from the rods and cones to the ganglion cells. The axons of the rods and cones make extremely intimate contact with the dendrites of both the bipolar cells and the horizontal cells. Recordings of intracellular signals in the horizontal cells indicate that these cells, like the rods and cones, perhaps also transmit no action potentials but instead transmit signals of the electrotonic variety. On the other hand, action potentials have been recorded from neurons in the inner nuclear layer that are believed to be the bipolar cells.

STIMULATION OF THE GANGLION CELLS

Spontaneous, Continuous Discharge of the Ganglion Cells. Though what happens in the transmission of visual signals from the rods and cones to the ganglion cells is not clear, it is known that the visual signals have already been changed tremendously by the time they reach the ganglion cells. The first important effect is that the ganglion cells transmit continuous nerve impulses at a rate of about 5 per second even when the retina is in total darkness. Then, the visual signal is superimposed onto this basic level of ganglion cell stimulation.

Effect of Light on Ganglion Cell Discharge. Figure 35–3 illustrates a minute spot of light shining on the very center of the *retinal field* of a peripheral

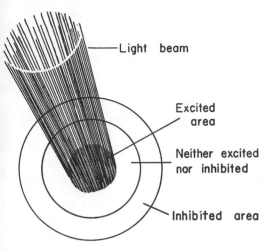

Light beam

Excited area

Neither excited nor inhibited

Inhibited area

FIGURE 35-3 Excitation and inhibition of a retinal area caused by a small beam of light.

ganglion cell. The light stimulates the rods and cones in this center, and the signals from these converge on the ganglion cell. The top record of Figure 35–4 illustrates the effect of turning the spot of light on. This record shows to the left the normal background rate of discharge of the ganglion cell and then shows that the rate of discharge increases markedly a fraction of a second after the light is turned on. Within another fraction of a second, the rate of impulses decreases almost back to the background rate of discharge; there is a slight excess rate of discharge as long as the light remains on.

When the light is turned off, exactly

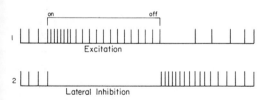

FIGURE 35-4 Responses of ganglion cells to light in (1) an area excited by a spot of light and (2) an area immediately adjacent to the excited spot; the ganglion cells in this area are inhibited by the mechanism of lateral inhibition. (Modified from Granit: Receptors and Sensory Perception. Yale University Press.)

opposite effects occur. For a fraction of a second, the ganglion cell transmits no impulses at all, but the rate gradually picks back up over another few seconds until the normal background rate of discharge returns.

Lateral Inhibition of Surrounding Ganglion Cells. Referring once again to Figure 35–3, one sees a concentric area surrounding the spot of light on the retina labeled "inhibited area." The lower record in Figure 35–4 shows the effect of the spot of light on discharge of a ganglion cell whose field center is in this area. Note that when the light is turned on, these ganglion cells become completely inhibited and stop transmitting even their normal background impulses. Then, when the light is turned off, the impulse rate increases at first to a level considerably greater than the normal rate of discharge, this effect lasting for a fraction of a second until the cells readapt to their normal level of discharge.

Mechanism of lateral inhibition. It is believed that lateral inhibition is caused by signals carried in the *horizontal cells* that have extensive horizontal connections in the outer plexiform layer of the retina. One of the reasons for believing this is that these cells have far more extensive fiber networks in the peripheral retina than in the central retina, and it is found experimentally that the area of inhibition is great in the periphery but circumscribed in the central retina, which is consonant with the more discrete visual functions of the central retina than of the peripheral retina.

Significance of lateral inhibition — enhancement of "contrast" in the visual signal. In the above description of the neural elements of the retina we noted that the dendrites in the retina spread laterally a moderate amount within the plexiform layers. This obviously causes the signals from adjacent receptors of the retina to spread into each other. Counteracting this lateral mixing of signals is the phenomenon of lateral inhibition. That is, the excitatory signals that spread laterally are opposed and nullified by the lateral inhibitory signals. This obviously helps to maintain sharp boundaries between the signals from excited and nonexcited areas of the retina, which enhances the *contrast* of the visual signal.

FUNCTION OF THE LATERAL GENICULATE BODY

Anatomical Organization of the Lateral Geniculate Nuclei. Each lateral geniculate body is composed of six nuclear layers. Layers 2, 3, and 5 (from the surface inward) receive signals from the temporal portion of the ipsilateral retina, while layers 1, 4, and 6 receive signals from the nasal retina of the opposite eye.

The pairing of layers from the two eyes probably plays a major role in *fusion of vision*, because corresponding retinal fields in the two eyes connect with respective neurons that are approximately superimposed over each other in the successive layers. Also, with a little imagination, one can postulate that interaction between the successive layers could be part of the mechanism by which stereoscopic visual depth perception occurs, because this depends on comparing the visual images of the two eyes and determining their slight differences, as was discussed in Chapter 34.

Characteristics of the Visual Signal in the Lateral Geniculate Body. The signals recorded in the relay neurons of the lateral geniculate body are almost the same as those recorded in the ganglion cells of the retinae except for one difference: the boundaries between excited and nonexcited areas of the visual signals are considerably sharper in the lateral geniculate neurons than in the ganglion cells of the retina. This indicates that additional lateral inhibition probably occurs in the lateral geniculate body in addition to that which occurs in

the retina, this lateral inhibition further enhancing the degree of contrast in the visual pattern.

FUNCTION OF THE PRIMARY VISUAL CORTEX

The ability of the visual system to detect spatial organization of the visual scene—that is, to detect the forms of objects, brightness of the individual parts of the objects, shading, and so forth—is dependent on function of the *primary visual cortex*, which is illustrated in Figure 35–5. This area lies mainly in the calcarine fissure located on the medial aspect of each occipital cortex. Specific points of the retina connect with specific points of the visual cortex, the right halves of the two respective retinae connecting with the right visual cortex and the left halves connecting with the left visual cortex. The macula is represented at the occipital pole of the visual cortex, and the peripheral regions of the retina are represented in concentric circles farther and farther forward from the occipital pole. Note the large area of cortex receiving signals from the macular region of the retina. It is in the center of this region that the fovea is represented, which gives the highest degree of visual acuity.

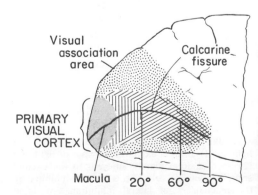

FIGURE 35–5 The visual cortex.

DETECTION OF LINES AND BORDERS BY THE PRIMARY VISUAL CORTEX

If a person looks at a blank wall, very little stimulation occurs in the neurons of the primary visual cortex, whether the illumination of the wall be bright or weak. Therefore, the question must be asked: What does the visual cortex do? To answer this question, let us now place on the wall a large black cross such as that illustrated to the left in Figure 35–6. To the right is illustrated the spatial pattern of excitation that one finds in the visual cortex. *Note that the areas of excitation occur along the sharp borders of the visual pattern.* Thus, by the time the visual signal is recorded in the primary visual cortex, it is concerned almost entirely with the *contrasts* in the visual scene rather than with the flat areas. At each point in the visual scene where there is a change from dark to light or light to dark, the corresponding area of the primary visual cortex becomes stimulated. The intensity of stimulation is determined by the *gradient of contrast*. That is, the greater the sharpness in the contrast border and the greater the difference in intensities between the light and dark areas, the greater is the degree of stimulation.

Thus, the *pattern of contrasts* in the visual scene is impressed upon the neurons of the visual cortex, and this pattern has a spatial orientation roughly the same as that of the retinal image.

Detection of Orientation of Lines and Borders. Not only does the visual cortex detect the existence of lines and borders in the different areas of the retinal image, but it also detects the orientation of each line or border—that is, whether it is a vertical or horizontal border, or lies at some degree of inclination. The way in which the visual cortex performs this function is the following:

The visual cortex is composed of thousands of minute columns of neurons, each column having a diameter of about

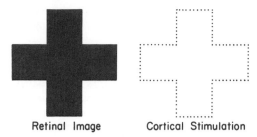

Retinal Image   Cortical Stimulation

FIGURE 35–6 Pattern of excitation occurring in the visual cortex in response to a retinal image of a black cross.

0.5 mm. and extending from the surface of the cortex down through its six layers. Each column of cells is excited by a border or line that is oriented only in one particular direction. Thus, a column of cells that will respond to a horizontal border will not respond to a vertical border. However, immediately adjacent to this horizontal column of cells is another column that responds to a border slightly inclined from the horizontal, then another adjacent column that responds to a border with still more inclination, and still another column that responds to a vertical border. Thus, each minute area of the visual cortex has multiple neuron columns, each of which responds specifically to a border or line oriented in a specific direction.

NEURAL ANALYSIS OF COLOR

Almost nothing is known at present about the mechanisms by which the visual nervous system decodes color information from the retina. However, decoding seems to begin in the retina itself by interaction among signals from different cones. A few ganglion cells transmit signals from blue receptive cones, others from green cones, and still others from red cones, representing the three primary colors. But most ganglion cells transmit combined color information from several types of cones. Two of the combinations are the following:

1. Some ganglion cells combine information from all three types of cones,

supposedly giving no color differentiation at all but only an overall luminosity signal.

2. Some ganglion cells give positive responses to blue and negative responses to yellow, whereas others give positive responses to yellow and negative to blue. Still other cells give positive responses to red and negative to blue, and so forth.

It is believed that further interactions of a similar nature but of a higher order occur at higher levels in the visual nervous system—in the lateral geniculate body and the visual cortex. That such interactions can occur at these higher levels is proved by the fact that looking at an object through a red filter with one eye and a green filter with the other eye will give the impression of seeing an image that has neither a red nor green color but some intermediate color such as yellow or orange.

PERCEPTION OF LUMINOSITY

One of the most important features of any visual image is its level of light intensity, or its *luminosity*. Yet, it has been very difficult to understand how luminosity is perceived by the visual neural apparatus because most ganglion cells in the retina respond primarily to *changes* in light intensities or to contrast borders in the retinal image. Yet, ganglion cells do seem to respond weakly at continuous rates proportional to the overall luminosity rather than simply to changes or to contrast. These signals are believed to carry luminosity information to the brain.

Yet, once the information reaches the brain, it is still a question which parts of the brain are responsible for perception of luminosity. In the human being, destruction of the primary visual cortex causes total blindness, including loss of ability to determine luminosity. However, some lower animals can still detect changes in luminosity almost as well as previously, and this suggests that lower centers could be very important in this function of the brain.

Perception of Rapid Changes in Light Intensity and of Movement. The primary visual cortex is very strongly excited by (1) sudden changes in light intensity, particularly when these changes occur along sharp borders, and (2) movement of an image across the retina. Both of these effects cause rapid discharge of ganglion cells for a fraction of a second after the initial stimulation. These effects emphasize again that the visual cortex is especially well adapted for perception of contrast signals in contradistinction to perception of steady signals.

TRANSMISSION OF VISUAL INFORMATION INTO OTHER REGIONS OF THE CEREBRAL CORTEX

After visual information reaches the primary visual cortex (area 17 in Fig. 35–7), it is further processed in adjacent areas called the visual association areas (areas 18 and 19 in Fig. 35–7).

The retinal image is spatially oriented in both area 18 and area 19, just as occurs in the primary visual cortex. Also, the visual projection images are organized into columns of cells in the same manner as that described earlier for the primary visual cortex. However, these columns of cells respond to more complex patterns than do those in the primary visual cortex. For instance, some columns of cells respond to lines or borders only of particular lengths

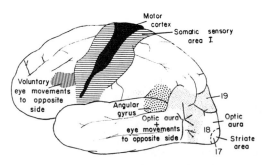

FIGURE 35–7 The visual association fields and the cortical areas for control of eye movements.

while other lengths fail to stimulate. Other columns are stimulated by simple geometric patterns, such as right angles, curving borders, or so forth, and others by specific combinations of colors. It is presumably these progressively more complex interpretations of form, colors, and probably also temporal changes that eventually decode the visual information, giving the person his overall impression of the visual scene that he is observing.

Effects of Destruction of Visual Cortical Areas in the Human Being. Destruction of the primary visual cortex causes blindness, but destruction in areas 18 and 19 only decreases one's ability to interpret the shapes of objects, their sizes, or their meanings, and can cause particularly an abnormality known as *alexia*, or *word blindness*, which means that the person cannot identify the meanings of words that he sees. Destruction of the cerebral cortex in the angular gyrus region where the temporal, parietal, and occipital lobes all come together usually makes it difficult for a person to correlate visual images with the motor functions. For instance, he is able to see his plate of food perfectly well but is unable to utilize the visual information to direct his fork toward the food. Other aspects of visual interpretation in relation to overall function of the cerebral cortex will be discussed in Chapter 39, and later in this chapter we will return to the cortical control of eye movements.

EYE MOVEMENTS AND THEIR CONTROL

To make use of the abilities of the eye, it is almost as important to direct the eyes toward the object to be viewed as to interpret the visual signals from the eyes.

Control of Eye Movements. The eye movements are controlled by three separate pairs of muscles shown in Figure 35–8: (1) the medial and lateral recti,

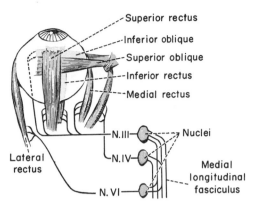

FIGURE 35–8 The extraocular muscles of the eye and their innervation.

(2) the superior and inferior recti, and (3) the superior and inferior obliques. The medial and lateral recti contract reciprocally to move the eyes from side to side. The superior and inferior recti contract reciprocally to move the eyes upward or downward. And the oblique muscles function mainly to rotate the eyeballs to keep the visual fields in the upright position.

Figure 35–8 also illustrates the innervation of the ocular muscles by the third, fourth, and sixth cranial nerves.

Simultaneous movement of both eyes in the same direction is called *conjugate movement* of the eyes. Though the precise brain centers that control conjugate movements are still unknown, it is believed that most of these are controlled by the *superior colliculi*. In turn, the association areas of the visual cortex, areas 18 and 19 in particular, transmit impulses through the *occipitocollicular tracts* to control the activities of the superior colliculi, and it is believed that these nuclei then transmit signals through the medial longitudinal fasciculus to the third, fourth, and sixth nuclei that control the eye muscles.

FIXATION MOVEMENTS OF THE EYES

Perhaps the most important movements of the eyes are those that cause the eyes to "fix" on a discrete portion of the field of vision.

Fixation movements are controlled by two entirely different neuronal mechanisms. The first of these allows the person to move his eyes voluntarily to find the object upon which he wishes to fix his vision; this is called the *voluntary fixation mechanism.* The second is an involuntary mechanism that holds the eyes firmly on the object once it has been found; this is called the *involuntary fixation mechanism.*

The voluntary fixation movements are controlled by a small cortical field located bilaterally in the premotor cortical regions of the frontal lobes, as illustrated in Figure 35–7. Bilateral dysfunction or destruction of these areas makes it difficult or almost impossible for the person to "unlock" his eyes from one point of fixation and then move them to another point. It is usually necessary for him to blink his eyes or put his hand over his eyes for a short time, which then allows him to move the eyes.

On the other hand, the fixation mechanism that causes the eyes to "lock" on the object of attention once it is found is controlled by the *eye fields of the occipital cortex* – mainly area 19 – which are also illustrated in Figure 35–7. When these areas are destroyed bilaterally, the person becomes completely unable to keep his eyes directed toward a given fixation point.

To summarize, the posterior eye fields automatically "lock" the eyes on a given spot of vision and thereby prevent movement of the image across the retina. To unlock this visual fixation, voluntary impulses must be transmitted from the "voluntary" eye fields located in the frontal areas.

Mechanism of Fixation. Visual fixation results from a negative feedback mechanism that prevents the object of attention from leaving the foveal portion of the retina. When a spot of light has become fixed on the foveal region of the retina, slow drifting movements of the eyes cause the spot to drift slowly across the cones. However, each time the spot of light approaches the edge of the fovea, a sudden flicking movement moves the spot away from this edge back toward the center. This is an automatic response to move the image back toward the central portion of the fovea.

Fixation on Moving Objects – "Pursuit Movements." The eyes can also remain fixed on a moving object, which is called *pursuit movement.* A highly developed cortical mechanism automatically detects the course of movement of an object and then gradually develops a similar course of movement of the eyes. For instance, if an object is moving up and down in a wavelike form at a rate of several times per second, the eyes at first may be completely unable to fixate on it. However, after a second or so the eyes begin to jump coarsely in approximately the same pattern of movement as that of the object. Then after a few more seconds, the eyes develop progressively smoother and smoother movements and finally follow the course of movement almost exactly. This represents a high degree of calculating ability by the cerebral cortex.

Optikokinetic nystagmus. A particularly important type of pursuit movement is that called *nystagmus,* which allows the eyes to fixate on successive points in a continuously moving scene. For instance, if a person is looking out the window of a train, his eyes fix on successive points in the visual scene. To do this, the eyes fixate on some object and move slowly backward as the object also moves backward. When the eyes have moved far to the side, they automatically jump forward to fix on a new object, which is followed again by slow movement in the backward direction. This type of movement is called *optikokinetic nystagmus.*

FUSION OF THE VISUAL IMAGES

To make the visual perceptions more meaningful and also to aid in depth perception by the mechanism of stereopsis, which was discussed earlier, the visual images in the two eyes normally *fuse* with each other on "corresponding points" of the two retinae. Furthermore, three different types of fusion are required: lateral fusion, vertical fusion, and torsional fusion (same rotation of the two eyes about their optical axes).

The lateral geniculate body possibly or probably plays a major role in fusion of visual images, because corresponding points of the two retinae transmit visual signals respectively to the successive nuclear layers of the lateral geniculate body. When the eyes are appropriately oriented, the retinal images will be appropriately superimposed upon each other in the respective layers.

Strabismus. Strasbismus, which is also called *squint* or *cross-eyedness*, means lack of fusion of the eyes in one or more of the coordinates described above.

Strabismus is believed to be caused by an abnormal "set" of the fusion mechanism of the visual system. That is, in the early efforts of the child to fixate the two eyes on the same object, one of the eyes fixates satisfactorily while the other fails to fixate, or they both fixate satisfactorily but never simultaneously. Soon, the patterns of conjugate movements of the eyes become abnormally "set" so that the eyes never fuse.

Suppression of visual image from a repressed eye. In most patients with strabismus, the eyes alternate in fixing on the object of attention. However, in some patients, one eye alone is used all the time while the other eye becomes repressed and is never used for vision. The vision in the repressed eye develops only slightly, usually remaining 20/400 or less. If the dominant eye then becomes blinded, vision in the repressed eye can develop partially in the adult and far more rapidly and completely in children. This illustrates that visual acuity is highly dependent on the proper development of the central synaptic connections from the eyes.

AUTONOMIC CONTROL OF ACCOMMODATION AND PUPILLARY APERTURE

The Parasympathetic Nerves to the Eyes. The eye is innervated mainly by parasympathetic nerve fibers. These arise in the· *Edinger-Westphal nucleus* and then pass in the *third nerve* to the *ciliary ganglion*, which lies about 1 cm. behind the eye. Here the fibers synapse with postganglionic parasympathetic neurons that pass through the *ciliary nerves* into the eyeball. These nerves excite the ciliary muscle and the sphincter of the iris.

CONTROL OF ACCOMMODATION

The accommodation mechanism — that is, the mechanism which focuses the lens system of the eye — is essential to a high degree of visual acuity. Accommodation results from contraction or relaxation of the ciliary muscle, contraction causing increased strength of the lens system, as explained earlier, and relaxation causing decreased strength.

Accommodation of the lens is regulated by a negative feedback mechanism that automatically adjusts the focal power of the lens for the highest degree of visual acuity. If the visual image is not in focus, the ciliary muscle automatically begins to contract or relax. If the focus improves, the accommodation continues to change in the same direction, but, if it worsens, the accommodation automatically turns around to the other direction and changes until the proper focus is attained.

Also, because the center of the fovea is depressed in relation to the remainder of the retina, an image might be in better focus in the center of the fovea than along the edges. From this information, the visual system theoretically could detect that the focal plane of the lens system lies too far posteriorly. On the other hand, if the focus on the edges of the fovea is better than in the very center, then it could be detected that the focal plane lies too far anteriorly. Therefore, appropriate integrative impulses could adjust the degree of accommodation to maintain a continual state of precise focus halfway between the depth and the sides of the fovea.

It is presumed that the cortical areas that control accommodation closely parallel those that control fixation movements of the eyes, with integration of visual signals in areas 18 and 19 and transmission of motor signals to the ciliary muscle through the pretectal area and Edinger-Westphal nucleus.

CONTROL OF THE PUPILLARY APERTURE

Stimulation of the parasympathetic nerves excites the pupillary sphincter, thereby decreasing the pupillary aperture; this is called *miosis*. On the other hand, stimulation of the sympathetic nerves excites the radial fibers of the iris and causes pupillary dilatation, which is called *mydriasis*.

The Pupillary Light Reflex. When light is shone into the eyes the pupils constrict, a reaction that is called the pupillary light reflex. The neuronal pathway for this reflex is illustrated in Figure 35–9. When light impinges on the retina, the resulting impulses pass through the optic nerves and optic tracts to the pretectal nuclei. From here, impulses pass to the *Edinger-Westphal nucleus* and finally back through the *parasympathetic nerves* to constrict the sphincter of the iris. In darkness, the Edinger-Westphal nucleus becomes inhibited, which results in dilatation of the pupil.

The function of the light reflex is to help the eye adapt extremely rapidly to changing light conditions, the importance of which was explained earlier.

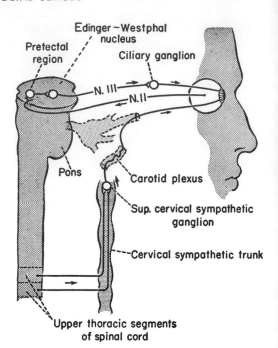

FIGURE 35–9 Autonomic innervation of the eye, showing also the reflex arc of the light reflex. (Modified from Ranson and Clark: Anatomy of the Nervous System.)

REFERENCES

Brindley, G. S.: Central pathways of vision. *Ann. Rev. Physiol., 32*:259, 1970.

Creutzfeldt, O., and Sakmann, B.: Neurophysiology of vision. *Ann. Rev. Physiol., 31*:499, 1969.

Hubel, D. H.: Eleventh Bowditch Lecture. *Physiologist, 10*:17, 1967.

Hubel, D. H.: The visual cortex of the brain. *Sci. Amer., 209(5)*:54, 1963.

Hubel, D. H., and Wiesel, T. N.: Receptive fields and functional architecture in two nonstriate visual areas (18 and 19) of the cat. *J. Neurophysiol., 28*:229, 1965.

Ogle, K. M., Martens, T. G., and Dyer, T. A.: Oculomotor Imbalance in Binocular Vision and Fixation Disparity. Philadelpha, Lea & Febiger, 1967.

Thomas, E. L.: Movements of the eye. *Sci. Amer., 219*:88(2), 1968.

Walsh, F. B.: Clinical Neuro-ophthalmology. 3rd ed., Baltimore, The Williams & Wilkins Co., 1968.

Wolbarsht, M. L., and Yeandle, S. S.: Visual processes in the Limulus eye. *Ann. Rev. Physiol., 29*:513, 1967.

CHAPTER 36

THE SENSE OF HEARING; AND THE CHEMICAL SENSES– TASTE AND SMELL

Hearing, like many somatic senses, is a *mechanoreceptive sense*, for the ear responds to mechanical vibration of the sound waves in the air. On the other hand, taste and smell are *chemical senses*, because the stimuli in both instances are chemicals. The purpose of the present chapter is to describe and explain the mechanisms by which the ear, the taste, and the smell organs transmit information into the central nervous system.

THE TYMPANIC MEMBRANE AND THE OSSICULAR SYSTEM

TRANSMISSION OF SOUND FROM THE TYMPANIC MEMBRANE TO THE COCHLEA

Figure 36–1 illustrates the *tympanic membrane*, commonly called the *eardrum*, and the *ossicular system*, which

transmits sound through the middle ear. The tympanic membrane is cone shaped, with its concavity facing downward toward the auditory canal. Attached to the very center of the tympanic membrane is the *handle* of the *malleus*. The malleus is tightly bound by ligaments at its other end with the *incus* so that whenever the malleus moves, the incus generally moves in unison with it. The opposite end of the incus in turn articulates with the stem of the *stapes*, and the *faceplate* of the stapes lies in the opening of

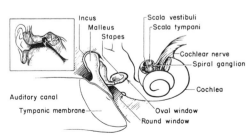

FIGURE 36–1 The tympanic membrane, the ossicular system of the middle ear, and the inner ear.

441

the oval window where sound vibrations are transmitted into the inner ear, the *cochlea*.

The ossicles of the middle ear are suspended by ligaments in such a way that the combined malleus and incus act as a single lever having its fulcrum approximately at the border of the tympanic membrane. The large *head* of the malleus, which is on the opposite side of the fulcrum from the handle, almost exactly balances the other end of the lever so that changes in position of the body will not increase nor decrease the tension on the tympanic membrane.

The handle of the malleus is constantly pulled inward by ligaments and by the tensor tympani muscle, which keeps the tympanic membrane tensed. This allows sound vibrations on *any* portion of the tympanic membrane to be transmitted to the malleus, which would not be true if the membrane were lax.

Impedance Matching by the Ossicular System. The amplitude of movement of the stapes faceplate with each sound vibration is only three-fourths as much as the amplitude of the handle of the malleus. Therefore, the ossicular lever system increases the *force* of movement about 1.3 times. Also, the surface area of the tympanic membrane is approximately 55 sq. mm., while the surface area of the stapes averages 3.2 sq. mm. This 17-fold difference times the 1.3-fold ratio of the lever system, allows all the energy of a sound wave impinging on the tympanic membrane to be applied to the small faceplate of the stapes, causing approximately 22 times as much pressure on the fluid of the cochlea as is exerted by the sound wave against the tympanic membrane. Since fluid has far greater inertia than air, it is easily understood that increased amounts of pressure are needed to cause vibration in the fluid. Therefore, the tympanic membrane and ossicular system provide *impedance matching* between the sound waves in air and the sound vibrations in the fluid of the cochlea. In the absence of this impedance matching, it is

calculated that 97 per cent of the sound energy would be reflected from the tympanic membrane back into the air and, therefore, would be lost, thus decreasing the sensitivity of hearing about 13 decibels (a decrease from a loud voice level to a very soft voice level).

Attenuation of Sound by Contraction of the Stapedius and Tensor Tympani Muscles. When loud sounds are transmitted through the ossicular system into the central nervous system, a reflex occurs after a latent period of only 10 milliseconds to cause contraction of both the stapedius and tensor tympani muscles. The tensor tympani muscle pulls the handle of the malleus inward while the stapedius muscle pulls the stapes outward. These two forces oppose each other and thereby cause the entire ossicular system to develop a high degree of rigidity, thus greatly reducing the transmission of sound, especially frequencies below 1000 cycles per second, to the cochlea.

This *attenuation reflex* can reduce the intensity of sound transmission by as much as 30 to 40 decibels, which is about the same difference as that between a whisper and the sound emitted by a loud radio speaker. The function of this mechanism is partly to allow adaptation of the ear to sounds of different intensities, but even more, to protect the cochlea from damaging vibrations caused by excessively loud sounds. It is mainly low frequency sounds, the ones that are attenuated, that frequently are loud enough to damage the basilar membrane of the cochlea. Also, when a person talks, collateral signals are sent to his attenuation muscles to decrease his own hearing sensitivity so that his speech will not overstimulate the hearing mechanism.

TRANSMISSION OF SOUND THROUGH THE SKULL

Because the inner ear, the *cochlea*, is embedded in a bony cavity in the temporal bone, vibrations of the entire skull can cause fluid vibrations in the cochlea itself. Therefore, under appropriate conditions, a tuning fork or an electronic vibrator placed on any bony protuberance of the skull causes the person to hear the sound if it is intense enough.

THE COCHLEA

FUNCTIONAL ANATOMY OF THE COCHLEA

The cochlea is a system of coiled tubes, shown in Figure 36-2, with three different tubes coiled side by side, the *scala vestibuli*, the *scala media*, and the *scala tympani*. The scala vestibuli and scala media are separated from each other by the *vestibular membrane*, and the scala tympani and scala media are separated from each other by the *basilar membrane*. On the surface of the basilar membrane lies a structure, the *organ of Corti*, which contains a series of mechanically sensitive cells, the *hair cells*. These are the receptive end-organs that generate nerve impulses in response to sound vibrations.

Figure 36-3 illustrates schematically the functional parts of the uncoiled

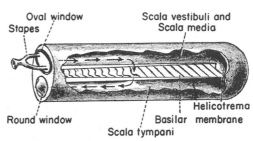

FIGURE 36-3   Movement of fluid in the cochlea following forward thrust of the stapes.

cochlea for transmission of sound vibrations. First, note that the vestibular membrane is missing from this figure. This membrane is so thin and so easily moved that it does not obstruct the passage of sound vibrations from the scala vestibuli into the scala media at all. Therefore, so far as the transmission of sound is concerned, the scala vestibuli and scala media are considered to be a single chamber. The importance of the vestibular membrane is to maintain a special fluid in the scala media that is required for normal function of the sound receptive hair cells, as discussed later in the chapter.

Sound vibrations enter the scala vestibuli from the faceplate of the stapes at the oval window. The faceplate covers this window and is connected with the window's edges by a relatively loose annular ligament so that it can move inward and outward with the sound vibrations. Inward movement causes the fluid to move into the scala vestibuli and scala media, which immediately increases the pressure in the entire cochlea and causes the round window to bulge outward.

Note from Figure 36-3 that the distal end of the scala vestibuli and scala tympani are continuous with each other by way of the *helicotrema*. If the stapes moves inward *very slowly*, fluid from the scala vestibuli is pushed through the helicotrema into the scala tympani, and this causes the round window to bulge outward. However, if the stapes vibrates

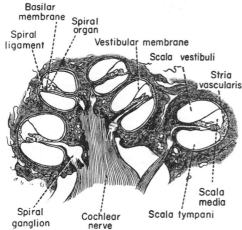

FIGURE 36-2   The cochlea. (From Goss. C. M.: Gray's Anatomy of the Human Body. Lea & Febiger.)

inward and outward rapidly, the fluid simply does not have time to pass all the way to the helicotrema, then to the round window, and back again to the oval window between each two successive vibrations. Instead, the fluid wave takes a shortcut through the basilar membrane, causing it to bulge back and forth with each sound vibration. We shall see later that each frequency of sound causes a different "pattern" of vibration in the basilar membrane and that this is one of the important means by which the sound frequencies are discriminated from each other.

The Basilar Membrane and Its Resonating Qualities. The basilar membrane contains about 20,000 or more *basilar fibers* that project from the bony center of the cochlea, the *modiolus*, toward the outer wall. These fibers are stiff hairlike structures that are not fixed at their distal ends except that they are embedded in the membrane. Because they are stiff and because they are free at one end, they can vibrate like reeds of a harmonica.

The lengths of the basilar fibers increase progressively from the base of the cochlea to the helicotrema, from approximately 0.04 mm. at the base to 0.5 mm. at the helicotrema, a 12-fold increase in length. This increase in length helps the shorter fibers near the base of the cochlea to vibrate at a high frequency, while those near the helicotrema vibrate more easily at a low frequency.

In addition to the differences in lengths of the basilar fibers, they are also differently "loaded" by the fluid of the cochlea. That is, when a fiber vibrates back and forth, all the fluid between the vibrating fiber and the oval and round windows must also vibrate back and forth at the same time. For a fiber vibrating near the base of the cochlea, the total mass of the fluid is slight in comparison with that for a fiber vibrating near the helicotrema. This difference, too, favors high frequency vibration near the windows and low frequency vibration near the tip of the cochlea.

Thus, high frequency resonance of the basilar fibers occurs near the base and low frequency resonance occurs near the apex because of (1) difference in lengths of the fibers and (2) difference in "loading." The product of these two factors multiplied together is called the *volume elasticity* of the fibers. The volume elasticity of the fibers near the helicotrema is at least 100 times as great as that of the fibers near the stapes, which corresponds to a difference in resonating frequencies between the two extremes of the cochlea of about 7 octaves.

Amplitude Patterns of Vibration of the Basilar Membrane for Different Sound Frequencies. The dashed curves of Figure 36–4A show, respectively, the displacement of the basilar membrane when the stapes (*a*) is all the way inward, (*b*) has moved back to the neutral point, (*c*) is all the way outward, and (*d*) has moved back again to the neutral point but is moving inward. The shaded area around these different waves shows the maximum extent of vibration of the

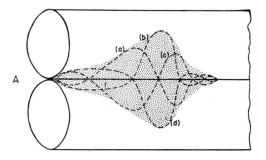

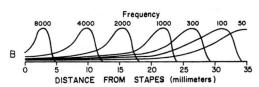

FIGURE 36–4 (*A*) Amplitude pattern of vibration of the basilar membrane for a medium frequency sound. (*B*) Amplitude patterns for sounds of all frequencies between 50 and 8000 per second, showing the points of maximum amplitude (the resonance points) on the basilar membrane for the different frequencies.

basilar membrane during a complete vibratory cycle. This is the amplitude pattern of vibration of the basilar membrane for this particular sound frequency. The area of maximal resonance is also demonstrated.

Figure 36–4*B* shows the amplitude patterns of vibration for different frequencies. The maximum amplitude (the resonant point) for 8000 cycles occurs near the base of the cochlea, while that for frequencies of 50 to 100 cycles per second occurs near the helicotrema.

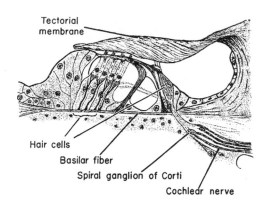

FIGURE 36–5 The organ of Corti, showing especially the hair cells and the tectorial membrane against the projecting hairs.

FUNCTION OF THE ORGAN OF CORTI

The organ of Corti, illustrated in Figure 36–5, is the receptor organ that generates nerve impulses in response to vibration of the basilar membrane. Note that the organ of Corti lies on the surface of the basilar fibers and that the actual sensory receptors in the organ of Corti are the *hair cells*: a single row of *internal hair cells* and three to four rows of *external hair cells*. The bases of the hair cells are enmeshed by a network of cochlear nerve endings. These lead to the *spiral ganglion of Corti*, which lies in the modiolus of the cochlea. The spiral ganglion in turn sends axons into the *cochlear nerve* and thence into the central nervous system at the level of the upper medulla. The relationship of the organ of Corti to the spiral ganglion is illustrated in Figure 36–2.

Excitation of the Hair Cells. Note in Figure 36–5 that minute hairs project upward from the hair cells and are embedded in the surface of the *tectorial membrane*, which lies above in the scala media. These hair cells are almost identical with the hair cells found in the macula and cristae ampullaris of the vestibular apparatus, which will be discussed in Chapter 37. Bending of the hairs excites the hair cells, and this in turn excites the nerve fibers enmeshing their bases.

Figure 36–6 illustrates the mechanism

by which vibration of the basilar membrane excites the hair endings. This shows that the upper ends of the hair cells are fixed tightly in a structure called the *reticular lamina*. Furthermore, the reticular lamina is very rigid and is continuous with a rigid triangular structure called the *rods of Corti* that rests on each basilar fiber. Therefore, the basilar fiber, the rods of Corti, and the reticular lamina all move as a unit.

Upward movement of the basilar fiber rocks the reticular lamina upward and *inward*. Then, when the basilar membrane moves downward, the reticular lamina rocks downward and *outward*. The inward and outward motion causes the hairs to sheer back and forth in the tectorial membrane, thus exciting the

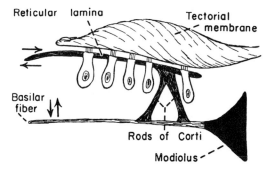

FIGURE 36–6 Stimulation of the hair cells by the to-and-fro movement of the hairs in the tectorial membrane.

hair cells and thence the cochlear nerve fibers whenever the basilar membrane vibrates.

Mechanism by which the Hair Cells Excite the Nerve Fibers — Receptor Potentials. Back-and-forth bending of the hairs causes alternate changes in the electrical potential across the surface of the hair cells. This alternating potential is the *receptor potential* of the hair cell. Most physiologists believe that the receptor potential stimulates the cochlear nerve filaments, which entwine the hair cells, by direct electrical excitation.

DETERMINATION OF PITCH — THE "PLACE" PRINCIPLE

From earlier discussions in this chapter it is already apparent that low pitch (or low frequency) sounds cause maximal resonance of the basilar membrane near the apex of the cochlea, and sounds of high pitch (or high frequency) activate the basilar membrane near the base of the cochlea. Furthermore, the cochlear nerve fibers from the cochlea to the brain are spatially organized so that excitation of a particular point on the basilar membrane excites a corresponding locus in the auditory cortex. Therefore, the primary method used by the nervous system to detect different pitches is to determine the position along the basilar membrane that is most stimulated. This is called the *place principle* for determination of pitch.

DETERMINATION OF LOUDNESS

Loudness is determined by the auditory system in at least two different ways: First, as the sound becomes louder, the amplitude of vibration of the basilar membrane and hair cells also increases so that the hair cells excite the nerve endings at faster rates. Second, as the amplitude of vibration increases, it causes more and more of the hair cells on the fringes of the vibrating portion of the basilar membrane to become stimulated, thus causing *spatial summation* of impulses — that is, transmission through many nerve fibers rather than through a few. The combination of these two mechanisms allows the ear to discriminate changes in sound intensity from the softest whisper to the loudest possible noise, a total change in sound intensity of *approximately one trillion times*.

The Decibel Unit. Because of the extreme changes in sound intensities that the ear can detect and discriminate, sound intensities are usually expressed in terms of the logarithm of their actual intensities. A 10-fold increase in intensity is called 1 *bel*, and one-tenth bel is called 1 *decibel*. One decibel represents an actual increase in intensity of 1.26 times.

Another reason for using the decibel system in expressing changes in loudness is that, in the usual sound range for communication, the ears can detect approximately a 1 decibel change in sound intensity.

CENTRAL AUDITORY MECHANISMS

THE AUDITORY PATHWAY

Figure 36–7 illustrates the major auditory pathways. It shows that neve fibers from the *spiral ganglion of the organ of Corti* enter the *dorsal* and *ventral cochlear nuclei* located in the upper part of the medulla. At this point, all the fibers synapse, and second order neurons pass mainly to the opposite side of the brain stem through the *trapezoid body* to end in the *superior olivary nucleus*. However, some of the second order fibers pass ipsilaterally to the superior olivary nucleus on the same side. From there the auditory pathway passes upward through the *lateral lemniscus* to the *inferior colliculus*, thence to the *medial geniculate nucleus*, and finally to the *auditory cortex* located mainly in the superior temporal gyrus.

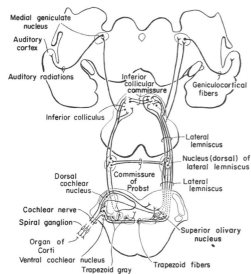

FIGURE 36–7 The auditory pathway. (Modified from Crosby, Humphrey, and Lauer: Correlative Anatomy of the Nervous System. The Macmillian Co.)

Integration of Auditory Information in the Relay Nuclei. Very little is known about the function of the different nuclei in the auditory pathway. However, some lower animals can still hear and can even detect normal threshold auditory signals when the cerebral cortex is removed bilaterally, which indicates that the nuclei in the brain stem and thalamus can perform many auditory functions even without the cerebral cortex. In man, *bilateral* destruction of the cortical auditory centers gives a different picture: it causes almost total deafness. This indicates that in man the conscious elements of the lower centers are almost completely suppressed.

One of the important features of auditory transmission through the relay nuclei is the spatial orientation of the pathways for sounds of different frequencies. For instance, in the dorsal cochlear nucleus, high frequencies ˙are represented along the medial edge while low frequencies are represented along the lateral edge, and a similar type of spatial orientation occurs throughout the auditory pathway as it travels upward to the cortex.

FUNCTION OF THE CEREBRAL CORTEX IN HEARING

The projection of the auditory pathway to the cerebral cortex is illustrated in Figure 36–8, which shows that the auditory cortex lies principally on the *supratemporal plane of the superior temporal gyrus.*

Locus of Sound Frequency Perception in the Auditory Cortex. Certain parts of the auditory cortex are known to respond to high frequencies and other parts to low frequencies. In monkeys, and presumably in man, the posterior part of the supratemporal plane responds to high frequencies, while the anterior part responds to low frequencies.

In the human being, direct electrical stimulation in an area of the auditory cortex called the *primary auditory cortex,* causes the person to hear (a) a tingling sound, (b) a loud roar, or (c) some other single discrete type of sound; whereas stimulation lower down on the side of the temporal lobe, an area called

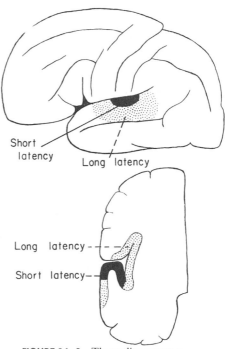

FIGURE 36–8 The auditory cortex.

the *auditory association cortex*, occasionally causes the person to perceive intelligible sounds that even include whole sentences. Thus, it would be possible to divide the auditory cortex into a *primary receptive area* and an *interpretive area*.

DISCRIMINATION OF DIRECTION FROM WHICH SOUND EMANATES

A person determines the direction from which sound emanates by at least two different mechanisms: (1) by the time lag between the entry of sound into one ear and into the opposite ear and (2) by the difference between the intensities of the sounds in the two ears. The time lag mechanism is believed to function in the following manner.

When sound enters one ear slightly before it enters the other ear, it excites the neurons in the contralateral superior olivary nucleus but, at the same time, *inhibits* the neurons in the ipsilateral superior olivary nucleus, this inhibition lasting for a fraction of a millisecond. Furthermore, certain portions of the superior olivary nuclei seem to have longer time lags of inhibition than do other portions. Therefore, when the sound hits the second ear and its lagging signal enters the inhibited superior olivary nucleus, the signal will pass up the auditory pathway through some of the neurons but not through others. Thus, a spatial pattern of neuronal stimulation develops, with the short lagging sounds stimulating one set of neurons maximally and the long lagging sounds stimulating another set of neurons maximally. This spatial orientation of signals is then transmitted all the way to the auditory cortex where sound direction is determined by the locus in the cortex that is stimulated maximally.

This mechanism for detection of sound direction indicates again how sensory signals are dissected as they pass through different levels of neuronal activity. In this case, the quality of sound

direction is believed to be separated from the quality of sound tones at the level of the superior olivary nuclei.

HEARING ABNORMALITIES

TYPES OF DEAFNESS

Deafness is usually divided into two types: first, that caused by impairment of the cochlea or auditory nerve, which is usually classed under the heading "nerve deafness," and, second, that caused by impairment of the middle ear mechanisms for transmitting sound into the cochlea, which is usually called "conduction deafness." Obviously, if either the cochlea or the auditory nerve is completely destroyed the person is permanently deaf. However, if the cochlea and nerve are still intact but the ossicular system has been destroyed or ankylosed, sound waves can still be conducted into the cochlea by means of bone conduction.

The Audiometer. To determine the nature of hearing disabilities, the "audiometer" is used. This is simply an earphone connected to an electronic oscillator capable of emitting pure tones ranging from low frequencies to high frequencies. Based on previous studies of normal persons, the instrument is calibrated so that the zero intensity level of sound at each frequency is the loudness that can barely be heard by the normal person. However, the calibrated volume control can be changed to increase or decrease the loudness of each tone above or below the zero level. If the loudness of a tone must be increased to 30 decibels above normal before the subject can hear it, he is said to have a *hearing loss* of 30 decibels for that particular tone.

In performing a hearing test using an audiometer, one tests the ear for approximately 8 to 10 tones covering the auditory spectrum, and the hearing loss is determined for each of these tones. Then the so-called "audiogram" is plotted as shown in Figures 36–9 and 36–10, depicting the hearing loss for each of the tones in the auditory spectrum.

The audiometer, in addition to being equipped with an earphone for testing air conduction by the ear, is also equipped with an electronic vibrator for testing bone conduction from the mastoid process into the cochlea.

The audiogram in nerve deafness. If a person has nerve deafness—this term including damage to the cochlea, to the auditory nerve, or to the central nervous system circuits from the ear—he has lost the ability to hear sound as tested by both the air conduction apparatus and the bone conduction apparatus. An audiogram depicting nerve deafness is illustrated in Figure 36–9. In this figure the deafness is mainly for high frequency sound. Such deafness could be caused by damage to the base of the cochlea. This type of deafness occurs to some extent in almost all older persons.

The audiogram in conduction deafness. A second and frequent type of deafness is that caused by fibrosis of the middle ear following infection in the middle ear. In this instance the sound waves cannot be transmitted easily to the oval window. Figure 36–10 illustrates an audiogram from a person with "middle ear deafness" of this type. In

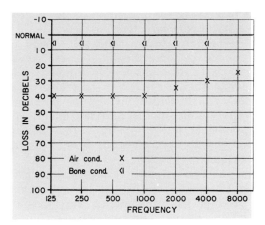

FIGURE 36–10 Audiogram of deafness resulting from middle ear sclerosis.

this case the bone conduction is essentially normal, but air conduction is greatly depressed at all frequencies, though more so at the low frequencies. In this type of deafness, the faceplate of the stapes frequently becomes "ankylosed" by bony overgrowth, to the edges of the oval window. In this case, the person becomes totally deaf for air conduction; but he can be made to hear again, with only 15 decibels hearing loss, by removing the stapes so that sound waves can strike the oval window directly.

THE SENSE OF TASTE

Taste is a function of the *taste buds* in the mouth, and its importance lies in th fact that it allows the person to select his food in accord with his desires and perhaps also in accord with the needs of the tissues for specific nutritive substances.

THE PRIMARY SENSATIONS OF TASTE

On the basis of psychologic studies, there are generally believed to be four *primary* sensations of taste, as follows:

The Sour Taste. The sour taste is caused by acids, and the intensity of the taste sensation is approximately pro-

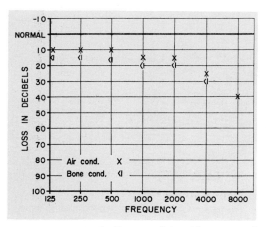

FIGURE 36–9 Audiogram of the old age type of nerve deafness.

portional to the logarithm of the *hydrogen ion concentration*.

The Salty Taste. The salty taste is elicited by ionized salts. The quality of the taste varies somewhat from one salt to another because the salts also elicit other taste sensations besides saltiness. The cations of the salts are mainly responsible for the salty taste, but the anions also contribute at least to some extent.

The Sweet Taste. The sweet taste is not caused by any single class of chemicals. A list of some of the types of chemicals that cause this taste includes: sugars, glycols, alcohols, aldehydes, ketones, amides, esters, amino acids, sulfonic acids, halogenated acids, and inorganic salts of lead and beryllium.

If *sucrose*, which is common table sugar, is considered to have a sweetness index of 1, then *saccharin*, a common sweetening agent, has an index of about 600.

The Bitter Taste. Two particular classes of substances are especially likely to cause bitter taste sensations, (1) long chain organic substances and (2) alkaloids. The alkaloids include many of the drugs used in medicines, such as quinine, caffeine, strychnine, and nicotine.

The bitter taste, when it occurs in high intensity, usually causes the person or animal to reject the food. This is undoubtedly an important purposive function of the bitter taste sensation because many of the deadly toxins found in poisonous plants are alkaloids, and these all cause an intensely bitter taste.

THRESHOLD FOR TASTE

The threshold for stimulation of the sour taste by hydrochloric acid averages 0.0009 N; for stimulation of the salty taste by sodium chloride: 0.01 M; for the sweet taste by sucrose: 0.01 M; and for the bitter taste by quinine: 0.000008 M. Note especially how much more sensitive is the bitter taste sense to stimuli than all the others, which would be ex-

pected since this sensation provides an important protective function.

Taste Blindness. Many persons are taste blind for certain substances, especially for different types of thiourea compounds. A substance used frequently by psychologists for demonstrating taste blindness is *phenylthiocarbamide*, for which approximately 20 per cent of all people exhibit taste blindness.

THE TASTE BUD AND ITS FUNCTION

Figure 36–11 illustrates a taste bud, which has a diameter of about $1/30$ millimeter and a length of about $1/16$ millimeter. The taste bud is composed of about 20 modified epithelial cells called *taste cells*. These cells are continually being replaced so that some are young cells and others mature cells; the mature cells lie toward the center of the bud. The life span of each taste cell is about seven days.

The outer tips of the taste cells are arranged around a minute *taste pore*, shown in Figure 36–11. From the tip of each taste cell, several *microvilli*, or *taste hairs*, about 3 microns in length and 0.2 micron in width, protrude outward through the taste pore into the cavity of the mouth. These microvilli are believed to provide the receptor surface for taste.

Interweaving among the taste cells is

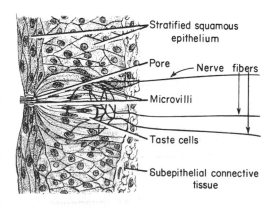

FIGURE 36–11 The taste bud.

a branching terminal network of two or three *taste nerve fibers* that are stimulated by the taste cells. These fibers invaginate deeply into folds of the taste cell membranes, so that there is extremely initimate contact between the taste cells and the nerves.

An interesting feature of the taste buds is that they completely degenerate when the taste nerve fibers are destroyed. Then, if the taste fibers regrow to the epithelial surface of the mouth, the local epithelial cells regroup themselves to form new taste buds. This illustrates the important principle of "trophic" function of nerve fibers in certain parts of the body.

Location of the Taste Buds. The taste buds are found mainly on the anterior and lateral surfaces of the tongue, but a few are located on the tonsillar pillars and at other points around the nasopharynx.

Specificity of Taste Buds for the Primary Taste Stimuli. In the foregoing paragraphs we have discussed taste buds as if each responded to a particular type of taste stimulus and not to other taste stimuli. In a statistical sense this is true, but so far as any single taste bud is concerned, it is not true, for most taste buds respond to varying extents to three and sometimes to all four of the primary taste stimuli.

Figure 36–12 illustrates the responsiveness of four different taste buds to the different primary tastes. Figure 36–12*A* illustrates a bud responsive to all four types of taste stimuli, but especially to saltiness. Figure 36–12*B* illustrates a taste bud strongly responsive to sourness and saltiness but also responsive to a moderate degree to both bitterness and sweetness. Likewise, Figures 36–12*C* and 36–12*D* illustrate response characteristics of two other types of taste buds.

Detection of different sensations of taste by the taste buds. Since each taste bud responds to multiple primary taste stimuli, it is difficult to understand how a person perceives the different primary taste sensations independently

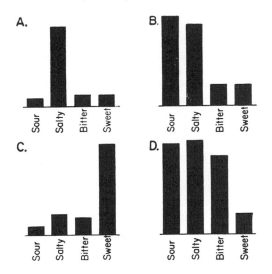

FIGURE 36–12 Specific responsiveness of four different types of taste buds, showing multiple stimulation by the different primary sensations of taste in the case of each of the taste buds.

of each other. However, a theory attempting to explain this is the following: Some area in the brain presumably is capable of detecting the *ratios* of stimulation of the different types of taste buds. For the sweet taste, a taste bud that responds strongly to sweet stimuli will transmit a stronger signal to the brain than will a taste bud that responds only weakly to the sweet taste. Thus, the ratio of the signal strength from the first bud to the signal strength from the second is high, and it is this *ratio* that elicits the sweet taste. Let us now assume that the second taste bud has a strong response to the bitter taste while the first has a weak response. The ratio will be exactly the reverse when a bitter substance is placed in the mouth.

Tastes besides the four primary tastes obviously would give still other ratios of stimulation of the different taste buds and thus give all the different gradations of taste sensations that are known to occur.

Mechanism of Stimulation of Taste Buds. *The receptor potential.* The membrane of the taste cell, like that of other sensory receptor cells, normally is negatively charged on the inside with respect to the outside. Application of a taste substance to the taste hairs causes partial loss of this negative potential, and the decrease in potential is approximately proportional to the logarithm of concentration of the stimulating substance. This change in potential in the taste cell is the *receptor potential* for taste. The receptor potential in turn stimulates the taste nerves.

The mechanism by which the stimulating substance reacts with the taste hairs to initiate the receptor potential is unknown. It is believed by most physiologists that the substance combines chemically with a specific receptor substance in the taste hair and that the combination alters the membrane permeability. Supposedly, the type of receptor substance or substances in each taste hair determines the type of taste substance that will elicit a response.

TRANSMISSION OF TASTE SIGNALS INTO THE CENTRAL NERVOUS SYSTEM

Figure 36–13 shows that taste signals are transmitted through the fifth, seventh, ninth, and tenth (vagus) cranial nerves, and thence into the *tractus solitarius* in the brain stem. From here, second order

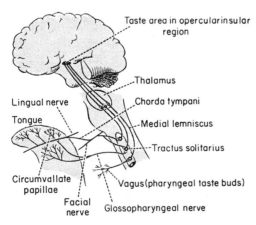

FIGURE 36–13 Transmission of taste impulses into the central nervous system.

neurons pass to a small area of the *thalamus* located slightly medial to the thalamic terminations of the facial regions of the dorsal column-medial lemniscal system. From the thalamus, third order neurons are believed to be transmitted to the *parietal opercular-insular area* of the cerebral cortex. This lies at the very lateral margin of the postcentral gyrus in the sylvian fissure in close association with, or even superimposed on, the tongue area of somatic sensory area I.

Taste Reflexes. From the tractus solitarius a large number of impulses are transmitted directly into the *superior* and *inferior salivatory nuclei*, and these in turn transmit impulses to the submaxillary and parotid glands to help control the secretion of saliva during the ingestion of food.

Adaptation of Taste. Everyone is familiar with the fact that taste sensations adapt rapidly. Electrophysiological studies of taste nerve fibers indicate that the taste buds do not adapt enough by themselves to account for all or most of the taste adaptation. Therefore, the extreme adaptation that occurs in the sensation of taste has been postulated to occur in the central nervous system itself, though the mechanism and site of this are not known.

THE SENSE OF SMELL

THE OLFACTORY MEMBRANE

The olfactory membrane lies in the superior part of each nostril, as illustrated in Figure 36–15. Medially it folds downward over the surface of the septum, and laterally it folds over the superior turbinate and even over a small portion of the upper surface of the middle turbinate. In each nostril the olfactory membrane has a surface area of approximately 2.4 square centimeters.

The Olfactory Cells. The receptor cells for the smell sensation are the olfactory cells, which are actually bipolar nerve cells derived originally from the central nervous system itself. There are on the order of 100 million of these cells in the olfactory epithelium interspersed among *sustentacular cells*, as shown in Figure 36–14. The mucosal end of the olfactory cell forms a *knob* from which large numbers of *olfactory hairs*, or cilia, 0.3 micron in diameter and 3 microns in length, project into the mucus that coats the inner surface of the nasal cavity. Specific chemical substances in these projecting olfactory hairs are believed to react with odors in the air and then to stimulate the olfactory cells.

The Necessary Stimulus for Smell. We do not know what it takes chemically to stimulate the olfactory cells. Yet we do know the physical characteristics of the substances that cause olfactory stimulation: First, the substance must be volatile so that it can be sniffed into the nostrils. Second, it must be at least slightly water soluble so that it can pass through the mucus to the olfactory cells. And, third, it must also be lipid soluble, presumably because the olfactory hairs and outer tips of the olfactory cells are composed principally of lipid materials.

Adaptation. The olfactory receptors adapt approximately 50 per cent in the first second or so after stimulation. Thereafter, they adapt further very slowly. Yet we all know from our own experience that smell sensations adapt almost to extinction within a minute or more after one enters a strongly odorous atmosphere. Since the psychological adaptation seems to be more rapid than the adaptation of the receptors, it has been suggested that at least part of this adaptation occurs in the central nervous system, as has also been postulated for adaptation of taste sensations.

Search for the Primary Sensations of Smell. All physiologists are convinced that the many sensations of smell are subserved by a few rather discrete primary sensations in the same way that taste is subserved by sour, sweet, bitter, and salty sensations. Thus far, only minor success has been achieved in classifying the primary sensations of smell. Yet, on the basis of psychological tests and on the basis of action potential studies from individual olfactory cells as well as from various points in the olfactory nerve pathways, it is believed that about seven different primary classes of olfactory stimulants selectively excite separate olfactory cells. One such classification is the following:

1. Camphoraceous   5. Ethereal
2. Musky   6. Pungent
3. Floral   7. Putrid
4. Pepperminty

It is unlikely that this list actually represents the primary sensations of smell, but it does illustrate one of the many attempts to classify them.

Threshold for Smell. One of the principal characteristics of smell is the minute quantity of the stimulating agent in the air required to effect a smell sensation. For instance, the substance *methyl mercaptan* can be smelled when only 1/25,000,000,000 milligram is present in each milliliter of air. Because of this low threshold, this substance is mixed with natural gas to give it an odor that can be detected when it leaks from a gas pipe.

Gradations of smell intensities. Though the threshold concentrations of substances that evoke smell are ex-

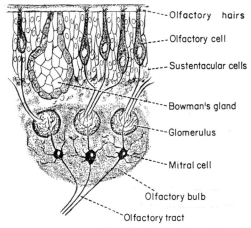

Olfactory hairs

Olfactory cell

Sustentacular cells

Bowman's gland

Glomerulus

Mitral cell

Olfactory bulb

Olfactory tract

FIGURE 36–14 Organization of the olfactory membrane. (From Maximow and Bloom: A Textbook of Histology.)

tremely slight, concentrations only 10 to 50 times above the threshold values evoke maximum intensity of smell. This is in contrast to most other sensory systems of the body, in which the ranges of detection are tremendous—for instance 500,000 to 1 in the case of an eye and almost a trillion to 1 in the case of the ears. It seems, therefore, that the smell sense is concerned more with detecting the presence or absence of odors than with quantitative detection of their intensities.

TRANSMISSION OF SMELL SENSATIONS INTO THE CENTRAL NERVOUS SYSTEM

The function of the central nervous system in olfaction is almost as vague as the function of the peripheral receptors. However, Figure 34–14 shows a number of separate *olfactory cells* sending axons into the *olfactory bulb* to end on *dendrites from mitral cells* in a structure called the *glomerulus*. Approximately 25,000 axons enter each glomerulus and synapse with only 25 mitral cells; these then send signals on into the brain.

It has been claimed that each glomerulus becomes excited in response to only a single type of smell.

Figure 36–15 shows the principal pathways for transmission of olfactory signals from the mitral cells into the brain. The fibers from the mitral cells travel through the olfactory tract and terminate in two principal areas of the brain called the *medial olfactory area* and the *lateral olfactory area*, respectively.

Secondary olfactory tracts pass from the nuclei of both the medial olfactory area and the lateral olfactory area into the *hypothalamus, thalamus, hippocampus*, and *brain stem nuclei*. These secondary areas control the automatic responses of the body to olfactory stimuli, including automatic feeding activities

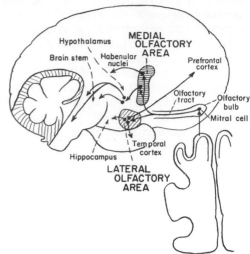

FIGURE 36–15 Neural connections of the olfactory system.

and also emotional responses, such as fear, excitement, and pleasure.

Secondary olfactory tracts also spread from the lateral olfactory area into the temporal cortex and prefrontal cortex. It is probably in this lateral olfactory area, especially in the amygdala and its associated cortical regions, that the more complex aspects of olfaction are integrated. Complete removal of the lateral olfactory area hardly affects the primitive responses to olfaction, such as licking the lips, salivation, and other feeding responses caused by the smell of food, or such as the various emotions associated with smell. On the other hand, its removal does abolish the more complicated conditioned reflexes depending on olfactory stimuli. Therefore, this region is considered to be the primary *cerebral* area for smell. In human beings, tumors in the region of the uncus and amygdala frequently cause the person to perceive abnormal smells.

Mechanism of Function of the Olfactory Tracts. Electrophysiological studies of the olfactory system show that the mitral cells are continually active, and superimposed on this background activity are evoked potentials caused by olfactory stimuli. Thus, the olfactory stimuli presumably *modulate* the rate of rhythmic impulses in the olfactory system and in this way transmit the olfactory information.

REFERENCES

Davis, H.: Biophysics and physiology of the inner ear. *Physiol. Rev., 37*:1, 1957.

Grinnell, A. D.: Comparative physiology of hearing. *Ann. Rev. Physiol., 31*:545, 1969.

Hawkins, J. E., Jr.: Hearing. *Ann. Rev. Physiol., 26*:453, 1964.

Katsuki, Y.: Comparative neurophysiology of hearing. *Physiol. Rev., 45*:380, 1965.

Moulton, D. G., and Beidler, L. M.: Structure and function in the peripheral olfactory system. *Physiol. Rev., 47*:1, 1967.

Oakley, B., and Benjamin, R. M.: Neural mechanisms of taste. *Physiol. Rev., 46*:173, 1966.

Schwartzkopff, J.: Hearing. *Ann. Rev. Physiol., 29*:485, 1967.

Wenzel, B. M., and Sieck, M. H.: Olfaction. *Ann. Rev. Physiol., 28*:381, 1966.

Wever, E. G.: Electrical potentials of the cochlea. *Physiol. Rev., 46*:102, 1966.

Whitfield, I. C.: Auditory Pathway. Baltimore, The Williams & Wilkins Co., 1968.

Zotterman, Y.: Olfaction and Taste. New York, The Macmillan Company, 1963.

CHAPTER 37

THE CORD AND BRAIN STEM REFLEXES; AND FUNCTION OF THE VESTIBULAR APPARATUS

In the discussion of the central nervous system thus far, we have considered principally the input of sensory information. In the following chapters we will discuss the origin and output of motor signals, the signals that cause muscular and other effects throughout the body. Sensory information is integrated at all levels of the nervous system and causes corresponding motor responses, beginning in the spinal cord with relatively simple reflexes and extending into the brain with still more complicated responses.

ORGANIZATION OF THE SPINAL CORD FOR REFLEX FUNCTIONS

The cord gray matter is the integrative area for the cord reflexes. Figure 37–1

shows the typical organization of the cord gray matter in a single cord segment. Sensory signals enter the cord through the sensory roots; and once in the cord, they travel to the brain for sensory perception. They also spread through the cord itself to control the cord reflexes, the motor signals for which leave by way of the motor roots. Between these two roots are two major types of cells concerned with reflex functions of the cord, the *anterior motoneurons* and the *internuncial cells* (also called *intermediate cells* or *interneurons*).

Function of the Anterior Motoneurons. The anterior motoneurons give rise to large type A alpha nerve fibers ranging from 9 to 16 microns in diameter and passing through the spinal nerves to innervate the skeletal muscle fibers. In addition, much smaller motoneurons

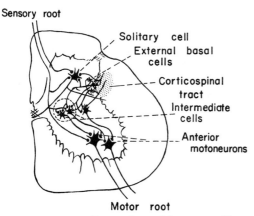

Sensory root

Solitary cell
External basal cells
Corticospinal tract
Intermediate cells
Anterior motoneurons

Motor root

FIGURE 37–1 Connections of the sensory fibers and corticospinal fibers with the internuncial cells ("intermediate cells") and anterior motoneurons of the spinal cord. (Modified from Lloyd: *J. Neurophysiol.,* 4:525, 1941.)

transmit impulses through type A gamma fibers, averaging 5 microns in diameter, to special skeletal muscle fibers called *intrafusal fibers.* These are part of the *muscle spindle,* which is discussed later in the chapter.

The Internuncial Cells. The internuncial cells are present in the base of the dorsal horns, are spread diffusely in the anterior horn, and are also in the intermediate areas between these two. These cells are numerous — approximately 30 times as numerous as the anterior motoneurons. They have many interconnections one with the other, and many of them directly innervate the anterior motoneurons as illustrated in Figure 37–1. The interconnections between the internuncial cells and anterior motoneurons are responsible for many of the integrative functions of the spinal cord that are discussed during the remainder of this chapter.

MUSCLE AND TENDON RECEPTORS

In addition to a few pain and deep pressure sensory endings, muscles and tendons have an abundance of two special types of receptors: (1) *muscle*

spindles, which detect momentary lengths of muscle fibers and rate of change of these lengths, and (2) *Golgi tendon organs,* which detect the tension applied to the tendon fibers during muscel contraction.

Signals transmitted to the central nervous system from these two receptors operate entirely at a subconscious level, causing no sensory perception at all. Instead, they transmit tremendous amounts of information from the muscles and tendons to (1) the motor control systems of the spinal cord and (2) the motor control systems of the cerebellum. They cause reflexes associated with "damping" of muscle movements, with equilibrium, and with posture.

THE MUSCLE SPINDLE AND ITS EXCITATION

The physiologic organization of the muscle spindle is illustrated in Figure 37–2*A.* Each spindle is built around 3 to 10 small *intrafusal muscle fibers* that are pointed at their ends and attach to the sheaths of the surrounding skeletal muscle fibers. Midway between the

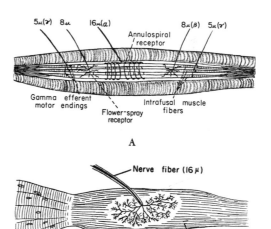

A

B

FIGURE 37–2 (*A*) Arrangement of the muscle spindle with respect to the surrounding skeletal muscle fibers. (*B*) Anatomy of Golgi tendon organ.

ends of each intrafusal fiber is a heavily nucleated area that has lost all cross striations and cannot contract, but instead becomes stretched whenever the muscle is stretched or whenever the ends of the intrafusal fibers contract. Entwined around the central areas of the intrafusal fibers is a nerve ending called the *primary* or *annulospiral receptor*, and from this a large type A nerve fiber, averaging 17 microns in diameter, passes into the sensory roots of the spinal cord. On each side of the annulospiral receptor is usually found another ending called the *secondary* or *flower-spray receptor* that excites another sensory nerve fiber that is much smaller than that excited by the annulospiral receptor, averaging approximately 8 microns in diameter.

The intrafusal muscle fibers of the spindle are innervated by small *gamma* motor nerve fibers, averaging 5 microns in diameter, which are distinct from the large alpha motor fibers to the extrafusal skeletal muscle fibers.

Stimulation of the Muscle Spindle. The muscle spindle can be stimulated by (1) stretch of the entire muscle belly, which also stretches the muscle spindle, or (2) contraction of the intrafusal fibers of the spindle, which contracts the two ends of the spindle but stretches the middle. In other words, the stimulus that normally excites the muscle spindle is stretch of its middle portion, caused either by stretching the entire muscle or by stimulating the intrafusal fibers.

THE GOLGI TENDON ORGAN AND ITS EXCITATION

Golgi tendon organs, illustrated in Figure 37–2*B*, lie within muscle tendons immediately beyond their attachments to the muscle fibers. On the average, 10 to 15 muscle fibers are usually connected in series with each Golgi tendon organ, and the organ is stimulated by the tension produced by this small bundle of muscle fibers.

THE STRETCH REFLEX

The stretch reflex, also called the myotatic reflex, employs the fewest number of neurons of any cord reflex: a single sensory neuron stimulated by a muscle spindle, and a single motoneuron, as illustrated in Figure 37–3. This reflex is initiated by stretch of the muscle, and it causes the stretched muscle to contract.

Note again in Figure 37–2*A* that the ends of the muscle spindle are attached to the connective tissue between the skeletal muscle fibers. Because of this arrangement, stretch of the skeletal muscle fibers also stretches the muscle spindle and thereby excites the receptor. And impulses are transmitted directly to the anterior motoneurons to excite these and thereby to contract reflexly the stretched muscle. The net effect of this reflex, therefore, is *to oppose any stretch of the muscle* beyond its present length.

Static versus Dynamic Stretch Reflexes. The muscle spindle is stimulated many times more strongly by sudden increase in the degree of stretch than by continuous stretch. Therefore, as would be expected, the intensity of the reflex caused by sudden increase in stretch is extremely powerful in comparison with the reflex caused by continuous stretch. The relatively weak reflex caused by continu-

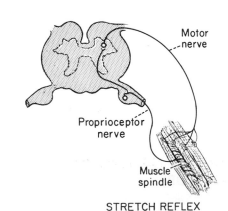

STRETCH REFLEX

FIGURE 37–3 Neuronal circuit of the stretch reflex.

ous stretch is called the *static stretch reflex*, while the strong reflex that occurs following sudden increase in stretch is called the *dynamic stretch reflex*.

The Negative Stretch Reflex. Even under normal resting conditions the primary receptor of the muscle spindle transmits a continuous discharge of nerve impulses. Therefore, the signal output from the muscle spindle can be *decreased* as well as increased. Thus, when a muscle is shortened, the rate of impulses becomes reduced, particularly so when the muscle is shortened rapidly. This decrease in impulses immediately inhibits muscle tone in the shortened muscle. Obviously, therefore, this negative stretch reflex opposes the shortening of the muscle length, particularly if the shortening occurs rapidly.

Damping Function of the Stretch Reflex. The powerful dynamic stretch reflex has a very important function to "damp" muscle contractions; that is, to remove jerkiness in the contractions. For instance, impulses transmitted down the corticospinal tract to the anterior motoneurons often arrive in irregular packets, and in animals whose posterior roots to a respective muscle have been sectioned and the stretch reflex thereby made nonfunctional, the elicited muscle contraction is jerky rather than smooth. However, with the dynamic stretch reflex intact, the movements are smooth, and no jerkiness is perceptible.

CONTROL OF MUSCLE CONTRACTION BY THE GAMMA EFFERENT MOTOR FIBERS — A SERVO CONTROL SYSTEM

Until recently it was believed that skeletal muscle contraction was controlled almost entirely by signals passing from the brain and thence through the large anterior motoneurons to the skeletal muscle fibers. It is now known that impulses transmitted to the muscle through the small *gamma nerve fibers*

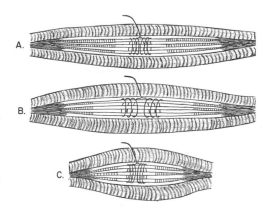

FIGURE 37-4 (*A*) The resting muscle spindle. (*B*) Contraction of the intrafusal fibers of the muscle spindle, resulting in stretch of the central receptor portion of the spindle. (*C*) Reflex contraction of the skeletal muscle fibers until the central portion of the spindle returns to its resting length.

can also control muscle contraction. However, they control muscle contraction in an indirect manner, as follows:

Figure 37-4*A* shows a normal muscle with an inlying muscle spindle. Figure 37-4*B* shows the same spindle after contraction of the intrafusal fibers. It will be recalled that the intrafusal fibers contract only at the two ends, and contraction of these ends *stretches the primary receptor* in the central portion of the spindle. Therefore, in Figure 37-4*B* the receptor transmits impulses that excite the anterior motoneurons of this same muscle, and these in turn cause the large muscle fibers surrounding the spindle to contract as shown in Figure 37-4*C* until the central portion of the spindle returns to its normal length. When this has occurred, the receptor stops sending excess impulses.

From the above sequence, one can see that contraction of the spindle *causes a secondary contraction of the surrounding muscle* until its contraction is equal to the contraction of the spindle. Furthermore, this effect is accomplished without any nerve impulses being transmitted directly from the brain to the anterior motoneurons. Such a mechanism as this is called a *servo mechanism*, the principles of which were discussed in Chapter 1.

Advantages of the Servo Type of Muscle Contraction. On first thought, it is difficult to understand why the nervous system would have two different types of muscle control mechanisms. However, these two mechanisms control two entirely different types of muscle functions. Direct stimulation of the muscle through the large anterior motoneurons causes instantaneous and rapid contraction of the muscle without any limit to the degree of shortening that will occur. On the other hand, the gamma system causes the muscle to contract to a *predetermined length*. The muscle will continue to contract until it has shortened to equal the degree of contraction of the spindle. Then it will stop contracting at that point.

But even more important than simply contracting the muscle to a fixed length is the fact that any extraneous force that tries to lengthen or shorten the muscle from the desired position will be automatically opposed by the muscle spindle servo feedback.

Function of the Gamma Efferent System in Postural Muscle Contractions. Many physiologists believe that the usual postural contractions of the body are controlled primarily by the gamma efferent servo control mechanism instead of by the alpha control mechanism. The advantage of the servo control mechanism would be that one could set the different parts of the body to exact positions, and these parts would remain in the same positions regardless of outside disturbing forces.

Brain Control of the Gamma Efferent System. The gamma efferent system is excited mainly by the *bulboreticular facilitatory* region of the brain stem and secondarily by impulses transmitted into the bulboreticular facilitatory area from (a) the *cerebellum*, (b) the *basal ganglia*, and (c) even the *cerebral cortex*. The bulboreticular facilitatory area is particularly concerned with postural contrac-

tions, which emphasizes the possible or probable important role of the gamma efferent mechanism in the control of muscular contraction for positioning the different parts of the body.

To emphasize the importance of the gamma system for control of muscles, it is particularly interesting that 31 per cent of all motor fibers to the muscles are gamma efferents rather than alpha efferents.

CLINICAL APPLICATIONS OF THE STRETCH REFLEX—THE KNEE JERK AND OTHER MUSCLE JERKS

Clinically, a method used to determine the functional integrity of the stretch reflexes is to elicit the knee jerk and other muscle jerks. The knee jerk can be elicited by simply striking the patellar tendon with a reflex hammer; this stretches the quadriceps muscle and initiates a dynamic stretch reflex to cause the lower leg to jerk forward.

Similar reflexes can be obtained from almost any muscle of the body either by striking the tendon of the muscle or by striking the belly of the muscle itself. In other words, sudden stretch of muscle spindles is all that is required to elicit a stretch reflex.

The muscle jerks are used by neurologists to assess the degree of facilitation of spinal cord centers. When large numbers of facilitatory impulses are being transmitted from the upper regions of the central nervous system into the cord, the muscle jerks are greatly exacerbated. On the other hand, if the facilitatory muscles are depressed or abrogated, the muscle jerks are considerably weakened or completely absent. These reflexes are used most frequently to determine the presence or absence of muscle spasticity following lesions in the motor areas of the brain. Ordinarily, diffuse lesions in the contralateral motor areas of the cerebral cortex cause greatly exacerbated muscle jerks.

THE TENDON REFLEX

Inhibitory Nature of the Tendon Reflex. The Golgi tendon organ detects tension applied to the tendon by muscle contraction, as was discussed earlier in the chapter. Signals from the tendon organ are then transmitted into the spinal cord to cause reflex effects in the respective muscle. However, this reflex is entirely inhibitory, the exact opposite to the muscle spindle reflex. The signal from the tendon organ supposedly excites inhibitory internuncial cells, and these in turn inhibit the anterior motoneurons to the respective muscle.

When tension on the muscle and, therefore, on the tendon becomes extreme, the inhibitory effect from the tendon organ can be so great that it causes sudden relaxation of the entire muscle. This effect is called the *lengthening reaction;* it is probably a protective mechanism to prevent tearing of the muscle or avulsion of the tendon from its attachments to the bone.

However, much more important than this protective reaction is probably the function of the tendon reflex as a part of the overall servo control of muscle contraction in the following manner:

The Tendon Reflex as a Servo Control Mechanism for Muscle Tension. In the same way that the stretch reflex operates as a feedback mechanism to control length of a muscle, the tendon reflex theoretically can operate as a servo feedback mechanism to control muscle tension. That is, if the tension on the muscle becomes too great, inhibition from the tendon organ decreases this tension back to a lower value. On the other hand, if the tension becomes too little, impulses from the tendon organ cease; and this loss of inhibition allows the anterior motoneurons to become active again, thus increasing muscle tension back toward a higher level.

Very little is known at present about the function of or control of this tension servo feedback mechanism, but it is postulated that signals from the brain to the cord centers set the gain of the tendon feedback system. This can be done by changing the degree of facilitation of the neurons in the feedback loop. In this way, control signals from higher nervous centers could automatically set the level of tension at which the muscle would be maintained. If the required tension is high, then the muscle tension would be set by the servo feedback mechanism to this high level of tension. On the other hand, if the desired tension level is low, the muscle tension would be set to this low level.

This description of the tension servo system is obviously very incomplete and almost entirely theoretical at the present time, but it describes the speculation on this effect.

Value of a Servomechanism for Control of Tension. An obvious value of a mechanism for setting degree of muscle tension would be to allow the different muscles to apply a certain amount of force irrespective of how far the muscles contract. An example of this would be paddling a boat, during which a person sets the amount of force that he pulls backwards on the paddle and maintains that degree of force throughout the entire movement. One can imagine hundreds of different patterns of muscle function that require maintenance of constant tension rather than maintenance of constant lengths of the muscles.

THE FLEXOR REFLEX

In the spinal or decerebrate animal (whose cord reflexes have been sensitized thereby), almost any type of sensory stimulus to a limb is likely to cause the flexor muscles of the limb to contract strongly, thereby withdrawing the limb from the stimulus. This is called the flexor reflex.

In its classical form the flexor reflex is elicited most frequently by stimulation of pain endings, such as by pinprick, heat, or some other painful stimulus, for which reason it is frequently called a *nociceptive reflex.* However, stimulation of the light touch receptors can also elicit a weaker and less prolonged flexor reflex.

If some other part of the body besides one of the limbs is stimulated, this part also will be withdrawn from the stimulus, but the reflex may not be confined entirely to flexor muscles even though it is basically the same reflex. Therefore, the reflex is frequently called a *withdrawal reflex*, too.

Neural Mechanism of the Flexor Reflex. The left-hand portion of Figure 37–5 illustrates the neuronal pathways for the flexor reflex. In this instance, a painful stimulus is applied to the hand; as a result, the flexor muscles of the upper arm become reflexly excited, thus withdrawing the hand from the painful stimulus.

The pathways for eliciting the flexor reflex do not pass directly to the anterior motoneurons but, instead, pass first into the internuncial pool of neurons and involve the following basic types of circuits: (1) diverging circuits to spread the reflex to the necessary muscles for withdrawal, (2) circuits to inhibit the antago-

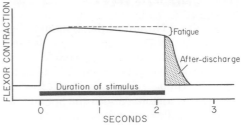

FIGURE 37–6   Myogram of a flexor reflex, showing rapid onset of the reflex, an interval of fatigue, and finally after-discharge after the stimulus is over.

nist muscles, called *reciprocal inhibition circuits,* and (3) circuits to cause a prolonged repetitive after-discharge even after the stimulus is over.

Figure 37–6 illustrates a typical myogram from a flexor muscle during a flexor reflex. Within a few milliseconds after a pain nerve is stimulated, the flexor response occurs. Then, in the next few seconds the reflex begins to *fatigue,* which is characteristic of essentially all integrative reflexes of the spinal cord. Then, soon after the stimulus is over, the contraction of the muscle begins to return toward the base line, but, because of *after-discharge,* will not return all the way for many milliseconds. The duration of the after-discharge depends on the intensity of the sensory stimulus that had elicited the reflex; a weak stimulus causes almost no after-discharge, in contrast to an after-discharge lasting for several seconds following a strong stimulus. Furthermore, a flexor reflex initiated by nonpainful stimuli and transmitted through the large sensory fibers causes essentially no after-discharge, whereas nociceptive impulses transmitted through the small type A fibers and type C fibers cause prolonged after-discharge.

The prolonged after-discharge that occurs following strong pain stimuli almost certainly involves reverberating circuits in the interneurons, these transmitting impulses to the anterior motoneurons sometimes for several seconds after the incoming sensory signal is completely over.

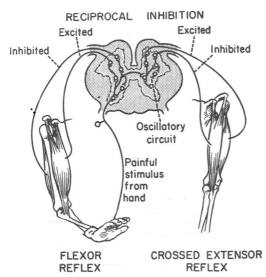

FIGURE 37–5   The flexor reflex, the crossed extensor reflex, and reciprocal inhibition.

Thus, the flexor reflex is appropriately organized to withdraw a pained or otherwise irritated part of the body away from the stimulus. Furthermore, because of the after-discharge it will hold the irritated part away from the stimulus for as long as 1 to 3 seconds even after the irritation is over. During this time, other reflexes and other actions of the central nervous system can move the entire body away from the painful stimulus.

The Pattern of Withdrawal. The pattern of withdrawal that results when the flexor (or withdrawal) reflex is elicited depends on the sensory nerve that is stimulated. Thus, a painful stimulus on the inside of the arm not only elicits a flexor reflex in the arm but also contracts the abductor muscles to pull the arm outward. In other words, the integrative centers of the cord cause those muscles to contract that can most effectively remove the pained part of the body from the object that causes pain. This same principle applies to any part of the body but especially to the limbs, for they have highly developed flexor reflexes.

THE CROSSED EXTENSOR REFLEX

Approximately 0.2 to 0.5 second after a stimulus elicits a flexor reflex in one limb, the opposite limb begins to extend. This is called the crossed extensor reflex. Extension of the opposite limb obviously can push the entire body away from the object causing the painful stimulus.

Neuronal Mechanism of the Crossed Extensor Reflex. The right-hand portion of Figure 37–5 illustrates the neuronal circuit responsible for the crossed extensor reflex, showing that signals from the sensory nerves cross to the opposite side of the cord to cause exactly opposite reactions to those that cause the flexor reflex. Because the crossed extensor reflex usually does not begin until 200 to 500 milliseconds following the initial pain stimulus, it is certain that many internuncial neurons are in the circuit between the incoming sensory neuron and the motoneurons of the opposite side of the cord responsible for the crossed extension. Furthermore, after the painful stimulus is removed, the crossed extensor reflex continues for an even longer period of after-discharge than that for the flexor reflex. Therefore, again, it is almost certain that this prolonged after-discharge results from reverberatory circuits among the internuncial cells. This after-discharge obviously would be of benefit in holding the entire body away from a painful object until other neurogenic reactions should cause the body to move away.

REFLEXES OF POSTURE AND LOCOMOTION

POSTURAL AND LOCOMOTIVE REFLEXES OF THE CORD

The Positive Supportive Reaction. Pressure on the footpad of a decerebrate animal causes the limb to extend against the pressure that is being applied to the foot. Indeed, this reflex is so strong in a decerebrate animal that the animal can be placed on its feet, and the pressure on the footpads will reflexly stiffen the limbs sufficiently to support the weight of the body — and the animal will stand in a rigid position. This reflex is called the positive supportive reaction. The positive supportive reaction involves a complex circuit in the internuncial cells similar to those responsible for the flexor and the crossed-extensor reflexes.

The Cord "Righting" Reflexes. When a spinal cat or even a well-recovered young spinal dog is laid on its side, it will make incoordinate movements that indicate that it is trying to raise itself to the standing position. This is called a cord righting reflex, and it illustrates that relatively complicated reflexes associated with posture are at least partially integrated in the spinal cord.

In the opossum with a transection of the thoracic cord, the walking movements of the hindlimbs are hardly different from those in the normal opossum — except that the hindlimb movements are not synchronized with those of the forelimbs as is normally the case.

The Rhythmic Stepping Reflex of the Single Limb. Rhythmic stepping movements are frequently observed in the limbs of spinal animals. Indeed, even when the lumbar portion of the spinal cord is separated from the remainder of the cord and a longitudinal section is made down the center of the cord to block neuronal connections between the two limbs, each hindlimb can still perform stepping functions.

The oscillation between the flexor and extensor muscles seems to result mainly from "reciprocal inhibition" and "rebound." That is, the forward flexion of the limb causes reciprocal inhibition of extensor muscles, but shortly thereafter the flexion begins to die out; as it does so, *rebound* excitation of the extensors causes the leg to move downward and backward. After extension has continued for a time, it, too, begins to die and is followed by rebound excitation of the flexor muscles.

Reciprocal Stepping of Opposite Limbs. If the lumbar spinal cord is not sectioned down its center, every time stepping occurs in the forward direction in one limb, the opposite limb ordinarily steps backward. This effect results from reciprocal innervation between the two limbs.

Diagonal Stepping of All Four Limbs — The "Mark Time" Reflex. If a decerebrate animal is held up from the table and its legs are allowed to fall downward, the stretch on the limbs occasionally elicits stepping reflexes that involve all four limbs. In general, stepping occurs diagonally between the fore- and hindlimbs. That is, the right hindlimb and the left forelimb move backward together while the right forelimb and left hindlimb move forward. This diagonal response is another manifestation of reciprocal innervation, this time occurring the entire distance up and down the cord between the fore- and hindlimbs. Such a walking pattern is often called a mark time reflex.

THE SPINAL CORD REFLEXES THAT CAUSE MUSCLE SPASM

In human beings, local muscle spasm is often observed. The mechanism of this has not been elucidated to complete satisfaction even in experimental animals, but it is known that pain stimuli can cause reflex spasm of local muscles, which presumably is the cause of much if not most of the muscle spasm observed at localized regions of the human body.

Muscle Spasm Resulting from a Broken Bone. One type of clinically important spasm occurs in muscles surrounding a broken bone. Pain impulses initiated from the broken edges of the bone cause the muscles surrounding the area to contract powerfully and tonically. Relief of the pain by injection of a local anesthetic relieves the spasm; a general anesthetic also relieves the spasm. One of these procedures is often necessary before the spasm can be overcome sufficiently for the two ends of the bone to be set back into appropriate positions.

Abdominal Spasm in Peritonitis. Another type of local spasm caused by a cord reflex is the abdominal spasm resulting from irritation of the parietal peritoneum by peritonitis. Almost the same type of spasm occurs during surgical operations; pain impulses from the parietal peritoneum cause the abdominal muscles to contract extensively and sometimes actually to extrude the gut through the surgical wound. For this reason deep surgical anesthesia is usually required for intra-abdominal operations.

Muscle Cramps. Still another type of local spasm is the typical muscle cramp. Electromyographic studies indicate that the cause of a muscle cramp is the following:

Any local irritating factor or metabolic abnormality of a muscle — such as severe cold, lack of blood flow to the muscle, or overexercise of the muscle — can elicit pain or other types of sensory impulses that are transmitted from the muscle to the spinal cord, thus causing reflex muscle contraction. The contraction in turn stimulates the same sensory receptors still more, which causes the spinal cord to increase the intensity of contraction still further. Thus, a positive feedback mechanism occurs so that a small amount of initial irritation causes more and more contraction until a full-blown muscle cramp ensues.

THE MASS REFLEX

In a spinal animal or human being, the spinal cord sometimes suddenly becomes excessively active, causing massive discharge of large portions of the cord. The usual stimulus that causes this is a strong nociceptive stimulus to the skin or excessive filling of a viscus, such as overdistention of the bladder or of the gut. Regardless of the type of stimulus, the resulting reflex, called the mass reflex, involves large portions or even all of the cord, and its pattern of reaction is the same. The effects of the mass reflex are: (1) a major portion of the body goes into strong flexor spasm, (2) the colon and bladder are likely to evacuate, (3) the arterial pressure often rises to maximal values — sometimes to a mean pressure well over 200 mm. Hg, and (4) large areas of the body break out into profuse sweating.

The precise neuronal mechanism of the mass reflex is unknown. However, since it can last for many minutes, it presumably results from great masses of reverberating circuits that excite large areas of the cord at once.

SPINAL CORD TRANSECTION AND SPINAL SHOCK

When the spinal cord is suddenly transected, essentially all cord functions immediately become depressed, a reaction called *spinal shock*. The reason for this is that normal activity of the cord neurons depends to a great extent on continual facilitatory signals from higher centers, particularly discharges transmitted through the vestibulospinal tract and the excitatory portion of the reticulospinal tracts.

After a few weeks of spinal shock, the spinal neurons usually regain their excitability. This seems to be a natural characteristic of neurons everywhere in the nervous system — that is, after they lose their source of facilitatory impulses, they usually increase their own natural degree of excitability to make up for the loss.

Some of the spinal functions specifically affected during or following spinal shock are: (1) The arterial blood pressure falls immediately — sometimes to as low as 40 mm. Hg — thus illustrating that sympathetic activity becomes blocked almost to extinction. However, the pressure ordinarily returns to normal within a few days. (2) All skeletal muscle reflexes integrated in the spinal cord are completely blocked during the initial stages of shock. In lower animals, a few hours to 3 weeks are required for these reflexes to return to normal; in man they rarely return to normal. Yet sometimes, both in animals and man, some reflexes eventually become hyperexcitable, particularly if a few facilitatory pathways remain intact between the brain and the cord while the remainder of the spinal cord is transected. (3) The sacral reflexes for control of bladder and colon evacuation are completely suppressed in man for the first few weeks following cord transection, but they eventually return.

THE BRAIN STEM

The brain stem is a complex extension of the spinal cord. Collected in it are numerous neuronal circuits to control respiration, cardiovascular function, gastrointestinal function, eye movement, equilibrium, support of the body against gravity, and many special stereotyped movements of the body. Some of these functions—such as control of respiration and cardiovascular functions—are described in special sections of this text. The remainder of this chapter deals primarily with the motor functions of the brain stem and with the functioning of the vestibular apparatus to maintain equilibrium.

THE RETICULAR FORMATION, AND SUPPORT OF THE BODY AGAINST GRAVITY

As illustrated in Figure 37–7, throughout the entire extent of the brain stem— in the medulla, pons, mesencephalon, and even in portions of the diencephalon —are areas of diffuse neurons collectively known as the *reticular formation*. By far the majority of the reticular formation is excitatory, especially the reticular formation in the pons, mesencephalon, and diencephalon. The superior and

lateral portions of the reticular formation in the medulla are also excitatory. Together, these areas are known as the *bulboreticular facilitatory area*. Diffuse stimulation in this facilitatory area causes a general increase in muscle tone either throughout the body or in localized areas.

A small portion of the reticular formation, located in the lower ventromedial portion of the medulla, has mainly inhibitory functions and is known as the *bulboreticular inhibitory area*. Diffuse stimulation in this area causes decreased tone of the musculature throughout the body.

SUPPORT OF THE BODY AGAINST GRAVITY

When a person or an animal is in a standing position, continuous impulses are transmitted from the reticular formation and from closely allied nuclei, particularly from the vestibular nuclei, into the spinal cord and thence to the extensor muscles to stiffen the limbs. This allows the limbs to support the body against gravity. These impulses are transmitted mainly by way of the *reticulospinal* and *vestibulospinal tracts*.

The normal excitatory nature of the upper portion of the reticular formation provides much of the intrinsic excitation required to maintain tone in the extensor muscles. However, the degree of activity in the individual extensor muscles is determined by the equilibrium mechanisms, which will be discussed later.

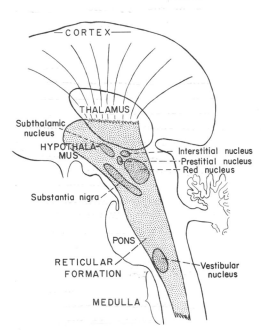

FIGURE 37–7 The reticular formation and associated nuclei.

VESTIBULAR SENSATIONS AND THE MAINTENANCE OF EQUILIBRIUM

THE VESTIBULAR APPARATUS

The vestibular apparatus is the sensory organ that detects sensations concerned with equilibrium. Its functional part is the membranous labyrinth, which is illustrated at the top of Figure 37–8. Its important parts for maintenance of equilibrium are the *utricle* and the *semicircular canals*.

The Utricle. On the wall of the utricle is the *macula*, which is covered by a gelatinous layer in which many small calcified masses known as *otoconia* are embedded. Also, in the macula, as illustrated in the lower right of Figure 37–8, are many *hair cells*, which project *cilia* or *hairs* up into the gelatinous layer. Around these hair cells are entwined sensory axons of the vestibular nerve. Bending the hairs to one side transmits impulses to apprise the central nervous system of the relative position of the otoconia in the gelatinous mass over the macula. Therefore, the weight of the otoconia compressing, bending, or pulling the hairs provides signals that are transmitted by appropriate nerve tracts to the brain to control equilibrium.

The Semicircular Canals. The three semicircular canals in each vestibular apparatus, known respectively as the *superior, posterior,* and *external* (or *horizontal*) *semicircular canals*, are arranged at right angles to each other so that they represent all three planes in space.

In the *ampullae* of the semicircular canals, as illustrated in Figure 37–8, are small *crests*, each called a *crista ampullaris*, and on top of the crest is a gelatinous mass similar to that in the utricle and known as the *cupula*. Into the cupula are projected hairs from hair cells located along the ampullary crest, and these hair cells in turn are connected to sensory nerve fibers that pass into the *vestibular nerve*. Bending the cupula to one side, caused by flow of fluid in the canal, stimulates the hair cells, while bending in the opposite direction inhibits them. Thus, appropriate signals are sent through the vestibular nerve to apprise the central nervous system of fluid movement in the respective canal.

Directional Sensitivity of the Hair Cells —The Kinocilium. Each hair cell has a large number of very small cilia plus one very large cilium called the *kinocilium* located always to one side on the hair cell. This is the cause of the directional sensitivity of the hair cells.

Neuronal Connections of the Vestibular Apparatus with the Central Nervous System. As illustrated in Figure 37–9, the primary pathway for the reflexes of equilibrium begins in the *vestibular nerves*, passes next to the *vestibular nuclei*, and then to the *reticular nuclei of the brain stem*. The reticular nuclei in turn control the interplay between facilitation and inhibition of the extensor muscles, thus automatically controlling equilibrium.

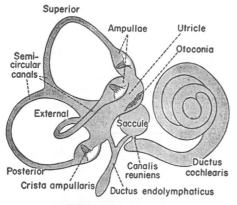

Superior
Ampullae
Utricle
Semicircular canals
Otoconia
External
Saccule
Posterior
Canalis reuniens
Ductus cochlearis
Crista ampullaris
Ductus endolymphaticus

MEMBRANOUS LABYRINTH

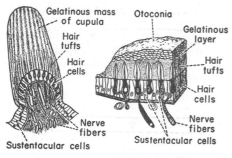

Gelatinous mass of cupula
Otoconia
Hair tufts
Gelatinous layer
Hair cells
Hair tufts
Hair cells
Nerve fibers
Nerve fibers
Sustentacular cells
Sustentacular cells

CRISTA AMPULLARIS AND MACULA

FIGURE 37–8 The membranous labyrinth, and organization of the crista ampullaris and the macula. (From Goss: Gray's Anatomy of the Human Body. Lea & Febiger; and modified from Kolmer by Buchanan: Functional Neuroanatomy. Lea & Febiger.)

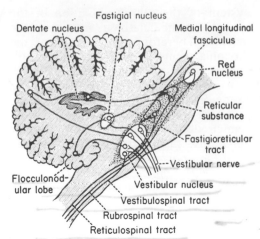

Dentate nucleus
Fastigial nucleus
Medial longitudinal fasciculus
Red nucleus
Reticular substance
Fastigioreticular tract
Vestibular nerve
Flocculonod-ular lobe
Vestibular nucleus
Vestibulospinal tract
Rubrospinal tract
Reticulospinal tract

FIGURE 37–9 Connections of vestibular nerves in the central nervous system.

Vestibular signals are also transmitted into the *flocculonodular lobes* of the cerebellum. This is in keeping with the usual function of the cerebellum to co-ordinate rapid body movements, for the semicircular canals are also concerned with equilibrium during *rapid changes in direction* of motion and not with equilibrium under static conditions, as is pointed out in subsequent sections.

FUNCTION OF THE UTRICLE IN THE MAINTENANCE OF STATIC EQUILIBRIUM

The density of the *otoconia* is almost three times that of the surrounding fluid and tissues, and because of this they bend the hair tufts forward when the head leans forward, backward when the head leans backward, and to one side when the head leans to that side.

It is especially important that the different hair cells are oriented in several different directions in the maculae so that at different positions of the head, different hair cells become stimulated. The "patterns" of stimulation of the different hair cells apprise the nervous system of the position of the head with respect to the pull of gravity. In turn, the reticular nuclei reflexly excite the appropriate muscles to maintain proper equilibrium.

Detection of Linear Acceleration by the Utricle. When the body is suddenly thrust forward, the otoconia, which have greater inertia than the surrounding fluids, fall backward on the hair tufts, and information of mal-equilibrium is sent into the nervous centers, causing the individual to feel as if he were falling backward. This automatically causes him to lean his body forward to correct the loss of equilibrium.

THE SEMICIRCULAR CANALS AND THEIR DETECTION OF ANGULAR ACCELERATION

When the head suddenly *begins* to rotate in any direction, the endolymph in one or more of the membranous semicircular canals, because of its inertia, tends to remain stationary while the semicircular canals themselves turn. This causes relative fluid flow in the canals in a direction opposite to the rotation of the head, and the hair cells are appropriately stimulated. However, with continued rotation, the signal from the hair cells subsides in about 20 seconds, because the fluid in the canals begins to turn with the head.

Thus, the semicircular canals transmit a signal when the head *begins* to rotate and a signal of opposite polarity when it *stops* rotating. Furthermore, this organ responds to rotation in any plane—horizontal, sagittal, or coronal—for fluid movement always occurs in at least one semicircular canal. In summary, the semicircular canals detect the *rate of change* of rotation of the head in any plane, which is called *angular acceleration.*

Rate of Angular Acceleration Required to Stimulate the Semicircular Canals. The angular acceleration required to stimulate the semicircular canals in the human being averages about 1 degree per second per second. In other words, when one begins to rotate, his velocity of rotation must be as much as 1 degree per second by the end of the first second, 2 degrees per second by the end of the second second, 3 degrees per second by the end of the third second, and so forth, in order for him barely to detect that his rate of rotation is increasing.

"Predictive" Function of the Semicircular Canals in the Maintenance of Equilibrium. We can explain the function of the semicircular canals best by the following illustration. If a person is running forward rapidly, and then suddenly begins to turn to one side, he falls off balance a second or so later unless appropriate corrections are made *ahead of time*. But, unfortunately, the utricle cannot detect that he is off balance until *after* this has occurred. On the other hand, the semicircular canals will have already detected that the person is beginning to turn, and this information can easily apprise the central nervous system of the fact that the person *will* fall off balance within the next second or so unless some correction is made. In other words, the semicircular canal mechanism *predicts ahead of time* that mal-equilibrium is going to occur even before it occurs and thereby causes the equilibrium centers to make appropriate adjustments.

Removal of the flocculonodular lobes of the cerebellum prevents normal function of the semicircular canals but does not prevent normal function of the macular receptors. It is especially interesting in this connection that the cerebellum serves as a "predictive" organ for most of the other rapid movements of the body as well as for those having to do with equilibrium. These other functions of the cerebellum are discussed in the following chapter.

NYSTAGMUS CAUSED BY STIMULATION OF THE SEMICIRCULAR CANALS

When a person rotates his head, his eyes must rotate in the opposite direction if they are to remain "fixed" on any one object long enough to gain a clear image. After they have rotated far to one side, they jump suddenly in the direction of rotation of the head to "fix" on a new object and then rotate slowly backward again, a process called nystagmus. The sudden jumping motion forward is called the *fast component* of the nystagmus, and the slow movement is called the *slow component*.

Nystagmus always occurs automatically when the semicircular canals are stimulated. For instance, if a person's head begins to rotate to the right, backward movement of fluid in the left horizontal canal and forward movement in the right horizontal canal cause the eyes to move slowly to the left; thus, the slow component of nystagmus is initiated by the vestibular apparatuses. But, when the eyes have moved as far to the left as they reasonably can, centers located in the brain stem in close approximation with the nuclei of the abducens nerves cause the eyes to jump suddenly to the right; then the vestibular apparatuses take over once more to move the eyes again slowly to the left.

OTHER FACTORS CONCERNED WITH EQUILIBRIUM

The Neck Proprioceptors. The vestibular apparatus detects the orientation and movements *only of the head*. Therefore, it is essential that the nervous centers also receive appropriate information depicting the orientation of the head with respect to the body as well as the orientation of the different parts of the body with respect to each other. This information is transmitted from the proprioceptors of the neck and body either directly into the reticular nuclei of the brain stem or by way of the cerebellum and thence into the reticular nuclei.

By far the most important proprioceptive information needed for the maintenance of equilibrium is that derived from the *joint receptors of the neck*, for this apprises the nervous system of the orientation of the head with respect to the body. When the head is bent in one direction or the other, impulses from the neck proprioceptors keep the vestibular apparatuses from giving the person a sense of mal-equilibrium. They do this by exactly opposing the impulses transmitted from the vestibular apparatuses. However, *when the entire body* is changed to a new position with respect to gravity, the impulses from the vestibular apparatuses *are not opposed* by the neck proprioceptors; therefore, the person in this instance does perceive a change in equilibrium status.

Proprioceptive and Exteroceptive Information from Other Parts of the Body. Proprioceptive information from other parts of the body besides the neck is also necessary for maintenance of equilibrium because appropriate equilibrium adjustments must be made whenever the body is angulated in the chest or abdomen region or elsewhere. Presumably, all this information is algebraically added in the reticular substance of the brain stem, thus causing appropriate adjustments in the postural muscles.

Importance of Visual Information in the Maintenance of Equilibrium. After complete destruction of the vestibular apparatuses, and even after loss of most proprioceptive information from the body, a person can still use his visual mechanisms effectively for maintaining equilibrium. Many persons with complete destruction of the vestibular apparatus have almost normal equilibrium as long as their eyes are open and as long as they perform all motions slowly.

Conscious Perception of Equilibrium. A cortical center for conscious perception of the state of equilibrium has been found in man to lie in the upper portion of the temporal lobe in close association with the primary cortical area for hearing. The sensations from the vestibular apparatuses, from the neck proprioceptors, and from most of the other proprioceptors are probably first integrated in the equilibrium centers of the brain stem before being transmitted to the cerebral cortex.

FUNCTIONS OF THE RETICULAR FORMATION AND SPECIFIC BRAIN STEM NUCLEI IN CONTROLLING SUBCONSCIOUS, STEREOTYPED MOVEMENTS

Rarely, a child is born without brain structures above the mesencephalic region, and some of these children have been kept alive for many months. Such a child is able to perform essentially all the functions of feeding, such as sucking, extrusion of unpleasant food from the mouth, and moving his hands to his mouth to suck his fingers. In addition, he can yawn and stretch. He can cry and follow objects with his eyes and by movements of his head. Also, placing pressure on the upper anterior parts of his legs will cause him to pull to the sitting position.

Therefore, it is obvious that many of the stereotyped motor functions of the human being are integrated in the brain stem. Unfortunately, the loci of most of these different motor control systems have not been found. However, a few control centers are the following:

1. The *interstitial nucleus*, illustrated in Figure 37–7, controls rotational movements of the head and eyes.

2. The *prestitial nucleus* causes raising of the head and body.

3. The *nucleus precommissuralis* causes flexing of the head and body.

4. The *magnocellular* portion of the *red nucleus* functions with the *reticular nuclei* to cause some of the gross body movements.

5. The *substantia nigra* seems to activate strongly the gamma efferent nerves to the muscle spindles and therefore perhaps provides a purposeful function in controlling muscle tone.

6. The *subthalamus* seems to be im-

portant in making the animal walk in a forward direction, a phenomenon called *forward progression.*

SUMMARY OF THE DIFFERENT FUNCTIONS OF THE CENTRAL NERVOUS SYSTEM IN POSTURE AND LOCOMOTION

From the discussions thus far in the chapter, we can see that almost all the discrete "patterns" of muscle movement required for posture and locomotion can be elicited by the spinal cord alone. However, coordination of these patterns to provide equilibrium, progression, and purposefulness of movement requires neuronal function at progressively higher levels of the central nervous system. Centers in the brain stem provide most of the nervous energy required to maintain postural tone and therefore to support the body against gravity. But, in addition, the brain stem centers provide especially the equilibrium reflexes. Then moving still higher we find a control center for "progression" in the subthalamic region, a center that makes the animal move forward in a normal rhythmic pattern of walking and also in a straight line. Even so, this animal still does not have purposefulness in his motion. For this the basal ganglia, the thalamus, and the cerebral cortex are required.

With this background, we can now describe the characteristics of animals with transections through their nervous systems at the different levels.

The Classic Decerebrate Animal — Decerebrate Rigidity. The classical decerebrate animal has a transection between the superior and the inferior colliculi. In this animal the inhibitory signals from the basal ganglia and cerebral cortex to the reticular facilitatory area are removed, which allows the bulboreticular facilitatory area to become intrinsically very active, as explained

earlier in the chapter. Impulses are transmitted downward to cause muscle rigidity called *decerebrate rigidity*. Especially the extensor muscles, but to some extent the flexor muscles as well, are maintained in a state of strong tonic contraction. The animal can occasionally stand, supporting his body against gravity. He also can make crude equilibrium adjustments; but for full expression of equilibrium, he needs a completely intact mesencephalon.

A major part of the increased postural tone in decerebrate rigidity results from stimulation of the *gamma activating system of the muscle spindles* and not from direct activation of the alpha motor neurons.

The Subthalamic Animal. With the transection immediately above the superior colliculi, preserving the subthalamus as well as the lower portions of the hypothalamus, lower animals, such as the cat, no longer have severe rigidity and are capable of walking with normal progression.

The Decorticate Animal. In a decorticate animal the cortex has been removed but the thalamus and basal ganglia are still intact. The capabilities of this type of animal depend on its degree of cerebral development. The cat, for instance, is capable of almost any type of motion, even to the extent of going around obstructions in his way, but, on the other hand, he still lacks much of the purposefulness of locomotion and frequently sits very still for hours at a time. The human being loses essentially all purposeful or voluntary motor functions on decortication.

FUNCTION OF THE CORD AND BRAIN STEM REFLEXES IN MAN

Though the foregoing discussions have concerned principally reflexes in animals rather than man, practically all the same principles of cord and brain stem function apply almost as well to man. Yet there are quantitative differences that must always be remembered.

In man with a spinal cord transection, most of the same reflexes as those that occur in lower animals can be elicited either wholly or partially. The only reflexes that are rarely if ever elicited in spinal man are the rhythmic reflexes; yet the reciprocal innervation that is required for rhythmic reflexes can be demonstrated. For instance, movement of an arm in one direction often causes simultaneous and directionally similar movement of the opposite leg, thus illustrating that appropriate reciprocal innervation is still present in the cord of man to cause at least crude types of diagonal movements necessary for the mark time reflex.

Also, stimulation in the subthalamic region in man causes rhythmic walking movements similar to those that occur in animals, illustrating that this same region of the brain subserves similar functions in man as in lower animals. However, a man who has lost large portions of the motor areas of his cortex cannot use these lower regions of the brain stem effectively enough actually to provide locomotion. This is a manifestation of the process of *encephalization* that has occurred in higher orders of animals, which suppresses many functions of the lower centers while the same functions are taken over, usually in a more complicated manner, by the higher centers. Thus, in man, the motor areas of the cortex are needed to provide progression in the act of locomotion and also to provide purposefulness of locomotion.

Nevertheless, in our subsequent discussions of the motor functions of the brain in man, we must always keep in mind the basic activities of the lower centers. The cord patterns of response are integrated into the overall control of muscular activity even when this control is initiated in the cerebral cortex. It is still the anterior motoneurons that have the final decisive power in the contraction of given muscles, and it is still the pattern of organization of the internuncial cells in the gray matter of the cord that provides many of the patterns of muscular contraction. Impulses from above ordinarily control only the sequence in which these patterns of contraction occur rather than the contractions of the individual muscles.

REFERENCES

Creed, R. S., Denny-Brown, D., Eccles, J. C., Liddell, E. G. T., and Sherrington, C. S.: Reflex Activity of the Spinal Cord. New York, Oxford University Press, Inc., 1932.

Hunt, C. C., and Perl, E. R.: Spinal reflex mechanisms concerned with skeletal muscle. *Physiol. Rev.,* 40:538, 1960.

Liddell, E. G. T.: The Discovery of Reflexes. New York, Oxford University Press, Inc., 1960.

Roberts, Tristan D. M.: Neurophysiology of Postural Mechanisms. New York, Plenum Publishing Corporation, 1967.

Sherrington, C. S.: The Integrative Action of the Nervous System. New Haven, Yale University Press, 1911.

Valdman, A. V. (ed.): Pharmacology and Physiology of the Reticular Formation. Progress in Brain Research. New York, American Elsevier Publishing Co., Inc., 1967, Vol. 20.

Wolfson, R. J.: The Vestibular System and Its Diseases. Philadelphia, University of Pennsylvania Press, 1966.

CHAPTER 38

CONTROL OF MOTOR FUNCTIONS BY THE MOTOR CORTEX, CEREBELLUM, AND BASAL GANGLIA

In the preceding chapter we were concerned with many of the subconscious motor activities integrated in the spinal cord and brain stem, especially those responsible for locomotion. In the present chapter we will discuss the control of motor function by the cerebral cortex and cerebellum, much of which is "voluntary" control in contradistinction to the subconscious control effected by the lower centers.

PHYSIOLOGIC ANATOMY OF THE MOTOR AREAS OF THE CORTEX AND THEIR PATHWAYS TO THE CORD

Those parts of the cortex which, when stimulated electrically, cause movements somewhere in the body are considered to be motor areas. Figure 38–1 illustrates the principal motor area of the cortex from which most motor responses are elicited, showing that this area is located in the posterior part of the frontal lobe. The area immediately in front of the central sulcus, designated in the figure by the darkest shading, contains large numbers of *giant Betz cells*, or *pyramidal cells*, for which reason it is

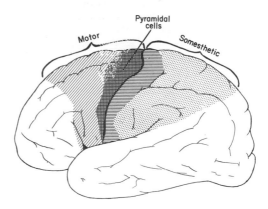

FIGURE 38–1 Relationship of the motor cortex to the somesthetic cortex.

473

called the *area pyramidalis*. This area causes motor movements following the least amount of electrical excitation and therefore is also called the *primary motor cortex*.

The Corticospinal Tract (Pyramidal Tract). One of the major pathways by which motor signals are transmitted from the motor areas of the cortex to the anterior motoneurons of the spinal cord is through the corticospinal tract, or pyramidal tract, which is illustrated in Figure 38–2. This tract originates in all the shaded areas in Figure 38–1, including both the motor and the somesthetic areas, about three-quarters from the motor area and about one-quarter from the somesthetic regions posterior to the central sulcus. The function of the fibers from the somesthetic cortex is not

clear, for only rarely does stimulation of this area cause motor movements.

The most impressive fibers in the corticospinal tract are the large myelinated fibers that originate in the giant Betz cells of the motor area. These account for approximately 34,000 large fibers (mean diameter of about 16 microns) in the corticospinal tract from each side of the cortex. However, since each corticospinal tract contains more than a million fibers, only 3 per cent of the total number of corticospinal fibers are of this large type. The other 97 per cent are mainly fibers smaller than 4 microns diameter, 60 per cent of which are myelinated and 40 per cent unmyelinated.

Even before the corticospinal tract leaves the brain, many collaterals are given off to:

1. Other adjacent cortical areas.
2. The basal ganglia.
3. The reticular substance of the brain stem.
4. The cerebellum.

The Extracorticospinal Tracts (Extrapyramidal Tracts). The extracorticospinal tracts, or extrapyramidal tracts, are collectively all the tracts, besides the corticospinal tract itself, that transmit motor signals from the cortex to the spinal cord. For instance, large numbers of neurons project from other motor areas of the cortex, as well as from the area pyramidalis, into the *caudate nucleus* and then through the *putamen, globus pallidus, subthalamic nucleus, red nucleus, substantia nigra,* and *reticular substance of the brain stem* before passing into the spinal cord. The multiplicity of connections within these intermediate nuclei of the basal ganglia and reticular substance will be presented later in the chapter.

The pathways for transmission of extracorticospinal signals into the cord are the *reticulospinal tracts*, which lie in both the ventral and lateral columns of the cord; and, to a less extent, the *rubrospinal, tectospinal,* and *vestibulospinal tracts*.

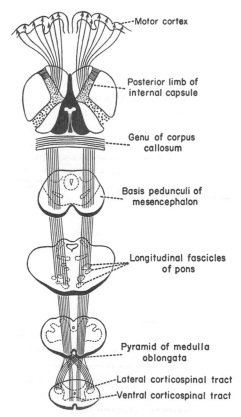

Motor cortex

Posterior limb of internal capsule

Genu of corpus callosum

Basis pedunculi of mesencephalon

Longitudinal fascicles of pons

Pyramid of medulla oblongata

Lateral corticospinal tract

Ventral corticospinal tract

FIGURE 38–2 The corticospinal tract. (From Ranson and Clark: Anatomy of the Nervous System.)

SPATIAL ORGANIZATION OF THE MOTOR CORTEX

Figure 38–3 shows the areas of the body in which muscle contractions occur when different parts of the primary motor cortex are stimulated. Thus, stimulation of the most lateral portions of the cortex causes swallowing or chewing movements, while stimulation of the midline portion of the motor cortex, where the motor cortex bends over into the longitudinal sulcus, causes contraction in the legs, feet, or toes. The spatial organization is similar to that of the sensory cortex, which was presented in Figure 32–8 of Chapter 32.

Degree of Representation of Different Muscle Groups in the Motor Cortex. The different muscle groups of the body are not represented equally in the motor cortex. In general, the degree of representation is proportional to the discreteness of movement required of the respective part of the body. Thus, the thumb and fingers have large representations, as is true also of the lips, tongue, and vocal cords.

When using barely threshold stimuli, only small segments of the peripheral musculature ordinarily contract at a time. In the "finger" and "thumb" regions, which have tremendous representation in the cerebral cortex, threshold stimuli can sometimes cause single muscles or, at times, even single fasciculi of muscles to contract, thus illustrating that a high degree of control is exercised by this portion of the motor cortex over discrete muscular movement.

On the other hand, threshold stimuli in the trunk region of the body might cause as many as 30 to 50 small trunk muscles to contract simultaneously, thus illustrating that the motor cortex does not control discrete trunk muscles but instead controls *groups* of muscles.

Coordinate Movements Elicited by Stimulation of the Motor Cortex — The Principle of "Final Position." Even though bare threshold stimulation of discrete points in the motor cortex can cause relatively minute portions of the musculature to contract individually, more moderate stimulation almost always causes coordinate contraction of several different muscles simultaneously. Furthermore, stimulation of an area of the motor cortex often causes a limb or other part of the body to move to essentially the same "final position" regardless of its original position. This has been demonstrated in normal unanesthetized monkeys in which electrodes had been implanted previously in a limb area of the motor cortex. Each time a stimulus was applied, regardless of the initial position of the limb, it would always move to a given position.

Though the mechanism by which the different "final positions" are achieved is unknown, it is reasonable to believe that they are accomplished by sending appropriate stimuli, perhaps by way of the basal ganglia, into the spinal cord to excite given groups of internuncial cells, each of which in turn is organized to give a particular final position.

Sequential Movements and Rhythmic Movements Elicited by Stimulating the Motor Cortex. Stimulation of a given area of the human cortex occasionally also causes rhythmic types of contraction, with alternate contractions between

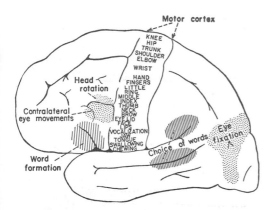

FIGURE 38–3 Representation of the different muscles of the body in the motor cortex and location of other cortical areas responsible for certain types of motor movements.

agonist and antagonist muscles. From the mouth region of the motor cortex, normal chewing or swallowing movements can be elicited. In lower animals stimulation of the foreleg region can occasionally cause rhythmic forward and backward movement of the limb in a stepping fashion, and stimulation of another immediately adjacent area might cause the limb to perform a batting type of movement instead of the stepping movement.

SKILLED MOVEMENTS, AND THE CONCEPT OF A "PREMOTOR AREA"

As one proceeds farther forward in the motor cortex, particularly into the area 1 to 3 centimeters anterior to the area pyramidalis, the patterns of elicited motor movements change somewhat from those observed upon stimulation of the pyramidal cells in the area pyramidalis. Stimulation in this forward region requires considerably greater electrical current and longer trains of stimuli to elicit responses. When responses do occur, they are likely to be slower to develop and also to be gross movements involving many muscles in contradistinction to more discrete movements elicited from the area pyramidalis. Therefore, some physiologists have termed this anterior portion of the motor cortex the premotor area and attribute to it special characteristics that are different from those of the posterior part of the motor cortex called the primary motor area.

The premotor area performs its functions by sending signals through two main routes: (1) through the area pyramidalis and thence down the corticospinal tract, and (2) through the extracorticospinal tracts. For instance, removal of the finger portion of the area pyramidalis prevents fine movements of the fingers when the premotor area is stimulated, indicating that the premotor area must work through the area pyramidalis to give fine movements. Yet, stimulation of the premotor area, even after removal of its corresponding area pyramidalis, can still cause postural movements such as movements of an entire arm, head rotation, movement of the trunk, or "fixation" of the upper arms or legs—all of which are functions generally ascribed to the extracorticospinal system.

Control of Skilled Movements. In human beings in whom small areas of the premotor region have been removed, the coordination of skilled movements becomes impaired. The discrete movements of the hands are still intact, but the organized movements needed for performance of skilled acts are gone. Therefore, it is believed that the premotor region is specifically concerned with the acquisition of specialized motor skills.

In addition, complex eye movements for fixation of the eyes and complex movements of the larynx and mouth for formation of words can be elicited by stimulating the eye, laryngeal, and mouth portions of the premotor regions, which is discussed elsewhere in the text.

EFFECTS OF DESTROYING PORTIONS OF THE MOTOR SYSTEM

Ablation of the Area Pyramidalis. Removal of a very small portion of the area pyramidalis in a monkey usually causes temporary flaccid paralysis of the represented muscles. If the ablated area is large, and particularly if the "premotor" area and underlying caudate nucleus are removed at the same time, the paralysis is permanent. If the caudate nucleus is left intact, gross postural and limb "fixation" movements can still be performed, but discrete movements of the distal segments of the limbs—the hands and fingers, for instance—are lost.

Ablation of the Premotor Area. If the premotor area is removed without destroying any of the adjacent area pyramidalis, the affected region of the body is likely to become spastic, but control of the discrete movements from the area is not affected. On the other hand, some of the skilled movements that had previously been performed by the involved muscles seem no longer to be performed in a coordinate manner.

Transection of the Corticospinal Tract. Transection of a corticospinal tract on one side causes *loss of all discrete, fine movements on the opposite side of the body*, which illustrates that the corticospinal tract, and not the extracorticospinal tracts, controls the fine movements—especially in the hands, fingers, feet, and toes.

The motor functions not lost after transection of the corticospinal tract are mainly the postural contractions and other gross movements of the body. For instance, the different movements required for equilibrium and support of the body against gravity are still completely intact. Furthermore, stimulation of motor cortical regions still causes gross movements of the trunk and upper portions of the limbs. These movements are sometimes called "fixation movements," and their function is the following: when a discrete motor function is to be performed by a hand, for instance, the trunk and upper arm must be "fixed" in appropriate positions before the hand can perform its required function.

FUNCTION OF THE INTERNUNCIAL SYSTEM OF THE SPINAL CORD

As illustrated in Figure 38–4, almost all corticospinal fibers end on small *internuncial cells in the base of the dorsal horns*. From these, impulses are transmitted mainly through *intermediate internuncial cells* to the anterior motoneurons. Thus, usually, at least two internuncial cells must relay the signals from the corticospinal tract to the an-

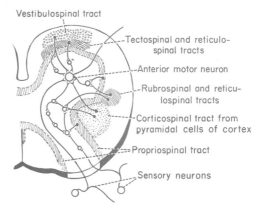

Vestibulospinal tract

Tectospinal and reticulo-
spinal tracts

Anterior motor neuron

Rubrospinal and reticu-
lospinal tracts

Corticospinal tract from
pyramidal cells of cortex

Propriospinal tract

Sensory neurons

FIGURE 38–4 Convergence of all the different motor pathways on the anterior motoneurons.

terior motoneurons, though rare corticospinal fibers do end directly on the anterior motoneurons.

Functions of the Internuncial Cells in the Motor Transmission System. In Chapter 37 we noted that the internuncial cells are the loci of "patterns of movement" in the spinal cord. These can be either fixed movements with "final positions" or rhythmic movements, such as stepping or scratching movements. Therefore, we must presume that these same patterns of movement can be activated by impulses from the corticospinal and extracorticospinal tracts. Thus, spread of excitation across a minute area of the motor cortex could be expected to cause a sequential movement of a limb, perhaps involving 100 or more successive cord-integrated patterns of movement.

Convergence of Signals in the Internuncial System. Another important function of the internuncial system is convergence of motor stimulatory signals from many different sources. For instance, sensory stimuli from the peripheral nerves can cause certain specific patterns of movement as described in relation to the cord reflexes in Chapter 37. Then the motor cortex can elicit its patterns of movement, and stimulation of other cortical motor areas as well as areas of the basal ganglia and the brain stem elicits still other patterns of movement mainly concerned with posture, equilibrium, and fixation of limbs. Since all these different types of patterns of movement might be elicited simultaneously, the internuncial cells and anterior motoneurons must algebraically add all the different stimuli and come out with the net result. Therefore, the internuncial system, along with the motoneurons themselves, constitutes a final level for integration of motor signals from all sources.

MOTOR FUNCTIONS OF THE BASAL GANGLIA

Physiologic Anatomy of the Basal Ganglia. The anatomy of the basal ganglia is so complex and is so poorly known in its details that it would be pointless to attempt a complete description at this time. However, Figure 38–5 illustrates the principal structures of the basal ganglia and their neural connections with other parts of the nervous system. Anatomically, the basal ganglia are the *caudate*

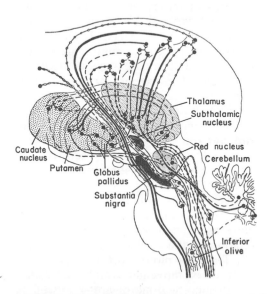

FIGURE 38–5 Pathways through the basal ganglia and related structures of brain stem, the thalamus, and cerebral cortex. (From Jung and Hassler: Handbook of Physiology, Sec. I, Vol. II, 1960. Baltimore, The Williams & Wilkins Co.)

nucleus, putamen, globus pallidus, amygdaloid nucleus, and *claustrum*. The amygdaloid nucleus, which will be discussed in Chapter 41, and the claustrum are not concerned directly with motor functions of the central nervous system. On the other hand, the *thalamus, subthalamus, substantia nigra*, and *red nucleus* all operate in close association with the caudate nucleus, putamen, and globus pallidus and are considered to be part of the basal ganglia system for motor control.

Two important features of the different pathways illustrated in Figure 38–5 are the following:

1. Numerous nerve pathways pass from the motor portions of the cerebral cortex to the caudate nucleus and putamen. In turn, the caudate nucleus and putamen send numerous fibers to the globus pallidus, the globus pallidus to the ventrolateral nucleus of the thalamus, and this nucleus back to the motor areas of the cerebral cortex. Thus, circular pathways are established from the motor cortical regions to the basal ganglia, the thalamus, and back to the same motor regions from which the pathways begin.

2. The basal ganglia have numerous short neuronal connections among themselves. Also, the lower basal ganglia, such as the globus pallidus, send tremendous numbers of nerve fibers into the lower brain stem, projecting especially onto the reticular nuclei of the mesencephalon and onto the red nucleus.

FUNCTIONS OF THE BASAL GANGLIA

Inhibition of Motor Tone by the Basal Ganglia. Though it is wrong to ascribe a single function to all the basal ganglia, nevertheless, one of the general effects of basal ganglial excitation is to inhibit muscle tone throughout the body. This effect results from inhibitory signals transmitted from the basal ganglia to the bulboreticular facilitatory area and excitatory signals to the bulboreticular inhibitory area. Therefore, whenever widespread destruction of the basal ganglia occurs, the facilitatory area becomes overactive and the inhibitory area underactive, and this causes muscle rigidity throughout the body.

Yet, despite this general inhibitory effect of the basal ganglia, stimulation of specific areas within the basal ganglia can at times elicit positive muscle contractions and at times even complex patterns of movements.

Function of the Caudate Nucleus and Putamen—The Striate Body. The caudate nucleus and putamen, because of their gross appearance on sections of the brain, are together called the *striate body*. They seem to function together to initiate and regulate gross intentional movements of the body. To perform this function they transmit impulses through two different pathways: (1) into the *globus pallidus*, thence by way of the *thalamus* to the *cerebral cortex*, and finally downward into the spinal cord through the *corticospinal* and *extracorticospinal pathways*; (2) downward through the *globus pallidus* by way of short pathways into the *reticular formation* and finally into the spinal cord mainly through the *reticulospinal tracts*.

In summary, the striate body helps to control gross intentional movements that we normally perform unconsciously. However, this control usually also involves the motor cortex, with which the striate body is very closely connected.

Function of the Globus Pallidus. It has been suggested that the principal function of the globus pallidus is to provide background muscle tone for intended movements, whether these be initiated by impulses from the cerebral cortex or from the striate body. That is, if a person wishes to perform an exact function with one hand, he positions his body appropriately and then tenses the muscles of the upper arm. These associated tonic contractions are supposedly initiated by a circuit that strongly involves the globus pallidus. Destruction of the globus pal-

lidus removes these associated movements and, therefore, makes it difficult for the distal portions of the limbs to perform their more discrete activities.

The globus pallidus is believed to function through two pathways: first, through feedback circuits to the thalamus, to the cerebral cortex, and thence by way of corticospinal and extracorticospinal tracts to the spinal cord; and, second, by way of short pathways to the reticular formation of the brain stem and thence mainly by way of the reticulospinal tracts into the spinal cord.

Electrical stimulation of the globus pallidus while an animal is performing a gross body movement will often stop the movement in a static position, the animal holding that position for many seconds while the stimulation continues. This reaction fits with the concept that the globus pallidus is involved in some type of servo feedback motor control system that is capable of locking the different parts of the body into specific positions. Obviously, such a circuit could be extremely important in providing the background body movements and upper limb movements when a person performs delicate tasks with his hands.

CLINICAL SYNDROMES RESULTING FROM DAMAGE TO THE BASAL GANGLIA

Much of what we know about the function of the basal ganglia comes from study of patients with basal ganglia lesions whose brains have undergone pathologic studies after death. Among the different clinical syndromes are:

Chorea. Chorea is a disease in which random uncontrolled contractions of different muscle groups of the body occur continuously. Normal progression of movements cannot occur; instead, the person may perform a normal sequence of movements for a few seconds and then suddenly begin another sequence of movements; then still another sequence begins after a few seconds. Because of this peculiar progression of movements, one type of chorea is frequently called *St. Vitus' dance*.

Pathologically, chorea results from *diffuse and widespread damage of the striate body*, but the manner in which this causes the choreiform movements is not known.

Athetosis. In this disease, slow, writhing movements of the peripheral parts of one or more limbs occur continually. The movements are likely to be worm-like, first, with overextension of the hands and fingers, then flexion, then rotatory or twisting to the side—all these continuing in a slow, rhythmic pattern. The contracting muscles exhibit a high degree of spasm, and the movements are enhanced by emotions or by excessive signals from the sensory organs. Furthermore, voluntary movements in the affected area are greatly impaired or sometimes even impossible.

The damage in athetosis is usually found in the *outer portion of the globus pallidus* or in this area and the striate body. Athetosis is usually attributed to the interruption of feedback circuits among the basal ganglia, thalamus, and cerebral cortex.

Parkinson's Disease. Parkinson's disease, which is also known as *paralysis agitans*, results almost invariably from *widespread destruction of the substantia nigra*. It is characterized by (1) *rigidity* of the musculature in either widespread areas of the body or in isolated areas, (2) *tremor at rest* of the involved areas in most but not all instances, and (3) *loss of involuntary and associated movements*. Also, the gamma activating system for exciting the muscle spindles becomes completely or almost completely inactivated, presumably because the substantia nigra is one of the major areas of the brain stem for controlling the gamma efferent system.

Tremor usually, though not always, occurs in Parkinson's disease. Its frequency normally is 4 to 8 cycles per second. When the person performs voluntary movements, weak tremors become temporarily blocked. The mechanism of the tremor is not known, but one theory that deserves serious consideration is that loss of the damping function of the gamma efferent muscle spindle system allows oscillation between the neuronal centers that control antagonistic muscle groups.

Coagulation of the Ventrolateral Nucleus of the Thalamus for Treatment of Parkinson's Disease. In recent years neurosurgeons have treated Parkinson's disease patients, with varying success, by destroying portions of the basal ganglia or of the thalamus. The most prevalent treatment at present is widespread destruction of the ventrolateral nucleus of the thalamus, either by electrocoagulation or by injecting a sclerosing chemical. Most fiber pathways from the basal ganglia and cerebellum to the cerebral cortex pass through this nucleus so that its destruction blocks many or most of the feedback functions of the basal ganglia and cerebellum. It is presumed that blockage of some of these feedbacks limits the functions of at least certain basal ganglia and thereby removes the factors that cause the rigidity and tremor of Parkinson's disease.

Treatment with L-Dopa. The neurons that have cell bodies in the substantia nigra and that have nerve endings terminating mainly in the lower basal ganglia are believed to secrete dopamine as their transmitter substance. Therefore, destruction of the substantia nigra eliminates the dopamine stimulation of the basal ganglia, and is supposedly the cause of Parkinson's disease. Administration of L-dopa to substitute for the lost dopamine greatly ameliorates the symptoms in over two-thirds of the patients.

MOTOR FUNCTIONS OF THE CEREBELLUM

The cerebellum has long been called a silent area of the brain, principally because electrical excitation of the cerebellum does not cause any sensation and rarely any motor movement. However, as we shall see, removal of the cerebellum does make motor movements highly abnormal.

Physiologic Anatomy. The basic *afferent pathways* to the cerebellum are illustrated in Figure 38–6. These include:

1. The *pontocerebellar tract*, which transmits signals mainly from the motor cortex.

2. The *olivocerebellar tract*, which transmits signals from the basal ganglia.

3. The *vestibulocerebellar tract*, which transmits signals of equilibrium from the vestibular system.

4. The *reticulocerebellar tract*, which transmits signals from the brain stem.

5. The *spinocerebellar tracts*, which transmit signals from the muscle spindles and the Golgi tendon organs.

6. The *tract from dorsal column nuclei to cerebellum*, which transmits signals from body proprioceptors.

The *efferent tracts* that leave the cerebellum are illustrated in Figure 38–7. These include mainly:

1. The *cerebellothalamocortical tract*, which transmits signals to the motor cortex and basal ganglia.

2. The *cerebellorubral* and *cerebello-*

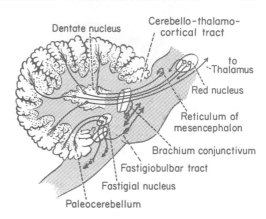

FIGURE 38–7 Principal efferent tracts from the cerebellum.

reticular tracts, which transmit signals mainly into the brain stem.

FUNCTION OF THE CEREBELLUM IN VOLUNTARY MOVEMENTS

The cerebellum functions only in association with motor activities initiated elsewhere in the central nervous system. These activities may be initiated originally in the spinal cord, in the reticular formation, in the basal ganglia, or in the motor areas of the cerebral cortex. We will discuss, first, the operation of the cerebellum in association with the motor cortex for the control of voluntary movements.

Figure 38–8 illustrates the basic cerebellar pathways involved in cerebellar control of voluntary movements. When motor impulses are transmitted from the cerebral cortex downward through the corticospinal tract to excite the voluntary muscles, collateral impulses are transmitted simultaneously into the cerebellum through the pontocerebellar tracts. Therefore, for every motor movement that is initiated, signals are sent not only to the muscles but to the cerebellum at the same time.

When the muscles respond to the motor signals, the muscle spindles, Golgi tendon organs, joint receptors,

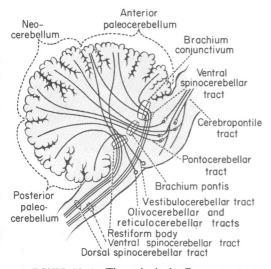

FIGURE 38–6 The principal afferent tracts to the cerebellum.

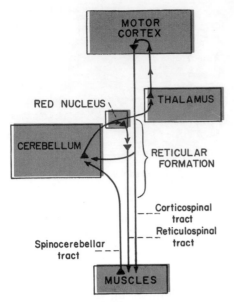

FIGURE 38-8 Pathways for cerebellar "error" control of voluntary movements.

and other peripheral receptors transmit impulses upward mainly through the spinocerebellar and dorsal column pathways to terminate in the anterior cerebellum. Furthermore, these impulses go to exactly the same portion of the anterior cerebellum that is stimulated by the downcoming impulses from the motor cortex.

After the signals from the periphery and those from the motor cortex are integrated, efferent impulses are transmitted from the cerebellar cortex to the dentate nuclei and thence upward through the ventrolateral nuclei of the thalamus back to the motor cortex where the stimulus first originated.

"Error Control" by the Cerebellum. One will readily recognize that the circuit described above represents a complicated feedback circuit beginning in the motor cortex and also returning to the motor cortex. Furthermore, experiments have shown that the cerebellum operates in much the same manner as a servomechanism such as that used in (a) industrial control systems, (b) the control system of an anti-aircraft gun, or

(c) the control system of an automatic pilot. That is, the cerebellum seems to compare the "intentions" of the cortex with the "performance" by the parts of the body, and, if the parts have not moved according to the intentions of the cortex, the "error" between these two is calculated by the cerebellum so that appropriate and immediate corrections can be made. Thus, if the cortex has transmitted a signal intending the limb to move to a particular point, but the limb begins to move too fast and obviously is going to overshoot the point of intention, the cerebellum can initiate "braking impulses" to slow down the movement of the limb and stop it at the point of intention.

The "Damping" Function of the Cerebellum—The Intention Tremor. One of the by-products of the cerebellar feedback mechanism is its ability to "damp" muscular movements. To explain the meaning of "damping" we must first point out that essentially all movements of the body are "pendular." For instance, when an arm is moved, momentum develops, and the momentum must be overcome before the movement can be stopped. And, because of the momentum, all pendular movements have a tendency to overshoot. The cerebellum ordinarily prevents this, but in a person whose cerebellum has been destroyed, the conscious centers of the cerebrum eventually recognize the overshoot and initiate a movement in the opposite direction to bring the arm to its intended position. But again the arm, by virtue of its momentum, overshoots, and appropriate corrective signals must again be instituted. Thus the arm oscillates back and forth past its intended point for several cycles before it finally reaches its mark. This effect is called an *action tremor*, or *intention tremor*.

Function of the Cerebellum in Prediction—Dysmetria. Another important by-product of the cerebellar feedback mechanism is its ability to help the central nervous system predict the future positions of moving parts of the body. Without the cerebellum a person "loses" his limbs when they move rapidly, indicating that feedback information from the periphery must be analyzed by the cerebellum if the brain is to keep track of the motor movements. Thus, somewhere in the circuit of the cerebellum there seems to be an integrative system that detects from the incoming proprioceptive signals the rapidity with which the limb is moving and then predicts from this the projected time course of movement. This allows the cerebellum, operating through the cerebral cortex, to inhibit the agonist muscles and to excite the antagonist muscles when the movement approaches the point of intention.

Without the cerebellum this predictive function is so deficient that moving parts of the body move much farther than the point of intention. This failure to control the distance that the parts of the body move is called *dysmetria*, which means simply poor control of the distance of movement. As would be expected, dysmetria becomes greatly enhanced in rapid movements in comparison with slow movements.

Extramotor Predictive Functions of the Cerebellum. The cerebellum also plays a role in predicting other events besides movements of the body. For instance, the rates of progression of both auditory and visual phenomena can be predicted. As an example, a person can predict from the changing visual scene how rapidly he is approaching an object. A striking experiment that demonstrates the importance of the cerebellum in this ability is the effect of midline lesions in the paleocerebellum of monkeys. Such a monkey occasionally charges the wall of a corridor and literally bashes its brains out because it is unable to predict when it will reach the wall.

Function of the Cerebellum in Involuntary Movements. The cerebellum functions in involuntary movements in almost exactly the same manner as it does in voluntary movements except that different pathways are involved, utilizing feedback to the basal ganglia and brain stem instead of to the motor cortex, and also utilizing the extracorticospinal tracts instead of the corticospinal tract.

Function of the Cerebellum in Equilibrium. In the discussion of the vestibular apparatus and of the mechanism of equilibrium in Chapter 37, it was pointed out that the flocculonodular lobes of the cerebellum are required for proper integration of equilibratory impulses from the semicircular canals. These areas of the cerebellum are particularly important for integration of *changes in direction of motion* as detected by the semicircular canals, but not so important for integrating static signals of equilibrium as detected by the maculae in the utricles. This is in keeping with the other functions of the cerebellum; that is, the semicircular canals allow the central nervous system to *predict* ahead of time that rotational movements of the body are going to cause mal-equilibrium.

CLINICAL ABNORMALITIES OF THE CEREBELLUM

Dysmetria and Ataxia. Two of the most important symptoms of cerebellar disease are dysmetria and ataxia. It was pointed out above that in the absence of the cerebellum a person cannot predict ahead of time how far his movements will go. Therefore, the movements ordinarily overshoot their intended mark. This effect is called *dysmetria*, and it results in incoordinate movements which are called *ataxia*.

Dysmetria and ataxia can also result from lesions in the spinocerebellar tracts, for the feedback information from the moving parts of the body is essential for accurate control of the muscular movements.

Failure of Progression. *Dysdiadochokinesia.* When the motor control system fails to predict ahead of time where the different parts of the body will be at a given time, it temporarily "loses" the parts during rapid motor movements.

As a result, the succeeding movement may begin much too early or much too late so that no orderly "progression of movement" can occur. One can demonstrate this readily in a patient with cerebellar damage by having him turn one of his hands upward and downward at a rapid rate. He rapidly "loses" his hand and does not know its position during any portion of the movements. As a result, a series of jumbled movements occurs instead of the normal coordinate upward and downward motions. This is called dysdiadochokinesia.

Dysarthria. Another instance in which failure of progression occurs is in talking, for the formation of words depends on rapid and orderly succession of individual muscular movements in the larynx. Lack of coordination between these and inability to predict either the intensity of the sound or the duration of each successive sound cause jumbled vocalization, with some syllables loud, some weak, some held long, some held for a short interval, and resultant speech that is almost completely unintelligible. This is called dysarthria.

Intention Tremor. When a person who has lost his cerebellum performs a voluntary act, his muscular movements are jerky; this reaction is called an *intention tremor* or an *action tremor*, as was discussed earlier.

Loss of Equilibrium. Destruction of the flocculonodular portion of the cerebellum causes loss of equilibrium, particularly when the person attempts to perform rapid movements. This loss of equilibrium is similar to that which occurs when the semicircular canals are destroyed, as was discussed in Chapter 38.

SENSORY FEEDBACK CONTROL OF MOTOR FUNCTIONS

The Sensory "Engram" for Motor Movements. It is primarily in the sensory and sensory association areas that a person experiences effects of motor movements and records "memories" of the different patterns of motor movements. These are called *sensory engrams* of the motor movements. When he wishes to achieve some purposeful act, he calls forth one of these engrams and then sets the motor system of the brain into action to reproduce the sensory pattern that is laid down in the engram. The mechanism of this is the following:

The Proprioceptor Feedback Servomechanism for Reproducing the Sensory Engram. In addition to feedback pathways through the cerebellum, more slowly acting feedback pathways also pass from proprioceptors to the sensory areas of the cerebral cortex and thence back to the motor cortex. (Also, some pathways feed directly back to the motor cortex by way of the thalamus.) Each of these feedback pathways is capable of modifying the motor response. For instance, if a person learns to cut with scissors, the movements involved in this process cause a particular sequential pattern of proprioceptive impulses to pass to the somatic sensory area. Once this pattern has been "learned" by the sensory cortex, the memory engram of the pattern can be used to control the motor system to perform the same sequential pattern whenever it is required.

To do this, the proprioceptor signals from the fingers, hands, and arms are compared with the engram, and if the two do not match each other, the difference, called the "error," supposedly initiates additional motor signals that automatically activate appropriate muscles to bring the fingers, hands, and arms into the necessary attitudes for performance of the task. Each successive portion of the engram presumably is projected according to a time sequence, and the motor control system automatically follows from one point to the next so that the fingers go through the precise motions necessary to duplicate exactly the sensory engram of the motor activity.

Thus, one can see that the motor system in this case actually acts as a servomechanism, for it is not the motor cortex itself that controls the pattern of activity to be accomplished. Instead, the pattern is located in the sensory part of the brain, and the motor system merely "follows" the pattern, which is the definition of a servomechanism.

An extremely interesting experiment that demonstrates the importance of the sensory engram for control of motor movements is one in which a monkey has been trained to perform some skilled activity and then various portions of his cortex are removed. Removal of small portions of the motor cortex that control the muscles normally used for the skilled activity does not prevent the monkey from performing the activity. Instead he automatically uses other muscles in place of the paralyzed ones to perform the same skilled activity. On the other hand, if the corresponding somesthetic cortex is removed but the motor cortex is left intact, the monkey loses all ability to perform the skilled activity. Thus, this experiment demonstrates that the motor system acts automatically as a servomechanism to use whatever muscles are available to follow the pattern of the sensory engram, and if some muscles are missing, other muscles are substituted automatically. The experiment also demonstrates forcefully that the somesthetic cortex is essential to the motor performance.

"MOTOR ENGRAMS" FOR MOTOR MOVEMENTS

Many motor activities are performed so rapidly that there is insufficient time for sensory feedback signals to control these activities. For instance, the movements of the fingers during typing occur much too rapidly for somesthetic sensory signals to be transmitted back to the cortex and for these then to control each discrete movement. The patterns for control of these rapid coordinate muscular movements are established in the motor areas of the frontal lobe, and these patterns are called "motor engrams" for motor movements, in contradistinction to the sensory engrams just discussed.

Role of Sensory Feedback During Establishment of the Rapid Motor Patterns. Even a highly skilled motor activity can be performed the very first time if it is performed extremely slowly — slowly enough for sensory feedback to guide the movements through each step. However, to be really useful, many skilled motor activities must be performed rapidly. This is achieved by successive performance of the same skilled activity until finally an engram of the skilled activity is laid down in the motor regions of the cortex as well as in the sensory region. This engram causes a precise set of muscles to go through a specific sequence of movements required to perform the skilled activity. Therefore, such an engram is called a *pattern of skilled motor function*, and the motor areas are primarily concerned with this. After a person has performed a skilled activity many times, the motor pattern of this activity can thereafter cause the hand or arm or other part of the body to go through the same pattern of activity again and again, now entirely *without* sensory feedback control. However, even though sensory feedback control is no longer present, the sensory system still determines whether or not the act has been performed correctly. This determination is made in retrospect rather than while the act is being performed. If the pattern has not been performed correctly, information from the sensory system supposedly can help to correct the pattern the next time it is performed.

Thus, eventually, hundreds of patterns of different coordinate movements are laid down in the motor areas of the cortex, and these can be called upon one at a time in different sequential orders to perform literally thousands of complex motor activities.

REFERENCES

Denny-Brown, D.: The Basal Ganglia: And Their Relation to Disorders of Movement. New York, Oxford University Press, Inc., 1962.

Evarts, E. V., and Thach, W. T.: Motor mechanisms of the CNS: cerebrocerebellar interrelations. *Ann. Rev. Physiol., 31*:451, 1969.

Fox, C. A., and Snider, R. S.: The Cerebellum. Progress in Brain Research. New York, American Elsevier Publishing Co., Inc., 1967, Vol. 25.

Marchiafava, P. L.: Activities of the central nervous system: motor. *Ann. Rev. Physiol., 30*: 359, 1968.

Martin, J. P.: Basal Ganglia and Posture. Philadelphia, J. B. Lippincott Co., 1967.

Massion, J.: The mammalian red nucleus. *Physiol. Rev., 47*:383, 1967.

Oscarsson, O.: Functional organization of the spino- and cuneocerebellar tracts. *Physiol. Rev., 45*:495, 1965.

Penfield, W., and Rasmussen, T.: The Cerebral Cortex of Man. New York, The Macmillan Company, 1950.

CHAPTER 39

THE CEREBRAL CORTEX AND INTELLECTUAL FUNCTIONS OF THE BRAIN

It is ironic that we know least about the mechanisms of the cerebral cortex of all parts of the brain, even though it is by far the largest portion of the nervous system. Nevertheless, in the early part of the present chapter the known facts about cortical functions are discussed and then some basic theories of the neuronal mechanisms involved in thought processes, memory, analysis of sensory information, and so forth, are presented briefly.

PHYSIOLOGIC ANATOMY OF THE CEREBRAL CORTEX

The functional part of the cerebral cortex is a thin layer of neurons 1.5 to 4 mm. in thickness, covering the surface of all the convolutions of the cerebrum and having a total area of about 2300 sq. cm. The total cerebral cortex contains approximately 9.3 billion neurons.

Figure 39-1 illustrates the typical structure of the cerebral cortex, showing six successive layers of different types of cells. Most of the cells are either *granule cells* or *pyramidal cells,* the latter named for their characteristic pyramidal shape. To the right in Figure 39-1 is illustrated the typical organization of nerve fibers within the different layers of the cortex. Note particularly the large number of horizontal fibers extending between adjacent areas of the cortex, but note also the vertical fibers that extend to and from the cortex to lower areas of the brain stem or to distant regions of the cerebral cortex through long association bundles of fibers.

Anatomical Relationship of the Cerebral Cortex to the Thalamus and Other Lower Centers. All areas of the cerebral cortex have direct afferent and efferent connections with the thalamus. Figure 39-2 shows the areas of the cerebral cortex and the names of the specific thalamic nuclei with which they connect. These connections are always in *two*

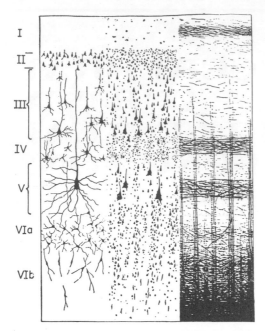

FIGURE 39–1 Structure of the cerebral cortex, illustrating: *I*, molecular layer; *II*, external granular layer; *III*, layer of pyramidal cells; *IV*, internal granular layer; *V*, ganglionic layer; *VI*, layer of fusiform or polymorphic cells. (From Ranson and Clark (after Brodmann): Anatomy of the Nervous System.)

directions, both from the thalamus to the cortex and then from the cortex back to the same area of the thalamus. Furthermore, when the thalamic connections are cut, the functions of the corresponding cortical area become entirely

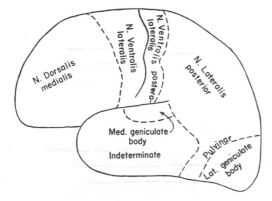

FIGURE 39–2 Areas of the cerebral cortex that connect with specific portions of the thalamus. (Modified from Elliott: Textbook of the Nervous System. J. B. Lippincott Co.)

or almost entirely abrogated. Therefore, the cortex operates in close association with the thalamus and can almost be considered both anatomically and functionally as a large outgrowth of the thalamus.

FUNCTIONS OF SPECIFIC CORTICAL AREAS

Studies in human beings by neurosurgeons have shown that specific functions are localized to certain general areas of the cerebral cortex. Figure 39–3 gives a map of these areas as determined by Penfield and Rasmussen from electrical stimulation of the cortex or by neurological examination of patients after portions of the cortex had been removed. The lightly shaded areas are primary sensory areas, while the darkly shaded area is the primary motor area from which muscular movements can be elicited with relatively weak electrical stimuli. These primary sensory and motor areas have highly specific functions, but all the other areas of the cortex perform the more general function that we commonly call cerebration.

FUNCTIONS OF THE PRIMARY SENSORY AREAS

The primary sensory areas all have certain functions in common. For instance, electrical stimulation of the primary visual cortex causes the person to see flashes of light, bright lines, colors, or other simple visions; and the visual images are localized to specific regions of the visual fields in accord with the portion of the primary visual cortex stimulated, as described in Chapter 35. But the visual cortex alone is not capable of analyzing complicated visual patterns; for this, the visual cortex must operate in association both with other regions of the cerebral cortex and with lower brain centers.

Similar effects occur in all the other

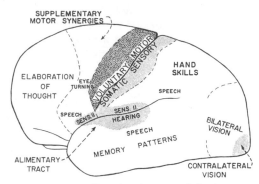

FIGURE 39-3 Functional areas of the human cerebral cortex as determined by electrical stimulation of the cortex during neurosurgical operations and by neurological examinations of patients with destroyed cortical regions. (From Penfield and Rasmussen: The Cerebral Cortex of Man. The Macmillan Co.)

primary sensory areas, which have been discussed in previous chapters. In the somatic area, electrical stimulation gives one the impression of "feeling" simple effects such as tingling and pinpricks. Stimulation in the auditory cortex causes one to "hear" a simple sound such as a loud or weak tone, a squeak, or a click; but never are words "heard." Therefore, for the information entering the primary sensory areas to become fully intelligible, it must be further analyzed elsewhere in the brain.

THE SECONDARY SENSORY AREAS

Around the borders of the primary sensory areas are regions which are variously called *secondary sensory areas*, or *sensory association areas*. In general these areas extend 1 to 5 cm. in all directions from the primary sensory areas; and each time a primary area receives a sensory signal, secondary signals immediately spread into these secondary areas as well.

Destruction of these secondary sensory areas greatly reduces the capability of the brain to analyze different characteristics of sensory experiences. For instance, damage in the superior temporal gyrus below and behind the primary auditory area in the "dominant hemis-

phere" of the brain often causes a person to lose all understanding of the meanings of words or of other auditory experiences.

Likewise destruction of areas 18 and 19 or the presence of a brain tumor or other lesion in these areas, which represent the secondary visual areas, does not cause blindness or prevent normal activation of the primary visual cortex but greatly reduces the person's ability to interpret what he sees. Such a person often loses his ability to recognize the meanings of words, a condition that is called *word blindness*.

Finally, destruction of the secondary somatic sensory area in the parietal cortex posterior to primary somatic area I causes the person to lose his spatial perception for location of the different parts of his body. In the case of the hand that has been "lost," the skills of the hand are greatly reduced. Thus, this area of the cortex seems to be necessary for interpretation of somatic sensory experiences.

INTERPRETATIVE FUNCTIONS OF THE ANGULAR GYRUS AND POSTERIOR TEMPORAL LOBE REGIONS

The posterior temporal lobe and the angular gyrus—the gyrus that lies at the posterior tip of the lateral fissure where the temporal, parietal, and occipital lobes all come together—are especially important for many of the intellectual functions of the cerebral cortex, particularly for interpretation of the different sensory experiences, whether these be visual, auditory, or somatic. Following severe damage in these regions, a person might hear perfectly well and even recognize different words but still be unable to arrange these words into a coherent thought. Likewise, he may be able to read words from a printed page but again be unable to recognize the thought that is conveyed.

Electrical stimulation in widespread areas of the posterior temporal lobe and angular gyrus of a conscious patient causes highly complex thoughts, including memories of complicated visual scenes or auditory hallucinations, such as specific musical pieces or a discourse by a specific person. For this reason it is believed that complicated memory patterns are stored here. This is in accord with the importance of these areas in the interpretation of complicated meanings of different sensory experiences.

The Dominant Hemisphere. The interpretative functions of the temporal lobe and angular gyrus are usually highly developed in only one cerebral hemisphere, which is called the dominant hemisphere. At birth these regions in both of the two hemispheres have almost the same capability of developing; we know this because removal of either hemisphere in early childhood causes the opposite to develop full dominant characteristics. Yet a theory that might explain dominance is the following: The attention of the "mind" is usually directed to one portion of the brain at a time, as will be explained in the next chapter. Presumably, one angular gyrus region begins to be used to a greater extent than the other, and, thenceforth, because of the tendency to direct one's attention to the better developed region, the rate of learning in the cerebral hemisphere that gains the first start increases rapidly while that in the opposite side remains slight. Therefore, in the normal human being, one side becomes dominant over the other.

In more than 9 out of 10 persons the left temporal lobe and angular gyrus become dominant, and in the remaining one tenth of the population either both sides develop simultaneously to have dual dominance, or, more rarely, the right side alone becomes highly developed.

Usually associated with the dominant temporal lobe and angular gyrus is dominance of certain portions of the somesthetic cortex and motor cortex for control of voluntary motor functions. For instance, as is discussed later in the

chapter, the premotor speech area, located far laterally in the intermediate frontal area, is almost always dominant on the same side of the brain as the dominant temporal lobe and angular gyrus. This speech area causes the formation of words by exciting simultaneously the laryngeal muscles, the respiratory muscles, and the muscles of the mouth.

Though the interpretative areas of the temporal lobe and angular gyrus, as well as many of the motor areas, are highly developed in only a single hemisphere, they are capable of receiving sensory information from both hemispheres and are also capable of controlling motor activities in both hemispheres. This unitary, cross-feeding organization prevents interference between the two sides of the brain; such interference, obviously, could create havoc with both thoughts and motor responses.

Effect of temporal lobe or angular gyrus destruction in the dominant hemisphere. Destruction of the dominant temporal lobe and angular gyrus in an adult person normally leaves a great void in his intellect because of his inability to interpret the meanings of sensory experiences. Therefore, the neurosurgeon assiduously avoids surgery in this region. The adult can only gradually develop the interpretative functions of the nondominant temporal lobe and angular gyrus, and even then only to a slight extent. Thus, only a small amount of intellect can return. However, if this area is destroyed in a child under six years of age, the opposite side can usually develop to full extent, thus eventually returning the capabilities of the child essentially to normal.

THE PREFRONTAL AREAS

Prevention of Distractibility by the Prefrontal Areas. The prefrontal areas are those portions of the frontal lobes that lie anterior to the motor regions. One of the outstanding characteristics of a person who has lost his prefrontal areas is the ease with which he can be *distracted* from a sequence of thoughts. Likewise, in lower animals whose prefrontal areas have been removed, the ability to concentrate on psychological tests is almost completely lost. The human being without prefrontal areas is still capable of performing many intellectual tasks, such as answering short questions and performing simple arithmetic computations (such as $9 \times 6 = 54$), thus illustrating that the basic intellectual activities of the cerebral cortex are still intact without the prefrontal areas. Yet if concerted *sequences* of cerebral functions are required of the person, he becomes completely disorganized. Therefore, the prefrontal areas seem to be important in keeping the mental functions directed toward goals.

Elaboration of Thought by the Prefrontal Areas. Another function that has been ascribed to the prefrontal areas by psychologists is called *elaboration of thought*. This means simply an increase in depth and abstractness of the different thoughts. It can be shown by psychological tests that lower animals presented with successive bits of sensory information forget these bits, however important they might be, within a second or more if the prefrontal areas have been removed. Yet if the prefrontal areas are intact, many successive bits of information can be remembered for many seconds. Therefore, there is much reason to believe that the prefrontal areas are especially capable of storing many bits of information *temporarily* and then correlating these into thoughts of a high order. This ability to hold and correlate many types of information simultaneously in the prefrontal areas could well explain the many functions of the brain that we associate with higher intelligence, such as abilities to (1) plan for the future, (2) delay action in response to incoming sensory signals so that the sensory information can be weighed until the best course of response is decided, (3) consider the consequences of motor actions even before these are performed, (4) solve complicated mathematical, legal, or philosophical problems, (5) correlate all avenues of information in diagnosing rare diseases, and (6) control one's activities in accord with moral laws.

Effects of Destruction of the Prefrontal Areas. The person without prefrontal areas ordinarily acts precipitously in response to incoming sensory impulses, such as striking an adversary too large to be beaten instead of pursuing the more judicious course of running away. Also, he is likely to lose many or most of his morals; he has little embarrassment in relation to his excretory, sexual, and social activities; and he is prone to quickly changing moods of sweetness, hate, joy, sadness, exhilaration, and rage. In short, he is a highly *distractible* person with lack of ability to pursue long and complicated thoughts.

THOUGHTS AND MEMORY

A thought probably results from the momentary "pattern" of stimulation of many different parts of the nervous system at the same time, probably involving most importantly the cerebral cortex, the thalamus, the rhinencephalon, and the upper reticular formation of the brain stem. The stimulated areas of the rhinencephalon, thalamus, and reticular formation probably determine the crude nature of the thought, giving it such qualities as pleasure, displeasure, pain, comfort, crude modalities of sensation, localization to general parts of the body, and other gross characteristics. On the other hand, the stimulated area of the cortex probably determines the discrete characteristics of the thought (such as specific localization of sensations in the body and of objects in the fields of vision), discrete patterns of sensation (such as the rectangular pattern of a concrete block wall or the texture of a rug), and other individual characteristics that enter into the overall awareness of a particular instant.

THE BASIS OF MEMORY

If we accept the above approximation of what constitutes a thought, we can see immediately that the mechanism of memory must be equally as complex as the mechanism of a thought, for, to provide memory, the nervous system must create the same spatial and temporal pattern of stimulation in the central nervous system at some future date. Though we cannot explain in detail what a memory is, we do know some of the basic neuronal processes that probably lead to the process of memory.

Instantaneous Memory. Instantaneous memory refers to one's ability to recall tremendous amounts of information from one second to the next or from minute to minute. For instance, if one is walking through a large industrial plant, literally thousands or perhaps millions of small bits of information are passed into the nervous system from the sensory receptors. Most of these bits are stored without one's thinking about them, but for periods of seconds or minutes thereafter, the information can still be recalled. Unfortunately, we do not know the basic mechanism of this type of memory, but the following theory is representative of possible mechanisms.

Reverberating circuit theory of instantaneous memory. When a tetanizing electrical stimulus is applied directly to the surface of the cerebral cortex and then removed after a second or more, the local area excited by this stimulus continues to emit rhythmic action potentials for minutes or, under favorable conditions, for as long as one hour. This effect results from local reverberating circuits, the impulses passing through a multiconnected circuit of neurons in the local area of the cortex itself

or back and forth between the cortex and the thalamus.

It is presumed that sensory signals reaching the cerebral cortex can set up similar reverberating oscillations that continue for many minutes and provide memory of the initiating sensation. Then gradually, as the reverberating circuits fatigue and the oscillation ceases, the instantaneous or temporary memory fades away.

One of the principal observations in support of this theory of instantaneous or temporary memory is that any factor that causes a general disturbance of brain function, such as sudden concussion of the brain, immediately erases all instantaneous or temporary memory, and the memory cannot be recalled when the disturbance is over.

Fixed Memory—The Time Factor in the Fixation of Memories. If a memory is to last in the brain so that it can be recalled days later, it must become "fixed" in the neuronal circuits. This process requires 15 to 20 minutes for minimal fixation and an hour or more for maximal fixation. For instance, if a strong sensory impression is made on the brain but is then followed by strong electrical stimulation of the brain, the sensory experience will not be remembered at all. However, if the same sensory stimulus is impressed on the brain and the strong electrical shock is delayed for as long as an hour, the memory "engram" will have become established. If we hold to the reverberation theory of instantaneous or temporary memory, one could explain these differences in the following way: If the reverberation is disrupted soon after it begins by strong electrical stimulation, significant physical or chemical changes in the synapses will not occur to develop a fixed memory. But if the thought pattern is not disrupted, sufficient reverberation can take place to establish a deeply trodden memory pathway.

Similarly, if many strong sensory experiences occur one after another, it usually will not be possible to remember any one of them well, presumably because each successive experience disrupts the previous reverberating circuit before an engram of the memory can be established.

Mechanism of fixed memory — alteration of transmission facility at the synapses. Fixed memory means the ability of the nervous system to recall thoughts long after initial excitation of the thoughts is over. It is assumed that fixed memory must result from some actual alterations of the synapses, either physical or chemical. Many different theories have been offered to explain the synaptic changes that cause fixed memory. Perhaps the most important of these is one dealing with *anatomical changes in the synapses.* Cajal, more than half a century ago, discovered that the number of terminal fibrils ending on neuronal cells and dendrites in the cerebral cortex increases with age. Also, some neuroanatomists believe that they can observe electronmicrographic changes in presynaptic terminals that have been subjected to intense and prolonged activity. These observations have led to a widely held belief that fixation of memories in the brain results from physical changes in the synapses themselves: perhaps changes in numbers of presynaptic terminals, perhaps in sizes of the terminals, or perhaps in their chemical compositions. Such physical changes could cause permanent or semipermanent increase in the degree of facilitation of the synapses, thus allowing signals to pass through the synapses with progressive ease the more often the memory engram is used. This obviously would explain the tendency for memories to become more and more deeply fixed in the nervous system the more often they are recalled or the more often the person repeats the sensory experience that leads to the memory engram.

Short-term versus long-term fixed memory. Physiologists also differentiate between memories that can be remembered for only a few days versus those that can be recalled many months or years later. There are several reasons for making this differentiation, as follows: (1) Destruction of the temporal lobes in a person causes him to lose his memory for recent events that have occurred over a period of about two to six days prior to the destruction. However, this destruction does not cause loss of long-term, well-ingrained memories. (2) Electroshock therapy in mental patients, or the occurrence of convulsions in epileptic patients, frequently causes loss of memories of the past few days but not loss of long-term memories.

Therefore, it is believed that certain areas of the brain can support memories for much longer periods of time than can other areas, some areas perhaps being able to store memories for only a few seconds, a few minutes, or a few days, while others can store memories for years or decades. It is also possible that memories are first stored in temporary areas, such as in the temporal lobes, and then are transferred over a period of time to other more permanent storage areas. This process of transferring memory storage has been proved to occur, for memories that utilize primarily one cerebral cortex can be transferred to the other cerebral cortex through the corpus callosum, as will be discussed in the following paragraph.

FUNCTION OF THE CORPUS CALLOSUM AND ANTERIOR COMMISSURE TO TRANSFER MEMORIES TO THE OPPOSITE HEMISPHERE

Fibers in the *corpus callosum* connect the respective cortical areas of the two hemispheres with each other except for the cortices of the anterior portions of the temporal lobes; these temporal cortical areas are interconnected by fibers that pass through the *anterior commissure.*

The function of the corpus callosum and anterior commissure can be explained best by recounting the following experiment. A monkey is first prepared by cutting the corpus callosum and splitting the optic chiasm longitudinally. Then he is taught to recognize different types of objects with his right eye while his left eye is covered. Then the right eye is covered and he is tested to determine whether or not his left eye can recognize the same object. The answer to this is that the left eye *can not* recognize the object. Yet, on repeating the same experiment with the optic chiasm still split but the corpus callosum intact, it is found invariably that recognition in one hemisphere of the brain creates recognition also in the opposite hemisphere.

Thus, the function of the corpus callosum and the anterior commissure is to associate respective cortical areas of the two hemispheres in such a way that a memory engram impressed on one hemisphere will become simultaneously impressed on the opposite hemisphere. Yet, it should be recalled once again that many other types of correlation between different cortical areas normally takes place through lower brain centers, particularly through centers in the reticular formation and thalamus. This explains why it is possible to transect the corpus callosum and other major association bundles in the cerebrum without affecting most of our sensory and motor functions of the brain.

INTELLECTUAL ASPECTS OF MOTOR CONTROL

Control of muscular movements by the cerebral cortex and other areas of the brain was discussed in Chapter 38, but we still need to consider the manner in which the intellectual operations of the brain elicit motor events. Almost all sensory and even abstract experiences of the mind are eventually expressed in some type of motor activity, such as actual muscular movements of a directed nature, tenseness of the muscles, total relaxation of the muscles, attainment of certain postures, tapping of the fingers, grimaces of the face, or speech.

Psychological tests show that the analytical portions of the brain control motor activities in the following sequence of three stages: (1) origin of the thought of the motor activity to be performed, (2) determination of the sequence of movements necessary to perform the overall task, and (3) control of the muscular movements themselves.

Origin of the Thought of a Motor Activity. A large number of motor movements, which we call subconscious movements, are conceived in the lower regions of the brain, but complex learned tasks can be performed only when the cerebral cortex is present. The most important cortical areas for this purpose are the posterior part of the superior temporal gyrus and the angular gyrus of the dominant hemisphere. The student will immediately recognize this area to be the major *interpretive region* of the brain. Thus, the sensory interactions that take place in this all-important sensory part of the dominant hemisphere not only play a major role in interpreting sensory experiences but also in determining the person's course of motor activities.

Determination of Sequences of Movement. In our earlier discussion of cortical control of motor activities, it was pointed out that the somatic association area is necessary for performance of complex learned movements; destruction of this area both in animals and human beings causes the motor activities to lose much of their purposefulness. Therefore, it is postulated that once the interpretive area of the brain has determined the motor activity to be performed, the actual sequences of movement are determined in the somesthetic areas as discussed in Chapter 38. Signals are transmitted from here to the motor portion of the brain to control the actual motor movements, as also discussed in Chapter 38.

FUNCTION OF THE BRAIN
IN COMMUNICATION

One of the most important differences between the human being and lower animals is the facility with which human beings can communicate with one another. Furthermore, because neurological tests can easily assess the ability of a person to communicate with others, we know perhaps more about the sensory and motor systems related to communication than about any other segment of cortical function. Therefore, we will briefly review the function of the cortex in communication, and from this one can see immediately how the principles of sensory analysis and motor control that were just discussed apply to this art.

There are two aspects to communication: first, the *sensory aspect*, involving the ears and eyes, and, second, the *motor aspect*, involving vocalization and its control.

Sensory Aspects of Communication. We noted earlier in the chapter that destruction of portions of the *secondary auditory* and *visual areas* of the cortex can result in inability to understand the spoken word or the written word. These effects are called respectively *auditory agnosia* and *visual agnosia* or, more commonly, *word deafness* and *word blindness*. On the other hand, some persons are perfectly capable of understanding either the spoken word or the written word but are unable to understand their meanings when used to express a thought. This results most frequently when the *angular gyrus* or *posterior portion of the superior temporal gyrus* in the dominant hemisphere is damaged or destroyed. This is considered to be *general agnosia*.

Motor Aspects of Communication — Aphasia. *Sensory aphasias.* The process of speech involves (a) formation in the mind of thoughts to be expressed and choice of the words to be used and (b) the actual act of vocalization. The formation of thoughts and choice of words is principally the function of the sensory areas of the brain, for we find that a person who has a destructive lesion in the same temporal and angular gyrus region that is involved in general agnosia also has inability to formulate intelligible thoughts to be communicated. And at other times the thoughts can be formulated, but the person is unable to put together the appropriate sequence of words to express the thought; this results most frequently from destruction of the lowermost tip of the somesthetic area where it dips into the sylvian fissure. These inabilities to formulate thoughts and word sequences are called *sensory aphasias*.

Motor aphasia. Often a person is perfectly capable of deciding what he wishes to say, and he is capable of vocalizing, but he simply cannot make his vocal system emit words instead of noises. This effect, called *motor aphasia*, almost always results from damage to *Broca's area*, which lies in the *anterior facial region of the motor cortex in the dominant hemisphere. Therefore, we assume that the *patterns* for control of

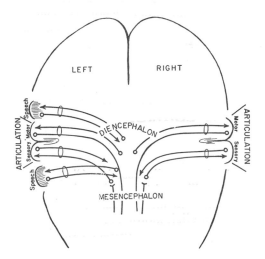

FIGURE 39-4 Centers of the cortex for speech control. (From Penfield and Rasmussen: The Cerebral Cortex of Man. The Macmillan Co.)

the larynx, lips, mouth, respiratory system, and other accessory muscles of articulation are all controlled in this area.

Finally, we have the act of *articulation* itself, which means the muscular movements of the mouth, tongue, larynx, and so forth, that are responsible for the actual emission of sound. The *posterior* facial and laryngeal regions of the motor cortex activate these muscles, and the sensory cortex helps to control the muscle contractions by feedback servomechanisms described in Chapter 38. Destruction of these regions can cause either total or partial inability to speak distinctly.

Note that the pathways from one part of the cortex to another that control communication pass mainly through the thalamic and mesencephalic areas, as illustrated in Figure 39–4, rather than directly from one area of the cortex to the next.

REFERENCES

Bitterman, M. E.: The evolution of intelligence. *Sci. Amer., 212:*92(1), 1965.

Bogoch, S.: The Biochemistry of Memory: With an Inquiry into the Function of the Brain Mucoids. New York, Oxford University Press, Inc., 1968.

Boycott, B. B.: Learning in the octopus. *Sci. Amer., 212:*42(3), 1965.

Darley, F. L.: Brain Mechanisms Underlying Speech and Language. New York, Grune & Stratton, Inc., 1967.

Kandel, E. R., and Spencer, W. A.: Cellular neurophysiological approaches in the study of learning. *Physiol. Rev., 48:*65, 1968.

Laursen, A. M.: Higher functions of the central nervous system. *Ann. Rev. Physiol., 29:*543, 1967.

Sperry, R. W.: Mental unity following surgical disconnection of the cerebral hemispheres. *Harvey Lectures,* 1966–1967, p. 293.

Young, J. Z.: The Memory System of the Brain. Berkeley, University of California Press, 1966.

CHAPTER 40

THE AUTONOMIC NERVOUS SYSTEM

The portion of the nervous system that controls the visceral functions of the body is called the *autonomic nervous system*. This system helps to control arterial pressure, gastrointestinal motility and secretion, urinary output, sweating, body temperature, and many other activities, some of which are controlled almost entirely by the autonomic nervous system and some only partially.

GENERAL ORGANIZATION OF THE AUTONOMIC NERVOUS SYSTEM

The autonomic nervous system is activated mainly by centers located in the *spinal cord, brain stem,* and *hypothalamus.* The autonomic signals, in turn, are transmitted to the body through two major subdivisions called the *sympathetic* and *parasympathetic systems,* the characteristics and functions of which follow.

PHYSIOLOGIC ANATOMY OF THE SYMPATHETIC NERVOUS SYSTEM

Figure 40–1 illustrates the general organization of the sympathetic nervous system, showing one of the two *sympathetic chains* to the side of the spinal column and nerves extending to the different internal organs. The sympathetic nerves originate in the spinal cord between the segments T-1 and L-2. Each sympathetic pathway is comprised of a *preganglionic neuron* and a *postganglionic neuron.* The cell body of the preganglionic neuron lies in the *intermediolateral horn* of the spinal cord, and its fiber passes through an *anterior root* of the cord into a *spinal nerve* and finally through the *white ramus* from the spinal nerve to a *ganglion* of the *sympathetic chain.* Here the fiber either synapses immediately with postganglionic neurons or often passes on through the chain into one of its radiating nerves to synapse with postganglionic neurons in an outlying sympathetic ganglion. The fiber of each postganglionic neuron then travels to its destination in one of the organs.

Many of the fibers from the postganglionic neurons in the sympathetic chain pass back into the spinal nerves through *gray rami* at all levels of the cord. These are type C fibers that extend to all parts of the body in the skeletal nerves. They control the blood vessels, sweat glands, and pilo-erector muscles

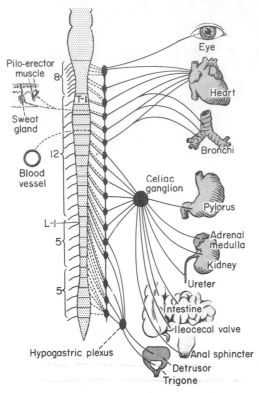

FIGURE 40–1 The sympathetic nervous system. Dashed lines represent postganglionic fibers in the gray rami leading into the spinal nerves for distribution to blood vessels, sweat glands, and pilo-erector muscles.

of the hairs. Approximately 8 per cent of the fibers in the average skeletal nerve are sympathetic fibers, a fact that indicates their importance.

Segmental Distribution of Sympathetic Nerves. The sympathetic pathways originating in the different segments of the spinal cord are not necessarily distributed to the same part of the body as the spinal nerve fibers from the same segments. Instead, the *sympathetic fibers from T-1 generally pass up the sympathetic chain into the head; from T-2 into the neck; T-3, T-4, T-5, and T-6 into the thorax; T-7, T-8, T-9, T-10, and T-11 into the abdomen; T-12, L-1, and L-2 into the legs.* This distribution is only approximate and overlaps greatly.

PHYSIOLOGIC ANATOMY OF THE PARASYMPATHETIC NERVOUS SYSTEM

The parasympathetic nervous system is illustrated in Figure 40–2, showing that parasympathetic fibers leave the central nervous system through several of the cranial nerves, the second and third sacral spinal nerves, and occasionally the first and fourth sacral nerves. Probably 80 per cent or more of all parasympathetic nerve fibers are in the vagus nerves, passing to the entire thoracic and abdominal regions of the body. The vagus nerves supply parasympathetic nerves to the heart, the lungs, the esophagus, the stomach, the small intestine, the proximal half of the colon, the liver, the gallbladder, the pancreas, and the upper portions of the ureters.

Parasympathetic fibers in the *third*

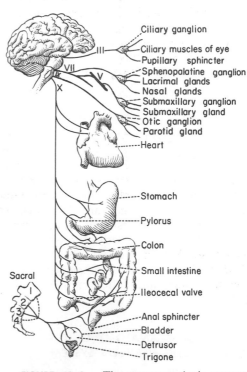

FIGURE 40–2 The parasympathetic nervous system.

nerve pass to the pupillary sphincters and ciliary muscles of the eye. Fibers from the *seventh nerve* pass to the lacrimal, nasal, and submaxillary glands, and fibers from the *ninth nerve* pass to the parotid gland.

The sacral parasympathetic fibers congregate in the form of the two *nervi erigentes*, which leave the sacral plexus on each side of the cord and distribute their peripheral fibers to the descending colon, rectum, bladder and lower portions of the ureters. Also, this sacral group of parasympathetics supplies fibers to the external genitalia to cause various sexual reactions.

Preganglionic and Postganglionic Parasympathetic Neurons. The parasympathetic system, like the sympathetic, has both preganglionic and postganglionic neurons, but, except in the case of a few cranial parasympathetic nerves, the preganglionic fibers pass uninterrupted to the organ that is to be excited by parasympathetic impulses. In the wall of the organ are located the *postganglionic neurons* of the parasympathetic system. The preganglionic fibers synapse with these; and then short postganglionic fibers, 1 millimeter to several centimeters in length, leave the neurons to spread in the substance of the organ.

CHOLINERGIC AND ADRENERGIC FIBERS — SECRETION OF ACETYLCHOLINE AND NOREPINEPHRINE BY THE POSTGANGLIONIC NEURONS

It will be recalled from Chapter 6 that skeletal nerve endings secrete acetylcholine. This is also true of the preganglionic neurons of both the sympathetic and parasympathetic system, and it is true, too, of the parasympathetic postganglionic neurons. Therefore, all these fibers are said to be *cholinergic because they secrete acetylcholine at their nerve endings.*

A few of the postganglionic endings of the sympathetic nervous system also secrete acetylcholine, and these fibers, too, are cholinergic; but by far the majority of the sympathetic postganglionic endings secrete *norepinephrine*. These fibers are said to be *adrenergic*, a term derived from *noradrenalin*, which is another name for norepinephrine. The acetylcholine and norepinephrine secreted by the postganglionic neurons act on the different organs to cause the respective parasympathetic or sympathetic effects. Therefore, these substances are called *parasympathetic* and *sympathetic mediators*, respectively, or sometimes *cholinergic* and *adrenergic* mediators.

Destruction of Acetylcholine, and Duration of its Action. The acetylcholine secreted by the parasympathetic nerve endings, like that also secreted at skeletal nerve endings, is destroyed by the enzyme *cholinesterase*, which is present in all the effector organs or the surrounding fluids. However, this destruction does not take place nearly so rapidly as it does in skeletal muscle fibers. The acetylcholine sometimes persists for as long as several seconds after its release and therefore has this period of action.

Removal of Norepinephrine from the Tissues, and Duration of its Action. Norepinephrine secreted by sympathetic nerve endings is removed from the tissues by (1) reabsorption into the sympathetic nerve endings themselves and (2) methylation caused by o-*methyl transferase*, an enzyme analogous to cholinesterase.

Ordinarily, the action of norepinephrine secreted in the tissues lasts for only a few seconds, illustrating that its destruction or removal from the tissues is rapid. However, norepinephrine and epinephrine secreted into the blood by the adrenal medullae are not removed or destroyed significantly until they diffuse into tissues. Therefore, when secreted into the blood, these two hormones remain active for one to several minutes.

TABLE 40–1   AUTONOMIC EFFECTS ON VARIOUS ORGANS OF THE BODY

Organ	Effect of Sympathetic Stimulation	Effect of Parasympathetic Stimulation
Eye: Pupil	Dilated	Contracted
Ciliary muscle	None	Excited
Glands: Nasal	Vasoconstriction	Stimulation of thin, copious secretion containing many enzymes
Lacrimal		
Parotid		
Submaxillary		
Gastric		
Pancreatic		
Sweat glands	Copious sweating (cholinergic)	None
Apocrine glands	Thick, odoriferous secretion	None
Heart: Muscle	Increased rate	Slowed rate
	Increased force of beat	Decreased force of atrial beat
Coronaries	Vasodilated	Constricted
Lungs: Bronchi	Dilated	Constricted
Blood vessels	Mildly constricted	None
Gut: Lumen	Decreased peristalsis and tone	Increased peristalsis and tone
Sphincter	Increased tone	Decreased tone
Liver	Glucose released	None
Gallbladder and bile ducts	Inhibited	Excited
Kidney	Decreased output	None
Ureter	Inhibited	Excited
Bladder: Detrusor	Inhibited	Excited
Trigone	Excited	Inhibited
Penis	Ejaculation	Erection
Systemic blood vessels:		
Abdominal	Constricted	None
Muscle	Constricted (adrenergic)	None
	Dilated (cholinergic)	
Skin	Constricted (adrenergic)	Dilated
	Dilated (cholinergic)	
Blood: Coagulation	Increased	None
Glucose	Increased	None
Basal metabolism	Increased up to 150%	None
Adrenal cortical secretion	Increased	None
Mental activity	Increased	None
Piloerector muscles	Excited	None

EXCITATORY AND INHIBITORY ACTIONS OF SYMPATHETIC AND PARASYMPATHETIC STIMULATION

Table 40–1 gives the effects on different visceral functions of the body of stimulating the parasympathetic and sympathetic nerves. From this table it can readily be seen that *sympathetic stimulation causes excitatory effects in some organs but inhibitory effects in others. Likewise, parasympathetic stimulation causes excitation in some organs but inhibition in others.* Also, when sympathetic stimulation excites a particular organ, parasympathetic stimulation often inhibits it, illustrating that the two systems occasionally act reciprocally to each other. However, most organs are dominantly controlled by one or the other of the two systems, so that, except in a few instances, the two systems do not actively oppose each other.

Though the specific functions of the autonomic nervous system are presented in detail at appropriate points in this text, some of the more important of these functions are the following:

EFFECTS OF SYMPATHETIC AND PARASYMPATHETIC STIMULATION ON SPECIFIC ORGANS

The Eye. Two functions of the eye are controlled by the autonomic nervous system: These are the pupillary opening and the focus of the lens. Sympathetic stimulation contracts the meridional *fibers of the iris* and, therefore, dilates the pupil, whereas parasympathetic stimulation contracts the *circular muscle of the iris* to constrict the pupil. The parasympathetics that control the pupil are reflexly stimulated when excess light enters the eyes; this reflex reduces the pupillary opening and decreases the amount of light that strikes the retina. On the other hand, the sympathetics become stimulated during periods of excitement and, therefore, increase the pupillary opening at these times.

Focusing of the lens is controlled entirely by the parasympathetic nervous system. The lens is normally held in a flattened state by tension of its radial ligaments. Parasympathetic excitation contracts the *ciliary muscle*, which releases this tension and allows the lens to become more convex. This causes the eye to focus on objects near at hand. The focusing mechanism was discussed in Chapter 35 in relation to function of the eyes.

The Gastrointestinal System. The gastrointestinal system has its own intrinsic set of nerves known as the *intramural plexus.* However, both parasympathetic and sympathetic stimulation can affect gastrointestinal activity — parasympathetic especially. Parasympathetic stimulation, in general, increases the overall degree of activity of the gastrointestinal tract by promoting peristalsis, thus allowing rapid propulsion of contents along the tract. This propulsive effect is associated with simultaneous increases in rates of secretion by many of the gastrointestinal glands.

Normal function of the gastrointestinal tract is not very dependent on sympathetic stimulation. However, in some diseases, strong sympathetic stimulation inhibits peristalsis and increases the tone of the sphincters. The net result is greatly slowed propulsion of food through the tract.

The Heart. In general, sympathetic stimulation increases the overall activity of the heart. This is accomplished by increasing both the rate and force of heartbeat. Parasympathetic stimulation causes mainly the opposite effects, decreasing the overall activity of the heart. To express these effects in another way, sympathetic stimulation increases the effectiveness of the heart as a pump, whereas parasympathetic stimulation decreases its effectiveness.

Systemic Blood Vessels. Most blood vessels, especially those of the abdominal viscera and the skin of the limbs, are constricted by sympathetic stimulation, whereas parasympathetic stimulation dilates vessels in certain restricted areas. In muscles, there are two types of sympathetic nerve fibers, adrenergic and cholinergic. The adrenergic fibers cause mild constriction of the vessels. On the other hand, the cholinergic fibers can strongly dilate the vessels during the early phases of muscular exercise. This was discussed in detail in Chapter 23.

Effect of sympathetic and parasympathetic stimulation on arterial pressure. The arterial pressure in the circulatory system is caused by propulsion of blood by the heart and by resistance to flow of this blood through the vascular system. In general, sympathetic stimulation increases both propulsion by the heart and resistance to flow, which can cause the pressure to increase greatly.

On the other hand, parasympathetic stimulation decreases the pumping effectiveness of the heart, which lowers the pressure a moderate amount, though not nearly so much as the sympathetics can increase the pressure.

Effects of Sympathetic and Parasympathetic Stimulation on Other Functions of the Body. Because of the great importance of the sympathetic and parasympathetic control systems, these are discussed many times in this text in relation to a myriad of body functions that are not considered in detail here. In general, most of the entodermal structures, such as the ducts of the liver, the gallbladder, the ureter, and the bladder, are inhibited by sympathetic stimulation but excited by parasympathetic stimulation. Sympathetic stimulation also has metabolic effects, causing release of glucose from the liver, increase in blood glucose concentrations, increase in basal metabolic rate, and increase in mental activity. Finally, the sympathetics and parasympathetics are involved in regulating the male and female sexual acts, as will be explained in Chapters 54 and 55.

FUNCTION OF THE ADRENAL MEDULLAE

Stimulation of the sympathetic nerves to the adrenal medullae causes large quantities of epinephrine and norepinephrine to be released into the circulating blood, and these two hormones in turn are carried in the blood to all tissues of the body. On the average, approximately 75 per cent of the secretion is epinephrine and 25 per cent norepinephrine, though the relative proportions of these change considerably under different physiological conditions.

The circulating hormones have almost the same effects on different organs as are caused by direct sympathetic stimulation, except that the *effects last about 10 times as long* because the hormones are removed from the blood slowly. For instance, epinephrine and norepinephrine cause constriction of essentially all the blood vessels of the body; they cause increased activity of the heart, inhibition of the gastrointestinal tract, dilatation of the pupil of the eye, and so forth.

The only significant differences caused by the adrenal hormones result from specific effects of epinephrine, which is not secreted at the sympathetic endings. This hormone causes an increase in the rate of metabolism and cardiac output perhaps 10 to 30 per cent more than the increase caused by direct sympathetic stimulation.

Value of the Adrenal Medullae to the Function of the Sympathetic Nervous System. Usually, when the sympathetic nervous system is stimulated, epinephrine and norepinephrine are almost always released by the adrenal medullae at the same time that the different organs are being stimulated directly by the sympathetic nerves. Therefore, the organs are actually stimulated in two different ways simultaneously, directly by the sympathetic nerves and indirectly by the medullary hormones. The two means of stimulation support each other, and either can actually substitute for the other. For instance, destruction of the direct sympathetic pathways to the organs does not abrogate excitation of the organs because norepinephrine and epinephrine are still released into the circulating fluids and indirectly cause stimulation. Likewise, total loss of the two adrenal medullae usually has little significant effect on the operation of the sympathetic nervous system.

Another important value of the adrenal medullae is the capability of epinephrine and norepinephrine to stimulate structures of the body that are not innervated by direct sympathetic fibers. For instance, the metabolic rate of every cell of the body is increased by these hormones, especially by epinephrine, even though only a small proportion of all the cells in the body are innervated by sympathetic fibers.

SYMPATHETIC AND PARASYMPATHETIC "TONE"

The sympathetic and parasympathetic systems are continually active, and the basal rates of activity are known, respectively, as *sympathetic tone* or *parasympathetic tone*.

The value of tone is that it allows a single nervous system to increase or to decrease the activity of a stimulated organ. For instance, sympathetic tone normally keeps almost all the blood vessels of the body constricted to approximately half their maximum diameter. By increasing the degree of sympathetic stimulation, the vessels can be constricted even more; but, on the other hand, by inhibiting the normal tone, the vessels can be dilated. If it were not for the continual sympathetic tone, the sympathetic system could cause only vasoconstriction, never vasodilatation.

Effect of Loss of Sympathetic or Parasympathetic Tone Following Denervation. Immediately after a sympthetic or parasympathetic nerve is cut, the innervated organ loses its sympathetic or parasympathetic tone. In the case of the blood vessels, for instance, cutting the sympathetic nerves results immediately in almost maximal vasodilatation. However, over several days or weeks, the *intrinsic tone* in the smooth muscle of the vessels increases, usually restoring almost normal vasoconstriction.

Essentially the same events occur in any organ whenever sympathetic or parasympathetic tone is lost. That is, compensation soon develops to return the function of the organ almost to its normal basal level. However, the compensation sometimes requires many months. For instance, loss of parasympathetic tone to the heart increases the heart rate from 90 to about 160 beats per minute in a dog, and this will still be about 120 six months later.

DENERVATION HYPERSENSITIVITY OF SYMPATHETIC AND PARASYMPATHETIC ORGANS FOLLOWING DENERVATION

During the first week or so after a sympathetic or parasympathetic nerve is destroyed, the innervated organ becomes more and more sensitive to injected norepinephrine or acetylcholine, respectively, the response often increasing

as much as 10-fold. The cause of this *denervation hypersensitivity* is not known, though one suggestion is that prolonged lack of nerve stimulation allows the quantity of "receptor substances" in the membranes of the effector cells to increase, thereby increasing the responsiveness of the cells to the circulating norepinephrine, epinephrine, or acetylcholine.

CONTROL OF THE AUTONOMIC NERVOUS SYSTEM BY THE BRAIN

Many areas in the reticular substance of the medulla, pons, and mesencephalon, as well as many special nuclei of the brain stem (see Fig. 40–3) control different autonomic functions such as arterial pressure, heart rate, glandular secretion in the upper part of the gastrointestinal tract, gastrointestinal peristalsis, the degree of contraction of the urinary bladder, and many others. The control of each of these is discussed in detail at appropriate points in this text. Suffice it to point out here that almost all portions of the autonomic nervous system are controlled to at least some degree by the brain stem.

In addition, signals from the hypothalamus and even from the cerebrum can affect the activities of almost all the lower brain stem autonomic control centers. For instance, stimulation in ap-

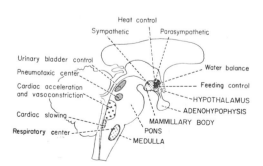

FIGURE 40–3 Autonomic control centers of the brain stem.

propriate areas of the hypothalamus can activate the medullary cardiovascular control centers strongly enough to increase the arterial pressure to more than double normal. Likewise, higher centers can transmit signals into the lower centers to increase or decrease salivation, gastrointestinal activity, or urinary bladder contraction. Therefore, to a great extent the autonomic centers in the lower brain stem are relay stations for control activities integrated at higher levels of the brain.

AUTONOMIC REFLEXES

It is mainly by means of autonomic reflexes that the autonomic nervous system regulates visceral functions. Throughout this text the functions of these reflexes are discussed in detail in relation to individual organs, but, to illustrate their importance, a few are presented here briefly.

Cardiovascular Autonomic Reflexes. Several reflexes in the cardiovascular system help to control the arterial blood pressure, cardiac output, and heart rate. One of these is the *baroreceptor reflex*, which was described in Chapter 19 along with other cardiovascular reflexes. Briefly, stretch receptors called *baroreceptors* are located in the walls of the major arteries, including the carotid arteries and the aorta. When these become stretched by high pressure, impulses are transmitted to the brain stem, where they inhibit the sympathetic centers. This results in decreased sympathetic impulses to the heart and blood vessels, which allows the arterial pressure to fall back toward normal.

The Gastrointestinal Autonomic Reflexes. The uppermost part of the gastrointestinal tract and the rectum are controlled principally by autonomic reflexes. For instance, the smell of appetizing food initiates impulses from the nose to the vagal, glossopharyngeal, and salivary nuclei of the brain stem. These in turn transmit impulses through the parasympathetic nerves to the secretory glands of the mouth and stomach, causing secretion of digestive juices even before food enters the mouth.

On the other hand, when fecal matter fills the rectum at the other end of the alimentary canal, sensory impulses initiated by stretching the rectum are sent to the sacral portion of the spinal cord, and a reflex signal is retransmitted through the parasympathetics to the distal parts of the colon; these result in strong peristaltic contractions that empty the bowel.

Other Autonomic Reflexes. Emptying of the bladder is also controlled in the same way as emptying of the rectum; stretching the bladder sends impulses to the sacral cord, and this in turn causes contraction of the bladder as well as relaxation of the urinary sphincters, thereby promoting micturition.

Also important are the sexual reflexes which are initiated both by psychic stimuli from the brain and stimuli from the sexual organs. Impulses from these sources converge on the sacral cord and, in the male, result, first, in erection and then in ejaculation.

Other autonomic reflexes include reflex regulation of pancreatic secretion, gallbladder emptying, urinary excretion, sweating, blood glucose concentration, and many other visceral functions, all of which are discussed in detail at many points throughout this text.

MASS DISCHARGE OF SYMPATHETIC SYSTEMS VERSUS DISCRETE CHARACTERISTICS OF PARASYMPATHETIC REFLEXES

In general, large portions of the sympathetic nervous system often become stimulated simultaneously, a phenomenon called *mass discharge*. This characteristic of sympathetic action is in keeping with the usually diffuse nature of sympathetic function, such as overall regulation of arterial pressure or of metabolic rate.

In contrast to the sympathetic system, most reflexes of the parasympathetic system are very specific. For instance, parasympathetic cardiovascular reflexes usually act only on the heart to increase or decrease its activity. Likewise, parasympathetic reflexes frequently cause secretion only in the mouth or, in other instances, secretion only by the stomach glands. Finally, the rectal emptying reflex does not affect other parts of the bowel to a major extent.

"ALARM" OR "STRESS" FUNCTION OF THE SYMPATHETIC NERVOUS SYSTEM

From the preceding discussion of the functions of the sympathetic nervous system one can readily understand that mass sympathetic discharge results in increased activity of many functions of the body, including increase in arterial pressure, increase in blood supply to the tissues, in rate of cellular metabolism throughout the body, blood glucose concentration, rate of blood coagulation,

and even mental activity. The sum of these effects permits the person to perform strenuous physical activity to a far greater extent than would otherwise be possible. And, since it is physical stress that usually excites the sympathetic system, it is frequently said that the purpose of the sympathetics is to provide extra energy for the body in states of stress, and this is often called the sympathetic *alarm reaction*, or *stress reaction*.

REFERENCES

Burn, J. H.: The Autonomic Nervous System: For Students of Physiology and of Pharmacology. 2nd ed., Philadelphia, F. A. Davis Co., 1965.
Hess, W. R.: The Functional Organization of the Diencephalon. New York, Grune & Stratton, Inc., 1958.
Malmejac, J.: Activity of the adrenal medulla and its regulation. *Physiol. Rev., 44*:186, 1964.
Root, W. S., and Hofmann, F. G. (eds.): Physiological Pharmacology. New York, Academic Press, Inc., 1967, Vol. 3, Pts. C and D.
Soderberg, U.: Neurophysiological aspects of homeostasis. *Ann. Rev. Physiol., 26*:271, 1964.
von Euler, U. S.: Neurotransmission in the adrenergic nervous system. *Harvey Lect., 55*: 43, 1961.
von Euler, U. S.: Noradrenaline. Springfield, Illinois, Charles C Thomas, Publisher, 1956.

CHAPTER 41

WAKEFULNESS; SLEEP; AND BEHAVIORAL FUNCTIONS OF THE BRAIN—ROLES OF THE RETICULAR ACTIVATING SYSTEM AND THE HYPOTHALAMUS

In the previous chapters of this section on neurophysiology we have discussed, first, the sensory mechanisms of the nervous system and, second, the motor mechanisms. In the present chapter we will consider the control of the level of activity in the brain itself—that is, the control of sleep, wakefulness, and attentiveness, and the special attributes of emotions, drives, and other aspects of behavior. To begin this discussion, we need first to consider the functions of the *reticular activating system*, a system that controls the overall degree of central nervous system activity, including control of wakefulness and sleep, and control of at least part of our ability to direct attention toward specific areas of our conscious minds.

Figure 41–1*A* illustrates the extent of this system, showing that it begins in the lower brain stem and extends upward through the *mesencephalon* and *thalamus* to be distributed throughout the cerebral cortex.

FUNCTION OF THE RETICULAR ACTIVATING SYSTEM IN WAKEFULNESS

Diffuse electrical stimulation in the *mesencephalic and upper pontile portion of the reticular formation*—an area also called the *bulboreticular facilitatory area* and discussed in Chapter 37 in relation to the motor functions of the nervous system—causes immediate and marked activation of the cerebral cortex and even causes a sleeping animal to awaken instantaneously. Furthermore, when this brain stem portion of the reticular formation is damaged severely, as occurs (a) when a *brain tumor* develops in this region, (b) when serious *hemorrhage* occurs, or (c) in diseases such as *encephalitis lethargica* (sleeping sickness), the person passes into coma

and is completely nonsusceptible to normal awakening stimuli.

Function of the Mesencephalic Portion of the Reticular Activating System. Electrical stimuli applied to different portions of the reticular activating system have shown that the mesencephalic portion functions quite differently from the thalamic portion. Electrical stimulation of the mesencephalic portion causes generalized activation of the entire brain, including activation of the cerebral cortex, thalamic nuclei, basal ganglia, hypothalamus, other portions of the brain stem, and even the spinal cord.

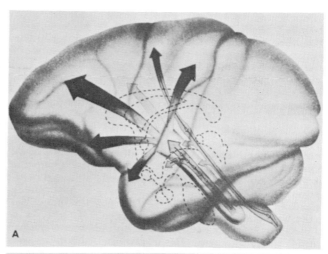

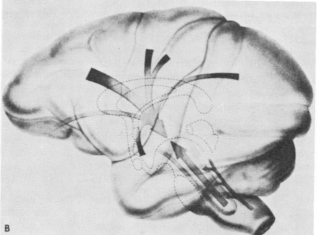

FIGURE 41-1 (*A*) The ascending reticular activating system schematically projected on a monkey brain. (From Lindsley: Reticular Formation of the Brain. Little, Brown and Co.) (*B*) Convergence of pathways from the cerebral cortex and from the spinal afferent systems on the reticular activating system. (From French, Hernandez-Peon, and Livingston: *J. Neurophysiol., 18*:74, 1955.)

Furthermore, once the mesencephalic portion is stimulated, the degree of activation throughout the nervous system remains high for as long as a half minute or more after the stimulation is over. Therefore, *it is believed that the mesencephalic portion of the reticular activating system is basically responsible for normal wakefulness of the brain.*

The mesencephalic portion of the reticular activating system can transmit signals to the cerebral cortex through two separate pathways: (1) a more direct pathway through the region of the subthalamus, and (2) an indirect pathway through many successive synapses in the thalamus. It is assumed that the direct pathway allows immediate activation of cortical function. On the other hand, transmission of signals through the indirect thalamic pathway is considerably delayed, and the signals are believed to stimulate some areas of the cortex more than others because of intrathalamic selection of pathways, as is explained in the following paragraphs.

Function of the Thalamic Portion of the Reticular Activating System. Electrical stimulation in different areas of the thalamic portion of the reticular activating system activates specific regions of the cerebral cortex more than others. This is distinctly different from stimulation in the mesencephalic portion which activates all the brain at the same time. Therefore, it is believed that the thalamic portion of the reticular activating system has two specific functions: first, it relays some of the diffuse facilitatory signals from the mesencephalic portion to all parts of the cerebral cortex to cause generalized activation of the cerebrum, and, second, stimulation of selected points in the thalamic activating system causes specific activation of certain areas of the cerebral cortex in distinction to the other areas. This selective activation of specific cortical areas probably plays an important role in our ability to direct our attention to certain parts of our mental activity, which is discussed later in the chapter.

THE AROUSAL REACTION — SENSORY ACTIVATION OF THE RETICULAR ACTIVATING SYSTEM

When an animal is asleep, the reticular activating system is mainly dormant. Yet almost any type of sensory signal can immediately activate the system. For instance, proprioceptive signals from the joints, pain impulses from the skin, visual signals from the eyes, auditory signals from the ears, or even visceral sensations from the gut can all cause sudden activation of the reticular activating system and therefore arouse the animal. This is called the *arousal reaction.*

An animal falls asleep permanently when the brain stem is transected slightly above the point of entrance of the sensory impulses from the fifth nerve into the pontine region of the reticular formation. Such a transection causes all somatic sensory impulses to be lost. However, even after this procedure, electrical stimulation in the remaining reticular activating system above the transection can still awaken the animal. This experiment illustrates the importance of the arousal reaction and of somatic impulses in maintaining the wakefulness state of the animal.

CEREBRAL ACTIVATION OF THE RETICULAR ACTIVATING SYSTEM

In addition to activation of the reticular activating system by sensory impulses, the cerebral cortex can also stimulate this system. Direct fiber pathways pass into the reticular activating system, as shown in Figure 41–1B, from almost all parts of the cerebrum. Because of an exceedingly large number of nerve fibers that pass from the motor regions of the cerebral cortex to the reticular formation, motor activity in particular is associated with a high degree of wakefulness, which partially explains the importance of movement to keep a person awake. However, intense activity of any other part of the cerebrum can also activate the reticular activating system and consequently can cause a high degree of wakefulness.

THE FEEDBACK THEORY OF WAKEFULNESS AND SLEEP

From the above discussion we can see that activation of the reticular activating system greatly intensifies the degree of activity in the cerebral cortex. But, in turn, increased activity in the cerebral cortex increases the degree of activity of the reticular activating system. Thus, there exists a *positive* "feedback" that helps to keep the ascending reticular activating system active once it becomes excited.

Likewise, increased activity of this system increases the degree of muscle tone throughout the body, which was discussed in Chapter 37 in relation to the gamma efferent system. Also, stimulation of the reticular activating system increases many autonomic activities throughout the body. In turn, these peripheral effects cause increased somatic impulses to be transmitted into the central nervous system—particularly important are the proprioceptor impulses, for these have a high degree of arousal activity. Thus, here again, we find another *positive* "feedback." That is, activity in the reticular activating system causes increased peripheral activity and this in turn feeds back to the reticular activating system to promote further increase in excitation there.

On the basis of these two positive feedback loops—(1) to the cerebral cortex and back to the reticular formation, and (2) to the peripheral muscles and back—a feedback theory of wakefulness and sleep can be formulated as follows:

It is assumed that once the reticular activating system becomes activated, the feedback impulses both from the cerebral cortex and from the periphery tend to maintain its excitation. Thus, after the

person has become awakened, he tends to remain awake at least for the time being.

After prolonged wakefulness, it can also be assumed that many of the neuronal cells of the feedback loops, particularly the neuronal cells in the reticular activating system itself, gradually become fatigued or less excitable for other reasons. When this happens, the degree of activation of the entire system begins to decrease. As a result, the intensity of the cortical and peripheral feedbacks also decreases, which results in further depression of the reticular activating system. The cortex and peripheral functions are further depressed, followed by still further depression of the reticular activating system. Thus, once the degree of excitability of the neurons falls to a critical level, a vicious cycle of depression ensues until all or most components of the feedback loops fade into decreased activity, and this then represents the state of sleep.

The next element of the feedback theory of wakefulness and sleep assumes that, after the reticular activating system has been dormant for a time, the neuronal cells involved in the feedback loops gradually regain excitability. Yet wakefulness still does not occur until some arousal signal initiates activity in the system. Once such has occurred, the feedback loops immediately come into play, and the person passes from the state of sleep into the state of wakefulness.

One might immediately ask: How is it that a person can also have various degrees of wakefulness and sleep? To answer this, we need only realize that the feedback loops contain literally millions of parallel pathways. If the feedback system is operating only through a few of these, a person would theoretically have a slight degree of wakefulness, but if the feedback is operating through tremendous numbers of pathways simultaneously, the degree of wakefulness obviously would be very great.

Active Production of Sleep by Some Neuronal Centers. Though sleep is generally associated with decreased activity of the reticular activating system and, therefore, also with decreased activity of the cerebrum as well, electrical stimulation of a few centers in the brain can actively cause sleep. The most important of these is located in the upper medullary and lower pontine area of the reticular formation, an area closely associated with the upper regions of the solitary fasciculus. Electrical stimulation in this region causes changes in the electroencephalogram that are typical of those changes in the early stages of sleep. On the contrary, when this area is destroyed bilaterally, the animal develops insomnia.

REM Sleep and the Process of Dreaming. Periodically during sleep, a person usually passes through a state of dreaming associated with mild involuntary muscle jerks and with rapid eye movements. The rapid eye movements have given this stage of sleep the name *REM sleep*. Electroencephalograms during these periods are typical of those of light sleep but not of deep sleep. Yet, strangely enough, muscle tone throughout the body is diminished almost to zero, the heart rate may be as low as 20 beats below normal, and the arterial pressure may be as low as 30 mm. Hg below normal. Thus, the person seems physiologically to be in very deep sleep, rather than light sleep, despite the fact that he is dreaming and has uncontrolled muscle movements.

REM sleep usually occurs three to four times during each night at intervals of 80 to 120 minutes, each occurrence lasting from 5 minutes to more than an hour. As much as 50 per cent of the infant's sleep cycle is composed of REM sleep; and in the adult, approximately 20 per cent is REM sleep.

The causation of REM sleep and its cyclic appearance is still unknown. However, it seems to be especially important, because the degree of restfulness of sleep is associated with the amount of REM sleep that a person receives during the night. For instance, in persons who have been awakened each time that the electroencephalogram demonstrated the onset of REM sleep, extreme tiredness and serious neurotic tendencies have developed.

PHYSIOLOGICAL EFFECTS OF SLEEP

Prolonged wakefulness is often associated with progressive malfunction of the mind and behavioral activities of the nervous system. We are all familiar with the increased sluggishness of thought that occurs toward the end of a prolonged wakeful period; but in addition, a person can become irritable or even psychotic following forced wakefulness for prolonged periods of time or even following loss of the REM portion of his sleep. Therefore, we can assume that sleep restores normal "balance" between the different parts of the central nervous system.

Even though wakefulness and sleep have not been shown to be necessary for somatic functions of the body, the cycle of enhanced and depressed nervous excitability that follows along with the cycle of wakefulness and sleep does have moderate effects on the peripheral body. For instance, there is enhanced sympathetic activity during wakefulness and also enhanced numbers of impulses to the skeletal musculature to increase muscular tone. Conversely, during sleep sympathetic activity decreases while parasympathetic activity occasionally increases, and the muscular tone becomes almost nil. Therefore, during sleep, arterial blood pressure falls, pulse rate decreases, skin vessels dilate, activity of the gastrointestinal tract sometimes increases, muscles fall into a completely relaxed state, and overall basal metabolic rate of the body falls by about 10 to 20 per cent.

ATTENTION

So long as a person is awake, he has the ability to direct his attention to specific aspects of his mental environment. Control of the general level of attentiveness is probably exerted by the same mechanism that controls wakefulness and sleep, the control center for which is located in the mesencephalon and upper pons.

On the other hand, one can surmise that the ability of specific thalamic areas to excite specific cortical regions might be one of the mechanisms by which a person can direct his attention to specific aspects of his mental environment, whether these be immediate sensory experiences or stored memories.

Also, nervous pathways extend centrifugally from almost all sensory areas of the brain toward the lower centers to control the intensity of sensory input to the brain. For instance, the auditory cortex can either inhibit or facilitate signals from the cochlea, the visual cortex can control the signals from the retina in the same way, and the somesthetic cortex can control the intensities of signals from the somatic areas of the body.

Thus, it is likely that activated regions of the cortex control their own sensory input under some conditions. This is another means by which the brain might direct its attention to specific phases of its mental activity.

COORDINATION OF DIFFERENT CORTICAL FUNCTIONS BY THE RETICULAR ACTIVATING SYSTEM

Though many fiber pathways interconnect different areas of the cerebral cortex directly, these have only limited function in coordinating the activities of different parts of the cerebral cortex with each other. For instance, the entire corpus callosum can be cut, which removes almost all direct pathways between one cerebral hemisphere and the other, and

yet it is difficult, when using usual psychological tests, to determine that there is any decrease in cerebral capability. Likewise, a deep transection can be made in the cerebral cortex in the bottom of the central sulcus, separating the parietal cortex from the frontal cortex, and yet the sensory information stored in the parietal cortex can still be used to control motor activities of the frontal cortex in such a way that the experimental observer cannot tell the difference. Thus, it is evident that the more important routes of communication from one major part of the cortex to another are mainly through deep areas of the brain.

Anatomically, many pathways exist between the different thalamic nuclei, which suggests that much of the coordination between the different cortical areas could occur in the thalamus. However, even in the thalamus only insignificant pathways exist *from one side of the brain to the other*. Therefore, it is likely that most of the coordination between one cerebral hemisphere and the other is controlled by the brain stem reticular activating area and that many of the important correlations even within a single hemisphere are also controlled through this area.

BRAIN WAVES

Electrical recordings from the surface of the brain or from the outer surface of the head demonstrate continuous electrical activity in the brain. Both the intensity and patterns of this electrical activity are determined to a great extent by the overall excitation of the brain resulting from functions in the reticular activating system. The undulations in the recorded electrical potentials, shown in Figure 41-2, are called *brain waves*, and the entire record is called an *electroencephalogram* (EEG).

Much of the time, the brain waves are irregular and no general pattern can be discerned in the EEG. However, at other times, distinct patterns do appear. Some

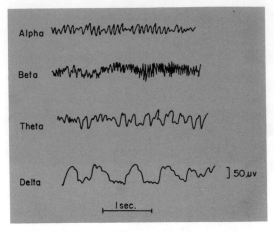

FIGURE 41–2 Different types of normal electroencephalographic waves.

sometimes as high as 50 cycles per second. These usually appear during activation of the central nervous system.

Theta waves have frequencies between 4 and 7 cycles per second. These occur mainly in the parietal and temporal regions in children, but they also occur during emotional stress in some adults, particularly during disappointment and frustration.

Delta waves include all the waves of the EEG below 3½ cycles per second and sometimes as low as 1 cycle every 2 to 3 seconds. These occur in deep sleep, in infancy, and in serious organic brain disease.

of these are characteristic of specific abnormalities of the brain, such as epilepsy, which is discussed later. Others occur even in normal persons and can be classified into *alpha, beta, theta,* and *delta waves,* which are all illustrated in Figure 41–2.

Alpha waves are rhythmic waves occurring at a frequency between 8 and 13 per second and are found in the EEG's of almost all normal persons when they are awake in a quiet, resting state of cerebration. During sleep the alpha waves disappear entirely, and when the awake person's attention is directed to some specific type of mental activity, the alpha waves are replaced by asynchronous, higher frequency but lower voltage waves. Figure 41–3 illustrates the effect on the alpha waves of simply opening the eyes in bright light and then closing the eyes again. Note that the visual sensations cause immediate cessation of the alpha waves and that these are replaced by low voltage, asynchronous waves.

Beta waves occur at frequencies of more than 14 cycles per second and

ORIGIN OF THE DIFFERENT TYPES OF BRAIN WAVES

The surface of the cerebral cortex is composed almost entirely of a mat of dendrites from neuronal cells in layers one and two of the cortex (there are four more layers of cortex below these). When signals impinge on these dendrites, they become partially discharged, emitting negative potentials characteristic of excitatory postsynaptic potentials as discussed in Chapter 6. This partially discharged state makes the neurons of the outer layers of the cortex highly excitable, and the negative potential is simultaneously recorded from the surface of the scalp. It is the undulating excitation of this outer cortex that gives rise to the brain waves.

One of the important sources of signals to excite the outer dendritic layer of the cerebral cortex is the ascending reticular activating system. Therefore, brain wave intensity is closely related to the degree of activity in either the brain stem or the thalamic portions of the reticular activating system.

Origin of Delta Waves. Transection of the fiber tracts from the thalamus to the cortex, which blocks the reticular activating system fibers, causes delta waves in the cortex. This indicates that some synchronizing mechanism can occur in the cortical neurons themselves

Eyes open   Eyes closed

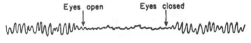

FIGURE 41–3 Replacement of the alpha rhythm by an asynchronous discharge on opening the eyes.

—entirely independently of lower structures in the brain—to cause the delta waves.

Delta waves also occur in very deep non-REM sleep; and this suggests that the cortex is then released from the activating influences of the reticular activating system, as was explained earlier in the chapter.

Origin of the Alpha Waves. Alpha waves will *not* occur in the cortex without connections with the reticular activating system. It is possible that this system enters into the genesis of the alpha waves. In support of this is the fact that strong electrical stimulation of the cortex causes reverberation of impulses among the nuclei of the thalamus at frequencies ranging between 8 and 13 per second, the frequency of the alpha waves.

EFFECT OF VARYING DEGREES OF CEREBRAL ACTIVITY ON THE BASIC RHYTHM OF THE ELECTROENCEPHALOGRAM

There is a general relationship between the degree of cerebral activity and the average frequency of the electroencephalographic rhythm, the frequency increasing progressively with higher and higher degrees of activity. This is illustrated in Figure 41–4, which shows the existence of delta waves in stupor, surgical anesthesia, and sleep; theta waves in psychomotor states and in infants; alpha waves during relaxed states; and beta waves during periods of intense mental activity. However, during periods of mental activity the waves usually become asynchronous rather than synchronous, so that the voltage falls considerably, despite increased cortical activity, as illustrated in Figure 41–3.

EPILEPSY

Epilepsy is characterized by uncontrolled excessive activity of either a part of the central nervous system or all of it. A person who is predisposed to epilepsy has attacks when the basal level of excitability of his nervous system (or of the part that is susceptible to the epileptic state) rises above a certain critical threshold. But, as long as the degree of excitability is held below this threshold, no attack occurs.

Two of the most common types of epilepsy are *grand mal epilepsy* and *petit mal epilepsy.*

Grand Mal Epilepsy. Grand mal epilepsy is characterized by extreme neuronal discharges originating in the brain stem portion of the reticular activating system. These then spread throughout the entire central nervous system, to the cortex, to the deeper parts of the brain, and even into the spinal cord to cause generalized *tonic convulsions* of the entire body followed toward the end of the attack by alternating muscular contractions called *clonic convulsions.* The grand mal seizure lasts from a few seconds to as long as three to four minutes and is characterized by post-seizure depression of the entire nervous system; the person might remain in a stupor for one minute to as long as a day or more after the attack is over.

The middle recording of Figure 41–5 illustrates a typical electroencephalogram from almost any region of the cortex during a grand mal attack. This illustrates that high voltage, spiking discharges occur over the entire cortex. Furthermore, the same type of discharge occurs on both sides of the brain at the same time, illustrating that the origin of the abnormality is in the lower centers of the brain that control the ac-

STUPOR SLEEP PSYCHOMOTOR INFANTS RELAXATION ATTENTION GRAND MAL
SURGICAL ANESTHESIA SLOW COMPONENT DETERIORATED EPILEPTICS FRIGHT FAST COMPONENT OF PETIT MAL
OF PETIT MAL LIGHT ETHER COMPULSION

⊢———— 1 SECOND ————⊣

FIGURE 41–4 Effect of varying degrees of cerebral activity on the basic rhythm of the EEG. (From Gibbs and Gibbs: Atlas of Encephalography. The Authors.)

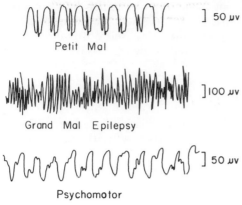

FIGURE 41–5 Electroencephalograms in different types of epilepsy.

tivity of the cerebral cortex and not in the cerebral cortex itself.

Presumably, therefore, a grand mal attack is caused by intrinsic overexcitability of the neurons that make up the reticular activating structures or from some abnormality of the local neuronal pathways. The synchronous discharges from this region could result from local reverberating circuits.

One might ask: What stops the grand mal attack after a given time? This is believed to result primarily from *fatigue of the neurons* involved in precipitating the attack. And the stupor that lasts for a few minutes to many hours after a grand mal seizure is over is believed to result from the intense fatigue of the neurons following their intensive activity during the grand mal attack.

Petit Mal Epilepsy. One of the most common kinds of epilepsy is the *absence type of petit mal epilepsy,* which is characterized by 5 to 20 seconds of unconsciousness during which the person has several twitchlike contractions of the muscles, usually in the head region— especially blinking of the eyes; this is followed by return of consciousness and continuation of previous activities.

This type of epilepsy is closely akin to grand mal epilepsy, and, in rare instances, it initiates grand mal attacks. Furthermore, persons who have petit mal attacks during early life may be-

come progressively disposed to grand mal attacks in later life. The first recording of Figure 41–5 illustrates a typical *spike and dome* pattern that is recorded during the absence type of petit mal epilepsy. The spike portion of this recording is almost identical to the spikes that occur in grand mal epilepsy, but the dome portion is distinctly different. The spike and dome can be recorded over the entire cerebral cortex, illustrating that the seizure originates in the reticular activating system of the brain.

BEHAVIOR — THE LIMBIC SYSTEM AND THE HYPOTHALAMUS

Behavior is a function of the entire nervous system, not of any particular portion. However, in this section we will deal with those special types of behavior associated with emotions, subconscious motor and sensory drives, and the intrinsic feelings of pain and pleasure. These functions of the nervous system are performed mainly by subcortical structures, such as the hypothalamus and adjacent areas, but older portions of the cerebral cortex that are located on the medial and ventral portions of the cerebral hemispheres also play a role. This overall group of brain structures is called the *limbic system.*

The Limbic System. Figure 41–6 gives a diagram of the important structures of the limbic system, showing the *hypothalamus* as one of the central elements of the system; surrounding it are other subcortical structures including the *preoptic area,* the *septum,* the *paraolfactory area,* the *epithalamus,* the *anterior nuclei of the thalamus, portions of the basal ganglia,* the *hippocampus,* and the *amygdala.* Surrounding the subcortical areas is the *limbic cortex* composed of a ring of cerebral cortex beginning (a) in the *orbitofrontal area* on the ventral surface of the frontal lobes, extending (b) upward in front of and over the corpus callosum onto the medial aspect of

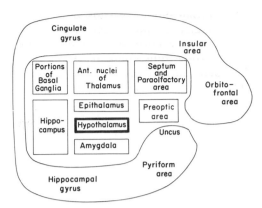

FIGURE 41-6 The limbic system.

the cerebral hemisphere to the *cingulate gyrus,* and finally (c) downward posterior to the corpus callosum onto the ventromedial surface of the temporal lobe in the *hippocampal gyrus, pyriform area,* and *uncus.*

BEHAVIORAL FUNCTIONS OF THE HYPOTHALAMUS AND RELATED AREAS

The vegetative functions of the hypothalamus—that is, the subconscious control of many of the body's internal activities by the hypothalamus—are discussed throughout this text. Some of these functions include regulation of arterial pressure, body fluid balance, electrolyte content of the body fluids, feeding, gastrointestinal activity, and many internal secretions of the endocrine glands.

In addition to these vegetative functions, the hypothalamus and closely associated areas of the diencephalon and mesencephalon also control many behavioral functions, as follows:

PLEASURE AND PAIN; REWARD AND PUNISHMENT

In the past several years, it has been learned that many hypothalamic and closely related structures are particularly concerned with the affective nature of sensory sensations—that is, with whether the sensations are *pleasant* or *painful.* These affective qualities are also called *reward* and *punishment.* Electrical stimulation of certain regions soothes the animal, whereas electrical stimulation of other regions causes extreme pain, fear, defense, escape reactions, and all the other elements of punishment. Obviously, these two oppositely responding systems greatly affect the behavior of the animal.

Figure 41-7 illustrates a technique that has been used for localizing the specific reward and punishment areas of the brain. In this figure a lever is placed at the side of the cage and is arranged so that depressing the lever makes electrical contact with a stimulator. Electrodes are placed successively in different areas in the brain so that the animal can stimulate the area by pressing the lever. If stimulating the particular area gives the animal a sense of reward, then he will press the lever again and again, sometimes as much as 5000 times per hour. Furthermore, when offered the choice of eating some de-

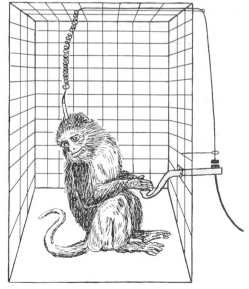

FIGURE 41-7 Technique for localizing reward and punishment centers in the brain of a monkey.

lectable food as opposed to the opportunity to stimulate the reward center, he often chooses the electrical stimulation.

By using this procedure, the major reward centers have been found to be located in the hypothalamus, primarily *in the ventromedial nuclei of the hypothalamus.* Less potent reward centers, which are probably secondary to the major ones in the hypothalamus, are found in the amygdala, septum, certain areas of the thalamus and basal ganglia, and finally extending downward into the basal tegmentum of the mesencephalon.

The apparatus illustrated in Figure 41–7 can also be connected so that pressing the lever turns off rather than turns on an electrical stimulus. In this case, the animal will not turn the stimulus off when the electrode is in one of the reward areas, but when it is in certain other areas he immediately learns to turn it off. Stimulation in these areas causes the animal to show all the signs of pain and displeasure. Furthermore, prolonged stimulation for 24 hours or more causes the animal to become severely sick and actually leads to death.

By means of this technique, the principal centers for pain, punishment, and escape tendencies have been found in the *perifornical nucleus* of the hypothalamus and in the central gray area of the mesencephalon.

Importance of Reward and Punishment in Behavior. Almost everything that we do depends on reward and punishment. If we are doing something that is rewarding, we continue to do it; if it is punishing, we cease to do it. Therefore, the reward and punishment centers undoubtedly constitute one of the most important of all the controllers of our bodily activities, our motivations, and so forth.

Importance of Reward and Punishment in Learning — Habituation and Reinforcement. It has been found in animal experiments that a sensory experience that causes neither reward nor punishment is remembered hardly at all. Also, electrical recordings have shown that new and novel sensory stimuli always excite the cerebral cortex. But repetition of the stimulus over a period of time leads to almost complete extinction of the cortical response if this stimulus does not excite either reward or punishment centers. Thus, the animal becomes *habituated* to the sensory stimulus. But if the stimulus causes either reward or punishment rather than indifference, the cortical response becomes progressively more and more intense with repetitive stimulation instead of fading away, and the response is said to be *reinforced.* Therefore, an animal builds up strong memory engrams for sensations which are either rewarding or punishing, but, on the other hand, develops complete habituation to indifferent sensory stimuli. Therefore, it is evident that the reward and punishment centers of the midbrain have much to do with controlling the type of information that we learn.

Stimulation of the *perifornical nuclei of the hypothalamus,* which are also the hypothalamic regions that give the most intense sensation of punishment, causes the animal to (1) develop a defense posture, (2) extend his claws, (3) lift his tail, (4) hiss, (5) spit, (6) growl, and (7) develop pilo-erection, wide-open eyes, and dilated pupils. Furthermore, even the slightest provocation causes an immediate savage attack. This is approximately the behavior that one would expect from an animal being severely punished, and it is a pattern of behavior that has also been called simply *rage.* It can even occur in decorticated animals, illustrating that the basic behavioral patterns for protective activities are present in the lower regions of the brain.

FUNCTIONS OF THE AMYGDALA

The amygdala is a complex of nuclei located immediately beneath the ventral surface of the cerebral cortex in the pole of each temporal lobe. In lower animals, this complex is concerned primarily with association of olfactory stimuli with stimuli from other parts of the brain. However, in the human being, a new portion of the amygdala, the *basolateral nuclei,* has become much more highly developed than the olfactory portion and plays very important roles in many behavioral activities not in any way associated with olfactory stimuli. The amygdala transmits signals especially into the hypothalamus.

Effects of Stimulating the Amygdala. In general, stimulation in the amygdala can cause almost all the same effects as those elicited by stimulation of the hypothalamus, plus still other effects. The effects that are mediated at least partially through the hypothalamus include (1) increases or decreases in arterial pressure, (2) increases or decreases in heart rate, (3) increases or decreases in gastrointestinal motility and secretion, (4) defecation and micturition, (5) pupillary dilatation or, rarely, constriction, (6) pilo-erection, (7) secretion of the various anterior pituitary hormones, including especially the gonadotropins and corticotropin.

In addition, stimulation of certain amygdaloid nuclei can cause a pattern of rage, escape, punishment, and pain similar to the affective-defense pattern elicited from the hypothalamus as described above. And stimulation of other nuclei can give reactions of reward and pleasure.

Finally, excitation of still other portions of the amygdala can cause sexual activities that include erection, copulatory movements, ejaculation, ovulation, uterine activity, and premature labor.

In short, stimulation of appropriate portions of the amygdaloid nuclei can give almost any pattern of behavior. It is believed that the normal function of the amygdaloid nuclei is to help control the overall pattern of behavior demanded for each occasion.

FUNCTION OF THE LIMBIC CORTEX

Probably the most poorly understood portion of the entire limbic system is the ring of cerebral cortex, called the *limbic cortex,* that surrounds the subcortical limbic structures. The limbic cortex is among the oldest of all parts of the cerebral cortex. In lower animals it plays a major role in various olfactory, gustatory, and feeding phenomena. However, in the human being, these functions of the limbic cortex are of minor importance. Instead, the limbic cortex of the human being is believed to be the cerebral association cortex for control of the lower centers that have to do primarily with behavior. For instance, the hippocampal gyrus has close interconnections with the hippocampus and thence through the fornix with almost all areas of the diencephalon, especially with the hypothalamus. The cingulate gyrus likewise connects directly with the hippocampus and also connects through the thalamus with almost all diencephalic areas. Finally, the orbitofrontal cortex interconnects profusely with the septal area, hypothalamus, preoptic area, and other associated regions.

Effect of Ablation of Different Portions of the Limbic Cortex. *Ablation of the anterior portion of the cingulate gyri.* Bilateral removal of the anterior portion of cingulate gyri causes increased tameness of an animal to the extent that he no longer has fear of a man even though this fear might have existed preoperatively. Furthermore, previous rage reactions are suppressed, and the animal seems to lack "social consciousness."

Ablation of the posterior orbital cortex. Bilateral removal of the posterior portion of the fronto-orbital cortex often causes an animal to develop an intense degree of motor restlessness, becoming unable to sit still, but instead moving continually.

Ablation of the anterior temporal region—the Kluver-Bucy syndrome. When the anterior temporal cortex is removed, many important subcortical structures are also frequently damaged or removed, including especially the amygdala. Therefore, it is doubtful that what has been learned from anterior temporal ablation is of particular value in determining the function of the temporal limbic *cortex*. However, removal of the entire anterior tip of the temporal lobe in the dominant hemisphere of a monkey, and sometimes in man, causes intense changes in behavior, including (1) excessive tendency to examine objects, (2) loss of fear, (3) decreased aggressiveness, (4) tameness, (5) changes in dietary habits such that a herbivorous animal sometimes even becomes carnivorous, and (6) sometimes psychic blindness. This combination of effects is called the Kluver-Bucy syndrome.

Overall Function of the Hippocampal Gyrus, Cingulate Gyrus, and Orbitofrontal Cortex. After the above discussion, we find ourselves immediately perplexed regarding the function of the cortical regions of the limbic system. The reason for this probably is that these regions are actually association areas that correlate information from many other regions of the brain but cause no direct, overt effects that can be observed objectively. Therefore, until further information is available, it is perhaps best to state that the cortical regions of the limbic system occupy intermediate associative positions between the functions of the remainder of the cerebral cortex and the functions of the lower centers for control of behavioral patterns.

REFERENCES

Adey, W. R., and Tokizane, T.: Structure and Function of the Limbic System. Progress in Brain Research. New York, American Elsevier Publishing Co., Inc., 1967, Vol. 27.

Brutkowski, S.: Functions of prefrontal cortex in animals. *Physiol. Rev.,* 45:721, 1965.

Gellhorn, E., and Loofbourrow, G. N.: Emotions and Emotional Disorders. New York, Paul B. Hoeber, 1962.

Jouvet, M.: Neurophysiology of the states of sleep. *Physiol. Rev.,* 47:117, 1967.

Kleitman, N.: Sleep and Wakefulness. Chicago, University of Chicago Press, 1963.

Laidlaw, J. P., and Stanton, J. B.: The EEG in Clinical Practice. Baltimore, The Williams & Wilkins Co., 1967.

Levi, L.: Emotional Stress. New York, American Elsevier Publishing Co., Inc., 1967.

Magoun, H. W.: The Waking Brain. 2nd ed., Springfield, Illinois, Charles C Thomas, Publisher, 1962.

Valdman, A. V.: Pharmacology and Physiology of the Reticular Formation: Progress in Brain Research. New York, American Elsevier Publishing Co., Inc., 1967, Vol. 20.

PART VI

GASTROINTESTINAL PHYSIOLOGY AND METABOLISM

CHAPTER 42

MOVEMENT OF FOOD THROUGH THE ALIMENTARY TRACT

Figure 42–1 illustrates the entire alimentary tract, showing separate parts adapted for specific functions, such as: (1) simple passage of food from one point to another, as in the esophagus, (2) storage of food or fecal matter in the body of the stomach and in the descending colon, respectively, (3) digestion of food in the stomach, duodenum, jejunum, and ileum, and (4) absorption of the digestive end-products in the entire small intestine and proximal half of the colon. The present chapter discusses the myriad of autoregulatory processes in the gut that keeps the food moving through the alimentary tract at a rate slow enough for digestion and absorption to take place but fast enough to provide the nutrients needed by the body.

Characteristics of the Intestinal Wall. Figure 42-2 illustrates a typical section of the intestinal wall, showing the following layers from the outside inward: (1) the *serosa*, (2) a *longitudinal muscle layer*, (3) a *circular muscle layer*, (4) the *submucosa*, and (5) the *mucosa*. In addition, a sparse layer of smooth muscle fibers, the *muscularis mucosae*, lies in the deeper layers of the mucosa.

INNERVATION OF THE GUT – THE INTRAMURAL PLEXUS

Beginning in the esophagus and extending all the way to the anus is an *intramural nerve plexus*. This is composed principally of two layers of neu-

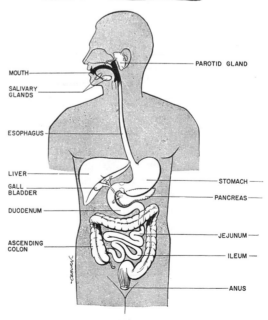

FIGURE 42–1 The alimentary tract.

521

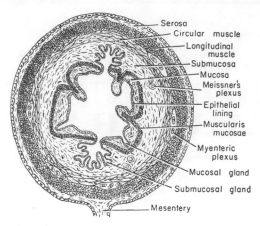

FIGURE 42-2 Typical cross-section of the gut.

rons and appropriate connecting fibers: the outer layer, called the *myenteric plexus* or *Auerbach's plexus*, lies between the longitudinal and circular muscular layers; and the inner layer, called the *submucosal plexus* or *Meissner's plexus*, lies in the submucosa. The myenteric plexus is far more extensive than the submucosal plexus, for which reason some physiologists often refer to the entire intramural plexus as simply the myenteric plexus.

In general, stimulation of the intramural plexus increases the activity of the gut, causing four principal effects: (1) increased tonic contraction, or "tone," of the gut wall, (2) increased intensity of rhythmic contractions, (3) increased rate of rhythmic contraction, and (4) increased velocity of conduction of excitatory waves along the gut wall.

Autonomic Control of the Gastrointestinal Tract. The entire gastrointestinal tract receives extensive parasympathetic and sympathetic innervation that is capable of altering the overall activity of the entire gut or of specific parts of it. The anatomy of this innervation was discussed in Chapter 40.

Parasympathetic signals to the upper gastrointestinal tract are carried almost entirely in the *vagus nerves*, while those to the distal colon pass through the *nervi erigentes*. The postganglionic neurons of the parasympathetic system lie in the myenteric plexus and are part of it, so that stimulation of the parasympathetic nerves causes a general increase in activity of this plexus. This in turn excites the gut wall and facilitates most of the intrinsic nervous reflexes of the gastrointestinal tract.

Sympathetic signals originate in the spinal cord between the segments T-8 and L-3 and pass through the sympathetic chains and outlying sympathetic ganglia to all parts of the gut. In general, stimulation of the sympathetic nervous system inhibits activity in the gastrointestinal tract, causing effects essentially opposite to those of the parasympathetic system. However, the sympathetic system excites (1) the ileocecal sphincter and (2) the internal anal sphincter. Thus, strong stimulation of the sympathetic system can totally block movement of food through the gastrointestinal tract by inhibition of the gut wall and by excitation of at least two major sphincters of the gastrointestinal tract.

PROPULSION OF FOOD IN THE GASTROINTESTINAL TRACT — PERISTALSIS

The basic propulsive movement of the gastrointestinal tract is peristalsis, which is illustrated in Figure 42-3. A contractile ring appears around the gut and then moves forward. Obviously, any

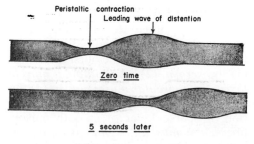

FIGURE 42-3 Peristalsis.

material in front of the contractile ring is moved forward.

The usual stimulus for peristalsis is *distention*. That is, if a large amount of food collects at any point in the gut, the distention stimulates the gut wall 2 to 3 cm. above this point, and a contractile ring appears and initiates a peristaltic movement.

Function of the Intramural Nerve Plexus in Peristalsis. Even though peristalsis is a basic characteristic of all tubular smooth muscle structures, it occurs only weakly or not at all in portions of the gastrointestinal tract that have congenital absence of the intramural nerve plexus. Also, the intramural plexus causes the velocity of peristalsis along the gut to be 1 to 10 cm. per second, which is some 3 to 25 times as great as the natural conductive velocity of impulses in the smooth muscle itself (4 mm./second). Furthermore, since the intramural plexus is principally under the control of the parasympathetic nerves, the intensity of peristalsis and its velocity of conduction can be altered by parasympathetic stimulation.

Analward Peristaltic Movements — Receptive Relaxation and the "Law of the Gut." Peristalsis, theoretically, can occur in either direction from a stimulated point, but it normally dies out rapidly in the orad direction while continuing for a considerable distance analward. This directional movement of peristalsis is believed to be caused by special organization of the intramural nerve plexus (principally the myenteric portion) which allows preferential transmission of analward signals. One reason for believing this is that peristaltic contraction at one point in the gut usually causes relaxation, called "receptive relaxation," several centimeters down the gut toward the anus. It is believed that this relaxation could occur only as a result of conduction in the myenteric plexus. Obviously, a leading wave of receptive relaxation could allow food to be propelled more easily analward than in the orad direction.

This response is also called the "law of the gut" or sometimes simply the "myenteric reflex." It can be particularly well demonstrated in the esophagus and in the pyloric region of the stomach.

SWALLOWING (DEGLUTITION)

In general, swallowing can be divided into: (1) the *voluntary stage*, which initiates the swallowing process, (2) the *pharyngeal stage*, which is involuntary and constitutes the passage of food through the pharynx into the esophagus, and (3) the *esophageal stage*, another involuntary phase which promotes passage of food from the pharynx to the stomach.

Voluntary Stage of Swallowing. When the food is ready for swallowing, it is "voluntarily" squeezed or rolled posteriorly by pressure of the tongue upward and backward against the palate, as shown in Figure 42–4. Thus, the tongue forces the bolus of food into the pharynx. From here on, the process of swallowing becomes entirely, or almost

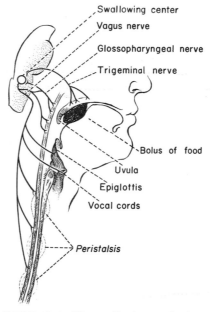

FIGURE 42–4 The swallowing mechanism.

entirely, automatic and ordinarily cannot be stopped.

Pharyngeal Stage of Swallowing. When the bolus of food is pushed backward in the mouth, it stimulates *swallowing receptor areas* all around the opening of the pharynx, especially on the tonsillar pillars, and impulses from these pass to the brain stem to initiate a series of automatic pharyngeal muscular contractions as follows:

1. The soft palate is pulled upward to close the posterior nares, in this way preventing reflux of food into the nasal cavities.

2. The pharynx narrows to form a sagittal slit through which the food must pass into the posterior pharynx. This slit performs a selective action, allowing food that has been masticated properly to pass with ease while impeding the passage of large objects.

3. The vocal cords of the larynx are strongly approximated, and the epiglottis swings backward over the superior opening of the larynx. Both of these effects prevent passage of food into the trachea.

4. The entire larynx is pulled upward and forward by muscles attached to the hyoid bone; this movement of the larynx stretches the opening of the esophagus. At the same time, the *hypopharyngeal sphincter* around the esophageal entrance, which normally prevents air from going into the esophagus during respiration, relaxes, thus allowing food to move easily and freely from the posterior pharynx into the upper esophagus. The upward movement of the larynx also lifts the glottis out of the main stream of food flow so that the food usually passes on either side of the epiglottis rather than over its surface; this adds still another protection against passage of food into the trachea.

5. At the same time that the larynx is raised and that the hypopharyngeal sphincter is relaxed, the superior constrictor muscle of the pharynx contracts, giving rise to a rapid peristaltic wave passing downward over the pharyngeal muscles and into the esophagus.

To summarize the mechanics of the pharyngeal state of swallowing—the trachea is closed, the esophagus is opened, and a fast peristaltic wave originating in the pharynx then forces the bolus of food into the upper esophagus, the entire process occurring in 1 to 2 seconds.

Nervous control of the pharyngeal stage of swallowing. The most sensitive tactile areas of the pharynx for initiation of the pharyngeal stage of swallowing lie in a ring around the pharyngeal opening, with greatest sensitivity in the tonsillar pillars. Impulses are transmitted from these areas through the sensory portions of the trigeminal and glossopharyngeal nerves into a region of the medulla oblongata closely associated with the *tractus solitarius,* which receives essentially all sensory impulses from the mouth.

The successive stages of the swallowing process are then automatically controlled in orderly sequence by neuronal areas distributed throughout the reticular substance of the medulla and lower portion of the pons. The sequence of the swallowing reflex remains the same from one swallow to the next, and the timing of the entire cycle also remains constant from one swallow to the next. The areas in the medulla and lower pons that control swallowing are collectively called the *deglutition* or *swallowing center.*

The motor impulses from the swallowing center to the pharynx and upper esophagus that cause swallowing are transmitted by the 5th, 9th, 11th, and 12th cranial nerves and even a few of the superior cervical nerves.

In summary, the pharyngeal stage of swallowing is principally a reflex act. It is initiated by voluntary movement of food into the pharynx, which in turn elicits the swallowing reflex.

Esophageal Stage of Swallowing. The esophagus functions primarily to conduct food from the pharynx to the stomach, and its movements are organized specifically for this function.

Normally the esophagus exhibits two

types of peristaltic movements — *primary peristalsis* and *secondary peristalsis*. Primary peristalsis is simply continuation of the peristaltic wave that begins in the pharynx and spreads into the esophagus during the pharyngeal stage of swallowing. This wave passes all the way from the pharynx to the stomach in approximately 5 to 10 seconds. However, food swallowed by a person who is in the upright position is usually transmitted to the lower end of the esophagus even more rapidly than the peristaltic wave itself, in about 4 to 8 seconds, because of the additional effect of gravity pulling the food downward. If the primary peristaltic wave fails to move all the food that has entered the esophagus on into the stomach, secondary peristaltic waves result from distention of the esophagus by the retained food. These waves are essentially the same as the primary peristaltic waves, except that they originate in the esophagus itself rather than in the pharynx. Secondary peristaltic waves continue to be initiated until all the food has emptied into the stomach.

The peristaltic waves of the esophagus are controlled almost entirely by vagal reflexes that are part of the overall swallowing mechanism. These reflexes are initiated by *vagal afferent fibers* from the esophagus to the medulla and then are transmitted back again to the esophagus through *vagal efferent fibers*.

logically, it remains tonically constricted in contrast to the remainder of the esophagus which normally remains completely relaxed. Then, when a peristaltic wave of swallowing passes down the esophagus, "receptive relaxation" relaxes the gastroesophageal sphincter ahead of the peristaltic wave, and allows propulsion of the swallowed food on into the stomach after a delay of only 1 to 3 seconds. Rarely, the gastroesophageal constrictor does not relax satisfactorily, resulting in the condition called *achalasia*, which will be discussed later in the chapter.

A principal function of the gastroesophageal constrictor is to prevent reflux of stomach contents into the upper esophagus. The stomach contents are highly acidic and contain many proteolytic enzymes. The esophageal mucosa, except in the lower eighth of the esophagus, is not capable of resisting for long the digestive action of gastric secretions in the esophagus.

Another factor that prevents reflux is a valvelike mechanism of that portion of the esophagus that lies immediately beneath the diaphragm. Increased intraabdominal pressure caves the esophagus inward at this point and therefore prevents the forcing of stomach contents into the esophagus. Otherwise, every time we walked, coughed, or breathed hard, we would expel acid into the esophagus.

FUNCTION OF THE GASTROESOPHAGEAL CONSTRICTOR

At the lower end of the esophagus, about 5 cm. above its juncture with the stomach at the *cardia*, the circular muscle is slightly hypertrophied to form a *gastroesophageal constrictor*. Anatomically this constrictor is no different from the remainder of the esophagus except for the hypertrophy. However, physio-

MOTOR FUNCTIONS OF THE STOMACH

The motor functions of the stomach are three-fold: (1) storage of large quantities of food until it can be accommodated in the lower portions of the gastrointestinal tract, (2) mixing of this food with gastric secretions until it forms a semifluid mixture called *chyme*, and (3) slow emptying of the food from the

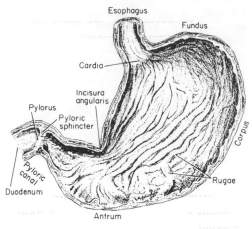

FIGURE 42–5 Physiologic anatomy of the stomach.

stomach into the small intestine at a rate suitable for proper digestion and absorption by the small intestine.

Figure 42–5 illustrates the basic anatomy of the stomach. Physiologically, the stomach can be divided into two major parts: (1) the *corpus*, or *body*, and (2) the *antrum*. The *fundus*, located at the upper end of the body of the stomach, is often considered by anatomists to be a separate entity from the body, but from a physiological point of view, the fundus is actually a functional part of the body.

STORAGE FUNCTION OF THE STOMACH

As food enters the stomach, it forms concentric circles in the body of the stomach, the newest food lying closest to the *cardia* (where the esophagus empties into the stomach) and the oldest food lying nearest the wall of the stomach. Normally the body of the stomach has relatively little tone in its muscular wall so that it can bulge progressively outward, thereby accommodating greater and greater quantities of food up to a limit of about 1 liter. The pressure in the stomach remains low until this limit is approached.

MIXING IN THE STOMACH

The digestive juices of the stomach are secreted by the *gastric glands*, which cover almost the entire wall of the body and fundus of the stomach. These secretions come immediately into contact with the stored food lying against the inner surface of the stomach; and when the stomach is filled, weak waves called *tonus waves*, or *mixing waves*, move along the stomach wall approximately once every 20 seconds. These waves begin near the cardia and spread for variable distances toward the pylorus. In general, they become more intense as they approach the antral portion of the stomach.

The mixing waves tend to move the gastric secretions and the outermost layer of food gradually toward the antral part of the stomach. On entering the antrum the waves become stronger, and the food and gastric secretions become progressively mixed to a greater and greater degree of fluidity.

Chyme. After the food has become mixed with the stomach secretions, the resulting mixture that passes on down the gut is called chyme. The degree of fluidity of chyme depends on the relative amounts of food and stomach secretions and on the degree of digestion that has occurred. The appearance of chyme is that of a murky, milky semifluid or paste.

EMPTYING OF THE STOMACH

Propulsion of Food Through the Stomach —The Pyloric Pump. Strong peristaltic waves occur about 20 per cent of the time in the antrum of the stomach. These waves, like the mixing waves, occur about once every 20 seconds. In fact, they are probably extensions of the mixing waves, which become potentiated as they spread from the body of the stomach into the antrum. They become intense approximately at the incisura angularis, from which they spread through the antrum, then over the pylorus, and often even into the duodenum. Each wave "pumps" 1 to 3 ml. of chyme into the duodenum, for which reason the mechanism is called the *pyloric pump.* As the stomach becomes progressively more empty, these waves begin farther and farther up the body of the stomach, gradually pinching off the lowermost portions of stored food, adding this food to the chyme in the antrum.

Regulation of Emptying. The rate at which food is emptied from the stomach into the duodenum by the pyloric pump is regulated by two principal factors: (1) the fluidity of the chyme in the stomach and (2) the degree of activity of the pyloric pump.

The *degree of fluidity of the chyme* is determined by (1) the type of food that has been eaten, (2) the degree of mastication of the food, (3) the length of time the food has been digested and mixed with secretions in the stomach, and (4) the intensity of the mixing and peristaltic waves of the stomach.

On the other hand, the activity of the pyloric pump is regulated principally by two feedback effects from the duodenum, as follows:

The enterogastric reflex and its effect on gastric emptying. When large quantities of chyme have been emptied into the duodenum, thereby raising the pressure in the duodenum, or whenever an obstruction in the small intestine prevents forward movement of the intestinal contents, the high pressure in the upper portion of the small bowel elicits the *enterogastric reflex,* which inhibits gastric peristalsis. Various aspects of this reflex are transmitted by three separate routes to the stomach, (1) by the myenteric plexus, (2) by vagal afferents to the medulla to inhibit vagal efferents, and (3) possibly by way of the celiac ganglion and then through sympathetic nerves to the stomach. Thus, the small intestine protects itself from becoming

overloaded because of too rapid emptying of the stomach contents.

Irritants in the small intestine, excessive *acidity* of the chyme, excessive *protein* breakdown products, and even *hypo-* or *hypertonicity* of the chyme can also elicit the enterogastric reflex. In other words, the duodenum prevents excess emptying of chyme that has chemical or physical characteristics which the duodenal secretions are not able to cope with.

Enterogastrone and its inhibition of stomach emptying. When fatty foods, especially fatty acids, are present in the chyme, the hormone *enterogastrone* is extracted by the chyme from the mucosa of the duodenum and jejunum. This in turn is absorbed into the blood and within a few minutes inhibits the motility of the stomach, thereby slowing the rate of stomach emptying to as little as half to one-third the previous rate. This allows prolonged time for digestion of fats in the small intestine, an important effect because fats are the slowest of all foods to be digested.

To a much less extent, carbohydrates and acid in the small intestine extract small quantities of enterogastrone and therefore slightly inhibit the rate of stomach emptying.

Summary. The duodenum and other portions of the upper intestine control the rate of stomach emptying by (1) the enterogastric reflex and (2) the enterogastrone mechanism. These two work together to slow down the rate of emptying when (a) too much food is already in the small intestine or (b) when the chyme is excessively acid, contains too much fat, is hypotonic or hypertonic, or is irritating. In this way, the rate of stomach emptying is limited to that amount of chyme that the small intestine can process.

Among the different foods that affect the rate of emptying, fats inhibit emptying the greatest, sometimes delaying the emptying of a fatty meal for as long as three to six hours; proteins have an intermediate effect, and carbohydrates only a mild delaying effect.

MOVEMENTS OF THE SMALL INTESTINE

MIXING CONTRACTIONS (SEGMENTATION CONTRACTIONS)

Small ringlike contractions occur most of the time throughout the entire small intestine. These often occur irregularly, but at other times they are rhythmic, occurring at rates of 8 to 9 per minute in the duodenum and at slower rates progressively farther down the small intestine. These contractions cause "segmentation" of the small intestine as illustrated in Figure 42–6, dividing the intestine at times into regularly spaced segments that have the appearance of a chain of sausages. The next contractions occur at new points. Therefore, the segmentation contractions "chop" the chyme many times per minute, in this way keeping the solid suspension of food mixed with the secretions of the small intestine.

PROPULSIVE MOVEMENTS

Chyme is propelled through the small intestine by *peristaltic waves*. These occur in any part of the small intestine, and they move analward at a velocity of 1 to 2 cm. per second. However, they are normally weak and usually die out after

traveling only a few centimeters. As a result, the movement of the chyme along the small intestine is slow, averaging only 1 cm. per minute. This means that 3 to 10 hours are normally required for passage of chyme from the pylorus to the ileocecal valve.

Peristaltic activity of the small intestine is greatly increased immediately after a meal. This is caused by a *gastroenteric reflex* that is initiated by distention of the stomach and conducted principally through the myenteric plexus from the stomach down along the wall of the small intestine. This reflex increases the overall degree of excitability of the small intestine, including both increased motility and secretion.

Also, irritation of the intestinal mucosa or distention of the intestine increases the intensity of peristalsis.

Movements Caused by the Muscularis Mucosae and Movements of the Villi. The muscularis mucosae, which is stimulated by local nervous reflexes in the intramural plexus and by the sympathetic nervous system, can cause short or long folds to appear in the intestinal mucosa. Also, individual fibers from this muscle extend upward into the intestinal villi and cause them to contract intermittently. The mucosal folds increase the surface area exposed to the chyme, thereby increasing the rate of absorption. The contractions of the villi—shortening, elongating, and shortening again—"milk" the villi so that lymph flows freely from the central lacteals into the lymphatic system. It has been claimed, though not proved, that villus contractions can also be stimulated by a hormone called *villikinin*. Supposedly, the chyme in the intestine extracts villikinin from the mucosa, and this in turn is absorbed into the blood to excite the villi.

FUNCTION OF THE ILEOCECAL VALVE

A principal function of the ileocecal valve is to prevent backflow of fecal con-

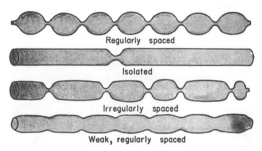

Regularly spaced

Isolated

Irregularly spaced

Weak, regularly spaced

FIGURE 42–6 Segmentation movements of the small intestine.

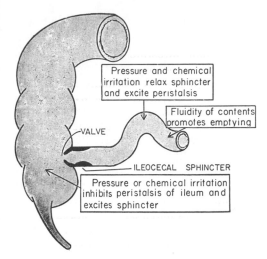

Pressure and chemical irritation relax sphincter and excite peristalsis

Fluidity of contents promotes emptying

VALVE

ILEOCECAL SPHINCTER

Pressure or chemical irritation inhibits peristalsis of ileum and excites sphincter

FIGURE 42–7 Emptying at the ileocecal valve.

tents from the colon into the small intestine. As illustrated in Figure 42–7, the lips of the ileocecal valve, which protrude into the lumen of the cecum, are admirably adapted for this function. Usually the valve can resist reverse pressure of as much as 50 to 60 cm. water.

The wall of the ileum for several centimeters immediately preceding the ileocecal valve has a thickened muscular coat called the *ileocecal sphincter.* This normally remains mildly constricted and prevents emptying of the ileal contents into the cecum except immediately following a meal when a gastroileal reflex (described above) intensifies the peristalsis in the ileum. When this occurs, about 4 ml. of chyme empties from the ileum into the cecum with each peristaltic wave. Yet, even so, only about 450 ml. of chyme empties into the cecum each day. The resistance to emptying at the ileocecal valve prolongs the stay of chyme in the ileum and, therefore, facilitates absorption.

Control of the Ileocecal Sphincter. The degree of contraction of the ileocecal sphincter is controlled primarily by reflexes from the cecum. Whenever the cecum is distended or irritated, the de-

gree of contraction of the ileocecal sphincter is intensified, which greatly delays emptying of additional chyme from the ileum.

MOVEMENTS OF THE COLON

The functions of the colon are (1) absorption of water and electrolytes from the chyme and (2) storage of fecal matter until it can be expelled. The proximal half of the colon, illustrated in Figure 42–8, is concerned principally with absorption, and the distal half with storage; since intense movements are not required for these functions, the movements of the colon are normally sluggish. Yet in a sluggish manner, the movements still have characteristics similar to those of the small intestine and can be divided once again into mixing movements and propulsive movements.

Mixing Movements — Haustrations. In the same manner that segmentation movements occur in the small intestine, large circular constrictions also occur in the large intestine. At each of these constriction points, about 2.5 cm. of the circular muscle contracts, sometimes constricting the lumen of the colon to almost complete occlusion. At the same time, the longitudinal muscle of the

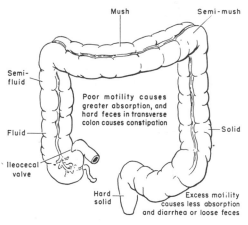

Mush

Semi-mush

Semi-fluid

Poor motility causes greater absorption, and hard feces in transverse colon causes constipation

Solid

Fluid

Ileocecal valve

Hard solid

Excess motility causes less absorption and diarrhea or loose feces

FIGURE 42–8 Absorptive and storage functions of the large intestine.

colon, which is aggregated into three longitudinal stripes called the *tineae coli*, contracts. These combined contractions of the circular and longitudinal smooth muscle cause the unstimulated portion of the large intestine to bulge outward into baglike sacs called *haustrations*. The haustral contractions, once initiated, usually reach peak intensity in about 30 seconds and then disappear during the next 60 seconds. They at times also move slowly analward during their period of contraction. After another few minutes, new haustral contractions occur in nearby areas but not in the same areas. Therefore, the fecal material in the large intestine is slowly "dug" into and rolled over in much the same manner that one spades the earth. In this way, all the fecal material is gradually exposed to the surface of the large intestine, and fluid is progressively absorbed until only 80 ml. of the 450 ml. daily load of chyme is lost in the feces.

Propulsive Movements—"Mass Movements." Peristaltic waves of the type seen in the small intestine do not occur in the colon. Instead, another type of movement, called a mass movement, propels the fecal contents toward the anus. These movements usually occur only a few times each day, most abundantly for about 10 minutes during the first hour or so after eating breakfast.

A mass movement is characterized by the following sequence of events: First, a constrictive point occurs at a distended or irritated point in the colon, and then rapidly thereafter the 20 or more cm. of colon *distal* to the constriction contracts almost as a unit, forcing the fecal material in this segment *en masse* down the colon. The initiation of contraction is complete in about 30 seconds, and relaxation then occurs during the next two to three minutes. Mass movements can occur in any part of the colon, though most often they occur in the transverse or descending colon. When they have forced a mass of feces into the rectum, the desire for defecation is felt.

The appearance of mass movements after meals is caused principally by the *duodenocolic reflex*. This results from filling of the duodenum, which initiates a reflex from the duodenum to the colon and thereby increases the excitability of the entire colon. To a less extent, a *gastrocolic reflex* also occurs.

Irritation or distension of the colon can also initiate intense mass movements.

DEFECATION

Continual dribble of fecal matter through the anus is prevented by tonic constriction of (1) the *internal anal sphincter*, a circular mass of smooth muscle that lies immediately inside the anus, and (2) the *external anal sphincter*, composed of striated voluntary muscle and controlled by the somatic nervous system.

Ordinarily, defecation results from the *defecation reflex*, which can be described as follows: When the feces enter the rectum, distention of the rectal wall initiates afferent signals that spread through the *myenteric plexus* to initiate peristaltic waves in the descending colon and sigmoid, forcing feces toward the anus. As the peristaltic wave approaches the anus, the internal anal sphincter is inhibited by the usual phenomenon of "receptive relaxation," and if the external anal sphincter is relaxed, defecation will occur. This overall effect is called the defecation reflex.

However, the defecation reflex itself is extremely weak, and to be effective in causing defecation it must be fortified by another reflex that involves the sacral segments of the spinal cord, as illustrated in Figure 42–9. When the afferent fibers in the rectum are stimulated, signals are transmitted into the spinal cord and thence, reflexly, back to the descending colon, sigmoid, rectum, and anus by way of parasympathetic nerve fibers in the *nervi erigentes*. These parasympathetic

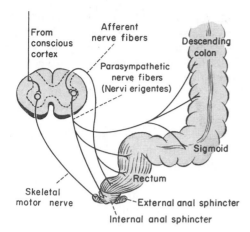

FIGURE 42-9 The afferent and efferent pathways of the parasympathetic mechanism for enhancing the defecation reflex.

signals greatly intensify the peristaltic waves and convert the defecation reflex from an ineffectual weak movement into a powerful process of defecation that is sometimes effective in emptying the large bowel all the way from the splenic flexure to the anus.

However, the normal person prevents defecation until a socially acceptable time presents itself. This is achieved by voluntary contraction of the external sphincter, a skeletal muscle that is kept tonically contracted when a person wishes to prevent defecation. When this is done, the defecation reflex dies out after a few minutes and usually will not return until many hours thereafter.

When it becomes convenient for the person to defecate, defecation reflexes can usually be initiated by taking a deep breath to move the diaphragm downward and then contracting the abdominal muscles to increase the pressure in the abdomen, thus forcing fecal contents into the rectum to elicit new reflexes. Unfortunately, reflexes initiated in this way are never as effective as those that arise naturally, for which reason people who inhibit their natural reflexes too often become severely constipated.

DISORDERS OF MOTOR FUNCTION IN THE GASTROINTESTINAL TRACT

ACHALASIA (FAILURE OF THE ESOPHAGUS TO EMPTY)

Achalasia is a condition in which the lower few centimeters of the esophagus fail to relax during the swallowing mechanism. As a result, food transmission from the esophagus into the stomach is impeded or is sometimes completely blocked. Pathological studies have shown the cause of this condition to be either damaged or absent myenteric plexus in the lower portion of the esophagus. The musculature of the lower end of the esophagus is still capable of contracting, and even exhibits incoordinate movements, but it has lost the ability to conduct a peristaltic wave and has lost the ability of "receptive relaxation" of the gastroesophageal sphincter as food approaches this area during the swallowing process.

Food is dammed up behind the aperistaltic portion of the esophagus, causing the middle and upper esophagus to dilate until it sometimes becomes 3 to 4 inches in diameter. In severe cases, food sometimes stays in the esophagus as long as 24 hours before passing into the stomach. This prolonged storage of food allows extreme putrefaction of the food and often also allows bacterial penetration of the esophageal mucosa, which results in ulceration of the esophageal wall and, possibly, death.

CONSTIPATION

Constipation means slow movement of feces through the large intestine, and it is usually associated with large quantities of dry, hard feces in the descending colon which accumulate because of the long time allowed for absorption of fluid.

The most frequent cause of constipation is irregular bowel habits that have developed through a lifetime of inhibition of the normal defecation reflexes. The newborn child is rarely constipated, but part of his training in the early years of life requires that he learn to control defecation, and this control is effected by inhibiting the natural defecation reflexes. Clinical experience shows that if one fails to allow defecation to occur when the defecation reflexes are excited, the reflexes themselves become progressively less strong over a period of time. For this reason, if a person establishes regular bowel habits early in life, usually defecating in the morning after breakfast when the gastrocolic and duodenocolic reflexes cause mass movements in the large intestine, he can generally prevent the development of constipation in later life.

DIARRHEA

Diarrhea, which is the opposite of constipation, results from rapid movement of fecal matter through the large intestine. The major causes of diarrhea are: (1) an infection in the gastrointestinal tract called *gastroenteritis* and (2) excessive parasympathetic stimulation of the large intestine, which is called *neurogenic diarrhea.*

Gastroenteritis. Gastroenteritis means infection at any point in the gastrointestinal tract or throughout the entire extent of the tract. In usual infectious diarrhea, the infection is most extensive in the large intestine, though sometimes it extends into the small intestine as well. Everywhere that the infection is present, the mucosa becomes extensively irritated, and its rate of secretion becomes greatly enhanced for reasons discussed in the following chapter. In addition, the motility of the intestinal wall usually increases many-fold. As a result, large quantities of fluid are made available for washing the infectious agent toward the anus, and at the same time strong mass movements propel this fluid forward. Obviously, this is an important mechanism for ridding the intestinal tract of the debilitating infection.

Neurogenic Diarrhea. Everyone is familiar with the diarrhea that accompanies periods of nervous tension, such as during examination time or when a soldier is about to go into battle. This type of diarrhea, called neurogenic diarrhea, is caused by excessive stimulation of the parasympathetic nervous system, which greatly excites both motility and secretion of mucus in the colon. These two effects added together can cause marked diarrhea. Indeed, in tense states, even the chyme in the small intestine can be flushed into the large intestine and then rapidly on through to the anus, causing at times loss of large quantities of water and electrolytes.

VOMITING

Vomiting is the means by which the upper gastrointestinal tract rids itself of its contents when the gut becomes excessively irritated, overdistended, or even overexcitable. The stimuli that cause vomiting can occur in any part of the gastrointestinal tract, though distention or irritation of the stomach or duodenum provides the strongest stimulus. Impulses are transmitted, by both vagal and sympathetic afferents to the bilateral *vomiting center* of the medulla, which lies near the tractus solitarius at approximately the level of the dorsal motor nucleus of the vagus. Once the vomiting center has been sufficiently stimulated and the vomiting act instituted, the first effects are (1) a deep breath, (2) raising the hyoid bone and the larynx to pull the cricoesophageal sphincter open, (3) closing the glottis, and (4) lifting the soft palate to close the posterior nares. Next comes a strong downward contraction of the diaphragm along with simultaneous contraction of all the abdominal muscles. This obviously squeezes the stomach between the two sets of muscles, building the intragastric pressure to a high level. At the same time, strong reverse peristalsis begins in the antral region of the stomach and passes over the body of the stomach, thus forcing the stomach contents toward the esophagus. Finally, the gastroesophageal sphincter relaxes, allowing expulsion of the gastric contents upward through the esophagus.

Thus, the vomiting act results from a squeezing action of the muscles of the stomach and abdomen in association with opening of the esophageal sphincters, so that the gastric contents can be expelled.

GASES IN THE GASTROINTESTINAL TRACT (FLATUS)

Gases can enter the gastrointestinal tract from three different sources: (1) swallowed air, (2) gases formed as a result of bacterial action, and (3) gases that diffuse from the blood into the gastrointestinal tract.

Most gases in the stomach are nitrogen and oxygen derived from swallowed air, and a large proportion of these are expelled by belching.

Only small amounts of gas are usually present in the small intestine, and these are composed principally of air that passes from the stomach into the intestinal tract.

In the large intestine, the greater proportion of the gases is derived from bacterial action; these include especially carbon dioxide, methane, and *hydrogen*. When the methane and hydrogen become suitably mixed with oxygen from swallowed air, an actual explosive mixture is occasionally formed.

The amount of gases entering or forming in the large intestine each day averages 7 to 10 liters, whereas the average amount expelled is usually only about 0.5 liter. The remainder is absorbed through the intestinal mucosa. Most often, a person expels large quantities of gases not because of excessive bacterial activity but because of excessive motility of the large intestine, the gases being moved on through the large intestine before they can be absorbed.

REFERENCES

Bosma, J. F.: Deglutition: pharyngeal stage. *Physiol. Rev., 37*:275, 1957.

Code, C. F., and Carlson, H. C.: Motor activity of the stomach. *In* Handbook of Physiology, The Williams & Wilkins Co., 1968, Sec. VI, Vol. IV, p. 1903.

Daniel, E. E.: Digestion: motor function. *Ann. Rev. Physiol., 31*:203, 1969.

Davenport, H.: Physiology of the Digestive Tract. 2nd ed., Chicago, Year Book Publishers, Inc., 1966.

Hunt, J. N., and Knox, M. T.: Regulation of gastric emptying. *In* Handbook of Physiology. Baltimore, The Williams & Wilkins Co., 1968, Sec. VI, Vol. IV, p. 1917.

Jenkins, G. N.: The Physiology of the Mouth. 3rd ed., Philadelphia, F. A. Davis Co., 1965.

Truelove, S. C.: Movements of the large intestine. *Physiol. Rev., 46*:457, 1966.

Youmans, W. B.: Innervation of the gastrointestinal tract. *In* Handbook of Physiology. Baltimore, The Williams & Wilkins Co., 1968, Sec. VI, Vol. IV, p. 1655.

CHAPTER 43

SECRETORY FUNCTIONS OF THE ALIMENTARY TRACT

The gastrointestinal secretions subserve two primary functions: (1) They provide the digestive enzymes, and (2) they provide mucus for lubrication and protection of all parts of the alimentary tract. The purpose of the present chapter is to describe the different alimentary secretions, their functions, and regulation of their production.

GENERAL PRINCIPLES OF GASTROINTESTINAL SECRETION

ANATOMICAL TYPES OF GLANDS

Several types of glands provide the different types of secretions in the gastrointestinal tract. First, on the surface of the epithelium in most parts of the gastrointestinal tract are literally billions of *single cell mucous glands*. These function to secrete much of the mucus that protects the mucosa.

Second, the most abundant type of gland in the gastrointestinal tract is the tubular gland, two examples of which

are illustrated in Figure 43–1A and 43–1B. Figure 43–1A shows a *crypt of Lieberkühn*, found throughout the small intestine and in modified form in the large intestine. These are simple pits lined with goblet cells to produce mucus, and epithelial cells that produce mainly serous fluids but also a small amount of enzymes. Figure 43–1B illustrates a *gastric gland* in the main body of the stomach. This is a considerably deeper tubular gland, and it is occasionally branched.

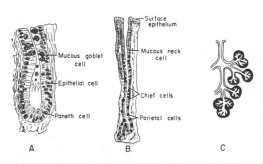

FIGURE 43–1 The anatomical types of gastrointestinal glands: (*A*) A simple tubular gland represented by a crypt of Lieberkühn. (*B*) A more elongated tubular gland represented by a gastric gland. (*C*) A compound acinous gland represented by the pancreas.

Third, associated with the gastrointestinal tract are several complex glands — the *salivary glands*, the *pancreas*, and the *liver* — which provide secretions for digestion or emulsification of food. The liver has a highly specialized structure that will be discussed later in this chapter. The salivary glands and the pancreas are compound acinous glands of the type illustrated in Figure 43-1C. These glands lie completely outside the walls of the gastrointestinal tract and, in this, differ from all other gastrointestinal glands. The secreting structures are lined with secreting glandular cells; these feed into a system of ducts that finally empty through one or more portals into the intestinal tract itself.

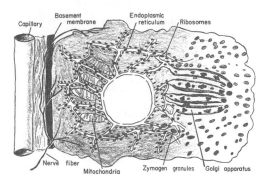

FIGURE 43-2 Basic mechanism of secretion by a glandular cell.

BASIC MECHANISM OF SECRETION BY GLANDULAR CELLS

Though all the basic mechanisms by which glandular cells form different secretions and then extrude these to the exterior are not known, experimental evidence points to the following basic principles of secretion by glandular cells, as illustrated in Figure 43-2. (1) Energy from adenosine triphosphate (see Chap. 45), along with appropriate nutrients, is used for synthesis of the organic substances, this synthesis occurring almost entirely in or on the surface of the *endoplasmic reticulum*. The *ribosomes* adherent to this reticulum are specifically responsible for formation of proteins that are to be secreted. (2) The secretory materials flow through the tubules of the endoplasmic reticulum into the vesicles of the Golgi apparatus, which lies near the secretory ends of the cells. (3) The materials then are concentrated and discharged into the cytoplasm in the form of *secretory granules*. (4) Active transport of electrolytes into the base of the cell is followed by osmosis of water, causing both water and electrolytes to flow through the cell and thereby to wash the organic substances (mainly

enzymes) through the secretory surface into the lumen of the gland.

LUBRICATING AND PROTECTIVE PROPERTIES OF MUCUS AND ITS IMPORTANCE IN THE GASTROINTESTINAL TRACT

Mucus is a thick secretion composed of water, electrolytes, and a mixture of several mucopolysaccharides. It has several important characteristics that make it both an excellent lubricant and a protectant for the wall of the gut. *First*, mucus has adherent qualities that make it adhere tightly to the food or other particles and spread as a thin film over the surfaces. *Second*, it has sufficient *body* that it coats the wall of the gut and prevents actual contact of food particles with the mucosa. *Third*, mucus has a low resistance to shear or slippage so that the particles can slide along the epithelium with great ease. *Fourth*, mucus causes fecal particles to adhere to each other to form the fecal masses that are expelled during a bowel movement. *Fifth*, mucus is strongly resistant to digestion by the gastrointestinal enzymes. And, *sixth*, the mucopolysaccharides of mucus have amphoteric properties (as is true of all proteins) and are therefore capable of buffering small amounts of either acids or alkalies.

In summary, mucus has the ability to allow easy slippage of food along the gastrointestinal tract and also to prevent excoriative or chemical damage to the epithelium.

SECRETION OF SALIVA

The Salivary Glands; Characteristics of Saliva. The principal glands of salivation are the *parotid, submaxillary,* and *sublingual glands,* but in addition to these are many small *buccal glands.* The daily secretion of saliva normally ranges between 1000 and 1500 milliliters, as shown in Table 43–1.

Saliva contains two different types of secretion: (1) a *serous secretion* containing *ptyalin* (an α-amylase), which is an enzyme for digesting starches, and (2) *mucous secretion* for lubricating purposes. It has a pH between 6.0 and 7.0, a favorable range for the digestive action of ptyalin.

Nervous Regulation of Salivary Secretion. Figure 43–3 illustrates the nervous pathways for regulation of salivation, showing the salivary glands to be controlled by nerve impulses from the *salivatory nuclei* located approximately at the junction of the medulla and pons. These nuclei are excited by both taste and tactile stimuli from the tongue and other areas of the mouth. Stimuli that are agreeable result in copious secretion of saliva, whereas disagreeable stimuli cause less salivation and occasionally even inhibit salivation.

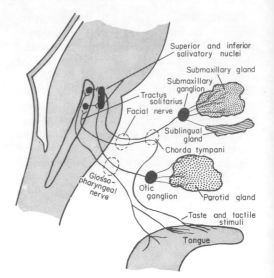

FIGURE 43–3 Nervous regulation of salivary secretion.

Salivation can also be stimulated or inhibited by impulses arriving in the salivatory nuclei from higher centers of the central nervous system. For instance, when a person eats food that he particularly likes, salivation is far greater than when he eats food that he detests. The *appetite area* of the brain that partially regulates these effects is located in close proximity to the parasympathetic centers of the anterior hypothalamus, and it functions to a great extent in response to signals from the taste and smell areas of the cerebral cortex or amygdala.

ESOPHAGEAL SECRETION

Esophageal secretions are entirely mucoid in character and function principally to provide lubrication for swallowing. Also, the mucus secreted by compound glands at the lower end of the esophagus protects the esophageal wall from being digested by gastric juices that reflux into the esophagus. Despite this protection, a peptic ulcer at times still occurs at the gastric end of the esophagus.

TABLE 43–1 DAILY SECRETION OF INTESTINAL JUICES

	Daily volume (ml.)	pH
Saliva	1200	6.0–7.0
Gastric secretion	2000	1.0–3.5
Pancreatic secretion	1200	8.0–8.3
Bile	700	7.8
Succus entericus	3000	7.8–8.0
Brunner's gland secretion	50(?)	8.0–8.9
Large intestinal secretion	60	7.5–8.0
Total	8210	

GASTRIC SECRETION

CHARACTERISTICS OF THE GASTRIC SECRETIONS

In addition to mucus-secreting cells that line the surface of the stomach, the stomach mucosa has two other types of tubular glands: the *gastric* or *fundic* glands and the *pyloric glands*. The gastric glands secrete the digestive juices, and the pyloric glands secrete almost entirely mucus for the protection of the pyloric mucosa. The gastric glands are located everywhere in the mucosa of the body and fundus of the stomach, and the pyloric glands are located in the antral portion of the stomach.

The Digestive Secretions from the Gastric Glands. The gastric glands contain three different types of cells: *mucous neck cells*, which secrete mucus; *chief cells*, which secrete digestive enzymes (*pepsin* in particular); and *parietal cells*, which secrete hydrochloric acid and which lie mainly behind the mucous neck cells or, less often, behind the chief cells. A postulated mechanism for secretion of the enzymes was given earlier in the chapter and was depicted in Figure 43–2. However, secretion of hydrochloric acid by the parietal cells involves particular problems that require further consideration as follows:

Basic mechanism of hydrochloric acid secretion. The parietal cells secrete a solution containing 150 millimols of hydrochloric acid per liter, having a pH of approximately 0.8, an extremely acidic solution. At this pH the hydrogen ion concentration is about four million times that of the arterial blood.

The parietal cell contains a system of *intracellular canaliculi*, as illustrated in Figure 43–4. It is believed that both the hydrogen ion and the chloride ion are transported actively through the wall of the canaliculus and into its lumen. Water is then pulled by osmosis into the cana-

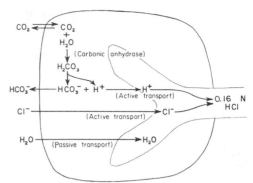

FIGURE 43–4 Postulated mechanism for the secretion of hydrochloric acid.

liculus to make the solution of hydrochloric acid.

Secretion of pepsin. The principal enzyme secreted by the chief cells is pepsin. This is formed inside the cells in the form of *pepsinogen*, which has no digestive activity. However, once pepsinogen is secreted and comes in contact with previously formed pepsin in the presence of hydrochloric acid, it is immediately activated to form pepsin.

Pepsin is an active proteolytic enzyme in a highly acid medium (optimum pH = 2.0), but above a pH of about 5 it has little proteolytic activity and soon becomes completely inactivated. Therefore, hydrochloric acid secretion is just as necessary as pepsin secretion for protein digestion in the stomach.

Secretion of Mucus in the Stomach. The pyloric glands are structurally similar to the gastric glands but contain almost no chief and parietal cells. Instead, they contain almost entirely mucous cells that are identical with the mucous neck cells of the gastric glands. All these mucous cells secrete a thin mucus, which protects the stomach wall from digestion by the gastric enzymes.

In addition, the surface of the stomach mucosa between glands has a continuous layer of mucous cells that secrete large quantities of a far more *viscid and alkaline mucus* that coats the mucosa with a mucous gel layer over 1 mm.

thick, thus providing a major shell of protection for the stomach wall.

REGULATION OF GASTRIC SECRETION BY NERVOUS AND HORMONAL MECHANISMS

Vagal Stimulation of Gastric Secretion. Nervous signals to cause gastric secretion originate in the dorsal motor nuclei of the vagi and pass via the vagus nerves to the myenteric plexus of the stomach and thence to the gastric glands. In response, these glands secrete vast quantities of both pepsin and acid. Also, vagal signals to the pyloric glands and the mucous neck cells of the gastric glands cause greatly increased secretion of mucus.

Stimulation of Gastric Secretion by Gastrin. When food enters the stomach, it causes the antral portion of the stomach mucosa to secrete the hormone gastrin, a heptadecapeptide. The food causes release of this hormone in two ways: (1) The actual bulk of the food distends the stomach, and this in some way not yet understood causes the hormone gastrin to be released from the antral mucosa. (2) Certain substances called secretagogues — such as food extractives, partially digested proteins, alcohol (in low concentration), caffeine, and so forth — also cause gastrin to be liberated from the antral mucosa.

Gastrin is absorbed into the blood and carried to the gastric glands where it stimulates the parietal cells to increase their rate of hydrochloric acid secretion as much as eight-fold, and the chief cells to increase their rate of enzyme secretion two- to four-fold.

The rate of secretion in response to gastrin is somewhat less than to vagal stimulation, 200 ml. per hour in contrast to about 500 ml. per hour, indicating that the gastrin mechanism is a less potent mechanism for stimulation of stomach secretion than is vagal stimulation. However, the gastrin mechanism usually continues for several hours in contrast to a much shorter period of time for vagal stimulation. Therefore, as a whole, it is likely that the gastrin mechanism is equally as important as, if not more important than, the vagal mechanism for control of gastric secretion.

Feedback inhibition of gastric acid secretion. When the acidity of the gastric juices increases to a pH of 2.0, the gastrin mechanism for stimulating gastric secretion becomes totally blocked. This effect probably results from two different factors. *First,* greatly enhanced acidity depresses or blocks the extraction of gastrin itself from the antral mucosa. *Second,* the acid seems to extract an inhibitory hormone from the gastric mucosa or to cause an inhibitory reflex that inhibits gastric acid secretion.

Obviously, this feedback inhibition of the gastric glands protects the stomach against excessively acid secretions, which would readily cause peptic ulceration.

PANCREATIC SECRETION

Characteristics of Pancreatic Juice. Pancreatic juice contains enzymes for digesting all three major types of food: proteins, carbohydrates, and fats.

The proteolytic enzymes are *trypsin,* two different *chymotrypsins, carboxypolypeptidase, ribonuclease,* and *deoxyribonuclease.* The first two of these split whole and partially digested proteins, while carboxypolypeptidase splits a particular peptide linkage of small peptides. The nucleases split the two types of nucleic acids: ribonucleic and deoxyribonucleic acids.

The digestive enzyme for carbohydrates is *pancreatic amylase,* which hydrolyzes starches, glycogen and most other carbohydrates, except cellulose, to form disaccharides.

The enzyme for fat digestion, *pancreatic lipase,* is capable of hydrolyzing neutral fat into glycerol and fatty acids.

The proteolytic enzymes as synthesized in the pancreatic cells are in the inactive forms *trypsinogen, chymotrypsinogen,* and *procarboxypolypeptidase,* which are all enzymatically inactive. These become activated only after they are secreted into the intestinal tract. Trypsinogen is activated by an enzyme called *enterokinase,* which is released from the intestinal mucosa whenever chyme comes in contact with the mucosa. Also, trypsinogen can be activated to a slight extent by trypsin that has already been formed. Chymotrypsinogen is activated by trypsin to form chymotrypsin, and procarboxypolypeptidase is presumably activated in some similar manner.

Secretion of Trypsin Inhibitor. Fortunately, the same cells that secrete the proteolytic enzymes into the acini of the pancreas secrete simultaneously another substance called *trypsin inhibitor.* This substance prevents activation of trypsin both inside the secretory cells and in the acini and ducts of the pancreas. Since it is trypsin that activates the other pancreatic proteolytic enzymes, trypsin inhibitor also prevents the subsequent activation of all these.

However, when the pancreas becomes severely damaged or when a duct becomes blocked, large quantities of pancreatic secretion become pooled in the damaged areas of the pancreas. Under these conditions, the effect of trypsin inhibitor is sometimes overwhelmed, in which case the pancreatic secretions rapidly become activated and literally digest the entire pancreas within a few hours, giving rise to the condition called *acute pancreatitis.* This usually is lethal because of the accompanying shock, and even if not lethal it leads to a lifetime of pancreatic insufficiency.

REGULATION OF PANCREATIC SECRETION

Pancreatic secretion, like gastric secretion, is regulated by both nervous and hormonal mechanisms. However, in this case, hormonal regulation is by far the more important of the two.

Nervous Regulation. When the cephalic and gastric phases of stomach secreation occur, parasympathetic impulses are simultaneously transmitted along the vagus nerves to the pancreas, resulting in secretion of moderate amounts of enzymes into the pancreatic acini. However, little secretion flows through the pancreatic ducts to the intestine because little water and electrolytes are secreted along with the enzymes. Therefore, most of the enzymes are temporarily stored in the acini.

Hormonal Regulation. After food enters the small intestine, pancreatic secretion becomes copious, mainly in response to the hormone *secretin.* In addition, a second hormone, *pancreozymin,* causes greatly increased secretion of enzymes.

Secretin stimulation of pancreatic secretion. Secretin is a small polypeptide that is present in the mucosa of the upper small intestine in the inactive form *prosecretin.* When chyme enters the intestine, it causes release and activation of secretin, which is subsequently absorbed into the blood. The one constituent of chyme that causes greatest secretin release is hydrochloric acid, though almost any type of food will cause at least some release.

Secretin causes the pancreas to secrete large quantities of fluid containing a high concentration of sodium bicarbonate (up to 145 mEq./liter). This immediately reacts with the hydrochloric acid from the stomach, raising the pH of the chyme up to 7.0 to 8.0. Since the mucosa of the small intestine cannot withstand the intense digestive properties of gastric juice, this is a highly important protective mechanism against the development of duodenal ulcers.

A *second* importance of this secretion by the pancreas is to provide an appropriate pH for action of the pancreatic enzymes. All of these function optimally in a slightly alkaline or neutral medium. The pH of the secretion averages 8.0.

Pancreozymin—"Ecbolic" secretion by the pancreas. The presence of food in the upper small intestine also causes a second hormone, pancreozymin, to be released from the mucosa. This results especially from the presence of proteoses and peptones, which are products of partial protein digestion. Pancreozymin, like secretin, passes by way of the blood to the pancreas but, instead of causing copious sodium bicarbonate secretion, causes secretion of large quantities of digestive enzymes, which is similar to the effect of vagal stimulation.

Figure 43–5 summarizes the overall regulation of pancreatic secretion. The total amount secreted each day is about 1200 ml.

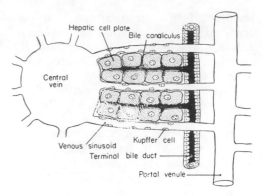

FIGURE 43–6 Basic structure of the liver lobule, showing the hepatic cellular plates, the blood vessels, and the bile ducts.

SECRETION OF BILE BY THE LIVER

PHYSIOLOGIC ANATOMY OF THE LIVER

The basic functional unit of the liver is the liver lobule, a cylindrical structure constructed around a *central vein* that empties into the hepatic veins and thence into the vena cava. The lobule itself is composed principally of many *hepatic cellular plates* (two of which are illustrated in Fig. 43–6) that radiate centrifugally from the central vein like spokes in a wheel. Each hepatic plate is usually two cells thick, and between the adjacent cells lie small *bile canaliculi* that empty into *terminal bile ducts* lying in the septa between the adjacent liver lobules.

Also in the septa are small *portal*

venules that receive their blood from the portal veins. From these venules blood flows into flat, branching *hepatic sinusoids* that lie between the hepatic plates, and thence into the central vein. Thus, the hepatic cells are exposed on one side to portal venous blood and on the other side to bile canaliculi.

In addition to the portal venules, *hepatic arterioles* are also present in the interlobular septa. These arterioles supply arterial blood to the septal tissues, and many of the small arterioles also empty directly into the hepatic sinusoids.

The venous sinusoids are lined with two types of cells: (1) typical *endothelial cells* and (2) large *Kupffer cells*, which are reticuloendothelial cells capable of phagocytizing bacteria and other foreign matter in the blood. The endothelial lining of the venous sinusoids has extremely large pores, some of which are almost 1 micron in diameter.

SECRETION OF BILE AND FUNCTIONS OF THE BILIARY TREE

Bile is secreted by the hepatic cells into the minute *bile canaliculi* that lie between the double layer of cells in the hepatic plates. The bile then flows peripherally toward the interlobular septa, where the canaliculi empty into *terminal*

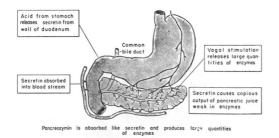

FIGURE 43–5 Regulation of pancreatic secretion.

bile ducts, then into progressively larger ducts, finally reaching the *hepatic duct* and *common bile duct* from which the bile either empties directly into the duodenum or is diverted into the gallbladder.

Bile contains no digestive enzyme and is important for digestion only because of the presence of bile salts which (1) help to emulsify fat globules so that they can be digested by the intestinal lipases and (2) help render the end-products of fat digestion soluble, so that they can be absorbed through the gastrointestinal mucosa into the lymphatics.

Bile is secreted continually by the liver rather than intermittently as in the case of most other gastrointestinal secretions, but the bile is stored in the gallbladder until it is needed in the gut. The gallbladder then empties the bile into the intestine in response to *cholecystokinin*, a hormone extracted from the intestinal mucosa by fats in the intestine. Therefore, bile becomes available to aid in the processes of fat digestion.

The rate of bile secretion can be altered in response to four different effects: (1) vagal stimulation can sometimes more than double the secretion, (2) secretin has a moderate effect on liver secretion, causing as much as 80 per cent increase in bile production (but without any increase in bile acid production) at the same time that it also causes the hydrelatic response of the pancreas, (3) the greater the liver blood flow (up to a point), the greater the secretion, and (4) the presence of large amounts of bile salts in the blood increases the rate of liver secretion proportionately. Most of the bile salts secreted in the bile are reabsorbed by the intestines and then resecreted by the liver. When the salts are lost to the exterior through a bile fistula rather than being reabsorbed, the rate of bile secretion becomes reduced several fold.

Storage of Bile in the Gallbladder. Bile is normally stored in the gallbladder until needed in the duodenum. The total secretion each day is some 600 to 700 ml., and the maximum volume of the gallbladder is only 40 to 70 ml. Nevertheless, at least 12 hours' bile secretion can be stored in the gallbladder because water, sodium, chloride, and most other small electrolytes are continually absorbed by the gallbladder mucosa, concentrating the other bile constituents, including the bile salts, cholesterol, and bilirubin. Bile is normally concentrated about five-fold, but it can be concentrated up to a maximum of about 10-fold.

Emptying of the gallbladder — Function of cholecystokinin. Two basic conditions are necessary for the gallbladder to empty: (1) The sphincter of Oddi must relax to allow bile to flow from the common bile duct into the duodenum and (2) the gallbladder itself must contract to provide the force required to move the bile along the common duct. After a meal, particularly one that contains a high concentration of fat, both these effects take place in the following manner:

First, the fat in the food entering the small intestine extracts a hormone called *cholecystokinin* from the intestinal mucosa, especially from the upper regions of the small intestine. The cholecystokinin in turn is absorbed into the blood and on passing to the gallbladder causes specific contraction of the gallbladder muscle. This provides the pressure that forces bile toward the duodenum.

Second, vagal stimulation associated with the cephalic phase of gastric secretion or with various other reflexes causes additional weak contraction of the gallbladder, which helps to force the bile from the gallbladder into the duodenum.

Third, when the gallbladder contracts, the sphincter of Oddi becomes inhibited, this effect resulting from either a neurogenic or a myogenic reflex from the gallbladder to the sphincter of Oddi.

Fourth, the presence of food in the duodenum causes the degree of peristalsis in the duodenal wall to increase. Each time a peristaltic wave travels toward the sphincter of Oddi, this sphincter, along with the adjacent intestinal wall, momentarily relaxes because of the phenomenon of "receptive relaxation," and, if the bile in the common bile duct is under sufficient pressure, a small quantity of the bile squirts into the duodenum.

In summary, the gallbladder empties its store of concentrated bile into the duodenum mainly in response to the cholecystokinin stimulus. When fat is not in the meal, the gallbladder empties poorly, but when adequate quantities of fat are present, the gallbladder empties completely in about one hour.

Figure 43–7 summarizes the secretion of bile, its storage in the gallbladder, and its release from the bladder to the gut.

Composition of Bile. The most abundant substance secreted in the bile is the *bile salts*, but also secreted or excreted in large concentrations are *bilirubin, cholesterol, fatty acids,* and the usual *electrolytes* of plasma. In the concentrating process in the gallbladder, water and large portions of the electrolytes are reabsorbed by the gallbladder mucosa, but essentially all the other constituents, including especially the bile salts and lipid substances such as cholesterol, are not reabsorbed and therefore become highly concentrated in the gallbladder bile.

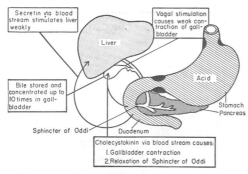

FIGURE 43–7 Mechanisms of liver secretion and gallbladder emptying.

THE BILE SALTS AND THEIR FUNCTION

The liver cells form about 1 gram of *bile salts* daily. The precursor of the bile salts is *cholesterol*, which is either supplied in the diet or synthesized in the liver cells during the course of fat metabolism and then is converted to *glycocholic acid* and, to a lesser extent, *taurocholic acid.* The salts of these acids are secreted in the bile.

The bile salts have two important actions in the intestinal tract. First, they have a detergent action on the fat particles in the food, which decreases the surface tension of the particles and allows the agitation in the intestinal tract to break the fat globules into minute sizes. This is called the *emulsifying* or *detergent function* of bile salts. Second, and even more important than the emulsifying function, bile salts help in the absorption of fatty acids, monoglycerides, cholesterol, and other lipids from the intestinal tract, and this is called their *hydrotropic function.* It is believed that the bile salt ions become physically adsorbed to the lipids, and the electrical charges of these ions presumably then increase the solubility of the lipids, thereby allowing ready passage through the intestinal mucosa. Without the presence of bile salts in the intestinal tract, up to 40 per cent of the lipids are lost into the stools, and the person often develops a metabolic deficit due to this nutrient loss.

Enterohepatic Circulation of Bile Salts. Approximately 94 per cent of the bile salts are reabsorbed by the intestinal mucosa along with absorption of the fats. After passing through the mucosa, the bile salts separate from the fats and are resecreted by the liver into the bile. In this way about 94 per cent of all the bile salts are recirculated into the bile, so that on the average these salts make the entire circuit some 18 times before being carried out in the feces. The small quantities of bile salts lost into the feces are replaced by new amounts formed continually by the liver cells. This recirculation of the bile salts is called the *enterohepatic circulation*.

SECRETION OF CHOLESTEROL; GALLSTONE FORMATION

In the process of secreting the bile salts, about one-tenth as much cholesterol as salts is also secreted into the bile. No specific function is known for the cholesterol in the bile, and it is presumed that it is simply a by-product of bile salt formation and secretion.

Cholesterol is almost insoluble in pure water, but the bile salts, fatty acids, and lecithin in bile exhibit a hydrotropic action on the cholesterol to make it soluble. When the bile becomes concentrated in the gallbladder, all the hydrotropic substances become concentrated along with the cholesterol, which keeps the cholesterol in solution. Under abnormal conditions, however, the cholesterol may precipitate, resulting in the formation of *gallstones*. The different conditions that can cause cholesterol precipitation are: (1) too much absorption of water from the bile, (2) too much absorption of bile salts, fatty acids, and lecithin from the bile, (3) too much secretion of cholesterol in the bile, and (4) inflammation of the epithelium of the gallbladder. The latter two of these require special explanation, as follows:

The amount of cholesterol in the bile is determined partly by the quantity of fat that the person eats, for the hepatic cells synthesize cholesterol as one of the products of fat metabolism in the body. For this reason, persons on a high fat diet over a period of many years are prone to develop gallstones.

Inflammation of the gallbladder epithelium often results from low grade chronic infection; this changes the absorptive characteristics of the gallbladder mucosa, sometimes allowing excessive absorption of water, bile salts, or other substances that are necessary to keep the cholesterol in solution. As a result, cholesterol begins to precipitate, usually forming many small crystals of cholesterol on the surface of the inflamed mucosa. These, in turn, act as nidi for further precipitation of cholesterol, and the crystals grow larger and larger. Occasionally tremendous numbers of sand-like stones develop, but much more frequently they coalesce to form a few large gallstones, or even a single stone that fills the entire gallbladder.

SECRETIONS OF THE SMALL INTESTINE

SECRETION OF MUCUS BY BRUNNER'S GLANDS AND BY MUCOUS CELLS OF THE INTESTINAL SURFACE

An extensive array of compound mucous glands, called *Brunner's glands*, is located in the first few centimeters of the duodenum, mainly between the pylorus and the papilla of Vater where the pancreatic juices and bile empty into the duodenum. These glands secrete large quantities of mucus mainly in response to direct tactile stimuli or irritating stimuli of the overlying mucosa. The function of the mucus secreted by Brunner's glands is to protect the duodenal wall from digestion by the gastric juice.

Mucus is also secreted in large quantities by goblet cells located extensively over the surface of the intestinal mucosa. This secretion results principally from direct tactile or chemical stimulation of the mucosa by the chyme or from local nervous reflexes.

SECRETION OF THE INTESTINAL DIGESTIVE JUICES — THE CRYPTS OF LIEBERKÜHN

Located on the entire surface of the small intestine, with the exception of the Brunner's gland area of the duodenum, are small crypts called crypts of Lieberkühn. The intestinal secretions are formed by the epithelial cells in these crypts at a rate of about 3000 ml. per day. The secretions are almost pure extracellular fluid, and they have a neutral pH in the range of 6.5 to 7.5. The secretions are rapidly reabsorbed by the villi. This circulation of fluid from the crypts to the villi supplies a watery vehicle for absorption of substances from the small intestine, which is one of the primary functions of the small intestine.

Enzymes in the Small Intestinal Secretion. When secretions of the small intestine are collected without cellular debris, they have almost no enzymes. However, the epithelial cells of the mucosa contain large quantities of digestive enzymes and presumably digest food substances *while* they are being absorbed through the epithelium. These enzymes are the following: (1) several different *peptidases* for splitting polypeptides into amino acids, (2) four enzymes for splitting disaccharides into monosaccharides — *sucrase, maltase, isomaltase*, and *lactase*, (3) *intestinal lipase* for splitting neutral fats into glycerol and fatty acids, and (4) a small amount of intestinal amylase for splitting carbohydrates into disaccharides.

The epithelial cells deep in the crypts of Lieberkühn continually undergo mitosis, and the new cells gradually migrate along the basement membrane upward out of the crypts toward the tips of the villi where they are finally shed into the intestinal secretions. The life cycle of an intestinal epithelial cell is approximately 48 hours. This rapid growth of new cells allows immediate repair of any excoriation that occurs in the mucosa.

Regulation of Small Intestinal Secretion. By far the most important means for regulating small intestinal secretion is various local intramural nervous reflexes or direct stimuli. Especially important is distention of the small intestine, which causes copious secretion from the crypts of Lieberkühn. In addition, tactile or irritative stimuli can result in intense secretion. Therefore, for the most part, secretion in the small intestine occurs simply in response to the presence of food in the intestine — the greater the amount, the greater the secretion.

SECRETIONS OF THE LARGE INTESTINE

Mucus Secretion. The mucosa of the large intestine, like that of the small intestine, is lined with crypts of Lieberkühn, but the epithelial cells contain almost no enzymes. Instead, they are lined almost entirely by goblet cells. Also, on the surface epithelium of the large intestine are large numbers of goblet cells dispersed among the other epithelial cells.

Therefore, the only significant secretion in the large intestine is mucus. Its rate of secretion is regulated principally by direct, tactile stimulation of the goblet cells on the surface of the mucosa and perhaps by local nervous reflexes. Also, stimulation of the *nervi erigentes*, which carry the parasympathetic innervation to the distal half of the large intestine, causes marked increase in the secretion of mucus. This occurs along with an increase in motility. Therefore, during extreme parasympathetic stimulation, often caused by severe emotional disturbances, so much mucus may be secreted into the large intestine that the person has a bowel movement of ropy mucus as often as every 30 minutes; the mucus contains little or no fecal material.

Mucus in the large intestine obviously protects the wall against excoriation, but, in addition, it provides the adherent qualities for holding fecal matter together. Furthermore, it protects the intestinal wall from the great amount of bacterial activity that takes place inside the feces.

Secretion of Water and Electrolytes in Response to Irritation. Whenever a segment of the large intestine becomes intensely irritated, such as occurs when bacterial infection becomes rampant during *gastroenteritis*, the mucosa then secretes large quantities of water and electrolytes in addition to the normal viscid solution of mucus. This acts to dilute the irritating factors and to cause rapid movement of the feces toward the anus. The usual result is *diarrhea*, with loss of large quantities of water and electrolytes, but also earlier recovery from the disease than would otherwise occur.

REFERENCES

Ciba Foundation Symposium: The Exocrine Pancreas: Normal and Abnormal Function. Boston, Little, Brown and Company, 1962.

Davenport, H. W.: Physiological structure of the gastric mucosa. *In* Handbook of Physiology. Baltimore, The Williams & Wilkins Co., 1967, Sec. VI, Vol. II, p. 759.

Farrar, G. E., Jr., and Bower, R. J.: Gastric juice and secretion: physiology and variations in disease. *Ann. Rev. Physiol.,* 29:141, 1967.

Gregory, R. A.: Secretory mechanisms of the digestive tract. *Ann. Rev. Physiol.,* 27:395, 1965.

Grossman, M. I.: Neuronal and hormonal stimulation of gastric secretion of acid. *In* Handbook of Physiology. Baltimore, The Williams & Wilkins Co., 1967, Sec. VI, Vol. II, p. 835.

Taylor, W. H.: Proteinases of the stomach in health and disease. *Physiol. Rev.,* 42:519, 1962.

Uvnas, B.: Gastrin and vagus. *Proc. Int. Union of Physiol. Sciences,* 6:189, 1968.

CHAPTER 44

DIGESTION AND ABSORPTION IN THE GASTROINTESTINAL TRACT; AND DISORDERS OF DIGESTION AND ABSORPTION

The foods on which the body lives, with the exception of small quantities of vitamins and minerals, can be classified as carbohydrates, fats, and proteins. However, these generally cannot be absorbed in their natural form through the gastrointestinal mucosa and, for this reason, are useless as nutrients without the preliminary process of digestion. Therefore, the present chapter discusses, first, the processes by which carbohydrates, fats, and proteins are digested into small enough compounds for absorption and, second, the mechanisms by which the digestive end-products as well as water, electrolytes, and other substances are absorbed.

DIGESTION OF THE VARIOUS FOODS

Hydrolysis as the Basic Process of Digestion. Almost all the carbohydrates of the diet are large *polysaccharides* or *disaccharides*, which are combinations of many *monosaccharides* bound to each other by the process of *condensation.* This means that a hydrogen ion is removed from one of the monosaccharides while a hydroxyl ion is removed from the next one; the two monosaccharides then combine with each other at these sites of removal, and the hydrogen and hydroxyl ions combine to form water. When the carbohydrates are digested back into monosaccharides, specific enzymes return the hydrogen and hydroxyl ions to the polysaccharides and thereby separate the monosaccharides from each other. This process, called hydrolysis, is the following:

$$R_1R_2 + H_2O \rightarrow R_1OH + R_2H$$

Almost the entire fat portion of the diet consists of triglycerides (neutral fats), which are combinations of three *fatty acid* molecules condensed with a single *glycerol* molecule. In the process of condensation, three molecules of water had been removed. Digestion of the triglycerides consists of the reverse process, the fat-digesting enzymes returning the three molecules of water to each molecule of neutral fat and thereby splitting the fatty acid molecules away from the glycerol. Here again, the process is one of hydrolysis.

Finally, proteins are formed from *amino acids* that are bound together by means of *peptide linkages*, which are also processes of condensation. Digestion of proteins, therefore, also involves hydrolysis, the proteolytic enzymes returning the water to the protein molecules to split them into their constituent amino acids.

Therefore, the chemistry of digestion is really simple, for in the case of all three major types of food, the same basic process of *hydrolysis is involved*. The only difference lies in the enzymes required to promote the reactions for each type of food.

DIGESTION OF CARBOHYDRATES

Only three major sources of carbohydrates exist in the normal human diet. These are sucrose, which is the disaccharide known popularly as cane sugar; lactose, which is a disaccharide in milk; and starches, which are large polysaccharides present in almost all foods and particularly in the grains.

Figure 44–1 gives a schema for digestion of the principal carbohydrates. This shows that the starches are first

hydrolyzed to maltose (or isomaltose), which is a disaccharide. Then this disaccharide, along with the other major disaccharides, lactose and sucrose, are hydrolyzed into the monosaccharides *glucose, galactose,* and *fructose.*

Hydrolysis of starches begins in the mouth under the influence of the enzyme *ptyalin,* which is secreted mainly in the saliva from the parotid gland. The hydrochloric acid of the stomach provides a slight amount of additional hydrolysis. Finally, the major share of hydrolysis occurs in the upper part of the small intestine under the influence of the enzyme *pancreatic amylase.*

The enzymes (*lactase, sucrase, maltase,* and *isomaltase*) for splitting the disaccharides are located in the brush border of the epithelial cells. The disaccharides are digested into monosaccharides as they come in contact with this border or as they diffuse into the microvilli that constitute the brush. The digestive products, the monosaccharides glucose, galactose, and fructose, are then immediately absorbed into the portal blood.

DIGESTION OF FATS

By far the most common fats of the diet are the neutral fats, also known as *triglycerides,* each molecule of which is composed of a glycerol nucleus and three fatty acids. Neutral fat is found in food of both animal origin and plant origin.

In the usual diet are also small quantities of phospholipids, cholesterol, and cholesterol esters. The phospholipids

FIGURE 44–1 Digestion of carbohydrates.

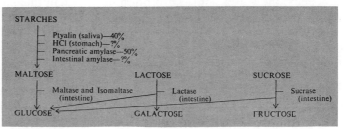

and cholesterol esters contain fatty acid and, therefore, can be considered to be fats themselves. Cholesterol, on the other hand, is a sterol compound containing no fatty acid, but it does exhibit some of the physical and chemical characteristics of fats; it is derived from fats, and it is metabolized similarly to fats. Therefore, cholesterol is considered from a dietary point of view to be a fat.

Though a minute amount of fat can be digested in the stomach under the influence of gastric lipase, 95 to 99 per cent of all fat digestion occurs in the small intestine mainly under the influence of *pancreatic lipase.*

Emulsification of Fat by Bile Salts. The first step in fat digestion is to break the fat globules into small sizes so that the digestive enzymes, which are not fat soluble, can act on the globule surfaces. This process is called emulsification of the fat, and it is achieved under the influence of bile salts that are secreted in the bile by the liver. The bile salts act as a detergent, greatly decreasing the interfacial tension of the fat. With a low interfacial tension, the gastrointestinal mixing movements can gradually break the globules of fat into finer and finer particles, with the total surface area of the fat increasing by a factor of two every time the diameters of the fat globules are decreased by a factor of two.

Digestion of Fat by Pancreatic Lipase and Enteric Lipase. Under the influence of *pancreatic lipase, most of the* fat is split into *fatty acids* and *glycerol.* Though a smaller portion of the fat is digested only to the mono- and diglyceride stages, as shown in Figure 44–2, the further the process of fat hydrolysis proceeds, the better is the absorption of the fats.

The epithelial cells of the small intestine contain a small quantity of lipase known as *enteric lipase.* This probably causes a very slight additional amount of fat digestion.

DIGESTION OF PROTEINS

The dietary proteins are derived almost entirely from meats and vegetables, and they are digested primarily in the stomach and upper part of the small intestine.

As illustrated in Figure 44–3, protein digestion begins in the stomach, the enzyme *pepsin* (and a smaller quantity of a similar enzyme called *gastricsin*) splitting the proteins into *proteoses, peptones,* and *large polypeptides.* These enzymes function only in a highly acid medium; they function best at a pH of 2 to 3. Therefore, the hydrochloric acid secreted in the stomach is essential for this digestive process.

Pepsin is especially important for its ability to digest collagen, an albuminoid that is affected little by other digestive enzymes. Since collagen is a major constituent of the fibrous tissue in meat, it is essential that this be digested so that the remainder of the meat can be attacked by the digestive enzymes.

The proteins are further digested in the upper part of the small intestine under the influence of *trypsin, chymotrypsin,* and *carboxypolypeptidases.* The final product of this digestion is mainly *small polypeptides* plus a few *amino acids.*

Finally, the small polypeptides are digested into amino acids when they come in contact with the epithelial cells of the small intestine. These cells contain several enzymes (*amino polypeptidase* and *dipeptidases*) that convert the remaining protein products into *amino acids.*

When food has been properly masticated and is not eaten in too large a quantity at any one time, about 98 per cent of all the protein finally becomes amino acids.

Fat	(Bile + Agitation)	Emulsified fat
Emulsified fat	(Pancreatic lipase) / (Enteric lipase)	Fatty acids, Glycerol } 60% (?) / Glycerides 40% (?)

FIGURE 44–2 Digestion of fats.

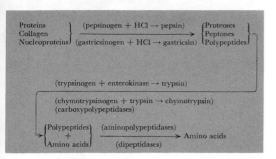

FIGURE 44–3 Digestion of proteins.

BASIC PRINCIPLES OF GASTROINTESTINAL ABSORPTION

ANATOMICAL BASIS OF ABSORPTION

The total quantity of fluid that must be absorbed each day is equal to the ingested fluid (about 1.5 liters) plus that secreted in the various gastrointestinal secretions (about 8.5 liters), approximately 10 liters. About 9.5 liters of this is absorbed in the small intestine, leaving only 0.5 liter to pass through the ileocecal valve into the colon each day.

The stomach is a poor absorptive area of the gastrointestinal tract. Only a few highly lipid-soluble substances, such as alcohol and some drugs, can be absorbed in small quantities.

The Absorptive Surface of the Intestinal Mucosa — The Villi. The absorptive surface of the intestinal mucosa has many folds called *valvulae conniventes,* which increase the surface area of the absorptive mucosa about three-fold. Also, located over the entire surface of the small intestine, from approximately the point at which the common bile duct empties into the duodenum down to the ileocecal valve, are literally millions of small *villi,* illustrated in Figure 44–4, which project about 1 mm. from the surface of the mucosa. These enhance the absorptive area another 10-fold. And, finally, the intestinal epithelial cells are characterized by a brush border, consisting of about 600 *microvilli* 1μ in length and 0.1μ in diameter, protruding from each cell; these are illustrated in the electron micrograph in Figure 44–5.

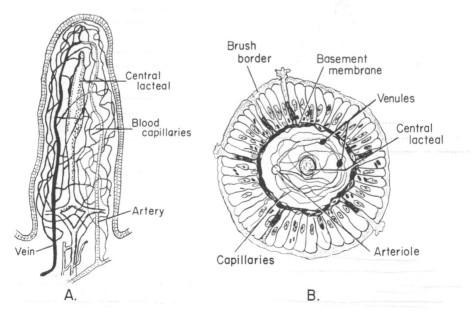

FIGURE 44–4 Functional organization of the villus: (*A*) Longitudinal section. (*B*) Cross-section showing the epithelial cells and basement membrane.

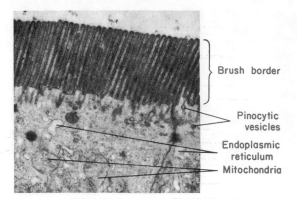

Brush border

Pinocytic
vesicles

Endoplasmic
reticulum

Mitochondria

FIGURE 44–5 Brush border of the gastro-intestinal epithelial cell, showing, also, pinocytic vesicles, mitochondria, and endoplasmic reticulum lying immediately beneath the brush border. (Courtesy of Dr. William Lockwood.)

This increases the surface area exposed to the intestinal materials another 20-fold. Thus, the combination of these factors increases the absorptive area of the mucosa about 600-fold, in all making a tremendous absorptive area of about 550 square meters for the entire small intestine.

Figure 44–4 illustrates the general organization of a villus, emphasizing especially the advantageous arrangement of the vascular system for absorption of fluid and dissolved material into the portal blood, and the arrangement of the *central lacteal* for absorption into the lymph.

BASIC MECHANISMS OF ABSORPTION

Absorption through the gastrointestinal mucosa occurs by *active transport* and by *diffusion*, as is also true for other membranes. The physical principles of these processes were explained in Chapter 4.

Briefly, active transport imparts energy to the substance as it is being transported for the purpose of concentrating it on the other side of the membrane or for moving it against an electrical potential. On the other hand, the term "diffusion" means simply transport of substances through the membrane as a result of molecular movement *along*, rather than against, an electrochemical gradient.

ABSORPTION IN THE SMALL INTESTINE

ABSORPTION OF NUTRIENTS

Absorption of Carbohydrates. Essentially all the carbohydrates are absorbed in the form of monosaccharides, and transport of these is an active process that is capable of transporting about one gram of glucose per minute. The order of preference for transporting different monosaccharides, and their relative rates of transport in comparison with glucose are:

Galactose	1.1
Glucose	1.0
Fructose	0.4
Mannose	0.2
Xylose	0.15
Arabinose	0.1

Mechanism of monosaccharide absorption. We still do not know the precise mechanism of monosaccharide absorption, but we do know that glucose transport becomes blocked whenever sodium transport is blocked. Therefore, it is assumed that the energy required for glucose transport is actually provided by the sodium transport system. That is, sodium is actively transported through the mucosal cells into the subepithelial spaces, and some coupled reaction between sodium transport and glucose transport provides the energy for moving glucose through the intestinal wall. Because of this dependence of glucose transport on sodium transport, the mechanism for glucose transport is said to be *secondary active transport,* in contradistinction to *primary active transport,* the term applied to the sodium type of transport mechanism.

Absorption of Proteins. Essentially all proteins are absorbed in the form of amino acids. However, minute quantities of dipeptides are also absorbed, and extremely minute quantities of whole proteins can be absorbed at times, probably by the process of pinocytosis described in Chapter 4.

Absorption of amino acids, like absorption of glucose, is an active process; certain types of amino acids are absorbed selectively and certain ones interfere with the absorption of others, illustrating that common carrier systems exist. Finally, metabolic poisons block the absorption of amino acids in the same way that they block the absorption of glucose.

Absorption of amino acids through the intestinal mucosa can occur far more rapidly than can protein digestion in the lumen of the intestine. As a result, the normal rate of absorption is determined not by the rate at which they can be absorbed but by the rate at which they can be released from the proteins during digestion. For these reasons, essentially no free amino acids can be found in the intestine during digestion — that is, they are absorbed as rapidly as they are formed.

Basic mechanisms of amino acid transport. As is true for monosaccharide absorption, very little is known about the basic mechanisms of amino acid transport. However, at least three different carrier systems transport different amino acids — one transports *neutral amino acids,* a second transports *basic amino acids,* and a third has

specificity for the two amino acids *proline* and *hydroxyproline.* Amino acid transport, like monosaccharide transport, ceases when sodium transport is blocked. Therefore, amino acid transport is also a type of secondary active transport, the energy for which is provided by the transport of sodium through the membrane.

Absorption of Fats. Fats are believed to be absorbed through the intestinal membrane principally in the form of fatty acids and monoglycerides, though a few diglycerides and triglycerides are also absorbed.

The mechanism for absorption of fatty acids appears to be the following: First, the fatty acid molecule, being highly lipid-soluble, becomes dissolved in the membrane of the brush border of the epithelial cell and diffuses to the interior of the cell. There it comes in contact with the endoplasmic reticulum, which uses the fatty acid molecules to resynthesize triglycerides. These triglycerides are then conducted through the tubules of the endoplasmic reticulum to the sides of the epithelial cell, where they are discharged through the intercellular channel into the submucosal fluids.

One of the important features of this mechanism for transporting fatty acids and glycerides is that the active chemical process occurs at the membrane of the endoplasmic reticulum rather than at the cell membrane itself where active transport of monosaccharides and amino acids occurs.

Chylomicrons, and transport of fat in the lymph. The triglycerides (and other fats) are discharged by the epithelial cells into the submucosal fluids in the form of minute fat droplets about 0.5 micron in diameter. Small quantities of phospholipids and cholesterol are present in these droplets along with the triglycerides, and the surfaces of the droplets are coated with a layer of protein, which makes them hydrophilic, allowing a reasonable degree of suspension stability in the extracellular fluids.

These small particles are called chylomicrons.

From beneath the epithelial cells the chylomicrons wend their way into the central lacteals of the villi and from here are pumped by the lymphatic pump upward through the thoracic duct to be emptied into the great veins of the neck. Between 80 and 90 per cent of all fat absorbed from the gut is absorbed in this manner and transported to the blood by way of the thoracic lymph in the form of chylomicrons.

Much smaller quantities of fatty acids, not over 10 to 20 per cent of the total, are absorbed directly into the portal blood rather than being converted into triglycerides and absorbed into the lymphatics. In general, the shorter the fatty acid chain, the greater is the proportion absorbed in this way.

Hydrotropic action of bile salts. Bile salts are known to increase the rate of absorption of fats. This effect results from the ability of bile salts to promote emulsification of the fatty acid and glyceride particles to extremely minute sizes so that they can become entrapped by the membrane of the brush border of the epithelial cells.

ABSORPTION OF WATER AND ELECTROLYTES

Absorption of Electrolytes. The monovalent electrolytes—sodium, potassium, chloride, nitrate, and bicarbonate—are easily absorbed through the intestinal membrane, while most of the polyvalent electrolytes, such as calcium, magnesium, and sulfate, are poorly absorbed.

Active absorption of sodium, chloride, and other ions. Sodium is actively absorbed in large quantity by the mucosa of the small intestine, and at least some chloride is also actively absorbed. However, in addition, transport of sodium across the epithelial membrane creates an electrical potential that pulls additional chloride ions through the membrane simultaneously. The basic mechanisms of this transport were discussed in Chapter 4.

Other ions that are actively absorbed include calcium, iron, hydrogen, potassium, magnesium, phosphate, and bicarbonate. Calcium absorption is discussed in Chapter 53, and iron absorption is explained in Chapter 8.

Absorption of Water. Water is absorbed by simple diffusion—by osmosis. That is, as the electrolytes and nutrients are absorbed, the intestinal fluids become hypotonic, which then causes water to be absorbed by osmosis into the more concentrated submucosal fluids.

ABSORPTION IN THE LARGE INTESTINE; FORMATION OF THE FECES

Approximately 500 ml. of chyme pass through the ileocecal valve into the large intestine each day. Most of the water and electrolytes in this are absorbed in the colon, leaving only about 100 ml. of fluid to be excreted in the feces.

Essentially all the absorption in the large intestine occurs in the proximal half of the colon, giving this portion the name *absorbing colon,* while the distal colon functions principally for storage and is therefore called the *storage colon.*

The mucosa of the large intestine is capable of absorbing sodium and some other ions actively. Because of this, almost no ions (except excesses of the divalent positive ions) are lost in the feces. Also, absorption of the electrolytes causes water to "follow" by the process of osmosis, as discussed above in relation to absorption in the small intestine.

Composition of the Feces. The feces normally are about three-fourths water and one-fourth solid matter composed of about 30 per cent dead bacteria, 10 to 20 per cent fat, 10 to 20 per cent inorganic matter, 2 to 3 per cent protein, and 30 per cent undigested roughage of the food and dried constituents of digestive juices, such as bile pigment and sloughed epithelial cells. The large amount of fat derives from unabsorbed fatty acids from the diet, fat formed by bacteria, and fat in the sloughed epithelial cells.

The brown color of feces is caused by *stercobilin* and *urobilin*, which are derivatives of bilirubin. The odor is caused principally by the products of bacterial action; these vary from one person to another, depending on each person's colonic bacterial flora and on the type of food he has eaten. The actual odoriferous products include *indole, skatole, mercaptans,* and *hydrogen sulfide.*

DISORDERS OF DIGESTION AND ABSORPTION

PEPTIC ULCER

A peptic ulcer is an excoriated area of the mucosa caused by the digestive action of gastric juice. Figure 44–6 illustrates the points in the gastrointestinal tract at which peptic ulcers frequently occur, showing that by far the most frequent site of peptic ulcers is in the pyloric region of the duodenum. In addition, peptic ulcers frequently occur along the lesser curvature of the stomach or, more rarely, in the lower end of the esophagus where stomach juices frequently reflux.

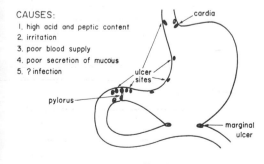

CAUSES:
1. high acid and peptic content
2. irritation
3. poor blood supply
4. poor secretion of mucous
5. ? infection

cardia

ulcer sites

pylorus

marginal ulcer

FIGURE 44–6 Peptic ulcer.

Basic Cause of Peptic Ulceration. The usual cause of peptic ulceration is too much secretion of gastric juice in relation to the degree of protection afforded the mucosa by the mucus that is also secreted. It will be recalled that all areas normally exposed to gastric juice are well supplied with mucous glands, beginning with the compound mucous glands of the lower esophagus, then including the mucous cell coating of the stomach mucosa, the mucous neck cells of the gastric glands, the deep pyloric glands that secrete almost nothing but mucus, and, finally, the coiled tubular glands of Brunner of the upper duodenum, which secrete a highly alkaline mucus.

Causes of Peptic Ulcer in the Human Being. Peptic ulcer occurs much more frequently in the white collar worker than in the laborer, and persons subjected to extreme anxiety for a long time seem particularly prone to peptic ulcer. For instance, the number of persons who developed peptic ulcer increased greatly during the air raids of London. Therefore, it is believed that most instances of peptic ulcer in the human being result from excessive secretion of gastric juices caused by stimulation of the dorsal motor nucleus of the vagus by impulses originating in the cerebrum. Supporting this theory is the fact that ulcer patients have a high rate of gastric secretion during the interdigestive period between meals when the stomach is empty. The normal stomach secretes only 18 milliequivalents of hydrochloric acid during the 12-hour interdigestive period through the night, while ulcer patients occasionally secrete as much as 300 milliequivalents of hydrochloric acid during this same time.

Physiology of Treatment. A usual medical treatment for peptic ulcer is a diet of six or more small meals per day rather than the normal three large meals. In this way food is kept in the stomach most of the time, and the food neutralizes much if not most of the acid and dilutes the gastric juice so that its digestive action on the mucosa is minimized. Also, the diet is bland and contains large quantities of fat. On entering the small intestine, the fat activates the entero-gastrone mechanism that inhibits gastric secretion.

Surgical treatment of peptic ulcer can be effected by one or both of two procedures: (1) removal of a large portion of the stomach or (2) vagotomy. When a peptic ulcer is surgically removed, usually at least the lower three-fourths to four-fifths of the stomach is also removed and the upper stump of the stomach then anastomosed to the jejunum. If less of the stomach than this is removed, far too much gastric juice continues to be secreted, and a marginal ulcer soon develops where the stomach is anastomosed to the intestine.

Vagotomy temporarily blocks almost all secretion by the stomach and usually cures a peptic ulcer within less than a week after the operation is performed. Unfortunately, though, a large amount of basal stomach secretion returns three to six months after the vagotomy, and in many patients the ulcer itself also returns. Even more distressing is the *gastric atony* that usually follows vagotomy, for the motility of the stomach is often reduced to such a low level that almost no gastric emptying occurs after vagotomy.

ABNORMAL DIGESTION OF FOOD IN THE SMALL INTESTINE; PANCREATIC FAILURE

Perhaps the commonest cause of abnormal digestion is failure of the pancreas to secrete its juice into the small intestine. Lack of pancreatic secretion frequently occurs (a) in *pancreatitis*, which is discussed below, (b) when the *pancreatic duct is blocked* by a gallstone at the papilla of Vater, or (c) after the *head of the pancreas has been removed* because of malignancy. Loss of pancreatic juice means loss of trypsin, chymotrypsin, carboxypolypeptidase,

pancreatic amylase, pancreatic lipase, and still a few other digestive enzymes. Without these enzymes, as much as three quarters of the fat entering the small intestine may go undigested and as much as a third to one half of the proteins and starches. As a result, large portions of the ingested food are not utilized for nutrition; and copious, fatty feces are excreted.

Pancreatitis. Pancreatitis means inflammation of the pancreas, and this can occur either in the form of *acute pancreatitis* or *chronic pancreatitis*. The commonest cause of acute pancreatitis is blockage of the papilla of Vater by a gallstone; this blocks the main secretory duct from the pancreas as well as the common bile duct. The pancreatic enzymes are then dammed up in the ducts and acini of the pancreas. Eventually, so much trypsinogen accumulates that it overcomes the *trypsin inhibitor* in the secretions, and a small quantity of trypsinogen becomes activated to form trypsin. Once this happens the trypsin activates still greater quantities of trypsinogen as well as chymotrypsinogen and carboxypolypeptidase, resulting in a vicious cycle until all the proteolytic enzymes in the pancreatic ducts and acini become activated. Rapidly these digest large portions of the pancreas itself, sometimes completely and permanently destroying the ability of the pancreas to secrete digestive enzymes.

MALABSORPTION FROM THE SMALL INTESTINE—"SPRUE"

Occasionally, nutrients are not adequately absorbed from the small intestine even though the food is well digested. Several different diseases can cause decreased absorbability by the mucosa; these are often classified together under the general heading of sprue, or in young children the condition is frequently called *celiac disease*. Obviously, also, malabsorption can occur when large portions of the small intestine have been removed.

The cause of celiac disease in children or so-called "idiopathic sprue" in adults is not known, though removal of wheat and rye flour from the food—that is, removal of the protein *gliadin* (a type of gluten) from the diet—will in many instances effect a cure or at least aid in the absorption of nutrients from the intestinal tract. It is believed that gliadin is not fully digested and that remaining glutamine-containing polypeptides are toxic to the mucosa. The newly forming epithelial cells in the crypts of Lieberkühn die before they can migrate upward onto the villi. As a result, the villi become blunted or disappear altogether, thus greatly reducing the absorptive area of the gut.

A different type of sprue called "tropical sprue" frequently occurs in the tropics and can often be treated with antibacterial agents. Even though no specific bacterium has ever been implicated as the cause of this tropical sprue, it is believed that this variety is often caused by inflammation of the intestinal mucosa resulting from a yet unidentified infectious agent.

In the early stages of sprue, the absorption of fats is more impaired than the absorption of other digestive products. The fat appearing in the stools is almost entirely in the form of soaps rather than undigested neutral fat, illustrating that the problem is one of absorption and not of digestion. In this stage of sprue, the condition is frequently called *idiopathic steatorrhea*, which means simply excess fats in the stools as a result of unknown causes.

REFERENCES

Benson, J. A., Jr., and Rampone, A. J.: Gastrointestinal absorption. *Ann. Rev. Physiol.,* 28:201, 1966.

Bockus, H. L. (ed.): Gastroenterology. 3 volumes. 2nd ed., Philadelphia, W. B. Saunders Company, 1963–1965.

Booth, C. C.: Absorption from the small intestine. *Scient. Basis Med. Ann. Rev., 171,* 1963.

Code, C. F.: The peptic ulcer problem. A physiologic appraisal. *Amer. J. Dig. Dis.,* 5:288, 1960.

Curran, P. F., and Schultz, S. G.: Transport across membranes: general principles. *In* Handbook of Physiology. Baltimore, The Williams & Wilkins Co., 1968, Sec. VI, Vol. III, p. 1217.

Jones, F. A., and Gummer, J. W.: Clinical Gastroenterology. 2nd ed., Philadelphia, F. A. Davis Co., 1965.

Palmer, E. D.: Functional Gastrointestinal Disease. Baltimore, The Williams & Wilkins Co., 1967.

Smyth, D. H.: Intestinal absorption. *Brit. Med. Bull.,* 23:205, 1967.

Wilson, T. H.: Intestinal Absorption. Philadelphia, W. B. Saunders Company, 1962.

CHAPTER 45

METABOLISM OF CARBOHYDRATES AND FORMATION OF ADENOSINE TRIPHOSPHATE

The next few chapters deal with metabolism in the body, which means the chemical processes that make it possible for the cells to continue living. It is not the purpose of this textbook, however, to present the chemical details of all the various cellular reactions, for this lies in the discipline of biochemistry. Instead, these chapters are devoted to: (1) review of the principal chemical processes of the cell and (2) analysis of their physiological implications, especially in relation to the manner in which they fit into the overall concept of homeostasis.

ROLE OF ADENOSINE TRIPHOSPHATE (ATP) IN METABOLISM

ATP is a labile chemical compound that is present in all cells and has the chemical structure shown below. *This substance is present everywhere in the cytoplasm and nucleoplasm of all cells, and essentially all the physiological mechanisms that require energy for operation obtain this directly from the stored ATP. In turn, the food in the cells is gradually oxidized, and the released energy is used to re-form the ATP, thus always maintaining a supply of this substance; all of these energy transfers take place by means of coupled reactions.*

The last two phosphate radicals of the ATP molecule are connected with the remainder of the molecule by *high energy bonds,* which are indicated by the symbol ~. Removal of each phosphate radical at this bond liberates 8000 calories of energy. After loss of one phosphate radical from ATP, the compound becomes *adenosine diphosphate* (ADP), and after loss of the second phosphate radical the compound becomes *adenosine monophosphate* (AMP). The interconversions between ATP, ADP, and AMP are the following:

$$\text{ATP} \underset{+8000 \text{ cal.}}{\overset{-8000 \text{ cal.}}{\rightleftarrows}} \begin{matrix}\text{ADP}\\+\\\text{PO}_4\end{matrix} \underset{+8000 \text{ cal.}}{\overset{-8000 \text{ cal.}}{\rightleftarrows}} \begin{matrix}\text{AMP}\\+\\2\text{PO}_4\end{matrix}$$

In summary, ATP is an intermediary compound that has the peculiar ability to enter into many coupled reactions — reactions with the food to extract energy, and reactions in many physiological mechanisms to provide energy for their operation. For this reason, ATP has frequently been called the energy *currency* of the body which can be gained and spent again and again.

The principal purpose of the present chapter is to explain how the energy from carbohydrates can be used to form ATP in the cells. At least 99 per cent of all the carbohydrates utilized by the body is used for this purpose.

TRANSPORT OF MONOSACCHARIDES THROUGH THE CELL MEMBRANE

From Chapter 44 it will be recalled that the final products of carbohydrate digestion that are absorbed into the blood are almot entirely *glucose, fructose,* and *galactose,* glucose representing by far the major share. These monosaccharides cannot diffuse through the pores of the cell membrane, but they do pass to the interior of cells with reasonable freedom by the process of *facilitated diffusion,* which was discussed in Chapter 4. That is, they combine with a carrier substance that makes them soluble in the lipid portion of the membrane, and then after passing through the membrane, they become dissociated from the carrier. The carrier is believed to be a protein of small molecular weight. The carrier mechanism works in both directions, both into and out of the cell. The rate of transport is approximately proportional to the concentration differences of the monosaccharides on the two sides of the membrane, and the monosaccharides are *not* transported through the usual cell membrane against a concentration difference.

Facilitation of Glucose Transport by Insulin. The rate of glucose transport and also transport of some other monosaccharides is greatly increased by insulin. When large amounts of insulin are secreted by the pancreas, the rate of glucose transport into some cells increases to as much as 10 times the rate of transport when no insulin at all is secreted. And, for practical considerations, the amounts of glucose that can diffuse to the insides of most cells of the body in the absence of insulin are far too little to supply anywhere near the amount of glucose normally required for energy metabolism. Therefore, in effect, the rate of carbohydrate utilization by the cells is controlled by the rate of insulin secretion in the pancreas. The functions of insulin and its control of carbohydrate metabolism will be discussed in detail in Chapter 52.

Phosphorylation of the Monosaccharides. Immediately upon entry into the cells, the monosaccharides combine with a phosphate radical as illustrated in Figure 45–1. This phosphorylation is almost completely irreversible except

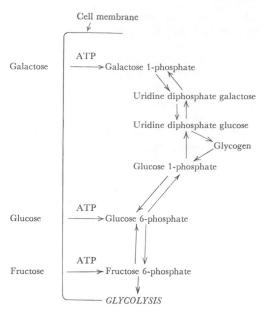

FIGURE 45-1 Interconversions of the three major monosaccharides — glucose, fructose, and galactose — in liver cells.

in the liver cells, the renal tubular epithelium, and the intestinal epithelial cells in which specific phosphatases are available for reversing the reaction. Therefore, in most tissues of the body phosphorylation serves to *capture* the monosaccharide in the cell — once *in* the cell the monosaccharide cannot diffuse back out except in special cells (such as those of the liver) that have the necessary phosphatases.

Conversion of Fructose and Galactose Into Glucose. In liver cells, appropriate enzymes are available to promote interconversions between the monosaccharides, as shown in Figure 45-1; and the dynamics of the reactions are such that the final product of these interconversions is almost entirely glucose. In essentially all other cells glucose and fructose can be reversibly interconverted, but some of the enzymes required for conversion of galactose into the other two monosaccharides are missing. Later in the chapter it is noted that the monosaccharides must all become either *glucose 6-phosphate* or *fructose 6-phosphate* before they can be used for

energy by the cells. For this reason, before galactose can be utilized by the tissues, it must be converted by the liver into glucose, which is then transported by the blood to the other cells.

Storage of Glycogen in Liver and Muscle. After absorption into the cells, glucose or the other monosaccharides can be used immediately for release of energy to the cells or it can be stored in the form of glycogen, which is a large polymer of glucose.

All cells of the body are capable of storing at least some glycogen, but certain cells can store large amounts, especially the liver cells, which can store up to 8 per cent of their weight as glycogen, and muscle cells, which can store up to 1 per cent glycogen. The glycogen molecular weight averages 5,000,000 or greater, and most of the glycogen precipitates in the form of solid granules.

Glycogenolysis. Glycogenolysis means the breakdown of glycogen to reform glucose in the cells. Glycogenolysis does not occur by reversal of the same chemical reactions that serve to form glycogen; instead, each succeeding glucose molecule in the glycogen polymer is split away by a process of *phosphorylation*, catalyzed by the enzyme *phosphorylase*.

Under resting conditions, the phosphorylase of most cells is in an inactive form so that glycogen can be stored but not reconverted into glucose. When it is necessary to re-form glucose from glycogen, therefore, the phosphorylase must first be activated. This is accomplished in the following two ways:

Activation of Phosphorylase by Epinephrine and Glucagon. Two hormones, epinephrine and glucagon, can specifically activate phosphorylase and thereby cause rapid glycogenolysis. The initial effect of each of these hormones is to increase the formation of cyclic adenylate in the cells, and it is this substance that activates the phosphorylase.

Epinephrine is released by the adrenal medullae when the sympathetic nervous system is stimulated. Therefore, one of the functions of the sympathetic nervous system is to promote glycogenolysis and thereby to increase the availability of glucose for immediate metabolism.

Glucagon is a hormone secreted by the *alpha cells* of the pancreas when the blood glucose concentration falls disastrously low. It stimulates the formation of cyclic adenylate mainly in the liver rather than elsewhere in the body. Therefore, its effect is primarily to dump glucose out of the liver into the blood, thereby elevating blood glucose concentration. The function of glucagon in blood glucose regulation is discussed in Chapter 52.

RELEASE OF ENERGY FROM THE GLUCOSE MOLECULE BY THE GLYCOLYTIC PATHWAY

GLYCOLYSIS AND THE FORMATION OF PYRUVIC ACID

By far the most important means by which energy is released from the glucose molecule is by the processes of *glycolysis* and then *oxidation of the end-products of glycolysis*. Glycolysis means splitting of the glucose molecule to form two molecules of pyruvic acid. This occurs by 10 successive steps of chemical reactions, as illustrated in Figure 45–2. Each step is catalyzed by at least one specific protein enzyme. Note that glucose is first converted into fructose 1,6-phosphate and then is split into two three-carbon atom molecules, each of which is then converted through five successive steps into pyruvic acid.

Formation of Adenosine Triphosphate (ATP) During Glycolysis. Despite the many chemical reactions in the glycolytic series, little energy is released, and most of the energy that is released simply becomes heat and is lost to the metabolic systems of the cells. However, as illustrated in the figure, 4 mols

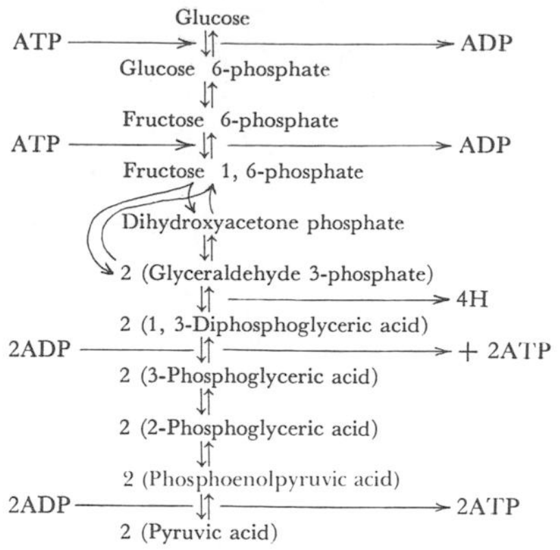

Net reaction:

Glucose $+ 2ADP + 2PO_4^{---} \longrightarrow$ 2 Pyruvic acid $+ 2ATP + 4H$

FIGURE 45–2 The sequence of chemical reactions responsible for glycolysis.

of ATP are formed for each mol of fructose 1,6-phosphate that is split to pyruvic acid, while 2 mols of ATP had been required to phosphorylate the original glucose to form fructose 1,6-phosphate before glycolysis could begin. Therefore, the net gain in ATP molecules by the entire glycolytic process is only two mols for each mol of glucose utilized. This amounts to 16,000 calories of energy stored in the form of ATP, but during glycolysis a total of 56,000 calories of energy is lost from the original glucose, giving an overall *efficiency* for ATP formation of 29 per cent. The remaining energy is lost in the form of heat.

CONVERSION OF PYRUVIC ACID TO ACETYL COENZYME A

The next stage in the degradation of glucose is conversion of the two pyruvic acid molecules into two molecules of *acetyl coenzyme A* (acetyl Co-A).

THE CITRIC ACID CYCLE

The next stage is called the *citric acid cycle* (also known as the *tricarboxylic acid cycle* or *Krebs cycle*). This is a sequence of chemical reactions in which the acetyl portion of acetyl Co-A is degraded to carbon dioxide and hydrogen atoms. Then the hydrogen atoms are subsequently oxidized, releasing still more energy to form ATP.

The enzymes responsible for the citric acid cycle and also for the subsequent oxidation of the hydrogen to form ATP are all contained in the *mitochondria*. These enzymes are attached to the surfaces of the intramitochondrial "shelves" and are arranged in sequential order so that successive products of the reactions can be shuttled from enzyme to enzyme. This obviously greatly enhances the rates of the chemical reactions and promotes increased rate of ATP formation. Because of the special function of the mitochondria in providing ATP to the remainder of the cell, they are frequently called the "power-houses" of the cell.

Figure 45–3 shows the different stages of the chemical reactions in the citric acid cycle. The substances to the left are added during the chemical reactions, and the products of the chemical reactions are shown to the right. Note at the top of the column that the cycle begins with *oxaloacetic acid*, and then at the bottom of the chain of reactions *oxaloacetic acid* is formed once more. Thus, the cycle can continue over and over again.

In the initial stage of the citric acid cycle, *acetyl Co-A* combines with *oxaloacetic acid* to form *citric acid*. The coenzyme A portion of the acetyl Co-A is released and can be used again and again for the formation of still more quantities of acetyl Co-A from pyruvic acid. The acetyl portion, however, becomes an integral part of the citric acid molecule. During the successive stages of the citric acid cycle, several molecules of water are added; and *carbon dioxide* and *hydrogen atoms* are released at various stages in the cycle, as shown on the right in the figure.

The net results of the entire citric

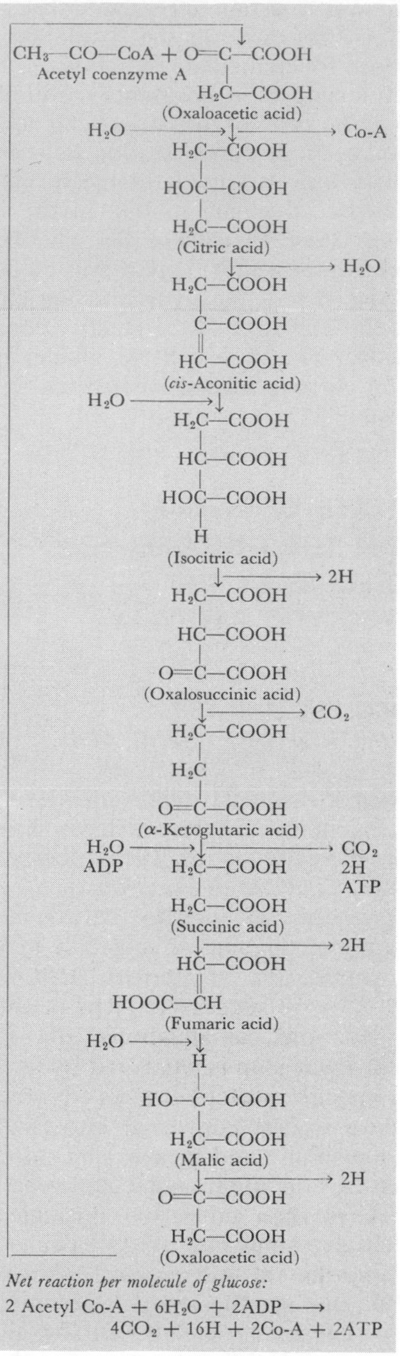

$CH_3-CO-CoA + O=C-COOH$
Acetyl coenzyme A
$$H_2C-COOH$$
(Oxaloacetic acid)
$H_2O \longrightarrow \downarrow \longrightarrow Co\text{-}A$
$$H_2C-COOH$$
$$HOC-COOH$$
$$H_2C-COOH$$
(Citric acid)
$\longrightarrow \downarrow \longrightarrow H_2O$
$$H_2C-COOH$$
$$C-COOH$$
$$HC-COOH$$
(*cis*-Aconitic acid)
$H_2O \longrightarrow \downarrow$
$$H_2C-COOH$$
$$HC-COOH$$
$$HOC-COOH$$
$$H$$
(Isocitric acid)
$\longrightarrow \downarrow \longrightarrow 2H$
$$H_2C-COOH$$
$$HC-COOH$$
$$O=C-COOH$$
(Oxalosuccinic acid)
$\longrightarrow \downarrow \longrightarrow CO_2$
$$H_2C-COOH$$
$$H_2C$$
$$O=C-COOH$$
(α-Ketoglutaric acid)
$H_2O \longrightarrow \downarrow \longrightarrow CO_2$
$ADP \qquad H_2C-COOH \qquad 2H$
ATP
$$H_2C-COOH$$
(Succinic acid)
$\longrightarrow \downarrow \longrightarrow 2H$
$$HC-COOH$$
$$HOOC-CH$$
(Fumaric acid)
$H_2O \longrightarrow \downarrow$
$$H$$
$$HO-C-COOH$$
$$H_2C-COOH$$
(Malic acid)
$\longrightarrow \downarrow \longrightarrow 2H$
$$O=C-COOH$$
$$H_2C-COOH$$
(Oxaloacetic acid)

Net reaction per molecule of glucose:
2 Acetyl Co-A $+ 6H_2O + 2ADP \longrightarrow$
$4CO_2 + 16H + 2Co\text{-}A + 2ATP$

FIGURE 45–3 The chemical reactions of the citric acid cycle, showing the release of carbon dioxide and an especially large number of hydrogen atoms during the cycle.

acid cycle are shown at the bottom of Figure 45–3, illustrating that for each molecule of glucose originally metabolized, 2 acetyl Co-A molecules enter into the citric acid cycle along with 6 molecules of water. These then are degraded into 4 carbon dioxide molecules, 16 hydrogen atoms, and 2 molecules of coenzyme A.

Formation of ATP in the Citric Acid Cycle. Not a large amount of energy is released during the citric acid cycle itself, and in only one of the chemical reactions — during the change from α-ketoglutaric acid to succinic acid — is a molecule of ATP formed. Thus, for each molecule of glucose metabolized, two acetyl Co-A molecules pass through the citric acid cycle, each forming a molecule of ATP; or, a total of two molecules of ATP is formed.

OXIDATION OF HYDROGEN ATOMS BY THE OXIDATIVE ENZYMES

Despite all the complexities of glycolysis and the citric acid cycle, pitifully small amounts of ATP are formed during all these processes. Instead, about 90 per cent of the final ATP is formed during subsequent oxidation of the hydrogen atoms that are released during these earlier stages of glucose degradation. Indeed, the principal function of all these earlier stages is to make the hydrogen of the glucose molecule available in a form that can be utilized for oxidation.

Oxidation of hydrogen is accomplished by a series of enzymatically catalyzed reactions that (a) change the hydrogen atoms into hydrogen ions and (b) change the dissolved oxygen of the fluids into hydroxyl ions. Then these two products combine with each other to form water. During the sequence of oxidative reactions, tremendous quantities of energy are released to form ATP. The oxidative enzymes, like the enzymes of the citric acid cycle, are believed to

be arranged in an orderly fashion on the inner surfaces of the mitochondria, thus allowing rapid procession of the chemical reactions.

Ionization of Hydrogen and Oxygen by the Oxidative Enzymes. Figure 45–4 illustrates a schema of the reactions that cause ionization of both hydrogen and oxygen, with subsequent formation of water. In this schema, the two hydrogens removed from each substrate during the degradation of glucose are shuttled from compound to compound until finally they are released into the body fluids at the cytochrome b stage in the form of two free hydrogen ions. To change a hydrogen atom into a hydrogen ion, an electron must be removed from it. This is performed by cytochrome b. Cytochrome b then passes the electron to cytochrome c, and cytochrome c in turn passes the electron to cytochrome a, which is also known as cytochrome oxidase. At this point, two electrons are transferred from two cytochrome a molecules to one-half molecule of dissolved oxygen plus a molecule of water

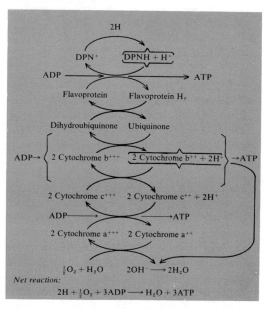

FIGURE 45–4 The chemical processes of oxidative phosphorylation, showing ionization of hydrogen and oxygen and formation of water and adenosine triphosphate.

to form two hydroxyl ions. Thus, hydrogen ions are formed by removing electrons from hydrogen atoms, and hydroxyl ions are formed by transferring these electrons to the dissolved oxygen in the water. Once both of these ions are present in the same fluid medium, they immediately react to form water, as shown in Figure 45–4.

Formation of ATP During Oxidation of Hydrogen — "Oxidative Phosphorylation." The net reaction for the oxidation of each two hydrogen atoms is shown at the bottom of Figure 45–4. This illustrates that three molecules of ATP are formed (but occasionally as many as four or five are formed) during this oxidation. This formation of ATP during the oxidation of hydrogen is called *oxidative phosphorylation*. The precise chemical means by which the oxidation of hydrogen is coupled with the conversion of adenosine diphosphate (ADP) into ATP is almost completely unknown, representing one of the most important voids in our understanding of the body's metabolic processes.

SUMMARY OF ATP FORMATION DURING THE BREAKDOWN OF GLUCOSE

During the entire schema of glucose breakdown, a total of 24 hydrogen atoms are released, and their oxidation gives a minimum production of 34 ATP molecules. These, plus 2 molecules formed during glycolysis and 2 more formed during the citric acid cycle, make at least *38 ATP molecules* formed for each molecule of glucose degraded to carbon dioxide and water. Thus, 304,000 calories of energy are stored in the form of ATP, while 686,000 calories are released during the complete oxidation of each gram-mol of glucose. This represents an overall *efficiency* of energy transfer of at least 44 per cent. The remaining 56 per cent of the energy becomes heat and therefore cannot be used by the cells to perform specific functions.

CONTROL OF GLYCOLYSIS AND OXIDATION BY ADENOSINE DIPHOSPHATE (ADP)

Referring back to the various chemical reactions of the glycolytic series, the citric acid cycle, and the oxidation of hydrogen, we see that at different stages *ADP* is converted into *ATP*. If *ADP is not available at each of these stages, the reactions cannot occur, thus stopping the degradation of the glucose molecule*. Therefore, once all the ADP in the cells has been converted to ATP, the entire glycolytic and oxidative process stops. Then, when more ATP is used to perform different physiological functions in the cell, new ADP is formed, which automatically allows glycolysis and oxidation to proceed once more. In this way, essentially a full store of ATP is automatically maintained all the time, except when the activity of the cell becomes so great that ATP is used up more rapidly than it can be re-formed.

GLUCOSE CONVERSION TO FAT

When glucose is not required for energy and the cells are already saturated with glycogen, the extra glucose that continually enters the cells is converted into fat and stored in this form. This is discussed in the following chapter.

ANAEROBIC RELEASE OF ENERGY — "ANAEROBIC GLYCOLYSIS"

Occasionally, oxygen becomes either unavailable or insufficient so that cellular oxidation of glucose cannot take place. Yet, even under these conditions, a small amount of energy can still be released to the cells by glycosis, for the chemical reactions in the glycolytic breakdown of glucose to pyruvic acid do not require oxygen. Unfortunately, this process is extremely wasteful of glucose because only 16,000 calories of energy are used to form ATP for each molecule of glucose utilized, which represents only a little over 2 per cent of the total energy in the glucose molecule. Nevertheless, this release of glycolytic energy to the cells can be a life-saving measure for a few minutes when oxygen becomes unavailable.

Formation of Lactic Acid During Anaerobic Glycolysis. The *law of mass action* states that as the end-products of a chemical reaction build up in the reacting medium, the rate of the reaction approaches zero. The two end-products of the glycolytic reactions (see Fig. 45–2) are (1) pyruvic acid and (2) hydrogen atoms, which are combined with DPN to form DPNH and H$^+$. The build-up of either or both of these would stop the glycolytic process and prevent further formation of ATP. Fortunately, when their quantities begin to be excessive these end-products react with each other to form lactic acid in accord with the following equation:

$$CH_3-\overset{O}{\overset{\|}{C}}-COOH + DPNH + H^+ \underset{}{\overset{\text{lactic}}{\underset{\text{dehydrogenase}}{\rightleftharpoons}}}$$
(Pyruvic acid)

$$CH_3-\overset{OH}{\underset{H}{\overset{|}{\underset{|}{C}}}}-COOH + DPN^+$$
(Lactic acid)

Thus, under anaerobic conditions, by far the major proportion of the pyruvic acid is converted into lactic acid, which diffuses readily out of the cells into the extracellular fluids and even into the intracellular fluids of other less active cells. Therefore, lactic acid represents a "sinkhole" into which the glycolytic end-products can disappear, thus allowing glycolysis to proceed for several minutes, supplying the body with considerable quantities of ATP even in the absence of respiratory oxygen.

Reconversion of Lactic Acid and Pyruvic Acid to Glucose in the Presence of Oxygen. When a person begins to breathe oxygen again after a period of anaerobic metabo-

lism, the lactic acid once again becomes pyruvic acid. Large portions of this are immediately utilized by the citric acid cycle to provide additional oxidative energy, and large quantities of ATP are formed. This excess ATP then causes as much as three-fourths of the remaining excess pyruvic acid to be converted back into glucose.

Thus, the great amount of lactic acid that forms during anaerobic glycolysis does not become lost to the body, for when oxygen is again available, the lactic acid can be either reconverted to glucose or used directly for energy.

RELEASE OF ENERGY FROM GLUCOSE BY THE PHOSPHOGLUCONATE PATHWAY

Though 95 per cent or more of all the carbohydrates utilized by the muscles are degraded to pyruvic acid by glycolysis and then oxidized, the glycolytic schema is not the only means by which glucose can be degraded and then oxidized to provide energy. A second important schema for breakdown and oxidation of glucose is called the *phosphogluconate pathway*, which is perhaps responsible for as much as 30 per cent of the glucose breakdown in the liver and even more than this in fat cells. It is especially important because it can provide energy independently of all the enzymes of the tricarboxylic acid cycle and therefore is an alternate pathway for energy metabolism in case of some enzymatic abnormality of the cells.

It is not possible to describe this pathway completely here, but the glucose molecule is attacked more directly than occurs in the glycolytic and citric acid systems; 24 hydrogen atoms are released from the glucose molecule and from water molecules that enter into the reactions, and these are oxidized in the usual manner by the process of oxidative phosphorylation to form ATP.

FORMATION OF CARBOHYDRATES FROM PROTEINS AND FATS— "GLUCONEOGENESIS"

When the body's stores of carbohydrates decrease below normal, moderate quantities of glucose can be formed from *amino acids* and from the *glycerol* portion of fat. This process is called *gluconeogenesis*. Approximately 60 per cent of the amino acids in the body proteins can be converted into carbohydrates, while the remaining 40 per cent have chemical configurations that make this difficult. Each amino acid is converted into glucose by a slightly different chemical process. For instance, alanine can be converted directly into pyruvic acid simply by deamination; the pyruvic acid then is converted into glucose, as explained previously. Several of the more complicated amino acids can be converted into different sugars containing three-, four-, five-, or seven-carbon atoms; these can then enter the phosphogluconate pathway and eventually form glucose. Similar interconversions can change glycerol into carbohydrates.

Regulation of Gluconeogenesis. Diminished carbohydrates in the cells and decreased blood sugar are the basic stimuli that set off an increase in the rate of gluconeogenesis. The diminished carbohydrates can directly cause reversal of many of the glycolytic and phosphogluconate reactions, thus allowing conversion of deaminated amino acids and glycerol into carbohydrates. However, in addition, several of the hormones secreted by the endocrine glands, especially the glucocorticoids, are especially important in this regulation.

Effect of glucocorticoids on gluconeogenesis. When normal quantities of carbohydrates are not available to the cells, it is believed that the adenohypophysis begins to secrete increased quantities of corticotropin, which stimulate the adrenal cortex to produce large quantities of *glucocorticoid hormones*, especially cortisol. In turn, cortisol mobilizes proteins from essentially all cells of the body, making these available in the form of amino acids in the body fluids. A high proportion of these immediately becomes deaminated in the liver and therefore provides ideal substrates for conversion into glucose. Thus, one of the most important means by which gluconeogenesis is promoted is through the release of glucocorticoids from the adrenal cortex.

BLOOD GLUCOSE

Except immediately after a meal, glucose is the only monosaccharide present in significant quantities in the blood and interstitial fluids, since the other monosaccharides have been converted to glucose. The normal blood glucose concentration in a person who has not eaten a meal within the past three to four hours is approximately 90 mg. per cent, and, even after a meal containing large amounts of carbohydrates, this rarely rises above 140 mg. per cent unless the person has diabetes mellitus.

The regulation of blood glucose concentration is so intimately related to insulin that this subject will be discussed in detail in Chapter 52 in relation to the function of insulin.

REFERENCES

Awapara, J., and Simpson, J. W.: Comparative Physiology: Metabolism. *Ann. Rev. Physiol.,* *29*:87, 1967.

Ciba Symposium on Control Glycogen Metabolism. Boston, Little, Brown and Company, 1964.

Bondy, P. K. (ed.): Duncan's Diseases of Metabolism. 6th ed., Philadelphia, W. B. Saunders Company, 1964.

Krebs, H. A.: The tricarboxylic acid cycle. *Harvey Lectures,* *44*:165, 1948–1949.

Lemberg, M. R.: Cytochrome oxidase. *Physiol. Rev.,* *49*:48, 1969.

Mommaerts, W. F. H. M.: Energetics of muscular contraction. *Physiol. Rev.,* *49*:427, 1969.

Pyke, D. (ed.): Disorders of Carbohydrate Metabolism. Philadelphia, J. B. Lippincott Co., 1963.

Stetten, D., Jr., and Stetten, M. R.: Glycogen metabolism. *Physiol. Rev.,* *40*:505, 1960.

CHAPTER 46

LIPID AND
PROTEIN METABOLISM

THE LIPIDS

A number of different chemical compounds of the food and body are classified together as *lipids*. These include (1) *neutral fat*, known also as *triglycerides*, (2) the *phospholipids*, (3) *cholesterol*, and (4) a few others of less importance. These substances have certain similar physical and chemical properties, especially the fact that they are miscible with each other. Chemically, the basic lipid moiety of both the triglycerides and the phospholipids is *fatty acids*, which are simply long chain hydrocarbon organic acids. Though cholesterol does not contain fatty acid, its sterol nucleus, as pointed out later in the chapter, is synthesized from degradation products of fatty acid molecules, thus giving it many of the physical and chemical properties of other lipid substances.

Basic Chemical Structure of Triglycerides (Neutral Fat). Since most of this chapter deals with utilization of triglycerides for energy, the following basic structure of the triglyceride molecule must be understood:

$$CH_3—(CH_2)_{16}—COO—CH_2$$
$$CH_3—(CH_2)_{16}—COO—CH$$
$$CH_3—(CH_2)_{16}—COO—CH_2$$
$$\text{Tri-stearin}$$

Note that three long chain fatty acid molecules are bound with one molecule of glycerol. In the human body, the three fatty acids most commonly present in neutral fat are (1) *stearic acid*, which has an 18-carbon chain and is fully saturated with hydrogen atoms, (2) *oleic acid*, which also has an 18-carbon chain but has one double bond in the middle of the chain, and (3) *palmitic acid*, which has 16-carbon atoms and is fully saturated. These or closely similar fatty acids are also the major constituents of the fats in the food.

TRANSPORT OF LIPIDS IN THE BLOOD

TRANSPORT FROM THE GASTROINTESTINAL TRACT— THE "CHYLOMICRONS"

It will be recalled from Chapter 44 that essentially all the fats of the diet are absorbed into the lymph. In the digestive tract, most triglycerides are split into glycerol and fatty acids or into monoglycerides and fatty acids. Then, on passing through the intestinal epithelial cells, these are resynthesized into new molecules of triglycerides which aggregate and enter the lymph as minute, dispersed droplets called *chylomicrons*, having a size about 0.5 micron. A small amount of protein adsorbs to the outer surfaces of the chylomicrons; this increases their suspension stability in the fluid of the lymph and prevents their adhering to the lymphatic vessel walls. The chylomicrons are then transported up the thoracic duct and emptied into the venous blood at the junction of the jugular and subclavian veins.

Removal of the Chylomicrons from the Blood. Immediately after a meal that contains large quantities of fat, the chylomicron concentration in the plasma may rise to as high as 1 to 2 per cent, and, because of the large sizes of the chylomicrons, the plasma appears turbid and sometimes yellow. However, the chylomicrons are gradually removed within two to three hours, and the plasma becomes clear once again.

Most of the triglycerides in the chylomicrons are probably hydrolyzed into glycerol and fatty acids under the influence of an enzyme in the blood called *lipoprotein lipase*. The fatty acid molecules can there be directly oxidized for energy, or they can be used by the fat cells of adipose tissue to resynthesize triglycerides, which are stored in the fat cells.

A second means by which chylomicrons are removed from the blood is by transport of whole chylomicrons through the capillary wall directly into liver cells, which then either use the lipids of the chylomicrons for energy or convert them into other lipid substances.

TRANSPORT OF FATTY ACIDS IN COMBINATION WITH ALBUMIN— "FREE FATTY ACID"

Fatty acids ionize strongly in water; therefore, immediately upon release of fatty acids into the blood, either by the action of lipoprotein lipase on chylomicrons or by the action of this same enzyme on triglycerides stored in the adipose tissue, the ionized fatty acid combines almost immediately with albumin of the plasma proteins. The fatty acid bound with proteins in this manner is called *free fatty acid* or *nonesterified fatty acid* (or simply *NEFA*) to distinguish it from essentially all the remainder of the fatty acid in the plasma which is transported in the form of esters of glycerol, cholesterol, or other substances.

The concentration of free fatty acid in the plasma under resting conditions is about 15 mg. per 100 ml. of blood, which is a total of only 0.75 gram of fatty acids in the entire circulatory system. Yet, strangely enough, even this small amount accounts for most of the transport of lipids from one part of the body to another for the following reasons:

(1) Despite the minute amount of free fatty acid in the blood, its rate of "turnover" is extremely rapid, *half the fatty acid being replaced by new fatty acid every two to three minutes.* One can calculate that at this rate over half of all the energy required by the body can be provided in this way even without increasing the free fatty acid concentration.

(2) All conditions that increase the rate of fat utilization for cellular energy also increase the free fatty acid concentration in the blood; this sometimes increases as much as 10- to 15-fold. Especially does this occur in starvation and in diabetes when a person is not or cannot be using carbohydrates for energy.

Under normal conditions about three molecules of fatty acid combine with each molecule of albumin, but as many as 30 fatty acid molecules can at times combine with a single molecule of albumin when the need for fatty acid transport is extreme. This shows how variable the rate of lipid transport can be under different physiological needs.

TRANSPORT IN LIPOPROTEINS

The plasma also contains large quantities of lipids in the form of lipoproteins, which are small suspended particles 0.01 to 0.05 micron in diameter. These contain mixtures of *triglycerides, phospholipids, cholesterol*, and *protein*. The protein in the mixture averages about one-fourth to one-third the total constituents, and the remainder is lipids. The total concentration of lipoproteins in the plasma averages about 700 mg. per 100 ml. This can be broken down into the following average concentrations of the individual constituents:

	mg./100 ml. of plasma
Cholesterol	180
Phospholipids	160
Triglycerides	160
Lipoprotein protein	200

Function of the Lipoproteins. The function of lipoproteins in the plasma is poorly known, though they are believed to be a means by which lipid substances can be transported from the liver to other parts of the body. For instance, the turnover of triglycerides in the lipoproteins is as much as 1.5 grams per hour, which could account for as much as 10 to 20 per cent of the total lipids utilized by the body for energy. It has especially been suggested that fats synthesized from carbohydrates in the liver are transported to the adipose tissue in lipoproteins because almost no free fatty acids are present in the blood when fat synthesis is occurring, while large quantities of lipoproteins are present. Even more important might be the transport of cholesterol and phospholipids by the lipoproteins, because these substances are not known to be transported to any significant extent in any other form.

The lipoproteins are formed either entirely or almost entirely in the liver, which is in keeping with the fact that most plasma phospholipids, cholesterol, and triglycerides (except those in the chylomicrons) are synthesized in the liver.

THE FAT DEPOSITS—ADIPOSE TISSUE

The Fat Cells. The fat cells of adipose tissue are modified fibroblasts that are capable of storing almost pure triglycerides in quantities equal to 80 to 95 per cent of their volume. The triglycerides are generally in a liquid form, and when the tissues of the skin are exposed to prolonged cold, the fatty acid chains of the triglycerides, over a period of weeks, become either shorter or more unsaturated to decrease their melting point, thereby always allowing the fat in the fat cells to remain in a liquid state. This is particularly important because only liquid fat can be hydrolyzed and then transported from the cells.

Fat cells can also synthesize fatty acids and triglycerides from carbohydrates, this function supplementing the synthesis of fat in the liver, as discussed later in the chapter.

Exchange of Fat Between the Adipose Tissue and the Blood—Tissue Lipase. Large quantities of lipase are present in adipose tissue. This enzyme catalyzes rapid exchange between the lipoprotein triglycerides, the free fatty acids of the plasma, and the triglycerides of the adipose tissue. Because of this rapid exchange, the triglycerides in the fat cells are renewed approximately once every two to three weeks, which means that the fat stored in the tissues one month, even of the obese person, is not the same fat that was stored the preceding month, thus emphasizing the dynamic state of the storage fat.

USE OF TRIGLYCERIDES FOR ENERGY, AND FORMATION OF ADENOSINE TRIPHOSPHATE (ATP)

Approximately 40 to 45 per cent of the calories in the normal American diet are derived from fats, which is about equal to the calories derived from carbohydrates. In addition, an average of 30 to 50 per cent of the carbohydrates ingested with each meal is converted into triglycerides, then stored, and later utilized as triglycerides for energy. Therefore, as much as two-thirds to three-quarters of all the energy derived directly by the cells might be supplied by triglycerides.

Hydrolysis of the Triglycerides. The first stage in the utilization of triglycerides for energy is hydrolysis into fatty acids and glycerol. The glycerol is immediately changed by intracellular enzymes into glyceraldehyde, which enters the phosphogluconate pathway of glucose breakdown and is used for energy. But before the fatty acids can be used for energy, they must be degraded further in the following way:

Degradation of Fatty Acid to Acetyl Coenzyme A by "Beta Oxidation." The fatty acid molecule is degraded by progressive release of two-carbon segments in the form of acetyl coenzyme A (acetyl Co-A). This process is illustrated in Figure 46–1, and it is called the *beta oxidation* process for degradation of fatty acids.

Oxidation of Acetyl Co-A. The acetyl Co-A molecules formed by beta oxidation of fatty acids enter into the citric acid cycle as explained in the preceding chapter, combining first with oxaloacetic acid to form succinic acid, which then is degraded into carbon dioxide and hydrogen atoms. The hydrogen is subsequently oxidized by the oxidative enzymes of the cells, which was also explained in the preceding chapter. The net reaction for each molecule of acetyl Co-A is the following:

$$CH_3COCo\text{-}A + \text{Oxaloacetic acid} + 3H_2O + ADP$$

$$\xrightarrow{\textit{Citric acid cycle}}$$

$$2CO_2 + 8H + HCo\text{-}A + ATP + \text{Oxaloacetic acid}$$

Thus, after the initial degradation of fatty acids to acetyl Co-A, their final breakdown is precisely the same as that of the acetyl Co-A formed from pyruvic acid during the metabolism of glucose.

ATP Formed by Oxidation of Fatty Acid. In Figure 46–1, note that 4 hydrogen atoms are released each time a molecule of acetyl Co-A is formed. In addition, for each acetyl Co-A degraded by the tricarboxylic acid cycle, 8 hydrogen atoms are removed, making a total of 104 hydrogen atoms that are released for each molecule of stearic acid that is split. The oxidation of these hydrogen atoms, plus the formation of ATP in the citric acid cycle, gives a total production of 156 or more molecules of ATP for each molecule of stearic acid.

FORMATION OF ACETOACETIC ACID IN THE LIVER AND ITS TRANSPORT IN THE BLOOD

Over 50 per cent of the initial degradation of fatty acids occurs in the liver. However, the liver cannot use anything like this amount of fatty acids for its own intrinsic metabolic processes. Instead, when the fatty acid chains have been split into acetyl Co-A, two molecules of acetyl Co-A condense to form one molecule of acetoacetic acid as follows:

$$2CH_3COCo\text{-}A + H_2O \underset{\textit{other cells}}{\overset{\textit{liver cells}}{\rightleftharpoons}}$$

Acetyl Co-A

$$CH_3COCH_2COOH + 2\,HCo\text{-}A$$

Acetoacetic acid

(1) $\underset{\text{(Fatty acid)}}{RCH_2CH_2CH_2COOH} + Co\text{-}A + ATP \underset{}{\overset{\text{thiokinase}}{\rightleftharpoons}} \underset{\text{(Fatty acyl Co-A)}}{RCH_2CH_2CH_2COCo\text{-}A} + AMP + \text{Pyrophosphate}$

(2) $\underset{\text{(Fatty acyl Co-A)}}{RCH_2CH_2CH_2COCo\text{-}A} + FAD \xrightarrow{\text{acyl dehydrogenase}} RCH_2CH{=}CHCOCo\text{-}A + FADH_2$

(3) $RCH_2CH{=}CHCOCo\text{-}A + H_2O \overset{\text{enoyl hydrase}}{\rightleftharpoons} RCH_2CHOHCH_2COCo\text{-}A$

(4) $RCH_2CHOHCH_2COCo\text{-}A + DPN^+ \underset{\text{dehydrogenase}}{\overset{\beta\text{-hydroxyacyl}}{\xrightarrow{\hspace{1cm}}}} RCH_2COCH_2COCo\text{-}A + DPNH + H^+$

(5) $RCH_2COCH_2COCo\text{-}A + Co\text{-}A \overset{\text{thiolase}}{\rightleftharpoons} \underset{\text{(Fatty acyl Co-A)}}{RCH_2COCo\text{-}A} + \underset{\text{(Acetyl Co-A)}}{CH_3COCo\text{-}A}$

FIGURE 46–1 Beta oxidation of fatty acids to yield acetyl coenzyme A.

The acetoacetic acid then diffuses freely through the liver cell membranes and is transported by the blood to the peripheral tissues. Here it again diffuses freely into the cells where exactly the reverse reactions occur and two acetyl Co-A molecules are formed. These in turn enter the citric acid cycle and are oxidized for energy as explained above.

Normally, the acetoacetic acid that enters the blood is transported so rapidly to the tissues that its concentration in the plasma rarely rises above 3 mg. per cent. Yet despite the small quantities in the blood, tremendous amounts are actually transported; this is analogous to the tremendous rate of fatty acid transport in the free form. The rapid transport of both these substances probably depends on their high degree of lipid solubility, which allows rapid diffusion through the cell membranes.

"Ketosis" and its Occurrence in Starvation, Diabetes, and Other Diseases. Large quantities of acetoacetic acid occasionally accumulate in the blood and interstitial fluids; this condition is called *ketosis* because acetoacetic acid is a keto acid. It occurs especially in starvation, in diabetes mellitus, or even when a person's diet is composed almost entirely of fat. In all these states, essentially no carbohydrates are metabolized — in starvation and following a high fat diet because carbohydrates are not available and in diabetes because insulin is not available to cause glucose transport into the cells.

When carbohydrates are not utilized for energy, essentially all the energy of the body must come from metabolism of fats. We shall see later in the chapter that lack of availability of carbohydrates automatically increases the rate of removal of fatty acids from adipose tissues, and in addition, several hormonal responses — such as increased secretion of corticotropin by the adenohypophysis, increased secretion of glucocorticoids by the adrenal cortex, and decreased secretion of insulin by the pancreas — all further enhance the removal of fatty acids from the fat tissues. As a result, tremendous quantities of fat become available to the liver for degradation into acetoacetic acid. The acetoacetic acid in turn pours out of the liver to be carried to the cells. Yet the cells are limited in the amount of acetoacetic acid that can be oxidized, for which reason the blood concentration of acetoacetic acid sometimes rises to as high as 30 times normal, thus leading to extreme acidosis, as explained in Chapter 26.

SYNTHESIS OF TRIGLYCERIDES FROM CARBOHYDRATES

Whenever a greater quantity of carbohydrates enters the body than can be used immediately for energy or stored in the form of glycogen, the excess is rapidly converted into triglycerides and then stored in this form in the adipose tissue. Most of this fat synthesis occurs in the adipose tissue itself, though a smaller portion occurs in the liver.

Conversion of Acetyl Co-A into Fatty Acids. The first step is conversion of carbohydrates into acetyl Co-A. It will be recalled from the preceding chapter that this occurs during the normal degradation of glucose by the glycolytic system. It will also be remembered from earlier in this chapter that fatty acids are actually large polymers of acetic acid. Therefore, it is easy to understand how acetyl Co-A could be converted into fatty acids.

However, the synthesis of triglycerides from acetyl Co-A is not simply the reverse of the oxidative degradation that was described above. Instead, its first step is conversion of acetyl Co-A into *malonyl Co-A* in accordance with step 1 of Figure 46–2. A large amount of energy is transferred from ATP to malonyl Co-A, and it is this energy that is utilized to cause the subsequent reactions that are required in the formation of the fatty acid molecule. Step 2 in Figure 46–2 gives the net reaction in the

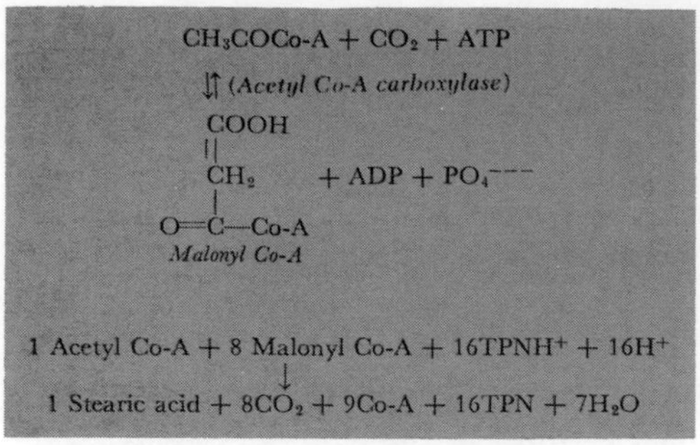

$$CH_3COCo\text{-}A + CO_2 + ATP$$

$$\Updownarrow (Acetyl\ Co\text{-}A\ carboxylase)$$

$$\underset{Malonyl\ Co\text{-}A}{\overset{\displaystyle COOH}{\underset{\displaystyle O=C-Co\text{-}A}{\overset{\displaystyle |}{\overset{\displaystyle CH_2}{|}}}}} + ADP + PO_4^{---}$$

1 Acetyl Co-A + 8 Malonyl Co-A + 16TPNH$^+$ + 16H$^+$
↓
1 Stearic acid + 8CO$_2$ + 9Co-A + 16TPN + 7H$_2$O

FIGURE 46–2 Synthesis of fatty acids.

formation of a stearic acid molecule, showing that one acetyl Co-A molecule and eight malonyl Co-A molecules combine with TPNH and hydrogen ions to form the fatty acid molecule.

The acetyl Co-A that is converted into fatty acid molecules is derived mainly from the *glycolytic* breakdown of glucose, and the TPNH required for fatty acid synthesis is derived from the *phosphogluconate pathway* of glucose degradation, which emphasizes the importance of both these pathways in fat synthesis. These two mechanisms of glucose degradation occur side-by-side in the fat cell, contributing the appropriate proportions of acetyl Co-A and TPNH required for fatty acid synthesis.

Combination of Fatty Acids with α-Glycerophosphate to Form Triglycerides. Once the synthesized fatty acid chains have grown to contain 14 to 18 carbon atoms, they are converted to triglycerides. This occurs because the enzymes that cause this conversion are highly specific for fatty acids with chain lengths of 14 carbon atoms or greater. The glycerol portion of the triglyceride is furnished by α-glycerophosphate, which is a product derived from the glycolytic schema of glucose degradation.

SYNTHESIS OF TRIGLYCERIDES FROM PROTEINS

Many amino acids can be converted into acetyl Co-A, as will be discussed later. Obviously, this can be synthesized into triglycerides. Therefore, when a person has more proteins in his diet than his tissues can use as proteins, a large share of the excess is stored as fat.

REGULATION OF ENERGY RELEASE FROM TRIGLYCERIDES

Regulation of Energy Release by Formation of Adenosine Diphosphate (ADP) in the Tissues. The primary factor that causes energy release from all foodstuffs is the rate of formation of ADP in the tissues. As explained at many points in this text, essentially all functions of the body are energized by the high energy phosphate bonds of ATP. In the process of liberating this energy the ATP becomes ADP, which in turn is a necessary substrate for the oxidation reactions responsible for energy release from essentially all foods. For instance, in the absence of ADP, acetyl Co-A cannot be oxidized, and it accumulates in the tissues. Then, in accordance with the *law of mass action*, this accumulation of acetyl Co-A brings the degradation of fatty acids to a halt. Likewise, failure to oxidize DPNH builds up this end-product of fat degradation, which likewise slows fatty acid degradation.

Conversely, when the activity of the tissues accelerates so that increased quantities of ADP are formed, all the oxidative processes accelerate, and the degradation and utilization of fatty acids and other foodstuffs for energy proceeds apace.

Hormonal Regulation of Fat Metabolism. At least seven of the hormones secreted by the endocrine glands also have marked effects on fat metabolism as follows:

Insulin lack causes depressed glucose utilization, which in turn decreases fat synthesis and promotes fat mobilization from the tissues as well as increased rate of fat utilization. In severe diabetes, a person can become extremely emaciated because of depletion of the fat stores. Conversely, an excess of insulin greatly enhances the availability of glucose to the cells, which inhibits fat utilization and enhances fat synthesis. In addition, insulin directly enhances the entry of triglycerides into fat cells, thus further increasing the fat stores while decreasing the use of fats for energy.

Glucocorticoids secreted by the adrenal cortex have a direct effect on the fat cells to increase the rate of fat mobilization. It has been postulated that this results from an effect of glucocorticoids to increase the cell membrane permeability of fat cells. In the absence of glucocorticoids, mobilization of fat is depressed, causing considerable depression in fat utilization. Furthermore, glucocorticoids must be available before sufficient fat can ever be mobilized to cause ketosis. Therefore, glucocorticoids are frequently said to have a *ketogenic effect.*

Corticotropin, especially, and *growth hormone*, to a lesser extent, from the adenohypophysis both have a fat-mobilizing effect similar to that of the adrenocortical glucocorticoids, though the precise causes of the mobilization are not known. The corticotropin also stimulates secretion of glucocorticoids which mobilize still more fat, as explained above.

Thyroid hormone causes rapid mobilization of fats, which is believed to result indirectly from an increased rate of energy metabolism in all cells of the body under the influence of this hormone. The resulting reduction in acetyl Co-A in the cells would then be a stimulus to cause fat mobilization.

Finally, *epinephrine* and *norepinephrine* have direct effects on fat cells to increase their rate of fat mobilization. In times of stress, the release of epinephrine from the adrenal medullae is an important means by which fatty acids are made available for metabolism, sometimes increasing the nonesterified fatty acids in the blood as much as 10- to 15-fold.

The effects of the different hormones on metabolism are discussed further in the chapters dealing with each of them.

PHOSPHOLIPIDS AND CHOLESTEROL

PHOSPHOLIPIDS

The three major types of body phospholipids are the *lecithins*, the *cephalins*, and the *sphingomyelins*. Typical examples of these are shown in Figure 46–3.

Phospholipids always contain one or more fatty acid molecules and one phosphoric acid radical, and they usually contain a nitrogenous base. Phospholipids are formed in essentially all cells of the body, though probably 90 per cent or more of the phospholipids that enter the blood are formed in the liver cells.

Specific Uses of Phospholipids. Several isolated functions of the phospholipids are the following: (1) Phospholipids possibly help to transport fatty acids through the intestinal mucosa into the lymph, possibly by making the fat particles more miscible with water. (2) Thromboplastin, which is necessary to initiate the clotting process, is composed mainly of one of the cephalins. (3) Large quantities of sphingomyelin are present in the nervous system; this substance acts as an insulator in the myelin sheath around nerve fibers. (4) Phospholipids are donors of phosphate radicals when

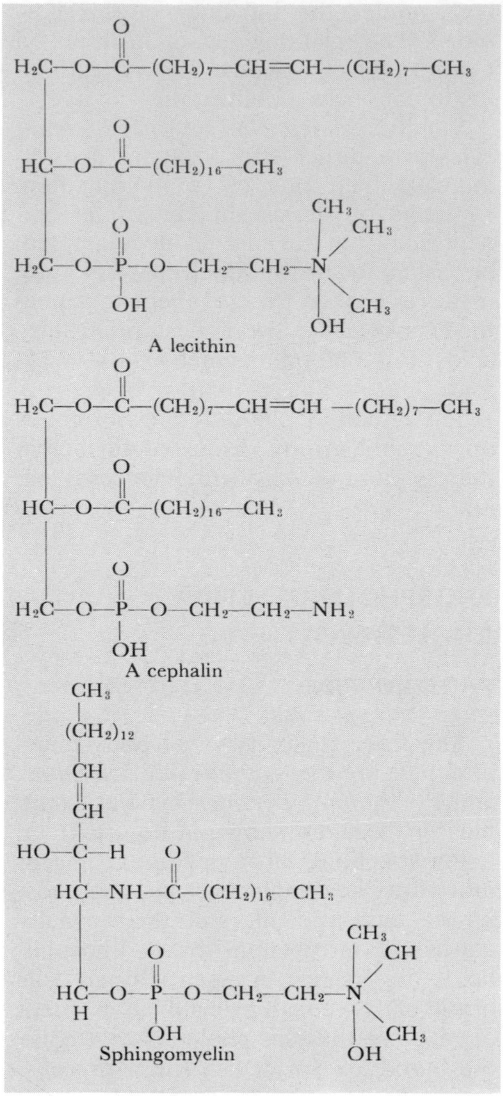

$$H_2C-O-\overset{\overset{\displaystyle O}{\|}}{C}-(CH_2)_7-CH=CH-(CH_2)_7-CH_3$$

$$HC-O-\overset{\overset{\displaystyle O}{\|}}{C}-(CH_2)_{16}-CH_3$$

$$H_2C-O-\overset{\overset{\displaystyle O}{\|}}{P}-O-CH_2-CH_2-\overset{CH_3}{\underset{\underset{\displaystyle OH}{|}}{N}}\overset{CH_3}{\diagdown}CH_3$$
$$\overset{|}{OH}$$

A lecithin

$$H_2C-O-\overset{\overset{\displaystyle O}{\|}}{C}-(CH_2)_7-CH=CH-(CH_2)_7-CH_3$$

$$HC-O-\overset{\overset{\displaystyle O}{\|}}{C}-(CH_2)_{16}-CH_3$$

$$H_2C-O-\overset{\overset{\displaystyle O}{\|}}{P}-O-CH_2-CH_2-NH_2$$
$$\overset{|}{OH}$$

A cephalin

$$CH_3$$
$$(CH_2)_{12}$$
$$CH$$
$$CH$$
$$HO-C-H$$
$$HC-NH-\overset{\overset{\displaystyle O}{\|}}{C}-(CH_2)_{16}-CH_3$$
$$HC-O-\overset{\overset{\displaystyle O}{\|}}{P}-O-CH_2-CH_2-\overset{CH_3}{\underset{\underset{\displaystyle OH}{|}}{N}}\overset{CH_3}{\diagup}CH$$
$$H \qquad OH$$

Sphingomyelin

FIGURE 46–3 Typical phospholipids.

these are needed for different chemical reactions in the tissues. (5) Perhaps the most important of all the functions of phospholipids is the formation of structural elements within cells throughout the body, as is discussed below in connection with cholesterol.

CHOLESTEROL

Cholesterol, the formula of which is illustrated in Figure 46–4, is present in

the diet of all persons, and it can be absorbed slowly from the gastrointestinal tract into the intestinal lymph without any previous digestion. It is highly fat soluble, but only slightly soluble in water. Approximately 70 per cent of the cholesterol of the plasma is in the form of cholesterol esters.

Formation of Cholesterol. Besides the cholesterol absorbed each day from the gastrointestinal tract, which is called *exogenous cholesterol*, a large quantity, called *endogenous cholesterol*, is formed in the cells of the body. Essentially all the endogenous cholesterol that circulates in the lipoproteins of the plasma is formed by the liver, but all the other cells of the body form at least some cholesterol, which is consistent with the fact that many of the membranous structures of all cells are partially composed of this substance.

As illustrated by the formula of cholesterol, its basic structure is a sterol nucleus. This is synthesized entirely from acetyl Co-A. In turn, the sterol nucleus can be modified by means of various side chains to form (a) cholesterol, (b) cholic acid, which is the basis of the bile acids formed in the liver, and (c) several important steroid hormones secreted by the adrenal cortex, the ovaries, and the testes (these are discussed in later chapters).

As much as 80 per cent of the cholesterol is converted into cholic acid in the liver. This, in turn, is conjugated with other substances to form bile salts, which have already been discussed in connection with fat digestion.

A large amount of cholesterol is precipitated in the corneum of the skin. This, along with other lipids, makes the skin highly resistant to the absorption of water-soluble substances and also to the action of many chemical agents. Also, these lipid substances help to prevent water evaporation from the skin; without this protection the amount of evaporation would probably be 15 to 20 liters per day instead of the usual 300 to 400 ml.

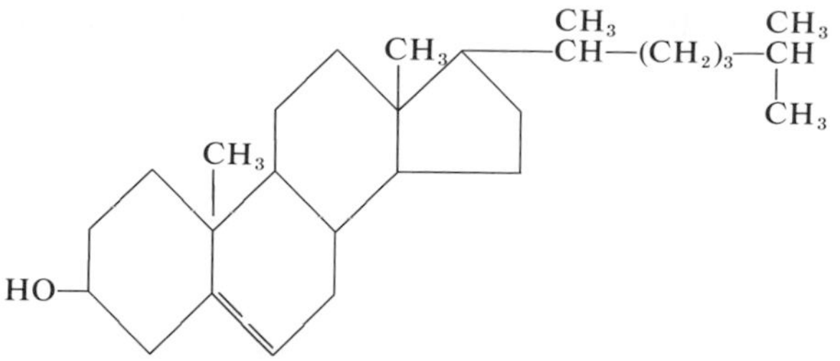

Cholesterol

FIGURE 46–4 Cholesterol.

STRUCTURAL FUNCTIONS OF PHOSPHOLIPIDS AND CHOLESTEROL

Large quantities of phospholipids and cholesterol are present in the cell membrane and in the membranes of the internal organelles of all cells. It has even been claimed that cholesterol and possibly also the phospholipids have a controlling effect over the permeability of cell membranes.

For membranes to develop, substances that are not soluble in water must be available, and, in general, the only substances in the body that are not soluble in water (besides the inorganic substances of bone) are the lipids and some proteins. Thus, the physical integrity of cells throughout the body is based mainly on phospholipids, triglycerides, cholesterol, and certain insoluble proteins. Some phospholipids are somewhat water soluble as well as lipid soluble, which gives them the important property of helping to decrease the interfacial tension between the membranes and the surrounding fluids.

ATHEROSCLEROSIS

Atherosclerosis is principally a disease of the large arteries in which lipid deposits called *atheromatous plaques* appear in the subintimal layer of the arteries. These plaques contain an espe-cially large amount of cholesterol and often are simply called cholesterol deposits. They usually are also associated with degenerative changes in the arterial wall. In a later stage of the disease, fibroblasts infiltrate the degenerative areas and cause progressive sclerosis of the arteries. In addition, calcium often precipitates with the lipids to develop *calcified plaques*. When these two reactions occur, the arteries become extremely hard, and the disease is then called *arteriosclerosis*, or simply "hardening of the arteries."

Obviously, arteriosclerotic arteries lose most of their distensibility, and, because of the degenerative areas, they are easily ruptured. Also, the atheromatous plaques often protrude through the intima into the flowing blood, and the roughness of their surfaces causes blood clots to develop, with resultant thrombus or embolus formation (see Chap. 10). Almost half of all human beings die of arteriosclerosis; approximately two-thirds of the deaths are caused by thrombosis of one or more coronary arteries and the remaining one-third by thrombosis or hemorrhage of vessels in other organs of the body—especially the brain, kidneys, liver, gastrointestinal tract, limbs, and so forth.

Effect of Age, Sex, and Heredity on Atherosclerosis. Atherosclerosis is mainly a disease of old age, but small atheromatous plaques can almost always be found in the arteries of young adults. Therefore, the full-blown disease seems to culminate from a lifetime of lipid deposition rather than deposition over a few years.

Far more men die of atherosclerotic heart disease than do women. This is especially true of men younger than 50. For this reason, it is possible that the male sex hormone accelerates the development of atherosclerosis, or that the female sex hormone *protects* a person from atherosclerosis. Indeed, administration of estrogens to men who have already had coronary thromboses seems to decrease the number of secondary coronary attacks.

Atherosclerosis and atherosclerotic heart disease are highly hereditary in some families. In some instances, this is related to an inherited hypercholesterolemia, but in other instances the blood cholesterol level is completely normal.

Relationship of Dietary Fat to Atherosclerosis in the Human Being. A high fat diet, especially one containing cholesterol and saturated fats, greatly increases one's chances of developing atherosclerosis. Therefore, decreasing the fat can help greatly in protecting against atherosclerosis, and some experiments indicate that this can benefit even patients who have already had coronary heart attacks. Also, life insurance statistics show that the rate of mortality — mainly from coronary disease — of normal weight older age persons is about half the mortality rate of overweight subjects of the same age.

THE BODY PROTEINS

About three-quarters of the body solids are proteins. These include *structural proteins, enzymes, genes, proteins that transport oxygen, proteins of the muscle that cause contraction*, and many other types that perform specific functions both intracellularly and extracellularly throughout the body.

The basic chemical properties of proteins that explain their diverse functions are so extensive that they are a major portion of the entire discipline of biochemistry. For this reason, the present discussion is confined to the general aspects of protein metabolism.

THE AMINO ACIDS

The principal constituents of proteins are amino acids, 21 of which are present in the body in significant quantities. Figure 46–5 illustrates the chemical formulas of these 21 amino acids, showing that they all have two features in common: Each amino acid has an acidic group (—COOH) and a nitrogen radical that lies in close association with the acidic radical, usually represented by the amino group (—NH$_2$).

Peptide Linkages and Peptide Chains. In proteins, the amino acids are aggregated into long chains by means of so-called *peptide linkages*, one of which is illustrated by the following reaction:

$$
\begin{array}{ccc}
\text{NH}_2 & & \text{NH}_2 \\
| & & | \\
\text{R—CH—COOH} + \text{R —CH—COOH} & \longrightarrow
\end{array}
$$

$$
\begin{array}{c}
\text{NH}_2 \\
| \\
\text{R—CH—CO} \\
| \\
\text{NH} \quad\quad + \text{H}_2\text{O} \\
| \\
\text{R —CH—COOH}
\end{array}
$$

Note in this reaction that the amino radical of one amino acid combines with the carboxyl radical of the other amino acid. A hydrogen atom is released from the amino radical, and a hydroxyl radical is released from the carboxyl radical; and these two combine to form a molecule of water. Note that after the peptide linkage has been formed, an amino radical and a carboxyl radical are still in the new molecule, both of which are capable of combining with additional amino acids.

PHYSICAL CHARACTERISTICS OF PROTEINS

With the exception of the few fibrous proteins, which are discussed subse-

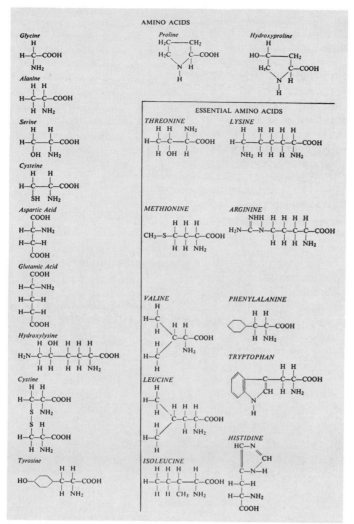

FIGURE 46–5 The amino acids, showing the 10 essential amino acids, which cannot be synthesized at all or in sufficient quantity in the body.

quently, most proteins of the body assume either a globular or an elliptical shape and are called *globular proteins*. These, in general, are soluble in water or salt solutions, and they are held in a globular shape by continuous coiling of the peptide chains.

The following are some of the important types of globular proteins:

1. *Albumins*, which are low molecular weight, simple proteins.

2. *Globulins*, which are medium mo-lecular weight, simple proteins found in many enzymes.

3. *Histones* and *protamines*, which are basic proteins present in nucleoproteins.

Many of the highly complex proteins are fibrillar and are called fibrous proteins. In these the peptide chains are elongated, and many chains are held together in parallel bundles by cross-linkages. There are three major types of fibrous proteins:

1. *Collagens*, which are the basic

structural proteins of connective tissue, tendons, cartilage, and bone.

2. *Elastins*, which are the elastic fibers of tendons, arteries, and connective tissue.

3. *Keratins*, which are the structural proteins of hair and nails.

TRANSPORT AND STORAGE OF AMINO ACIDS

THE BLOOD AMINO ACIDS

The normal concentration of amino acids in the blood is between 35 and 65 mg. per cent. This is an average of about 2 mg. per cent for each of the 21 amino acids, though some are present in far greater concentrations than others. Since the amino acids are relatively strong acids, they exist in the blood principally in the ionized state and account for 2 to 3 milliequivalents of the negative ions in the blood.

Fate of Amino Acids Absorbed from the Gastrointestinal Tract. After the amino acids are absorbed from the gastrointestinal tract into the blood, most of them are absorbed within 5 to 10 minutes by cells throughout the entire body. Therefore, almost never do large concentrations of amino acids accumulate in the blood. The turnover rate of the amino acids is so rapid that many grams of proteins can be carried from one part of the body to another in the form of amino acids each hour.

Active Transport of Amino Acids into the Cells. The molecules of essentially all the amino acids are much too large to diffuse through the pores of the cell membranes. Small quantities perhaps can dissolve in the matrix of the cell membrane and diffuse to the interior of the cells in this manner, but large quantities of amino acids can be transported through the membrane only by active transport utilizing carrier mechanisms. The nature of the carrier mechanisms is still poorly understood, but some of the theories are discussed in Chapter 4.

Renal threshold for amino acids. One of the special functions of carrier transport of amino acids is to prevent loss of these in the urine. All the different amino acids can be *actively transported* through the proximal tubular epithelium, thus removing them from the glomerular filtrate and returning them to the blood. However, as is true of other active transport mechanisms in the renal tubules, there is an upper limit to the rate at which each type of amino acid can be transported. For this reason, when a particular type of amino acid rises to too high a concentration in the plasma and glomerular filtrate, the excess above that which can be actively reabsorbed is lost into the urine.

STORAGE OF AMINO ACIDS AS PROTEINS IN THE CELLS

Almost immediately after entry into the cells, amino acids are conjugated under the influence of intracellular enzymes into cellular proteins so that the concentration of amino acids inside the cells always remains low. Yet many intracellular proteins can be rapidly decomposed again into amino acids under the influence of intracellular enzymes called *kathepsins*, and these amino acids in turn can be transported back out of the cell into the blood. The proteins that can be thus decomposed include many cellular enzymes as well as some other functioning proteins. However, the genes of the nucleus and structural proteins, such as collagen and muscle contractile proteins, do not participate significantly in this reversible storage of amino acids. Some tissues of the body participate in the storage of amino acids to a greater extent than others, particularly the liver.

Reversible Equilibrium Between the Proteins of Different Parts of the Body. Since cellular proteins can be synthesized rapidly from plasma amino acids and many of these in turn can be degraded and returned to the plasma almost equally as rapidly, there is constant equilibrium between the plasma amino acids and most of the proteins in the cells of the body. Therefore, it follows that there is also equilibrium between the proteins from one type of cell to the next. These effects are particularly noticeable in relation to protein synthesis in cancer cells. Cancer cells are prolific users of amino acids, and, simultaneously, the proteins of the other tissues become markedly depleted.

THE PLASMA PROTEINS

The three major types of protein present in the plasma are *albumin, globulin*, and *fibrinogen*. The principal function of albumin is to provide *colloid osmotic pressure*, which in turn prevents plasma loss from the capillaries, as discussed in Chapter 16. The globulins perform a number of enzymatic functions in the plasma itself, but, more important than this, they are also the antibodies responsible for both the natural and acquired immunity that a person has against invading organisms, which was discussed in Chapter 9. The fibrinogen polymerizes into long fibrin threads during blood coagulation, thereby forming blood clots that help to repair leaks in the circulatory system, which was discussed in Chapter 10.

Formation of the Plasma Proteins. Essentially all the albumin and fibrinogen of the plasma proteins, as well as 50 per cent or more of the globulins, are formed in the liver. The remainder of the globulins, the antibodies, are formed in the lymphoid tissues and other cells of the reticuloendothelial system.

The rate of plasma protein formation by the liver can be extremely high, as great as 4 grams per hour or as much as 100 grams per day. Certain disease conditions often cause rapid loss of plasma proteins; severe burns that denude large surface areas cause loss of many liters of plasma through the denuded areas each day. The rapid production of plasma proteins by the liver is obvi-

ously valuable in preventing death in such states. Furthermore, occasionally, a person with severe renal disease loses as much as 20 to 30 grams of plasma protein in the urine each day for years. In some of these patients the plasma protein concentration may remain normal or almost normal throughout the entire illness.

Reversible Equilibrium Between the Plasma Proteins and the Tissue Proteins. The rate of synthesis of plasma proteins by the liver is dependent on the concentration of amino acids in the blood, which means that the concentration of plasma proteins becomes reduced whenever an appropriate supply of amino acids is not available. On the other hand, whenever excess proteins are available in the plasma but insufficient proteins are present in the cells, the plasma proteins are used to form tissue proteins. Thus, there is a constant state of equilibrium, as illustrated in Figure 46–6, between the plasma proteins, the amino acids of the blood, and the tissue proteins. It has been estimated from radioactive tracer studies that about 400 gms. of body protein are synthesized and degraded each day as part of the continual state of flux of amino acids. This illustrates once again the general principle of reversible exchange of amino acids

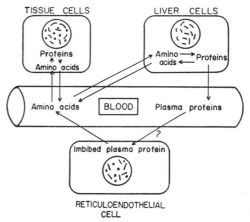

FIGURE 46–6 Reversible equilibrium between the tissue proteins, plasma proteins, and plasma amino acids.

among the different proteins of the body.

Because of this reversible equilibrium between plasma proteins and the other proteins of the body, one of the most effective of all therapies for severe acute protein deficiency is intravenous administration of plasma protein. Within hours, the protein becomes distributed wherever it is needed throughout the cells of the body.

USE OF PROTEINS FOR ENERGY

Once the cells are filled to their limits with protein, any additional amino acids in the body fluids are degraded and used for energy or stored as fat. This degradation occurs almost entirely in the liver, and it begins with the process known as *deamination*, which means removal of the amino radical in the form of ammonia.

Urea formation by the liver. The ammonia released during deamination is removed from the blood almost entirely by conversion into urea, two molecules of ammonia and one molecule of carbon dioxide combining in accordance with the following net reaction:

$$2NH_3 + CO_2 \rightarrow H_2N-\underset{\underset{O}{\|}}{C}-NH_2 + H_2O$$

Essentially all urea formed in the human body is synthesized in the liver. In the absence of the liver or in serious liver disease, ammonia accumulates in the blood. This in turn is extremely toxic, especially to the brain, often leading to a state called *hepatic coma.*

Oxidation of Deaminated Amino Acids. Once amino acids have been deaminated, the resulting keto acid products can in most instances be oxidized to release energy for metabolic purposes. This usually involves two processes: (1) the keto acid is changed into an appropriate chemical substance that can combine with one or more of the active compounds of the tricarboxylic acid cycle and (2) this substance is then degraded by this cycle in the same manner that acetyl Co-A derived from carbohydrate and lipid metabolism is degraded.

In general, the amount of adenosine triphosphate formed for each gram of protein that is oxidized is approximately the same as that formed for each gram of glucose oxidized.

Gluconeogenesis and Ketogenesis. Certain deaminated amino acids are similar to the breakdown products that result from glucose and fatty acid metabolism. For instance, deaminated alanine is pyruvic acid. Obviously, this can be converted into glycose or glycogen. Or it can be converted into acetyl Co-A, which can then be polymerized into fatty acids. Also, two molecules of acetyl Co-A can condense to form acetoacetic acid, which is one of the ketone bodies, as explained earlier.

The conversion of amino acids into glucose or glycogen is called *gluconeogenesis*, and the conversion of amino acids into keto acids (or fatty acids) is called *ketogenesis*. Eighteen out of 21 of the deaminated amino acids have chemical structures that allow them to be converted into glucose, and 19 can be converted into fats.

HORMONAL REGULATION OF PROTEIN METABOLISM

Growth Hormone. Growth hormone increases the rate of synthesis of cellular proteins, causing the tissue proteins to increase while the plasma amino acid concentration falls. Growth hormone enhances the transport of some amino acids through the cell membranes, and it probably also accelerates the actual chemical processes of protein synthesis.

Insulin. Lack of insulin reduces protein synthesis almost to zero. Insulin slightly accelerates amino acid transport into the cells, which could be a stimulus to protein synthesis. Also, insulin increases the availability of glucose to the cells so that the use of amino acids for energy becomes correspondingly reduced.

Glucocorticoids. The glucocorticoids secreted by the adrenal cortex *decrease* the quantity of protein in most tissues while increasing the amino acid concentration in the plasma. However, contrary to elsewhere in the body, these hormones *increase* both the liver proteins and the plasma proteins. It is believed that the glucocorticoids act by increasing the rate of breakdown of extrahepatic proteins, thereby making increased quantities of amino acids available in the body fluids. This in turn supposedly allows the liver to synthesize increased quantities of liver and plasma proteins.

Testosterone. Testosterone, the male sex hormone, causes increased deposition of protein in the tissues throughout the body, including especially an increase in the contractile proteins of the muscles. The mechanism of this effect is unknown.

Thyroxine. Thyroxine increases the rate of metabolism of all cells, and, as a result, indirectly affects protein metabolism. If insufficient carbohydrates and fats are available for energy, thyroxine causes rapid degradation of proteins to be used for energy. On the other hand, if adequate quantities of carbohydrates and fats are available and excesses of amino acids are also available in the extracellular fluid, thyroxine can actually increase the rate of protein synthesis.

REFERENCES

Adipose tissue and metabolism (conference). *Ann. N.Y. Acad. Sci., 131*:1, 1965.

Goodman, DeW. S.: Cholesterol ester metabolism. *Physiol. Rev., 45*:747, 1965.

Greenstein, J. P., and Winitz, M.: Chemistry of the Amino Acids. 3 volumes. New York, John Wiley & Sons, Inc., 1961.

Kirk, J. E.: Arteriosclerosis and arterial metabolism. *In* Bittar, E. E., and Bittar, N. (eds.): The Biological Basis of Medicine. New York, Academic Press, Inc., 1968, Vol. I, p. 493.

Munro, H., and Allison, J. B.: Mammalian Protein Metabolism. New York, Academic Press, Inc., 1967, Vols. 1 and 2.

Neurath, H. (ed.): The Proteins: Composition, Structure, and Function. 2nd ed., New York, Academic Press, Inc., 1963, Vol. 1.

Spain, D. M.: Atherosclerosis. *Sci. Amer., 215*: 48(2), 1966.

Tubbs, P. K.: Membranes and fatty acid metabolism. *Brit. Med. Bull., 24*:158, 1968.

CHAPTER 47

ENERGETICS; METABOLIC RATE; AND REGULATION OF BODY TEMPERATURE

IMPORTANCE OF ADENOSINE TRIPHOSPHATE (ATP) IN METABOLISM

The last few chapters have pointed out that carbohydrates, fats, and proteins can all be used by the cells to synthesize large quantities of ATP, and in turn the ATP can be used as an energy source for many other cellular functions, most of which have been discussed in Chapter 2 and at many other points in this text. To recapitulate, energy derived from ATP can be used for:

1. Synthesis of cellular components, such as synthesis of proteins.
2. Muscle contraction.
3. Transport across membranes as occurs (a) for glandular secretion, (b) for active absorption in both the gut and the renal tubules, and (c) in nerve fibers to provide the sodium and potassium concentration gradients necessary for nerve transmission.

For these reasons, ATP has been called an energy "currency" that can be created and expended.

CREATINE PHOSPHATE AS A STORAGE DEPOT FOR ENERGY

Despite the paramount importance of ATP as a coupling agent for energy transfer, this substance is not the most abundant store of high energy phosphate bonds in the cells. On the contrary, *creatine phosphate*, which also contains high energy phosphate bonds, is many times more abundant. The high energy bond of creatine phosphate contains about 8500 calories per mol under standard conditions, or 9500 calories per mol under conditions in the body (38° C. and low concentrations of the reactants). This is not greatly different from the 8000 calories per mol in each of the two high energy phosphate bonds of ATP. The formula for creatine phosphate is the following:

$$\underset{\substack{}}{HOOC-CH_2-N-\overset{\overset{\displaystyle NH}{\|}}{C}-\overset{\overset{\displaystyle H}{|}}{N} \sim \overset{\overset{\displaystyle O}{\|}}{\underset{\underset{\displaystyle H}{\|}}{P}}-OH}$$

CH₃

Creatine phosphate cannot act in the same manner as ATP as a coupling agent for transfer of energy between the foods and the functional cellular systems. But it can transfer energy interchangeably with ATP. When extra amounts of ATP are available in the cell, much of its energy is utilized to synthesize creatine phosphate, thus building up this storehouse of energy. Then when the ATP begins to be used up, the energy in the creatine phosphate is transferred rapidly back to ATP and from this to the functional systems of the cells. This reversible interrelationship between ATP and creatine phosphate is illustrated by the following equation:

$$\text{Creatine phosphate} + \text{ADP}$$
$$\Updownarrow$$
$$\text{ATP} + \text{Creatine}$$

Note particularly that the higher energy level of the high energy phosphate bond in creatine phosphate, 9500 in comparison with 8000 calories per mol, causes the reaction between creatine phosphate and ATP to proceed to an equilibrium state very much in favor of ATP. Therefore, the slightest utilization of ATP by the cells calls forth the energy from the creatine phosphate to synthesize new ATP. This keeps the concentration of ATP at an almost constant level.

ANAEROBIC VERSUS AEROBIC ENERGY

Anaerobic energy means energy that can be derived from the foods without the simultaneous utilization of oxygen; *aerobic energy* means energy that can be derived from the foods only by oxidative metabolism. In the discussions in the preceding three chapters it was noted that carbohydrates, fats, and proteins can all be oxidized to cause synthesis of ATP. However, carbohydrates are the only significant foods that can be utilized to provide energy without utilization of oxygen; this energy release occurs during glycolytic breakdown of glycogen to pyruvic acid. When glycogen is split to pyruvic acid, each mol of glucose in the glycogen gives rise to 3 mols of ATP.

Anaerobic Energy Usage in Strenuous Bursts of Activity. It is common knowledge that muscles can perform extreme feats of strength for a few seconds but are much less capable during prolonged activity. The energy used during strenuous activity is derived from: (1) ATP already present in the muscle cells, (2) stored creatine phosphate in the cells, (3) anaerobic energy released by glycolytic breakdown of glycogen to lactic acid, and (4) oxidative energy released continuously by oxidative processes in the cells.

The amount of ATP in cells is only about 5 millimols per liter of intracellular fluid, and this amount can maintain maximum muscle contraction for not more than a few seconds. The amount of creatine phosphate in the cells may be as much as 10 times this amount, but even by utilization of all the creatine phosphate, the amount of time that maximum contraction can be maintained is still only 20 to 30 seconds. Release of energy by glycolysis can occur much more rapidly than can oxidative release of energy. Consequently, most of the extra energy required during strenuous activity that lasts for more than 30 seconds but less than two minutes is probably derived from anaerobic glycolysis. As a result, the glycogen content of muscles after strenuous bouts of exercise becomes almost zero, while the lactic acid concentration of the blood rises.

Oxygen Debt. After a period of strenuous exercise, the oxidative metabolic processes continue to operate at a high level of activity for many minutes to (1) reconvert the lactic acid into glucose and (2) reconvert the decomposed ATP and creatine phosphate to their original states. The extra oxygen that must be used to rebuild these substances plus that required to replenish the stores of oxygen in the tissue fluids is called the *oxygen debt*. This debt can be as great as four liters or more. It is usually repaid within three to four minutes, but occasionally as much as 10 minutes are required.

SUMMARY OF ENERGY UTILIZATION BY THE CELLS

With the background of the past few chapters and of the preceding discussion, we can now synthesize a composite picture of overall energy utilization by the cells as illustrated in Figure 47–1. This figure shows the anaerobic utilization of glycogen and glucose to form ATP and also the aerobic utilization of compounds derived from carbohydrates, fats, proteins, and other substances for the formation of still additional ATP. In turn, ATP is in reversible equilibrium with creatine phosphate in the cells, and, since large quantities of creatine phosphate are present in the cell, much of the stored energy of the cell is in this energy storehouse. Energy from ATP can be utilized by the different functioning systems of the cells to provide for synthesis and growth, muscular contraction, glandular secretion, impulse conduction, active absorption, and other cellular activities.

Adenosine Diphosphate (ADP) Concentration as the Rate Controlling Factor in Energy Release. Under normal resting conditions, the concentration of ADP in the cells is extremely slight so that the chemical reactions that depend on ADP as one of the substrates likewise are very slow. These include all the oxidative metabolic pathways as well as essentially all other pathways for release of energy in the body. Thus, *ADP is the major rate limiting factor* for almost all energy metabolism of the body.

When the cells become active, regardless of the type of activity, ATP is converted into ADP, increasing the concentration of ADP in direct proportion to the degree of activity of the cell. This automatically increases the rates of reactions of the ADP-rate-limiting steps in the metabolic release of energy. Thus, by this simple process, the amount of energy released in the cell is controlled by the degree of activity of the cell.

THE METABOLIC RATE

The *metabolism* of the body means simply all the chemical reactions in all the cells of the body, and the *metabolic rate* is normally expressed in terms of the rate of heat liberation during the chemical reactions.

Heat as the Common Denominator of All the Energy Released in the Body. In discussing many of the metabolic reactions of the preceding chapters, we have noted that a large portion of the energy in the foods becomes heat before it can be transferred to the functional systems of the cells, so that not more than about 25 per cent of all the energy from food is finally utilized by the systems. Then,

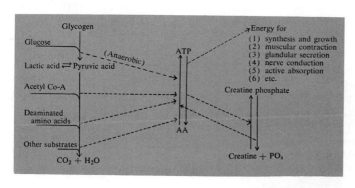

FIGURE 47–1 Overall schema of energy transfer from foods to the adenylic acid system and then to the functional elements of the cells. (Modified from Soskin and Levine: Carbohydrate Metabolism. University of Chicago Press.)

even this 25 per cent usually also becomes heat after it performs its function. For instance, energy is used to form proteins, but later the proteins are degraded in the tissues, and the energy is then lost as heat. And energy is expended in pumping blood, but as the blood flows through the peripheral vessels, the friction of the different layers of blood flowing over each other and the friction of the blood against the walls of the vessels turns this energy also into heat.

Therefore, essentially all the energy expended by the body is converted into heat. The only exception to this occurs when the muscles are used to perform some form of work outside the body. For instance, when the muscles elevate an object to a height or carry the person's body up steps, a type of potential energy is thus created by raising a mass against gravity.

The Calorie. To discuss the metabolic rate and related subjects intelligently, it is necessary to use some unit for expressing the quantity of energy released from the different foods or expended by the different functional processes of the body. In general, the *Calorie* is the unit used for this purpose. It will be recalled that 1 *calorie*, spelled with a small "c," is the quantity of heat required to raise the temperature of 1 gram of water 1° C. The calorie is much too small a unit for ease of expression in speaking of energy in the body. Consequently the large Calorie, spelled with a capital "C," which is equivalent to 1000 calories, is the unit ordinarily used in discussing energy metabolism.

MEASUREMENT OF THE METABOLIC RATE — INDIRECT CALORIMETRY

Indirect Calorimetry. Since essentially all the energy expended in the body is derived from reaction of oxygen with the different foods, the metabolic rate can be calculated with a high degree of accuracy from the rate of oxygen utilization. When 1 liter of oxygen is metabolized with glucose, 5.01 Calories of energy are released; when metabolized with starches, 5.06 Calories are released; with fat, 4.70 Calories; and with protein, 4.60 Calories.

From these figures it is striking how nearly equivalent are the quantities of energy liberated regardless of the type of food that is being burned. For the average diet, the *quantity of energy liberated per liter of oxygen utilized in the body averages approximately 4.825 Calories*. This is called the *energy equivalent* of oxygen. Using this value, one can calculate, within plus or minus 4 per cent accuracy, the rate of heat liberated in the body from the quantity of oxygen utilized in a given period of time.

The metabolator. Figure 47–2 illustrates the metabolator usually used for indirect calorimetry. This apparatus contains a floating drum, under which is an oxygen chamber connected to a mouthpiece through two rubber tubes. A valve in one of these rubber tubes allows air to pass from the oxygen chamber into the mouth, while air passing from the mouth back to the chamber is directed by means of another valve through the second tube. Before the expired air enters the upper portion of the oxygen chamber, it flows through a canister containing pellets of soda lime, which combine chemically with the carbon dioxide in the expired air. Therefore, as oxygen is used by the person's body and the

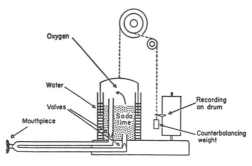

FIGURE 47–2 The metabolator.

carbon dioxide is absorbed by the soda lime, the floating oxygen chamber, which is precisely balanced by a weight, gradually sinks in the water, owing to the oxygen loss. This chamber is appropriately coupled to a pen that records on a moving drum the rate at which the chamber sinks in the water and thereby records the rate at which the body utilizes oxygen.

FACTORS THAT AFFECT THE METABOLIC RATE

Any factor that increases the chemical activity in the cells increases the metabolic rate. Some of these are the following:

Exercise. The factor that causes by far the most dramatic effect on metabolic rate is exercise. Table 47–1 illustrates the rates of energy utilization during different types of activities. Note that walking up stairs requires approximately 17 times as much energy as lying in bed asleep. And, in general, a laborer can, over 24 hours, average a rate of energy utilization as great as 6000 to 7000 Calories, or, in other words, as much as 3½ times the basal rate of metabolism.

Specific Dynamic Action of Protein. After a meal is ingested, the metabolic rate increases. This results mainly from

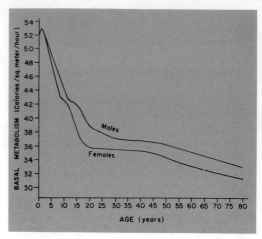

FIGURE 47–3 Normal basal metabolic rates at different ages for each sex.

direct stimulation of cellular chemical processes by products of absorbed amino acids. After a meal containing large quantities of proteins, the metabolic rate usually begins rising within one to two hours, reaches a maximum usually about 30 per cent above normal (*but sometimes as high as 50 to 70 per cent above normal*) an hour or so later, and then fades back to normal in 10 to 12 hours. This effect on the metabolic rate is called the *specific dynamic action of protein.*

Age. The metabolic rate of the newborn child in relation to his body surface area is almost two times that of an aged person. This is illustrated in Figure 47–3, which shows the metabolic rates of both males and females from birth until very old age.

Thyroid Hormone. When the thyroid gland secretes maximal quantities of thyroxine, the metabolic rate sometimes rises to as much as 100 per cent above normal. On the other hand, total loss of thyroid secretion decreases the metabolic rate to as low as 50 per cent of normal. These effects can readily be explained by the basic function of thyroxine to increase the rates of activity of almost all the chemical reactions in all cells of the body. This relationship between thyroxine and metabolic rate will

TABLE 47–1   ENERGY EXPENDITURE PER HOUR DURING DIFFERENT TYPES OF ACTIVITY FOR A 70 KILOGRAM MAN

Form of Activity	Calories per hour
Sleeping	65
Sitting at rest	100
Dressing and undressing	118
Typewriting rapidly	140
"Light" exercise	170
Walking slowly (2.6 miles per hour)	200
Carpentry, metal working, industrial painting	240
"Severe" exercise	450
Swimming	500
Running (5.3 miles per hour)	570
"Very severe" exercise	600
Walking up stairs	1100

Extracted from data compiled by Professor M. S. Rose.

be discussed in much greater detail in Chapter 51 in relation to thyroid function.

Sympathetic Stimulation. Stimulation of the sympathetic nervous system with liberation of norepinephrine and epinephrine increases the metabolic rates of essentially all the tissues of the body. These hormones have a direct effect on cells to cause glycogenolysis, and this, probably along with other intracellular effects of these hormones, increases cellular activity.

Maximal stimulation of the sympathetic nervous system can increase the metabolic rate as much as 30 to 80 per cent.

Fever. Fever, regardless of its cause, increases the metabolic rate. This is because all chemical reactions, either in the body or in the test tube, increase their rates of reaction approximately 130 per cent for every 10° C. rise in temperature. An increase in body temperature to 110° F. increases the metabolic rate almost 100 per cent.

Sleep. The metabolic rate falls approximately 10 to 15 per cent below normal during sleep. This fall is presumably due to two principal factors: (1) decreased tone of the skeletal musculature during sleep and (2) decreased activity of the sympathetic nervous system.

THE BASAL METABOLIC RATE

Basal Conditions. The basal metabolic rate means the rate of energy utilization in the body during absolute rest but while the person is awake. The following basal conditions are necessary for measuring the basal metabolic rate:

1. The person must not have eaten any food for at least 12 hours because of the specific dynamic action of food.

2. The basal metabolic rate is determined after a night of restful sleep, for rest reduces the activity of the sympathetic nervous system and of other metabolic excitants to their minimal level.

3. No strenuous exercise is performed after the night of restful sleep, and the person must remain at complete rest in a reclining position for at least 30 minutes prior to actual determination of the metabolic rate. This is perhaps the most important of all the conditions for attaining the basal state because of the extreme effect of exercise on metabolism.

4. All psychic and physical factors that cause excitement must be eliminated, and the subject must be made as comfortable as possible. These conditions, obviously, help to reduce the degree of sympathetic activity to as little as possible.

5. The temperature of the air must be comfortable and be somewhere between the limits of 65° and 80° F. Below 65° F., the sympathetic nervous system becomes progressively more activated to help maintain body heat, and above 80° F., discomfort, sweating, and other factors increase the metabolic rate.

Usual Technique for Determining the Basal Metabolic Rate. The usual method for determining basal metabolic rate is to measure the rate of oxygen utilization using a metabolator of the type illustrated in Figure 47–2. Then the basal metabolic rate is calculated from the energy equivalent of oxygen.

Expressing the Basal Metabolic Rate in Terms of Surface Area. Obviously, if one subject is much larger than another, the total amount of energy utilized by the two subjects will be considerably different simply because of differences in body size. Experimentally, among normal persons the average basal metabolic rate varies approximately *in proportion to the body surface area*.

For instance, if a person utilizes 15 liters of oxygen per hour, this equals 15 $\times$ 4.825, or 72.4 Calories. If his surface area is 1.5 square meters (as determined from a chart based on height and weight), then his basal metabolic rate is stated to be 48.3 Calories per square meter per hour. Finally, if this person's normal BMR (as determined from the chart in Fig. 47–3 based on age and sex) is 38.5 Calories per hour, the value for a 20-year-old male, then one can calculate that the BMR is 25.5 per cent above normal. This is expressed as a BMR of plus 25.5. In severe hyperthyroidism, the BMR can, rarely, be as high as plus 100; in severe hypothyroidism, it can fall to as low as minus 50.

Constancy of the Basal Metabolic Rate from Person to Person. When the basal metabolic rate is measured in a wide variety of different persons and comparisons are made within single age, weight, and sex groups, 85 per cent of normal persons have been found to have basal metabolic rates within 10 per cent of the mean. Thus, it is obvious that measurements of metabolic rates performed under basal conditions offer an excellent means for comparing the rates of metabolism from one person to another.

BODY TEMPERATURE

The temperature of the inside of the body remains almost exactly constant, within ± 1° F., day in and day out except when a person develops a febrile illness. Indeed, the nude person can be exposed to temperatures as low as 55° to 60° F. or as high as 150° F. and still maintain an almost constant internal body temperature. Therefore, it is obvious that the mechanisms for control of body temperature represent a beautifully designed control system. It is the purpose, here, to discuss this system as it operates in health and in disease.

The Normal Body Temperature. No single temperature level can be considered to be normal, for measurements on many normal persons have shown a *range* of normal temperatures, as illustrated in Figure 47–4, from approximately 97° F. to over 99° F. When measured by rectum, the values are approximately 1° F. greater than the oral temperatures. The average normal temperature is generally considered to be 98.6° F. (37° C.) when measured orally and approximately 1° F. or 0.6° C. higher when measured rectally.

The body temperature varies considerably with exercise and with the temperature of the surroundings, for the temperature regulatory mechanisms are not 100 per cent effective. When excessive heat is produced in the body by strenuous exercise, the rectal temperature can rise to as high as 101° to 104° F. On the other hand, when the body is exposed to extreme cold, the rectal temperature can often fall to values considerably below 98° F.

BALANCE BETWEEN HEAT PRODUCTION AND HEAT LOSS

Heat is continually being produced in the body as a by-product of metabolism, and body heat is also continually being lost to the surroundings. When the rate of heat production is exactly equal to the rate of loss, the person is said to be

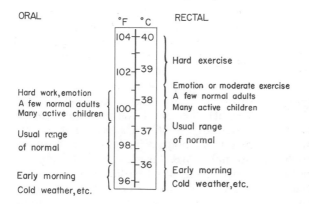

FIGURE 47–4 Estimated range of body temperature in normal prsons. (From DuBois: Fever. Charles C Thomas.)

in heat balance. But when the two are out of equilibrium, the body heat, and the body temperature as well, will be either increasing or decreasing.

The important factors that play major roles in determining the rate of heat production may be listed as follows: (1) basal rate of metabolism of all the cells of the body; (2) increase in rate of metabolism caused by muscle activity, including that caused by shivering; (3) increase in metabolism caused by the effect of thyroxine on cells; (4) increase in metabolism caused by the effect of epinephrine and sympathetic stimulation on cells; and (5) increase in metabolism caused by increased temperature of the body cells.

HEAT LOSS

The various methods by which heat is lost from the body include *radiation, conduction*, and *evaporation*. However, the amount of heat loss by each of these different mechanisms varies with atmospheric conditions.

Radiation. A nude person in a room at normal room temperature loses about 60 per cent of his heat by radiation.

Loss of heat by radiation means loss in the form of infrared heat rays, a type of electromagnetic wave. If the temperature of the body is greater than the temperature of the surroundings, a greater quantity of heat is radiated from the body than is radiated to the body. This is the usual situation. Yet, at times, especially in the summer, the surroundings become hotter than the human body, under which circumstances more radiant heat is transmitted to the body than from the body.

The surface of the human body is extremely absorbent for heat rays. It absorbs approximately 97 per cent of the rays that hit it. This absorption is approximately equal for human beings with either white or black skin; for at infrared wavelengths, the different colors of the skin have no effect on absorption.

On the other hand, the energy from the sun is transmitted mainly in the form of light rays rather than infrared rays. Approximately 35 per cent of these waves are reflected from the white skin but only a small amount from the dark skin. Consequently, in sunlight, dark skin does absorb more heat than white skin.

Conduction. Usually, only minute quantities of heat are lost from the body by direct conduction from the surface of the body to other objects, such as a chair or a bed. On the other hand, loss of heat by *conduction to air* represents a sizeable proportion of the body's heat loss even under normal conditions. However, once the temperature of the air immediately adjacent to the skin equals the temperature of the skin, no further exchange of heat from the body to the air can occur. Therefore, conduction of heat from the body to the air is self-limited unless the heated air moves away from the skin so that new, unheated air is continually brought in contact with the skin.

Convection. Movement of air is known as convection, and the removal of heat from the body by convection air currents is commonly called "heat loss by convection." Actually, the heat must first be *conducted* to the air and then carried away by the convection currents.

A small amount of convection almost always occurs around the body because of the tendency for the air adjacent to the skin to rise as it becomes heated. Therefore, a nude person seated in a comfortable room without gross air movement still loses about 12 per cent of his heat by conduction to the air and then by convection away from the body.

Evaporation. When water evaporates from the body surface, 0.58 Calorie of heat is lost for each gram of water that evaporates, and water evaporates *insensibly* from the skin and lungs at a rate of about 600 ml. per day. This causes continual heat loss at a rate of 12 to 18 Calories per hour.

When the temperature of the surroundings is greater than that of the skin, instead of losing heat, the body gains heat by radiation and conduction from the surroundings. Under these conditions, *the only means by which the body can rid itself of heat is by evaporation.* Therefore, any factor that prevents adequate evaporation when the surrounding temperatures are higher than body temperature permits the body temperature to rise. This occurs occasionally in human beings who are born with congenital absence of sweat glands. These persons can withstand cold temperatures as well as normal persons, but they are likely to die of heat stroke in tropical zones.

SWEATING AND ITS REGULATION BY THE AUTONOMIC NERVOUS SYSTEM

When the body becomes overheated, large quantities of sweat are secreted onto the surface of the skin by the eccrine sweat glands to provide rapid *evaporative cooling* of the body. Stimulation of the preoptic area immediately anterior to the hypothalamus excites sweating. The impulses from this area that cause sweating are transmitted in the autonomic pathways to the cord and thence through cholinergic nerves of the sympathetic outflow to the skin everywhere in the body.

Mechanism of Sweat Secretion. The sweat glands are tubular structures consisting of two parts: (1) a deep *coiled portion* that secretes the sweat, and (2) a *duct portion* passing outward through the dermis of the skin. The secretory portion of the sweat gland secretes a fluid called the *precursor secretion*; then certain constituents of the fluid are reabsorbed as it flows outward through the duct.

When the rate of sweat secretion is very low, the sodium and chloride concentrations of the sweat are also very low, sometimes as low as 5 mEq./liter each, because these ions are reabsorbed from the precursor secretion before it reaches the surface of the body. On the

other hand, when the rate of secretion becomes progressively greater, the rate of sodium chloride reabsorption does not increase commensurately, so that then the concentration of sodium chloride in the sweat can rise almost to the level in plasma.

Acclimatization of the Sweating Mechanism. A person exposed to hot weather for several weeks sweats progressively more profusely, sweating an average maximum of about 1.5 liters per hour at first, which rises to about double this value within 10 days and to about 2½ times as much within six weeks, as illustrated in Figure 47–5. This increased effectiveness of the sweating mechanism is caused by a direct increase in sweating capability of the sweat glands themselves. Associated with the increased sweating is usually decreased concentration of sodium chloride in the sweat, which allows progressively better conservation of salt. This adaptation for conserving salt is caused mainly by increased secretion of aldosterone. A person who sweats profusely may lose as much as 15 to 20 grams of sodium chloride each day until he becomes acclimatized. On the other hand, after four to six weeks of accli-

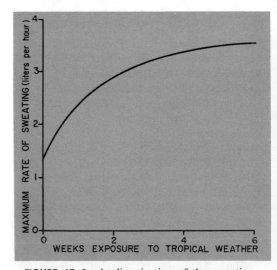

FIGURE 47–5 Acclimatization of the sweating mechanism, showing progressive increase in the maximum rate of sweating during the first few weeks of exposure to tropical weather.

matization the loss of sodium chloride may be as little as 3 to 5 grams per day.

Long-term Acclimatization of the Sweat Apparatus. Each person is born with considerable excess of sweat glands, but if he lives in a temperate zone, many of these become permanently inactivated during childhood. However, if he lives in the tropics they remain functional throughout life. Therefore, a person who has spent his childhood in the tropics usually possesses a much more effective sweating mechanism than does a person reared elsewhere.

THE INSULATOR SYSTEM OF THE BODY

The skin, the subcutaneous tissues, and especially the fat of the subcutaneous tissues are heat insulators for the body. The fat is especially important because it conducts heat only *one-fourth* as rapidly as do the other tissues. When no blood is flowing from the heated internal organs to the skin, the insulating properties of the male body are approximately equal to three-quarters the insulating properties of a usual suit of clothes. In women this insulation is still better.

FLOW OF BLOOD TO THE SKIN— THE "RADIATOR" SYSTEM OF THE BODY

Immediately beneath the skin is a continuous venous plexus that is supplied by inflow of arterial blood. Indeed, in the most exposed areas of the body—the hands, feet, and ears—blood is supplied through direct arteriovenous shunts from the small arteries to the veins. The rate of blood flow into this venous plexus can vary tremendously—from barely above zero to as great as 30 per cent of the total cardiac output. A high rate of blood flow causes heat to be conducted from the internal portions of the body to the skin with great efficiency, whereas

reduction in the rate of blood flow decreases the efficiency of heat conduction from the internal portions of the body.

Obviously, therefore, the skin is an effective radiator system, and the flow of blood to the skin is the principal mechanism of heat transfer from the body "core" to the skin. If blood flow from the internal structures to the skin is depressed, the only means by which heat produced internally can be lost to the exterior is by heat diffusion through the insulator tissues of the skin and subcutaneous areas.

Control of Heat Conduction to the Skin. Ordinarily, the sympathetics remain tonically active, causing continual constriction of the arterioles supplying the skin. When the sympathetic centers of the posterior hypothalamus are stimulated, the blood vessels are constricted even more, and blood flow to the skin may almost cease, but, when these centers of the hypothalamus are inhibited, decreased numbers of sympathetic impulses are transmitted to the periphery, and the blood vessels dilate.

REGULATION OF BODY TEMPERATURE—FUNCTION OF THE HYPOTHALAMUS

In general, between approximately 60° and 130° F. in dry air, the nude body is capable of maintaining indefinitely a normal body core temperature somewhere between 98° and 100° F.

The temperature of the body is regulated almost entirely by nervous feedback mechanisms, and almost all of these operate through a *temperature regulating center* located in the hypothalamus. However, for these feedback mechanisms to operate, there must also exist temperature detectors to determine when the body temperature becomes either too hot or too cold. Probably the most important temperature receptors for control of body temperature are many special *heat sensitive neurons* located in the preoptic area of the anterior hypothalamus. These neurons increase their impulse output as the temperature rises and decrease their output when the temperature decreases. The firing rate sometimes increases as much as 10-fold with an increase in body temperature of 10° C.

In addition to these heat sensitive neurons of the preoptic area, *skin temperature receptors*, including both *warmth* and *cold receptors*, transmit nerve impulses into the spinal cord and thence to the hypothalamic region of the brain to help control body temperature, as will be discussed later.

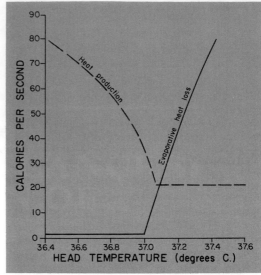

FIGURE 47–6 Effect of hypothalamic temperature on: (1) evaporative heat loss from the body and (2) heat production caused primarily by muscular activity and shivering. This figure demonstrates the extremely critical temperature level at which increased heat loss begins and increased heat production stops. (Drawn from data in Benzinger, Kitzinger, and Pratt: Temperature, Part 3. Ed. Hardy, p. 637. Reinhold Publishing Corp.)

THERMOSTATIC CENTER IN THE ANTERIOR HYPOTHALAMUS

The principal area in the brain where heat or cold affects body temperature control is the preoptic area and adjacent regions of the anterior hypothalamus, which is also where the heat sensitive neurons are found. Therefore, this small area is called the *thermostatic center*. This center controls body temperature by altering both the rate of heat loss from the body and the rate of heat production, as will be explained later.

Figure 47–6 demonstrates the extreme precision with which the hypothalamic thermostat can control heat loss and heat production. The solid curve shows that almost precisely at 37° C. (98.4° F.), sweating begins and then increases rapidly as the temperature rises above this value; on the other hand, it ceases at any temperature below this critical level.

Likewise, the thermostat controls the rate of heat production, which is illustrated by the dashed curve. At any temperature above 37.1° C., the heat production remains almost exactly constant, but whenever the temperature falls below this level, the various mechanisms for increasing heat production become markedly activated, especially the increase in muscular activity which culminates in shivering.

MECHANISMS OF INCREASED HEAT LOSS WHEN THE BODY BECOMES OVERHEATED

Overheating the preoptic thermostatic area increases the rate of heat loss from the body in two principal ways: (1) by stimulating the sweat glands to cause evaporative heat loss from the body and (2) by inhibiting sympathetic centers in the posterior hypothalamus; this removes the normal vasoconstrictor tone to the skin vessels, thereby allowing vasodilatation and loss of heat from the skin.

MECHANISMS OF HEAT CONSERVATION AND INCREASED HEAT PRODUCTION WHEN THE BODY BECOMES COOLED

When the preoptic area of the hypothalamus is cooled below approximately 37° C., special mechanisms are set into play to conserve the heat that is already in the body, and still others are set into play to increase the rate of heat production. These mechanisms include (1) vasoconstriction, to decrease the flow of heat to the skin; (2) pilo-erection in lower animals, to trap more "insulator" air in the fur; (3) abolition of sweating; (4) shivering, to increase heat production by the muscles; and (5) secretion of norepinephrine, epinephrine, and thyroxine, to increase heat production.

Hypothalamic Stimulation of Shivering. Located in the dorsomedial portion of the posterior hypothalamus near the wall of the third ventricle is an area called the *primary motor center for shivering.* This area is normally inhibited by signals from the preoptic thermostatic area. However, when the preoptic thermostat becomes cooled, the normal inhibition of the primary motor center no longer exists. The self-excitation property of this center then causes it to transmit impulses to the spinal cord to increase the tone of the skeletal muscles throughout the body. When the tone of the muscles rises above a certain critical level, shivering begins, which probably results from feedback oscillation of the muscle spindle stretch reflex mechanism. When shivering occurs, body heat production rises as high as 100 to 300 per cent above normal.

Sympathetic "Chemical" Excitation of Heat Production. It was pointed out earlier that either sympathetic stimulation or circulating epinephrine in the blood can cause an immediate increase in the rate of cellular metabolism; this effect is called *chemical thermogenesis.*

The process of acclimatization greatly affects the intensity of chemical thermogenesis; an animal that has been exposed for several weeks to a very cold environment can increase its heat production as much as 100 to 200 per cent by this mechanism. However, in man, who rarely becomes completely acclimatized to cold environments, it is rare that chemical thermogenesis increases the rate of heat production more than 25 to 50 per cent, but even this amount is important in the control of body temperature.

Increased Thyroxine Output as a Cause of Increased Heat Production. Cooling the preoptic area of the hypothalamus also increases the production of *thyrotropin-releasing factor, thyrotropin,* and *thyroxine,* as will be explained in detail in Chapter 51. The increased thyroxine then increases the rate of cellular metabolism throughout the body. However, this mechanism requires several weeks to reach full effect.

Exposure of animals to extreme cold for several weeks can cause their thyroid gland to increase in size as much as 20 to 40 per cent. Also, military personnel residing for several months in the Arctic develop basal metabolic rates 16 to 20 per cent above normal, and Eskimos have basal metabolic rates even higher than this.

EFFECT OF THE SKIN COLD AND WARMTH RECEPTORS ON BODY TEMPERATURE REGULATION

Nerve receptors for both cold and heat are present everywhere in the skin. The function of these in transmitting cold and warmth sensations into the nervous system was discussed in Chapter 33. However, in addition to their effect in providing these sensations, these receptors play roles in regulating internal body temperature in three different ways: they (1) cause a psychic desire for warmer surroundings and therefore make the person seek appropriate shelter or clothing, (2) transmit nerve impulses into the central nervous system to alter the "setting" of the hypothalamic thermostat, and (3) elicit local cord reflexes that affect skin blood flow or sweating and thereby help to maintain normal body temperature.

Shift in "Setting" of the Hypothalamic Thermostat by Cold or Warmth Sensations from the Skin. When the temperature receptors of the skin are strongly stimulated, impulses from these interact with the hypothalamic temperature regulatory mechanism to reset the control level of the hypothalamic thermostat slightly. For instance, when the skin becomes overheated, the thermostatic setting is reduced a few tenths of a degree below 37.0° C. That is, sweating begins to occur at perhaps 36.5° C. instead of at the normal 37.0° C. level, and increased heat production begins to occur at levels below 36.6° C. instead of below the normal level of 37.1° C.

Thus, signals from the skin temperature receptors make the thermostatic mechanisms more effective than usual, because through their auspices overheating of the skin causes the heat-losing activities of the thermostatic center to come into play long before the excessive heat can be transmitted from the skin to the interior of the body. Likewise, excessive cooling of the skin causes heat conservation and increased heat production to begin even before the cold can be transmitted to the interior of the body.

ABNORMALITIES OF BODY TEMPERATURE REGULATION

FEVER

Fever, which means a body temperature above the usual range of normal, may be caused by abnormalities in the brain itself or by toxic substances that affect the temperature regulating centers. Some causes of fever are bacterial diseases, brain tumors, and deterioration of body tissues.

Resetting the Hypothalamic Thermostat in Febrile Diseases—Effect of Pyrogens. Many proteins, breakdown products of proteins, and certain other substances, such as lipopolysaccharide toxins secreted by bacteria, can cause the "setting" of the hypothalamic thermostat to rise. Substances that cause this effect are called *pyrogens*. It is pyrogens secreted by toxic bacteria or pyrogens released from degenerating tissues of the body that cause fever during disease conditions. When the setting of the hypothalamic thermostat becomes increased to a higher level than normal, all the mechanisms for raising the body temperature are set into play, including heat conservation and increased heat production. Simultaneously, the person experiences a severe chill because of a cold skin caused by intense cutaneous vasoconstriction. Also, he begins to shiver. Within a few hours after the thermostat has been set to a higher level, the body temperature also approaches this level.

To give an idea of the extremely powerful effect of pyrogens in resetting the hypothalamic thermostat, as little as 1 *microgram* of lipopolysaccharide pyrogen derived from *E. coli* can cause severe fever when injected into man.

Antipyretics. Aspirin, antipyrine, aminopyrine, and a number of other substances known as "antipyretics" have an effect on the hypothalamic thermostat opposite to that of the pyrogens. In other words, they cause the setting of the thermostat to be lowered so that the body temperature falls, though usually not more than a degree or so. Aspirin is especially effective in lowering the hypothalamic setting when pyrogens have raised the setting, but aspirin will not lower the normal temperature.

HEAT STROKE

The limits of extreme heat that one can stand depend almost entirely on whether the heat is dry or wet. If the air is completely dry and sufficient convection air currents are flowing to promote rapid evaporation from the body, a person can withstand air temperature at 200° F. with no apparent ill effects. On the other hand, if the air is 100 per cent humidified and evaporation cannot occur or if the body is in water, the body temperature begins to rise whenever the surrounding temperature rises above approximately 94° F. If the person is performing heavy work, this critical temperature level may fall to 85° to 90° F.

Also, as the body temperature rises, the basal metabolism increases about 6 per cent for each degree F. rise, or 10 per cent for each degree C. rise, because of the intrinsic effect of heat to increase the rates of chemical reactions. Once the body temperature rises to approximately 110° F., the rate of metabolism has doubled.

Unfortunately, there is a limit to the rate at which the body can lose heat even with maximal sweating. Furthermore, when the hypothalamus becomes excessively heated, its heat regulating ability becomes greatly depressed and sweating diminishes. As a result, a vicious cycle develops: high temperature causes still greater production of heat and a still higher body temperature, etc. Once the body temperature rises above 107° to 110° F., the heat regulating mechanisms often can no longer dissipate the excessive heat being produced. Therefore, the temperature may then rise abruptly until it causes death, unless this rise is checked artificially.

Harmful Effects of High Temperature. When the body temperature rises above approximately 106° F., the parenchyma of many cells usually begins to be damaged. The pathologic findings in a person who dies of hyperpyrexia are local hemorrhages and parenchymatous degeneration of cells throughout the entire body. The brain is especially likely to suffer, because neuronal cells once destroyed can never be replaced. When the body temperature rises to 110° F., the person usually has only a few hours to live unless his temperature is brought back within normal range rapidly by sponging his body with alcohol, which evaporates and cools the body, or by bathing him in ice water.

EXPOSURE OF THE BODY TO EXTREME COLD

A person exposed to ice water for approximately 20 to 30 minutes ordinarily dies unless treated immediately. By that time, the internal body temperature will have fallen to about 77° F. Yet if he is warmed rapidly by application of external heat, his life can often be saved.

Once the body temperature has fallen below 85° F., the ability of the hypothalamus to regulate temperature is completely lost, and it is greatly impaired even when the body temperature falls below approximately 94° F. Part of the reason for this loss of temperature regulation is that the rate of heat production in each cell is greatly depressed by the low temperature. Also, sleepiness and even coma are likely to develop, which depress the activity of the central nervous system heat-control mechanisms and prevent shivering.

REFERENCES

American Physiological Society: Adaptation to Environment. *In* Handbook of Physiology. Baltimore, The Williams & Wilkins Co., 1965, Sec. IV.

Atkins, E.: Pathogenesis of fever. *Physiol. Rev.,* 40:580, 1960.

Benzinger, T. H.: Heat regulation: homeostasis of central temperature in man. *Physiol. Rev.,* 49:671, 1969.

Consolazio, C. F.; Johnson, R.; and Pecora, L.: Physiological Measurements of Metabolic Functions in Man. New York, McGraw-Hill Book Company, 1963.

Cooper, K. E.: Temperature regulation and the hypothalamus. *Brit. Med. Bull.,* 22:238, 1966.

Hammel, H. T.: Regulation of internal body temperature. *Ann. Rev. Physiol.,* 30:641, 1968.

Hardy, J. D. (ed.): Temperature; Its Measurement and Control in Science and Industry. Biology Medicine Level. New York, Reinhold Publishing Corp., 1963, Pt. 3.

Hemingway, A.: Shivering. *Physiol. Rev., 43*: 397, 1963.

Kleiber, M.: Respiratory exchange and metabolic rate. *In* Handbook of Physiology. Baltimore, The Williams & Wilkins Co., 1965, Sec. III, Vol. II, p. 927.

Myant, N. B.: Some aspects of the control of cell metabolism. *In* Bittar, E. E., and Bittar, N. (eds.): The Biological Basis of Medicine. New York, Academic Press, Inc., 1968, Vol. 2, p. 133.

CHAPTER 48

DIETARY BALANCES; REGULATION OF FEEDING; OBESITY; AND VITAMINS

The intake of food must always be sufficient to supply the metabolic needs of the body and yet not enough to cause obesity. Also, since different foods contain different proportions of proteins, carbohydrates, and fats, appropriate balance must be maintained between these different types of food so that all segments of the body's metabolic systems can be supplied with the requisite materials. This chapter therefore discusses the problems of balance between the three major types of food, the mechanisms by which the intake of food is regulated in accordance with the metabolic needs of the body, and some aspects of vitamin metabolism.

DIETARY BALANCES

Energy Available in Foods. The energy liberated from each gram of carbohydrate as it is oxidized is 4.1 Calories; from fat, 9.3 Calories; and from protein, 4.35 Calories. However, these different substances vary in the average percentages absorbed from the gastrointestinal tract: approximately 98 per cent of the carbohydrate, 95 per cent of the fat, and 92 per cent of the protein. Therefore, in round figures the average *physiologically available energy* in each gram of the three different foodstuffs in the diet is:

	Calories
Carbohydrates	4.0
Fat	9.0
Protein	4.0

Average Composition of the Diet. The average American receives approximately 15 per cent of his energy from protein, about 40 per cent from fat, and 45 per cent from carbohydrates. In most other parts of the world the quantity of energy derived from carbohydrates far exceeds that derived from both proteins and fats.

Daily requirement for protein.
About 30 grams of the body proteins are
degraded and used for energy daily.
Therefore, an average man can maintain
his normal stores of protein provided
that his *daily intake is above 30 to 45
grams.*

Another factor that must be considered in analyzing the proteins of the
diet is whether the dietary proteins are
complete proteins or *partial* proteins. A
partial protein is one that is deficient in
one or more of the essential amino acids,
so that it cannot be used fully to maintain protein balance. A particular example of this occurs in the diet of many
African natives who subsist primarily on
a corn meal diet. The protein of corn is
almost totally lacking in tryptophan; and
this means that this diet, in effect, is
almost completely protein deficient. As a
result, the natives, especially the children, develop the protein deficiency
syndrome called *kwashiorkor,* which
consists of failure to grow, lethargy, depressed mentality, and hypoprotein
edema.

**Relative Utilization of Fat and Carbo-
hydrates—The Respiratory Quotient.**
When one molecule of glucose is oxidized, the number of molecules of carbon dioxide liberated is exactly equal to
the number of oxygen molecules necessary for the oxidative process. Therefore, the *respiratory quotient,* which is
defined as the *ratio of carbon dioxide
output to oxygen usage,* is 1.00. On the
other hand, during the metabolism of fat
the ratio of carbon dioxide output to
oxygen usage falls to about 0.71, and
pure protein metabolism gives a respiratory quotient of 0.80 to 0.85.

Shortly after a meal, almost all the
food metabolized is carbohydrates.
Consequently, the respiratory quotient
approaches 1.00 at this time. Approximately 8 to 10 hours following a meal,
the quantity of carbohydrate being
metabolized is relatively slight, and the
respiratory quotient approaches that
for fat metabolism; that is, approximately 0.71.

In diabetes mellitus, little carbohydrate is utilized by the body, and consequently, most of the energy is derived
from fat. Therefore, most persons with
severe diabetes have respiratory quo-
tients approaching the value for fat
metabolism, 0.71.

REGULATION OF FOOD INTAKE

Hunger. The term "hunger" means a
craving for food, and it is associated
with a number of objective sensations.
For instance, in a person who has not
had food for many hours, the stomach
undergoes intense rhythmic contractions called *hunger contractions.* These
cause a tight or a gnawing feeling in the
pit of the stomach and sometimes actually cause pain called *hunger pangs.* In
addition to the hunger pangs, the hungry
person also becomes more tense and
restless than usual, and often has a
strange feeling throughout his entire
body that might be described by the non-
physiological term "twitterness."

Satiety. Satiety is the opposite of
hunger. It means a feeling of fulfillment
in the quest for food. Satiety usually
results from a filling meal, particularly
when the person's nutritional storage
depots, the adipose tissue and the glycogen stores, are already filled.

NEURAL CENTERS FOR REGULATION OF FOOD INTAKE

Stimulation of the *lateral hypothalamus* causes an animal to eat voraciously,
while stimulation of the *ventromedial
nuclei of the hypothalamus* causes complete satiety, and, even in the presence
of highly appetizing food, the animal
will still refuse to eat. Therefore, we can
label the lateral nuclei of the hypothalamus as the *hunger center* or the *feeding
center,* and the ventromedial nuclei of
the hypothalamus as a *satiety center.*

The feeding center operates by directly exciting the emotional drive to search for food. On the other hand, it is believed that the satiety center operates primarily by inhibiting the feeding center.

Other Neural Centers that Enter into Feeding. Higher centers than the hypothalamus also play important roles in the control of feeding, particularly in the control of appetite. These centers include especially the amygdala and the cortical areas of the limbic system. Destruction of the amygdala on both sides of the brain causes "psychic blindness" in the choice of foods. In other words, the person loses or at least partially loses his appetite control of the type and quality of food he eats.

The cortical regions of the limbic system, including the infraorbital regions, the hippocampal gyrus, and the cingulate gyrus, all have areas that can either increase or decrease feeding activities. These areas seem especially to play a role in the animal's drive to search for food when he is hungry. It is presumed that these centers are also responsible for determining the quality of food that is eaten. For instance, a previous unpleasant experience with almost any type of food often kills a person's appetite for that food thenceforth.

FACTORS THAT REGULATE FOOD INTAKE

We can divide the regulation of food intake into *long-term regulation*, which means regulation over weeks, and *short-term regulation*, which means regulation of hunger and feeding from minute to minute or hour to hour. Long-term regulation is concerned primarily with maintenance of normal quantities of nutrient stores in the body. Short-term regulation is concerned primarily with the immediate effects of feeding on the alimentary tract.

Long-term Regulation (Metabolic Regulation). An animal that has been starved for a long time and is then presented with unlimited food eats a far greater quantity than does an animal that has been on a regular diet. Thus, the feeding center in the hypothalamus is geared to the nutritional status of the body. Some of the nutritional factors that control the degree of activity of the feeding center are the following:

The glucostatic theory of hunger and of feeding regulation. It has long been known that a decrease in blood glucose concentration is associated with development of hunger, which has led to the so-called glucostatic theory of hunger and of feeding regulation, as follows: When the blood glucose level falls too low, this automatically causes the animal to increase his feeding, which eventually returns the glucose concentration back toward normal. There are two other observations that also support the glucostatic theory: (1) An increase in blood glucose level increases the measured electrical activity in the satiety center in the ventromedial nuclei of the hypothalamus and simultaneously decreases the electrical activity in the feeding center of the lateral nuclei. (2) Chemical studies show that the ventromedial nuclei (the satiety center) concentrates glucose while other areas of the hypothalamus fail to concentrate glucose; therefore, it is assumed that glucose acts by increasing the degree of satiety.

Effect of blood amino acid concentration on feeding. An increase in amino acid concentration in the blood also reduces feeding, and a decrease enhances feeding. In general, though, this effect is not as powerful as the glucostatic mechanism.

Effect of fat metabolites on feeding. The overall degree of feeding varies almost inversely with the amount of adipose tissue in the body. Therefore, many physiologists believe that *long-term regulation* of feeding is controlled mainly by fat metabolites — perhaps by the concentration of free fatty acids in the blood, which is known to be directly proportional to the quantity of adipose tissue in the body. Therefore, it is likely that fats act in the same manner as glucose and amino acids to cause a negative feedback regulatory effect on feeding. It is also possible that this is by far the most important long-term regulator of feeding.

Summary of long-term regulation. Even though our information on the different feedback factors in long-term feeding regulation is not precise, we can make the following general statement: When the nutrient stores of the body fall below normal, the feeding center of the hypothalamus becomes highly active, and the person exhibits increased hunger; on the other hand, when the nutrient stores are abundant, the person loses his hunger and develops a state of satiety.

Short-term Regulation (Nonmetabolic Factors). The degree of hunger or satiety can be temporarily increased or decreased by daily habits. For instance, the normal person has the habit of eating three meals a day, and, if he misses one, he is likely to develop a state of hunger at mealtime despite completely adequate nutritional stores in his tissues. But, in addition to habit, several other short-term physiological stimuli can alter one's desire for food for several hours at a time as follows:

Gastrointestinal distention. When the gastrointestinal tract becomes distended, especially the stomach, inhibitory signals suppress the feeding center, thereby reducing the desire for food. Obviously, this mechanism is of particular importance in bringing one's feeding to a halt during a heavy meal.

Metering of food by head receptors. It is postulated that various "head factors" relating to feeding, such as chewing, salivation, swallowing, and tasting, "meter" the food as it passes through the mouth, and after a certain amount has passed through, the hypothalamic feeding center becomes inhibited. For instance, when a person with an esophageal fistula is fed large quantities of food, even though this food is immediately lost again to the exterior, his appetite is decreased after a reasonable quantity of food has passed through his mouth. This effect occurs despite the fact that the gastrointestinal tract does not become the least bit filled.

Importance of Having Both Long- and Short-term Regulatory Systems for Feeding. The long-term regulatory system obviously helps an animal to maintain constant stores of nutrients in his tissues, preventing these from becoming too low or too high. On the other hand, the short-term regulatory stimuli make the animal eat in a rhythmic pattern, so that food passes through his gastrointestinal tract fairly continuously and so that his digestive, absorptive, and storage mechanisms can all work at a steady pace rather than only when the animal needs food for energy.

It is important, then, that feeding occur rather continuously (but at a rate that the gastrointestinal tract can accommodate), regulated principally by the short-term regulatory mechanism. However, it is also important that the intensity of the daily rhythmic feeding habits be modulated up or down by the long-term regulatory system, based principally on the level of nutrient stores in the body.

OBESITY

Energy Input versus Energy Output. When greater quantities of energy (in the form of food) enter the body than are expended, the body weight increases. Therefore, obesity is obviously caused by excess energy input over energy output. For each 9.3 Calories excess energy entering the body, 1 gram of fat is stored.

Effect of Muscular Activity on Energy Output. About half the energy used each day by the normal person goes into muscular activity, and in the laborer as much as three-fourths. Therefore, muscular activity is by far the most important means by which energy is expended.

Consequently, it is frequently said that obesity results from *too high a ratio of food intake to daily exercise.*

ABNORMAL FEEDING REGULATION AS A PATHOLOGIC CAUSE OF OBESITY

Obesity is often caused by an abnormality of the feeding regulatory mechanism. This can result from either psychogenic factors that affect the regulation or actual abnormalities of the hypothalamus itself.

Psychogenic Obesity. Studies of obese patients show that a large proportion of obesity results from psychogenic factors. Perhaps the most common psychogenic factor contributing to obesity is the prevalent idea that healthy eating habits require three meals a day and that each meal must be filling. Many children are forced into this habit by overly solicitous parents, and the children continue to practice it throughout life. In addition, persons are known often to gain large amounts of weight during or following stressful situations, such as the death of a parent, a severe illness, or even mental depression. It seems that eating is often a means of release from tension.

Hypothalamic Abnormalities as a Cause of Obesity. Many persons with hypophyseal tumors that encroach on the hypothalamus develop progressive obesity, illustrating that obesity often results from damage to the hypothalamus. Though in the normal obese person hypothalamic damage is almost never found, it is possible that function of the feeding center in the obese person is different from that in the nonobese person. For instance, an obese person who has made himself reduce to normal weight usually develops hunger that is demonstrably far greater than that of the normal person. This indicates that the "setting" of his feeding center is at a much higher level of nutrient storage than is that of the normal person.

Genetic Factors in Obesity. Obesity definitely runs in families. For instance, identical twins usually maintain weight levels within 2 pounds of each other throughout life if they live under similar conditions.

The genes can direct the degree of feeding in several different ways, including (1) a genetic abnormality of the feeding center to set the level of nutrient storage high or low and (2) abnormal hereditary psychic factors that either whet the appetite or cause the person to eat as a "release" mechanism.

A genetic abnormality in the *chemistry of fat storage* is also known to cause obesity in a certain strain of rats. In these rats, fat is easily stored in the adipose tissue, but the quantity of lipoprotein lipase formed in the adipose tissue is greatly reduced, so that little of the fat can be removed. This obviously results in a one-way path, the fat continually being deposited but never released. This, too, is another possible mechanism of obesity in some human beings.

STARVATION

Use of Food Stores During Starvation. During starvation the body preferentially uses carbohydrate first, fat second, and protein last. However, because the carbohydrate stores are very slight (only

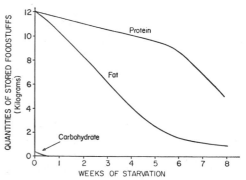

FIGURE 48–1 Effect of starvation on the food stores of the body.

a few hundred grams of glycogen in the liver and muscles), they are used up almost entirely within 24 to 48 hours, as shown in Figure 48–1. Indeed, the carbohydrate stores, if utilized alone without utilization of either fat or protein, would be sufficient to maintain life for only about 13 hours.

Figure 48–1 also illustrates the rates of fat and protein depletion throughout the weeks of starvation, assuming that the person has approximately 15 per cent body fat at the beginning of starvation. Under these conditions essentially all the fat is removed within approximately five to six weeks, and thereafter no fat remains to act as a protein sparer.

Unfortunately, starvation occurs more frequently in unhealthy than in healthy people, such as in cancer patients or patients with fever. In these, the basal metabolic rate is usually considerably above the mean value, and, therefore, the food stores are depleted at a much more rapid rate than would otherwise occur.

Premortem depletion of proteins. In the early stages of starvation, the only utilization of protein each day is the "obligatory" utilization due simply to degradation of intracellular protein as a result of their participation in cell metabolism. However, after most of the fat is gone—that is, after five to six weeks—the proteins must also be used for energy. Consequently, the quantity of protein thereafter decreases rapidly. Obviously, the disappearance of protein from the cells decreases the ability of the cells to perform normal cellular functions, and death soon ensues.

Vitamin Deficiencies in Starvation. The stores of some of the vitamins, especially the water-soluble vitamins—the vitamin B group and vitamin C—do not last long during starvation. Consequently, after a week or more of starvation mild vitamin deficiencies usually begin to appear, and over several weeks severe vitamin deficiencies can occur. Obviously, these can add to the debility that leads to death.

VITAMINS

The study of vitamin and mineral metabolism rightfully falls in the province of biochemistry. However, a few vitamins are very important to the physiology of the body and therefore deserve special comment.

A vitamin is an organic compound that is needed in small quantities for operation of normal bodily metabolism and which cannot be manufactured in the cells of the body. Table 48–1 illustrates the usually recommended daily requirements of the more important vitamins.

VITAMIN A

The basic chemical function of vitamin A in the metabolism of the body is not known except in relation to its use in the formation of retinal pigments, which was discussed in Chapter 34. Nevertheless, some of the other physiologic results of vitamin A lack have been well documented. Vitamin A is necessary for normal growth of most cells of the body and especially for normal growth and proliferation of epithelial cells. When vitamin A is lacking, the epithelial structures of the body tend to become stratified and keratinized. Therefore, vitamin A deficiency manifests itself by (1) scaliness of the skin and sometimes acne, (2) failure of growth of young

TABLE 48–1   DAILY REQUIREMENTS OF
THE VITAMINS

Vitamin	Daily Requirement
A	3.1 mg.
Thiamine	1.3 mg.
Riboflavin	1.8 mg.
Niacin	18 mg.
Ascorbic acid	80 mg.
D (children and during pregnancy)	11 μg.
E	unknown
K	none
Folic acid	unknown
B_{12}	unknown
Pyridoxine	unknown
Pantothenic acid	unknown

animals, (3) failure of reproduction in many animals, associated especially with atrophy of the germinal epithelium of the testes and sometimes with interruption of the female sexual cycle, and (4) keratinization of the cornea with resultant corneal opacity and blindness.

Also, the damaged epithelial structures often become infected; for example, in the eyes, the kidneys, or the respiratory passages. Therefore, vitamin A has been called an "anti-infection" vitamin. Vitamin A deficiency also frequently causes kidney stones, probably owing to infection in the renal pelvis.

THIAMINE (VITAMIN B₁)

Thiamine operates in the metabolic systems of the body principally as *thiamine pyrophosphate;* this compound functions as a *cocarboxylase,* operating mainly in conjunction with a protein decarboxylase for decarboxylation of pyruvic acid and other α-keto acids.

Thiamine deficiency causes decreased utilization of pyruvic acid and some amino acids by the tissues but increased utilization of fats. Thus, thiamine is specifically needed for final metabolism of carbohydrates and proteins.

Thiamine Deficiency and the Nervous System. In thiamine deficiency the utilization of glucose by the central nervous system may be decreased as much as 50 to 60 per cent. Therefore, it is readily understandable how thiamine deficiency could greatly impair function of the central nervous system. The neuronal cells of the central nervous system frequently show chromatolysis and swelling during thiamine deficiency, changes that are characteristic of neuronal cells with poor nutrition.

Also, thiamine deficiency can cause *degeneration of myelin sheaths* of nerve fibers both in the peripheral nerves and in the central nervous system. In severe thiamine deficiency, the peripheral nerve fibers and fiber tracts in the cord can degenerate to such an extent that *paraly-*

sis occasionally results; and, even in the absence of paralysis, the muscles atrophy, with resultant severe weakness.

Thiamine Deficiency and the Cardiovascular System. Thiamine deficiency also weakens the heart muscle, so that a person with severe thiamine deficiency sometimes develops *cardiac failure* with associated *peripheral edema* and *ascites.* Also, thiamine deficiency causes *peripheral vasodilation* throughout the circulatory system, possibly because of the result of metabolic deficiency in the smooth muscle of the vascular system. As a result, the cardiac output is as great as two times normal despite the weak heart.

Thiamine Deficiency and the Gastrointestinal Tract. Among the gastrointestinal symptoms of thiamine deficiency are indigestion, severe constipation, anorexia, gastric atony, and hypochlorhydria. All these effects possibly result from failure of the smooth muscle and glands of the gastrointestinal tract to derive sufficient energy from carbohydrate metabolism.

The overall picture of thiamine deficiency, including polyneuritis, cardiovascular symptoms, and gastrointestinal disorders, is frequently referred to as "beriberi" — especially when the cardiovascular symptoms predominate.

NIACIN

Niacin, also called *nicotinic acid,* functions in the body in the forms of *diphosphopyridine nucleotide* (DPN) and *triphosphopyridine nucleotide* (TPN). These are hydrogen acceptors that combine with hydrogen atoms as they are removed from food substrates by many different types of dehydrogenases. Typical operation of both of them is presented in Chapter 45. When deficiency of niacin exists, the normal rate of dehydrogenation presumably cannot be maintained, and, therefore, oxidative delivery of energy from the foodstuffs to the functioning elements of the cells likewise cannot occur at normal rates. Clinically, niacin deficiency causes mainly gastrointestinal symptoms, neurologic symptoms, and a characteristic dermatitis. However, it is probably much more proper to say that essentially all functions of the body are depressed. Pathologic lesions often appear in many parts of the central nervous system, and permanent dementia or any of many different types of psychoses may result. The skin develops a cracked, pigmented scaliness in areas that are exposed to mechanical irritation or sun irradiation; thus, it seems as if the skin were unable to repair the different types of irritative damage. Also, niacin deficiency causes intense irritation and inflammation of the mucous membranes of the mouth and other portions of the gastrointestinal tract, thus instituting many digestive abnormalities.

The clinical entity called "pellagra" and the canine disease called "black tongue" are caused mainly by niacin deficiency. Pellagra is greatly exacerbated in persons on a corn diet (such as many of the natives of Africa) because corn is very deficient in the amino acid tryptophan, which can be converted in limited quantities to niacin in the body.

RIBOFLAVIN (VITAMIN B₂)

Riboflavin normally combines in the tissues with phosphoric acid to form two coenzymes, *flavin mononucleotide* (*FMN*), and *flavin adenine dinucleotide* (*FAD*). These in turn operate as hydrogen carriers in several of the important oxidative systems of the body. Usually, DPN, operating in association with specific dehydrogenases, accepts hydrogen removed from various food substrates and then passes the hydrogen to FMN or FAD; finally, the hydrogen is released as an ion into the surrounding fluids to become oxidized by nascent oxygen, the system for which is described in Chapter 45.

Deficiency of riboflavin in lower animals causes severe *dermatitis, vomiting, diarrhea, muscular spasticity* which finally becomes muscular weakness, and then *death* preceded by coma and decline in body temperature. Thus, severe riboflavin deficiency can cause many of the same effects as lack of niacin in the diet; presumably the debilities that result in each instance are due to generally depressed oxidative processes within the cells.

In the human being riboflavin deficiency has never been known to be severe enough to cause the marked debilities noted in animal experiments, but mild riboflavin deficiency is probably common.

VITAMIN B₁₂

Several different *cobalamin* compounds exhibit so-called "vitamin B₁₂" activity. These compounds contain cobalt that has four coordination bonds similar to those found in the hemoglobin molecule. It is likely that the cobalt atom functions in much the same way that the iron atom functions in hemoglobin.

Almost nothing is known about the manner in which vitamin B₁₂ enters into the metabolic organization of cells, though experiments indicate that it acts as a coenzyme in the conversion of amino acids and similar compounds into other substances. It also seems to be important for reducing ribonucleotides to deoxyribonucleotides, a step that is important in the formation of genes. This could explain the two major functions of vitamin B₁₂, (1) promotion of growth and (2) red blood cell maturation. This latter function was described in Chapter 8.

FOLIC ACID (PTEROYLGLUTAMIC ACID)

Several different pteroylglutamic acids exhibit the "folic acid effect." Folic acid enters into the chemical reactions for conversion of glycine into serine and also for formation of methionine, purines, and thymine. These latter two are required for formation of deoxyribonucleic acid. Therefore, folic acid is required for reproduction of the cellular genes. This perhaps explains one of the most important functions of folic acid — that is, to promote growth.

Folic acid is an even more potent growth promotor than vitamin B_{12} and, like vitamin B_{12}, is also important for the maturation of red blood cells, as discussed in Chapter 8.

PYRIDOXINE (VITAMIN B₆)

Pyridoxine exists in the form of *pyridoxal phosphate* in the cells and functions as a coenzyme for many different chemical reactions relating to amino acid and protein metabolism. For instance, it is required as a coenzyme in (a) transamination; (b) decarboxylation; (c) synthesis of at least one amino acid, tryptophan; (d) conversion of a number of the amino acids into other substances needed by the cells; and (e) desaturation of fats in the liver to form unsaturated fats. As a result of this myriad of chemical functions, pyridoxine plays many key roles in metabolism — especially in protein metabolism. Also, it is believed to act in the transport of some amino acids across cell membranes.

In the human being, pyridoxine deficiency has been known to cause convulsions, dermatitis, and gastrointestinal disturbances such as nausea and vomiting. However, this deficiency is rare.

PANTOTHENIC ACID

Pantothenic acid mainly is incorporated in the body into coenzyme A, which has many metabolic roles in the cells. Two of these are (1) acetylation of decarboxylated pyruvic acid to form acetyl Co-A prior to its entry into the tricarboxylic acid cycle and (2) degradation of fatty acid molecules into multiple molecules of acetyl Co-A. Thus, lack of pantothenic acid can lead to depressed metabolism of both carbohydrates and fats.

Deficiency of pantothenic acid in lower animals can cause retarded growth, failure of reproduction, graying of the hair, dermatitis, fatty liver, and hemorrhagic adrenal cortical necrosis. In the human being, no definite deficiency syndrome has been proved, presumably because of the wide occurrence of this vitamin in almost all foods.

ASCORBIC ACID (VITAMIN C)

Ascorbic acid is a strong reducing compound, and it probably can be reversibly oxidized and reduced within the body. It is believed that ascorbic acid functions either as a reducing agent in specific metabolic processes or as an oxidation-reduction system, reversibly exchanging electrons with other oxidation-reduction systems. However, the precise · function of ascorbic acid is almost totally unknown, though many of the physiologic effects of ascorbic acid deficiency have been well catalogued.

Physiologically, the major function of ascorbic acid appears to be maintenance of normal intercellular substances throughout the body. This includes the formation of collagen, probably because of the action of ascorbic acid in synthesis of hydroxyproline. It also enhances the intercellular cement substance between the cells, the formation of bone matrix, and the formation of tooth dentin.

Deficiency of ascorbic acid for 20 to 30 weeks, as occurred during long sailing voyages in olden days, causes *scurvy*, some effects of which are (1) *failure of wounds to heal*, caused by failure of the cells to deposit collagen fibrils and intercellular cement substances; (2) *cessation of bone growth*, because no new matrix is laid down between the cells; and (3) *fragile blood vessel walls*, with resultant small hemorrhagic points throughout the body.

In extreme scurvy, lesions of the gums with loosening of the teeth occur; infections of the mouth develop; vomiting of blood, bloody stools, and cerebral hemorrhage can all occur; and, finally, high fever often develops.

VITAMIN D

Vitamin D increases calcium absorption from the gastrointestinal tract and also helps to control calcium deposition in the bone. These effects are discussed in Chapter 53.

VITAMIN E

Several related compounds exhibit so-called "vitamin E activity." It is doubtful whether vitamin E deficiency occurs significantly in human beings. In lower animals, lack of vitamin E can cause degeneration of the germinal epithelium in the testis and therefore can cause male sterility. Lack of vitamin E can also cause resorption of a fetus after conception in the female. Because of these effects of vitamin E deficiency, vitamin E is sometimes called the "anti-sterility vitamin."

Vitamin E deficiency in animals can also cause paralysis of the hindquarters, and pathologic changes occur in the muscles similar to those found in the disease entity "muscular dystrophy" of the human being. However, administration of vitamin E to patients with muscular dystrophy has not proved to be of any benefit.

Vitamin E is believed to prevent oxidation of the unsaturated fats. In the absence of vitamin E, the quantity of unsaturated fats in the cells becomes diminished, causing abnormal structure and function of such cellular organelles as the mitochondria and the lysosomes. Indeed, the muscular dystrophy-like syndrome that occurs in the vitamin E deficiency probably results from continual rupture of lysosomes and subsequent autodigestion of the muscle.

VITAMIN K

Vitamin K is necessary for formation by the liver of prothrombin and factor VII (proconvertin), both of which are important in blood coagulation. These were discussed in Chapter 10 in relation to blood clotting.

REFERENCES

Adolescent obesity (symposium): *Fed. Proc.*, *25*:1, 1966.
Anand, B. K.: Central chemosensitive mechanisms related to feeding. *In* Handbook of Physiology. Baltimore, The Williams & Wilkins Co., 1967, Sec. VI, Vol. I, p. 249.
Ciba Foundation Study Group No. 11: The Mechanism of Action of Water-Soluble Vitamins. Boston, Little, Brown and Company, 1962.
Ciba Symposium on Nutrition and Infection. Boston, Little, Brown and Company, 1967.
Ciba Symposium on Thiamine Deficiency. Boston, Little, Brown and Company, 1967.
Hunter, S. H., and Provasoli, L.: Comparative physiology: nutrition. *Ann. Rev. Physiol.*, *27*:19, 1965.
Kennedy, G. C.: Food intake, energy balance and growth. *Brit. Med. Bull.*, *22*:216, 1966.
Mayer, J.: General characteristics of the regulation of food intake. *In* Handbook of Physiology. Baltimore, The Williams & Wilkins Co., 1967, Sec. VI, Vol. I, p. 3.
Pike, R. L., and Brown, M. L.: Nutrition: An Integrated Approach. New York, John Wiley & Sons, Inc., 1967.
Robinson, F. A.: The Vitamin Co-factors of Enzyme Systems. New York, Pergamon Press, Inc., 1967.
Wohl, M. G., and Goodhart, R. S.: Modern Nutrition in Health & Disease. 4th ed., Philadelphia, Lea & Febiger, 1968.

ENDOCRINOLOGY AND REPRODUCTION

CHAPTER 49

INTRODUCTION TO ENDOCRINOLOGY, AND THE HYPOPHYSEAL HORMONES

The functions of the body are regulated by two major control systems: (1) the nervous system, which has been discussed, and (2) the hormonal, or endocrine, system. In general, the hormonal system is concerned principally with the different metabolic functions of the body, such as controlling the rates of chemical reactions in the cells, controlling the transport of substances through cell membranes, or taking part in other aspects of cellular metabolism.

A hormone is a chemical substance that is secreted into the body fluids by one cell or a group of cells and that exerts a physiological effect on other cells of the body. The following hormones have proved to be of major significance and are discussed in detail in this and the following chapters:

Adenohypophyseal hormones: *growth hormone, corticotropin, thyrotropin, follicle-stimulating hormone, luteinizing hormone, luteotropic hormone,* and *melanocyte-stimulating hormone.*
Neurohypophyseal hormones: *antidiuretic hormone* and *oxytocin.*

Adrenocortical hormones: especially *cortisol* and *aldosterone.*
Thyroid hormones: *thyroxine* and *calcitonin.*
Pancreatic hormones: *insulin* and *glucagon.*
Ovarian hormones: *estrogens* and *progesterone.*
Parathyroid hormone: *parathormone* and *calcitonin.*
Placental hormones: *chorionic gonadotropin, estrogens,* and *progesterone.*

INTRACELLULAR HORMONAL MEDIATOR—CYCLIC AMP

Many hormones probably exert their effects on cells by first causing the substance *cyclic 3',5'-adenosine monophosphate* (cyclic AMP) to be formed in the cell. It is then the cyclic AMP and not the original hormone that causes the hormonal effects inside the cell. Thus, cyclic AMP is an intracellular hormonal mediator. Figure 49–1 illustrates the function of cyclic AMP in more detail. It is believed that the stimulating hormone acts at the membrane of the target cell to combine with a specific receptor

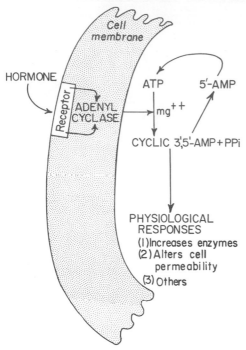

FIGURE 49–1 The cyclic AMP mechanism by which many hormones exert their control of cell function.

for that particular type of hormone. After binding with the receptor, the combination of hormone and receptor activates the enzyme *adenyl cyclase* in the membrane, and the portion of the adenyl cyclase that is exposed to the cytoplasm causes immediate conversion of cytoplasmic ATP into cyclic AMP. The cyclic AMP then initiates any number of cellular functions before it itself is destroyed—functions such as increasing the number of enzymes in the cell, altering the cell permeability, initiating synthesis of specific intracellular chemicals, and so forth. The types of effects that will occur inside the cell are determined by the character of the cell itself. Thus, a thyroid cell stimulated by cyclic AMP synthesizes thyroxine, whereas an adrenocortical cell forms adrenocortical hormones. On the other hand, cyclic AMP affects epithelial cells of the renal tubules by increasing their permeability to water.

The cyclic AMP mechanism has been proposed as an intracellular hormonal mediator for at least the following hormones (and perhaps there are more): thyrotropin, corticotropin, melanocyte-stimulating hormone, antidiuretic hormone, glucocorticoids, thyroid hormone, luteinizing hormone, insulin, glucagon, parathyroid hormone, catecholamines, seratonin, and estrogens.

THE HYPOPHYSIS AND ITS RELATIONSHIP TO THE HYPOTHALAMUS

The hypophysis (Fig. 49–2), also called the *pituitary gland,* is a small gland—about 1 cm. in diameter—that lies in the *sella turcica* at the base of the brain and is connected with the hypothalamus by the *hypophyseal,* or *pituitary, stalk.* Physiologically, the hypophysis is divisible into two distinct portions: the *adenohypophysis,* also known as the *anterior pituitary gland,* and the *neurohypophysis,* also known as the *posterior pituitary gland.* Between these is a small, relatively avascular zone called the *pars intermedia,* which is almost absent in the human being but

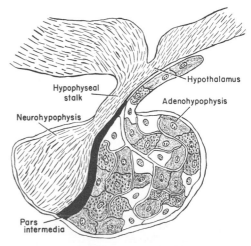

FIGURE 49–2 The hypophysis.

is much larger and much more functional in some lower animals.

Six important hormones plus several less important ones are secreted by the *adeno*-hypophysis, and two important hormones are secreted by the *neuro*-hypophysis. The hormones of the adeno-hypophysis play major roles in the control of metabolic functions throughout the body: (1) *Growth hormone* promotes growth of the animal by affecting many metabolic functions throughout the body, especially protein formation. (2) *Corticotropin* controls the secretion of some of the adrenocortical hormones, which in turn affect the metabolism of glucose, proteins, and fats. (3) *Thyrotropin* controls the rate of secretion of thyroxine by the thyroid gland, and thyroxine in turn controls the rates of most chemical reactions of the entire body. Three separate gonadotropic hormones, (4) *follicle-stimulating hormone,* (5) *luteinizing hormone,* and (6) *luteotropic hormone,* control the growth of the gonads as well as their reproductive activities.

The two hormones secreted by the neurohypophysis play other roles: (1) *Antidiuretic hormone* controls the rate of water excretion into the urine and in this way helps to control the concentration of water in the body fluids. (2) *Oxytocin* (a) helps to deliver milk from the glands of the breast to the nipples during suckling; (b) probably helps to promote fertilization of the ovum following discharge of sperm into the vagina; and (c) probably helps in the delivery of the baby at the end of gestation.

CELL TYPES OF THE ADENOHYPOPHYSIS

The adenohypophysis is composed of three major types of cells: (1) the *chromophobes,* (2) the *acidophils,* and (3) the *basophils.*

The *acidophils* are believed to produce *growth hormone* and *luteotropic*

hormone, whereas the *basophils* are believed to produce *luteinizing hormone, follicle-stimulating hormone, thyrotropin,* and probably *corticotropin.* The *chromophobes* have been supposed to secrete none of the adenohypophyseal hormones but instead simply to be precursor cells of either the acidophils or basophils.

THE HYPOTHALAMIC-HYPOPHYSEAL PORTAL SYSTEM

The adenohypophysis is a highly vascular gland with extensive sinuses among the glandular cells. Almost all of the blood that enters these sinuses passes first through the tissue of the lower hypothalamus and then through small *hypothalamic-hypophyseal portal vessels* into the sinuses. Thus, Figure 49–3 illustrates a small artery supplying the lowermost portion of the hypothalamus called the *median eminence.* Small vascular tufts project into the substance of the median eminence and then return to its surface, coalescing to form the hypothalamic-hypophyseal portal vessels. These in turn pass downward along the hypophyseal stalk to supply the adenohypophyseal sinuses.

Secretion of "Releasing Factors" in the Hypothalamus. Neurons originating in various parts of the hypothalamus send

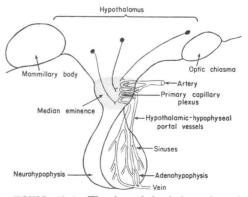

FIGURE 49–3 The hypothalamic-hypophyseal portal system.

nerve fibers into the *median eminence* and into the *tuber cinereum,* which is the hypothalamic tissue that surrounds the hypophyseal stalk. These nerve endings are different from most endings in the central nervous system because their function is not to transmit signals from one neuron to another but merely to secrete small polypeptide hormones called *releasing factors.* These are then transmitted in the portal system to the adenohypophysis to control release of adenohypophyseal hormones from the adenohypophyseal cells.

For each type of adenohypophyseal hormone, with the possible exception of luteotropic hormone, there is a corresponding hypothalamic-releasing factor. Thus, the releasing factor that causes secretion of thyrotropin is called *thyrotropin-releasing factor (TRF)*, the factor that causes release of corticotropin is called *corticotropin-releasing factor (CRF)*, and the factor that causes release of somatotropin (growth hormone) is called *somatotropin-releasing factor (SRF)*. Similarly, releasing factors exist for follicle-stimulating and luteinizing hormones. In addition, a luteotropin *inhibitory* factor has been postulated; this will be discussed in Chapter 55.

PHYSIOLOGIC FUNCTIONS OF GROWTH HORMONE

The functions of most adenohypophyseal hormones are so intimately concerned with the functions of the target glands they stimulate that they will subsequently be discussed along with the functions of the target gland hormones. On the other hand, *growth hormone,* also called *somatotropin,* does not function through a target gland but acts directly on all or almost all tissues of the body. It promotes both increased sizes of the cells and increased mitosis, with development of increased numbers of cells.

Basic Metabolic Effects of Growth Hormone. Growth hormone is known to have the following basic effects on the metabolic processes of the body:

1. Increased rate of protein synthesis in all the cells of the body.

2. Decreased rate of carbohydrate utilization throughout the body.

3. Increased mobilization of fats and use of fats for energy.

Thus, in effect, growth hormone enhances the body proteins, conserves carbohydrates, and uses up the fat stores. It is probable that the increased rate of growth results mainly from the increased rate of protein synthesis.

Enhancement of Amino Acid Transport Through the Cell Membrane as the Basic Mechanism for Promotion of Growth. Since the earliest studies on growth hormone, it has been known that its important effect is to cause protein accumulation in all or almost all cells of the body. Recently, it has been learned that growth hormone directly enhances transport of some of the amino acids through the cell membrane to the interior of the cell. This increases the concentration of amino acids in the cells, which is possibly the factor that sets off the cycle for protein building.

Possible Effects of Growth Hormone on RNA. Some physiologists do not believe that all or even most of the effects of growth hormone in increasing protein synthesis can be ascribed to the promotion of amino acid transport into cells. There are two reasons for this: (1) In the presence of growth hormone, the quantity of messenger RNA in the cells is increased, which could in itself increase the rate of protein synthesis. (2) Protein synthesis can be enhanced by growth hormone even in cell homogenates in which the cell membranes have been totally disrupted. Therefore, it has been suggested that growth hormone might increase the rate of protein synthesis either by increasing the rate of formation of RNA in the nucleus or by increasing the activity of RNA after it has been formed.

Mobilization of Fatty Acids—Ketogenic Effect of Growth Hormone. Growth hormone also increases the release of fatty acids from the adipose tissue and therefore increases the fatty acid concentration in the body fluids. This in turn increases the use of fatty acids for supplying energy to the body. Likewise, far greater quantities of acetoacetic acid are formed in the liver and transported into the blood. Therefore, a person exposed to excessive growth hormone is much more likely to develop ketosis than is the normal person. This is called the *ketogenic effect* of growth hormone.

Effect of Carbohydrate Metabolism—Diabetogenic Effect of Growth Hormone. Growth hormone causes a simultaneous decrease in utilization of carbohydrate for energy. Instead, glycogen is stored in the cells until the cells become saturated; thereafter, the uptake of glucose by the cells becomes greatly depressed. As a result, the blood glucose concentration rises above normal, for which reason it is said that growth hormone has a *diabetogenic effect*. In addition, growth hormone diminishes the transport of glucose into the cells, which is an effect opposite to that of insulin on the transport of glucose. Unfortunately, the causes of these effects are unknown.

Regulation of Growth Hormone Secretion. The rate of growth hormone secretion increases and decreases extremely rapidly in relation to the person's state of nutrition. For instance, extremely high levels occur during wasting diseases or during starvation; the hormone concentration is mainly an inverse reflection of the degree of protein depletion. Figure 49–4 illustrates this relationship: the first block shows growth hormone levels in children with extreme protein deficiency during the malnutrition disease called *kwashiorkor;* the second block shows the levels in the same children after three days of treatment with more than adequate quantities of carbohydrates in their diets; block three shows the levels after treatment with protein supplement in their diet for three days; and block four

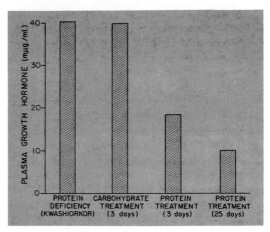

FIGURE 49–4 Effect of extreme protein deficiency on the concentration of growth hormone in the plasma in the disease kwashiorkor. The figure also shows the failure of carbohydrate treatment but the effectiveness of protein treatment in lowering growth hormone concentration. (Drawn from data in Pimstone: *Amer. J. Clin. Nutr., 21*: 482, 1968.)

shows the levels after treatment with protein supplement for 25 days. These results demonstrate that under very severe conditions of protein malnutrition, adequate calories alone are not sufficient to correct 'the excess production of growth hormone. Instead, the protein deficiency must also be corrected before the growth hormone concentration will return to normal.

Thus, it is almost certain that growth hormone secretion is controlled moment by moment by the nutritional status of the body, and it seems that the most important nutrient in the control of growth hormone secretion is protein, though changes in blood glucose concentration can also cause extremely rapid and dramatic alterations in growth hormone secretion. Consequently, it can be postulated that growth hormone operates in a feedback control system as follows: When the tissues begin to suffer from malnutrition, especially from poor protein nutrition, large quantities of growth hormone are secreted. Growth hormone, in turn, promotes the transport of amino acids into the cells and perhaps also promotes increased activity of the RNA system for formation of proteins, thus enhancing the protein nutritional status of the cells and thereby correcting the malnutrition.

ABNORMALITIES OF GROWTH HORMONE SECRETION

Dwarfism. Some instances of dwarfism result from deficiency of adenohypophyseal secretion during childhood — especially deficiency of growth hormone. In general, the features of the body develop in appropriate proportion to each other, but the rate of development is greatly decreased. A child who has reached the age of 10 may have the bodily development of a child of 4 to 5, whereas the same person on reaching the age of 20 may have the bodily development of a child of 7 to 10.

Giantism. Occasionally, the acidophilic cells of the adenohypophysis become excessively active, and some-

times even acidophilic tumors occur in the gland. As a result, large quantities of growth hormone and probably, to a less extent, some of the other adenohypophyseal hormones are produced. All body tissues grow rapidly, including the bones, and if the epiphyses of the long bones have not become fused with the shafts — that is, if the person has not reached adolescence — his height increases so that he becomes a giant. The giant ordinarily has hyperglycemia, and the beta cells of the islets of Langerhans in the pancreas may eventually degenerate, partially because they become overactive owing to the hyperglycemia and partially because of a direct overstimulating effect of growth hormone on the islet cells. Consequently, many giants finally develop full-blown diabetes mellitus.

Acromegaly. If adenohypophyseal acidophilia occurs after adolescence — that is, after the epiphyses of the long bones have fused with the shafts — the

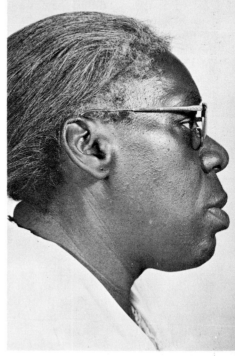

FIGURE 49–5   An acromegalic. (Courtesy of Dr. Herbert Langford.)

person cannot grow taller; but his soft tissues can continue to grow, and the bones can grow in thickness. This condition is known as "acromegaly." Enlargement is especially marked in the small bones of the hands and feet and the *membranous bones,* including the cranium, the nose, the bosses on the forehead, the supra-orbital ridges, the lower jawbone, and portions of the vertebrae, for their growth does not cease at adolescence anyway. Consequently, the jaw protrudes forward, sometimes as much as a half inch or more, the forehead slants forward because of excess development of the supra-orbital ridges, the nose increases to as much as twice normal size, the foot requires a size 14 or larger shoe, and the fingers become extremely thickened so that the hand develops a size almost twice normal. A typical acromegalic is shown in Figure 49–5.

CHEMISTRY OF THE ADENOHYPOPHYSEAL HORMONES

All the adenohypophyseal hormones thus far isolated are small proteins or polypeptides. The molecular weights of these differ for the same hormone in different types of animals, but they usually range between 25,000 and 100,000. Corticotropin, for instance, is a polypeptide containing 39 amino acids.

The molecular weight of human growth hormone is approximately 21,500. That of growth hormone secreted by most lower animals is about twice this. It is particularly significant that bovine growth hormone, or that from other animals below the level of primates, has no significant effect in the human being. For this reason, therapeutic use of growth hormone, or that from other animals below the level of primates, has no significant effect in the human being. Because of this, therapeutic use of growth hormone has been long impeded.

MELANOCYTE-STIMULATING HORMONE

Another hormone normally secreted in small quantities, and perhaps in large quantities under abnormal conditions, is melanocyte-stimulating hormone (MSH), also known as *intermedin.* This hormone is secreted by the *pars intermedia* of the hypothalamus. MSH stimulates the *melanocytes,* which are cells that contain the black pigment melanin and that occur in abundance between the dermis and epidermis of the skin.

In some amphibia, melanin is collected in the melanocytes in small granules called *melanosomes.* When melanocyte-stimulating hormone is not available, the melanosomes become concentrated near the nuclei of the melanocytes in such a way that the melanocytes appear light colored. Then, when MSH is secreted, the melanosomes disperse throughout the cytoplasm of the melanocytes, and the entire cell becomes almost black within a few seconds to minutes. In this way, the color of the skin can be changed from very light in the absence of the hormone to very dark in its presence.

Exposure of the human being to melanocyte-stimulating hormone over a period of days causes intense darkening of the skin. It has a much greater effect in persons who have a genetically dark skin than in persons who have a genetically light skin. It is likely that this hormone acts in a different manner in the human being than in amphibia, for much of the melanin formed by human melanocytes actually leaves the melanocytes and becomes dispersed in the cells of the epidermis rather than remaining in the melanocytes.

Corticotropin, one of the major hormones secreted by the adenohypophysis, has about one-thirtieth the melanocyte-stimulating effect of MSH, and the two hormones are often secreted in high concentrations simultaneously. This frequently occurs in Addison's disease in which the adrenal cortex is damaged so much that it cannot secrete adreno-cortical hormones; lack of these hormones causes compensatory increase in corticotropin secretion and, along with it, increased MSH secretion. Consequently, the skin becomes considerably darkened. MSH has exactly the same chemical structure as the first 13 amino acid portion of the 39 amino acid corticotropin polypeptide chain.

THE NEUROHYPOPHYSIS AND ITS RELATION TO THE HYPOTHALAMUS

The neurohypophysis is composed mainly of glial-like cells called *pituicytes*. However, the pituicytes do not secrete hormones; they act simply as a supporting structure for large numbers of *terminal nerve fibers* and *terminal nerve endings* from nerve tracts that originate in the *supraoptic* and *paraventricular nuclei* of the hypothalamus, as shown in Figure 49–6. These tracts pass to the neurohypophysis through the *hypophyseal stalk* (pituitary stalk). It is the nerve endings that secrete the neurohypophyseal hormones: (1) *antidiuretic hormone* (ADH, also called *vasopressin*) and (2) *oxytocin*. Both of these hormones are small peptides, having eight amino acids in the peptide chain. If the hypophyseal stalk is cut near the hypophysis, leaving

the entire hypothalamus intact, the neurohypophyseal hormones continue to be secreted almost normally, but they are then secreted by the cut ends of the fibers within the hypothalamus and not by the nerve endings in the neurohypophysis. Therefore, it is almost certain that the neurohypophyseal hormones are formed in the cell bodies of the supraoptic and paraventricular nuclei and are then transported down the nerve fibers to the nerve endings in the neurohypophysis. *ADH is formed primarily in the supraoptic nuclei*, while *oxytocin is formed primarily in the paraventricular nuclei.*

Under resting conditions, large quantities of both ADH and oxytocin accumulate in the neurohypophysis, probably bound with a "carrier protein" in the terminal nerve fibers themselves. However, when nerve impulses are transmitted downward along the fibers from the supraoptic and paraventricular nuclei, the hormones are immediately released from the nerve endings and are absorbed into the local blood vessels.

PHYSIOLOGIC FUNCTIONS OF ANTIDIURETIC HORMONE (VASOPRESSIN)

Extremely minute quantities of antidiuretic hormone (ADH), as little as 2 millimicrograms, when injected into a person cause antidiuresis; that is, decreased excretion of water by the kidneys. This antidiuretic effect was discussed in detail in Chapter 25. Briefly, in the absence of ADH, the distal tubules, collecting tubules, and parts of the loops of Henle are almost totally impermeable to water, which prevents significant reabsorption of water and therefore allows extreme loss of water into the urine. On the other hand, in the presence of ADH, the permeability of these tubules to water increases greatly and allows most of the water to be reabsorbed, thereby conserving water in the body.

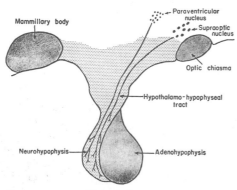

FIGURE 49–6 Hypothalamic control of the neurohypophysis.

When the body fluids become highly concentrated, the supraoptic nuclei become excited, impulses are transmitted to the neurohypophysis, and ADH is secreted. This passes by way of the blood to the kidneys, where it increases the permeability of the distal tubules and collecting tubules to water. As a result, most of the water is reabsorbed from the urine, while electrolytes continue to be lost. This effect dilutes the extracellular fluids, returning them to a reasonably normal osmotic composition.

Effect of Hemorrhage—Pressor Effect of ADH. The most powerful stimuli of all for increasing the secretion of ADH are severe hemorrhage and circulatory shock. It is most generally believed that the decreased pressure in the various baroreceptor areas of the body causes baroreceptor signals to be transmitted to the supraoptic nuclei to increase the output of ADH.

The *normal* rate of ADH secretion is much too small to cause any significant pressor effect. However, the quantities of ADH secreted following severe hemorrhage *are* large enough to cause a mild pressor effect in the circulatory system, thereby helping to return the arterial pressure toward normal.

Other Factors that Affect ADH Production. Other factors that frequently *increase* the output of ADH include *trauma* to the body, *pain, anxiety,* and drugs such as *morphine, nicotine, tranquilizers,* and some *anesthetics.* Each of these factors can cause retention of water in the body. This explains the frequent accumulation of water in many emotional states, and it also explains the diuresis that occurs when the state is over.

A substance that *inhibits* ADH secretion is *alcohol.* Therefore, during an alcoholic bout, lack of ADH allows marked diuresis. Alcohol probably also dilates the afferent arterioles of the nephrons, which adds to the diuretic effect.

Other Actions of ADH. Large doses of ADH can also cause contraction of almost any smooth muscle tissue in the body, including contraction of most of the intestinal musculature, the bile ducts, and the uterus. However, the concentrations required to cause these effects are far greater than that required to cause antidiuresis, and it is doubtful that these are significant physiologic effects for normal function of the body.

OXYTOCIC HORMONE

Effect on the Uterus. An "oxytocic" substance is one that causes contraction of the pregnant uterus. The hormone *oxytocin* powerfully stimulates the pregnant uterus, especially toward the end of gestation. Therefore, many obstetricians believe that this hormone is at least partially responsible for effecting birth of the baby. This is supported by the following facts: (1) In a hypophysectomized animal, the duration of labor is considerably prolonged, thus indicating a probable effect of oxytocin during delivery. (2) The amount of oxytocin in the plasma has been reported to increase during labor. (3) Stimulation of the cervix in a pregnant animal elicits nervous signals that pass to the hypothalamus and cause increased secretion of oxytocin.

Possible Effect of Oxytocin in Promoting Fertilization of the Ovum. Sexual stimulation of the female during intercourse increases the secretion of oxytocin, and the increased oxytocin is probably at least partially responsible for the uterine contractions that occur during the female orgasm. For these reasons, it has been proposed that oxytocin promotes fertilization of the ovum by causing uterine contractions that propel the male semen upward through the fallopian tubes.

Effect of Oxytocin on Milk Ejection. Oxytocin plays an especially important function in the process of lactation, for this hormone causes milk to be expressed from the alveoli into the ducts so that the baby can obtain it by suckling. This mechanism works as follows: The suckling stimuli on the nipple of the breast cause impulses to be transmitted through the sensory nerves to the brain, the impulses finally reaching the paraventricular nuclei in the anterior hypothalamus to cause release of oxytocin. The oxytocin then is carried by the blood to the breasts where it causes contraction of *myoepithelial cells,* which lie outside of and compress the alveoli of the mammary glands. In less than a minute after the beginning of suckling, milk begins to flow. This effect is frequently called *milk letdown* or *milk ejection.*

REFERENCES

Bajusz, E.: Physiology and Pathology of Adaptation Mechanisms: Neural-Neuro-Endocrine-Hormonal. New York, Pergamon Press, Inc., 1968.

Berson, S. A., and Yalow, R. S.: Peptide hormones in plasma. Harvey Lectures, 1966–1967, p. 107

Brown-Grant, K.: The action of hormones on the hypothalamus. *Brit. Med. Bull., 22:*273, 1966.

Bullough, W. S.: The control of tissue growth. *In* Bittar, E. E., and Bittar, N. (eds.): The Biological Basis of Medicine. New York, Academic Press, Inc., 1968, Vol. 1, p. 311.

Farrell, G.: Fabre, L. F.: and Rauschkolb, E. W.: The neurohypophysis. *Ann. Rev. Physiol., 30:*557, 1968.

Growth hormone (symposium). *Ann. N.Y. Acad. Sci., 148:*289, 1968.

Guillemin, R.: The adenohypophysis and its hypothalamic control. *Ann. Rev. Physiol., 29:*313, 1967.

Harris, G. W., and Donovan, B. T.: The Pituitary Gland. 3 volumes. Berkeley, University of California Press, 1966.

Knobil, E., and Hotchkiss, J.: Growth hormone. *Ann. Rev. Physiol., 26:*47, 1964.

McCann, S. M., and Porter, J. C.: Hypothalamic pituitary stimulating and inhibiting hormones. *Physiol. Rev., 49:*240, 1969.

Venning, E. H.: Adenohypophysis and adrenal cortex. *Ann. Rev. Physiol., 27:*107, 1965.

CHAPTER 50

THE ADRENOCORTICAL HORMONES

The adrenal glands, which lie at the superior poles of the two kidneys, are each composed of two distinct parts, the *adrenal medulla* and the *adrenal cortex*. The adrenal medulla is functionally related to the sympathetic nervous system, and it secretes the hormones *epinephrine* and *norepinephrine* in response to sympathetic stimulation. In turn, these hormones cause almost the same effects as does direct stimulation of the sympathetic nerves in all parts of the body. These effects were discussed in detail in Chapter 40 in relation to the sympathetic nervous system.

The adrenal cortex secretes an entirely different group of hormones called *corticosteroids*. As far as is known, there is no direct functional relationship between the adrenal medulla and the adrenal cortex, their anatomical association being simply a chance occurrence.

Two major types of hormones—the *mineralocorticoids* and the *glucocorticoids*—are secreted by the adrenal cortex. In addition to these, the cortex secretes small amounts of *androgenic hormones*, which exhibit the same ef-fects in the body as the male sex hormone testosterone.

FUNCTIONS OF THE MINERALOCORTICOIDS— ALDOSTERONE

Loss of adrenocortical secretion causes death within three days to a week unless the person receives extensive salt therapy or mineralocorticoid therapy. Without mineralocorticoids, the sodium and chloride concentrations of the extracellular fluid decrease markedly, and the total extracellular fluid volume and blood volume also become greatly reduced. The person soon develops diminished cardiac output, which proceeds to a shocklike state followed by death. This entire sequence can be prevented by the administration of aldosterone or some other mineralocorticoid. Therefore, the mineralocorticoids are said to be the "life-saving" portion of the adrenocortical hormones, while the

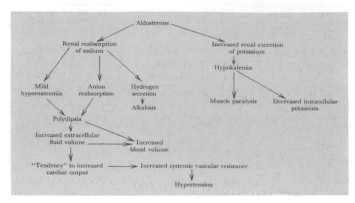

FIGURE 50-1 Effects of aldosterone.

glucocorticoids are of particular importance in helping the person resist different types of stresses, as is discussed later in the chapter.

Aldosterone exerts at least 95 per cent of the mineralocorticoid activity of the adrenocortical secretion, but two other hormones secreted by the adrenal cortex provide at least some significant mineralocorticoid activity. These are *corticosterone*, which also exerts glucocorticoid effects, and *deoxycorticosterone*, which is secreted in minute quantities and has almost the same effects as aldosterone, but a potency only one-thirtieth that of aldosterone.

The basic actions of the mineralocorticoids are illustrated in Figure 50-1, which shows two major basic effects as follows: (1) *increased renal tubular reabsorption of sodium* and (2) *increased renal excretion of potassium*. These two basic effects, however, cause many additional secondary effects also illustrated in the figure.

RENAL EFFECTS OF ALDOSTERONE

Effect on Tubular Reabsorption of Sodium. By far the most important effect of aldosterone is to increase the rate of tubular reabsorption of sodium. Aldosterone has an especially potent effect in the distal tubule, collecting tubule, and at least part of the loop of Henle. A high secretion rate of aldosterone can decrease the loss of sodium in the urine to as little as a few milligrams per day. Conversely, total lack of aldosterone secretion can cause loss of as much as 12 grams of sodium in the urine in a day, an amount equal to one-seventh of all the sodium in the body.

Effect on Tubular Reabsorption of Chloride. Aldosterone also increases the reabsorption of chloride ions from the tubules. This occurs secondarily to the increased sodium reabsorption as follows: Absorption of the positively charged sodium ions causes an electrical potential gradient to develop between the lumen and the outside of the tubules, with positivity on the outside. This positivity in turn attracts the negatively charged chloride ions through the membrane.

Alkalosis Caused by Aldosterone. As pointed out in the discussion of acid-base balance in Chapter 26, a large proportion of the sodium reabsorption from the tubules results from an exchange reaction in which hydrogen ions are secreted into the tubules to take the place of the sodium that is reabsorbed. Therefore, when the rate of sodium reabsorption is enhanced in response to aldosterone, the secreted hydrogen ions reduce the hydrogen ion content of the body fluids and therefore promote alkalosis.

Increased Renal Excretion of Potassium. At the same time that aldosterone causes increased tubular reabsorption of sodium, it also increases the loss of potassium into the urine. The probable cause of this is the interaction between sodium and potassium when sodium is actively absorbed by the tubular epithelium. That is, when aldosterone stimulates active reabsorption of sodium, much of the sodium is exchanged for potassium ions, similar to its exchange for hydrogen ions as discussed above. Therefore, increased numbers of potassium ions are lost in the urine in response to aldosterone.

The low potassium ion concentration in the body fluids (hypokalemia) sometimes leads to muscle paralysis; this is caused by hyperpolarization of the nerve and muscle fiber membranes (see Chap. 5), which prevents transmission of action potentials.

EFFECTS OF ALDOSTERONE ON FLUID VOLUMES AND CARDIOVASCULAR DYNAMICS

Effect on Extracellular Fluid Volume — Edema. From the preceding discussion, it is evident that mineralocorticoids greatly increase the quantities of sodium and chloride ions in the extracellular fluids. This in turn promotes increased water reabsorption from the tubules (a) by stimulating the hypothalamic antidiuretic hormone system as explained in the preceding chapter and (b) by creating an osmotic gradient across the tubular membrane when the electrolytes are absorbed, this in turn carrying water through the membrane in the wake of the electrolyte absorption. Also, the increased electrolyte concentration causes thirst, thereby making the person drink excessive amounts of water. The final result, therefore, is an increase in extracellular fluid volume that occasionally is enough to cause generalized edema.

Also, since the plasma portion of the blood is part of the extracellular fluid, one of the effects of aldosterone secretion is a mild to moderate increase in blood volume.

Effect on Cardiac Output. Lack of aldosterone secretion has a profound effect on cardiac output, for the extracellular fluid and blood volumes decrease so greatly that venous return becomes markedly compromised. As a result, cardiac output actually falls to shock levels.

Conversely, when aldosterone secretion is excessive the resultant increase in extracellular fluid volume and blood volume increases the cardiac output about 10 to 20 per cent.

Effect on Arterial Pressure. When aldosterone is not secreted, a shock-like state of hypotension occurs because of the greatly reduced cardiac output.

On the other hand, excess secretion of aldosterone usually causes moderate hypertension. The cause of this is probably the following. The increased cardiac output supplies each tissue with too much blood flow, and the local vessels constrict in an attempt to return the blood flow to normal levels. This response obviously increases the total peripheral resistance, which *multiplies* the effect of the increased cardiac output to cause hypertension.

MECHANISM OF ACTION OF ALDOSTERONE

Though we have known for many years the overall effects of mineralocorticoids on the body, the means by which aldosterone acts on the tubular cells to increase transport of sodium has until recently been a complete mystery. But several experiments now point toward the following mechanism.

Aldosterone probably stimulates the specific DNA molecules (the genes) that control sodium reabsorption. These set into motion synthesis of specific messenger RNA molecules that in turn cause formation of specific enzymes that catalyze one or more of the chemical reactions responsible for sodium transport through the cell membrane. About 45 minutes is required for these effects to take place before sodium transport begins to increase.

REGULATION OF ALDOSTERONE SECRETION

Basic Mechanisms for Stimulating Aldosterone Secretion. Though we know most of the effects of aldosterone secretion, we are much less sure of the mechanisms that control its secretion. In general, the theories fall into three different categories, as follows:

Direct stimulation of the adrenal cortex. When blood containing a *low concentration of sodium ions* is infused directly into the adrenal artery, the output of aldosterone often increases 1.5- to 3-fold almost immediately. Likewise, blood containing *excess potassium ions* causes an even greater effect. This illustrates that decreased blood sodium or increased blood potassium could have a direct effect on the adrenal glands to regulate the output of aldosterone. Indeed, the quantitative changes in aldosterone output are sufficient to account for much of the regulation of aldosterone secretion.

The renin-angiotensin mechanism. Infusion of angiotensin into the circulation causes marked increase in aldosterone secretion from the adrenal cortex. Also, either low blood sodium or diminished blood flow to the kidneys increases the formation of renin, which in turn leads to the formation of angiotensin (see Chap. 20). Therefore, it has been suggested that either low blood sodium or low renal blood flow increases aldosterone secretion through the mediation of angiotensin.

Neurosecretory control of adrenocortical secretion. We shall see later in the chapter that one of the adenohypophyseal hormones, corticotropin (ACTH), is the primary regulator of glucocorticoid secretion. This hormone also increases aldosterone secretion to a slight extent but perhaps not enough to have major significance.

Destruction of the diencephalic portion of the brain has, in the hands of some research workers, a depressant effect on adrenocortical secretion of aldosterone, and substances isolated from this area of the brain have been claimed to restore the secretion of aldosterone. Therefore, it has been suggested that decreased sodium, increased potassium, decreased blood flow to the tissues of the diencephalon, or signals from the neck baroreceptors cause the release of a substance (or substances) which in turn increases the output of aldosterone and returns the body fluids and circulatory conditions to normal.

Summary. The precise mechanism or mechanisms by which aldosterone secretion is regulated is unknown. Plasma sodium and potassium changes have direct effects on adrenal output of aldosterone, but most research workers doubt that these direct effects are quantitatively sufficient to explain all the known aspects of regulation of aldosterone secretion.

Unfortunately, both the nervous and renal mechanisms of regulation are still doubtful because not all research workers have been able to confirm all the postulations.

FUNCTIONS OF THE GLUCOCORTICOIDS

Even though mineralocorticoids usually save the life of an adrenalectomized animal, the animal still is not completely normal. Instead, its metabolic systems for utilization of carbohydrates, proteins, and fats are considerably deranged. Furthermore, the animal cannot resist different types of physical or even mental stress, and minor illnesses, such as respiratory tract infections, can lead to death. Therefore, the glucocorticoids have functions just as important to the life of the animal as those of the mineralocorticoids. These are explained in the following sections.

At least 95 per cent of the glucocorticoid activity of the adrenocortical secretions results from the secretion of *cortisol*, known also as *hydrocortisone*. In addition to this, a small amount of glucocorticoid activity is provided by *corticosterone* and a minute amount by *cortisone*.

EFFECT OF CORTISOL ON CARBOHYDRATE METABOLISM

Stimulation of Gluconeogenesis. By far the best-known metabolic effect of cortisol and other glucocorticoids on metabolism is their ability to stimulate gluconeogenesis by the liver, often increasing the rate of gluconeogenesis as much as 6- to 10-fold. The mechanism of this effect is unknown, but it could result from (a) mobilization of amino acids from peripheral tissues, these acids then being converted into glucose; (b) increased active transport of amino acids into the liver cells, where gluconeogenesis occurs; or (c) formation of liver cell enzymes that convert amino acids into glucose.

Decreased Glucose Utilization by the Cells. Cortisol also decreases the rate of glucose utilization by the cells, though this effect is slight. It is possible that the reduction in glucose utilization results secondarily from enhanced utilization of fats, which is discussed below. Also, it is known that glucocorticoids slightly depress glucose transport into the cells; this might be the primary means by which these hormones diminish glucose utilization.

Elevated Blood Glucose Concentration and Adrenal Diabetes. Both the increased rate of gluconeogenesis and the mild reduction in rate of glucose utilization by the cells cause the blood glucose concentration to rise. In addition, the enzyme *glucose 6-phosphatase* is increased in the liver in response to cortisol; this enzyme catalyzes the dephosphorylation of liver glucose and thereby promotes its transport into the

blood, thus further increasing the blood glucose concentration.

The increased blood glucose concentration is called *adrenal diabetes*, and it has many similarities to pituitary diabetes.

EFFECTS OF CORTISOL ON PROTEIN METABOLISM

Reduction in Cellular Protein. One of the principal effects of cortisol on the metabolic systems of the body is reduction of the protein stores in essentially all body cells except those of the liver.

Depressed Transport of Amino Acids into Peripheral Tissues, and Enhanced Hepatic Transport. Recent studies in isolated tissue have demonstrated that cortisol depresses amino acid transport into muscle cells and perhaps into other cells besides those of the liver. In contrast, cortisol enhances the transport of amino acids into liver cells.

Obviously, the decreased transport of amino acids into extrahepatic cells decreases their intracellular amino acid concentrations and as a consequence decreases the synthesis of protein. Yet catabolism of proteins in the cells continues to release amino acids from the already existing proteins, and these diffuse out of the cells to increase the plasma amino acid concentration. Therefore, it is said that *cortisol mobilizes amino acids from the tissues.*

Increased Blood Amino Acids. The failure of amino acids to be incorporated into proteins by the cells under the influence of glucocorticoids, plus the increased catalysis of cellular proteins, increases the blood amino acid concentration; and this perhaps accounts for many of the important functions of glucocorticoids. For instance, the different cells continually synthesize a multitude of chemical substances from amino acids. Therefore, mobilization of amino acids from the tissues might play a valuable function in providing an available supply of amino acids in time of need.

The increased plasma concentration of amino acids, plus the fact that cortisol enhances transport of amino acids into the hepatic cells, could also account for enhanced utilization of amino acids by the liver in the presence of cortisol — such effects as (1) increased rate of deamination of amino acids by the liver, (2) increased protein synthesis in the liver, (3) increased formation of plasma proteins by the liver, and (4) increased conversion of amino acids to glucose — that is, enhanced gluconeogenesis.

Effect on Cellular Enzymes. Cortisol also increases the enzymes required for protein metabolism. This possibly results from the formation of appropriate messenger RNA's.

EFFECTS OF CORTISOL ON FAT METABOLISM

Mobilization of Fatty Acids. In much the same manner that cortisol promotes amino acid mobilization from muscle, it also promotes mobilization of fatty acids from adipose tissue. This in turn increases the concentration of free fatty acids in the plasma and increases their utilization for energy. In this way, cortisol helps to shift the metabolic systems of the cells from utilization of glucose for energy to utilization of fatty acids instead.

REGULATION OF CORTISOL SECRETION — CORTICOTROPIN (ACTH)

Effect of Corticotropin on Adrenocortical Secretion of Cortisol. Corticotropin, also known as *corticotropic hormone* or *adrenocorticotropic hormone* (ACTH), is secreted by the adenohypophysis. It in turn stimulates the adrenal cortex to secrete greatly increased quantities of the glucocorticoids — cortisol and corticosterone — and slightly increased quantities of aldosterone and adrenal androgens. Within minutes, physiological stress can enhance corticotropin secretion as much as 20-fold. For instance, pain stimuli caused by stress are transmitted into the hypothalamus, and this in turn transmits impulses into the median eminence where a neurosecretory product, corticotropin-releasing factor (CRF) is secreted into the hypophyseal portal system and carried by this system to the sinuses of the adenohypophysis. CRF then excites the glandular cells to cause corticotropin secretion. This adenohypophyseal-adrenocortical response to stress is especially important in helping an animal survive under adverse conditions, which are discussed later in the chapter.

Inhibitory effect of cortisol on corticotropin secretion. An elevated concentration of cortisol in the blood causes negative feedback to the adenohypophysis to reduce the production of corticotropin. Though the exact means by which this feedback occurs is unknown, physiologists in general believe that cortisol acts both on the hypothalamus and directly on the adenohypophysis itself to inhibit adenohypophyseal secretion of corticotropin.

Function of Cortisol in Different Types of Stress. It is amazing that almost any type of stress, whether it be physical or neurogenic, will cause an immediate and marked increase in corticotropin secretion, followed within minutes by greatly increased adrenocortical secretion of cortisol. Some of the different types of stress that increase cortisol release are the following:

1. Trauma of almost any type
2. Intense heat
3. Intense cold
4. Injection of norepinephrine and other sympathomimetic drugs
5. Surgical operations
6. Injection of necrotizing substances beneath the skin
7. Restraining an animal so that he cannot move
8. Almost any debilitating disease

Yet we are not sure why enhanced cortisol secretion is of significant benefit to the animal. One guess, which is probably as good as any other, is that the glucocorticoids cause rapid mobilization of amino acids and fats from their cellular stores, making these available both for energy and for synthesis of other compounds needed by the different tissues of the body. It is possible that damaged tissues which are momentarily depleted of proteins can utilize the newly available amino acids to form new proteins that are essential to the lives of the cells. Or perhaps the amino acids are used to synthesize such essential intracellular substances as purines, pyrimidines, and creatine phosphate, which are necessary for maintenance of cellular life.

Effect of Cortisol in Inflammation: Stabilization of Lysosomes. Almost all tissues of the body respond to tissue damage by the process called *inflammation*. There are three stages in this process: (1) leakage of large quantities of plasma-like fluid out of the capillaries into the damaged area followed by clotting of the fluid, (2) infiltration of the area by leukocytes, (3) tissue healing, which is accomplished to a great extent by ingrowth of fibrous tissue.

The amount of cortisol normally secreted does not significantly affect inflammation and healing; but large amounts of cortisol or other glucocorticoids administered to a person block all stages of the inflammatory process, even blocking the initial leakage of plasma from the capillaries into the damaged area.

Unfortunately, we still know very little about the basic mechanism of cortisol's *anti-inflammatory effect*. However, it has recently been discovered that cortisol stabilizes the membranes of cellular lysosomes so that they rupture only with difficulty. It will be recalled that lysosomes contain large quantities of hydrolytic enzymes that are capable of digesting intracellular proteins. Therefore, this stabilizing effect on the lyso-some membrane could account for a major share of the anti-inflammatory effect of cortisol, because it would prevent the usual destruction of tissues that occurs in inflammation as a result of liberation of lysosomal enzymes.

The anti-inflammatory effect of cortisol plays a major role in combating certain types of diseases, such as rheumatoid arthritis, rheumatic fever, and acute glomerulonephritis. All these are characterized by severe local inflammation, and the harmful effects to the body are caused mainly by the inflammation itself and not by other aspects of the disease. When cortisol or other glucocorticoids are administered to patients with these diseases, almost invariably the inflammation subsides within 24 to 48 hours. And even though the cortisol does not correct the basic disease condition itself but merely prevents the damaging effects of the inflammatory response, this alone can often be a lifesaving measure.

CHEMISTRY OF THE ADRENOCORTICAL HORMONES

All the adrenocortical hormones are steroid compounds, and they are formed in the adrenal cortex from acetyl coenzyme A or from cholesterol preformed in the cortical cells in the body. Figure 50–2 illustrates the chemical formulas of cortisol and aldosterone.

Several steroids that do not naturally occur in the adrenal cortex have even more potent actions than the natural hormones. Thus, *prednisolone*, which is a slight modification of cortisol, has three times as much glucocorticoid potency as cortisol but no increase in mineralocorticoid potency. Likewise, *2-methyl-9α-fluorocortisol* has three times as much mineralocorticoid activity as aldosterone and therefore can be used in minute quantities, only 25 micrograms per day, to treat total deficiency of mineralocorticoid secretion.

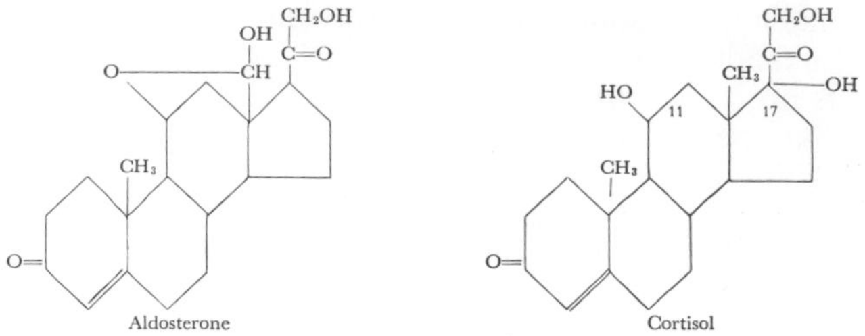

FIGURE 50–2 Two important corticosteroids.

THE ADRENAL ANDROGENS

Several moderately active male sex hormones called *adrenal androgens* are also secreted by the adrenal cortex. Also, progesterone and estrogens, which are female sex hormones, have been extracted from the adrenal cortex, though these are secreted in only minute quantities.

In normal physiology of the human being, the adrenal androgens have almost insignificant effects, but when the adrenal gland oversecretes adrenocortical hormones, the quantity of adrenal androgens then often has excessive masculinizing effects.

ABNORMALITIES OF ADRENOCORTICAL SECRETION

HYPOADRENALISM — ADDISON'S DISEASE

Addison's disease results from failure of the adrenal cortices to produce adrenocortical hormones. Basically, the disturbances in Addison's disease are:

Mineralocorticoid Deficiency. Lack of aldosterone secretion greatly decreases sodium reabsorption and consequently allows sodium ions, chloride ions, and water to be lost into urine in great profusion. The net result is greatly decreased extracellular fluid volume and plasma volume. As a result, the cardiac output decreases, and the patient dies in shock four to seven days after complete cessation of mineralocorticoid secretion.

Glucocorticoid Deficiency. Loss of cortisol secretion makes it impossible for the person with Addison's disease to maintain normal blood glucose concentration between meals because he cannot synthesize significant quantities of glucose by gluconeogenesis. Furthermore, lack of cortisol reduces the mobilization of both proteins and fats from the tissues, thereby depressing many other metabolic functions of the body. Indeed, lack of adequate glucocorticoid secretion makes the person with Addison's disease highly susceptible to the deteriorating effects of different types of stress, and even a mild respiratory infection can sometimes cause death.

Treatment of Persons with Addison's Disease. The untreated person with Addison's disease dies within a few days because of electrolyte disturbances. Yet such a person can usually live for years, though in a weakened condition, if small quantities of mineralocorticoids are administered — for instance, 25 mg. of 2-methyl-9α-fluorocortisol by mouth daily.

Unfortunately, persons treated with mineralocorticoids but without glucocorticoids are still subject to stress reactions that may cause rapid demise. Therefore, full treatment of Addison's disease includes administration of glucocorticoids — usually cortisone, cortisol, or prednisolone — as well as mineralocorticoids.

HYPERADRENALISM — CUSHING'S DISEASE

One type of hypersecretion by the adrenal cortex causes a complex of hormonal effects called *Cushing's disease*. This usually results from general hyperplasia of both adrenal cortices. The hyperplasia in turn is most often caused by increased secretion of corticotropin by the adenohypophysis.

Effects on Electrolytes, Body Fluids, and the Cardiovascular System. In Cushing's disease, sodium retention is moderately enhanced because of the moderately increased aldosterone secretion. This in turn leads to mildly increased extracellular fluid volume, increased plasma volume, mild degrees of hypokalemia, and mild alkalosis. The increase in extracellular fluid volume is usually not enough to cause frank edema, but it may cause slight puffiness of the skin, especially in the facial region, as illustrated in Figure 50–3.

The changes in fluid and electrolyte balance also usually lead to slightly increased cardiac output, as explained earlier in the chapter, and often to hypertension.

Effects on Carbohydrate and Protein Metabolism. The abundance of gluco-corticoids secreted in Cushing's disease causes increased blood glucose concentration, sometimes to values as high as 140 to 200 mg. per cent. This effect probably results mainly from enhanced gluconeogenesis. If this "adrenal diabetes" lasts for many months, the beta cells in the islets of Langerhans occasionally "burn out" because the high blood glucose greatly overstimulates them to secrete insulin. The destruction of these cells then causes frank diabetes mellitus, which is permanent for the remainder of life.

The effects of glucocorticoids on protein catabolism are often profound in Cushing's disease, causing greatly decreased tissue proteins almost everywhere in the body with the exceptions of the liver and the plasma proteins. For instance, decreased protein in the bones causes *osteoporosis* with consequent weakness of the bones.

Effects of the Adrenal Sex Hormones. In Cushing's disease the quantities of adrenal androgens are usually also increased to significant levels, and these

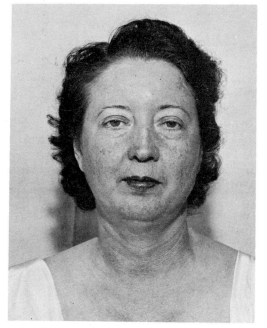 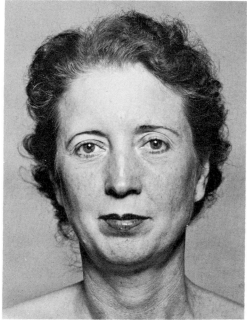

FIGURE 50–3 A person with Cushing's disease before subtotal adrenalectomy (*left*) and after subtotal adrenalectomy (*right*). (Courtesy of Dr. Leonard Posey.)

result in typical masculinizing effects in a female (or in a male child), such as growth of a beard, increased deepness of the voice, occasional baldness, growth of the clitoris to resemble a penis, or growth of a child's penis to adult proportions.

Treatment of Cushing's Disease. Treatment in Cushing's disease usually consists of decreasing the secretion of corticotropin, if this is possible. Hypertrophied pituitary glands or tumors that oversecrete corticotropin can be surgically removed, or destroyed by radiation.

PRIMARY ALDOSTERONISM

Occasionally a small tumor of the adrenal cortex produces extreme amounts of aldosterone but not the other adrenal hormones, giving rise to *primary aldosteronism*. The effects include *increased extracellular fluid volume, hypertension, hypokalemia, mild alkalosis,* and a *tendency toward hypernatremia.* Especially interesting in primary aldosteronism are occasional periods of *muscular paralysis caused by the hypokalemia*, the mechanism of which was explained in Chapter 5.

ADRENOGENITAL SYNDROME

Tumors of the adrenal cortex occasionally secrete large quantities of adrenal androgens (masculinizing hormones) without secreting the other adrenal hormones. The results in the female are growth of a beard, development of a much deeper voice, occasionally development of baldness if she also has the genetic inheritance for baldness, development of a masculine distribution of hair on the body and on the pubis, growth of the clitoris to resemble a penis, and deposition of proteins in the

skin and especially in the muscles to give typical masculine characteristics.

In the prepubertal male a virilizing adrenal tumor causes the same characteristics as in the female, plus rapid development of the male sexual organs and creation of male sexual desires. Typical development of the male sexual organs in a 4-year-old boy with the adrenogenital syndrome is shown in Figure 50–4.

In the adult male, the virilizing characteristics of the adrenogenital syndrome are usually completely obscured by the normal virilizing characteristics of the testosterone secreted by the testes.

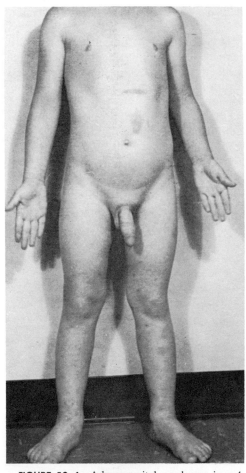

FIGURE 50–4 Adrenogenital syndrome in a 4-year-old boy. (Courtesy of Dr. Leonard Posey.)

REFERENCES

Bransome, E. D., Jr.: Adrenal cortex. *Ann. Rev. Physiol.*, *30*:171, 1968.

Ciba Foundation Study Group No. 27: The Human Adrenal Cortex. Boston, Little, Brown and Company, 1967.

Denton, D. A.: Evolutionary aspects of the emergence of aldosterone secretion and salt appetite. *Physiol. Rev.*, *45*:245, 1965.

Eisenstein, A. B.: The Adrenal Cortex. Boston, Little, Brown and Company, 1967.

Hechter, O., and Halkerston, I. D. K.: Effects of steroid hormones of gene regulation and cell metabolism. *Ann. Rev. Physiol.*, *27*:133, 1965.

McKerns, K. W. (ed.): Functions of the Adrenal Cortex. New York, Appleton-Century-Crofts, 1968, Vols. 1 and 2.

Sharp, G. W. G., and Leaf, A.: Mechanism of action of aldosterone. *Physiol. Rev.*, *46*:593, 1966.

Yates, F. E., and Urquhart, J.: Control of plasma concentrations of adrenocortical hormones. *Physiol. Rev.*, *42*:359, 1962.

CHAPTER 51

THE THYROID HORMONES

The thyroid gland, which is located immediately below the larynx on either side of and anterior to the trachea, secretes *thyroxine* and, also, much less quantities of several other closely related iodinated hormones that have a profound effect on the metabolic rate of the body. Complete lack of thyroid secretion usually causes the basal metabolic rate to fall to −40 to −50, and extreme excesses of thyroid secretion can cause the basal metabolic rate to rise as high as +60 to +100. Thyroid secretion is controlled primarily by *thyrotropin* secreted by the adenohypophysis.

The purpose of this chapter is to discuss the formation and secretion of the thyroid hormones, their functions in the metabolic schema of the body, and regulation of their secretion.

FORMATION AND SECRETION OF THE THYROID HORMONE

Physiologic Anatomy of the Thyroid Gland. The thyroid gland is composed, as shown in Figure 51–1, of large numbers of encysted *follicles* filled with a

substance called *colloid* and lined with *cuboidal epithelioid cells* that secrete into the interior of the follicles. Once the secretion has entered the follicles, it must be absorbed back through the follicular epithelium into the blood before it can function in the body.

Requirements of Iodine for Formation of Thyroxine. To form normal quantities of thyroxine, approximately 35 to 50 mg. of ingested iodine are required *each year,* or approximately *1 mg. per week.* To prevent iodine deficiency, common

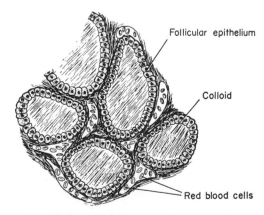

FIGURE 51–1 Microscopic appearance of the thyroid gland, showing the secretion of thyroglobulin into the follicles.

table salt is iodized with one part sodium iodide to every 100,000 parts sodium chloride.

The Iodide Pump (Iodide Trapping). As shown in Figure 51–2, the thyroid cell membranes have a specific ability to transport iodides actively to the interior of the cell and then into the follicle; this is called the *iodide pump,* or *iodide trapping.* In a normal gland, the iodide pump can concentrate iodide to 25 to 50 times its concentration in the blood. However, when the thyroid gland becomes maximally active, the concentration can rise to as high as 350 times that in the blood.

CHEMISTRY OF THYROXINE AND TRIIODOTHYRONINE FORMATION

Oxidation of the Iodides. The first stage in the formation of the thyroid hormones is believed to be oxidative conversion of iodides either to *elemental iodine* or to *some other oxidized form* of iodine that is then capable of combining with the amino acid *tyrosine* to form the thyroid hormone. The reason for believing this is that iodination of tyrosine can be effected readily even in vitro by elemental iodine but not by iodides. Furthermore, large amounts of the enzyme *peroxidase* are present in the thyroid glandular cells. This enzyme is capable of oxidizing iodides, and, when it is absent from the cells, the rate of formation of thyroid hormones is greatly decreased.

Iodination of Tyrosine and Formation of the Thyroid Hormones. Figure 51–3 illustrates the successive stages of iodination of tyrosine and final formation of the two most important thyroid hormones, *thyroxine* and *triiodothyronine.* Tyrosine is first iodized to *monoiodotyrosine* and then to *diiodotyrosine.* Two molecules of diiodotyrosine then conjugate, with the loss of the amino acid alanine, to form one molecule of *thyroxine.* Or one molecule of monoiodotyro-

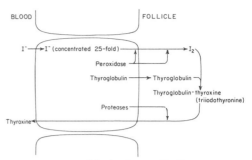

FIGURE 51–2 Mechanisms of iodine transport, thyroxine formation, and thyroxine release into the blood. (Triiodothyronine formation and release parallels that of thyroxine.)

sine combines with one molecule of diiodotyrosine to form *triiodothyronine.*

Storage of Thyroid Hormones in Thyroglobulin. In addition to synthesizing the thyroid hormones, the thyroid cells also secrete a large globulin called *thyroglobulin* (molecular weight 680,000) directly into the thyroid follicles and not into the circulating blood. Furthermore, it is the amino acid tyrosine in this molecule that is converted into monoiodotyrosine, diiodotyrosine, triiodothyronine, and thyroxine. This conversion of tyrosine probably can occur both before and after the thyroglobulin has been secreted into the follicles. It is in the form of this complex molecule thyroglobulin that thyroid hormones are stored in the thyroid gland for a period of several weeks before being released into circulating blood.

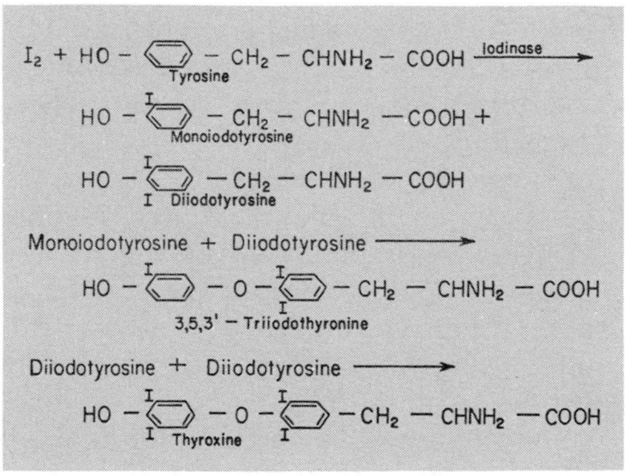

$$I_2 + HO - \langle \rangle - CH_2 - CHNH_2 - COOH \xrightarrow{\text{Iodinase}}$$
Tyrosine

$$HO - \overset{I}{\underset{\text{Monoiodotyrosine}}{\langle \rangle}} - CH_2 - CHNH_2 - COOH +$$

$$HO - \overset{I}{\underset{I \ \text{Diiodotyrosine}}{\langle \rangle}} - CH_2 - CHNH_2 - COOH$$

$$\text{Monoiodotyrosine} + \text{Diiodotyrosine} \longrightarrow$$

$$HO - \overset{I}{\langle \rangle} - O - \overset{I}{\underset{I}{\langle \rangle}} - CH_2 - CHNH_2 - COOH$$
3,5,3' − Triiodothyronine

$$\text{Diiodotyrosine} + \text{Diiodotyrosine} \longrightarrow$$

$$HO - \overset{I}{\underset{I \ \text{Thyroxine}}{\langle \rangle}} - O - \overset{I}{\underset{I}{\langle \rangle}} - CH_2 - CHNH_2 - COOH$$

FIGURE 51–3 Chemistry of thyroxine and triiodothyronine formation.

RELEASE OF THYROXINE FROM THE FOLLICLES

Thyroglobulin itself is not released into the circulating blood but instead is digested by *proteinases* that are also secreted into the follicles by the thyroid cells. The proteinases split thyroxine and minute quantities of triiodothyronine from the thyroglobulin complex and allow these to diffuse through the glandular cells into the blood outside the follicles. The rate of activity of the proteinases (and, therefore, also, the rate of thyroxine release into the circulating body fluids) is controlled by thyrotropin from the adenohypophysis, which is discussed in detail later in the chapter.

TRANSPORT OF THYROXINE AND TRIIODOTHYRONINE TO THE TISSUES

Quantitatively, at least 95 per cent, and probably closer to 100 per cent, of the active thyroid hormones entering the blood are thyroxine, and the remainder is almost entirely triiodothyronine.

Binding of Thyroxine and Triiodothyronine with the Plasma Proteins. On entering the blood, both thyroxine and triiodothyronine immediately combine with several of the plasma proteins, mainly with a globulin called *thyroxine-binding globulin.*

The affinity of thyroxine-binding globulin for thyroxine is approximately three times as great as its affinity for triiodothyronine. For this reason, whenever these two hormones are injected intravenously or secreted by the thyroid gland in equal quantities, the amount of *free* thyroxine that remains in the plasma is only one-third the amount of free triiodothyronine. This allows far more rapid entry of triiodothyronine into the cells than of thyroxine, which causes triiodothyronine to have a potency about three to five times that of thyroxine.

About half the thyroxine in the blood is released to the tissue cells every eight days. Since thyroxine constitutes the major bulk of the thyroid hormones, this slow release of thyroxine causes the very slow action of thyroxine, which is discussed later in the chapter.

FUNCTIONS OF THYROXINE IN THE TISSUES

GENERAL INCREASE IN METABOLIC RATE

The principal effect of thyroxine is to increase the metabolic activities of essentially all tissues of the body. The basal metabolic rate increases to as much as 60 to 100 per cent above normal when large quantities of thyroxine are secreted. The rate of utilization of foods for energy is greatly accelerated. The rate of protein synthesis is at times increased, while at the same time the rate of protein catabolism is also increased. The growth rate of young persons is greatly accelerated. The mental processes are excited, and activity of many other endocrine glands is often increased. Yet despite the fact that these many changes in metabolism are under the influence of thyroxine, the basic mechanism (or mechanisms) by which thyroxine acts is almost completely unknown. Some of the theories are the following:

Effect of Thyroxine on Mitochondria. When thyroxine is given to an animal, the mitochondria in most cells of the body increase in size and also in number. Furthermore, the total membrane surface of all the mitochondria increases almost directly in proportion to the increased metabolic rate of the whole animal. Therefore, it is an obvious deduction that the principal function of thyroxine might be simply to increase the number and activity of mitochondria, and that these in turn increase the rate of formation of ATP to energize cellular function. Unfortunately, though, the increase in number and activity of mito-chondria could just as well be the *result* of increased cell activity as the cause of the increase.

Effect of Thyroxine on Cellular Enzyme Systems. Vast numbers of intracellular enzymes increase in quantity a week or so following administration of thyroxine. For instance, one enzyme, α-glycero-phosphate dehydrogenase, is often in-creased to an activity six times its nor-mal level. Since this enzyme is particu-larly important in the degradation of carbohydrates, its increase could help to explain the rapid utilization of carbo-hydrates under the influence of thyrox-ine. Also, the oxidative enzymes and the elements of the electron transport sys-tem, both of which are normally found in mitochondria, are greatly increased.

Since the enzymes that are increased are mainly those found also in mito-chondria, it is possible that this effect is indirectly caused by the increase in mitochondria.

Latency and Duration of Thyroid Hor-mone Activity. After injection of a rela-tively large quantity of thyroxine into a human being, essentially no effect on the metabolic rate can be discerned during the first 24 hours, which illustrates that there is a long *latent period* before thy-roxine activity begins. Once activity does begin, it increases progressively and reaches a maximum in about 12 days in the human being. Thereafter, it de-creases with a half-time of about 15 days. Some of the activity still persists as long as six weeks to two months later.

The long time required for maximum effect of thyroxine to occur is believed to be caused by the time required for the cells to accumulate increased num-bers of mitochondria and to synthesize the enzymes whose development is promoted by the thyroid hormones. Once these elements have been formed, they presumably continue to act for many weeks, which would explain the long duration of thyroid hormone effect even after the thyroid hormone itself is gone.

PHYSIOLOGIC EFFECTS OF THYROXINE ON DIFFERENT BODILY MECHANISMS

Effect on Basal Metabolic Rate. Exces-sive quantities of thyroxine can occa-sionally increase the basal metabolic rate to as much as 100 per cent above normal. However, in most patients with severe hyperthyroidism the basal meta-bolic rate ranges between 40 and 60 per cent above normal. On the other hand, when no thyroid hormone is produced, the basal metabolic rate falls to −40 to −50, as discussed in Chapter 47. Figure 51–4 shows the approximate relation-ship between the daily supply of thy-roxine and the basal metabolic rate.

Effect on Body Weight. Greatly in-creased thyroxine production almost always decreases the body weight, and greatly decreased thyroxine production almost always increases the body weight; but these effects do not always occur, because thyroxine increases the appetite, which may overbalance the change in metabolic rate.

Effect on Growth. Because anabolism of proteins cannot occur normally in the absence of thyroxine, the growth effect of growth hormone from the pituitary gland is insignificant without concurrent presence of thyroxine in the body fluids. There are two conditions in which this effect on growth is especially evident: First, in growing children who are hypo-

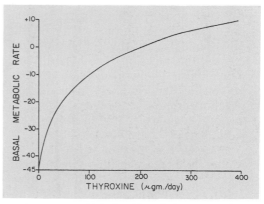

FIGURE 51–4 Relationship of thyroxine daily rate of secretion to the basal metabolic rate.

thyroid, the rate of growth is greatly retarded. Second, in growing children who are hyperthyroid, excessive skeletal growth often occurs, causing the child to become considerably taller than otherwise.

Effect on the Cardiovascular System. Because increased metabolism induced by thyroxine increases the demand of the tissues for nutrient substances, activity of the cardiovascular system is increased in all ways. Especially, there is increase in local blood flow in each tissue, and concomitant increase in cardiac output sometimes to two times normal. Other effects are an increase in heart rate, systolic arterial pressure, pulse pressure, and blood volume.

Effect on Respiration. The increased rate of metabolism caused by thyroid hormone increases the utilization of oxygen and the formation of carbon dioxide; these effects activate all the mechanisms that increase the rate and depth of respiration.

Effect on the Gastrointestinal Tract. In addition to increased rate of absorption of foodstuffs, which has been discussed, thyroxine increases both the rate of secretion of the digestive juices and the motility of the gastrointestinal tract. Often, diarrhea results. Also, associated with this increased secretion and motility is increased appetite, so that the food intake usually increases. Lack of thyroxine causes constipation.

Effect on the Central Nervous System. In general, thyroxine increases the rapidity of cerebration, while, on the other hand, lack of thyroid hormone decreases this function. The hyperthyroid individual is likely to develop extreme nervousness and is likely to have many psychoneurotic tendencies such as anxiety complexes, extreme worry, or paranoias.

Effect on the Function of the Muscles. Moderate increase in thyroxine usually makes the muscles react with considerable vigor, but when the quantity of thyroxine becomes extreme, the muscles become weakened because of excess protein catabolism. On the other hand, lack of thyroxine causes the muscles to become extremely sluggish; they contract readily but relax slowly after a contraction.

Muscle tremor. One of the most characteristic signs of hyperthyroidism is a fine muscle tremor. This is not the coarse tremor that occurs in Parkinson's disease or in shivering, for it occurs at the rapid frequency of 8 to 15 times per second. The tremor can be observed easily by placing a sheet of paper on the extended fingers and noting the degree of vibration of the paper. The cause of this tremor is not definitely known, but it is possibly due to increased activity in the areas of the cord that control muscle tone. The tremor is an excellent means for assessing the degree of thyroxine effect on the central nervous system.

Effect on Sleep. Because of the exhausting effect of thyroxine on the musculature and on the central nervous system, the hyperthyroid subject often has a feeling of constant tiredness; but, also, because of the excitable effects of thyroxine on the synapses, it is difficult for him to sleep. On the other hand, extreme somnolence is characteristic of hypothyroidism.

REGULATION OF THYROID SECRETION

To maintain a normal basal metabolic rate, precisely the right amount of thyroid hormone must be secreted all the time, and to provide this, a specific feedback mechanism operates through the hypothalamus and adenohypophysis to control the rate of thyroid secretion in proportion to the metabolic needs of the body. This system is the following:

Effects of Thyrotropin on Thyroid Secretion—Function of Cyclic AMP. Thyrotropin is an adenohypophyseal hormone, a polypeptide with a molecular weight of about 10,000, that was discussed in Chapter 49; it increases the secretion of thyroxine by the thyroid gland. Thyrotropin is also known as *thyrotropic hormone,* or *thyroid-stimulating hormone* (TSH). Its specific effects on the thyroid gland are: (1) increased proteolysis of the thyroglobulin in the follicles with resultant release of thyroxine into the circulating blood and diminishment of the follicular substance itself; (2) increased activity of the iodide pump, which increases the rate of "iodide trapping" in the glandular cells, often increasing the ratio of thyroid intracellular to extracellular iodide concentration to as great as 350:1 during maximal stimulation; (3) increased size and increased secretory activity of the thyroid cells; and (4) increased number of thyroid cells, plus a change from cuboidal to columnar cells and much infolding of the thyroid epithelium into the follicles. In summary, thyrotropin *increases all the known activities of the thyroid glandular cells.*

Recent experiments indicate that the primary effect of thyrotropin is to increase the quantity of cyclic AMP in the thyroid cell. This in turn has a stimulatory effect on many different cell functions such as increasing the trapping of iodine, increasing the secretion of proteases into the thyroid follicle, and even increasing the growth of the cellular elements. This method for control of thyroid cell activity is similar to the function of cyclic AMP in other target tissues of the body.

Hypothalamic Regulation of Thyrotropin Secretion by the Adenohypophysis. Electrical stimulation of the anterior hypothalamus slightly above the median eminence and immediately posterior to the optic chiasm increases the output of thyrotropin and thereby causes a corresponding increase in activity of the thyroid gland. Therefore, the hypothalamus has the ability to control the adenohypophyseal secretion of thyrotropin. This control is exerted by hypothalamic

secretion of *thyrotropin-releasing factor* (TRF) into the hypophyseal portal blood. TRF has been obtained in an almost pure form by Guillemin, and it is probably an octapeptide. This factor in turn acts on the adenohypophyseal glandular cells to increase the output of thyrotropin. When the portal system from the hypothalamus to the adenohypophysis is completely blocked, the output of thyrotropin is moderately decreased.

Effects of cold and other neurogenic stimuli on thyrotropin secretion. One of the best-known stimuli for increasing the rate of thyrotropin secretion by the adenohypophysis is exposure of an animal to cold. Exposure for several weeks increases the output of thyroxine sometimes more than 100 per cent and can increase the basal metabolic rate of the animal 20 to 30 per cent. Indeed, even human beings moving to arctic regions develop basal metabolic rates 15 to 20 per cent above normal.

Various emotional reactions can also affect the output of thyrotropin and can, therefore, indirectly affect the secretion of thyroxine. For instance, prolonged periods of anxiety, especially in young women soon after reaching adulthood, are frequently associated with the development of thyrotoxic goiter. On the basis of this, most clinicians believe that prolonged emotional states cause increased secretion of thyrotropin.

Neither these emotional effects nor the effect of cold is observed when the hypophyseal stalk is cut, illustrating that both of these effects are mediated by way of the hypothalamus.

Inverse Effect of Thyroid Hormones on Adenohypophyseal Secretion of Thyrotropin — Feedback Regulation of the Metabolic Rate. Increased thyroxine in the body fluids decreases the secretion of thyrotropin by the adenohypophysis. When the rate of thyroxine secretion rises to about 1.75 times normal, the rate of thyrotropin secretion falls essentially to zero. This feedback

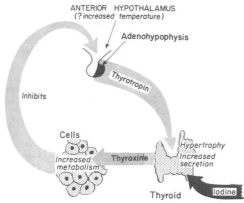

FIGURE 51-5 Regulation of thyroid secretion.

mechanism probably results from the effect of thyroxine on cellular metabolic rate. Thus, if the cellular metabolic rate falls too low, the thyrotropin-thyroxine system becomes activated until enough thyroxine becomes available to increase the metabolic activity back to normal.

One of the ways by which increased metabolic rate could exert its feedback control over the thyrotropin-thyroxine system would be by causing increased heat formation in the cells of the temperature regulating center of the anterior hypothalamus; the increased heat in this center would decrease the output of thyrotropin-releasing factor, which would decrease the output of thyrotropin and thyroxine, thus returning the metabolic activities of the cells back to normal. These possible events are summarized in Figure 51-5.

DISEASES OF THE THYROID

HYPERTHYROIDISM

At some time or other, several per cent of the population develop hyperthyroidism. Young adult women are especially prone to this. The basic cause of the disease is probably excess secretion of thyrotropin, or increased sensitivity of the thyroid gland to this hormone.

The cause of the excessive secretion of thyrotropin is usually unknown. Hyperthyroidism frequently follows severe physical or emotional disturbances; it occurs several times as frequently in cold climates as in hot climates; and it often occurs in successive generations of the same family.

Characteristics of Hyperthyroidism (Toxic Goiter, Thyrotoxicosis, Graves' Disease). In hyperthyroidism, the thyroid gland is usually increased to two to three times normal size. This increase in size is not an indication, however, of the actual increase in rate of thyroid hormone secretion, for each cell increases its rate of secretion tremendously. Studies with radioactive iodine indicate that such hyperplastic glands secrete thyroxine at a rate as great as 5 to 15 times normal.

The symptoms of toxic goiter—these are obvious from the preceding discussion of the physiology of the thyroid hormones—are intolerance to heat, increased sweating, mild to extreme weight loss (sometimes as much as 100 pounds), varying degrees of diarrhea, muscular weakness, nervousness or other psychic disorders, and tremor of the hands.

Exophthalmos and Exophthalmos-Producing Substance. Most, but not all, persons with hyperthyroidism develop some degree of protrusion of the eyeballs, as illustrated in Figure 51-6. This condition is called *exophthalmos*. A major degree of exophthalmos occurs in about one-third of the hyperthyroid patients, and the condition occasionally becomes severe enough that it causes blindness because the eyeball protrusion stretches the optic nerve. Much more often, the eyes are damaged because the eyelids do not close completely when the person blinks or when he is asleep. As a result, the dry epithelial surfaces of the eyes become irritated and often infected, resulting in ulceration of the cornea.

FIGURE 51–6 Patient with exophthalmic hyperthyroidism. Note protrusion of the eyes and retraction of the superior eyelids. The basal metabolic rate was +40. (Courtesy of Dr. Leonard Posey.)

The exophthalmos is probably caused by a substance called *exophthalmos-producing substance,* which is often secreted as a by-product of excessive thyrotropin secretion in hyperthyroidism. The effect of exophthalmos-producing substance on the retro-orbital tissues is mainly to cause edematous swelling and deposition of large quantities of mucopolysaccharides in the extracellular spaces, followed by secondary fibrosis. When the hyperthyroidism is finally brought under control, the exophthalmos usually does not regress significantly, partly because the secretion of exophthalmos-producing substance may continue even after thyroid hormone secretion has been brought to normal, but partly also because the fibrosis that has occurred in the retro-orbital tissues will not resorb.

HYPOTHYROIDISM

The effects of hypothyroidism in general are opposite to those of hyperthyroidism, but here again, a few physiological mechanisms peculiar to hypothyroidism alone are involved.

Hypothyroid Goiter. The term "goiter" means a greatly enlarged thyroid gland. As pointed out in the discussion of iodine metabolism, 35 to 50 mg. of iodine is necessary each year for the formation of adequate quantities of thyroxine. In certain areas of the world, notably in the Swiss Alps and in the Great Lakes region of the United States, insufficient iodine is present in the soil for the foodstuffs to contain even this minute quantity of iodine. Therefore, in days prior to iodized table salt, many persons living in these areas developed extremely large thyroid glands called *endemic goiters.*

The mechanism for development of the large endemic goiters is the following: Lack of iodine prevents production of thyroxine by the thyroid gland. As a result, no thyroxine is available to inhibit production of thyrotropin by the adenohypophysis; this allows the adenohypophysis to secrete excessively large quantities of thyrotropin. The thyrotropin then causes the thyroid gland to continue growing and growing even though it cannot form thyroxine. The gland sometimes grows to a size 25 times normal, causing a grapefruit-like protrusion of the anterior neck.

Similar goiters, called *idiopathic goiters,* often result from other factors that prevent the thyroid cells from forming thyroxine, factors such as (1) enzyme abnormalities of the gland or (2) the presence in foods of *goitrogenic substances,* which block thyroxine formation. These substances are found in some varieties of turnips and cabbages.

Characteristics of Hypothyroidism. Whether hypothyroidism is caused by endemic goiter, idiopathic goiter, destruction of the thyroid gland by irradiation, surgical removal of the thyroid gland, or destruction of the thyroid gland by various other diseases, the physiological effects are the same. These include extreme somnolence with sleeping 14 to 16 hours a day, extreme muscular sluggishness, slowed heart rate, decreased cardiac output, decreased blood volume, increased weight, constipation, mental sluggishness, failure of many trophic functions in the body evidenced by depressed growth of hair and scaliness of the skin, development of a frog-like husky voice, and, in severe cases, development of an edematous appearance throughout the body.

Myxedema. The patient with almost total lack of thyroid function develops an edematous appearance, and the condition is known as "myxedema." Figure 51–7 shows such a patient with bagginess under the eyes and swelling of the face. Greatly increased quantities of mucopolysaccharides, mainly hyaluronic acid, collect in the interstitial spaces, and this causes the total quantity of interstitial fluid also to increase. The fluid is adsorbed to the mucopolysaccharides, thus greatly increasing the quantity of "ground substance" gel in the tissues.

Arteriosclerosis in hypothyroidism. Lack of thyroxine increases the quantity of blood lipids and thereby causes atherosclerosis and arteriosclerosis. Therefore, many hypothyroid patients, particularly those with myxedema, develop severe arteriosclerosis, which results in peripheral vascular disease, deafness, and often extreme coronary sclerosis with early demise.

Cretinism. Cretinism is the condition caused by extreme hypothyroidism during infancy and childhood, and it is characterized especially by failure of growth. A newborn baby without a thyroid gland may have absolutely normal appearance and function because he has been supplied with thyroxine by the mother while *in utero,* but a few weeks after birth his movements become sluggish, and both his physical and mental growth are greatly retarded. Treatment of the cretin at any time usually causes normal return of physical growth, but, unless the cretin is treated within a few months after birth, his mental growth will be permanently retarded. This is probably due to the fact that physical development of the neuronal cells of the central nervous system is rapid during the first year of life so that any retardation at this point is extremely detrimental.

Skeletal growth in the cretin is characteristically more inhibited than is soft tissue growth, though both are inhibited to a certain extent. However, as a result of this disproportionate rate of growth, the soft tissues are likely to enlarge excessively, giving the cretin the appearance of a very obese and stocky, short child.

DIAGNOSIS AND TREATMENT OF HYPO- AND HYPERTHYROIDISM

The two procedures most widely used for diagnosing either hyper- or hypothyroidism are (1) measurement of the *protein-bound iodine* in the plasma and (2) measurement of the *basal metabolic rate.* The normal concentration of pro-

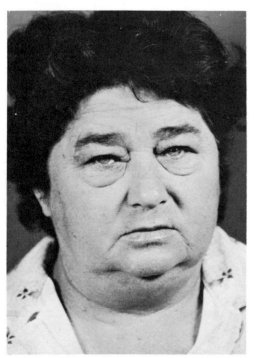

FIGURE 51–7 Patient with myxedema. (Courtesy of Dr. Herbert Langford.)

tein-bound iodine is between 4 and 7.5 μg. per 100 ml. plasma. In severe hypothyroidism this falls below 2 μg. per 100 ml., and in severe hyperthyroidism it rises to 15 to 20 μg per 100 ml. The basal metabolic rate in severe hypothyroidism falls to as low as −40 to −50, and it rises to as high as +40 to +100 in severe hyperthyroidism.

Treatment of hyperthyroidism is usually most successfully achieved by surgery. However, several different drugs can block the formation of thyroxine by the thyroid gland and, therefore, can be used to advantage in treating this disease. These drugs include especially *propylthiouracil* and *thiocyanate ions*.

Hypothyroidism can probably be treated more easily than any other endocrine disorder — simply by the ingestion each day of a single desiccated tablet of thyroid gland tissue removed from animals. Because of its slow activity, this hormone requires several days to a week to take effect.

REFERENCES

Ciba Foundation Study Group No. 18: Brain-Thyroid Relationships. Boston, Little, Brown and Company, 1964.

Danowski, T. S.: Clinical Endocrinology. Vol. II: Thyroid. Baltimore, The Williams & Wilkins Co., 1962.

Harrison, T. S.: Adrenal medullary and thyroid relationships. *Physiol. Rev., 44*:161, 1964.

McKenzie, J. M.: Humoral factors in the pathogenesis of Graves' disease. *Physiol. Rev., 48*:252, 1968.

Rawson, R. W.; Sonenberg, M.; and Money, W. L.: Diseases of the thyroid. *In* Duncan, G. G. (ed.): Diseases of Metabolism. 5th ed., Philadelphia, W. B. Saunders Company, 1964, p. 1159.

Rosenberg, I. S., and Bastomsky, C. H.: The thyroid. *Ann. Rev., Physiol., 27*:71, 1965.

Werner, S. C., and Nauman J. A.: The thyroid. *Ann. Rev. Physiol., 30*:213, 1968.

Wolff, J.: Transport of iodide and other anions in the thyroid gland. *Physiol. Rev., 44*:45, 1964.

CHAPTER 52

INSULIN, GLUCAGON, AND DIABETES MELLITUS

The pancreas, in addition to its digestive functions, secretes two hormones, *insulin* and *glucagon.* The purpose of this chapter is to discuss the functions of these hormones in regulating glucose, lipid, and protein metabolism, as well as to discuss the two diseases—*diabetes mellitus* and *hyperinsulinism*—caused, respectively, by hyposecretion of insulin and excess secretion of insulin.

Physiologic Anatomy of the Pancreas. The pancreas is composed of two major types of tissues, as shown in Figure 52–1: (1) the *acini,* which secrete diges-

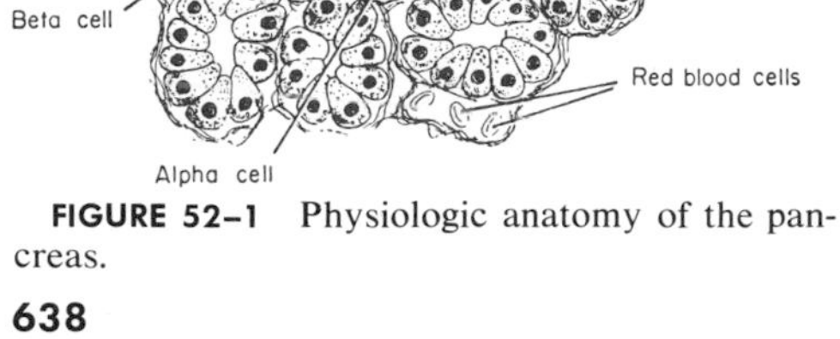

FIGURE 52-1 Physiologic anatomy of the pancreas.

tive juices into the duodenum, and (2) the *islets of Langerhans,* which do not have any means for emptying their secretions externally but instead secrete insulin and glucagon directly into the blood. The digestive secretions of the pancreas were discussed in Chapter 43.

The islets of Langerhans of the human being contain two major types of cells, the *alpha* and *beta* cells, which are distinguished from one another by their morphology and staining characteristics. The beta cells produce insulin, and glucagon is secreted by the alpha cells.

Chemistry of Insulin. Insulin is a small protein having a molecular weight of 5734. Before insulin can exert its function it must become "fixed" to the tissues, probably to the cell membranes, as explained later in the chapter.

EFFECT OF INSULIN ON CARBOHYDRATE METABOLISM

The earliest studies of the effect of insulin on carbohydrate metabolism showed three basic effects of the hormone: (1) enhanced rate of glucose metabolism, (2) decreased blood glucose concentration, and (3) increased glycogen stores in the tissues. All of these effects probably result from the following basic mechanisms.

FACILITATION OF GLUCOSE TRANSPORT THROUGH THE CELL MEMBRANE

The single basic effect of insulin that has been demonstrated many times is its ability to increase the rate of glucose transport through the membranes of most cells in the body. In the complete absence of insulin, the overall rate of glucose transport into the cells of the entire body becomes only about one-fourth the normal value. On the other hand, when great excesses of insulin are secreted, the rate of glucose transport into the cells may be as great as five times normal. This means that, between the limits of no insulin at all and great excesses of insulin, the rate of glucose transport can be altered as much as 20-fold.

Mechanism by which Insulin Accelerates Glucose Transport. As pointed out in the discussion of the cell membrane in Chapter 4, glucose cannot pass into the cell through the cell pores but instead must enter by some transport mechanism through the membrane matrix. Figure 52–2 depicts the generally believed method by which glucose enters the cell, showing that glucose probably combines with a carrier substance in the cell membrane and then is transported to the inside of the membrane where it is released to the interior of the cell. The carrier then returns to the outer surface of the membrane to transport additional quantities of glucose. This process can occur *in either direction,* as shown by the reversible arrows in the diagram.

Glucose transport through the cell membrane cannot occur against a con-

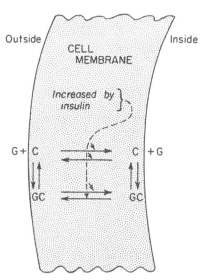

FIGURE 52–2 Effect of insulin in increasing glucose transport in either direction through the cell membrane.

centration gradient. That is, once the glucose concentration inside the cell rises to as high as the glucose concentration on the outside, further glucose will not be transported to the interior. Therefore, the glucose transport process is one of *facilitated diffusion,* which means simply that glucose diffuses through the membrane by means of a facilitating carrier mechanism.

In the complete absence of insulin, the carrier mechanism for the transport of glucose to the interior of the cell operates very poorly. Unfortunately, the manner in which insulin affects this mechanism is unknown. The insulin first becomes "fixed" in the cell membrane and presumably enhances the activity of one or more of the processes related to transport.

Tissues in which Insulin Is Effective. Enhanced transport of glucose through the cell membrane is particularly effective in skeletal muscle and adipose tissue. These two together make up approximately 65 per cent of the entire body by weight. In addition, insulin enhances glucose transport into the heart and at least into certain smooth muscle organs, such as the uterus.

On the other hand, insulin does not enhance glucose transport into the brain cells and does not enhance its transport through the intestinal mucosa or through the tubular epithelium of the kidney. However, these tissues together amount to less than 5 per cent of the total body mass. The effects of insulin in the remaining 30 per cent of the body are mainly unknown, but there is reason to believe that most of this tissue also responds at least partially to insulin.

In summary, insulin is extremely important for glucose transport into the cells of most tissues of the body. The most important exception to this is the brain; here, glucose transport is probably more dependent on *diffusion* through the blood-brain barrier (see Chap. 17) than through the cell membrane.

Effect of Insulin on Glucose Utilization and Glycogen Storage in Extrahepatic Tissues. Since insulin lack decreases glucose transport into most body cells to about one-fourth normal, without insulin most tissues of the body (with the major exception of the brain) must depend on other metabolic substrates for energy.

Another major effect of insulin is greatly enhanced storage of glycogen in the skeletal muscle cells throughout the body and moderate enhancement of glycogen storage in the skin and glandular tissues. This effect is undoubtedly caused by the availability of greatly increased intracellular glucose.

Action of Insulin on Carbohydrate Metabolism in the Liver. Knowledge of the action of insulin on carbohydrate metabolism in the liver is still somewhat confused because insulin does not cause immediate increase in glucose transport into liver cells similar to that which occurs in skeletal muscle and most other tissues of the body. Instead, insulin causes initial loss of glycogen from the liver and transfer of this to the skeletal muscle. This effect probably results from the following two factors: (1) The liver cells are naturally highly permeable to glucose, which means that glucose can go through the liver cell membranes in either direction with ease regardless of the presence of insulin. (2) Liver cells contain large quantities of *glucose 6-phosphatase,* an enzyme that continually dephosphorylates glucose that has previously entered the liver cell; therefore, glucose cannot become trapped inside the liver cell as it does in other cells. Because of these two factors, when insulin promotes glucose transport into muscle and other peripheral tissues, the blood glucose concentration falls immediately, which in turn causes glucose mobilization from the liver. Thus, in effect, insulin transfers glycogen from the liver to the peripheral cells.

Long-term effect of insulin on the liver. If insulin is administered continuously for days or weeks along with simultaneous availability of adequate carbohydrates, the quantity of glycogen in the liver becomes greater than normal rather than being decreased. This effect is caused by still another effect of insulin in liver cells: it markedly increases the quantity of *glucokinase,* and this accelerates the rate at which glucose can be processed for utilization in liver cells. Therefore, carbohydrate metabolism in the liver becomes greatly enhanced, causing storage of glycogen in the liver cells, which is opposite to the acute effect of insulin.

REGULATION OF BLOOD GLUCOSE CONCENTRATION, AND CONTROL OF INSULIN SECRETION

Glucose concentration in the plasma and other extracellular fluids is controlled by four separate mechanisms: (1) the ability of the liver to "buffer" the blood glucose concentration, (2) an automatic feedback mechanism involving insulin secretion by the pancreas, (3) an automatic feedback mechanism involving sympathetic stimulation, epinephrine, and norepinephrine to cause release of glucose from the liver, and (4) a similar feedback mechanism involving glucagon secreted by the alpha cells of the pancreas.

Glucose-Buffer Function of the Liver. As just discussed, the liver acts as a large storage vault for glucose, and because of this the liver also acts as a blood glucose buffer system. When excess quantities of glucose enter the blood, about two-thirds of this is stored almost immediately in the liver, and this prevents excessive increase in blood glucose concentration. Conversely, when the blood glucose concentration falls below normal, the stored glucose in the liver rapidly replenishes the blood glucose.

Insulin plays an important role in controlling the glucose-buffer function of the liver, because in the absence of insulin, glucokinase is greatly depressed in the liver cells, so that glucose is not readily stored in the liver in times of excess.

In addition to acting as a major storehouse for glucose, the liver also helps to regulate blood glucose concentration by increasing its rate of gluconeogenesis when glucose is needed in the blood.

The liver glucose-buffer system reduces the variation in blood glucose concentration to approximately one-third what it is without the liver.

Effect of Insulin on Blood Glucose Concentration. In the absence of insulin, little of the glucose absorbed from the gastrointestinal tract can be transported into the tissue cells. As a consequence, the blood glucose concentration rises very high—from a normal value of 90 mg./100 ml. sometimes to as high as 300 to 1200 mg./100 ml.

On the other hand, in the presence of great amounts of excess insulin, glucose is transported into the cells so rapidly that its concentration in the blood can fall to as low as 20 to 30 mg./100 ml. Therefore, the rate of insulin secretion by the pancreas must be regulated accurately so that the blood glucose concentration also will be regulated accurately at a constant and normal value.

The glucose concentration of the plasma has a direct and immediate effect on the islets of Langerhans to control their rate of insulin secretion. Even the isolated pancreas perfused with a solution containing a high concentration of glucose secretes greatly increased quantities of insulin. Therefore, the secretion of insulin by the pancreas affords an important feedback mechanism for continual regulation of blood glucose concentration.

Besides the immediate increase in insulin secretion that results from increased blood glucose concentration, still an additional gradual increase in insulin secretion occurs if the elevated glucose concentration persists. Over a period of one to three weeks, the islets of Langerhans hypertrophy, and secretion of insulin increases correspondingly. Therefore, if a person suddenly begins to eat a diet containing excessive amounts of carbohydrates, his normal pancreatic insulin secretion may not be capable of taking care of all the excess glucose, and his blood glucose concentration rises above normal. But after several weeks of pancreatic adaptation, an adequate amount of insulin is secreted, and the excess glucose is transported into the cells.

Regulation of Blood Glucose Concentration by the Sympathetic Nervous System and by Glucagon. Decreased blood glucose concentration excites the sympathetic nuclei of the hypothalamus, which then transmit impulses through the sympathetic nervous system to cause epinephrine release by the adrenal medullae and norepinephrine release both by the adrenals and by the sympathetic nerve endings. These hormones, especially epinephrine, exert a direct effect on the liver cells to increase the rate of glycogenolysis. They do this by activating the liver cell enzyme *adenyl cyclase*, which converts much ATP to cyclic AMP. The cyclic AMP in turn activates the normally inactive phosphorylase in the liver cells. This enzyme then causes glycogen to split into glucose phosphate, which is rapidly dephosphorylated by glucose 6-phosphatase and diffuses into the blood to elevate the blood glucose concentration back toward normal.

Later in the chapter we shall see that a decrease in blood glucose concentration also causes glucagon secretion by the pancreas; the glucagon then has the same effect as that of epinephrine to cause glucose release from the liver. In addition, glucagon and glucocorticoids from the adrenal gland enhance gluconeogenesis, which converts amino acids into glucose. These, then, are additional feedback mechanisms for regulation of blood glucose concentration.

Purpose of Blood Glucose Regulation. One might ask the question: Why is it important to maintain a constant blood glucose concentration, particularly since most tissues can shift to utilization of fats and proteins for energy in the absence of glucose? The answer is that glucose is the only nutrient that can be utilized by the brain, retina, and germinal epithelium in sufficient quantities to supply them with their required energy. Therefore, it is important to maintain a blood glucose concentration at a sufficiently high level to provide this necessary nutrition.

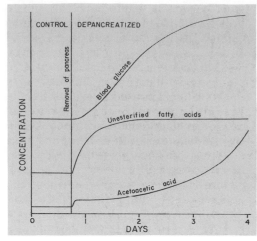

FIGURE 52-3 Effect of removing the pancreas on the concentrations of blood glucose, plasma free (unesterified) fatty acids, and acetoacetic acid.

OTHER METABOLIC EFFECTS OF INSULIN

EFFECTS ON FAT METABOLISM

As long as adequate amounts of insulin and glucose are available, glucose is preferentially metabolized by the cells to supply essentially all the energy required. On the other hand, in the absence of adequate glucose or insulin, the major share of the energy required by the body is then supplied by fats.

Mechanism of Fat Mobilization in Insulin Lack. Figure 52-3 illustrates the effect of sudden insulin lack on blood glucose and on the fat mobilization products, free fatty acids and acetoacetic acid, showing that all of these increase. The cause of this fat mobilization is two-fold.

First, in the absence of insulin, glucose cannot enter the fat cell with ease. Several products of glucose metabolism are very important for causing, and also for maintaining, fat storage, the most important being *α-glycerophosphate,* which supplies the glycerol nucleus that combines with fatty acids to form the neutral fats in adipose tissue. In the absence of these products of glucose, the process of storing triglycerides in fatty tissue is reversed, causing, instead, release of free fatty acids into the blood.

Second, lack of insulin—in some way not yet understood, but perhaps by acting directly on the fat cell membrane—causes direct enhancement of lipolysis with resultant release of free fatty acids into the circulating body fluids.

Also, additional fat mobilization results from the action of cortisol on the fat cell; this hormone is usually secreted in large quantities in conditions of insulin lack.

Effect of Insulin Lack on Blood Lipid Concentration. In addition to the increase in free fatty acids in the circulating blood, essentially all the other lipid components of the plasma also become increased in the absence of insulin. This includes an increase in the overall quantity of lipoproteins and of their constituents, *triglycerides, cholesterol,* and *phospholipids.* At times the blood lipids increase as much as five-fold, giving a total concentration of plasma lipids of several per cent rather than the normal 0.6 per cent. These high lipid concentrations, especially the high concentration of cholesterol, is the probable cause of extreme atherosclerosis in persons with serious diabetes.

Ketogenic Effect of Insulin Lack. Associated with increased fat mobilization from the tissues is usually also a ketogenic effect of insulin lack, which was discussed in Chapter 46 in relation to utilization of fats for energy. That is, acetoacetic acid, a product of fatty acid metabolism, is released by the liver into the blood in greatly increased quantities, causing the condition called *ketosis.*

The Glucose–Fatty Acid Cycle. From this discussion, it is clear that glucose and insulin have a direct inhibitory effect on fatty acid utilization by the body. Conversely, fatty acids have an inhibitory effect on the utilization of glucose, which results primarily from the following effect: When excess fatty acids are available, some of their degradation products inhibit the conversion of fructose 6-phosphate into fructose 1,6-phosphate, one of the early steps essential to the degradation of glucose before it can be utilized for energy.

This reciprocal inhibitory relationship between glucose and fatty acid utilization is called the *glucose–fatty acid cycle.* It plays a particularly important role in insulin lack, because in addition to the direct effect of insulin lack in depressing glucose utilization, the resultant buildup of fatty acids also indirectly depresses glucose metabolism.

This effect almost doubles the glucose metabolic defect in the presence of insulin deficiency.

EFFECT OF INSULIN ON PROTEIN METABOLISM AND ON GROWTH

Effect of Insulin on Protein Anabolism — Transport of Amino Acids Through the Cell Membranes. The total quantity of protein stored in the tissues of the body is increased by insulin and greatly decreased by insulin lack. Indeed, a person with severe diabetes mellitus can become incapacitated almost as much from protein lack as from failing glucose metabolism.

Most of the effect of insulin on protein metabolism is probably secondary to the action of insulin on carbohydrate metabolism, for utilization of carbohydrates for energy acts as a protein sparer. However, *insulin also increases the transport of amino acids through the cell membranes* in a manner similar to that of its increasing glucose transport. This transport is carrier mediated, as is also true of glucose transport. Insulin enhances the transport several-fold, but the effect is generally far less intense than the effect on glucose transport. Insulin is essential for growth of an animal, principally because of the protein-sparing effect of increased carbohydrate metabolism, but also because of the direct effect of insulin on protein anabolism.

DIABETES MELLITUS

Diabetes mellitus is caused by destruction, or a failure to function, of the beta cells in the islets of Langerhans of the pancreas. Though various factors, such as excess growth hormone from the adenohypophysis, can at times cause the beta cells of the islets of Langerhans to "burn out," diabetes is, in general, a hereditary disease, for almost all persons who develop diabetes, especially those who develop it in early years, can trace the disease to one or more forebears.

PATHOLOGIC PHYSIOLOGY OF DIABETES

Most of the pathology of diabetes mellitus can be attributed to one of the following three major effects of insulin lack: (1) decreased utilization of glucose by the body cells, with a resultant increase in blood glucose concentration to as high as 300 to 1200 mg. per 100 ml.; (2) markedly increased mobilization of fats from the fat storage areas, causing abnormal fat metabolism and especially deposition of lipids in vascular walls to cause atherosclerosis; and (3) diminished deposition of protein in the tissues of the body, caused partly by failure of glucose to be used as a protein sparer and partly by loss of the direct effect of insulin to promote protein anabolism.

However, in addition, some special pathologic physiological problems occur in diabetes mellitus, and these are not so readily apparent. These are as follows:

Loss of Glucose in the Urine of the Diabetic Person. Whenever the quantity of glucose entering the kidney tubules in the glomerular filtrate rises above approximately 225 mg. per minute, a significant proportion of the glucose begins to spill into the urine. If normal quantities of glomerular filtrate are formed per minute, 225 mg. of glucose will enter the tubules each minute when the blood glucose level rises to 180 mg. per cent. Consequently, it is frequently stated that the blood "threshold" for the appearance of glucose in the urine is approximately 180 mg. per cent. When the blood glucose level rises to 300 to 500 mg. per cent — common values in persons with diabetes — several hundred grams of glucose can be lost into the urine each day.

Acidosis in Diabetes. The shift from carbohydrate to fat metabolism in diabetes has already been discussed. When the body depends almost entirely on fats for metabolism, the level of acetoacetic acid and other keto acids in the body fluids may rise from 1 mEq./liter to as high as 30 mEq./liter. This, obviously, is likely to result in acidosis.

All the usual reactions that occur in metabolic acidosis take place in diabetic acidosis. These include *rapid and deep breathing* called "Kussmaul respiration," which causes excessive expiration of carbon dioxide and *marked decrease in bicarbonate content of the extracellular fluids.* Likewise, *large quantities of chloride ion are excreted by the kidneys* as an additional compensatory mechanism for correction of the acidosis. These extreme effects, however, occur only in the most severe degrees of untreated diabetes. The overall changes in the electrolytes of the blood as a result of severe diabetic acidosis are illustrated in Figure 52–4.

The Glucose Tolerance Test. An important test performed for the diagnosis

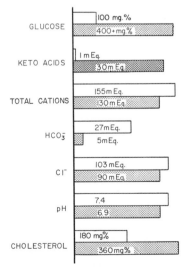

FIGURE 52–4 Changes in blood constituents in severe diabetes, showing normal values (light bars) and diabetic values (dark bars).

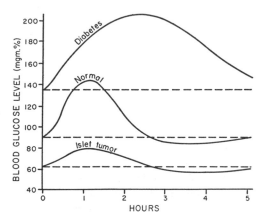

FIGURE 52–5 Glucose tolerance curve in the normal person, in a diabetic person, and in a person with an islet tumor (hyperinsulinism).

of diabetes mellitus is the glucose tolerance test illustrated in Figure 52–5. The middle curve is the normal one: when a normal fasting person ingests 50 grams of glucose, his blood glucose level rises from a normal value of approximately 90 mg. per cent to approximately 140 mg. per cent and falls back to below normal within three hours.

Though an occasional diabetic person has a normal fasting blood glucose concentration, it is usually above 120 mg. per cent, and his glucose tolerance test is almost always abnormal. On ingestion of 50 grams of glucose, these persons exhibit a progressive, slow rise in blood glucose level for two to three hours, as illustrated by the upper curve in Figure 52–5, and the glucose level falls back to the control value only after some five to six hours; it never falls below the control level. This slow fall of the curve and its failure to fall below the control level illustrates that the normal increase in insulin secretion following glucose ingestion does not occur in the diabetic, and a diagnosis of diabetes mellitus can usually be definitely established on the basis of such a curve.

Treatment of Diabetes. The theory of treatment in diabetes mellitus is to administer enough insulin so that the patient will have normal carbohydrate metabolism. If this is done, most conse-

quences of diabetes can be prevented. Ordinarily, the diabetic patient is given a single dose of one of the long-acting insulins each morning; this increases his overall carbohydrate metabolism throughout the day. Then additional quantities of rapidly acting regular insulin are given at those times of the day when his blood glucose level tends to rise too high, such as at meal times. Each patient must be established on an individualized routine of treatment.

Diet of the diabetic. The insulin requirements of a diabetic are established with the patient on a standard diet containing normal, well-controlled amounts of carbohydrates; any change in the quantity of carbohydrate intake changes the requirements for insulin. In the normal person, the pancreas has the ability to adjust the quantity of insulin produced to the intake of carbohydrate; but in the completely diabetic person, this control function is totally lost.

Relationship of treatment to arteriosclerosis. In the early days of treating diabetes it was the tendency to reduce the carbohydrates in the diet so that the insulin requirements would be minimized. This procedure kept the blood sugar level down to normal values and prevented the loss of glucose in the urine, but it did not prevent the abnormalities of fat metabolism, especially arteriosclerosis with early death from heart disease. Actually, it exacerbated these effects. Consequently, there is a tendency at present to allow the patient a normal carbohydrate diet and then to give large quantities of insulin to metabolize the carbohydrates. This depresses the rate of fat metabolism and also depresses the high level of blood cholesterol that occurs in diabetes as a result of abnormal fat metabolism.

DIABETIC COMA

If diabetes is not controlled satisfactorily, severe dehydration and acidosis may result; and sometimes, even when the person is receiving treatment, sporadic changes in metabolic rates of the cells, such as might occur during bouts of fever, can also precipitate dehydration and acidosis.

If the pH of the body fluids falls below approximately 7.0 to 6.9, the diabetic person develops coma. Also, in addition to the acidosis, dehydration is believed to exacerbate the coma. Once the diabetic person reaches this stage, the outcome is usually fatal unless he receives immediate treatment.

Treatment of Diabetic Coma. The patient with diabetic coma is extremely refractory to insulin because acidic plasma has an *insulin antagonist,* an alpha globulin, that opposes the action of the insulin. Also, the very high free fatty acid and acetoacetic acid levels in the blood inhibit the usage of glucose, as was discussed earlier. Therefore, instead of the usual 60 to 80 units of insulin per day, which is the dosage usually necessary for control of severe diabetes, as much as 1500 to 2000 units of insulin must often be given the first day of treatment of coma. Administration of insulin alone may be sufficient to reverse the abnormal physiology and to effect a cure. However, it is usually advantageous to correct also the dehydration and acidosis immediately.

HYPERINSULINISM

Though much more rare than diabetes, increased insulin production, which is known as hyperinsulinism, does occasionally occur. This usually results from an adenoma of an islet of Langerhans. To prevent hypoglycemia in some of these patients, more than 1000 grams of glucose have had to be administered each 24 hours.

Insulin Shock and Hypoglycemia. As already emphasized, the central nervous system derives essentially all its energy from glucose metabolism. If insulin causes the level of blood glucose to fall to low values, the metabolism of the central nervous system becomes depressed. Consequently, in patients with hyperinsulinism or in patients who administer too much insulin to themselves, the syndrome called "insulin shock" may occur as follows:

As the blood sugar level falls into the range of 50 to 70 per cent, the central nervous system usually becomes quite excitable, for this degree of hypoglycemia seems to facilitate neuronal activity. Sometimes various forms of hallucinations result, but more often the patient simply experiences extreme nervousness, and he may tremble all over. As the blood glucose level falls to 20 to 50 mg. per cent, clonic convulsions and loss of consciousness are likely to occur. As the glucose level falls still lower, the convulsions cease, and only a state of coma remains. Indeed, at times it is difficult to distinguish between diabetic coma as a result of insulin lack and coma due to hypoglycemia caused by excess insulin. If glucose is not administered immediately, permanent damage to the neuronal cells of the central nervous system occurs.

Depressed Glucose Tolerance Curve and Decreased Insulin Sensitivity in Hyperinsulinism. In addition to the usual signs of hypoglycemia that appear in hyperinsulinism, a definite diagnosis of the condition can usually be made by performing a glucose tolerance test. This glucose tolerance curve is usually depressed, as shown by the lower curve of Figure 52–5, illustrating that the initial glucose level is low, and the increase after ingestion of glucose is slight.

FUNCTIONS OF GLUCAGON

Glucagon, like insulin, is a small protein having a molecular weight of 3482. On injecting purified glucagon into an animal, profound *hyper*glycemia occurs, 1 microgram per kg. of glucagon elevating the blood glucose concentration approximately 20 mg./100 ml. of blood.

Glycogenolysis and Increased Blood Glucose Concentration Caused by Glucagon. The most dramatic effect of glucagon is its ability to cause glycogenolysis in the liver, which in turn increases the blood glucose concentration. This effect is similar to but many times more powerful than that of epinephrine in causing glycogenolysis. Like epinephrine, glucagon stimulates the formation of cyclic AMP, which in turn activates the normally inactive phosphorylase in the liver cells. The phosphorylase then causes rapid glycogenolysis followed by release of glucose into the blood.

Infusion of glucagon for about four hours can cause such intensive liver glycogenolysis that all the liver stores of glycogen will be totally depleted.

Prolonged hyperglycemic effect of glucagon. Even after all the glycogen in the liver has been exhausted under the influence of glucagon, continued infusion of this hormone causes continued hyperglycemia and glycosuria. This is due to glucagon's ability to increase the rate of gluconeogencsis.

Regulatory Function of Glucagon to Prevent Hypoglycemia. When the blood glucose concentration falls to as low as 70 mg. per cent, the pancreas secretes large quantities of glucagon, which rapidly mobilizes glucose from the liver.

Thus, glucagon protects against hypoglycemia.

It has been suggested that glucagon is secreted during muscular activity and that it helps to cause glucose mobilization from the liver for use by the muscles; and in starvation, glucagon is abundantly secreted, perhaps in an attempt to maintain a normal blood glucose concentration.

REFERENCES

Cahill, G. F., Jr.; Owen, O. E.; and Felig, P.: Insulin and fuel homeostasis. *Physiologist, 11*:97, 1968.

Diabetes mellitus and obesity. *Ann. N.Y. Acad. Sci., 148*:573, 1968.

Frohman, L. A.: The endocrine function of the pancreas. *Ann. Rev. Physiol., 31*:353, 1969.

Hales, C. N.: The glucose fatty acid cycle and diabetes mellitus. *In* Bittar, E. E., and Bittar, N. (eds.): The Biological Basis of Medicine. New York, Academic Press, Inc., 1968, Vol. I, p. 309.

Levine, R.: Concerning the mechanisms of insulin action. *Diabetes, 10*:421, 1961.

Park, C. R.; Reinwein, D.; Henderson, M. J.; Cadenas, E.; and Morgan, H. E.: The action of insulin on the transport of glucose through the cell membrane. *Amer. J. Med., 26*:674, 1959.

Rieser, P.: Insulin, Membranes and Metabolism. Baltimore, The Williams & Wilkins Co., 1968.

Sutherland, E. W., and Robinson, G. A.: The role of cyclic AMP in the control of carbohydrate metabolism. *Diabetes, 18*:797, 1969.

CHAPTER 53

PARATHYROID HORMONE, CALCITONIN, CALCIUM AND PHOSPHATE METABOLISM, VITAMIN D, BONE AND TEETH

The physiology of parathyroid hormone and of the hormone calcitonin is closely related to calcium and phosphate metabolism, the function of vitamin D, and the formation of bone and teeth. Therefore, these are discussed together in the present chapter.

CALCIUM AND PHOSPHATE IN THE EXTRACELLULAR FLUID AND PLASMA

Gastrointestinal Absorption of Calcium and Phosphate. Calcium is poorly absorbed from the gastrointestinal tract because of the relative insolubility of many of its compounds and also because bivalent cations are poorly absorbed through the gastrointestinal mucosa anyway. On the other hand, phosphate is absorbed exceedingly well most of the time except when excess calcium is in the diet; the calcium tends to form almost insoluble calcium phosphate compounds that fail to be absorbed but instead pass on through the bowels to be excreted in the feces. Therefore, if calcium is absorbed, phosphate is also absorbed.

Promotion of Calcium Absorption by Vitamin D. The most important function of vitamin D is its ability to increase the rate of calcium absorption from the gastrointestinal tract. This effect results from a direct action of vitamin D on the epithelial cells of the mucosa in the duodenum and jejunum to increase the amount of carrier substance required for active transport of calcium through the intestinal membrane.

Regulation of Calcium Absorption by Parathyroid Hormone. Parathyroid hormone has a direct effect on the gastrointestinal mucosa to increase the rate of calcium absorption. Also, decreased calcium ion concentration in the extracellular fluid causes marked parathyroid hormone secretion. Thus, a feedback mechanism operates whereby decreased calcium ion concentration increases parathyroid hormone secretion, which in turn increases calcium absorption from the gut. This is part of the overall parathyroid control of calcium ion concentration in the extracellular fluid, as explained later in the chapter.

Excretion of Calcium, and Function of Parathyroid Hormone To Limit Calcium Excretion. About seven-eighths of the daily intake of calcium is excreted in the feces and the remaining one-eighth is excreted in the urine. Excretion of calcium in the urine conforms to much the same principles as the excretion of sodium. When the calcium ion concentration of the extracellular fluids is low, the rate at which calcium is excreted by the kidneys also becomes low, while even a minute increase in calcium ion concentration increases the calcium ion secretion markedly. One of the factors that helps to regulate this excretion of calcium ion is parathyroid hormone, for increased parathyroid hormone — which occurs when the extracellular calcium ion concentration falls — increases the rate of calcium reabsorption from the tubules, while decreased parathyroid hormone allows a greater proportion of the calcium to pass on into the urine. Thus, here again, the parathyroid feedback mechanism operates to control the concentration of calcium in the extracellular fluids.

Calcium in the Plasma and Interstitial Fluid. The concentration of calcium in plasma is approximately 10 mg. per cent, normally varying between 9 and 11 mg. per cent. This is equivalent to approximately 5 mEq./liter. The calcium is in three different forms, as shown in Figure 53-1. (1) Approximately 50 per cent

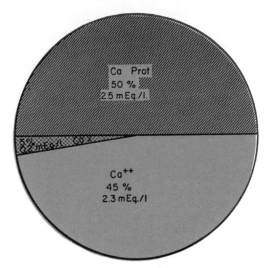

FIGURE 53-1 Distribution of ionic calcium (Ca^{++}), diffusible but un-ionized calcium ($Ca\ X$), and calcium proteinate ($Ca\ Prot$) in blood plasma.

is combined with plasma proteins and consequently is nondiffusible through the capillary membrane. (2) Approximately 5 per cent of the calcium is diffusible through the capillary membrane but is combined with other substances of the plasma and interstitial fluids (citrate, for instance) in such a manner that it is not ionized. (3) The remaining 45 per cent of the calcium in the plasma is both diffusible through the capillary membrane and ionized. Thus, the plasma and interstitial fluids have a normal *calcium ion concentration of approximately 2.3 mEq./liter.* This ionic calcium is important for most functions of calcium in the body, including the effect of calcium on the heart, on the nervous system, and on bone formation.

Tetany resulting from hypocalcemia. When the extracellular fluid concentration of calcium ions falls below normal, the nervous system becomes progressively more excitable because of increased neuronal membrane permeability. This increase in excitability occurs both in the central nervous system and in the peripheral nerves, though most symptoms are manifest peripherally. The nerve fibers become so excitable that they begin to discharge spontaneously, initiating nerve impulses that pass to the peripheral skeletal muscles where they elicit tetanic contraction. Consequently, hypocalcemia causes tetany.

Figure 53–2 illustrates tetany in the hand, which usually occurs before generalized tetany develops. This is called "carpopedal spasm."

Acute hypocalcemia in the human being ordinarily causes essentially no other serious effects besides tetany, because tetany kills the patient as a result of laryngospasm before other effects can develop. Tetany ordinarily occurs when the blood concentration of calcium reaches approximately 7 mg. per cent, which is only 30 per cent below the normal calcium concentration.

The Inorganic Phosphate in the Extracellular Fluids. Inorganic phosphate exists in the plasma mainly in two forms: HPO_4^{--} and $H_2PO_4^-$. However, because it is difficult to determine chemically the exact quantities of HPO_4^{--} and $H_2PO_4^-$ in the blood, ordinarily the total quantity of phosphate is expressed in terms of milligrams of *phosphorus* per 100 ml. of blood. The average total quantity of inorganic phosphorus represented by both phosphate ions is about 4 mg./100 ml.

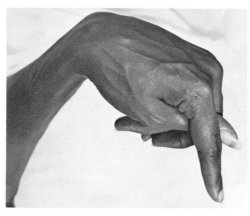

FIGURE 53–2 Hypocalcemic tetany in the hand, called "carpopedal spasm." (Courtesy Dr. Herbert Langford.)

BONE AND ITS RELATIONSHIPS WITH EXTRACELLULAR CALCIUM AND PHOSPHATES

Bone is composed of a tough *organic matrix* that is greatly strengthened by deposits of *calcium salts*. Average compact bone contains by weight approximately 25 per cent matrix and 75 per cent salts. However, *newly formed bone* may have a considerably higher percentage of matrix in relation to salts.

The Organic Matrix of Bone. The organic matrix of bone is approximately 97 per cent *collagen fibers,* and the remaining 3 per cent is a homogeneous medium called *ground substance.* The collagen fibers extend in all directions in the bone but to a great extent along the lines of tensional force. These fibers give bone its powerful tensile strength.

The ground substance is composed of extracellular fluid plus *mucoprotein, chondroitin sulfate,* and *hyaluronic acid.* The precise function of these is not known, though perhaps they help to provide a medium for deposition of calcium salts.

The Bone Salts. The crystalline salts deposited in the organic matrix of bone are composed principally of *calcium* and *phosphate,* and the formula for the major crystalline salts, known as *hydroxyapatites,* is the following:

$$Ca^{++}_{10-x}(H_3O^+)_{2x} \cdot (PO_4)_6(OH^-)_2$$

Magnesium, sodium, potassium, and *carbonate* ions are also present among the bone salts. These are believed to be adsorbed to the surfaces of the hydroxyapatite crystals rather than organized into distinct crystals of their own. This ability of many different types of ions to adsorb to bone crystals extends to many ions normally foreign to bone, such as *strontium, uranium, plutonium, the other transuranic elements, lead, gold, other heavy metals,* and *at least 9 out of 14 of the major radioactive products released by explosion of the hydrogen bomb.* Deposition of radioactive substances in the bone can cause prolonged irradiation of the bone tissues, and, if a sufficient amount is deposited, an osteogenic sarcoma almost invariably eventually develops.

Tensile and Compressional Strength of Bone. The collagen fibers of bone have cross-striations approximately every 64 millimicrons along their length; and hydroxyapatite crystals, standing end on end like blocks in pillars, lie adjacent to each segment of the fiber bound tightly to it. This intimate bonding prevents "shear" in the bone; that is, it prevents the crystals and collagen fibers from slipping out of place, which is essential in providing strength to the bone.

The collagen fibers of bone, like those of tendons, have great tensile strength, while the calcium salts, which are similar in physical properties to marble, have great compressional strength. These combined properties, plus the degree of bondage between the collagen fibers and the crystals, provide a bony structure that has both extreme tensile and compressional strength.

PRECIPITATION AND ABSORPTION OF CALCIUM AND PHOSPHATE IN BONE—EQUILIBRIUM WITH THE EXTRACELLULAR FLUIDS

Supersaturated State of Calcium and Phosphate Ions in Extracellular Fluids with Respect to Bone Salts. The concentrations of calcium and phosphate ions in extracellular fluid are considerably greater than those required to cause precipitation of hydroxyapatite. However, because of the large number of ions required to form a single molecule of hydroxyapatite, it is very difficult for all of these ions to come together simultaneously; therefore, the precipitation process is normally extremely slow; and, without the intermediation of some active process, it requires months to years to occur significantly. Furthermore, inhibitors are present in most tissues of the body, as well as in plasma, to prevent such precipitation; one such inhibitor is pyrophosphate. Therefore, hydroxyapatite crystals fail to precipitate in most normal tissues despite the state of supersaturation of the ions.

Mechanism for Causing Hydroxyapatite Precipitation in Bones. The initial stage in bone production is secretion by the osteoblasts of ground substance and of collagen. The collagen polymerizes rapidly to form collagen fibers, and the resultant tissue becomes cartilage. Shortly thereafter, hydroxyapatite crystals begin to precipitate in the cartilage, and most of the ground substance is resorbed. The crystals first appear within the substance of the collagen fibers, and then additional crystals form around the peripheries of these fibers until very little space remains for the ground substance. Thus, the final product, bone, is mainly collagen fibers and hydroxyapatite crystals.

It is believed that the collagen fibers of the cartilage are specially prepared in advance for causing precipitation of the hydroxyapatite crystals. Several theories have been offered as an explanation of this effect. One theory suggests that the osteoblasts secrete a substance into the cartilage to neutralize the inhibitor that normally prevents hydroxyapatite crystallization. Once this has occurred, a natural affinity of the collagen fibers for calcium salts supposedly causes their precipitation.

Precipitation of calcium in non-osseous tissues under abnormal conditions. Though calcium salts almost never precipitate in normal tissues besides bone, under abnormal conditions they do precipitate. For instance, they precipitate in arterial walls in the condition called arteriosclerosis and cause the arteries to become bone-like tubes. Likewise, calcium salts frequently deposit in degenerating tissues or in old blood clots. Presumably, in these instances, the inhibitor factors that normally prevent deposition of calcium salts disappear from the tissues, thereby allowing precipitation.

EXCHANGEABLE CALCIUM IN BONE

If soluble calcium salts are injected intravenously, the calcium ion concentration can be made to increase immediately to very high levels. However, within minutes to an hour or more, the calcium ion concentration returns to normal. Likewise, if large quantities of calcium ions are removed from the circulating body fluids, the calcium ion concentration again returns to normal within minutes to hours. These effects result from the fact that the bones contain a type of *exchangeable* calcium that is always in equilibrium with the calcium ions in the extracellular fluids. This calcium is probably deposited by the process of adsorption or in the form of some readily mobilizable salt such as $Ca\ HPO_4$.

The importance of exchangeable calcium to the body is that it provides a rapid buffering mechanism to keep the calcium ion concentration in the extracellular fluids from rising to excessive levels or falling to very low levels under transient conditions of excess or hypo-availability of calcium.

DEPOSITION AND ABSORPTION OF BONE

Bone is continually being deposited by *osteoblasts,* which are found on most surfaces of the bones and in many cavities. A small amount of osteoblastic activity occurs continually in all living bones so that at least some new bone is being formed constantly.

Bone is also being continually absorbed in the presence of *osteoclasts,* which are located in many cavities throughout all bones. Osteoclasts can form from osteocytes or osteoblasts, or

they can even form from fibroblasts in the bone marrow. Later in the chapter we will see that parathyroid hormone controls the bone absorptive activity of osteoclasts.

Histologically, bone absorption occurs immediately adjacent to the osteoclasts, as illustrated in Figure 53–3. Recent electron microscopic studies have shown the presence of a few minute bone crystals inside the osteoclasts, which indicates that the osteoclasts in some instances phagocytize and digest bony particles, and finally release calcium, phosphate, and the digestive end-products of the organic matrix into the extracellular fluids. However, this phagocytic process probably is not the major means by which bone is absorbed, for it is believed that most bone absorption results from osteoclastic secretion of enzymes or acids that digest or dissolute the organic matrix and simultaneously cause dissolution of the bone salts.

The continual deposition and absorption of bone has a number of physiologic-

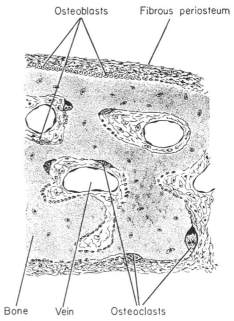

FIGURE 53–3 Osteoblastic and osteoclastic activity in the same bone.

ally important functions. For instance, because old bone becomes weak and brittle, new organic matrix is needed as the old organic matrix degenerates. In this manner the normal toughness of bone is maintained. Indeed, the bones of children, in whom the rate of deposition and absorption is rapid, show little brittleness in comparison with the bones of old age, at which time the rates of deposition and absorption are slow.

Control of the Rate of Bone Deposition by Bone "Stress." Bone is deposited in proportion to the compressional load that the bone must carry. For instance, the bones of athletes become considerably heavier than those of nonathletes. Also, if a person has one leg in a cast but continues to walk on the opposite leg, the bone of the leg in the cast becomes thin and decalcified, while the opposite bone remains thick and normally calcified. Therefore, continual physical stress stimulates osteoblastic deposition of bone.

Bone stress also determines the shape of bones under certain circumstances. For instance, if a long bone of the leg breaks in its center and then heals at an angle, the compression stress on the inside of the angle causes increased deposition of bone, while increased absorption occurs on the outer side of the angle where the bone is not compressed. After many years of increased deposition on the inner side of the angulated bone and absorption on the outer side, the bone becomes almost straight. This is especially true in children because of the rapid remodeling of bone at younger ages.

Repair of a Fracture. A fracture of a bone in some way maximally activates all the periosteal and intraosseous osteoblasts involved in the break. Indeed, even fibroblasts in some surrounding tissues become osteoblasts. Therefore, within a short time a large bulge of fibrous tissue, osteoblastic tissue, and new organic bone matrix, followed shortly by the deposition of calcium salts, develops between the two broken ends of the bone. This is called a *callus*. The callus is then reshaped over months or years to become normal bone.

PARATHYROID HORMONE

For many years it has been known that increased activity of the parathyroid gland causes rapid absorption of calcium salts from the bones with resultant hypercalcemia in the extracellular fluids; conversely, hypofunction of the parathyroid glands causes hypocalcemia, often with resultant tetany, as described earlier in the chapter.

Physiologic Anatomy of the Parathyroid Glands. Normally there are four parathyroid glands in the human being, located, respectively, behind each of the upper and each of the lower poles of the thyroid. Each parathyroid gland is approximately 6 mm. long, 3 mm. wide, and 2 mm. thick, and has a macroscopic appearance of dark brown fat; therefore, the parathyroid glands are difficult to locate during thyroid operations. For this reason total or subtotal thyroidectomy frequently resulted in total removal of the parathyroid glands before the importance of these glands was generally recognized.

The parathyroid gland of the adult human being, illustrated in Figure 53–4, contains mainly *chief cells* and *oxyphil*

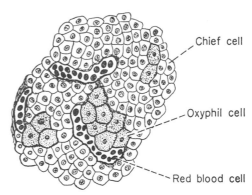

FIGURE 53–4 Histologic structure of a parathyroid gland.

cells, but oxyphil cells are absent in many animals and in young human beings. The chief cells secrete most of the parathyroid hormone. The function of the oxyphil cells is not certain; they are probably aged chief cells that still secrete some hormones.

Chemistry of Parathyroid Hormone. Parathyroid hormone is a small protein having a molecular weight of approximately 9000 and composed of 74 to 80 amino acids.

EFFECT OF PARATHYROID HORMONE ON CALCIUM AND PHOSPHATE CONCENTRATIONS IN THE EXTRACELLULAR FLUID

Figure 53–5 illustrates the effect of injecting parathyroid hormone subcutaneously into a human being, showing marked elevation in calcium ion concentration of the extracellular fluids and depression of phosphate concentration. The rise in calcium ion concentration is caused principally by a direct effect of parathyroid hormone to increase bone absorption. The decline in phosphate concentration, on the other hand, is caused principally by an effect of parathyroid hormone on the kidneys to increase the rate of phosphate excretion.

Bone Absorption Caused by Parathyroid Hormone. The absorptive effect of parathyroid hormone in bones is be-

lieved to result from the ability of this hormone to convert osteoblasts and osteocytes in bone into osteoclasts and, also, to increase their osteoclastic activity. In turn, the osteoclasts are believed to secrete either enzymes or acids that absorb the bone, as discussed earlier in the chapter.

Bone contains such great amounts of calcium in comparison with the total amount in all the extracellular fluids (about 1000 times as much) that even when parathyroid hormone causes a great rise in calcium concentration in the fluids, it is impossible to discern any immediate effect at all on the bones. Yet prolonged administration of parathyroid hormone finally results in evident absorption in all the bones with development of large cavities filled with very large, multinucleated osteoclasts.

Effect of Parathyroid Hormone on Phosphate and Calcium Excretion by the Kidneys. Administration of parathyroid hormone causes immediate and rapid loss of phosphate in the urine. This effect is caused by diminished renal tubule reabsorption of phosphate ions. It was long believed that this was the most important function of parathyroid hormone. However, it is now known that this function is of secondary importance to the effect of parathyroid hormone on bone absorption.

Parathyroid hormone causes increased renal tubular reabsorption of calcium at the same time that it diminishes the rate of phosphate reabsorption. Were it not for this effect of parathyroid hormone on the kidneys to increase calcium reabsorption, the continual loss of calcium into the urine would eventually deplete the bones of this mineral.

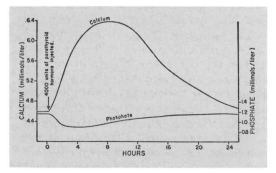

FIGURE 53–5 Effect on plasma calcium and phosphate concentrations of injecting 4000 units of parathyroid hormone into a human being.

REGULATION OF EXTRACELLULAR CALCIUM ION CONCENTRATION BY THE PARATHYROID GLANDS

The calcium ion concentration in the extracellular fluids under normal conditions rarely varies more than 5 per cent above or below the normal concentration. Therefore, calcium ion regulation is one of the most highly developed homeostatic mechanisms of the entire body. This regulation is effected almost entirely by the parathyroid glands in the following way:

Control of Parathyroid Secretion by Calcium Ion Concentration. Even the slightest decrease in calcium ion concentration in the extracellular fluid causes the parathyroid glands to increase their rate of secretion and to hypertrophy. For instance, the parathyroid glands become greatly enlarged in *rickets,* in which the level of calcium is usually depressed only a few per cent; also they become greatly enlarged in pregnancy, even though the decrease in calcium ion concentration in the mother's extracellular fluid is hardly measurable; and, they are greatly enlarged during lactation because calcium is used for milk formation.

Feedback Control of Calcium Ion Concentration. It can now be understood how the parathyroid glands control the calcium ion concentration in the extracellular fluids, for reduced calcium concentration increases the rate of parathyroid secretion; and this in turn increases bone absorption, thereby raising the calcium ion concentration to normal. Conversely, increased calcium ion concentration inhibits the parathyroid glands, which automatically decreases the calcium ion concentration to normal.

Note particularly that the regulatory properties of parathyroid hormone do not derive only from its effect on bone, for in addition it *increases the absorption of calcium from the gut* and *from the renal tubules,* which also helps to raise the calcium ion concentration. These are particularly valuable effects when the bone salts have been so depleted that parathyroid hormone can no longer cause additional absorption of calcium from the bone.

CALCITONIN

About ten years ago, a new hormone that has effects on blood calcium opposite to those of parathyroid hormone was discovered in several lower animals. This hormone was named *calcitonin* because it reduces the blood calcium ion concentration. In the human being calcitonin is secreted by the thyroid gland — by the so-called *parafollicular cells* in the interstitial tissue between the follicles.

Physiologic Effects of Calcitonin. Calcitonin inhibits the absorption of bone. It presumably acts at the same site in bone as does parathyroid hormone, because it even inhibits the absorption of bone after administration of large quantities of parathyroid hormone. Unfortunately, the mechanism of this inhibition is unknown, though it occurs extremely rapidly. It causes almost total cessation of bone salt reabsorption within minutes; as a result, the calcium ion concentration in the body fluids begins to fall also within minutes. Thus, the action of calcitonin on calcium ion concentration occurs many times as rapidly as the effect of parathyroid hormone.

Role of Calcitonin in Feedback Homeostasis of Calcium Ion Concentration. Perfusion of the thyroid gland with plasma containing even slightly elevated calcium ion concentration causes an immediate, several-fold increase in rate of secretion of calcitonin. Conversely, decreased plasma calcium ion concentration can decrease calcitonin secretion to as little as one-fifth normal. Therefore, the calcitonin mechanism, like the parathyroid mechanism, can also act in a feedback system to control calcium ion concentration.

However, there is one major difference between the calcitonin and the parathyroid feedback systems: the calcitonin mechanism operates extremely rapidly, reaching peak activity in less than an hour; this is in contrast to the many hours required for peak activity to be attained following parathyroid secretion.

PHYSIOLOGY OF PARATHYROID AND BONE DISEASES

HYPOPARATHYROIDISM

When the parathyroid glands do not secrete sufficient parathyroid homone, the osteoclasts of the bone become almost totally inactive, and the calcium level in the blood falls from the normal value of 10 mg. per cent to 7 mg. per cent within two to three days. When this level is reached, the usual signs of hypocalcemic tetany develop. Among the muscles of the body most sensitive to tetanic spasm are the laryngeal muscles. Spasm of these obstructs respiration, which is the usual cause of death in tetany unless appropriate treatment is applied.

Treatment of Hypoparathyroidism. *Parathyroid hormone (parathormone).* Parathyroid hormone is occasionally used for treating hypoparathyroidism. However, because of the expense of this hormone, because its effect lasts only about 24 to 36 hours, and because the tendency of the body to develop immune bodies against it makes it progressively less active in the body, treatment of hypoparathyroidism with parathyroid hormone is rare in present-day therapy.

Dihydrotachysterol and vitamin D. In addition to its ability to cause increased absorption of calcium from the gastrointestinal tract, vitamin D also causes a weak effect similar to that of parathyroid hormone in promoting calcium and phosphate absorption from bones. Therefore, a person with hypoparathyroidism can be treated satisfactorily by administration of *large quantities* of vitamin D. One of the vitamin D compounds, dihydrotachysterol (A.T. 10), has a more marked ability to cause bone absorption than do most of the other vitamin D compounds. Administration of calcium plus vitamin D three or more times a week can almost completely control the calcium level in the extracellular fluid of a hypoparathyroid person.

HYPERPARATHYROIDISM

The cause of hyperparathyroidism ordinarily is a tumor of one of the parathyroid glands, and such tumors occur much more frequently in women than in men or children. Consequently, it is believed that pregnancy, lactation, and perhaps other causes of prolonged low calcium levels, all of which stimulate the parathyroid gland, may predispose to the development of such a tumor. Treatment is always simply to remove the tumor, though this may be difficult to do because these tumors are sometimes only pea-sized and cannot be found at operation.

In hyperparathyroidism extreme osteoclastic activity occurs in the bones, and this elevates the calcium ion concentration in the extracellular fluid while usually (but not always) depressing slightly the concentration of phosphate ions.

Bone Disease in Hyperparathyroidism. In severe hyperparathyroidism, osteoclastic absorption of bone soon far outstrips osteoblastic deposition, and the bone may be eaten away almost entirely. Indeed, the reason a hyperparathyroid person comes to the doctor is often a bone that has broken without cause.

Effects of Hypercalcemia in Hyperparathyroidism. Hyperparathyroidism can at times cause the plasma calcium level to rise to as high as 15 to 17 mg. per cent. The effects of such elevated calcium levels are mild to moderate depression of the central and peripheral nervous systems, muscular weakness, constipation, abdominal pain, peptic ulcer, lack of appetite, and depressed relaxation of the heart during diastole.

RICKETS

Rickets occurs mainly in children as a result of calcium or phosphate deficiency in the extracellular fluids. Ordinarily, rickets is caused by lack of vitamin D rather than lack of calcium or phosphate in the diet. If the child is properly exposed to sunlight, the 7-dehydrocholesterol in his skin becomes activated by the ultraviolet rays and forms vitamin D_3, which prevents rickets by promoting calcium and phosphate absorption from the intestines as discussed earlier in the chapter.

Children who remain indoors through the winter generally do not receive adequate quantities of vitamin D without some supplementary therapy in the diet. Rickets tends to occur especially in the spring months because vitamin D formed during the preceding summer can be stored for several months in the liver, and, also, calcium and phosphorus absorption from the bones must take place for several months before clinical signs of rickets become apparent.

Effect of Rickets on the Bone. During prolonged deficiency of calcium and phosphate in the body fluids, increased parathyroid hormone secretion protects the body against hypocalcemia by causing osteoclastic absorption of the bone; however, this causes the bone to become progressively weaker and also imposes marked physical strain on the bone, resulting in rapid osteoblastic activity as well as osteoclastic activity. The osteoblasts lay down large quantities of organic bone matrix, which does not become calcified because the calcium and phosphate ions are insufficient to cause calcification. Consequently, the newly formed, uncalcified organic matrix gradually takes the place of other bone that is being reabsorbed.

Obviously, hyperplasia of the parathyroid glands is marked in rickets because of the decreased blood calcium level.

Tetany in Rickets. In the early stages of rickets, tetany almost never occurs because the parathyroid glands continually stimulate osteoclastic absorption of bone and therefore maintain an almost normal level of calcium in the body fluids. However, when the bones become exhausted of calcium, the level of calcium may fall rapidly. As the blood level of calcium falls below 7 mg. per cent, the usual signs of tetany develop, and the child may die of tetanic respiratory spasm unless intravenous calcium is administered, which relieves the tetany within seconds.

PHYSIOLOGY OF THE TEETH

The teeth cut, grind, and mix the food. To perform these functions the jaws have powerful muscles capable of providing an occlusive force between the front teeth of as much as 50 to 100 pounds and as much as 150 to 200 pounds for the jaw teeth. Also, the upper and lower teeth are provided with projections and facets which interdigitate so that each set of teeth fits with the other. This fitting is called *occlusion,* and it allows even small particles of food to be caught and ground between the tooth surfaces.

FUNCTION OF THE DIFFERENT PARTS OF THE TEETH

Figure 53–6 illustrates a sagittal section of a tooth, showing its major functional parts, the *enamel, dentine, cementum,* and *pulp.* The tooth can also be divided into the *crown,* which is the portion that protrudes out of the gum into the mouth, and the *root,* which is the portion that protrudes into the bony socket of the jaw. The collar between the crown and the root where the tooth is surrounded by the gum is called the *neck.*

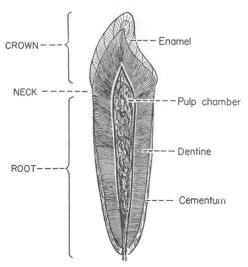

FIGURE 53-6 Functional parts of a tooth.

Dentine. The main body of the tooth is composed of dentine, which has a strong, bony structure. Dentine is made up principally of hydroxyapatite crystals similar to those in bone, but much more dense. These are embedded in a strong meshwork of collagen fibers. In other words, the principal constituents of dentine are very much the same as those of bone. The major difference is its histologic organization, for dentine does not contain any osteoblasts, osteoclasts, or spaces for blood vessels or nerves. Instead, it is deposited and nourished by a layer of cells called *odontoblasts,* which line its inner surface along the wall of the pulp cavity.

The calcium salts in dentine make it extremely resistant to compressional forces, while the collagen fibers make it tough and resistant to tensional forces that might result when the teeth are struck by solid objects.

Enamel. The outer surface of the tooth is covered by a layer of enamel that is formed prior to eruption of the tooth by special epithelial cells called *ameloblasts.* Once the tooth has erupted, no more enamel is formed. Enamel is composed of small crystals of calcium phosphate salts (hydroxyapatite) with adsorbed carbonate, magnesium, so-dium, potassium, and other ions embedded in a fine meshwork of very strong and almost completely insoluble protein fibers that are similar to (but not identical with) the keratin of hair. The smallness of the crystalline structure of the salts makes the enamel extremely hard, much harder than the dentine. Also, the protein fiber meshwork makes enamel very resistant to acids, enzymes, and other corrosive agents because this protein is one of the most insoluble and resistant proteins known.

Cementum. Cementum is a bony substance secreted by cells of the *periodontal membrane,* which lines the tooth socket. Many collagen fibers pass directly from the bone of the jaw, through the periodontal membrane, and then into the cementum. These collagen fibers and the cementum hold the tooth in place. When the teeth are exposed to excessive strain, the layer of cementum becomes thicker and stronger. Also, it increases in thickness and strength with age, causing the teeth to become progressively more firmly seated in the jaws as one reaches adulthood and older.

Pulp. The inside of each tooth is filled with pulp, which in turn is composed of connective tissue with an abundant supply of nerves, blood vessels, and lymphatics. The cells lining the surface of the pulp cavity are the odontoblasts, which, during the formative years of the tooth, lay down the dentine but at the same time encroach more and more on the pulp cavity, making it smaller. In later life the dentine stops growing and the pulp cavity remains essentially constant in size. However, the odontoblasts are still viable and send projections into small *dentinal tubules* that penetrate all the way through the dentine; these are of importance for providing nutrition.

DENTITION

Each human being and most other mammals develop two sets of teeth during a lifetime. The first teeth are called the *deciduous teeth,* or *milk teeth,* and they number 20 in the human being. These erupt between the seventh month and second year of life, and they last until the sixth to the thirteenth year. After each deciduous tooth is lost, a permanent tooth replaces it, and an additional 8 to 12 molars appear posteriorly in the jaw, making the total number of permanent teeth 28 to 32, depending on whether the four *wisdom teeth* finally appear, which does not occur in everyone.

Formation of the Teeth. Figure 53–7 illustrates the formation and eruption of teeth. Figure 53–7*A* shows invagination of the oral epithelium into the *dental lamina;* this is followed by the development of a tooth-producing organ. The epithelial cells above form ameloblasts, which secrete the enamel on the outside of the tooth. The epithelial cells below invaginate upward to form a pulp cavity and also to form the odontoblasts that secrete dentine. Thus, enamel is secreted on the inside, forming an early tooth, as illustrated in Figure 53–7*B.*

Eruption of teeth. During early childhood, the teeth begin to protrude upward from the jaw bone through the oral epithelium into the mouth. The cause of "eruption" is unknown, though several theories have been offered in an attempt to explain this phenomenon.

One of these assumes that an increase of the material inside the pulp cavity of the tooth causes much of it to be extruded through the root canal and that this pushes the tooth upward. However, a more likely theory is that the bone surrounding the tooth progressively hypertrophies and in so doing shoves the tooth forward.

Development of the permanent teeth. During embryonic life, a tooth-forming organ also develops in the dental lamina for each permanent tooth that will be needed after the deciduous teeth are gone. These tooth-producing organs slowly form the permanent teeth throughout the first 6 to 20 years of life. When each permanent tooth becomes fully formed, it, like the deciduous tooth, pushes upward through the bone of the jaw. In so doing it erodes the root of the deciduous tooth and eventually causes it to loosen and fall out. Soon thereafter, the permanent tooth erupts to take the place of the original one.

Metabolic factors in development of the teeth. The rate of development and the speed of eruption of teeth can be accelerated by both thyroid and growth hormones. Also, the deposition of salts in the early forming teeth is affected considerably by various factors of metabolism, such as the availability of calcium and phosphate in the diet, the amount of vitamin D present, and the rate of parathyroid hormone secretion. When all these factors are normal, the dentine and enamel will be correspondingly healthy, but, when they are deficient, the calcification of the teeth also may be defective so that the teeth will be abnormal throughout life.

MINERAL EXCHANGE IN TEETH

The salts of teeth, like those of bone, are composed basically of hydroxyapatite with adsorbed carbonates and various cations bound together in a hard crystalline substance. Also, new salts

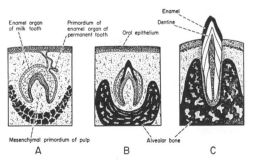

FIGURE 53–7 (*A*) Primordial tooth organ. (*B*) The developing tooth. (*C*) The erupting tooth. (Modified from Bloom and Fawcett: A Textbook of Histology, 8th Ed.)

are constantly being deposited while old salts are being reabsorbed from the teeth, as also occurs in bone. However, experiments indicate that deposition and reabsorption occur mainly in the dentine and cementum, while very little occurs in the enamel. Much of that which does occur in the enamel occurs by exchange of minerals with the saliva instead of with the fluids of the pulp cavity. The rate of absorption and deposition of minerals in the cementum is approximately equal to that in the surrounding bone of the jaw, while the rate of deposition and absorption of minerals in the dentine is only one-third that of bone. The cementum has characteristics almost identical with those of usual bone, including the presence of osteoblasts and osteoclasts, while dentine does not have these characteristics, as was explained above; this difference undoubtedly explains the different rates of mineral exchange.

The mechanism by which minerals are deposited and reabsorbed from the dentine is unknown. It is possible that the small processes of the odontoblasts protruding into the tubules of the dentine are capable of absorbing salts and then providing new salts to take the place of the old. It is also possible that the odontoblasts provide continuous replacement of the collagen fibers of the dentine. This would be comparable to the rejuvenation of the bone matrix by osteoblasts, which is necessary for adequate maintenance of bone strength.

In summary, rapid mineral exchange occurs in the dentine and cementum of teeth, though the mechanism of this exchange in dentine is unknown. On the other hand, enamel exhibits extremely slow mineral exchange so that it maintains much of its original mineral complement throughout life.

DENTAL ABNORMALITIES

The two most common dental abnormalities are *caries* and *malocclusion.*

Caries means erosions of the teeth, while malocclusion means failure of the projections of the upper and lower teeth to interdigitate properly.

Caries. Two major but differing theories have been proposed to explain the cause of caries. One of these postulates that acids formed in crevices of the teeth by acid-producing bacteria cause erosion and absorption of the protein matrix of the enamel and dentine. This theory assumes that the acids are formed by splitting carbohydrates into lactic acid. For this reason it has been taught that eating a diet high in carbohydrate content and, particularly, eating large quantities of sweets between meals can lead to excessive development of caries.

The second theory proposes that proteolytic enzymes secreted by bacteria in the crevices of the teeth or in plaques that develop on the surfaces of unbrushed teeth digest the keratin matrix of the enamel. Then the calcium salts, unprotected by their protein fibers, are slowly dissolved by the saliva.

It is not known which of the above theories correctly describes the primary process for development of caries. Indeed, both might be operative simultaneously to cause carious teeth.

Some teeth are more resistant to caries than others. Studies in recent years indicate that teeth formed in children who drink water containing small amounts of fluorine develop enamel that is more resistant to caries than the enamel in children who drink water not containing fluorine. Fluorine does not make the enamel harder than usual, but instead it is said to inactivate proteolytic enzymes before they digest the protein matrix of the enamel. Regardless of the precise means by which fluorine protects the teeth, it is known that small amounts of fluorine deposited in enamel make teeth about twice as resistant to caries as are teeth without fluorine.

Malocclusion. Malocclusion (failure of the facets of upper and lower teeth to fit properly) is usually caused by a heredity abnormality that causes the teeth of one jaw to grow in an abnormal direction. In malocclusion, the teeth cannot perform their normal grinding or cutting action adequately; occasionally malocclusion results in abnormal displacement of the lower jaw in relation to the upper jaw, causing such undesirable effects as pressure on the anterior portion of the ear or pain in the mandibular joint.

The orthodontist can often correct malocclusion by applying prolonged gentle pressure against the teeth with appropriate braces. The gentle pressure causes absorption of alveolar jaw bone on the compressed side of the tooth and deposition of new bone on the tensional side of the tooth. In this way the tooth gradually moves to a new position as directed by the applied pressure.

REFERENCES

Arnaud, C. D., Jr.; Tenenhouse, A. M.; and Rasmussen, H.: Parathyroid hormone. *Ann. Rev. Physiol,* 29:349, 1967.

Copp, D. H.: Endocrine regulation of calcium metabolism. *Ann. Rev. Physiol.,* 32:61, 1970.

Fourman, P.: Calcium Metabolism and the Bone. 2nd ed., Philadelphia, F. A. Davis Co., 1968.

Gaillard, P.; Talmage, R. V.; and Budy, A. M.: The Parathyroid Glands. Chicago, University of Chicago Press, 1965.

Hirsch, P. F., and Munson, P. L.: Thyrocalcitonin. *Physiol. Rev.,* 49:548, 1969.

Jackson, W. P. U.: Calcium Metabolism and Bone Disease. Baltimore, The Williams & Wilkins Co., 1967.

McLean, F. C. (ed.): Radioisotopes and Bone. Philadelphia, F. A. Davis Co., 1963.

Posner, A. S.: Crystal chemistry of bone mineral. *Physiol. Rev.,* 49:760, 1969.

CHAPTER 54

REPRODUCTIVE FUNCTIONS OF THE MALE, AND THE MALE SEX HORMONES

The reproductive functions of the male can be divided into three major subdivisions: first, spermatogenesis, which means simply the formation of sperm; second, performance of the male sexual act; and third, regulation of male sexual functions by the various hormones. Associated with these reproductive functions are the effects of the male sex hormones on the accessory sexual organs, on cellular metabolism, on growth, and on other functions of the body.

Physiologic Anatomy of the Male Sexual Organs. Figure 54–1 illustrates the various portions of the male reproductive system. Note that the testis is composed of a large number of *seminiferous tubules* where the sperm are formed. The sperm then empty into the *vasa recta* and from there into the *epididymis*. The epididymis leads into the *vas deferens,* which enlarges into the *ampulla of the vas deferens* immediately proximal to

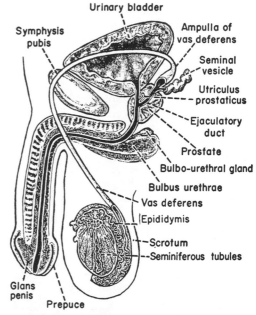

FIGURE 54–1 The male reproductive system. (Modified from Bloom and Fawcett: Textbook of Histology, 8th Ed.)

the prostate gland. A *seminal vesicle,* one located on each side of the prostate, empties into the prostatic end of the ampulla, and the contents from both the ampulla and the seminal vesicle pass into an *ejaculatory duct* leading through the body of the prostate gland to empty into the *internal urethra.* Finally, the *urethra* is the last connecting link from the testis to the exterior. The urethra is supplied with mucus derived from a large number of small *glands of Littré* located along its entire extent and also from large bilateral *bulbo-urethral glands* (Cowper's glands) located near the origin of the urethra.

SPERMATOGENESIS

Spermatogenesis occurs in all the seminiferous tubules during active sexual life, beginning at the age of approximately 13 as the result of stimulation by adenohypophyseal gonadotropic hormones and continuing throughout the remainder of life.

The seminiferous tubules, one of which is illustrated in Figure 54–2*A*, contain a large number of small to medium-sized cells called *spermatogonia* located in two to three layers along the outer border of the tubular epithelium. These continually proliferate and differentiate through definite stages of development to form sperm, as shown in Figure 54–2*B*. During this process, each sperm loses one member of each of its 23 pairs of chromosomes, leaving only 23 chromosomes, none of them paired.

The sex-determining pair of chromosomes in the spermatogonium is composed of one "X" chromosome, which is called the *female* chromosome, and one "Y" chromosome, the *male* chromosome. During sperm formation, however, the sex-determining chromosomes divide among the sperm, so that half of the sperm become *male sperm* containing the "Y" chromosome and the other

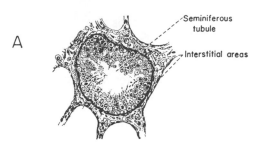

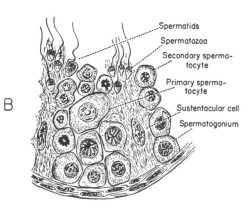

FIGURE 54–2 (*A*) Cross-section of a seminiferous tubule. (*B*) Spermatogenesis. (Modified from Arey: Developmental Anatomy. 7th Ed.)

half *female sperm* containing the "X" chromosome. The sex of the offspring is determined by which of these two types of sperm fertilizes the ovum. This will be discussed further in Chapter 56.

During formation of sperm from spermatogonia, most of the cytoplasm disappears and each cell begins to elongate into a spermatozoon, illustrated in 54–3, composed of a *head, neck, body,* and *tail.* To form the head, the nuclear material is rearranged into a compact mass, and the cell membrane contracts around the nucleus. It is this nuclear material that fertilizes the ovum. The tail has the same structure as a cilium, which was described in detail in Chapter 2. Upon release of sperm from the male genital tract into the female tract, the tail begins to wave back and forth, providing snake-like propulsion that moves the sperm forward at a

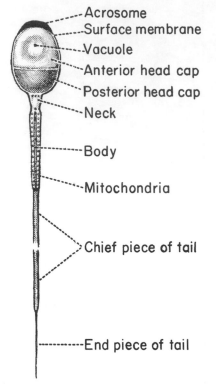

Acrosome
Surface membrane
Vacuole
Anterior head cap
Posterior head cap
Neck
Body
Mitochondria
Chief piece of tail
End piece of tail

FIGURE 54–3 Structure of the human spermatozoon.

they are stored. Sperm can maintain their fertility in the genital ducts for as long as 42 days.

FUNCTION OF THE PROSTATE GLAND

The prostate gland secretes a thin, milky, alkaline fluid containing citric acid, calcium, acid phosphate, and fibrinolysin. During ejaculation, the capsule of the prostate gland contracts simultaneously with the contractions of the vas deferens and seminal vesicles so that the thin, milky fluid of the prostate gland adds to the bulk of the semen. The alkaline characteristic of the prostatic fluid may be quite important for successful fertilization of the ovum, because the fluids of the vas deferens and of the female vagina are relatively acidic and, consequently, inhibit sperm fertility. Sperm do not become optimally motile until the pH of the surrounding fluids rises to approximately 6 to 6.5.

maximum velocity of about 1 foot per hour.

Maturation of Sperm in the Epididymis. Following formation in the seminiferous tubules, the sperm pass through the *vasa recta* into the *epididymis*. After the sperm have been in the epididymis for some 18 hours to 10 days, they mature, develop the power of motility, and become capable of fertilizing the ovum.

Storage of Sperm. A small quantity of sperm can be stored in the epididymis, but probably most sperm are stored in the vas deferens and possibly to some extent in the ampulla of the vas deferens. Though the sperm in these areas become motile if released to the exterior, they are relatively dormant as long as they are stored, probably because of the acidic nature of the fluid in which

SEMEN AND THE PHENOMENON OF CAPACITATION

Semen, which is ejaculated during the male sexual act, is composed of the fluids from the vas deferens, from the seminal vesicles, from the prostate gland, and from the mucous glands, especially the bulbo-urethral glands. The average pH of the combined semen is approximately 7.5, the alkaline prostatic fluid having neutralized the mild acidity of the other portions of the semen. The prostatic fluid gives the semen a milky appearance, while fluid from the seminal vesicles and from the mucous glands gives the semen a mucoid consistency. Within approximately one-half hour after ejaculation the mucoid consistency of semen disappears because of proteolytic enzymes in the fluid.

For reasons not yet understood, sperm that are washed clear of the other components of semen become rapidly infertile but regain this fertility in a few minutes to an hour or more when remixed with the other components. This phenomenon is called *capacitation*.

Though sperm can live for many weeks in the male genital ducts, once they are ejaculated in the semen their maximal life span is only 24 to 72 hours at body temperature. At lowered temperatures, however, semen may be stored for weeks or even years.

MALE FERTILITY

The seminiferous tubular epithelium can be destroyed by a number of different diseases. For instance, bilateral orchiditis resulting from mumps usually causes sufficient degeneration of tubular epithelium that sterility results in a large percentage of males so afflicted. Another disease that frequently localizes in the testes and can cause severe tubular damage is typhus fever. Also, many male infants are born with degenerate tubular epithelium as a result of strictures in the genital ducts or as a result of unknown causes. Finally, a cause of sterility seems to be excessive temperature of the testes.

Effect of Sperm Count on Fertility. An average total of 400,000,000 sperm are usually present in each ejaculate, in a total semen volume of about 3 ml. When the number of sperm falls below approximately 100,000,000, the person is likely to be almost completely infertile. Only a single sperm is necessary to fertilize the ovum; but, for reasons not yet completely understood, the ejaculate must contain a tremendous number of sperm for at least one to fertilize the ovum. It is believed that many sperm must be present to secrete the enzyme hyaluronidase, which removes a barrier of granulosa cells from around the ovum to be fertilized.

THE MALE SEXUAL ACT

NEURONAL STIMULUS FOR PERFORMANCE OF THE MALE SEXUAL ACT

The most important source of impulses for initiating the male sexual act is the *glans penis*. The massaging action of intercourse on the glans stimulates the sensory end-organs, and the sexual sensations in turn pass into the sacral portion of the spinal cord, and finally up the cord to undefined areas of the cerebrum. Impulses may also enter the spinal cord from areas adjacent to the penis to aid in stimulating the sexual act. For instance, stimulation of the anal epithelium, the scrotum, and perineal structures in general may all send into the cord impulses which add to the sexual sensation. Sexual sensations can even originate in internal structures, such as irritated areas of the urethra, the bladder, the prostate, the seminal vesicles, the testes, and the vas deferens. Indeed, one of the causes of "sexual drive" is probably overfilling of the sexual organs with secretions. Infection and inflammation of these sexual organs sometimes cause almost continual sexual desire, and "aphrodisiac" drugs, such as cantharides, increase the sexual desire by irritating the bladder and urethral mucosa.

The Psychic Element of Male Sexual Stimulation. Appropriate psychic stimuli can greatly enhance the ability of a person to perform the sexual act. Simply thinking sexual thoughts or even dreaming that the act of intercourse is being performed can cause the male sexual act to occur and to culminate in ejaculation. Indeed, *nocturnal emissions* during dreams occur in many males during some stages of sexual life, especially during the teens.

Integration of the Male Sexual Act in the Spinal Cord. Though psychic factors usually play an important part in the male sexual act and can actually initiate it, the cerebrum is probably not absolutely necessary for its performance, for appropriate genital stimulation can cause ejaculation in some animals and in an occasional human being after their spinal cords have been cut above the lumbar region. Therefore, the male sexual act results from inherent reflex mechanisms integrated in the sacral and lumbar spinal cord, and these mechanisms can be initiated by either psychic stimulation or actual sexual stimulation.

STAGES OF THE MALE SEXUAL ACT

Erection. Erection is the first effect of male sexual stimulation. It is caused by parasympathetic impulses that pass from the sacral portion of the spinal cord to the penis. These parasympathetic impulses dilate the arteries of the penis, thus allowing arterial blood to flow under high pressure into the *erectile tissue* of the penis, illustrated in Figure 54–4. This erectile tissue is nothing more than large, cavernous venous sinusoids, which are normally relatively empty but which become dilated tremendously when arterial blood flows into them under pressure, since the venous outflow is partially occluded. Also, the erectile bodies are surrounded by strong fibrous coats; therefore, high pressure within the sinusoids causes ballooning of the erectile tissue to such an extent that the penis becomes hard and elongated.

Lubrication. During sexual stimulation, parasympathetic impulses, in ad-

dition to promoting erection, cause the glands of Littré and the bulbo-urethral glands to secrete mucus. Thus mucus flows through the urethra during intercourse to aid in the lubrication of coitus. However, most of the lubrication of coitus is provided by the female sexual organs rather than by the male. Without satisfactory lubrication, the male sexual act is rarely successful because unlubricated intercourse causes pain impulses which inhibit rather than excite sexual sensations.

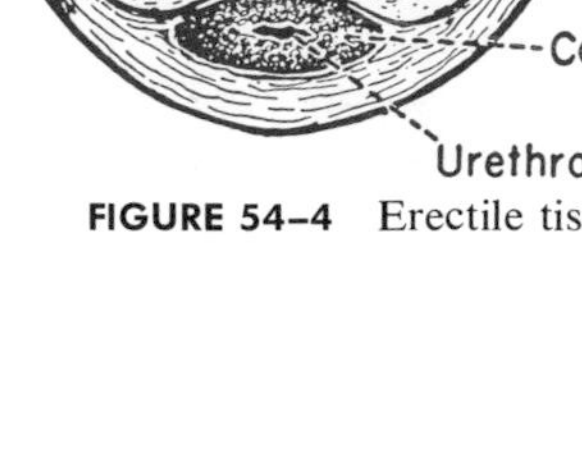

Ejaculation. Ejaculation is the culmination of the male sexual act. When the sexual stimulus becomes extremely intense, the reflex centers of the spinal cord begin to emit rhythmic sympathetic impulses that leave the cord at L-1 and L-2 and pass to the genital organs through the hypogastric plexus to initiate emission, which is the forerunner of ejaculation.

Emission begins with peristaltic contractions in the ducts of the testis, the epididymis, and the vas deferens to cause expulsion of sperm into the internal urethra. Simultaneously, rhythmic contractions in the seminal vesicles and the muscular coat of the prostate gland expel seminal fluid and prostatic fluid along with the sperm. These mix with the mucus already secreted by the bulbo-urethral glands, and all of these different types of fluid combine to form the semen. The process to this point is emission.

Then, rhythmic nerve impulses are sent from the cord through S-1 and S-2 and thence to skeletal muscles that encase the base of the erectile tissue, causing rhythmic increases in pressure in this tissue, which expresses the semen from the urethra to the exterior. This process is called *ejaculation* proper.

FIGURE 54–4 Erectile tissue of the penis.

Deep penile fascia

Corpus cavernosum

Central artery

Corpus spongiosum

Urethra

TESTOSTERONE: THE MALE SEX HORMONE

Secretion, Metabolism, and Chemistry of Testosterone. Two different steroid male sex hormones have been isolated from the venous blood draining from the testes, *testosterone* and *Δ4-andro-stene-3,17,-dione.* However, the quantity of testosterone is so much greater than that of the second hormone that one can consider testosterone to be the single significant hormone responsible for the male hormonal effects caused by the testes.

Testosterone is formed by the *interstitial cells of Leydig,* which lie in the interstices between the seminiferous tubules as illustrated in Figure 54–5. It is a steroid compound, as shown by the formula in Figure 54–6. After secretion by the testes, testosterone circulates in the blood for not over 15 to 30 minutes before it either becomes fixed to the tissues or is degraded into inactive products that are subsequently excreted.

FUNCTIONS OF TESTOSTERONE

In general, testosterone is responsible for the distinguishing characteristics of the masculine body. Even the testes of the fetus are stimulated by chorionic gonadotropin from the placenta to pro-

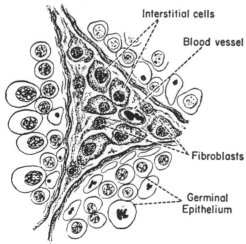

FIGURE 54–5 Interstitial cells located in the interstices between the seminiferous tubules. (Modified from Bloom and Fawcett: Textbook of Histology, 8th Ed.)

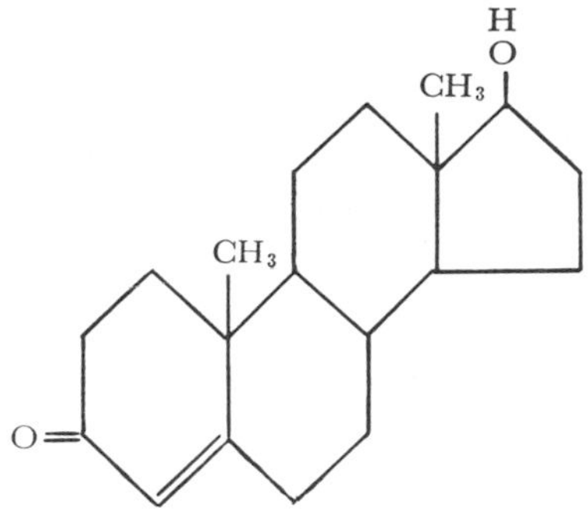

Testosterone

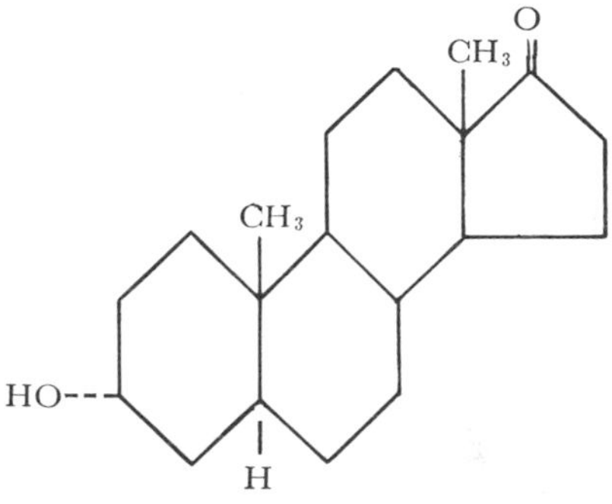

Androsterone

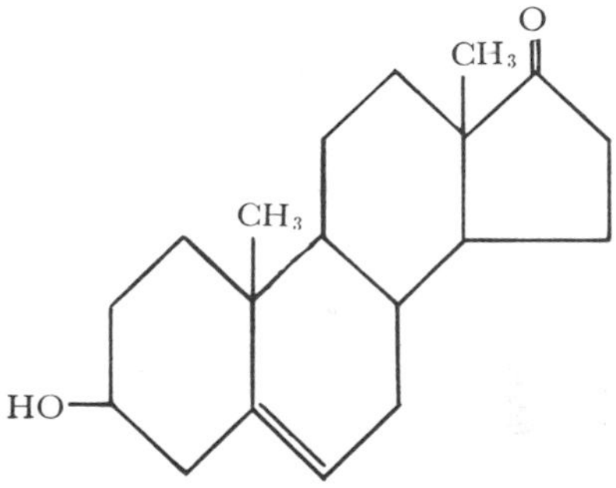

Dehydroepiandrosterone

FIGURE 54–6 Testosterone.

duce a small quantity of testosterone during fetal development, but essentially no testosterone is produced during childhood until approximately the age of 11 to 13. Then testosterone production increases rapidly at the onset of puberty and lasts throughout most of the remainder of life, dwindling rapidly beyond the age of 40 to become perhaps one-fifth the peak value by the age of 80.

Functions of Testosterone During Fetal Development. Testosterone begins to be elaborated by the male at about the second month of embryonic life. This testosterone is probably at least partly responsible for development of the male sex characteristics, including growth of a penis, a scrotum, a prostate gland, the seminal vesicles, and the genital ducts. If testosterone is very deficient, the male fetus is likely to become a *hermaphrodite*.

Effect on descent of the testes. The testes usually descend into the scrotum during the last three months of gestation when the testes are secreting reasonable quantities of testosterone. In the absence of testosterone, descent does not occur and the testes remain in the abdomen, a condition called *cryptorchidism*.

Effect of Testosterone on Development of Adult Primary and Secondary Sexual Characteristics. Testosterone secretion after puberty causes the penis, the scrotum, and the testes all to enlarge manyfold until about the age of 20. In addition, testosterone causes the "secondary sexual characteristics" of the male to develop at the same time, beginning at puberty and ending at maturity. These secondary sexual characteristics, in addition to the sexual organs themselves, distinguish the male from the female as follows:

Effect on body hair. Testosterone causes a masculine distribution of hair, including baldness in men who have a *genetic background* for the development of baldness.

Effect on the voice. Testosterone causes hypertrophy of the laryngeal mucosa and enlargement of the larynx. These effects cause at first a relatively hoarse voice, but this gradually changes into the typical masculine bass voice.

Effect on nitrogen retention and muscular development. One of the most important male characteristics is the development of increasing musculature following puberty. This is associated with increased protein in other parts of the body as well. Many of the changes in the skin are probably also due to deposition of proteins in the skin, and the changes in the voice could even result from this protein anabolic function of testosterone.

Testosterone has often been considered to be a "youth hormone" because of its effect on the musculature, and it is occasionally used for treatment of persons who have poorly developed muscles.

Effect on bone growth and calcium retention. Following puberty or following prolonged injection of testosterone, the bones grow considerably in thickness and also deposit considerable calcium salts. Thus, testosterone increases the total quantity of bone matrix, and it also causes calcium retention. The increase in bone matrix is believed to result from the general protein anabolic function of testosterone, and the deposition of calcium salts to result from increased bone matrix available to be calcified.

Testosterone also causes the epiphyses of the long bones to unite with the shafts of the bones at an early age in life. Therefore, despite rapidity of growth after puberty in the male, this early uniting of the epiphyses prevents the person from growing as tall as he would have grown if testosterone had not been secreted. Even in normal men the final adult height is slightly less than that which would have been attained had the person been castrated prior to puberty.

Other Effects of Testosterone. Testosterone also (1) increases the basal metabolic rate 5 to 15 per cent, (2) increases the red cell count about 15 per cent, and (3) slightly increases reabsorption of sodium by the distal tubules of the kidneys.

Basic Mechanism of Action of Testosterone. Though it is not known exactly how testosterone causes all the effects just listed, it is believed that they result mainly from increased rate of protein formation in cells. For instance, within 30 minutes after testosterone injection, the concentration of RNA in prostate gland cells begins to increase, and this is followed by progressive increase in cellular protein. After several days the quantity of DNA in the gland has also increased, and there has been a simultaneous increase in the number of prostatic cells. The obvious single type of action that would cause all these effects would be the activation of DNA molecules in the nucleus, but this is still speculation.

CONTROL OF MALE SEXUAL FUNCTIONS BY THE GONADOTROPIC HORMONES — FSH, LH, AND LTH

The adenohypophysis secretes three different gonadotropic hormones: (1) *follicle-stimulating hormone* (FSH); (2) *luteinizing hormone* (LH), also called *interstitial cell-stimulating hormone* (ICSH); and (3) *luteotropic hormone* (LTH), also called *lactogenic hormone,* or *prolactin.* FSH and LH play major roles in the control of male sexual functions. LTH is not known to play a significant role in male sexual function but does play significant roles in the normal monthly ovarian cycle of the female and in the secretion of milk.

Regulation of Testosterone Production by LH. Testosterone is produced by the interstitial cells of Leydig only when the testes are stimulated by LH from the adenohypophysis. Also, simultaneous administration of a small amount of FSH along with LH greatly potentiates the effect of LH in promoting testosterone production.

Effect of chorionic gonadotropin on the fetal testes. During gestation the placenta secretes large quantities of chorionic gonadotropin, a hormone that has almost the same properties as LH. This hormone stimulates the formation of interstitial cells in the testes of the fetus and causes testosterone secretion. As pointed out earlier in the chapter, the secretion of testosterone during fetal life is important for promoting formation of male sexual organs.

Regulation of Spermatogenesis by Follicle-Stimulating Hormone (FSH). The conversion of spermatogonia into sperm is stimulated by FSH; and in the absence of FSH, spermatogenesis will not proceed. However, FSH cannot by itself cause complete formation of spermatozoa. For spermatogenesis to proceed to completion, testosterone must be secreted simultaneously in small amounts by the interstitial cells. Thus, FSH seems to initiate the proliferative process of spermatogenesis, and testosterone apparently is necessary for final maturation of the spermatozoa. Because testosterone is secreted by the interstitial cells under the influence of LH, both FSH and LH must be secreted by the adenohypophysis if spermatogenesis is to occur.

Regulation of Gonadotropin Secretion by the Hypothalamus. The gonadotropins, like corticotropin and thyrotropin, are secreted by the adenohypophysis mainly in response to nervous activity in the hypothalamus. For instance, in the female rabbit, coitus with a male rabbit elicits nervous activity in the hypothalamus that in turn stimulates the adenohypophysis to secrete FSH and LH. These hormones then cause rapid ripening of follicles in the rabbit's ovaries, followed a few hours later by ovulation.

Many other types of nervous stimuli are also known to affect gonadotropin secretion. For instance, in sheep, goats, and deer, nervous stimuli in response to changes in weather and amount of light in the day increase the quantities of gonadotropins during the mating season of the year, thus allowing birth of the young during an appropriate period for survival. In the human being, it is known too that various psychogenic stimuli feeding into the hypothalamus can cause marked excitatory or inhibitory effects on gonadotropin secretion, in this way sometimes greatly altering the degree of fertility.

The hypothalamus controls gonadotropin secretion by way of the hypothalamic-hypophyseal portal system. Neurosecretory hormones called *gonadotropin-releasing factors* are carried in the hypophyseal portal blood to the adenohypophysis to control the secretion of both LH and FSH. These releasing factors are analogous to growth hormone-releasing factor, corticotropin-releasing factor, and thyrotropin-releasing factor, which increase the secretion of respective adenohypophyseal hormones.

Reciprocal Inhibition of the Hypothalamic-Anterior Pituitary Secretion of Gonadotropic Hormones by the Testicular Hormones. Injection of testosterone into either a male or a female animal inhibits the secretion of gonadotropins. This inhibition is dependent on normal function of the hypothalamus. Therefore, it is assumed that testosterone inhibits formation of the gonadotropin-releasing factors. One can readily see that this inhibitory effect of testosterone provides a feedback control system for maintaining testosterone secretion at a constant level; that is, excess testosterone secretion inhibits LH secretion, which in turn reduces the testosterone secretion back to normal level.

Puberty and Regulation of Its Onset. During the first 10 years of life, the male child secretes almost no gonadotropins and, consequently, almost no testosterone. Then, at the age of about 10, the adenohypophysis begins to secrete progressively increasing quantities of gonadotropins; and this is followed by a corresponding increase in testicular function. By approximately the age of 13, the male child reaches full adult sexual capability. This period of change is called *puberty*.

Initiation of the onset of puberty has long been a mystery. However, it is now almost certain that *during childhood the hypothalamus, not the two glands, is at fault; that is, the hypothalamus simply does not secrete gonadotropin-releasing factors*. Also, it is now known that even the minutest amount of testosterone inhibits the production of gonadotropin-releasing factors by the childhood hypothalamus. This leads to the theory that the childhood hypothalamus is so extremely sensitive to inhibition that the minutest amount of testosterone secreted by the testicles inhibits the entire system. For reasons yet unknown, the hypothalamus loses this inhibitory sensitivity at the time of puberty, which allows the secretory mechanisms to develop full activity.

ABNORMALITIES OF MALE SEXUAL FUNCTION

The Prostate Gland and Its Abnormalities. The prostate gland remains relatively small throughout childhood and begins to grow at puberty. This gland reaches an almost stationary size by the age of about 20 and remains this size up to the age of approximately 40 to 50. At that time in some men it begins to degenerate along with the decreased production of testosterone by the testes. However, a benign prostatic fibroadenoma frequently develops in the prostate in older men and causes urinary obstruction. This effect is not caused by testosterone.

Cancer of the prostate gland is an extremely common cause of death, resulting in approximately 2 to 3 per cent of all male deaths. Once cancer of the prostate gland does occur, the cancerous cells are usually stimulated to more rapid growth by testosterone and are inhibited by removal of the testes so that testosterone cannot be formed. Also, prostatic cancer can usually be inhibited by administration of estrogens.

Hypogonadism in the Male. Hypogonadism is caused by any of several abnormalities: First, the person may be born without functional testes. Second, he may have undeveloped testes, owing to failure of the adenohypophysis to secrete gonadotropic hormones. Third, cryptorchidism (undescended testes) may occur, resulting in partial or total degeneration of the testes. Fourth, he may lose his testes, which is called *castration.*

When a boy loses his testes prior to puberty, a state of *eunuchism* ensues in which he continues to have neuter sexual characteristics throughout life.

In the castrated adult male, sexual desires are decreased but not totally lost, provided that sexual activities have been practiced previously. Indeed erection can still occur as before though with less ease, and ejaculation can still take place. Therefore, the male sexual act is not dependent on production of male sex hormone nor on production of sperm, though male sex hormone does enhance the psychic factors that promote successful culmination of the act.

Hypergonadism in the Male. *Interstitial cell tumors* develop rarely in the testes, but when they do develop they sometimes produce as much as 100 times normal quantities of testosterone. When such tumors develop in young children, they cause rapid growth of the musculature and bones but also early uniting of the epiphyses so that the eventual adult height actually is less than that which would have been achieved otherwise. Obviously, such interstitial cell tumors cause excessive development of the sexual organs and of the secondary sexual characteristics. In the adult male, small interstitial cell tumors are difficult to diagnose because masculine features are already present. Diagnosis can be made, however, from urine tests that show greatly increased excretion of testosterone end-products such as 17-ketosteroids.

REFERENCES

Armstrong, D. T.: Reproduction. *Ann. Rev. Physiol., 32:*439, 1970.

Bacci, G.: Sex Determination. New York, Pergamon Press, Inc., 1966.

Beech, F. A.: Cerebral and hormonal control of reflexive mechanisms involved in copulatory behavior. *Physiol. Rev., 47:*289, 1967.

Donovan, B. T., and van der Werff ten Bosch, J. J.: Physiology of Puberty. Baltimore, The Williams & Wilkins Co., 1966.

Everett, J. W.: Neuroendocrine aspects of mammalian reproduction. *Ann. Rev. Physiol., 31:*383, 1969.

Kristen, B. E.: Effects of gonadotrophins on secretion of steroids by the testis and ovary. *Physiol. Rev., 44:*609, 1964.

Lloyd, C. W.: Human Reproduction and Sexual Behavior. Philadelphia, Lea & Febiger, 1964.

Segal, S. J.: Research in fertility regulation. *New Eng. J. Med., 279:*364, 1968.

CHAPTER 55

SEXUAL FUNCTIONS IN THE FEMALE, AND THE FEMALE SEX HORMONES

The sexual and reproductive functions in the female can be divided into two major phases: first, preparation of the body for conception and gestation, and second, the period of gestation itself. The present chapter is concerned with the preparation of the body for gestation, and the following chapter presents the physiology of pregnancy.

PHYSIOLOGIC ANATOMY OF THE FEMALE SEXUAL ORGANS

Figure 55–1 illustrates the principal organs of the human female reproductive tract, including the *ovaries,* the *fallopian tubes,* the *uterus,* and the *vagina.* Reproduction begins with the development of ova in the ovaries. A single ovum is expelled from an ovarian follicle into the abdominal cavity in the middle of each monthly sexual cycle. This ovum then passes through one of the fallopian tubes into the uterus, and,

if it has been fertilized by a sperm, it implants in the uterus where it develops into a fetus, a placenta, and fetal membranes.

The outer surface of the ovary is covered by a *germinal epithelium,* which embryologically is derived di-

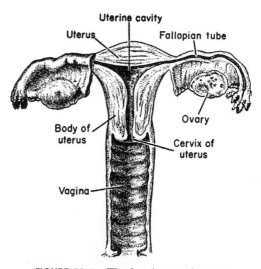

FIGURE 55–1 The female sexual organs.

672

rectly from the epithelium of the germinal ridges. During fetal life and to a lesser extent during childhood and early sexual life, *primordial ova* differentiate from the germinal epithelium and migrate into the substance of the ovarian cortex, carrying with them a layer of epithelioid granulosa cells. The ovum surrounded by a single layer of epithelioid granulosa cells is called a *primordial follicle*.

THE MONTHLY OVARIAN CYCLE AND FUNCTION OF THE GONADOTROPIC HORMONES

The normal sexual life of the female is characterized by monthly rhythmic changes in the rates of secretion of the sex hormones and corresponding changes in the sexual organs themselves. This rhythmic pattern is called the *female sexual cycle*. The duration of the cycle averages 28 days. It may be as short as 20 days or as long as 45 days in completely normal women, though abnormal cycle length is often associated with decreased fertility.

The two significant results of the female sexual cycle are: First, only a *single* mature ovum is normally released from the ovaries each month so that only a single fetus can begin to grow at a time. Second, the uterine endometrium is prepared for implantation of the fertilized ovum at the required time of the month.

The Gonadotropic Hormones. The sexual cycle is controlled entirely by the gonadotropic hormones secreted by the adenohypophysis. These are the same gonadotropic hormones that function in the male, FSH and LH, as well as an additional one, luteotropic hormone (LTH). Ovaries that are not stimulated by gonadotropic hormones remain completely inactive, which is essentially the case throughout child-

hood when almost no gonadotropic hormones are secreted. However, at the age of about eight, the pituitary begins secreting progressively more and more gonadotropic hormones, which culminates in the initiation of monthly sexual cycles between the ages of 11 and 15; this culmination is called *puberty,* as already discussed in relation to the male.

FSH and LH are small glycoproteins having molecular weights of 30,000 and 26,000, respectively, and LTH is a single long polypeptide chain having a molecular weight of about 24,000. The only significant effects of FSH and LH are on the ovaries in the female and the testes in the male. However, LTH, in addition to effects on the ovaries, helps to promote secretion of milk by the breasts and, therefore, is often called *lactogen, prolactin,* or *lactogenic hormone.*

FOLLICULAR GROWTH— FUNCTION OF FOLLICLE- STIMULATING HORMONE (FSH)

Figure 55–2 depicts the various stages of follicular growth in the ovaries, illustrating, first, the primordial follicle (primary follicle). Throughout childhood the primordial follicles do not grow, but at puberty, when FSH from the adenohypophysis begins to be secreted in large quantity, the entire ovaries and especially the follicles within them begin to grow. The first stage of follicular growth is enlargement of the ovum itself. This is followed by development of increasing numbers of granulosa cells and development of *layers of theca cells* around the outer surface of the granulosa cells.

The Vesicular Follicles—Function of Luteinizing Hormone (LH). The cells of the theca, and perhaps also the granulosa cells of the growing follicles, eventually begin secreting a *follicular fluid.* This causes an *antrum* to appear within the epithelioid mass of theca and granulosa cells. At this time the secretion of

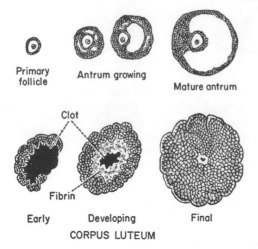

Primary follicle   Antrum growing   Mature antrum

Clot

Fibrin

Early   Developing   Final

CORPUS LUTEUM

FIGURE 55–2 Stages of follicular growth in the ovary, showing also formation of the corpus luteum. (Modified from Arey: Developmental Anatomy, 7th Ed.)

LH by the adenohypophysis begins to increase, and it acts synergistically with the FSH to promote greatly accelerated further growth of the follicular tissues and increased secretion into the follicle. Therefore, the antrum becomes a frank vesicle, as illustrated in Figure 55–2. During growth of the *vesicular follicles,* the granulosa and theca cells continue to proliferate, and a large mass of these cells develops at one pole of the vesicular follicle; in this mass is located the ovum. This mass of cells secretes large quantities of estrogens, as explained later in the chapter.

During each monthly cycle, perhaps 20 or more primordial follicles develop antra and begin the vesicular phase of follicular growth, but ordinarily one of the follicles outgrows all the others and reaches a size of approximately 1 to 1.5 cm. This follicle ovulates, that is, expels its ovum.

OVULATION

Shortly before ovulation, the protruding outer wall of the follicle swells rapidly and a small area in the center of the swelling, called the *stigma,* protrudes like a nipple. The pressure in

the follicle at this time is about 15 mm. Hg. Within another half hour or so, the stigma ruptures, and fluid begins to ooze from the follicle. About two minutes later, as the follicle becomes smaller because of loss of fluid, a more viscous fluid that has occupied the central portion of the follicle and that contains both the ovum and a mass of granulosa and theca cells is extruded into the abdomen. The cause of ovulation seems to be necrosis or weakening of the cells of the stigma or of the cement substance between them.

Necessity of Luteinizing Hormone (LH) for Ovulation; The Ovulatory Surge of LH. LH is necessary for final follicular growth and ovulation. Without this hormone, even though large quantities of FSH are available, the follicle will not progress to the stage of ovulation. LH acts synergistically with FSH to cause rapid swelling of the follicle shortly before ovulation. The swelling in itself undoubtedly contributes to the rupture, but it is likely that LH also has a direct effect to cause the stigma to break open at the time of ovulation.

It is significant that an especially large amount of LH, called the *ovulatory surge,* is secreted by the adenohypophysis during the day immediately preceding ovulation.

Atretic Follicles. In general only one ovum is expelled from the ovaries during each monthly sexual cycle. All the developing vesicular follicles that do not ovulate begin to become atretic either before or immediately after the single ovum is expelled. The ovum within each atretic follicle degenerates, and the follicular cells disappear completely, followed by ingrowth of connective tissue. This process of atresia is undoubtedly caused by a hormonal signal from the ovulating follicle, but the nature of this signal is unknown.

THE CORPUS LUTEUM

Within a few hours after expulsion of the ovum from the follicle, the granulosa and theca cells of the follicle undergo *luteinization,* and the mass of cells becomes a corpus luteum, which secretes progesterone and also large amounts of estrogens. That is, these cells become greatly enlarged and develop lipid inclusions that give the cells a distinctive yellowish color, whence is derived the term *luteum.*

In the normal female, the corpus luteum grows to approximately 1.5 cm., reaching this stage of development approximately seven or eight days following ovulation. After this, it begins to involute and loses its secretory function, as well as its lipid characteristics, approximately 12 days following ovulation.

Luteinizing Function of Luteinizing Hormone (LH). The change of follicular cells into lutein cells is completely dependent on the secretion of LH by the adenohypophysis. In fact, this function gave LH its name "luteinizing."

Secretion by the Corpus Luteum: Function of Luteotropic Hormone (LTH) and LH. The corpus luteum is a highly secretory organ, secreting large amounts of both *progesterone* and *estrogens.* Unfortunately, we do not know the precise role of the individual gonadotropic hormones in causing secretion by the corpus luteum. However, total abrogation of gonadotropin secretion prevents further function of the corpus luteum. It is believed, on the basis of animal experiments, that in the human being LH is responsible for initial development of the corpus luteum while LTH is responsible for secretion by the corpus luteum.

SUMMARY

Approximately each 28 days, new follicles begin to grow in the ovary, one of which finally ovulates. The secretory cells of the follicle then develop into a corpus luteum that continues to secrete estrogens and progesterone under the influence of appropriate gonadotropic hormones. After another two weeks, the gonadotropin stimulus is lost, whereupon the ovarian hormones decrease greatly, and menstruation begins. A new sexual cycle then follows.

FEMALE SEX HORMONES

The two types of female sex hormones are the *estrogens* and *progesterone.* The estrogens mainly promote proliferation of specific cells in the body and are responsible for growth of sexual organs and most secondary sexual characteristics of the female. On the other hand, progesterone is concerned almost entirely with final preparation of the uterus for pregnancy and of the breasts for lactation.

CHEMISTRY OF THE SEX HORMONES

The Estrogens. At least six different natural estrogens have been isolated from the plasma of the human female, but only two are secreted in significant quantities by the ovaries, β-estradiol and *estrone,* both of which are steroid compounds not greatly different from testosterone. Considering both the available quantities and their potencies, the total estrogenic effect of β-estradiol is about three times that of estrone. For this reason β-estradiol is considered to be the major estrogen secreted by the ovaries, though the estrogenic effects of estrone are certainly far from negligible.

Fate of the estrogens; function of the liver in estrogen degradation. Soon after they are secreted by the ovary, the estradiol and estrone that do not enter cells to perform physiologic functions are oxidized to estriol; this oxidation occurs principally in the liver but also to a slight extent elsewhere in the body. The liver also conjugates the estrogens to form glucuronides and sulfates, and about one-fifth of these conjugated products are excreted in the bile while larger quantities are excreted in the urine. The liver also combines estrogens loosely with a protein to form so-called *estroprotein,* and it is mainly in this form that the estrogens circulate in the extracellular fluids.

Progesterone. Progesterone is a steroid having a molecular structure not very different from those of the other steroid hormones, the estrogens, testosterone, and the corticosteroids.

Within a few minutes after secretion, almost all the progesterone is degraded to other steroids that have no progesteronic effect. Here, as is also true with the estrogens, the liver is especially important for this metabolic degradation.

A major end-product of progesterone degradation is *pregnanediol,* a large portion of which is excreted in the urine in this form. One can estimate the rate of progesterone formation in the body from the rate of this excretion.

FUNCTIONS OF THE ESTROGENS—EFFECTS ON THE PRIMARY AND SECONDARY SEXUAL CHARACTERISTICS

The principal function of the estrogens is to cause cellular proliferation and growth of the tissues of the sexual organs and of other tissues related to reproduction.

Effect on the Sexual Organs. During childhood, estrogens are secreted only in small quantities, but following puberty the quantity of estrogens secreted under the influence of adenohypophyseal gonadotropic hormones increases some 20-fold or more. At this time the female sexual organs change from those of a child to those of an adult. The fallopian tubes, uterus, and vagina all increase in size. Also, the external genitalia develop, with deposition of fat in the mons pubis and labia majora and with enlargement of the labia minora.

In addition to increase in size of the vagina, estrogens change the vaginal epithelium from a cuboidal into a stratified type, which is considerably more resistant to trauma and infection than is the prepubertal epithelium.

Estrogens also cause marked proliferation of the endometrium and development of glands that will later be used to aid in nutrition of the implanting ovum. These effects are discussed later in the chapter in connection with the endometrial cycle.

Effect on the Fallopian Tubes. The estrogens have an effect on the mucosal lining of the fallopian tubes similar to that on the uterine endometrium: They cause the glandular tissues to proliferate; and, especially important, they cause the number of ciliated epithelial cells that line the fallopian tubes to increase. The activity of the cilia is considerably enhanced, these always beating toward the uterus. This undoubtedly helps to propel the fertilized ovum toward the uterus.

Effect on the Breasts. The primordial breasts of both female and male are exactly alike, and under the influence of appropriate hormones, the masculine breast can develop sufficiently to produce milk in the same manner as the female breast.

Estrogens cause fat deposition in the breasts, development of the stromal tissues of the breasts, and growth of an extensive ductile system. The lobules and alveoli of the breast develop to a slight extent, but it is progesterone and prolactin that cause marked growth of these structures. In summary, the estrogens initiate growth of the breasts and the breasts' milk-producing apparatus, and they are also responsible for the characteristic external appearance of the mature female breast, but they do not complete the job of converting the breasts into milk-producing organs.

Effect on the Skeleton. Estrogens cause increased osteoblastic activity. Therefore, when the female enters her sexual life at puberty, her growth rate becomes rapid for several years. However, estrogens cause early uniting of the epiphyses with the shafts of the long bones. This effect is much stronger in the female than is the similar effect of testosterone in the male. As a result, growth of the female usually ceases several years earlier than growth of the male. The female eunuch who is completely devoid of estrogen production usually grows several inches taller than the normal mature female because her epiphyses do not unite early.

Effect on Metabolism and Fat Deposition. Estrogens increase the metabolic rate slightly but not so much as the male sex hormone testosterone. However, they cause deposition of increased quantities of fat in the subcutaneous tissues. As a result, the overall specific gravity of the female body, as judged by flotation in water, is considerably less than that of the male body, which contains more protein and less fat. In addition to deposition of fat in the breasts and subcutaneous tissues, estrogens cause especially marked deposition of fat in the buttocks and thighs, causing the broadening of the hips that is characteristic of the feminine figure.

Effect on Hair Distribution. Estrogens do not greatly affect hair distribution except in the pubic region. Here, they cause growth of the usual female pubic hair with a flat upper border rather than the triangular border characteristic of the male.

Intracellular Functions of Estrogens. Thus far we have discussed the gross effects of estrogens on the body. The precise functions of estrogens in causing these effects are not known other than for the following clues: Estrogens circulate in the blood for only a few minutes before they are delivered to the target cells. On entry into these cells, the estrogens combine within 10 to 15 seconds with a protein in the cytoplasm and then are slowly released from this combination to cause intracellular effects. These effects are believed to occur mainly in the nucleus, for RNA begins to be produced in about 30 minutes; and over a period of many hours even DNA is produced, resulting eventually in division of the cell. The RNA diffuses to the cytoplasm where it causes greatly increased protein formation.

FUNCTIONS OF PROGESTERONE

Effect on the Uterus. By far the most important function of progesterone is *to promote secretory changes in the endometrium,* thus preparing the uterus for implantation of the fertilized ovum. This function is discussed specifically in connection with the endometrial cycle of the uterus.

In addition to this effect on the endometrium, progesterone decreases the frequency of uterine contractions, thereby helping to prevent expulsion of the implanted ovum, an effect discussed in the following chapter.

Effect on the Fallopian Tubes. Progesterone also promotes secretory changes in the mucosal lining of the fallopian tubes. These secretions are important for nutrition of the fertilized, dividing ovum as it traverses the fallopian tube prior to implantation.

Effect on the Breasts. Progesterone promotes development of the lobules and alveoli of the breasts, causing the alveolar cells to proliferate, to enlarge, and to become secretory in nature. However, progesterone does not cause the alveoli actually to secrete milk, for, as discussed in the following chapter, milk is secreted only after the prepared breast is further stimulated by prolactin (luteotropic hormone) from the adenohypophysis.

REGULATION OF THE FEMALE RHYTHM — INTERPLAY BETWEEN THE OVARIAN HORMONES AND THE GONADOTROPIC HORMONES

The lower part of Figure 55-3 illustrates the cyclic changes in estrogens and progesterone in the circulating blood

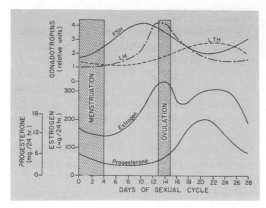

FIGURE 55-3 Approximate cyclic changes in ovarian and pituitary secretion of the different hormones involved in the female sexual cycle.

of the normal female during the normal sexual cycle, while the upper part of the figure shows the approximate cyclic changes of gonadotropic hormones secreted by the adenohypophysis.

A repetitive back-and-forth interplay is believed to occur between the ovarian hormones and the gonadotropins, causing a continual oscillatory increase and decrease of both the ovarian and gonadotropic hormones. The basis of this rhythmic oscillation can be explained as follows:

Effect of the Ovarian Hormones on Adenohypophyseal Secretion of Gonadotropic Hormones. Both estrogens and progesterone, in large enough quantity, inhibit the production of FSH and LH. Much larger quantities of progesterone than estrogen must be administered to have this effect, and even then the effect of progesterone is poor unless estrogens have been secreted or injected prior to the progesterone.

The feedback effects of both estrogens and progesterone seem to operate entirely, or almost entirely, by actions of these hormones on the hypothalamus rather than by direct actions on the adenohypophysis. That is, the quantities of *FSH-* and *LH-releasing factors* are diminished.

When FSH and LH are secreted in large quantities, the pituitary secretes diminished quantities of LTH. Thus,

there is a reciprocal relationship between the first two of these gonadotropins and the third, a relationship that probably plays a significant role in control of the monthly sexual cycle. The diminished secretion of LTH is believed to be caused by an *LTH-inhibiting factor* that is secreted along with the FSH- and LH-releasing factors.

FEEDBACK OSCILLATION OF THE HYPOTHALAMIC-PITUITARY-OVARIAN SYSTEM

It is clear from the preceding discussion that the gonadotropic hormones from the anterior pituitary gland are the basic cause of hormonal secretion by the ovaries; and, in turn, both the estrogens and progesterone of the ovaries, in large enough concentration, can decrease production of at least two of the gonadotropins, FSH and LH. Thus, conditions are appropriate for feedback oscillation to result, which can perhaps be explained as follows:

1. The adenohypophysis, when not affected by outside stimuli, secretes mainly FSH but also some LH. During the first part of the month, FSH begins to be secreted, and this causes the follicles of the ovaries to begin growing.

2. The adenohypophysis then begins to secrete LH, which, acting synergistically with FSH, causes the vesicular follicles of the ovaries to grow rapidly and to secrete progressively more estrogens. Finally, one of the follicles ruptures, causing ovulation.

3. Immediately before and after ovulation, under the influence of LH, the follicular cells of the follicle take on lutein characteristics and become the corpus luteum. Then, large amounts of estrogens and progesterone are secreted by the corpus luteum. These exert their normal *negative* feedback effect on the hypothalamic-adenohypophysis system to decrease FSH and LH, but they simultaneously increase the output of

LTH. Under the influence of LTH and the waning LH, the corpus luteum secretes large quantities of both progesterone and estrogens for 14 days, reaching peak outputs 7 to 8 days after ovulation.

4. From the 8th to the 12th day after ovulation, the corpus luteum involutes, probably because the output of LH has decreased greatly by this time. The secretion of estrogens and progesterone diminishes accordingly, and menstruation occurs for reasons to be discussed shortly. A few days later, the diminished production of estrogens and progesterone automatically allows the adenohypophysis to begin producing FSH once again in large quantities, followed soon by increasing LH.

5. The increasing FSH and LH then stimulate development of new follicles to start a new cycle.

THE MENOPAUSE

Figure 55–4 illustrates (a) the increasing levels of estrogen secretion at puberty, (b) the cyclic variation during the monthly sexual cycles, (c) the further increase in estrogen secretion during the first few years of sexual life, (d) then progressive decrease in estrogen secretion toward the end of sexual life, and (e) finally almost no estrogen secretion at a time of life called the *menopause*.

At an average age of approximately

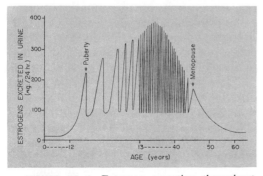

FIGURE 55–4 Estrogen secretion throughout sexual life.

45 to 50 years the sexual cycles usually become irregular, and ovulation fails to occur during many of the cycles. After a few months to a few years, the cycles cease altogether, as illustrated in the figure.

The cause of the menopause (cessation of the sexual cycles) apparently is "burning out" of the ovaries. In other words, throughout a woman's sexual life many of the primordial follicles grow into vesicular follicles with each sexual cycle, and eventually almost all the ova either degenerate or are ovulated. Therefore, at the age of about 45 only a few primordial follicles still remain to be stimulated by FSH and LH, and the production of estrogens by the ovary decreases as the number of primordial follicles approaches zero. When estrogen production falls below a critical value, the estrogens can no longer inhibit the production of FSH and LH sufficiently to cause oscillatory cycles. Consequently, FSH and LH (mainly FSH) are produced thereafter in large and continuous quantities.

The Female Climacteric. The term "female climacteric" means the entire time, lasting from several months to several years, during which the sexual cycles become irregular and gradually stop. In this period the woman must readjust her life from one that has been physiologically stimulated by estrogen and progesterone production to one devoid of these feminizing hormones. The secretion of estrogens decreases rapidly, and essentially no progesterone is secreted after the last ovulatory cycle. The loss of the estrogens often causes marked physiologic changes in the function of the body, including (1) "hot flashes" characterized by extreme flushing of the skin, (2) psychic sensations of dyspnea, (3) irritability, (4) fatigue, (5) anxiety, and (6) occasionally various psychotic states. These symptoms are of sufficient magnitude in approximately 15 per cent of women to warrant treatment. If psychotherapy fails, daily administration of an estrogen in small quantities will reverse the symptoms, and by gradually decreasing the dose the menopausal woman is likely to avoid severe symptoms; unfortunately, though, such treatment prolongs the symptoms.

THE ENDOMETRIAL CYCLE AND MENSTRUATION

Associated with the cyclic production of estrogens and progesterone by the ovary is an endometrial cycle operating through the following stages: first, proliferation of the uterine endometrium; second, secretory changes in the endometrium; and third, desquamation of the endometrium, which is known as *menstruation*. The various phases of the endometrial cycle are illustrated in Figure 55–5.

Proliferative Phase (Estrogen Phase) of the Endometrial Cycle. At the beginning of each menstrual cycle, most of the endometrium is desquamated by the process of menstruation, and only a thin layer of endometrial stroma remains at the base of the original endometrium. The only remaining epithelial cells are those located in the deep portions of the glands and crypts of the endometrium. *Under the influence of estrogens,* secreted in increasing quantities by the ovary during the first part of the ovarian cycle, the stromal cells and the epithelial cells proliferate rapidly. The endometrial surface is re-epithelialized within approximately four to five days after the beginning of menstruation. For the first two weeks of the sexual cycle — that is, until ovulation — the endometrium increases greatly in thickness, owing to increasing numbers of stromal cells and to progressive growth of the endometrial glands and blood vessels into the endometrium, all of which effects are promoted by the estrogens. At the time of ovulation the endometrium is approximately 2 to 3 mm. thick.

Secretory Phase (Progestational Phase) of the Endometrial Cycle. During the latter half of the sexual cycle, both estrogens and progesterone are secreted in large quantities by the corpus luteum. The progesterone, probably acting synergistically with the estrogens, causes considerable swelling of the endometrium. The glands increase in tortuosity, secretory substances develop in the glandular epithelial cells, and the glands secrete small quantities of endometrial fluid. Also, the cytoplasm and the lipid and glycogen deposits of the stromal cells increase. The thickness of the endometrium approximately doubles during the secretory phase, so that toward the end of the monthly cycle the endometrium has a thickness of 4 to 6 mm.

The whole purpose of all these endometrial changes is to produce a highly secretory endometrium containing large amounts of stored nutrients that can provide appropriate conditions for implantation of a fertilized ovum during the latter half of the monthly cycle.

Menstruation. Menstruation is caused by the sudden reduction in both estrogens and progesterone at the end of the monthly ovarian cycle. The first effect is decreased stimulation of the endometrial cells by these two hormones, followed rapidly by involution of the endometrium to about 65 per cent of its previous thickness. During the 24 hours preceding the onset of menstruation, the tortuous blood vessels leading to the mucosal layers of the endometrium become vasospastic, presumably be-

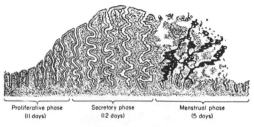

Proliferative phase (11 days)   Secretory phase (12 days)   Menstrual phase (5 days)

FIGURE 55–5 Phases of endometrial growth and menstruation during each monthly female sexual cycle.

cause of some effect of the involution, such as release of a vasoconstrictor material, or perhaps because of a direct effect of estrogen withdrawal, since estrogens are an endometrial vasodilator. The vasospasm and loss of hormonal stimulation cause beginning necrosis in the endometrium, especially of blood vessels. As a result, blood seeps into the deeper layers of the endometrium, the hemorrhagic areas growing over a period of approximately 24 to 36 hours. Gradually, the necrotic outer layers of the endometrium separate from the uterus at the site of the hemorrhages, until, at approximately 48 hours following the onset of menstruation, all the superficial layers of the endometrium have desquamated. The desquamated tissue and blood in the uterine vault initiate uterine contractions that expel the uterine contents.

During menstruation, approximately 35 ml. of blood and an additional 35 ml. of serous fluid are lost. Within approximately five days after menstruation starts, the loss of blood ceases, for by this time the endometrium has become completely re-epithelialized.

ABNORMALITIES OF SECRETION BY THE OVARIES

Hypogonadism. Less than normal secretion by the ovaries can result from poorly formed ovaries or lack of ovaries. When ovaries are absent from birth or when they never become functional, *female eunuchism* occurs. In this condition the usual secondary sexual characteristics do not appear, and the sexual organs remain infantile. Especially characteristic of this condition is excessive growth of the long bones because the epiphyses do not unite with the shafts of these bones at as early an age as in the normal adolescent woman. Consequently, the female eunuch is essentially as tall as, or perhaps even

slightly taller than, her male counterpart of similar genetic background.

When the ovaries of a fully developed woman are removed, the sexual organs regress to some extent so that the uterus becomes almost infantile in size, the vagina becomes smaller, and the vaginal epithelium becomes thin and easily damaged. The breasts atrophy and become pendulous, and the pubic hair becomes suddenly thinner. These are the same changes that occur in the woman after the menopause.

Irregularity of menses and amenorrhea due to hypogonadism. As pointed out in the preceding discussion of the menopause, the quantity of estrogens produced by the ovaries must rise above a critical value if they are to be able to inhibit the production of follicle-stimulating hormone sufficiently to cause an oscillatory sexual cycle. Consequently, in hypogonadism or when the gonads are secreting small quantities of estrogens as a result of other factors, the ovarian cycle likely will not occur normally. Instead, several months may elapse between menstrual periods, or menstruation may cease altogether (amenorrhea). Characteristically, prolonged ovarian cycles are frequently associated with failure of ovulation, presumably due to insufficient secretion of luteinizing hormone, which is necessary for ovulation.

Excessive menstrual bleeding. Contrary to what might be expected, excessive menstrual bleeding most commonly occurs in hypogonadism rather than in hypergonadism. Such bleeding is especially likely to result at the end of anovulatory cycles, for in the absence of preliminary progesterone stimulation, the endometrium does not slough away easily as occurs in normal menstruation; instead, it denudes slowly and bleeds severely.

Hypersecretion by the Ovaries. A rare granulosa-theca cell tumor occasionally develops in an ovary, occurring more often after menopause than before. These tumors secrete large quantities of estrogens which exert the usual estrogenic effects, including hypertrophy of the uterine endometrium and irregular bleeding from this endometrium. In fact, bleeding is often the first indication that such a tumor exists.

Endometriosis. Endometriosis is the development and growth of endometrium in the peritoneal cavity, this growth usually occurring in the pelvis closely associated with the sexual organs. A theory explaining the origin of intra-abdominal endometrial tissue is that contraction of the uterus during menstruation occasionally expels viable endometrium backward through the fallopian tubes into the abdominal cavity and that this endometrial tissue then implants on the peritoneum.

During each ovarian cycle the endometrium in the peritoneal cavity proliferates, secretes, and desquamates in the same manner that the intrauterine endometrium does. However, when desquamation occurs within the peritoneal cavity, the tissue and the hemorrhaging blood cannot be expelled to the exterior. Consequently, the quantity of endometrial tissue in the peritoneal cavity progressively increases with each subsequent menstrual cycle.

The presence of necrotic and hemorrhagic material in the abdominal cavity and also the swelling of the endometrial tissue during each ovarian cycle can cause considerable irritation of the peritoneum, sometimes producing severe abdominal pain. Also, fibrosis occurs in the areas of endometriosis, thereby promoting adhesions from one sexual organ to another and from the sexual organs to other intrapelvic and intra-abdominal structures.

THE FEMALE
SEXUAL ACT

Stimulation of the Female Sexual Act. As is true in the male sexual act, successful performance of the female sexual act depends on both psychic stimulation and local sexual stimulation.

The sex hormones seem to exert a direct influence on the woman to create such a sex drive, but, on the other hand, the growing female child in modern society is often taught that sex is something to be hidden and that it is immoral. As a result of this training, much of the natural sex drive is inhibited, and whether the woman will have little or no sex drive ("frigidity") or will be more highly sexed depends on a balance between natural factors and previous training.

Local sexual stimulation in women occurs in more or less the same manner as in men, for massage or other types of stimulating of the perineal region, sexual organs, and urinary tract create sexual sensations. The *clitoris* is especially sensitive for initiating sexual sensations. Once these sensations have entered the spinal cord, they are transmitted thence to the cerebrum. Local reflexes that are at least partly responsible for the female orgasm are integrated in the sacral and lumbar spinal cord.

Female Erection and Lubrication. Located around the introitus and extending into the clitoris is erectile tissue almost identical with the erectile tissue of the penis. This erectile tissue, like that of the penis, is controlled by the parasympathetic nerves that pass from the cord to the external genitalia. In the early phases of sexual stimulation, the parasympathetics dilate the arteries to the erectile tissues, and this allows rapid inflow of blood into the erectile tissue so that the introitus tightens around the penis; this aids the male greatly in his attainment of sufficient sexual stimulation for promoting ejaculation.

Parasympathetic impulses also pass to the bilateral Bartholin's glands located beneath the labia minora to cause secretion of mucus immediately inside the introitus. This mucus, along with large quantities of mucus secreted by the vaginal mucosa itself, is responsible for most of the lubrication during sexual intercourse. The lubrication in turn is necessary for establishing during intercourse a satisfactory massaging sensation rather than an irritative sensation, which may be provoked by a dry vagina. A massaging sensation constitutes the optimal type of sensation for evoking the appropriate reflexes that culminate in both the male and female climaxes.

The Female Orgasm. When local sexual stimulation reaches maximum intensity, and especially when the local sensations are supported by appropriate conditioning impulses from the cerebrum, reflexes are initiated that cause the female orgasm, also called the *female climax*. The female orgasm is analogous to ejaculation in the male, and it probably is important for fertilization of the ovum. Indeed, the human female is known to be somewhat more fertile when inseminated by normal sexual intercourse rather than by artificial methods, thus indicating an important function of the female orgasm. Possible effects that could result in this are:

First, during the orgasm the perineal muscles of the female contract rhythmically, which presumably results from spinal reflexes similar to those that cause ejaculation in the male. It is possible, also, that these same reflexes increase uterine and fallopian tube motility during the orgasm, thus helping to transport the sperm toward the ovum, but the information on this subject is scanty.

Second, in many lower animals, copulation causes the neurohypophysis to secrete oxytocin; this effect is probably mediated through the amygdaloid nuclei and then through the hypothalamus to the pituitary. The oxytocin in turn causes increased contractility of the uterus, which also is believed to cause rapid transport of the sperm. Sperm have been shown to traverse the entire length of the fallopian tube in the cow in approximately five minutes, a rate at least 10 times as fast as that which the sperm themselves could achieve.

In addition to the effects of the orgasm on fertilization, the intense sexual sensations that develop during the orgasm also pass into the cerebrum and in some manner satisfy the female sex drive.

FEMALE FERTILITY

The Fertile Period of Each Sexual Cycle. The time that the ovum remains viable and capable of being fertilized after it is expelled from the ovary is probably not over 24 hours. Therefore, sperm must be available soon after ovulation if fertilization is to take place. A few sperm can remain viable in the female reproductive tract for up to 72 hours, though probably most of them for not more than 24 hours. Therefore, for fertilization to take place, intercourse usually must occur some time between one day prior to ovulation up to one day after ovulation. Thus the period of female fertility during each sexual cycle is short.

One of the often practiced methods of contraception is to avoid intercourse near the time of ovulation. Since the interval from ovulation until the next succeeding onset of menstruation is almost always between 13 and 15 days, the time of ovulation averages 14 days prior to the next onset of menstruation. In other words, if the periodicity of the menstrual cycle is 28 days, ovulation usually occurs within one day of the 14th day of the cycle. If, on the other hand, the periodicity of the cycle is 40 days, ovulation usually occurs within one day of the 26th day of the cycle.

Hormonal Suppression of Fertility— "The Pill." It has long been known that administration of either estrogens or progesterone, if given in sufficient quantity, can inhibit ovulation. The mechanism of this effect is presumably inhibition of production of the gonadotropins required for ovulation.

The problem in devising methods for hormonal suppression of ovulation has been to develop appropriate combinations of estrogens and progestins that will suppress ovulation but that will not cause unwanted effects of these two hormones. For instance, too much of either of the hormones can cause abnormal menstrual bleeding patterns. However, use of a synthetic progestin in place of progesterone, especially the 19-nor-steroids, along with small amounts of estrogens will usually prevent ovulation and yet, also, allow almost a normal pattern of menstruation. Therefore, almost all "pills" used for control of fertility consist of some combination of estrogen and one of the synthetic progestins.

Abnormal Conditions Causing Female Sterility. Approximately one out of every 6 marriages is infertile; and in more than half of these, the infertility is due to female sterility rather than to male sterility.

Probably by far the most common cause of female sterility is failure to ovulate. This can result from either hyposecretion of gonadotropic hormones, in which case the intensity of the hormonal stimuli simply is not sufficient to cause ovulation, or it can result from abnormal ovaries which will not allow ovulation. For instance, thick capsules occasionally exist on the outside of the ovaries which prevent ovulation.

Lack of ovulation caused by hyposecretion of the gonadotropic hormones can be treated by administering *human chorionic gonadotropin,* a hormone that will be discussed in the following chapter and that is extracted from the human placenta. This hormone, though secreted by the placenta, has almost exactly the same effects as luteinizing hormone and, therefore, is a powerful stimulator of ovulation. However, excessive use of this hormone can cause ovulation from many follicles simultaneously; and this results in multiple births, an effect that has caused as many as six children to be born to mothers treated for infertility with this hormone.

One of the most common causes of female sterility is *endometriosis,* for, as described earlier, endometriosis causes fibrosis throughout the pelvis; and this fibrosis frequently so enshrouds the ovaries that an ovum cannot be released into the abdominal cavity. Often, also, endometriosis occludes the fallopian tubes either at the fimbriated ends or elsewhere along their extent. Another common cause of female infertility is salpingitis, that is, inflammation of the fallopian tubes; this causes fibrosis in the tubes, thereby occluding them. In past years, such inflammation was extremely common as a result of gonococcal infection, but with modern therapy this is becoming a less prevalent cause of female infertility.

REFERENCES

Behrman, S. J., and Kistner, R. W.: Progress in Infertility. Boston, Little, Brown and Company, 1968.

Drill, V.: Oral Contraceptives. New York. McGraw-Hill Book Company, 1966.

Greenblatt, R. B.: Ovulation: Stimulation, Suppression, Detection. Philadelphia, J. B. Lippincott Co., 1966.

Masters, W. H., and Johnson, V. E.: Human Sexual Response. Boston, Little, Brown and Company, 1966.

Pincus, G.: The Control of Fertility. New York, Academic Press, Inc., 1965.

Rogers, J.: Endocrine and Metabolic Aspects of Gynecology. Philadelphia, W. B. Saunders Company, 1963.

Schmidt-Matthiesen, H.: The Normal Human Endometrium. McGraw-Hill Book Company, New York, 1968.

CHAPTER 56

PREGNANCY, LACTATION, AND NEONATAL PHYSIOLOGY

In the preceding chapter the sexual functions of the female were described to the point of fertilization of the ovum. If the ovum becomes fertilized, a completely new sequence of events called *gestation,* or *pregnancy,* takes place, and the fertilized ovum eventually develops into a full-term fetus. The purpose of the present chapter is to discuss these events.

Shortly before the ovum is released from the follicle, each of the 23 pairs of chromosomes loses one of the partners so that 23 *unpaired* chromosomes remain in the mature ovum that is now capable of being fertilized. This process is called *maturation of the ovum.*

FERTILIZATION OF THE OVUM

Only one sperm is required for fertilization, the process of which is illustrated in Figure 56–1. Furthermore, almost never does more than one sperm enter the ovum for the following reason: The zona pellucida of the ovum has a

lattice-type structure, and once the ovum is punctured, some substance seems to diffuse out of the ovum into the lattice to prevent penetration by additional sperm. Indeed, microscopic studies

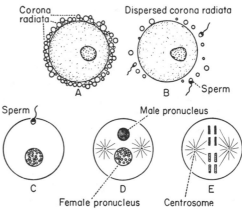

FIGURE 56–1 Fertilization of the ovum, showing (*A*) the mature ovum surrounded by the corona radiata, (*B*) dispersal of the corona radiata, (*C*) entry of the sperm, (*D*) formation of the male and female pronuclei, and (*E*) reorganization of a full complement of chromosomes and beginning division of the ovum. (Modified from Arey: Developmental Anatomy, 7th Ed.)

show that many sperm do attempt to penetrate the zona pellucida but become inactivated before traveling only part way through.

Once a sperm enters the ovum, its head swells rapidly to form a *male pronucleus,* which is also illustrated in Figure 56–1. Later, the 23 chromosomes of the male pronucleus and the 23 of the *female pronucleus* align themselves to re-form a complete complement of 46 chromosomes (23 pairs) in the fertilized ovum.

TRANSPORT AND IMPLANTATION OF THE DEVELOPING OVUM

Transport of the Ovum in the Fallopian Tube. When ovulation occurs, the ovum is expelled directly into the peritoneal cavity and must then enter one of the fallopian tubes. The fimbriated end of each fallopian tube falls naturally around the ovaries, and the inner surfaces of the fimbriated tentacles are lined with ciliated epithelium, the *cilia* of which continually beat toward the *abdominal ostium* of the fallopian tube. One can actually see a slow fluid current flowing toward the ostium. By this means the ovum enters one or the other fallopian tube.

Fertilization of the ovum normally takes place either before the ovum enters the fallopian tube or soon after entering. An additional three days is usually required for transport into the cavity of the uterus. Transport is caused mainly by the cilia, which beat toward the uterus. The slowness of transport allows several stages of division to occur in the fertilized ovum before it enters the uterus. During this time, large quantities of secretions are formed by the glands lining the fallopian tube, and these are believed to be important for nutrition of the developing ovum.

Implantation of the Ovum in the Uterus. After reaching the uterus, the developing ovum usually remains in the uterine cavity an additional four to five days before it implants in the endometrium, which means that implantation ordinarily occurs on the seventh to eighth day following fertilization. During this time the ovum obtains its nutrition from the endometrial secretions called the "uterine milk." Figure 56–2 shows a very early stage of implantation, illustrating an implanted ovum in the *blastocyst stage*.

Implantation results from the action of trophoblastic cells that develop over the surface of the blastocyst. These cells secrete proteolytic enzymes that digest and liquefy the cells of the endometrium. Simultaneously, much of the fluid and nutrients thus released is actively absorbed into the blastocyst as a result of phagocytosis by the trophoblastic cells. Also at the same time, additional trophoblastic cells form cords of cells that extend to and attach to the edges of the digested endometrium. Thus, the blastocyst eats a hole in the endometrium and attaches to it at the same time.

Once implantation has taken place, the trophoblastic and sub-lying cells proliferate rapidly; and these, along with cells from the mother's endometrium, form the placenta and the various membranes of pregnancy.

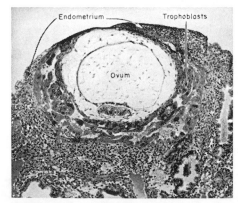

FIGURE 56–2 Implantation of the early human embryo, showing trophoblastic digestion and invasion of the endometrium. (Courtesy of Dr. Arthur Hertig.)

EARLY INTRA-UTERINE NUTRITION OF THE EMBRYO

As the trophoblastic cells invade the decidua (the thickened endometrium prepared for reception of the ovum), digesting and imbibing it, the stored nutrients in the decidua are used by the embryo for appropriate growth and development. The embryo continues to obtain a large measure of its total nutrition in this way for 8 to 12 weeks, though the placenta also begins to provide slight amounts of nutrition after approximately the sixteenth day beyond fertilization (a little over a week after implantation).

FUNCTION OF THE PLACENTA

DEVELOPMENTAL AND PHYSIOLOGIC ANATOMY OF THE PLACENTA

While the trophoblastic cords from the blastocyst are attaching to the uterus, blood capillaries grow into the cords from the vascular system of the embryo, and, by the sixteenth day after fertilization, blood begins to flow. Simultaneously, blood sinuses supplied with blood from the mother develop between the surface of the uterine endometrium and the trophoblastic cords. The trophoblastic cells then gradually send out more and more projections, which become the *placental villi* into which capillaries from the fetus grow. Thus, the villi, carrying fetal blood, are surrounded by sinuses containing maternal blood.

The final structure of the placenta is illustrated in Figure 56–3. The mother's blood flows from the *uterine arteries* into large *blood sinuses* surrounding the villi and then back into the *uterine veins* of the mother.

The mature placenta has a total surface area of approximately 16 square meters, which is about one-fourth the total area of the pulmonary membrane. However, remember that even at full maturity the placental membrane is still

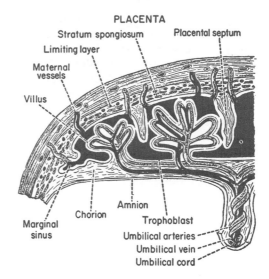

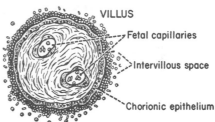

FIGURE 56–3 *Above:* Organization of the mature placenta. *Below:* Relationship of the fetal blood in the villus capillaries to the mother's blood in the intervillus spaces. (Modified from Gray and Gross: Anatomy of the Human Body. Lea & Febiger; and from Arey: Developmental Anatomy, 7th Ed.)

three cells thick, and the distance between the maternal blood and the fetal blood is many times the distance across the alveolar membranes of the lung. Nevertheless, many nutrients and other substances pass through the placental membrane by diffusion in very much the same manner as through the alveolar membranes and capillary membranes elsewhere in the body.

PERMEABILITY AND FUNCTION OF THE PLACENTAL MEMBRANE

The major function of the placenta is to allow diffusion of foodstuffs from the mother's blood into the fetus' blood and diffusion of excretory products from the fetus back into the mother. In the early months of development, placental permeability is relatively slight, but as the placenta becomes older, the permeability increases progressively until the last month or so of pregnancy when it begins to decrease again. The increase in permeability is caused by both progressive enlargement of the surface area of the placental membrane and progressive thinning of the layers of the villi. On the other hand, the decrease shortly before birth results from deterioration of the placenta caused by its age and sometimes from destruction of whole segments due to infarction of isolated areas.

Diffusion of Oxygen Through the Placental Membrane. Almost exactly the same principles are applicable for the diffusion of oxygen through the placental membrane as through the pulmonary membrane; these principles were discussed in Chapter 28. The dissolved oxygen in the blood of the large placental sinuses simply passes through the villus membrane into the fetal blood because of a pressure gradient of oxygen from the mother's blood to the fetus' blood. The mean P_{O_2} in the mother's blood in the placental sinuses is approximately 50 mm. Hg toward the end of pregnancy, and the mean P_{O_2} in the blood leaving the villi and returning to the fetus is about 30 mm. Hg. The mean pressure gradient for diffusion of oxygen through the placental membrane is therefore about 20 mm. Hg.

One might wonder how it is possible for a fetus to obtain sufficient oxygen when the fetal blood leaving the placenta has a P_{O_2} of only 30 mm. Hg. However, there are two major reasons why even this low P_{O_2} is capable of allowing the fetal blood to transmit almost as much oxygen to the fetal tissues as is transmitted by the mother's blood to her tissues:

First, the hemoglobin of the fetus is primarily *fetal hemoglobin,* a type of hemoglobin synthesized in the fetus prior to birth. This type of hemoglobin can carry as much as 20 to 30 per cent more oxygen at low blood P_{O_2}'s than can maternal hemoglobin.

Second, the hemoglobin concentration of the fetus is about 50 per cent greater than that of the mother; and this is an even more important factor than the first in enhancing the amount of oxygen transported to the fetal tissues.

Diffusion of Carbon Dioxide Through the Placental Membrane. Carbon dioxide is continually formed in the tissues of the fetus in the same way that it is formed in maternal tissues. And the only means for excreting the carbon dioxide is through the placenta. The P_{CO_2} builds up in the fetal blood until it is about 41 to 46 mm. Hg in contrast to about 40 to 45 mm. Hg in maternal blood. Thus, a low pressure gradient for carbon dioxide develops across the placental membrane, but this is sufficient to allow adequate diffusion of carbon dioxide from the fetal blood into the maternal blood, because carbon dioxide diffuses about 20 times as rapidly as oxygen.

Diffusion of Other Substances Through the Placental Membrane. Other metabolic substrates needed by the fetus diffuse into the fetal blood in the same manner as oxygen. For instance, the glucose level in the fetal blood ordinarily is approximately 20 to 30 per cent lower than the glucose level in the maternal blood, for glucose is being metabolized rapidly by the fetus. This in turn causes rapid diffusion of additional glucose from the maternal blood into the fetal blood.

Because of the small molecular size of the amino acids and because of the high solubility of fatty acids in cell membranes, both these substances also diffuse from the maternal blood into the fetal blood. And in the same manner that carbon dioxide diffuses from the fetal blood into the maternal blood, other excretory products formed within the fetus diffuse into the maternal blood. These include especially the nonprotein nitrogens such as urea, uric acid, and creatinine.

HORMONAL FACTORS IN PREGNANCY

In pregnancy, the placenta secretes large quantities of *chorionic gonadotropin, estrogens, progesterone,* and *human placental lactogen,* the first three of which, and perhaps the fourth as well, are essential to the continuance of pregnancy. The functions of these hormones are discussed in the following sections.

CHORIONIC GONADOTROPIN AND ITS EFFECT TO CAUSE PERSISTENCE OF THE CORPUS LUTEUM AND TO PREVENT MENSTRUATION

Menstruation normally occurs approximately 14 days after ovulation, at which time most of the secretory endometrium of the uterus sloughs away from the uterine wall and is expelled to the exterior. If this should happen after an ovum has implanted, the pregnancy would terminate. However, this is prevented by the secretion of chorionic gonadotropin in the following manner:

Coincidently with ovum implantation, the hormone *chorionic gonadotropin* is secreted by the trophoblastic cells into the fluids of the mother. As illustrated in Figure 56–4, the rate of secretion rises rapidly to reach a maximum approximately seven weeks after ovulation, then decreases to a relatively low value at 16 weeks after ovulation. The chorionic gonadotropin, a glycoprotein, then has very much the same function as luteinizing hormone secreted by the adenohypophysis. Most important, it prevents the normal involution of the corpus luteum at the end of the sexual month, and instead causes the corpus luteum to secrete even larger quantities of its

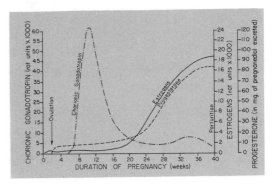

FIGURE 56–4 Rates of secretion of estrogens, progesterone, and chorionic gonadotropin at different stages of pregnancy.

usual hormones, progesterone and estrogens. These excess hormones cause the uterine endometrium to continue growing and to store large amounts of nutrients rather than to be passed in the menstruum. If the corpus luteum is removed before approximately the eleventh week of pregnancy, spontaneous abortion usually occurs, though after this time the placenta itself secretes sufficient quantities of progesterone and estrogens to maintain pregnancy for the remainder of the gestation period.

HUMAN PLACENTAL LACTOGEN

Recently, a new hormone called human placental lactogen has been discovered. This hormone begins to be secreted by the trophoblastic cells of the placenta about the fifth week of pregnancy, and its rate of secretion increases progressively throughout pregnancy. In lower animals it has a luteotropic function, causing the corpora lutea to secrete large quantities of estrogens and progesterone. Though this function has not yet been proved in human beings, it is likely that this and many other functions previously ascribed to chorionic gonadotropin are perhaps performed by human placental lactogen.

Human placental lactogen also has a prolactin-like effect on the breasts and probably plays a role in development of the mother's breasts prior to birth of the baby.

SECRETION OF ESTROGENS BY THE PLACENTA

The placenta, like the corpus luteum, secretes both estrogens and progesterone. Both histochemical and physiologic studies indicate that these two hormones are secreted by *trophoblastic cells* (that line the outsides of the villi), along with the secretion of chorionic gonadotropin and human placental lactogen.

Figure 56–4 shows that the daily production of placental estrogens increases markedly toward the end of pregnancy, to as much as 300 times the daily production in the middle of a normal monthly sexual cycle.

These hormones exert mainly a proliferative function on certain reproductive and associated organs. During pregnancy, the extreme quantities of estrogens cause (1) enlargement of the uterus, (2) enlargement of the breasts and growth of the breast glandular tissue, and (3) enlargement of the female external genitalia.

The estrogens also relax the various pelvic ligaments so that the sacroiliac joints become relatively limber and the symphysis pubis becomes elastic. These changes obviously make for easy passage of the fetus through the birth canal.

There is much reason to believe that estrogens also affect the development of the fetus during pregnancy, for example, by controlling the rate of cell reproduction in the early embryo.

SECRETION OF PROGESTERONE BY THE PLACENTA

In addition to being secreted in moderate quantities by the corpus luteum at the beginning of pregnancy, progesterone is secreted in tremendous quantities by the placenta, sometimes as much as 1 gram per day, toward the end of pregnancy. Indeed, the rate of progesterone secretion increases by as much as 10-fold during the course of pregnancy, as illustrated in Figure 56–4.

The special effects of progesterone that are essential for normal progression of pregnancy are the following:

1. Progesterone causes decidual cells to develop in the uterine endometrium, and these then play an important role in the nutrition of the early embryo.

2. Progesterone has a special effect to decrease the contractility of the gravid uterus, thus preventing uterine contractions from causing spontaneous abortion.

3. Progesterone also contributes to the development of the ovum even prior to implantation, for it specifically increases the secretions of the fallopian tubes and uterus to provide appropriate nutritive matter for the developing *morula* and *blastocyst*. There are some reasons to believe, too, that progesterone even helps to control cell cleavage in the early developing embryo.

4. The progesterone secreted during pregnancy also helps to prepare the breasts for lactation, which is discussed in detail later in the chapter.

RESPONSE OF THE MOTHER TO PREGNANCY

Obviously, the presence of a growing fetus in the uterus adds an extra physiologic load on the mother, and much of the response of the mother to pregnancy is due to this increased load. However, special effects include:

Blood Flow Through the Placenta and Cardiac Output. About 750 ml. of blood flows through the maternal circulation of the placenta each minute during the latter phases of gestation. Obviously, this also increases the cardiac output in the same manner that arteriovenous shunts increase the output. This factor, plus a general increase in metabolism, increases the cardiac output to 30 to 40 per cent above normal.

Blood Volume of the Mother. The maternal blood volume shortly before term is approximately 30 per cent above normal. This increase occurs mainly because of increased secretion during pregnancy of adrenocortical hormones, estrogens, and progesterone, all of which cause increased fluid retention by the kidneys.

At the time of birth of the baby, the mother has approximately 1 to 2 liters of extra blood in her circulatory system. Only about one-fourth of this amount is normally lost during delivery of the baby, thereby allowing a considerable safety factor for the mother.

Nutrition during Pregnancy. The supplemental food needed by the mother during pregnancy to supply the needs of the fetus and fetal membranes includes especially extra dietary quantities of the various minerals, vitamins, and proteins. The growing fetus assumes priority in regard to many of the nutritional elements in the mother's body, and many portions of the fetus continue to grow even though the mother does not eat a sufficient diet. For instance, lack of adequate nutrition in the mother hardly changes the rate of growth of the fetal nervous system, and the length of the fetus increases almost normally; on the other hand, lack of adequate nutrition can decrease the fetus' weight considerably, can decrease ossification of the bones, and can cause anemia, hypoprothrombinemia, decreased size of many bodily organs of the fetus, and a tendency to mental deficiency.

By far the greatest growth of the fetus occurs during the last trimester of pregnancy; the weight of the child almost doubles during the last two months of pregnancy. Ordinarily, the mother does not absorb sufficient protein, calcium, phosphorus, and iron from the gastrointestinal tract during the last month of pregnancy to supply the fetus. However, from the beginning of pregnancy the mother's body has been storing these substances to be used during the latter months of pregnancy. Some of this storage is in the placenta, but most of it is in the normal storage depots of the mother.

If appropriate nutritional elements are not present in the mother's diet, a number of maternal deficiencies can occur during pregnancy. Such deficiencies often occur for calcium, phosphates, iron, and the vitamins. For example, approximately 375 mg. of iron is needed by the fetus to form its blood and an additional 600 mg. is needed by the mother to form her own extra blood. The normal store of nonhemoglobin iron in the mother at the outset of pregnancy is often only 100 or so mg. and almost never over 700 mg. Consequently, without sufficient iron in the food the mother herself usually develops anemia. Therefore, in general, the obstetrician supplements the diet of the mother with the needed substances. It is especially important that the mother receive large quantities of vitamin D, for, even though the total quantity of calcium utilized by the fetus is small, calcium even normally is poorly absorbed by the gastrointestinal tract. Finally, shortly before birth of the baby vitamin K is often added to the diet so that the baby will have sufficient prothrombin to prevent postnatal hemorrhage.

The Amniotic Fluid and its Formation. Normally, the volume of amniotic fluid (the fluid that surrounds the fetus in the uterus) is between 500 ml. and 1 liter. Studies with isotopes of the rate of formation of amniotic fluid show that on the average the water in amniotic fluid is completely replaced once every three hours, and the electrolytes sodium and potassium are replaced once every 15 hours. Yet, strangely enough, the sources of the fluid and the points of reabsorption are mainly unknown. A small portion of the fluid is derived from renal excretion by the fetus. On the other hand, a certain amount of absorption occurs by way of the gastrointestinal tract and lungs of the fetus.

ABNORMAL RESPONSES OF THE MOTHER TO PREGNANCY

Hyperemesis Gravidarum. In the earlier months of pregnancy, the mother frequently develops hyperemesis gravidarum, a condition characterized by nausea and vomiting and commonly known as "morning sickness." Occasionally, the vomiting becomes so severe that the mother becomes greatly dehydrated, and in rare instances the condition even causes death.

The cause of the nausea and vomiting is unknown, but it occurs to its greatest extent during the same time that chorionic gonadotropin is secreted in large quantities by the placenta. Because of this coincidence, many clinicians believe that chorionic gonadotropin is in some way responsible for the nausea and vomiting. On the other hand, during the first few months of pregnancy rapid trophoblastic invasion of the endometrium also takes place, and, because the trophoblastic cells digest portions of the endometrium as they invade it, it is possible that degenerative products resulting from this invasion, instead of chorionic gonadotropin, are responsible for the nausea and vomiting. Indeed, degenerative processes in other parts of the body, such as occur following gamma ray irradiation and burns, can all cause similar nausea and vomiting.

Toxemia of Pregnancy. Approximately 7 per cent of all pregnant women experience rapid weight gain, edema, and often elevation of arterial pressure. This condition, known as toxemia of pregnancy, is characterized by inflammation and spasm of the arterioles in many parts of the body.

Various attempts have been made to prove that toxemia of pregnancy is caused by excessive secretion of placental or adrenal hormones, but proof of a hormonal basis for toxemia is yet completely lacking. Perhaps a more plausible theory is that toxemia of pregnancy results from some type of autoimmunity or allergy resulting from the presence of the fetus. Indeed, the acute symptoms disappear within a few days after birth of the baby.

Because of the diminished glomerular filtration rate in the kidneys, one of the major problems of toxemia is retention of salt and water. Therefore, it is a dictum among obstetricians that any pregnant woman who has a tendency toward toxemia must limit her salt intake. It usually is not necessary to limit the mother's water intake, for limitation of salt alone prevents excessive absorption of water by the kidney tubules for reasons discussed in Chapter 26.

Eclampsia. Eclampsia is a severe degree of toxemia that occurs in one out of several hundred pregnancies. It is characterized by everything that occurs in toxemia, plus clonic convulsions followed by coma. Usually, it occurs shortly before, or sometimes within a day after parturition.

Even with the best treatment, some 5 per cent of eclamptic mothers still die. However, injection of vasodilator drugs plus dehydration of the patient can often reverse the vascular spasm and lower the blood pressure, bringing the mother out of the eclamptic state.

PARTURITION

INCREASED UTERINE IRRITABILITY NEAR TERM

Parturition means simply the process by which the baby is born. At the termination of pregnancy the uterus becomes progressively more excitable until finally it begins strong rhythmic contractions with such force that the baby is expelled. The exact cause of the increased activity of the uterus is not known, but at least two major categories of effects lead up to the culminating contractions responsible for parturition; these are, first, progressive hormonal changes that cause increased excitability of the uterine musculature, and, second, progressive mechanical changes.

Hormonal Factors That Cause Increased Uterine Contractility. *Ratio of estrogens to progesterone.* Progesterone probably inhibits uterine contractility during pregnancy, thereby helping to prevent expulsion of the fetus. On the other hand, estrogens have a definite tendency to increase the degree of uterine contractility. Both these hormones are secreted in progressively greater quantities throughout pregnancy, but from the seventh month onward, estrogen secretion increases more than progesterone secretion. Therefore, it has been postulated that the *estrogen to progesterone ratio* increases sufficiently toward the end of pregnancy to be at least partly responsible for the increased contractility of the uterus.

Effect of oxytocin on the uterus. Oxytocin, secreted by the neurohypophysis, specifically causes uterine contraction (see Chap. 49), and this hormone might be particularly important in increasing the contractility of the uterus near term. The rate of oxytocin secretion seems to be considerably increased at the time of labor, possibly because of reflexes from the uterus that will be discussed later. Though hypophysectomized animals and human beings can still deliver their young at term, labor is prolonged.

Mechanical Factors that Increase Contractility of the Uterus. Simply stretching smooth muscle organs usually increases their contractility. Furthermore, intermittent stretch, as occurs repetitively in the body and cervix of the uterus because of movements of the fetus, can also elicit smooth muscle contraction.

Note especially that twins are born on the average *19 days* earlier than a single child, which emphasizes the importance of mechanical stretch in eliciting uterine contractions.

The obstetrician frequently induces labor by dilating the cervix or by rupturing the membranes so that the head of the baby stretches the cervix more forcefully than usual or irritates it in some other way. The mechanism by which cervical stretch excites the body of the uterus is not known, but the effect could result simply from myogenic transmission from the cervix to the body of the uterus.

ONSET OF LABOR – A POSITIVE FEEDBACK THEORY FOR ITS INITIATION

During the last few months of pregnancy the uterus undergoes periodic episodes of weak and slow rhythmic contractions called *Braxton-Hicks contractions.* These become progressively stronger toward the end of pregnancy; and they eventually change rather suddenly, within hours, to become exceptionally strong contractions that start stretching the cervix and later force the baby through the birth canal, thereby causing parturition. This process is called *labor,* and the strong contractions that result in final parturition are called *labor contractions.*

On the basis of experience during the past few years with other types of control systems, a theory has been proposed for explaining the onset of labor based on "positive feedback." This theory suggests that stretch of the cervix by the fetus' head finally becomes great enough to elicit a reflex increase in contractility of the uterine body. This contraction pushes the baby forward, which stretches the cervix some more and initiates a new cycle. Thus, the process repeats again and again until the baby is expelled. This theory is illustrated in Figure 56–5.

In addition to the mechanical factors that cause the vicious cycle of labor, oxytocin from the neurohypophysis probably also plays a role. On the basis of studies in cows, this results from the following sequence of events: (a) uterine contraction, (b) cervical stretch or other type of cervical stimulation, (c)

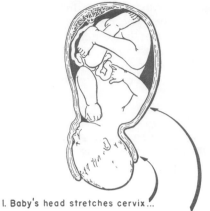

1. Baby's head stretches cervix...

2. Cervical stretch exc tes fundic contraction...

3. Fundic contraction pushes baby down and stretches cervix some more...

4. Cycle repeats over and over again...

FIGURE 56–5 Theory for the onset of intensely strong contractions during labor.

nerve signals from the cervix to the hypothalamic-neurohypophysis axis, (d) increased secretion of oxytocin, and (e) increased contractility of the uterus, which leads to (f) further intensification of the vicious cycle of increasing contraction.

To summarize the theory, we can assume that multiple factors increase the contractility of the uterus toward the end of pregnancy. These are additive in their effects, and they cause the Braxton-Hicks contractions to become progressively stronger. Eventually, one of these becomes strong enough that the contraction itself irritates the cervix of the uterus, thereby increasing the uterine contractility because of positive feedback and resulting in a second contraction stronger than the first, and a third stronger than the second, and so forth. Once these contractions become strong enough to cause this type of feedback, with each succeeding contraction greater than the one preceding, the process proceeds to completion — all simply *because positive feedback becomes a vicious cycle when the degree of feedback is greater than a critical level.*

ABDOMINAL CONTRACTION DURING LABOR

Once labor contractions become strong and painful, neurogenic reflexes, mainly from the birth canal to the spinal cord and thence back to the abdominal muscles, cause intense abdominal contraction. This additional contraction of the abdominal muscles and the reflexes causing it add greatly to the positive feedback that eventually causes expulsion of the fetus.

MECHANICS OF PARTURITION

In the early part of labor, the uterine contractions might occur only once every 30 minutes and last for about 1 minute. As labor progresses, the contractions finally appear as often as once every one to three minutes, and the intensity of contraction increases greatly with only a short period of relaxation between contractions.

The combined contractions of the uterine and abdominal musculature during delivery of the baby cause a downward force on the fetus of approximately 25 pounds during each strong contraction.

In 19 out of 20 births the head is the first part of the baby to be expelled. The head acts as a wedge to open the structures of the birth canal as the fetus is forced downward from above. The first major obstruction to expulsion of the fetus is the uterine cervix. Toward the end of pregnancy the cervix becomes soft, which allows it to stretch when labor pains cause the body of the uterus to contract. The so-called *first stage of labor* is the period of progressive cervical dilatation, lasting until the opening is as large as the head of the fetus. This stage usually lasts 8 to 24 hours in the first pregnancy but often only a few minutes after many pregnancies.

Once the cervix has dilated fully, the fetus' head moves rapidly into the birth canal, and, with additional force from above, continues to wedge its way through the canal until delivery is effected. This is called the *second stage of labor,* and it may last from as little as a minute after many pregnancies up to half an hour or more in the first pregnancy.

SEPARATION AND DELIVERY OF THE PLACENTA

During the succeeding 10 to 45 minutes after birth of the baby, the uterus contracts to a very small size, which causes a *shearing* effect between the walls of the uterus and the placenta, thus separating the placenta from its implantation site. Obviously, separation of the placenta opens the placental sinuses and causes bleeding. However, the amount of bleeding is limited to an average of 350 ml. by the following mechanism: The smooth muscle fibers of the uterine musculature are arranged in figures of 8 around the blood vessels as they pass through the uterine wall. Therefore, contraction of the uterus following delivery of the baby constricts the vessels that had previously supplied blood to the placenta.

LACTATION

DEVELOPMENT OF THE BREASTS

Growth of the Ductile System — Role of the Estrogens. All through pregnancy, the tremendous quantities of estrogens secreted by the placenta cause the ductile system of the breasts to grow and to branch. Simultaneously, the stroma of the breasts also increases in quantity, and large quantities of fat are laid down in the stroma.

Moderate quantities of growth hormone from the adenohypophysis, or *human placental lactogen* from the placenta are also required for the estrogens to produce their effect on the breasts. These latter two hormones both cause

protein deposition in the glandular cells, which is essential to their growth.

Development of the Lobule-Alveolar System — Role of Progesterone. The synergistic action of estrogens and growth hormone can cause only a primitive lobule-alveolar system to develop in the breasts at the same time that the ducts are growing, but the simultaneous action of progesterone causes growth of the lobules, budding of alveoli, and development of secretory characteristics in the cells of the alveoli.

Function of Prolactin and Other Hormones in Development of the Breasts. Prolactin is the hormone most probably concerned with causing milk secretion after birth of the baby, and it is probably the same hormone as luteotropic hormone, the importance of which for maintaining secretion by the corpus luteum was discussed in the preceding chapter. Prolactin has a powerful synergistic effect with estrogens and progesterone to stimulate development of the alveolar secretory system of the breast. It is possible that human placental lactogen from the placenta might also play a similar role during pregnancy. This hormone has physiologic properties almost the same as those of prolactin, though very little is yet known about it.

INITIATION OF LACTATION — PROLACTIN

By the end of pregnancy, the mother's breasts are fully developed for nursing, but only a few milliliters of fluid is secreted each day until after the baby is born. This fluid is called *colostrum;* it contains essentially the same amounts of proteins and lactose as milk but almost no fat, and its maximum rate of production is about $1/100$ the subsequent rate of milk production.

The absence of lactation during pregnancy is believed to be caused by suppressive effects of progesterone and estrogens on the milk secretory process of the breasts and also on the secretion of prolactin by the adenohypophysis.

However, immediately after the baby is born, the sudden loss of both estrogen and progesterone secretion by the placenta removes any inhibitory effects of these two hormones and allows marked increase in production of prolactin by the adenohypophysis, as illustrated in Figure 56–6. The prolactin stimulates synthesis of large quantities of fat, lactose, and casein by the mammary glandular cells, and within two to three days the breasts begin to secrete copious quantities of milk instead of colostrum.

THE EJECTION OR "LET-DOWN" PROCESS IN MILK SECRETION — FUNCTION OF OXYTOCIN

Milk is secreted continuously into the alveoli of the breasts, but milk does not flow easily from the alveoli into the ductile system and therefore does not continually leak from the breast nipples. Instead, the milk must be "ejected" or "let-down" from the alveoli to the ducts before the baby can obtain it. This process is caused by a combined neurogenic and hormonal reflex involving the hormone *oxytocin* as follows:

When the baby suckles the breast, sensory impulses are transmitted through somatic nerves to the spinal cord and then to the hypothalamus, there causing

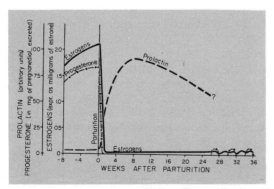

FIGURE 56–6 Changes in rates of secretion of estrogens, progesterone, and prolactin at parturition and during the succeeding weeks after parturition, showing especially the rapid increase in prolactin secretion immediately after parturition.

oxytocin and, to a lesser extent, *vasopressin* secretion, as described in Chapter 49. These two hormones, principally oxytocin, flow in the blood to the breasts where they cause the *myoepithelial cells* that surround the outer walls of the alveoli to contract, thereby expressing the milk from the alveoli into the ducts. Thus, within 30 seconds to a minute after a baby begins to suckle the breast, milk begins to flow. This process is called milk ejection, or milk let-down.

EFFECT OF LACTATION ON THE UTERUS AND ON THE SEXUAL CYCLE

The uterus involutes after parturition far more rapidly in women who lactate than in women who do not lactate. This difference probably results from greatly diminished estrogen secretion by the ovaries during the period of lactation, for estrogens are known to enlarge the uterus and presumably also to prevent rapid uterine involution.

Lactation usually prevents the sexual cycle for the first few months. Presumably this is caused by preoccupation of the adenohypophysis with production of prolactin, which reduces the rate of secretion of the other gonadotropic hormones. However, after several months of lactating, the adenohypophysis usually begins once again to produce sufficient quantities of follicle-stimulating hormone and luteinizing hormone to re-initiate the monthly sexual cycle.

GROWTH AND FUNCTIONAL DEVELOPMENT OF THE FETUS

During the first two to three weeks of intrauterine life the fetus remains almost microscopic in size, but thereafter the dimensions of the fetus increase almost in proportion to age. At 12 weeks the length of the fetus is approximately

10 cm.; at 20 weeks, approximately 25 cm.; and at term (40 weeks), approximately 53 cm. (about 21 inches). Because the weight of the fetus is proportional to the cube of the length, the weight increases approximately in proportion to the cube of the age of the fetus. The weight remains almost nothing during the first few months and reaches 1 pound only at five and a half months of gestation. Then, during the last trimester of pregnancy, the fetus gains tremendously so that two months prior to birth the weight averages 3 pounds, one month prior to birth 4.5 pounds, and at birth 7 pounds.

DEVELOPMENT OF THE ORGAN SYSTEMS

Within one month after fertilization of the ovum all the different organs of the fetus have already been at least partly formed, and during the next two to three months the minute details of the different organs are established. Beyond the fourth month, the organs of the fetus are grossly the same as those of the newborn child, even including most of the smaller structures of the organs. However, cellular development of these structures is usually far from complete at this time and requires the full remaining five months of pregnancy for complete development. Even at birth certain structures, particularly the nervous system, the kidneys, and the liver, still lack full development, as is discussed in more detail later in the chapter.

The Circulatory System. The human heart begins beating during the fourth week following fertilization, contracting at the rate of about 65 beats per minute. This increases steadily as the fetus grows and reaches a rate of approximately 140 per minute immediately before birth.

Formation of blood cells. Nucleated red blood cells begin to be formed in the yolk sac and mesothelial layers of the placenta at about the third week of fetal development. This is fol-lowed a week later by the formation of non-nucleated red blood cells by the fetal mesenchyme and by the endothelium of the fetal blood vessels. Then at approximately six weeks, the liver begins to form blood cells, and in the third month the spleen and other lymphoid tissues of the body also begin forming blood cells. Finally, from approximately the third month on, the bone marrow forms more and more red and white blood cells. During the midportion of fetal life, the liver, spleen, and lymph nodes are the major sources of the fetus' blood cells, but, during the latter three months of fetal life, the bone marrow gradually takes over while these other structures lose their ability completely to form blood cells.

The Respiratory System. Obviously, respiration cannot occur during fetal life. However, respiratory movements do take place beginning at the end of the first trimester of pregnancy. Tactile stimuli or fetal asphyxia especially cause respiratory movements.

The Nervous System. Most of the peripheral reflexes of the fetus are well formed by the third to fourth month of pregnancy. However, some of the more important higher functions of the central nervous system are still undeveloped even at birth. Indeed, myelinization of some major tracts of the central nervous system becomes complete only after approximately a year of postnatal life.

The Gastrointestinal Tract. Even in midpregnancy the fetus ingests and absorbs large quantities of amniotic fluid, and during the latter two to three months, gastrointestinal function approaches that of the normal newborn infant. Small quantities of *meconium* are continually formed in the gastrointestinal tract and excreted from the bowels into the amniotic fluid. Meconium is composed partly of unabsorbed residue of amniotic fluid and partly of excretory products from the gastrointestinal mucosa and glands.

The Kidneys. The fetal kidneys are capable of excreting urine during at least the latter half of pregnancy, and urination occurs normally *in utero.* However, the renal control systems for regulation of extracellular fluid electrolyte balances and acid-base balance are almost nonexistent until after midfetal life and do not reach full development until about a month after birth.

ADJUSTMENTS OF THE INFANT TO EXTRA-UTERINE LIFE

ONSET OF BREATHING

The most obvious effect of birth on the baby is loss of the placental connection with the mother, and therefore loss of this means for metabolic support. Especially important is loss of the placental oxygen supply and placental excretion of carbon dioxide. Therefore, by far the most important immediate adjustment of the infant to extra-uterine life is the onset of breathing.

Cause of Breathing at Birth. Following completely normal delivery from a mother who has not been depressed by anesthetics, the child ordinarily begins to breathe immediately and has a completely normal respiratory rhythm from the outset. The promptness with which the fetus begins to breathe indicates that breathing is initiated by sudden exposure to the exterior world, probably resulting from sensory impulses originating in the suddenly cooled skin. However, if the infant does not breathe immediately, his body becomes progressively more hypoxic and hypercapnic, which provides additional stimulus to the respiratory center and usually causes breathing within a few minutes after birth.

Delayed and Abnormal Breathing at Birth — Danger of Hypoxia. If the mother has been depressed by an anesthetic during delivery, respiration by the child is likely to be delayed for several minutes, thus illustrating the importance of using as little obstetrical anesthesia as feasible. Also, many infants who have traumatic deliveries are slow to breathe or sometimes will not breathe at all.

Degree of hypoxia that an infant can stand. In the adult, failure to breathe for only four minutes often causes death, but a newborn infant often survives as long as 15 minutes of failure to breathe after birth. Unfortunately, though, permanent mental impairment often ensues if breathing is delayed more than 7 to 10 minutes.

This ability of the neonatal infant to survive long periods of hypoxia probably results at least partly from special adaptation of its cellular respiratory enzymes to low Po_2. However, within a week or so after birth, the infant loses this resistance to hypoxia.

Expansion of the Lungs at Birth. At birth, the walls of the alveoli are held together by the surface tension of the viscid fluid that fills them. Therefore more than 25 mm. Hg of negative pressure is required to open the alveoli for the first time. But once the alveoli are open, further respiration can be effected with relatively weak respiratory movements. Fortunately, the first inspirations of the newborn infant are extremely powerful, usually capable of creating as much as 50 mm. Hg negative pressure in the intrapleural space.

Figure 56–7 illustrates the tremendous forces required to open the lungs at the onset of breathing. To the left is shown the pressure-volume curve (compliance curve) for the first breath after birth. Observe, first, the lowermost curve, which shows that the lungs expand essentially not at all until the negative pressure has reached −40 cm. water

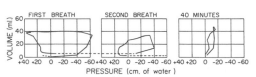

FIGURE 56–7 Pressure-volume curves of the lungs (compliance curves) of a newborn baby immediately after birth, showing (a) the extreme forces required for breathing during the first two breaths of life and (b) development of a nearly normal compliance within 40 minutes after birth. (From Smith: *Sci. Amer., 209*(Oct.):32, 1963.)

(−30 mm. Hg). As the negative pressure increases to −60 cm. water, about 40 ml. air enters the lungs. Then, to deflate the lungs, considerable positive pressure is required, probably because of the viscous resistance offered by the fluid in the bronchioles.

Note that the second breath is much easier. However, breathing does not become completely normal until about 40 minutes after birth, as shown by the third compliance curve.

Respiratory distress syndrome. A small number of infants breathe satisfactorily immediately after birth but gradually develop severe respiratory distress during the ensuing few hours to several days and frequently succumb within the next day or so. The alveoli of these infants at death is filled with fluid containing large quantities of protein, almost as if pure plasma had leaked out of the capillaries into the alveoli.

Unfortunately, the cause of the respiratory distress syndrome is not certain. However, the fluid from these lungs is deficient in *surfactant*, a substance secreted into the alveoli of normal lungs that decreases the surface tension of the alveolar fluid and therefore allows the alveoli to open easily. This substance was discussed in Chapter 27 in relation to the mechanics of respiration.

CIRCULATORY READJUSTMENTS AT BIRTH

As important as the onset of breathing at birth are the immediate circulatory readjustments that allow adequate blood flow through the lungs. Also, circulatory readjustments during the first week of life shunt more and more blood through the liver as well. To describe these readjustments we must first consider briefly the anatomic structure of the fetal circulation.

Structure of the Fetal Circulation. Because the lungs are mainly non-functional during fetal life and because the liver is only partially functional, it is not necessary for the fetal heart to pump much blood through either the lungs or the liver. On the other hand, the fetal heart must pump large quantities of blood through the placenta. Therefore, special anatomic arrangements cause the fetal circulatory system to operate considerably differently from that of the adult. First, as illustrated in Figure 56–8, blood returning from the placenta passes through the *ductus venosus,* mainly bypassing the liver. Then, most of the blood entering the right atrium from the inferior vena cava is directed in a straight pathway across the posterior aspect of the right atrium and thence through the *foramen ovale* directly into the left atrium. Thus, the well-oxygenated blood from the placenta enters the left side of the heart rather than the right side and is pumped mainly into the vessels of the head and forelimbs.

The blood entering the right atrium from the superior vena cava is directed downward through the tricuspid valve

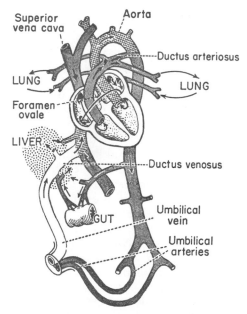

FIGURE 56–8 Organization of the fetal circulation. (Modified from Arey: Developmental Anatomy, 7th Ed.)

into the right ventricle. This blood is mainly deoxygenated blood from the head region of the fetus, and it is pumped by the right ventricle into the pulmonary artery, then mainly through the *ductus arteriosus* into the descending aorta and through the two umbilical arteries into the placenta. Thus, the deoxygenated blood becomes oxygenated.

Changes in Fetal Circulation at Birth. The basic changes in the fetal circulation at birth are the following:

Primary changes in pulmonary and systemic vascular resistance at birth. The primary changes in the circulation at birth are: First, loss of the tremendous blood flow through the placenta *approximately doubles the systemic vascular resistance at birth*. This obviously *increases the aortic pressure* as well as the pressures in the left ventricle and left atrium.

Second, the *pulmonary vascular resistance greatly decreases* as a result of expansion of the lungs. In the unexpanded fetal lungs, the blood vessels are compressed because of the small volume of the lungs. Immediately upon expansion, these vessels are no longer compressed, and the resistance to blood flow decreases several-fold. Also, in fetal life the hypoxia and hypercapnia of the lungs cause marked tonic vasoconstriction of the lung blood vessels, but vasodilation takes place when aeration of the lungs eliminates the hypoxia and hypercapnia. These changes reduce the resistance to blood flow through the lungs as much as 6-fold, which obviously *reduces the pulmonary arterial pressure,* the right ventricular pressure, and the right atrial pressure.

Secondary changes in the vascular system at birth—Closure of the foramen ovale and closure of the ductus arteriosus. The *low right atrial pressure* and the *high left atrial pressure* that occur secondarily to the changes, respectively, in pulmonary and systemic resistance at birth cause a tendency for blood to flow backward from the left atrium into the right atrium rather than in the other direction, as occurred during fetal life. Consequently, the small valve that lies over the foramen ovale on the left side of the atrial septum closes over this opening, thereby preventing further flow. In two-thirds of all persons the valve becomes adherent over the foramen ovale within a few months to a few years and forms a permanent closure. But, even if permanent closure does not occur, the left atrial pressure throughout life remains 3 to 4 mm. Hg greater than the right atrial pressure, and the back pressure keeps the valve closed.

Similar effects occur in relation to the ductus arteriosus, for the increased systemic resistance *elevates the aortic pressure* while the decreased pulmonary resistance *reduces the pulmonary arterial pressure.* As a consequence, shortly after birth, blood begins to flow backward from the aorta into the pulmonary artery rather than in the other direction as in fetal life. However, within three to four days after birth, the muscular wall of the ductus arteriosus gradually constricts to stop all blood flow. This is called *functional closure* of the ductus arteriosus. Then, sometime during the second month of life the ductus arteriosus ordinarily becomes anatomically *occluded* by growth of fibrous tissue into its lumen.

The causes of functional closure and anatomical closure of the ductus are not completely known. However, the most likely cause is increased oxygenation of the blood flowing through the ductus. In fetal life the P_{O_2} may be as low as 20 mm. Hg, but it increases to about 100 mm. Hg within a few minutes after birth. Furthermore, the degree of contraction of the ductus is highly related to the availability of oxygen.

In one out of several thousand infants, the ductus fails to close at all, resulting in a *patent ductus arteriosus,* the consequences of which were discussed in Chapter 22.

SPECIAL PROBLEMS OF PREMATURE BIRTH

All the problems of neonatal life are especially exacerbated in prematurity. These can be categorized under the following two headings: (1) immaturity of certain organ systems and (2) instability of the different homeostatic control systems. Because of these effects, a premature baby rarely lives if it is born more than two and a half to three months prior to term.

IMMATURE DEVELOPMENT

Almost all the organ systems of the body are immature in the premature infant, but some require particular attention if the life of the premature baby is to be saved.

Respiration. The respiratory system is especially likely to be underdeveloped in the premature infant. The vital capacity and the functional residual capacity of the lungs are especially small in relation to the size of the infant. As a consequence, respiratory distress is a common cause of death. Indeed, there is a very high incidence of the respiratory distress syndrome. Also, the low functional residual capacity in the premature infant is often associated with periodic breathing of the Cheyne-Stokes type.

Gastrointestinal Function. Another major problem of the premature infant is to ingest and absorb adequate food. If the infant is more than two months premature, the digestive and absorptive systems are almost always inadequate. The absorption of fat is so poor that the premature infant must have a low fat diet. Furthermore, the premature infant has unusual difficulty in absorbing calcium and therefore can develop severe rickets before one recognizes the difficulty. For this reason, special attention must be paid to adequate calcium and vitamin D intake.

Function of Other Organs. Immaturity of other organ systems that frequently causes serious difficulties in the premature infant includes: (a) immaturity of the liver, which results in *poor intermediary metabolism* and often also a *bleeding tendency* as a result of poor formation of blood coagulation factors; (b) immaturity of the kidneys, which are particularly deficient in their ability to rid the body of acids, thereby predisposing to *acidosis* as well as to many serious fluid balance abnormalities; and (c) immaturity of the blood-forming mechanism of the bone marrow, which allows rapid development of *anemia*.

INSTABILITY OF THE CONTROL SYSTEMS IN THE PREMATURE INFANT

Immaturity of the different organ systems in the premature infant creates a high degree of instability in the homeostatic systems of the body. For instance, the acid-base balance can vary tremendously, particularly when the food intake varies from time to time. Likewise, the blood protein concentration is usually low because of immature liver development, often leading to *hypoproteinemic edema*. And inability of the infant to regulate its calcium ion concentration frequently brings on hypocalcemic tetany. Also, the blood glucose concentration can vary between the extremely wide limits of 20 mg./100 ml. to over 100 mg./100 ml., depending principally on the regularity of feeding. It is no wonder, then, with these extreme variations in the internal environment of the premature infant, that mortality is high.

Instability of Body Temperature. One of the particular problems of the premature infant is inability to maintain normal body temperature. Its temperature tends to approach that of its surroundings. At normal room temperature the temperature may stabilize in the low 90's or even in the 80's. Statistical studies show that a body temperature maintained below 96° F. is associated with a particularly high incidence of death, which explains the common use of the incubator in the treatment of prematurity.

BEHAVIORAL GROWTH OF THE INFANT

Behavioral growth is principally a matter of nervous system maturity. However, it is extremely difficult to dissociate maturity of the anatomical structures of the nervous system from maturity caused by training. Anatomical studies show that certain major tracts in the central nervous system are not completely myelinated until the end of the first year of life. For this reason we frequently state that the nervous system is not fully functional at birth. On the other hand, we know that most reflexes of the fetus are fully developed by approximately the third to fourth month of intra-uterine life, and that unmyelinated nerve fibers can be just as functional as myelinated nerve fibers. Therefore, much of the functional immaturity of the nervous system at birth might well be caused by lack of training rather than actual immaturity of the anatomical structures.

Nevertheless, the brain weight of the child increases rapidly during the first year and less rapidly during the second year, reaching almost adult proportions by the end of the second year. This is also associated with closure of the fontanels and sutures of the skull, which prevents much additional growth of the cranium beyond the first two years of life.

Figure 56–9 illustrates a normal progress chart for the infant during the first year of life. Comparison of a baby's actual development in relation to such a chart is frequently used for clinical assessment of the child's mental and behavioral growth.

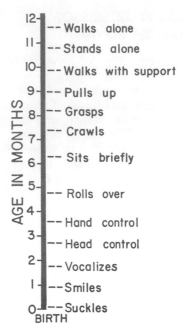

FIGURE 56–9 Behavioral development of the infant during the first year of life.

REFERENCES

Assali, N. S. (ed.): Biology of Gestation. 2 volumes. New York, Academic Press, Inc., 1968.

Barnes, A. C.: Intra-Uterine Development. Philadelphia, Lea & Febiger, 1968.

Ciba Foundation Study Group No. 23: Egg Implantation. Boston, Little, Brown and Company, 1966.

Dawes, G. S.: Foetal and Neonatal Physiology. Chicago, Year Book Medical Publishers, Inc., 1968.

Hadek, R.: Mammalian Fertilization: An Atlas of Ultrastructure. Academic Press, Inc., New York, 1969.

Locke, M.: Control Mechanisms in Developmental Processes. New York, Academic Press, Inc., 1968.

Meites, J., and Nicoll, C. S.: Adenohypophysis: prolactin. *Ann. Rev. Physiol.,* 28:57, 1966.

Metcalf, J.; Bartels, H.; and Moll, W.: Gas exchange in the pregnant uterus. *Physiol. Rev.,* 47:782, 1967.

Parkes, A. S.: Marshall's Physiology of Reproduction. 3rd ed., 3 volumes. Boston, Little, Brown and Company, 1956–1966.

Saunders, F. J.: Effects of sex steroids and related compounds on pregnancy and on development of the young. *Physiol. Rev.,* 48:601, 1968.

Wynn, R. M.: Fetal Homeostasis. New York, New York Academy of Sciences, 1968.

INDEX